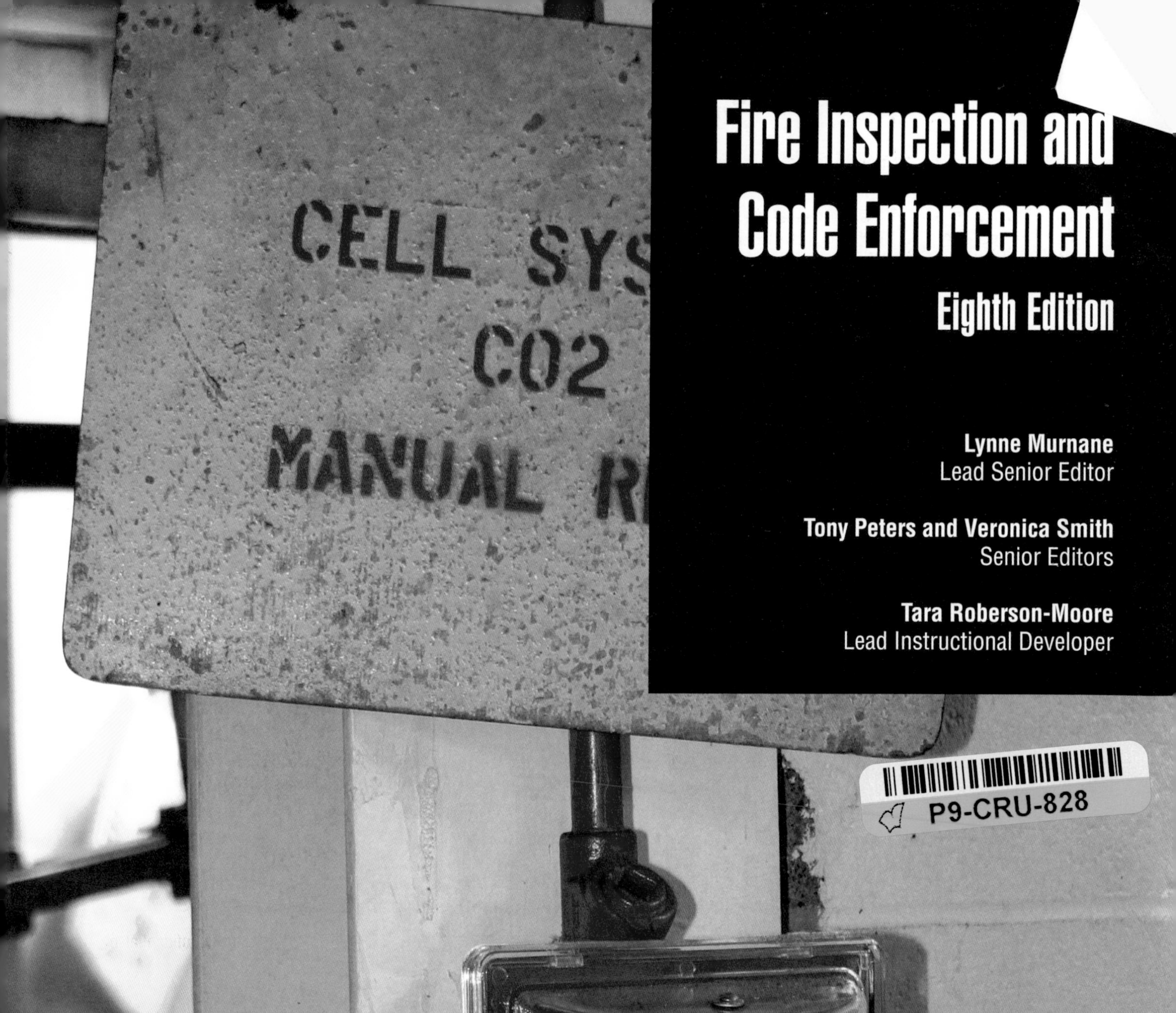

Fire Inspection and Code Enforcement

Eighth Edition

Lynne Murnane
Lead Senior Editor

Tony Peters and Veronica Smith
Senior Editors

Tara Roberson-Moore
Lead Instructional Developer

IFSTA

Validated by the International Fire Service Training Association

Published by
Fire Protection Publications • Oklahoma State University

RECYCLABLE

The International Fire Service Training Association (IFSTA) was established in 1934 as a *nonprofit educational association of fire fighting personnel who are dedicated to upgrading fire fighting techniques and safety through training.* To carry out the mission of IFSTA, Fire Protection Publications was established as an entity of Oklahoma State University. Fire Protection Publications' primary function is to publish and distribute training materials as proposed, developed, and validated by IFSTA. As a secondary function, Fire Protection Publications researches, acquires, produces, and markets high-quality learning and teaching aids consistent with IFSTA's mission.

IFSTA holds two meetings each year: the Winter Meeting in January and the Annual Validation Conference in July. During these meetings, committees of technical experts review draft materials and ensure that the professional qualifications of the National Fire Protection Association® standards are met. These conferences bring together individuals from several related and allied fields, such as:

- Key fire department executives, training officers, and personnel
- Educators from colleges and universities
- Representatives from governmental agencies
- Delegates of firefighter associations and industrial organizations

Committee members are not paid nor are they reimbursed for their expenses by IFSTA or Fire Protection Publications. They participate because of a commitment to the fire service and its future through training. Being on a committee is prestigious in the fire service community, and committee members are acknowledged leaders in their fields. This unique feature provides a close relationship between IFSTA and the fire service community.

IFSTA manuals have been adopted as the official teaching texts of many states and provinces of North America as well as numerous U.S. and Canadian government agencies. Besides the NFPA® requirements, IFSTA manuals are also written to meet the Fire and Emergency Services Higher Education (FESHE) course requirements. A number of the manuals have been translated into other languages to provide training for fire and emergency service personnel in Canada, Mexico, and outside of North America.

ISBN 978-0-87939-605-3   Library of Congress Control Number: 201595098

Eighth Edition, First Printing, April 2016   *Printed in the United States of America*

10 9 8 7 6 5 4

If you need additional information concerning the International Fire Service Training Association (IFSTA) or Fire Protection Publications, contact:

Customer Service, Fire Protection Publications, Oklahoma State University
930 North Willis, Stillwater, OK 74078-8045
800-654-4055   Fax: 405-744-8204

For assistance with training materials, to recommend material for inclusion in an IFSTA manual, or to ask questions or comment on manual content, contact:

Editorial Department, Fire Protection Publications, Oklahoma State University
930 North Willis, Stillwater, OK 74078-8045
405-744-4111   Fax: 405-744-4112   E-mail: editors@osufpp.org

Chapter Summary

Chapter

1 Duties and Authority 10
2 Codes, Standards, and Permits 38
3 Fire Behavior 70
4 Construction Types and Occupancy Classifications 112
5 Building Construction 150
6 Building Components 190
7 Means of Egress 238
8 Site Access 282
9 Fire Hazard Recognition 304
10 Hazardous Materials 376
11 Water Supply Distribution Systems 462
12 Water-Based Fire Suppression Systems 500
13 Special-Hazard Fire Extinguishing Systems and Portable Extinguishers 560
14 Fire Detection and Alarm Systems 604
15 Plans Review 648
16 Inspection Procedures 686

Appendices

A NFPA® 1031 Job Performance Requirements (JPRs) with Chapter and Page References 731
B Example of a Citation Program 733
C Adams vs. State of Alaska 736
D Sample Fire Lane Sign Requirements 737
E Tent Guidelines 741
F Hot Work Program 743
G Floating Roof Tanks 746
H Audible Warning Devices 747
I Automatic Sprinkler Systems Acceptance Checklist 749
J Inspection Report Checklist 752

Glossary **755**
Index **767**

Table of Contents

Acknowledgments ....
Introduction 1
Purpose and Scope 3
Book Organization 3
Terminology 3
Resources 4
Key Information 4
Metric Conversions 6

1 Duties and Authority 12
Case History 13
Inspector I
Duties 14
Public Organizations 15
Fire Department *15*
Building Department *15*
Code Enforcement Department *16*
Third-Party Inspectors 16
Private Fire Safety Inspectors 16
Training and Certification 17
Professional Development 17
Inspection Categories 18
Legal Guidelines for Inspections 19
Compliance Procedures 19
Civil Proceedings 20
Criminal Proceedings 20
Types of Criminal Violations *21*
Evidentiary Requirements *22*
Court Proceedings 22
Preliminary Hearings *22*
Felony Trial *22*
Testimony 23
Fact Witness *24*
Expert Testimony *24*
Inspector II
Authority 24
Enabling Legislation 24
Types of Laws 25
Federal Laws *25*
State/Provincial Laws *26*
Local Laws and Ordinances *27*
Legal Status of Inspectors 27
Public Sector *28*
Private Sector *29*
Conflicts Between Public and Private Requirements *29*
Liability 29
Right of Entry *29*
Warrants *30*
Duty to Inspect *31*
Civil Rights *31*
Indemnification *32*
Malfeasance and Negligence *33*
Outside Technical Assistance 33
Recommend Policies and Modifications 34
Developing Policies and Procedures 34
Maintaining and Revising Policies 35
Creating Forms and Job Aids 36
Chapter Summary 36
Review Questions 37

2 Codes, Standards, and Permits 38
Case History 41
Codes and Standards 41
Codes 42
Prescriptive/Model Codes *43*
Performance-Based Design *44*
Application of Codes *44*
Current Codes and Standards *46*
Consistent Codes and Standards *46*
Standards 47
National Fire Protection Association® (NFPA®) *48*
ASTM International *49*
Underwriters Laboratories (UL) *49*
American National Standards Institute *49*
Standards Council of Canada (SCC) *50*
Complaint Procedures 51
Permits 52
Operational Permits 52
Construction Permits 55
Inspector II
Local Code Development Process 56
Identify the Problem 56
Identify Affected Stakeholders 57
Form a Task Force 57
Draft the Proposed Code 58
Submit Code for Legal Review 59
Formally Adopt the Code 59
Prepare Formal Resolution *59*
Discuss Legislation and Adopt Code *61*
Introduce New Code *61*

Code Modification and Appeals Procedures........62
Code Modification 63
Appeals Procedures 65
Permit Process 65
Application 66
Research and Review 66
Issue Permit 66
Evaluate Status 66
Permit Issued in Error 67
Chapter Summary 67
Review Questions 68

3 Fire Behavior 70
Case History 73
I Inspector I
Science of Fire 74
Physical and Chemical Changes 74
Modes of Combustion 76
Models of Fire 76
Nonflaming Combustion 76
Flaming Combustion 77
Temperature, Energy, and Heat 77
Temperature 78
Energy 78
Heat 78
Forms of Ignition 79
Sources of Thermal Energy 80
Chemical Energy 80
Electrical Energy 81
Mechanical Energy 82
Heat Transfer 83
Conduction 83
Convection 84
Radiation 84
Passive Agents 87
Fuel 87
Gaseous Fuel 89
Liquid Fuel 90
Solid Fuel 91
Oxidizer 92
Self-Sustained Chemical Reaction 95
Classification of Fires 96
Class A Fires 96
Class B Fires 96
Class C Fires 98
Class D Fires 98
Class K Fires 98
Stages of Fire Development 98
Factors that Affect Fire Development 99
Fuel Type 99
Availability and Location of Additional Fuels 99
Compartment Volume and Ceiling Height 101
Ventilation 102
Thermal Properties of the Compartment 102
Fuel Load 102
Incipient Stage 103
Growth Stage 104
Thermal Layering 104
Isolated Flames 106
Rapid Transition 106
Fully Developed Stage 107
Decay Stage 108
Fuel Consumption 108
Limited Ventilation 108
II Inspector II
Fire Control 108
Temperature Reduction 109
Fuel Removal 110
Oxygen Exclusion 110
Chemical Flame Inhibition 111
Chapter Summary 111
Review Questions 111

4 Construction Types and Occupancy Classifications 112
Case History 115
I Inspector I
Construction Types 116
United States Construction 117
Type I 118
Type II 120
Type III 120
Type IV 122
Type V 122
Canadian Construction 125
Occupancy Classifications 125
Assembly Occupancies 132
Business Occupancies 133
Educational Occupancies 133
Factory/Industrial Occupancies 134
Institutional Occupancies 134
Health Care and Ambulatory Health Care Occupancies 136
Detention and Correctional Occupancies 137
Residential Board and Care Occupancies 137
Day Care Occupancies 138
Mercantile Occupancies 138

Residential Occupancies 139
One- and Two-Family Dwellings *139*
Lodging (Boarding) or Rooming Houses *140*
Hotels *141*
Dormitories *142*
Apartment Buildings *143*
Storage Occupancies 144
Utility/Miscellaneous Occupancies 145
Inspector II
Multiple-Use Occupancies 146
Incidental Use 147
Mixed-Use 147
Accessory *148*
Nonseparated *148*
Separated *148*
Chapter Summary 148
Review Questions 149

5 Building Construction 150
Case History 153
Inspector I
Construction Materials 153
Fire Protection and Resistance Terminology 154
Wood 154
Solid Lumber *155*
Engineered Wood Products *155*
Fire-Retardant Treated Wood *157*
Exterior Wall Materials *157*
Masonry 160
Brick *161*
Concrete Block *161*
Stone *161*
Clay Tile Block and Gypsum Block *162*
Concrete 163
Steel 163
Characteristics *164*
Fire Protection *164*
Membrane Ceilings *165*
Glazing 166
Gypsum Board 167
Plastics 168
Flammability *169*
Fire Hazards *169*
Thermal Barriers *170*
Exterior Insulation and Finish Systems (EIFS) *170*
Inspector II
Structural Systems 171
Bearing Wall Structures 172
Loads 172
Shell and Membrane Structures 174
Frame Structural Systems 176
Steel Stud Wall Framing *176*
Post and Beam Framing *176*
Truss Frames *176*
Slab and Column Frames *176*
Rigid Frames *178*
Wood Structural Systems 178
Wood Framing *179*
Heavy-Timber Framing *181*
Masonry Structural Systems 181
Masonry Walls *181*
Structural Stability *182*
Interior Structural Framing *183*
Fire Resistance *184*
Steel Structural Systems 184
Concrete Structural Systems 186
Cast-in-Place Concrete *186*
Precast Concrete *186*
Ordinary Reinforcing *187*
Pretensioning and Posttensioning *187*
Chapter Summary 188
Review Questions 189

6 Building Components 190
Case History 193
Inspector I
Walls 194
Fire Walls 194
Construction Types *194*
Openings *194*
Party Walls 195
Fire Partitions and Fire Barriers 196
Enclosure and Shaft Walls 196
Curtain Walls 197
Roofs 199
Floors 200
Construction Materials 200
Supports and Coverings 200
Penetrations and Openings 202
Ceilings 203
Stairs 203
Doors 204
Types 204
Swinging Doors *205*
Sliding Doors *205*
Folding Doors *205*
Vertical Doors *205*
Revolving Doors *206*

Styles and Construction Materials 206
Wood Panel and Flush Doors *206*
Glass Doors *206*
Metal Doors *207*
Fire Doors **207**
Classifications 209
Frames and Hardware 210
Construction and Operational Types 211
Rolling Steel Fire Doors *211*
Horizontal Sliding Fire Doors *212*
Swinging Fire Doors *213*
Special Types *213*
Glass Panels and Louvers *213*
Windows **213**
Components 213
Types 214
Security 215
Interior Finishes **216**
Building Services **217**
Elevator Hoistways and Doors 217
Utility Chases and Vertical Shafts 218
Refuse and Linen Chutes *218*
Grease Ducts *218*
Inspector II
Evaluating Fire Walls **220**
Evaluating Rooftop Photovoltaic Systems **221**
Firefighter Safety 221
Access Pathways and Smoke Ventilation 221
Evaluating Structural Stability **222**
Evaluating Interior Components **222**
Floor Finishes 222
Ceilings 223
Stairs 223
Evaluating Fire Doors **223**
Evaluating Interior Finishes **225**
Flame Spread Ratings 225
Smoke-Developed Index 226
Fire-Retardant Coatings 228
Evaluating Building Services **228**
Heating, Ventilating, and Air Conditioning Systems 229
Refrigeration Systems 231
Electrical Systems 232
Electrical Service Panels *233*
Switch Gear *233*
Generators *234*
Transformers *234*
Elevator Hoistways and Doors 234
Utility Chases and Vertical Shafts 235
Chapter Summary **236**
Review Questions **237**

7 Means of Egress **238**
Case History **241**
Inspector I
Means of Egress Systems **241**
Elements 246
Exit Access *246*
Exit *246*
Exit Discharge *250*
Building Components 251
Doors *251*
Walls *256*
Ceilings *257*
Floors *257*
Stairs *258*
Ramps *260*
Fire Escape Stairs, Ladders, and Slides *260*
Exit Illumination and Markings 262
Exit Illumination *262*
Emergency Lighting *263*
Markings *263*
Auxiliary Power *264*
Occupant Loads **265**
Inspector II
Multiuse Occupant Loads **268**
Means of Egress Determinations **268**
Capacity 268
Exit Capacity *269*
Total Exit Capacity *269*
Required Number of Exits *270*
Arrangement 270
Location of Exits *271*
Maximum Travel Distance to an Exit *272*
Effectiveness 273
Chapter Summary **275**
Calculations **276**
Inspector I Calculations 276
Inspector II Calculations 277
Review Questions **280**

8 Site Access **282**
Case History **285**
Inspector I
Fire Lanes and Fire Apparatus Access Roads **285**
Requirements 286
Dead-End Access Roads 288
Road Markings and Signs 289
Construction and Demolition Sites **291**

Structure Access Barriers **294**
Exterior Access 294
Weight Requirements *295*
Illegal Parking *296*
Overhead Obstructions *297*
Landscaping *298*
Topographical Conditions *298*
Seasonal Climate Conditions *298*
Interior Access 299
Chapter Summary **302**
Review Questions **302**

9 Fire Hazard Recognition **304**
Case History **307**
Inspector I
Unsafe Behaviors **308**
Inadequate Housekeeping 308
Unintentional Ignition Sources 308
Open Burning 309
Improper Use of Electrical Equipment 311
Improper Use and Storage of Flammable and Combustible Liquids 313
Dispensing *314*
Transporting *317*
Improper Use and Storage of Compressed/Liquefied Gases 317
Unsafe Conditions **319**
Electrical Hazards 320
Worn Electrical Equipment *320*
Improper Use of Electrical Equipment *320*
Defective or Improper Electrical Installations *321*
Material Storage Facilities 322
Storage Methods *323*
Inspection Guidelines *325*
Warehouses *327*
Lumberyards *328*
Tire Storage Facilities *330*
Pallet Storage Facilities *330*
Recycling Facilities 331
Waste-Handling Facilities *332*
Incinerators *333*
Inspector II
Building Systems **335**
Heating, Ventilating, and Air Conditioning Equipment/Systems 335
Furnaces *336*
Boilers *338*
Unit Heaters *340*
Room Heaters *340*
Temporary/Portable Heating Equipment *342*
Air Conditioning Systems *343*
Ventilation Systems *344*
Filtering Devices *344*
Cooking Equipment 344
Ventilation-Hood Systems *346*
Solid-Fuel Cooking Equipment *346*
Fire Suppression 347
Industrial Furnaces and Ovens 348
Classes *348*
Special Considerations *348*
Powered Industrial Trucks 349
Tents 350
Hazardous Processes **352**
Welding and Thermal Cutting Operations 352
Hot-Work Program *354*
Permits *354*
Fire Watch *355*
Flammable Finishing Operations 356
Spray Finishing *357*
Powder Coating *358*
Immersion Coating *358*
Quenching Operations 359
Dry Cleaning Operations 361
Dust Hazards 362
Process Hazard Analysis *365*
Classification and Grouping *367*
Dust Controls *368*
Fire Protection *368*
Grain Facilities *369*
Woodworking and Processing Facilities *371*
Machine Shops and Manufacturing Facilities *371*
Inspection Procedures *371*
Torch-Applied Roofing Materials 372
Asphalt and Tar Kettles 372
Distilleries 373
Chapter Summary **374**
Review Questions **375**

10 Hazardous Materials **376**
Case History **379**
Inspector I
Application of Hazardous Materials Regulations **379**
Product Containment 381
Pressure Relief 381
Fire Protection 382
Exemptions 382

Classification of Hazardous Materials................385
Physical Hazard Materials................................386
Flammable and Combustible Liquids.......386
Compressed and Liquefied Gases..............388
Cryogenic Fluids..388
Flammable Solids or Gases.........................390
Organic Peroxides......................................391
Oxidizers and Oxidizing Gases..................392
Pyrophorics..393
Unstable (Reactive) Materials...................394
Water-Reactive Materials...........................394
Explosives and Blasting Agents..................395
Health Hazard Materials.................................396
Highly Toxic and Toxic Materials..............396
Corrosives..398
Mixtures..399
Incompatible Materials.............................399
Identification of Hazardous Materials...............401
Safety Data Sheets 402
Transportation Placards, Labels, and Markings...402
UN Hazard Classes....................................404
UN Commodity Identification Numbers....404
DOT Placards, Labels, and Markings........405
Other Markings 411
Manufacturers' Warnings..........................414
Military Markings......................................414
Pipeline Markings......................................414
NFPA® 704 System....................................416
Resource Guidebooks 418
Emergency Response Guidebook..............418
NIOSH Pocket Guide to Chemical Hazards...420
Hazardous Materials Guide for First Responders...421
Hazardous Materials Information Resource System...421
Canadian Dangerous Goods System 421
Mexican Hazard Communication System 427
Piping Identification 430
Cylinder Markings 430
Permissible Amount of Hazardous Materials in a Building...430
Maximum Allowable Quantity (MAQ) per Control Area ...431
Control Areas...434
Nonbulk and Bulk Packaging............................435
Containers 435
Safety Cans..436
Intermediate Bulk Containers...................437
Cylinders...438
Bulk Packaging 438
Shop-Fabricated Aboveground Storage Tanks...440
Field-Erected Aboveground Storage Tanks....443
Underground Storage Tanks......................446
Pressure Vessels..446
Testing, Maintenance, and Operations..............449
II Inspector II
Process Control..449
Unauthorized Discharge 450
Hazardous Materials Piping, Valves, and Fittings...451
Hazardous/High-Hazard Occupancies 452
Categories...452
Explosives...453
Requirements..454
Engineering Controls for Hazardous Materials...455
Spill Control and Secondary Containment......456
Mechanical Ventilation System........................458
Automatic Sprinkler Protection 458
Summary...459
Review Questions..460

11 Water Supply Distribution Systems.......462
Case History..465
I Inspector I
Water Supply System Components 466
System Design 466
Water Supply Sources......................................468
Reservoirs..469
Suction Tanks...469
Pressure Tanks...469
Gravity Tanks...470
Means of Moving Water 470
Distribution Systems 472
Piping..472
Storage Tanks...474
Control Valves...474
Backflow Preventers..................................475
Fire Hydrants..476
Water Supply Testing..477
Fire Hydrant Inspections 480
Pitot Tube and Gauge 481
Fire Flow Test Computations 482
Required Residual Pressure 484
Fire Flow Test Procedures 485
Precautions...487
Obstructions..489

Available Fire Flow Test Results
Computations 490
Graphical Analysis *490*
Mathematical Method *493*
Chapter Summary 493
Calculations 493
Review Questions 498

12 Water-Based Fire Suppression Systems 500
Case History 503
Inspector I
Automatic Sprinkler Systems 504
Basic Types and Design 505
Components 508
Water Supplies *508*
Waterflow Control Valves *508*
Operating Valves *510*
Water Distribution Pipes *510*
Sprinklers *512*
Detection and Activation Devices *516*
Residential Systems 518
Water Supply and Flow Rate Requirements *519*
Spacing *519*
Automatic Fire Suppression Systems 519
Water-Spray Fixed System 520
Water-Mist Systems 521
Foam-Water Systems 523
Standpipe and Hose Systems 524
Components 526
Classifications 526
Class I: Firefighters *526*
Class II: Trained Building Occupants *526*
Class III: Combination *526*
Types 527
Water Supplies and Residual Pressure 529
Standpipe Hose Valves 530
Pressure-Regulating Devices 530
Fire Department Connections 531
Stationary Fire Pumps 532
Types 532
Horizontal Split-Case *532*
Vertical Split-Case *533*
Vertical Inline *534*
Vertical Turbine *534*
End Suction *534*
Pressure-Maintenance *535*
Pump Drivers 535
Electric Motor Driver *535*
Diesel Engine Driver *536*
Steam Turbine *536*
Controllers 536
Electric Motor *537*
Diesel Motor *537*
Inspector II
Evaluate System Components and Equipment 538
Plans Review 539
Preacceptance Inspections 539
Acceptance Testing 540
Routine Inspections and Testing 540
Automatic Sprinkler Systems *546*
Residential Sprinkler Systems *552*
Water-Spray Fixed Systems *554*
Foam-Water Systems *555*
Standpipe and Hose Systems *555*
Stationary Fire Pumps *557*
Chapter Summary 557
Review Questions 558

13 Special-Agent Fire Extinguishing Systems and Portable Extinguishers 560
Case History 563
Inspector I
Special-Agent Fire Extinguishing Systems 564
Fire Hazard Classification 565
Dry-Chemical Systems 565
Application Methods *566*
Agents *567*
Components *568*
Dry Powder Systems 569
Wet-Chemical Systems 571
Clean-Agent Systems 572
Agents *573*
Components *574*
Carbon Dioxide Systems 574
Personnel Safety *575*
Components *577*
Foam Systems 578
Types of Foam Systems *578*
Foam Generation *579*
Foam Proportioning Rates *580*
Foam Expansion Rates *581*
Foam Concentrate Types *582*
Proportioners *583*
Portable Fire Extinguishers 584
Classification Systems 585
Rating Systems 585
Agents 586

Types 590
Installation and Placement 592

Inspector II

Inspection and Maintenance of Special-Agent Fire Extinguishing Systems 594
Dry-Chemical Systems 594
Wet-Chemical Systems 595
Clean-Agent Systems 596
Carbon Dioxide Systems 596
Foam Systems 596
Fire Extinguishers: Selection, Location, Training, and Inspection 597
Nature of the Hazard 597
Extinguisher Size/Travel Distance 599
Training 600
Inspection and Maintenance 600
Chapter Summary 603
Review Questions 603

14 Fire Detection and Alarm Systems 604
Case History 607

Inspector I

Fire Alarm System Components 608
Fire Alarm Control Units 608
Primary Power Supply 609
Secondary Power Supply 609
Initiating Devices 610
Notification Appliances 610
Additional Alarm System Functions 612
Alarm Signaling Systems 613
Protected Premises Systems (Local) 614
Conventional Alarm Systems 615
Zoned Conventional Alarm Systems 616
Addressable Alarm Systems 617
Supervising Station Alarm Systems 617
Central Station System 618
Proprietary System 619
Remote Receiving System 619
Public Emergency Alarm Reporting Systems 621
Emergency Communications Systems 621
Voice Notification Systems 621
Two-Way Communication Systems 622
Mass Notification Systems (MNS) 622
Automatic Alarm-Initiating Devices 623
Fixed-Temperature Heat Detectors 624
Fusible Links/Frangible Bulbs 625
Bimetallic Heat Detector 625
Continuous-Line Heat Detector 626
Rate-of-Rise Heat Detectors 627
Pneumatic Rate-of-Rise Line Heat Detector 627
Pneumatic Rate-of Rise Spot Heat Detector 628
Rate-Compensation Heat Detector 628
Electronic Spot-Type Heat Detector 628
Smoke Detectors 629
Photoelectric Smoke Detectors 630
Ionization Smoke Detectors 631
Duct Smoke Detectors 632
Video-Based Detectors 633
Flame Detectors 633
Combination Detectors 634
Sprinkler Waterflow Alarm-Initating Devices 634
Manually Actuated Alarm-Initiating Devices 635
Inspection and Testing 637
Inspection Considerations for Fire Alarm Control Units 638
Inspection Considerations for Alarm-Initiating Devices 639
Inspection Considerations for Alarm Signaling Systems 641

Inspector II

Inspecting, Testing, and Evaluating Fire Detection and Alarm Systems 642
Acceptance Testing 642
Occupant Notification Devices 643
Audible Notification 644
Visual Notification 645
Required Detection and Alarm System Documentation 646
Chapter Summary 647
Review Questions 647

15 Plans Review 648
Case History 651

Inspector I

Overview of Plans Review 651
Benefits of Plans Review 652
Field Verification 653
Building Plans and Construction Drawings 654

Inspector II

Plans Review Process 656
Sequence 658
Fees 658
Construction and Support Documents 658
Building Construction Plans 662
Plan Views 663
Site Plan 663

Floor Plan 665
Elevation View 666
Sectional View 668
Detailed View 669
System Plans 669
Mechanical Systems 670
Electrical Systems 672
Plumbing Systems 672
Automatic Sprinkler Systems 673
Standpipe and Hose Systems 677
Special-Agent Fire Extinguishing Systems 677
Fire Detection and Alarm Systems 677
Systematic Plans Review 679
Overall Size of Building 680
Occupancy and Construction Classification 680
Occupant Load 681
Means of Egress 681
Exit Capacity 682
Building Compartmentation 682
Additional Concerns 682
Chapter Summary 683
Review Questions 685

16 Inspection Procedures 686
Case History 689
Inspector I
Interpersonal Communication 690
Communication Model Elements 690
Interpersonal Skills 692
Listening Skills 692
Conversing Skills 693
Persuading Skills 695
Administrative Duties 695
Written Communications 695
Memos 696
E-Mail Messages 696
Letters 696
Reports 696
Files and Records 697
Inspection Preparation 698
Personal Appearance 700
Equipment Lists 700
Inspection Scheduling 701
Inspection Records Review 702
Inspection Procedures 702
Guidelines for Inspections 703
General Inspection Practices 705
Code Requirements 706
Photographs 708
Inspection Checklists 708
Building Occupancy Changes 712
Results Interview 712
Violation Discussions 713
Educational Opportunities 714
Long-Term Relationships 714
Letters and Reports 714
Follow-Up Inspections 715
Emergency Planning and Preparedness 716
Educational Facilities 718
Health Care Facilities 719
Correctional Facilities 721
Hotels and Motels 721
Complaint Management 722
Inspector II
Complex Complaint Procedures 724
Evaluate Emergency Planning 725
Chapter Summary 727
Review Questions 727

Appendices
A NFPA 1031 Job Performance Requirements (JPRs) with Chapter and Page References 731
B Example of a Citation Program 733
C Adams vs. State of Alaska 736
D Sample Fire Lane Sign Requirements 737
E Tent Guidelines 741
F Hot Work Program 743
G Floating Roof Tanks 746
H Audible Warning Devices 747
I Automatic Sprinkler Systems Acceptance Checklist 749
J Inspection Report Checklist 752

Glossary 755

Index 767

List of Tables

Table 2.1 Model Code Activities Requiring Permits 53

Table 3.1 Spontaneous Heating Materials and Locations 81

Table 3.2 Thermal Conductivity of Common Substances 84

Table 3.3 Representative Peak Heat Release Rates (HRR) During Unconfined Burning....88

Table 3.4 Characteristics of Common Flammable Gases....90

Table 3.5 Common Oxidizers 94

Table 3.6 Flammable Ranges of Common Flammable Gases and Liquids (Vapor)....95

Table 3.7 Common Products of Combustion and Their Toxic Effects 97

Table 3.8 Factors Influencing Development of a Fuel-Controlled Fire 105

Table 4.1 Fire-Resistance Rating Requirements for Building Elements (Hours)....118

Table 4.2 Fire-Resistance Rating Requirements for Exterior Walls Based on Fire Separation Distance..... 119

Table 4.3 Comparative Overview of Occupancy Categories 127

Table 4.4 Canadian Codes: Factory/Industrial Occupancies 135

Table 4.5 Canadian Codes: Industrial Occupancies....136

Table 4.6 Required Separation of Occupancies (Hours) 149

Table 5.1 Plastics Commonly Used in Construction....170

Table 5.2 Minimum Uniformly Distributed Live Loads 174

Table 6.1 Interior Wall and Ceiling Finish Requirements by Occupancy 227

Table 7.1 Factors in Loss-of-Life Fires....242

Table 7.2 Maximum Floor Area Allowances per Occupant....267

Table 7.3 Egress Width per Occupant Load....269

Table 7.4 Common Path, Dead-End, and Travel Distance Limits (By Occupancy)....274

Table 9.1 Classifications of Dry Cleaning Plants and Solvents 361

Table 10.1 Model Fire Code Classification of Flammable and Combustible Liquids 387

Table 10.2 Hazard Classification of Organic Peroxides 392

Table 10.3 Hazard Classifications of Solid and Liquid Oxidizers....393

Table 10.4 Model Fire Code Hazard Classifications for Unstable (Reactive) Materials....395

Table 10.5 Model Fire Code Hazard Classifications for Water-Reactive Materials 395

Table 10.6 Chemical Compatibility Matrix....400

Table 10.7 Information Disclosed n a U.S. Material Safety Data Sheet 403

Table 10.8 Unique U.S. DOT Labels 409

Table 10.9	Unique U.S. DOT Markings	412
Table 10.10	U.S. and Canadian Military Symbols	415
Table 10.11	Emergency Response Guidebook Contents	419
Table 10.12	Canadian Transportation Placards, Labels, and Markings	423
Table 10.13	WHMIS Symbols and Hazard Classes	428
Table 10.14	Sample ISO-3864 Type Symbols	429
Table 10.15	Storage Tank Types by Pressure Rating	441
Table 10.16	UNDMGC/US DOT Explosives Hazard Classification	455
Table 11.1	Classifications and Markings of Municipal Fire Hydrants	479
Table 11.2	Correction Factors for Large Diameter Outlets	484
Table 11.3	Values for Computing Fire Flow Tests	497
Table 12.1	Sprinkler Color Coding, Temperature Classification, and Temperature Rating	517
Table 12.2	Acceptance Testing Procedures	541
Table 13.1	Portable Fire Extinguisher Ratings	586
Table 13.2	Pictorial and Letter Systems	587
Table 13.3	Portable Fire Extinguisher Requirements for Class A Fire Hazards	599
Table 14.1	Heat-Sensing Fire Detector Color Coding, Temperature Classification, and Temperature Rating	624
Table 15.1	Common Architectural and Fire Safety Symbols	655
Table 16.1	Occupancy Category Requirements	709
Table 16.2	Emergency Evacuation Plans and Drills per Occupancy Type	717

Acknowledgements

The eighth edition of the **Fire Inspection and Code Enforcement** is designed to address the Inspector I and Inspector II portions of NFPA® 1031, *Standard for Professional Qualifications for Fire Inspector and Plan Examiner.*

Acknowledgement and special thanks are extended to the members of the IFSTA validating committee who contributed their time, wisdom, and knowledge to the development of this manual.

IFSTA Fire Inspection and Code Enforcement Eighth Edition IFSTA Validation Committee

Chair

Brett T. Lacey
Fire Marshal
Colorado Springs Fire Department
Colorado Springs, Colorado

Vice-Chair

Scott Strassburg
Division Chief/ Fire Marshal
Blooming Grove Fire Department
Madison, WI

Secretary

Scott A. Stookey
Senior Technical Staff
International Code Council
Austin, TX

Committee Members

Glen K. Albright
Fire Prevention Specialist
Ventura City Fire Department
Ventura, CA

George W. Apple
Assistant Chief - Fire Marshal
Cosumnes Fire Department
Elk Grove, CA

Doug DeGraff
Deputy State Fire Marshal
Washington State Patrol / Fire Protection Bureau
Bellevue, WA

Joheida (Fred) Fister
Fire Marshal/Battalion Chief
Hilton Head Island Fire Rescue
Hilton Head Island, SC

Keith Flood
Fire Marshal
West Haven Fire Department
West Haven, CT

Ronald F. Hoelle II
Course Supervisor Fire Inspector Course
Department of Defense, Goodfellow AFB
301 Comanche Trail
San Angelo, TX

George Hollingsworth
Captain II
Fairfax County Fire and Rescue Department
Fairfax, VA
Representing: Virginia Fire Prevention Association

Tonya L. Hoover
State Fire Marshal
CAL FIRE – Office of the State Fire Marshal
Sacramento, CA

Martin M. King
Assistant Chief, Bureau of Fire Prevention, Urban Affairs and EMS
West Allis Fire Department
West Allis, WI

Mark Larson
State Fire Marshal
Boise, ID

IFSTA Fire Inspection and Code Enforcement Eighth Edition Validation Committee

Committee Members (cont.)

Richard Pippenger
Fire Marshal
Lower Valley Fire District
Fruita, CO

Jon Roberts
Lead Regulatory Engineer, Underwriters Laboratories LLC.
Oklahoma City, OK

Walter G. M. Schneider III, Ph.D., P.E., CBO
Agency Director
Centre Region Code Administration
State College, Pennsylvania
Department Chief
Bellefonte Fire Department
Bellefonte, PA

Tim Stemple, Fire Chief (Ret.)
Lockheed Martin
Fort Worth, TX

R. Paul Valentine
Senior Engineer
Nexus Engineering
Oakbrook Terrace, IL

Ray Wolf
Deputy Fire Marshal Fire Code Enforcement
New Mexico State Fire Marshal's Office
Santa Fe, NM

The following individuals and organizations contributed information, photographs, photography assistance, and other assistance that made the completion of this manual possible:

Rich Mahaney
Russell Chandler
Ron Jeffers
Oregon OSHA
Click2Enter, Inc.
FM Global

George Apple
Chemetron Fire Systems
Dave Coombs
Wil Dane
David DeStefano
Joe Elam
Ed Hartin
Greg Havel
Brett Lacey
Bill Lefkowitz, Central Fla Fire Academy
Dan Madryzkowski/NIST
Dennis Marx
Ron Moore
McKinney (TX) Fire Department
Ed Prendergast
Jon Roberts
Sand Springs (OK) Fire Department
Walt Schneider
Scott Stookey
Scott Strassburg
Ralph E. Tingley
Steve Toth
Tyco Products
Wayne State University

Fire Inspection and Code Enforcement Project Team

Project Manager
Lynne Murnane, Senior Editor

Director of Fire Protection Publications
Craig Hannan

Curriculum Manager
Leslie Miller

IFSTA Projects Coordinator
Clint Clausing

Production Manager
Ann Moffat

Senior Editors
Alex Abrams
Tony Peters

Technical Reviewers
Bill Peterson
Greg Rogers
Chris Conroy

Proofreaders
Alex Abrams
Veronica Smith

Curriculum Development
Tara Roberson-Moore
Beth Ann Fulgenzi
Lori Raborg

Library Researcher
Susan Walker, Librarian

Editorial Administrative Support Specialist
Tara Gladden

Photographers
Clint Parker
Veronica Smith

Illustrators and Layout Designers
Ben Brock
Ann Moffat

Indexer
Nancy Kopper

The IFSTA Executive Board at the time of validation of **Fire Inspection and Code Enforcement**, 8th Edition, was as follows:

Dedication

This manual is dedicated to the men and women who hold devotion to duty above personal risk, who count on sincerity of service above personal comfort and convenience, who strive unceasingly to find better and safer ways of protecting lives, homes, and property of their fellow citizens from the ravages of fire, medical emergencies, and other disasters

...The Firefighters of All Nations.

Introduction

Introduction Contents

Introduction....................................1
Purpose and Scope....................................3
Book Organization....................................3
Terminology....................................3
Resources....................................3
Key Information....................................4
Metric Conversions....................................6

Introduction

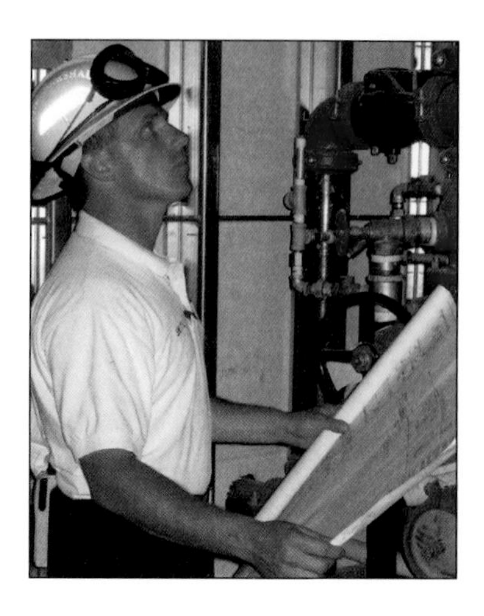

The number of structural fires in North America has decreased over the past half century, and the average fire loss has decreased. This reduction is due in part to the increase in modern, NFPA®-compliant building construction. At the same time, old buildings that have been modernized have been upgraded to meet current codes and standards. Local municipalities have adopted and enforced model building codes as well as fire and life safety codes.

No matter how safe the design of a building might be or how fire resistant construction materials are, however, unsafe acts on the part of occupants can still result in fires. A high-rise fire in Sao Paulo, Brazil, in 1972 cost many lives in a building that was supposed to be fireproof. The Station Nightclub fire in West Warwick, Rhode Island, on February 20, 2003, killed over 100 people due to the improper use of pyrotechnics and blocked exits.

To reduce the potential for life and property loss, most fire departments have established fire prevention bureaus or divisions. Reducing community risk requires trained individuals to verify that a locality adheres to locally adopted fire and life-safety standards structural safety components in construction and renovations. Staffed by both uniformed and non-uniformed inspectors, these units provide building inspection, perform plans review, issue permits, and perform code enforcement services. The employees of these units should be certified to the levels established by NFPA® 1031, *Standard for Professional Qualifications for Fire Inspector and Plan Examiner.*

NFPA® 1031 also contains the basic requirements for Plan Examiner Levels I and II. While Level II Fire Inspectors can perform plans review, this function is more technical and requires additional training. Depending on the size and organization of the local jurisdiction, plans review may be the responsibility of the building department or fire department. If the building department has the responsibility, the fire department may or may not be involved; and that involvement may be total or limited to a review of the fixed fire suppression, detection, and alarm systems. NFPA® 1031 provides the certification requirements of the Level III Fire Inspector. This level is generally that of the organization's fire marshal and that person may hold the rank of chief officer. Therefore, Level III is excluded from this manual.

Regardless of the type of organization, an inspector must possess a high level of expertise to verify that all fire and life safety hazards are identified and actions are taken to correct them. This level of expertise is gained through training, experience, and certification based on nationally recognized standards. This manual is intended to provide the information necessary for the inspector to certify to the requirements established by the National Fire Protection Association® (NFPA®) in its standard NFPA® 1031, *Standard for Professional Qualifications for Fire Inspector and Plan Examiner* (2014). NFPA® 1031 separates the inspector's duties into three levels. Only Levels I and II are included in this manual.

While duties vary between organizations and between the public and private sectors, a Level I Inspector must be able to perform the following:

- Handle citizen complaints related to fire and life safety.
- Perform fire and life safety inspections of new and existing structures.
- Determine occupancy classification and occupant loads for single-use buildings.
- Inspections of means of egress for code compliance.
- Determine the required type of construction for expansion or remodeling of a structure.
- Determine the operational readiness of existing fixed fire suppression systems, fire detection and alarm systems, and existing portable fire extinguishers.
- Recognize hazardous conditions involving equipment, processes, and operations governed by adopted codes and standards.
- Verify required emergency planning and preparedness measures are in place and have been practiced.
- Verify required emergency access is marked and available for use by fire apparatus.
- Verification of code compliance for incidental storage, handling and use of flammable and combustible liquids and gases, and hazardous materials.
- Recognize a hazardous fire growth potential in an occupancy or space.
- Verify water supply fire flow capacity to determine the ability of water supply systems to provide the required level of protection.
- Participate in legal proceedings involving fire and life safety code issues.

A Level II Inspector is expected to be able to perform additional duties:

- Conduct research.
- Interpret codes.
- Implement policies.
- Testify at legal proceedings.
- Create forms.
- Understand the local permit application process.
- Communicate orally and in writing.
- Understand the local plans review process.
- Apply local fire and life safety codes to complex situations.
- Understand the laws and ordinances that authorize the inspection of occupancies.
- Analyze and recommend modifications to local codes.

- Evaluate fire protection systems.
- Analyze egress elements of a structure.
- Evaluate hazardous conditions.
- Evaluate emergency planning and preparedness procedures.
- Evaluate code compliance in the storage, use, and manufacture of flammable and combustible liquids and gases and hazardous materials.
- Evaluate emergency access to sites.
- Review and evaluate the installation of fire protection systems.
- Identify building construction characteristics.

Purpose and Scope

The *purpose* of **Fire Inspection and Code Enforcement**, 8th Edition, is to provide fire and emergency services personnel and civilian inspectors with basic information necessary to meet the job performance requirements (JPRs) of NFPA® 1031 for Level I and Level II Fire Inspectors. Additional information that exceeds the standard requirements based on the experiences of the IFSTA validation committee and editors has also been included.

The *scope* of the manual addresses the basic duties assigned to a Level I or Level II Fire Inspector. Because NFPA® 1031 does not require the fire inspector to hold a certification as a firefighter or fire officer or have previous fire fighting experience, some basic information normally included in training for Firefighters I and II and Fire Officers I and II is included in this manual.

Book Organization

A close review of the NFPA® 1031 standard indicates that the knowledge, skills, and abilities required for each level are similar. In fact, most of the differences can be attributed to experience on the part of the inspector. For instance, a Level I Fire Inspector must be able to determine the occupant load for a single-use building, while the Level II Fire Inspector must be able to determine the occupant load for a multi-use building. To assist both readers and instructors, the two levels that are addressed in this manual are designated by specific icons.

Terminology

This manual is written with a global, international audience in mind. For this reason, it often uses general descriptive language in place of regional- or agency-specific terminology (often referred to as jargon). Additionally, in order to keep sentences uncluttered and easy to read, the word *state* is used to represent both state and provincial level governments (or their equivalent). This usage is applied to this manual for the purposes of brevity and is not intended to address or show preference for only one nation's method of identifying regional governments within its borders.

The glossary at the end of the manual will assist the reader in understanding words that may not have their roots in the fire and emergency services. The IFSTA **Fire Service Orientation and Terminology** manual is the source for the definitions of fire and emergency services-related terms in this manual.

Resources

To help you increase your knowledge of inspection and life safety issues, this manual contains references to additional materials, particularly in the Appendix materials. Other standards and information sources are included within each chapter.

Additional educational resources to supplement this manual are available from IFSTA and Fire Protection Publications (FPP). Of particular interest to the inspector are IFSTA **Fire Protection, Detection, and Suppression Systems, Essentials of Fire Fighting, Plans Review, and Fire Service Orientation and Terminology** manuals and the FPP *Plans Review* manual.

Key Information

Various types of information in this book are given in shaded boxes marked by symbols or icons. See the following definitions:

Case History

A case history analyzes an event. It can describe its development, action taken, investigation results, and lessons learned.

Safety Alert Icon

Safety alert boxes are used to highlight information that is important for safety reasons. (In the text, the title of safety alerts will change to reflect the content.)

Asbestos Components

Asbestos is no longer used as a building material because it is known to cause respiratory ailments. However, it was commonly used in the U.S. from the 1930s until the 1970s and is still found in many older buildings. Asbestos siding may be covered with another siding material, or, in some cases, still be exposed.

The inspector needs to be aware of the health hazards associated with asbestos, and take care not to disturb asbestos materials when conducting inspections. Exercise caution while in proximity to components that may potentially contain asbestos. Any activity near or affecting asbestos products must include proper PPE and decontamination procedures.

In addition to respiratory hazards, asbestos fibers are also difficult to remove from contaminated clothing and other resources. The AHJ may require contaminated resources to be discarded after use. For additional information, refer to NFPA® 1851, *Standard of Selection, Care and Maintenance of Protective Ensembles, Structural Firefighting and Proximity Firefighting.*

Information Box Icon

Information boxes give facts that are complete in themselves but belong with the text discussion. It is information that needs more emphasis or separation. In the text, the title of information boxes will change to reflect the content.

Automatic Notification

If fire inspections are an integral part of the procedure for obtaining business licenses and permits, inspectors may automatically receive notification when business and commercial changes occur in the community.

A **key term** is designed to emphasize key concepts, technical terms, or ideas that the inspector needs to know. They are listed at the beginning of each chapter and the definition is placed in the margin for easy reference. An example of a key term is:

Code — A collection of rules and regulations that has been enacted by law in a particular jurisdiction. Codes typically address a single subject area; examples include a mechanical, electrical, building, or fire code.

Three key signal words are found in the book: **WARNING**, **CAUTION**, and **NOTE**. Definitions and examples of each are as follows:

- **WARNING** indicates information that could result in death or serious injury to anyone on the scene. See the following example:

WARNING!

During inspections of facilities where toxic materials are being manufactured, stored, or shipped, an inspector must have appropriate personal protective equipment (PPE) and training in its use when there is any possibility of being exposed to these products.

- **CAUTION** indicates important information or data that inspectors need to be aware of in order to perform their duties safely. See the following example:

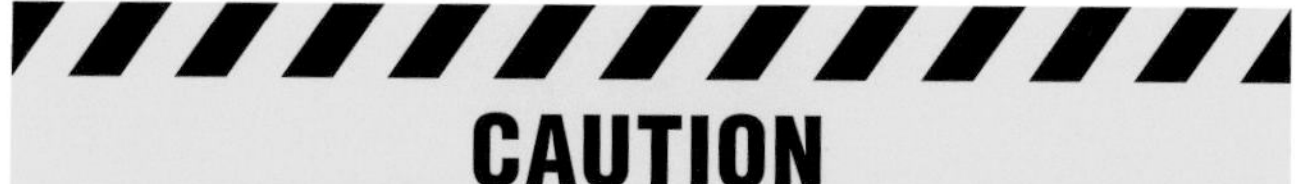

CAUTION

The U.S. military ships some hazardous materials and chemicals by common carrier. In these instances, materials are not required to be marked with U.S. Department of Transportation (DOT) or Transport Canada (TC) markings.

- **NOTE** indicates important operational information that helps explain why a particular recommendation is given or describes optional methods for certain procedures. See the following example:

NOTE: Unless inspectors are specifically given the authority, they cannot apply current building code requirements to existing structures.

Metric Conversions

Throughout this manual, U.S. units of measure are converted to metric units for the convenience of our international readers. Be advised that we use the Canadian metric system. It is very similar to the Standard International system, but may have some variation.

We adhere to the following guidelines for metric conversions in this manual:

- Metric conversions are approximated unless the number is used in mathematical equations.
- Centimeters are not used because they are not part of the Canadian metric standard.
- Exact conversions are used when an exact number is necessary, such as in construction measurements or hydraulic calculations.
- Set values, such as hose diameter, ladder length, and nozzle size, use their Canadian counterpart naming conventions and are not mathematically calculated. For example, 1½-inch hose is referred to as 38 mm hose.

The following two tables provide detailed information on IFSTA's conversion conventions. The first table includes examples of our conversion factors for a number of measurements used in the fire service. The second shows examples of exact conversions beside the approximated measurements you will see in this manual.

U.S. to Canadian Measurement Conversion

Measurements	Customary (U.S.)	Metric (Canada)	Conversion Factor
Length/Distance	Inch (in) Foot (ft) [3 or less feet] Foot (ft) [3 or more feet] Mile (mi)	Millimeter (mm) Millimeter (mm) Meter (m) Kilometer (km)	1 in = 25 mm 1 ft = 300 mm 1 ft = 0.3 m 1 mi = 1.6 km
Area	Square Foot (ft^2) Square Mile (mi^2)	Square Meter (m^2) Square Kilometer (km^2)	1 ft^2 = 0.09 m^2 1 mi^2 = 2.6 km^2
Mass/Weight	Dry Ounce (oz) Pound (lb) Ton (T)	gram Kilogram (kg) Ton (T)	1 oz = 28 g 1 lb = 0.5 kg 1 T = 0.9 T
Volume	Cubic Foot (ft^3) Fluid Ounce (fl oz) Quart (qt) Gallon (gal)	Cubic Meter (m^3) Milliliter (mL) Liter (L) Liter (L)	1 ft^3 = 0.03 m^3 1 fl oz = 30 mL 1 qt = 1 L 1 gal = 4 L
Flow	Gallons per Minute (gpm) Cubic Foot per Minute (ft^3/min)	Liters per Minute (L/min) Cubic Meter per Minute (m^3/min)	1 gpm = 4 L/min 1 ft^3/min = 0.03 m^3/min
Flow per Area	Gallons per Minute per Square Foot (gpm/ft^2)	Liters per Square Meters Minute (L/(m^2.min))	1 gpm/ft^2 = 40 L/(m^2.min)
Pressure	Pounds per Square Inch (psi) Pounds per Square Foot (psf) Inches of Mercury (in Hg)	Kilopascal (kPa) Kilopascal (kPa) Kilopascal (kPa)	1 psi = 7 kPa 1 psf = .05 kPa 1 in Hg = 3.4 kPa
Speed/Velocity	Miles per Hour (mph) Feet per Second (ft/sec)	Kilometers per Hour (km/h) Meter per Second (m/s)	1 mph = 1.6 km/h 1 ft/sec = 0.3 m/s
Heat	British Thermal Unit (Btu)	Kilojoule (kJ)	1 Btu = 1 kJ
Heat Flow	British Thermal Unit per Minute (BTU/min)	watt (W)	1 Btu/min = 18 W
Density	Pound per Cubic Foot (lb/ft^3)	Kilogram per Cubic Meter (kg/m^3)	1 lb/ft^3 = 16 kg/m^3
Force	Pound-Force (lbf)	Newton (N)	1 lbf = 0.5 N
Torque	Pound-Force Foot (lbf ft)	Newton Meter (N.m)	1 lbf ft = 1.4 N.m
Dynamic Viscosity	Pound per Foot-Second (lb/ft.s)	Pascal Second (Pa.s)	1 lb/ft.s = 1.5 Pa.s
Surface Tension	Pound per Foot (lb/ft)	Newton per Meter (N/m)	1 lb/ft = 15 N/m

Conversion and Approximation Examples

Measurement	U.S. Unit	Conversion Factor	Exact S.I. Unit	Rounded S.I. Unit
Length/Distance	10 in	1 in = 25 mm	250 mm	250 mm
	25 in	1 in = 25 mm	625 mm	625 mm
	2 ft	1 in = 25 mm	600 mm	600 mm
	17 ft	1 ft = 0.3 m	5.1 m	5 m
	3 mi	1 mi = 1.6 km	4.8 km	5 km
	10 mi	1 mi = 1.6 km	16 km	16 km
Area	36 ft^2	1 ft^2 = 0.09 m^2	3.24 m^2	3 m^2
	300 ft^2	1 ft^2 = 0.09 m^2	27 m^2	30 m^2
	5 mi^2	1 mi^2 = 2.6 km^2	13 km^2	13 km^2
	14 mi^2	1 mi^2 = 2.6 km^2	36.4 km^2	35 km^2
Mass/Weight	16 oz	1 oz = 28 g	448 g	450 g
	20 oz	1 oz = 28 g	560 g	560 g
	3.75 lb	1 lb = 0.5 kg	1.875 kg	2 kg
	2,000 lb	1 lb = 0.5 kg	1 000 kg	1 000 kg
	1 T	1 T = 0.9 T	900 kg	900 kg
	2.5 T	1 T = 0.9 T	2.25 T	2 T
Volume	55 ft^3	1 ft^3 = 0.03 m^3	1.65 m^3	1.5 m^3
	2,000 ft^3	1 ft^3 = 0.03 m^3	60 m^3	60 m^3
	8 fl oz	1 fl oz = 30 mL	240 mL	240 mL
	20 fl oz	1 fl oz = 30 mL	600 mL	600 mL
	10 qt	1 qt = 1 L	10 L	10 L
	22 gal	1 gal = 4 L	88 L	90 L
	500 gal	1 gal = 4 L	2 000 L	2 000 L
Flow	100 gpm	1 gpm = 4 L/min	400 L/min	400 L/min
	500 gpm	1 gpm = 4 L/min	2 000 L/min	2 000 L/min
	16 ft^3/min	1 ft^3/min = 0.03 m^3/min	0.48 m^3/min	0.5 m^3/min
	200 ft^3/min	1 ft^3/min = 0.03 m^3/min	6 m^3/min	6 m^3/min
Flow per Area	50 gpm/ft^2	1 gpm/ft^2 = 40 L/(m^2.min)	2 000 L/(m^2.min)	2 000 L/(m^2.min)
	326 gpm/ft^2	1 gpm/ft^2 = 40 L/(m^2.min)	13 040 L/(m^2.min)	13 000L/(m^2.min)
Pressure	100 psi	1 psi = 7 kPa	700 kPa	700 kPa
	175 psi	1 psi = 7 kPa	1225 kPa	1 200 kPa
	526 psf	1 psf = 0.05 kPa	26.3 kPa	25 kPa
	12,000 psf	1 psf = 0.05 kPa	600 kPa	600 kPa
	5 psi in Hg	1 psi = 3.4 kPa	17 kPa	17 kPa
	20 psi in Hg	1 psi = 3.4 kPa	68 kPa	70 kPa
Speed/Velocity	20 mph	1 mph = 1.6 km/h	32 km/h	30 km/h
	35 mph	1 mph = 1.6 km/h	56 km/h	55 km/h
	10 ft/sec	1 ft/sec = 0.3 m/s	3 m/s	3 m/s
	50 ft/sec	1 ft/sec = 0.3 m/s	15 m/s	15 m/s
Heat	1200 Btu	1 Btu = 1 kJ	1 200 kJ	1 200 kJ
Heat Flow	5 BTU/min	1 Btu/min = 18 W	90 W	90 W
	400 BTU/min	1 Btu/min = 18 W	7 200 W	7 200 W
Density	5 lb/ft^3	1 lb/ft^3 = 16 kg/m^3	80 kg/m^3	80 kg/m^3
	48 lb/ft^3	1 lb/ft^3 = 16 kg/m^3	768 kg/m^3	770 kg/m^3
Force	10 lbf	1 lbf = 0.5 N	5 N	5 N
	1,500 lbf	1 lbf = 0.5 N	750 N	750 N
Torque	100	1 lbf ft = 1.4 N.m	140 N.m	140 N.m
	500	1 lbf ft = 1.4 N.m	700 N.m	700 N.m
Dynamic Viscosity	20 lb/ft.s	1 lb/ft.s = 1.5 Pa.s	30 Pa.s	30 Pa.s
	35 lb/ft.s	1 lb/ft.s = 1.5 Pa.s	52.5 Pa.s	50 Pa.s
Surface Tension	6.5 lb/ft	1 lb/ft = 15 N/m	97.5 N/m	100 N/m
	10 lb/ft	1 lb/ft = 15 N/m	150 N/m	150 N/m

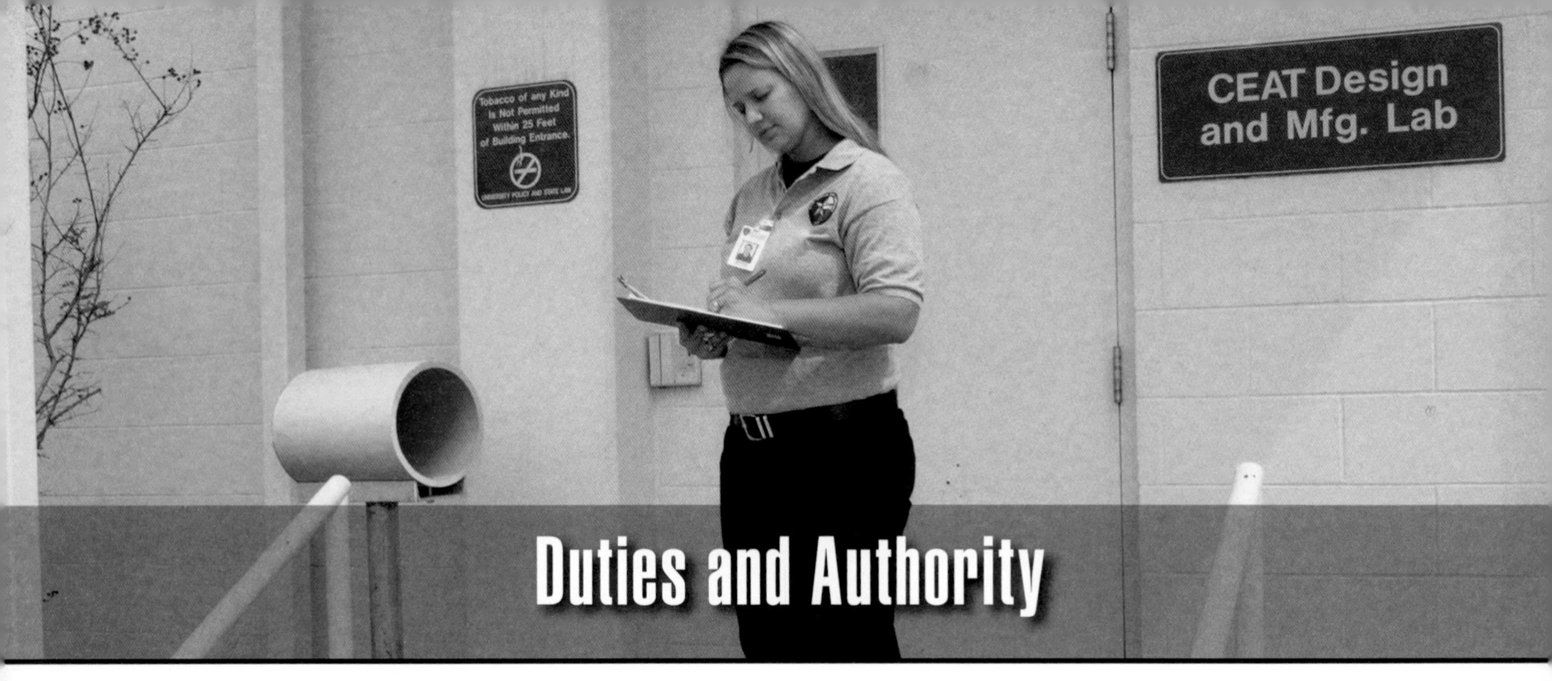

Duties and Authority

Chapter Contents

Case History **13**
I **Duties** **14**
Public Organizations 14
Third-Party Inspectors 16
Private Fire Safety Inspectors 16
Training and Certification 17
Professional Development 17
Inspection Categories 18
Legal Guidelines for Inspections **19**
Compliance Procedures 19
Civil Proceedings 20
Criminal Proceedings 20
Court Proceedings 22
Testimony 23
II **Authority** **24**
Enabling Legislation 24
Types of Laws 25
Legal Status of Inspectors 27
Liability 29
Outside Technical Assistance 33
Recommend Policies and Modifications ... **34**
Developing Policies and Procedures 34
Maintaining and Revising Policies 35
Creating Forms and Job Aids 36
Chapter Summary **36**
Review Questions **37**

chapter 1

Key Terms

Administrative Warrant 31
Authority Having Jurisdiction (AHJ) 17
Due Process 19
Due Process Clause 19
Enabling Legislation 14
Indemnify 32
Indict 21
Inspector 14
Liability 29
Malfeasance 33
Misfeasance 33
Mitigate 16
Negligence 33
Nonfeasance 33
Ordinance 14
Police Power 28
Public-Duty Doctrine 33
Regulations 14
Right of Entry 29
Sanction 20
Search Warrant 31
Special Duty 33
Statute or Statutory Law 14

NFPA® Job Performance Requirements

This chapter provides information that addresses the following job performance requirements of NFPA® 1031, *Standard for Professional Qualifications for Fire Inspector and Plan Examiner* (2014).

Fire Inspector I	Fire Inspector II
4.2.6	5.2.4
	5.2.5

Duties and Authority

Learning Objectives

After reading this chapter, students will be able to:

Inspector I

1. Describe the duties of an inspector.
2. Describe the legal guidelines for inspections. (4.2.6)

Inspector II

1. Explain the types of laws, legal status, and liabilities that impact inspections. (5.2.4)
2. Recognize an inspector's role in developing, maintaining, and revising policies or procedures and forms. (5.2.4, 5.2.5)

Chapter 1
Duties and Authority

Case History

Independence, Wis. – A fire in the Hotel Tap in 2009 resulted in the deaths of three tenants. When the families of the dead sued the building's owners for negligence, an investigation revealed that the hotel had failed three previous fire inspections. Each inspection had cited improperly maintained fire escapes and lack of fire extinguishers, exit lights, and smoke alarms.

The building's owner stated that deficiencies had been corrected at the time of the fire, but the three inspection reports noting deficiencies were the only ones on file at City Hall, where copies of all inspections are kept. Fire inspectors are also obligated to report violations to the city attorney, but the city attorney said he was never notified of the failed inspections.

Several fire extinguishers were found in the building's rubble, but investigators were unable to determine whether additional extinguishers or alarms had been present due to severe fire damage. The lack of knowledge about the safety features in place at the time of the fire hampered the police investigation into the blaze.

Lesson Learned: It is critical that inspectors make sure that reports are accurate, complete, and follow regulations for notifying other city departments. Follow-up inspections to verify corrections are vital.

The statement *a fire that does not occur is the one that is most easily controlled* accurately describes the ultimate goal of any fire and life safety inspection program. By identifying hazards or potential hazards and causing them to be corrected, the majority of uncontrolled fires can be avoided.

This chapter provides the inspector with a general description of the duties, legal guidelines, and authority of an inspector. It also describes recommendations for policies and procedures and creating forms and job aids.

The authority having jurisdiction (AHJ) is responsible for training and certifying inspection and code enforcement personnel to perform the functions assigned to them. The inspector is responsible for continuing professional development.

NOTE: Chapter 2, Standards, Codes, and Permits, provides an overview of the standards and codes that the inspector will enforce.

I

Duties

To accomplish the goal of protecting lives and property from uncontrolled fires and other hazards, all levels of government have created fire inspection and code enforcement organizations with the authority to manage these programs. Those persons responsible for inspection programs may be members of the fire department, building department, code enforcement department, or some other agency. Other agencies can include insurance underwriters, third-party code compliance certification, and so on. These individuals may be referred to as *fire and life safety inspectors, code enforcement officers*, or, as in this manual, simply **inspectors**.

Inspector — Person who is trained to perform fire prevention and life safety inspections of all types of new construction and existing occupancies; also called *Code Enforcement Officer* and *Fire and Life Safety Inspector*.

Statute or Statutory Law — Federal or state/provincial legislative act that becomes law; prescribes conduct, defines crimes, and promotes public good and welfare.

Regulations — Rules or directives of administrative agencies that have authorization to issue them.

Enabling Legislation — Legislation that gives appropriate officials the authority to implement or enforce the law.

Ordinance — Local or municipal law that applies to persons and things of the local jurisdiction; a local agency act that has the force of a statute; different from law that is enacted by federal or state/provincial legislatures.

To perform their jobs properly, inspectors must understand their assigned duties, know the applicable building and fire code provisions and how they relate to one another, and be aware of the authority they have for conducting their duties. Inspectors must become familiar with the **statutes**, codes, **regulations**, and permitting processes that are in effect in a particular jurisdiction, including the following:

- **Enabling legislation** that creates the inspection position or that designates individuals to perform fire and life safety inspections.
- State/provincial or local statutes that provide the legal basis for the inspections and establish the minimum requirements to perform fire prevention and related activities.
- State/provincial and local laws, codes, **ordinances**, and statutes that detail various fire and life safety requirements and establish an inspector's duties and responsibilities regarding the mitigation of any hazards or violations discovered.
- Statutes that set the limits of authority that may be part of state/provincial and local enabling legislation.
- Ways in which statutes, regulations, and ordinances can be altered or amended.

An inspector's duties will vary based on the inspection organization's type and size. While duties vary between organizations and between the public and private sectors, a Level I Inspector must be able to perform the following duties:

- Handle citizen complaints related to fire and life safety.
- Interpret and apply adopted codes and standards.
- Perform fire and life safety inspections of new and existing structures **(Figure 1.1)**.
- Determine occupancy loads for single-use buildings.
- Participate in legal proceedings involving fire and life safety code issues.
- Verify water supply fire flow capacity to determine the ability of water supply systems to provide the required level of protection.

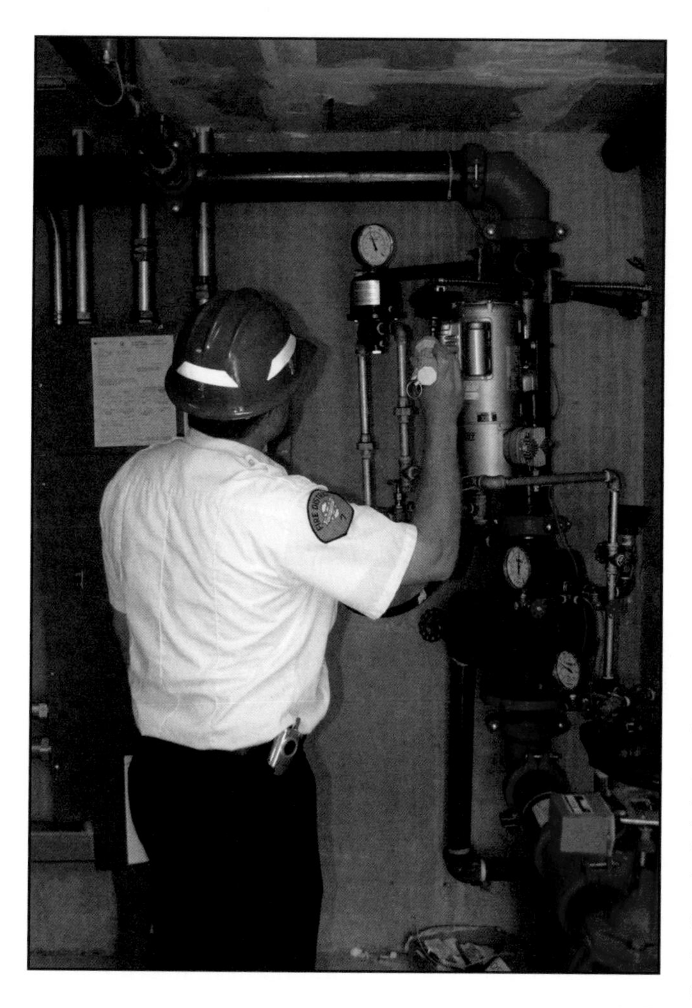

Figure 1.1 An inspector evaluating part of a sprinkler system in a new structure to see that it conforms to the locally adopted codes.

An inspector may be an employee of a public government agency, a private company, or an insurance company. This section provides a brief overview of both public and private inspection organizations; training, certification, and professional development; and categories of inspections.

Public Organizations

Within a public organization, fire and life safety inspection programs may be located in the fire, building, or code enforcement departments. The size and complexity of the local government may determine the location of the program.

Fire Department

Staffing for fire prevention divisions may include non-sworn personnel, sworn personnel, or a combination of both. Depending on the department's size, inspectors may be non-sworn employees who have been hired specifically as inspectors or plans examiners. In some departments, sworn firefighters or fire officers may be assigned to the prevention division. When the authority for inspections and code enforcement is not assigned to the fire department, the fire department should develop a strong working relationship with the department that performs the inspections.

Building Department

In some jurisdictions, fire inspection and code enforcement responsibility may be assigned to the building department. In this case, the inspectors are usually civilian employees. The building department is generally responsible for:

- Reviewing and approving all new construction and alterations to existing structures
- Conducting plans reviews **(Figure 1.2, p. 16)**

Figure 1.2 The building department reviews construction plans and documents and issues permits for all new construction in the jurisdiction.

- Issuing permits related to buildings and their use
- Performing field inspections to verify that approved plans are followed in the construction and alteration process

Code Enforcement Department

Some jurisdictions may have a separate code enforcement department. In addition to conducting fire inspections and in some cases responding to complaints, the code enforcement department may have additional responsibilities:

- Parking
- Weed abatement
- Vacant structures
- Water usage
- Pyrotechnics displays
- Tents

Third-Party Inspectors

The role of inspector is not limited to the public sector. Some jurisdictions use a private firm to perform plan reviews and conduct inspections. These inspectors are employed by the third-party inspection service to ensure that the minimum fire and life safety requirements are satisfied within the jurisdiction.

Mitigate — To make less harsh or intense; to alleviate.

Private Fire Safety Inspectors

Inspectors employed by an insurance or underwriting company function as risk assessors. These inspectors evaluate processes and facilities that are insured or underwritten by the insurance company in an attempt to identify hazards. This evaluative process is designed to **mitigate** risks before any loss occurs that affects both the insured and the insurance company.

The insurance inspector's scope and power to make improvements or enforce corrections varies widely. The insurance company's policies and procedures dictate the working relationship between the insured and the inspector. The insurance inspector must always follow appropriate insurance company policies and procedures for handling noted hazards.

Private inspectors should have good working relationships with the public sector inspectors within the jurisdiction because it may become necessary for them to work together. Private inspectors should also regularly communicate about facility and process changes that will affect fire and life safety at their facilities.

Authority Having Jurisdiction (AHJ) — Term used in codes and standards to identify the legal entity, such as a building or fire official, who has the statutory authority to enforce a code and to approve or require equipment. In the insurance industry it may refer to an insurance rating bureau or an insurance company inspection department.

Training and Certification

The **authority having jurisdiction (AHJ)** establishes the minimum level of training for inspectors and code officials and defines the certification requirements. The AHJ may be the national agency, state/province, the county, or the local municipality. Professional qualifications may be based on NFPA® 1031, *Standard for Professional Qualifications for Fire Inspector and Plan Examiner,* or individually developed standards.

Code development organizations, such as the International Code Council® (ICC®) and the National Fire Protection Association® (NFPA®), provide courses and training materials for all levels and types of code enforcement and interpretation. The U.S. National Fire Academy (NFA) as well as state/provincial and regional fire academies also provides courses in inspection, plans review, and code enforcement.

The AHJ, local colleges, or technical schools may provide training and certification testing. Because NFPA® 1031 does not require that inspectors meet Fire Fighter I, Fire Fighter II, or Fire Officer certification, initial inspection training may include information normally provided in these programs. The AHJ may also require continuing education hours and recertification on a two-year or greater cycle.

Professional Development

Once trained and certified, the inspector may pursue a professional development path established by the AHJ. In addition to NFPA® 1031 Fire Inspector I, II, and III and Plans Examiners I and II levels, there are other levels of advancement and certification:

- Residential inspection
- Commercial inspection **(Figure 1.3)**
- Code enforcement
- Special inspector
- General inspection

Figure 1.3 An inspector verifies that an extinguisher system is installed correctly in a commercial kitchen. *Courtesy of Sand Springs (OK) Fire Department.*

- Code official or building official
- Master code professional

Course topics may include a variety of information:

- Report writing
- Verbal communication skills
- Ethics
- Principles of code enforcement
- Personnel issues
- Public speaking
- Stress management
- Case development
- Evidentiary procedures
- Courtroom presentations

The programs and workshops provided by professional organizations are a final continuing education source. For example, the American Code Enforcement Association (ACEA) sponsors annual conferences and workshops that provide attendees with continuing education credits that many AHJs accept toward annual requirements.

Inspection Categories

An inspector may be authorized to perform several categories of fire and life safety inspections:

- **Routine** — Performed on a set basis, usually involving occupancy classifications, such as places of assembly, education, and health care.
- **Issuance of a permit** — Conducted when the owner/occupant is required to obtain a permit for a special event or use of the occupancy. Examples of special events are the limited-term use of a circus-type tent or use of pyrotechnics during a performance or show.
- **Response to a complaint** — Performed when a complaint has been filed against a business or occupancy, such as overcrowding or blocked exits.
- **Imminent hazard** — Conducted when it becomes obvious that an occupancy or process poses a hazard to life and property, such as blocked egress or inaccessibility to the means of egress.
- **New construction** — Conducted when a new structure is built, an addition is made to an existing structure, or an existing structure is altered in such a way that code compliance must be determined **(Figure 1.4)**. The demolition of a structure is also included in this type of inspection.
- **Change in occupancy** — Performed when a structure's use or occupancy classification changes and code compliance must be reviewed. An example of such a change would be a warehouse converted into a movie theater or other place of assembly.
- **Owner/occupant request** — Occasionally the owner/occupant may request an inspection. This inspection may be required by another government agency or an insurance company.

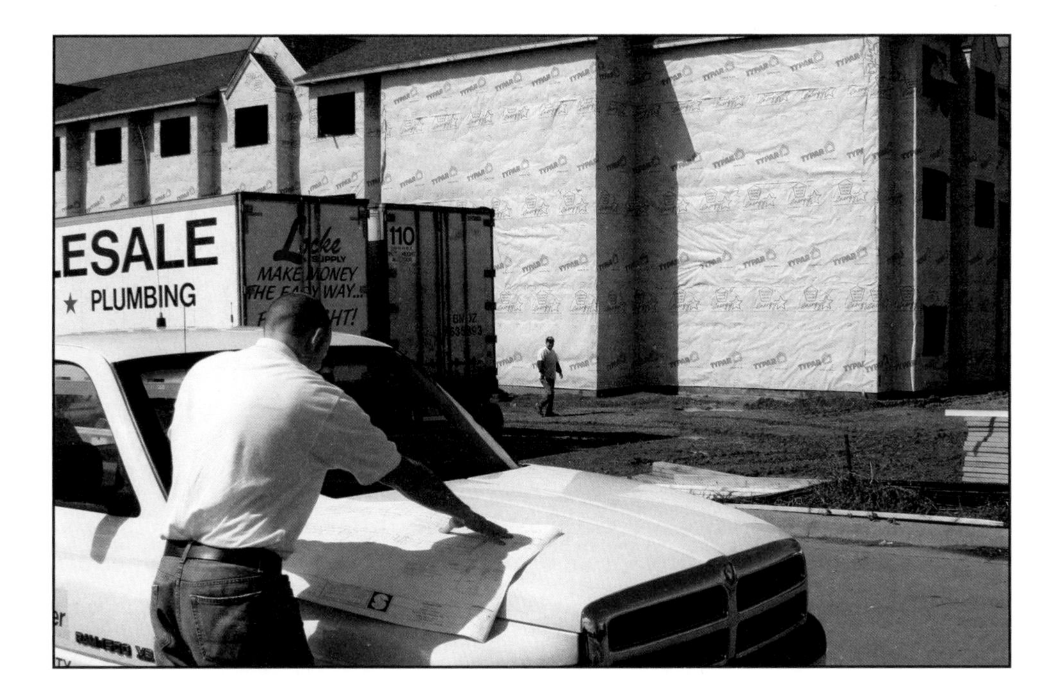

Figure 1.4 Inspectors are responsible for field inspections of new construction. Knowledge of both construction techniques and building codes is essential to this task.

As mentioned previously, the fire department may share inspection responsibilities with the building or code enforcement department. In cases of joint responsibility, the jurisdiction must establish clear inspection guidelines or Memorandums of Understanding (MOUs) regarding procedures.

Legal Guidelines for Inspections

In assessing the statutes of the local jurisdiction that dictate code compliance, the inspector should know the answers to the following questions:

- Is noncompliance with the code a violation of criminal or civil law?
- What process has been employed to attempt to achieve compliance?
- Have accurate records of all correspondence, communications, and personal contact been kept in accordance with local policy and law?
- Will the possible penalties induce compliance?

Inspectors must understand the steps in prosecuting a case for a fire code violation. Although compliance is the primary goal of all fire prevention code activities, it is occasionally necessary to use stronger measures to implement a fire code's requirements. When exercising this enforcement power, the inspector must ensure that the legal rights of the accused are protected. This concept is known as **due process**.

Due Process — Conduct of legal proceedings according to established rules and principles for the protection and enforcement of private rights, including notice and the right to a fair hearing before a tribunal with the power to decide the case; *also called due process of law or due course of law.*

Two **due process clauses** are in the U.S. Constitution: one in the 5th Amendment applying to the federal government and one in the 14th Amendment applying to the states. The 5th Amendment's Due Process Clause also applies to the states under the incorporation doctrine of the U.S. Constitution.

Due Process Clause — Constitutional provision that prohibits the government from unfairly or arbitrarily depriving a person of life, liberty, or property.

Compliance Procedures

Official actions to ensure code compliance vary from jurisdiction to jurisdiction and are dictated by local codes. Inspectors must understand the procedures involved and thoroughly and accurately document all actions taken. These actions are generally enacted in the following order:

Sanction — Notice or punishment attached to a violation for the purpose of enforcing a law or regulation.

1. **Notification** – Written inspection form provided to the responsible party after an initial inspection describing the fire code infraction(s).
2. **Follow-up inspection** – Conducted after a predetermined time allowed by a local code or ordinance.
3. **Sanction** – Issued by most jurisdictions if the violation is not corrected within a certain time. Sanctions may be in the form of the following items:
   — Citations
   — Complaints
   — Fines
   — Court summonses
   — Stop work orders

   NOTE: The procedures for issuing sanctions differ; however, each jurisdiction should have a written procedure that details the process. **Appendix B** contains an example of a citation program.

4. **Prosecution** – Evidence presented by the inspector or local prosecutor supporting the charges against the accused for fire code violations. A magistrate then weighs the evidence as well as defense arguments before rendering a decision.

   It is usually best to seek advice from legal counsel before taking action if there is a question of authority or meaning of the code.

Occasionally, it is necessary for the jurisdiction to prosecute a fire code violation in court. Prosecution usually occurs when the property owner/occupant has failed to correct an outstanding fire code violation. As a result, the inspector and any other inspection personnel who were involved during the inspection and subsequent action may find themselves involved in legal proceedings. Depending upon the local justice system, inspectors may be considered either witnesses or courtroom advisors to the prosecution.

NOTE: The inspector's role in case prosecution and court testimony is covered in the Inspector II portion of this chapter.

Civil Proceedings

A civil proceeding may be the result of a noncriminal violation but usually involves one party filing suit against another party. A suit may allege a breach of contract or may be for any other reason.

In some cases, the judge will determine the outcome of the case. Both parties must agree to this proceeding, which is known as a bench trial. Otherwise, the decision will be determined by a jury in a jury trial. During the trial phase, the judge makes determinations based on law as to what evidence is presented to the jury.

Criminal Proceedings

Criminal proceedings vary according to jurisdiction, and the prosecuting attorney's office should be consulted for complete information regarding the local process. Criminal proceedings are most often initiated as a result of action taken in the form of a "probable cause" arrest, service of an arrest warrant, or grand jury indictment. This action is the result of an investigation establishing that a crime has occurred and that sufficient probable cause is present to believe that the arrested or **indicted** party committed the crime. The potential punishment a party could receive if found guilty (responsible) most often dictates whether the crime is a misdemeanor or a felony and how the case proceeds through the justice system.

Indict — To present a formal, written accusation charging a defendant with a crime.

Types of Criminal Violations

In most jurisdictions, there are three types of criminal violations with which the inspector should be familiar:

- **Summary offenses** — The lowest form of offense in most legal systems. These are minor infractions of laws or local ordinances. The person who commits a summary offense is typically given a written citation of the violation when the violation occurs. The offender then has the choice of paying the fine that accompanies the citation or challenging the citation before a judge. The most common forms of summary offenses include speeding or parking violations and similar offenses. In some jurisdictions, unauthorized controlled burns (trash, leaves, etc.) are considered summary offenses, and the inspector may be involved in writing the citations, depending on local laws.
- **Misdemeanors** — Generally viewed as a "lesser" crime that is punishable by a fine or (generally) a term of less than two years in jail or prison. In the event of a misdemeanor, the arrestee is given an initial appearance, during which he or she is allowed to plead guilty or not guilty. If the plea is guilty, a sentence may be imposed or levied in the form of a fine, incarceration, or both. If the subject pleads not guilty, a trial before a judge or jury is set.
- **Felonies** — Generally viewed as a "serious" crime and is punishable by a fine, prison, and/or death, depending on the severity of the felony and the jurisdiction. Like the misdemeanor, the arrestee is given an initial appearance and the option of pleading guilty or not guilty. If the plea is guilty, a sentence may be imposed at that time or after a study outlining the circumstances and the arrestee's past actions has been completed. If the plea is not guilty, a preliminary hearing (also known as a "probable cause hearing") date is set. A person indicted by a grand jury and pleading not guilty at the initial appearance is immediately "bound over for trial" and the preliminary hearing stage is bypassed.

Sample Case and Prosecution

As an example, a city may have an inspection schedule that is formulated by statute. The inspector examines a business, finds serious deficiencies, and issues an inspection report. This report is sent as certified mail and also includes an Abatement Order. Within the report is contained a 30-day time limit for correction or a plan of action. The property owner has the right to appeal the inspector's findings.

After the time limit has passed and no correction has been made, the Fire Marshal or inspector then refers the case to Housing Court and an arrest warrant is made. From that point, the case is taken over by Housing Court.

Evidentiary Requirements

An important part of an inspector's job is to keep careful, accurate notes about the inspection process. Accurate record-keeping is not only important for an efficient program, these notes and photographs may someday be needed as evidence in a trial.

NOTE: Chapter 16, Inspection Procedures, contains more information about keeping records.

Whether the inspector records information digitally or on paper, inspection visits must include notes about conditions found on the premises and document any violations found. These notes, plus photographs of violations, provide a record for the inspection department and the owner. After the visit is complete, the inspector can print a copy of the inspection form for the owner to sign or e-mail it as a record of the inspector's visit.

Depending on their severity, violations are given several time frames for abatement, such as 30 days, 60 days, or immediately for severe hazards. Inspectors must know the legal requirements for entering premises. Hazards that are in plain sight during an inspection may be recorded, but evidence that is obtained without a scheduled visit or warrant may be inadmissible in court if the matter comes to a trial.

In all cases, the inspector is considered to be part of the prosecution on the side of the state or AHJ. In essence, the inspector is presenting the fire code as it is applied in that locality. In this role, the inspector assists the prosecuting attorney with information about fire ordinances, technical terms, and facts of the case. The inspector must, however, present unbiased testimony, always based upon fact.

Court Proceedings

Inspectors should be familiar with the various types of court proceedings with which they may become involved. This section highlights some common proceedings.

Preliminary Hearings

A preliminary hearing takes place before a judge. During this hearing, the prosecutor's office is required to establish that a crime has been committed and that it is more likely than not that the arrested party has committed the offense specified. If sufficient evidence is presented, the judge may "find in favor" of the prosecutor and "remand or bind the defendant (arrestee) over for trial." For example, during a preliminary hearing regarding an arson prosecution, most jurisdictions will require a fire origin and cause expert to establish that the fire in question was a result of a crime (typically intentionally and maliciously set) prior to "a finding of probable cause."

Felony Trial

In the event of a felony trial, the defendant (person charged with the crime) has the constitutional right to a trial before a jury. The defendant sometimes waives this right, and the trial takes place with only the judge (bench trial) making the determination as to the defendant's guilt or innocence. The burden of proof is higher at the criminal trial stage. The prosecutor's office must present evidence sufficient to convince the judge/jury that the defendant is guilty "beyond a reasonable doubt." A reasonable doubt standard does not require a 100 percent certainty; however, it does require a level of proof beyond that which a "reasonable" person would have after hearing all the evidence.

Because the prosecution has the burden of proving that a crime was committed and that the person standing trial was the one responsible for that crime, they present evidence first. The defendant and defense attorneys then present evidence in support of their not guilty position after the prosecutor finishes or "rests." Witnesses are often called upon to testify to facts indicating that a crime has taken place or that a person is responsible for the crime.

Testimony

Testimony is simply an affirmation or declaration made under oath, typically in a court of law. All verbal information provided in a trial is in the form of sworn or affirmed testimony **(Figure 1.5)**. It is generally divided into two categories: fact witness and expert witness.

Figure 1.5 During court testimony, inspectors must remain impartial and be familiar with the relevant facts of the case.

In order to preserve the case and be as helpful as possible, the inspector should note the following suggestions regarding courtroom procedure and behavior:

- Provide evidence that a follow-up inspection was conducted in the facility where the infraction is alleged to have occurred.
- Review all files and notes regarding the infractions with the prosecutor before entering the courtroom.
- Confine testimony to the facts of the case. Avoid hearsay information from a third party, biased opinions, or irrelevant statements.
- Remain impartial and do not give the impression that you have a personal opinion or prejudice regarding the trial.
- Limit the information provided to only that which is necessary to answer a given question. Never volunteer information or expand an answer to include information that has not been requested.
- Make responses as brief as possible while conveying all the information necessary. If the answer calls for a simple yes or no answer, answer yes or no. If it is necessary to explain an answer, request the court's permission to do so.
- Remember that if a question is asked beyond an inspector's ability to answer, the inspector should simply state that he or she is unable to answer that question.
- Verify that all physical evidence, exhibits, photographs, notes, reference materials, and other materials pertinent to the case have been reviewed by the prosecutor and have been brought to court.
- Answer all questions factually and truthfully.
- Anticipate personal attacks or challenges to credibility.
- Never become argumentative or unprofessional on the witness stand.

Fact Witness

A fact witness is a witness who answers questions posed by an attorney regarding what he or she saw, heard, touched, smelled, or tasted. The testimony presents information to the judge or jury regarding "facts" without the witness being allowed to provide his or her own interpretation of those facts.

An example of fact testimony (also known as *lay testimony*) would be a witness's description of seeing a structure built without a permit and with no regard for building and fire codes. Typically, that witness would not be allowed to provide further testimony; that is, which codes or code sections were violated, etc. This further type of testimony would be classified as *expert testimony*.

Expert Testimony

In contrast to the fact witness, an expert witness is allowed, at the court's discretion, to provide opinion testimony. An expert witness is generally defined as a person with sufficient skill, knowledge, or experience in a given field so as to be capable of drawing inferences or reaching conclusions or opinions that an average person would not be competent to reach. The evidence upon which an inspector bases an expert opinion must be relevant and reliable, thereby allowing it to be admissible.

Authority

An inspector is given the authority to perform his or her duties based on the laws adopted by the AHJ or regulations established by the private sector organization. In the case of the private sector, the regulations may be written to meet the legal requirements of the applicable jurisdiction. For instance, a private organization may require its inspectors to enforce Occupational Safety and Health Administration (OSHA) requirements for the use or storage of flammable liquids. Therefore, all inspectors must understand the laws upon which their authority is based.

The sections that follow describe how inspection duties are affected by enabling legislation, types of laws, and legal status of inspectors. Securing outside assistance for special situation is also described.

Enabling Legislation

One of the most important laws to the inspector is the enabling legislation to establish the organization that performs fire inspections. Enabling legislation is the foundation upon which the department and local fire and life safety codes are built. This legislation defines the scope of the department's authority.

The scope of powers granted to inspection organizations varies widely among jurisdictions. Some AHJs have almost unlimited power to develop, enact, and enforce fire codes. Others may be required to enforce a code established by the state/province, county, or regional government. Some organizations may not have any power to modify or amend the provisions of the fire code. These codes are usually the result of a cooperative effort among community stakeholders, the AHJ, and the legislative body. Legal counsel should provide information on the inspector's authority and liability as defined by the law.

Inspectors must not only be familiar with the AHJ's enabling legislation, they must also have a clear understanding of the building and fire codes they are responsible for enforcing. The fire marshal or building code official should provide building or fire code information to the inspector. This information can be in the form of an operating procedure, a directive, or a training program that clearly describes the building and fire codes.

Types of Laws

This section details the various types of federal, state/provincial, and local laws that inspectors may encounter. Inspectors may also be involved in issuing various types of permits for their jurisdictions. The types of permits and permit processes will be addressed in Chapter 2, Codes, Standards, and Permits.

Federal Laws

Many federal agencies have developed regulations designed to ensure the safety of the public. These regulations cover a broad spectrum of activities, including:

- Employee safety
- Transportation of hazardous materials
- Patient safety in health care facilities
- Accessibility for disabled citizens
- Minimum housing standards

The federal agency that sets standards is usually responsible for their enforcement. In some instances, such as workplace safety laws, the state or province may choose to enforce federal regulations.

In most cases, the local inspector is not responsible for enforcing federal regulations. However, the inspector who becomes aware of hazards or violations within a federal property or jurisdiction should know how to report them to the proper agency or authority to seek corrective action.

Federal and some state/provincial-owned buildings located within the local jurisdiction are not required to comply with local codes **(Figure 1.6)**. In the past, agencies that operated these buildings usually enforced their own fire protection regulations. In recent years, the U.S. govern-

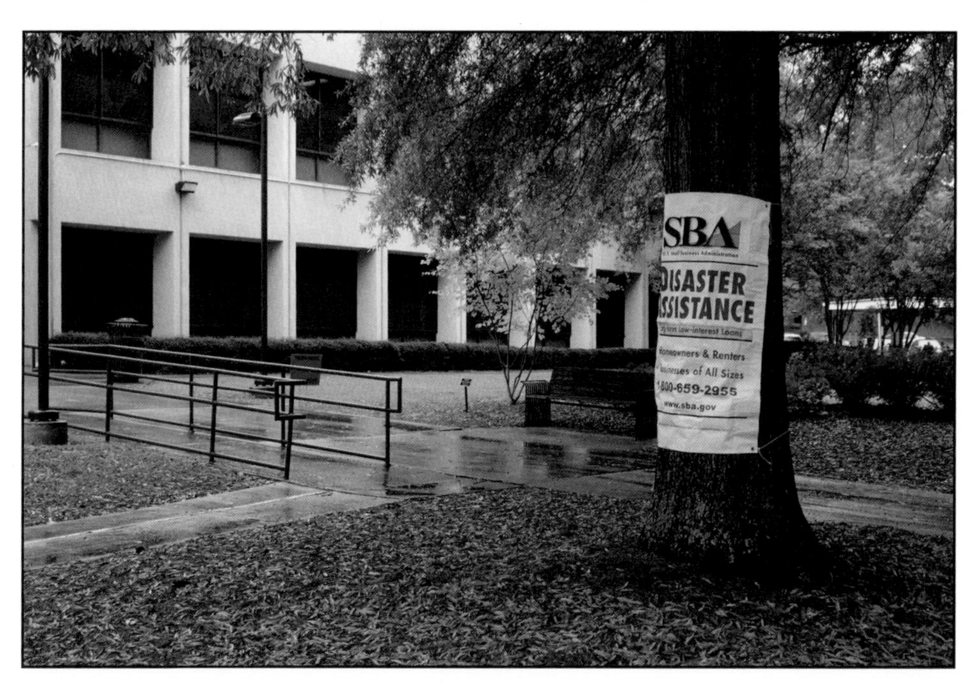

Figure 1.6 Federal buildings in a jurisdiction do not have to comply with locally adopted codes. *Courtesy of Federal Management Agency, George Armstrong, Photographer.*

ment has chosen to follow local codes for facilities such as post offices, Social Security buildings, and Armed Forces Reserve facilities protected by municipal fire departments.

Military bases that have their own fire departments and fire prevention staff follow the *Unified Facilities Criteria (UFC)* from the U.S. Department of Defense (DoD), portions of the *International Fire Code® (IFC®)*, and NFPA® standards. However, a combination of the *UFC*, NFPA® standards, and local code adaptations are often used in local federally owned buildings. Inspectors should know who is responsible for federal properties and work with them to ensure that fire protection is maintained at a high level, regardless of which code they follow.

Even if the local fire department does not have the authority to perform code compliance inspections in a federal facility, it still should perform pre-incident planning visits. These visits will prepare the department to handle any emergency that may occur within the facility as well as help department members prepare to assist federal authorities.

State/Provincial Laws

In addition to enforcing selected federal laws, state/provincial governments may also regulate certain fire and life safety inspection activities within their jurisdiction. For example, in some states/provinces, the duty to inspect nursing homes and day care centers is assigned to the state/provincial fire marshal's office. Similarly, authority is sometimes removed from the state/provincial level, and building and fire codes are enacted that are enforced by local jurisdictions. It is the inspector's duty to become familiar with the relationship between state/provincial and local laws in his or her state.

State/provincial laws can also define and specify building construction and maintenance details in terms of fire protection and empower agencies to issue regulations. State/provincial labor laws, insurance laws, and health laws also have a bearing on fire safety and sometimes encompass fire and life safety inspection responsibilities. It is the inspector's responsibility to communicate the local jurisdiction's authority to other agencies with fire-related concerns. This communication is necessary to form an effective working relationship between local and state or provincial agencies. Each inspector must be familiar with his or her state/provincial adoption regulations as well as know how oversight works.

Minimum/Maximum Laws

States may enact laws that would apply to local governments throughout the state; therefore, an inspector must apply these **minimum** standards when enforcing local regulations.

States can also enact laws that dictate the **maximum** level of enforcement capabilities permitted at the local level. These laws place certain jurisdictional restraints on inspectors at the local level.

Local Laws and Ordinances

Local laws and ordinances, although sometimes based on state/provincial laws, are more specific and tailored toward the needs of a particular county, municipality, or fire protection district. States or provinces may enact legislation that enables local jurisdictions to adopt state/provincial regulations. Two ways to do this are adopting by reference and by enabling acts.

Adopting by reference. Adoption by reference means that the local jurisdiction will follow the state/provincial laws exactly as written.

Enabling acts. Adopting laws in the form of enabling acts gives the local jurisdiction the use of state/provincial laws as their basis, but then allows additions or deletions to regulations or ordinances based on local needs or preferences **(Figure 1.7)**.

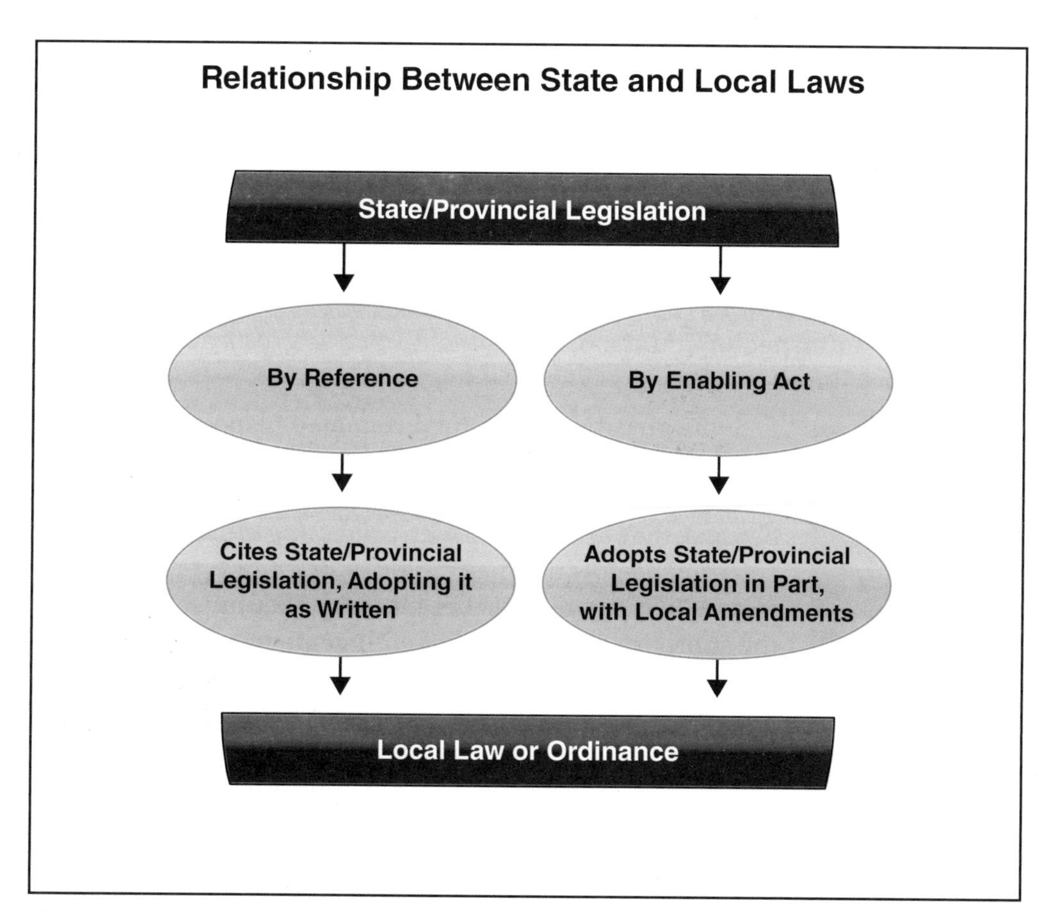

Figure 1.7 Local jurisdictions may adopt state/provincial laws by two methods: by reference or by enabling act.

Legal Status of Inspectors

The inspector's legal status defines the amount of authority granted to the inspector, the responsibility to act, and the protection against legal action that the jurisdiction provides. In the private sector, the employer's risk management program, job description, and fire and life safety policies determine an inspector's legal status.

Public inspection personnel often inspect facilities that insurance inspectors or private sector companies also inspect. Joint or overlapping inspections should increase fire and life safety. In cases where conflicts arise between insurance requirements and the adopted code, compliance will be determined by meeting or exceeding the minimum requirements of the code. The adopted code's requirements take precedence, unless a performance design or other means of equivalent protection is offered.

Public Sector

Individual state and provincial governments determine public sector employees' legal status. In addition to legal requirements for inspectors, the states may also have regulations regarding the organization of fire protection agencies, retirement systems, and civil service requirements. The state may also specify jurisdictional lines between the state/provincial fire marshal's office and other local agencies, such as those at the county and municipal levels. Inspectors may be authorized to do some or all of the following:

- Arrest or detain individuals
- Issue a summons
- Issue citations
- File complaints for code violations
- Obtain and execute warrant

Police Power — Constitutional right of the state government to make laws and regulations to protect the safety, health, welfare, and morals of the community. Police power is used as the basis for enacting a variety of laws including land use (zoning), fire and building codes, gambling, vehicle registration and parking, nuisances, schools and sanitation.

Fire codes authorize inspectors to interrupt business operations or to evacuate and close a building that presents imminently dangerous conditions. For example, an inspector could close a nightclub that is grossly over its legal occupancy limit.

Some jurisdictions grant designated inspectors different levels of authority based on the code they are enforcing. For example, fire officials usually have a unique level of authority because they are charged with protecting life safety. In other cases, inspectors may be granted **police power** to act in these roles. If inspectors have this authority, they must also receive the appropriate legal and law enforcement training. In many states, training and certification as a peace officer is required before an inspector can exercise any law enforcement-type powers. This situation is particularly true if the inspector can expect to be involved in the prosecution of a fire code violator.

The exact relationship between a municipal police department and an inspector must be clearly defined in communities where the inspector has limited police powers **(Figure 1.8)**. Understanding this relationship avoids conflicts when the inspector must exercise police powers that require the assistance of the police department. The jurisdiction's adopted fire and life safety codes should contain language that explains the relationship between the inspector and the police department.

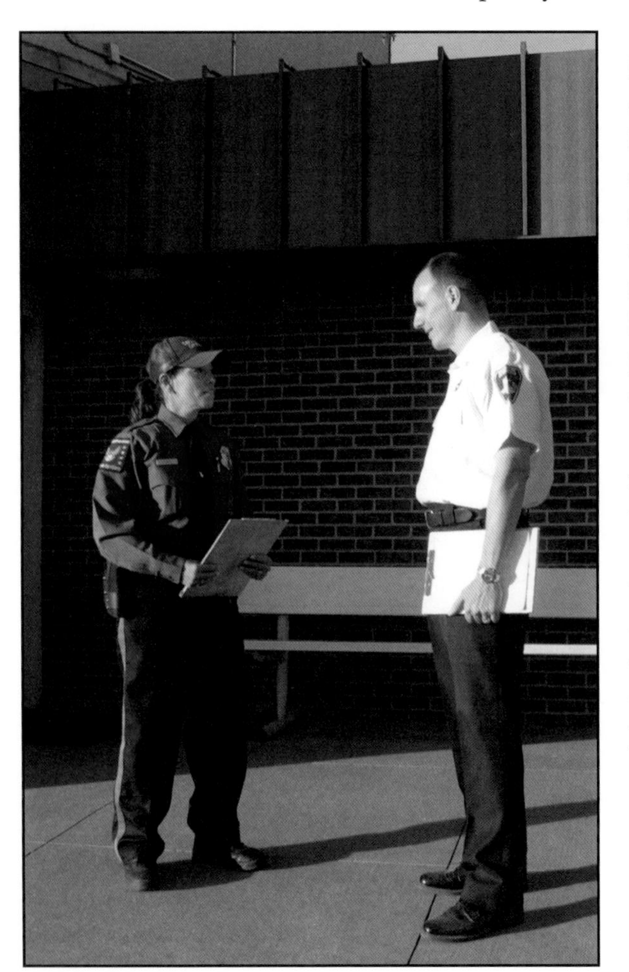

Figure 1.8 Inspectors often have the opportunity to work with law enforcement officials. Maintaining a productive relationship will help in enforcement matters.

Private Sector

In the private sector, an inspector's legal status may be less specific than it is in the public sector. The status may be included in the inspector's job description, the company's risk management plan, or in the company's fire and life safety policies. Private sector inspectors should be familiar with the legal basis for performing inspection and code enforcement within their companies.

Conflicts Between Public and Private Requirements

Insurance companies often dictate protection requirements that are more stringent than local fire codes or ordinances for their insured risks. Life safety code issues always have precedence over property conservation considerations. The public inspector must point out these issues to the building owner/occupant and the insurance inspector to ensure that life safety is paramount.

Insurance company inspectors may find that their employer's requirements bring them into conflict with either the company they are inspecting or local public inspectors, or both. Interpretations sometimes differ concerning the applicable fire code or its provisions. An insurance inspector should work with the municipal inspector to ensure that the insurance requirements meet or exceed the locally adopted fire code.

Liability

In general, inspectors are not held liable for discretionary acts, namely, the manner in which the fire inspector, acting for a superior, performs an act or enforces a policy. Because new cases resulting from lawsuits and administrative challenges can result in modification of laws, an inspector may need to be aware of changes in the limits of **liability**.

Liability — State of being legally obliged and responsible.

Although it is not possible for this manual to give specific information on liability due to differences in laws, there are some general liability considerations that are useful for all inspectors. This section will describe issues of right of entry, warrants, duty to inspect, civil rights, indemnification, and malfeasance and negligence.

Right of Entry

Right of Entry — Rights set forth by the administrative powers that allow the inspector to inspect buildings to ensure compliance with applicable codes.

Model fire codes grant the AHJ **right of entry**, namely the authority to enter any premises or building as often as necessary to perform an inspection or to execute any duties authorized by the fire code. These actions can include licensing inspections for health care occupancies, conducting an inspection for the approval of a regulated activity, such as an indoor pyrotechnics display, or responding to a complaint of a malfunctioning fire alarm system.

Even though inspectors have a right of entry, they must be aware that property owners also have certain rights under the 4th Amendment of the U.S. Constitution. The 4th Amendment is part of the Bill of Rights, which guards against unreasonable searches. Two U.S. Supreme Court cases, *Sea vs. City of Seattle* and *Camara vs. City and County of San Francisco*, set federally protected constitutional limits on the authority of fire inspectors to enter a business or a private dwelling for the purpose of conducting inspections.

In both cases, the Supreme Court clarified that an inspector always has the authority to enter a premises if an emergency exists. An *emergency* is a condition or event that creates reasonable belief that someone is in distress.

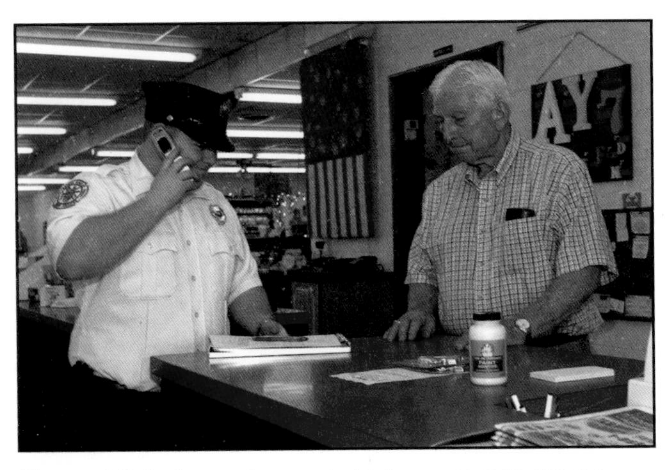

Figure 1.9 An inspector must obtain consent to enter a property when there is no emergency.

The Court has also ruled that if a hazard is obvious and in public view, an inspector has the right to enter the premises. For example, if an inspector smells a natural gas leak in a restaurant and the owner refuses entry, the Supreme Court has ruled that such an event allows the inspector to enter without the owner's permission. Entry is permitted because an emergency exists and the hazards of a natural gas leak are clearly identifiable due to the odorants required by Federal law to warn the public of its release.

The inspector must obtain consent to perform an inspection if there is no emergency or visible hazard. To obtain consent, the inspector should provide identification, state the purpose of the inspection, and seek verbal or written permission from the building owner, tenant, or owner's representative **(Figure 1.9)**. The inspector should invite the individual to participate in the inspection. The individual should be given a copy of the applicable code sections, which show that the inspector has the right of entry.

The AHJ may have available a refusal form when a property owner or occupant refuses the right to enter a property. The refusal form provides information for obtaining a warrant to enter the property at a later time.

Occasionally, it is practical to obtain written permission to enter a site before the inspection. The use of a consent-to-enter form will remove questions regarding the original authorization to enter and perform an inspection. Any form that is developed for these purposes should be approved by the jurisdiction's legal counsel and authorized by the fire chief.

Canadian Right of Entry

Canadian national, provincial, and municipal fire prevention laws are generally based on the inspection requirements outlined in the *National Fire Code of Canada (NFC).* Section 8 132-28 provides the authority for the right of entry to perform an inspection:

Every Fire Prevention Officer may, upon the complaint of a person interested, inspect any residential occupancies or dwelling units, and for such purpose may, at all reasonable hours and upon producing proper identification, enter into and upon the building or premises containing the dwelling units for the purpose of examination and ascertaining whether provisions of this Article have been obeyed and to enforce or carry into effect the Article.

Warrants

If an inspector is denied entry and believes hazards exist, he or she will need to obtain a warrant to enter the premises. Laws vary greatly across the U.S., Canada, and U.S. territories on the requirements for obtaining a warrant; therefore, the inspector should know the procedures for obtaining a warrant and work closely with legal counsel and the AHJ in these situations.

In two decisions, the U.S. Supreme Court has ruled that an inspector must obtain an administrative warrant to search portions of commercial premises that are not open to the public without the consent of the owner. The use of an administrative warrant, as opposed to a search warrant, to gain entry is necessary because of the legal protocols required for issuing these warrants.

Administrative Warrant — Written order that authorizes an inspector to enter a property for the purpose of conducting an inspection.

The **administrative warrant** is a demand to enter for the purpose of inspection. There is no requirement for the inspector to claim or list a probable cause in the same sense as in a criminal search warrant. The requirement for an administrative warrant protects the owner from continuous and sometimes frivolous inspections, while at the same time giving the inspector a legal means to demand entry with legal backing.

Search Warrant — Written order, in the name of the People, State, Province, Territory, or Commonwealth, signed by a magistrate, commanding a peace officer to search for personal property or other evidence and return it to the magistrate.

A **search warrant** is an authorization to enter for the purpose of gathering evidence. A judge issues a search warrant once probable cause is established that a crime has been committed and evidence supporting the accusation is on the property to be searched. The warrant restricts the search to seeking evidence for the suspected violation that is specifically listed on the warrant **(Figure 1.10)**.

Figure 1.10 When an inspector can demonstrate probable cause that a crime has been committed, a judge can issue a search warrant.

Duty to Inspect

Most model codes contain a duty-to-inspect clause. This clause is significant because it charges the inspector with total enforcement. Inspectors are not limited to certain buildings or occupancy classifications. The code must be applied equally, within reason, to all applicable occupancies in a given jurisdiction. Failure to follow this clause may subject the department or individual inspector to personal and professional liability.

From a liability standpoint, it is better for inspectors to conduct fewer but more thorough inspections and to follow up on all violations than to perform more frequent, less-thorough inspections. Failure to inspect a property does not impose a duty on the inspector unless laws or statutes impose such a duty or there is a known code violation present. Laws that single out a particular class occupancy for a predetermined number of inspections do establish a duty for the inspection staff.

Civil Rights

Another potential liability source for inspectors involves civil rights. For example, an inspector might perform inspections in a manner that discriminates against a certain group of people or classification of business. In such a case, the discrimination may indicate a civil rights violation. Other examples of questionable practices, also known as oppression, might include the following:

- Singling out certain groups or classes of people, including businesses, without justification.
- Targeting a specific business or industry but seldom or never inspecting similar businesses; for example, inspecting pool halls but ignoring bowling alleys.

- Conducting inspections according to different frequencies for certain businesses throughout a jurisdiction; for example, inspecting a tavern in a residential neighborhood more frequently than a tavern located in a commercial or industrial setting.
- Conducting inspections based upon the race, religion, or ethnic background of the owner or clientele.
- Conducting inspections as a means of retaliating against the owner of the business; for example, inspecting a business based on a complaint like, "The owner threw me out, so I want you to inspect his business."

Civil rights cases can become high-profile media events. These can be avoided through the consistent application and enforcement of the adopted local codes. Inspection procedures need to be reviewed regularly to ensure that deviations from acceptable practices are avoided.

Indemnify — To agree that one party will compensate another party for losses or damages that are incurred if specific actions or events occur.

Indemnification

Most jurisdictions **indemnify** their inspection personnel or provide liability insurance to protect them in the areas where they may be held liable. To indemnify the inspector means that the AHJ assumes the responsibility for any claims against the individual. The procedures for indemnification generally depend on prevailing state law. Inspectors need to determine whether they are indemnified or protected by liability insurance when they are performing their official duties. Liability insurance provided by the jurisdiction protects the inspector from costs involved with providing legal counsel or court judgments.

Decisions in court cases have established inspector liability precedence over the past 30 years. For instance, a 1976 court ruling (*Adams v. State of Alaska;* see **Appendix C**), held that fire inspectors, in conducting code inspections, had taken on a duty and must use reasonable care in the exercise of that duty.

If inspectors discover violations in a property but fail to perform a follow-up inspection to verify that the violations are corrected, they can be held liable if a fire related to the violations occurs. In addition, inspectors can be held liable for the resulting deaths or injuries if that can be attributed to any outstanding code violations **(Figure 1.11)**.

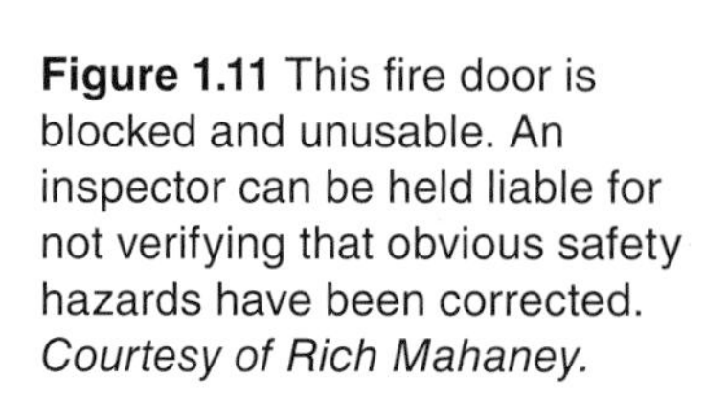

Figure 1.11 This fire door is blocked and unusable. An inspector can be held liable for not verifying that obvious safety hazards have been corrected. *Courtesy of Rich Mahaney.*

Likewise, inspectors who take on a **special duty** or obligation to a person can be held liable. This liability results from the inspector placing the person in potential harm or danger through incorrect or unsafe advice or who acts outside the scope of his or her authority.

Special Duty — Type of obligation that an inspector assumes by providing expert advice or assistance to a person; this obligation may make the inspector liable if it creates a situation in which a person moves from a position of safety to a position of danger by relying upon the expertise of the inspector.

Malfeasance and Negligence

Malfeasance is the commission of an unlawful act that is knowingly committed by a public official, but **negligence** consists of misfeasance or nonfeasance, which are unlawful acts that were not intentionally committed. One of the key questions in negligence is the adequacy of performance. There are two ways in which one can be judged negligent: wrongful performance (**misfeasance**) or the omission of performance (nonfeasance).

Misfeasance occurs when an individual has the knowledge, ability, and legal authority to act but performs the act incorrectly. For instance, an inspector who is trained to recognize a fire hazard but during an inspection inadvertently overlooks the presence of a violation that later results in injury or property loss.

Nonfeasance would occur if after recognizing a fire code violation or fire hazard, an inspector failed to order the correction of the violation or hazard, which later resulted in injury or property loss because proper inspection techniques were not used.

As states have modified their constitutions to eliminate the **public-duty doctrine,** there have been a number of court rulings against the immunity provisions contained in the model codes. The courts have ruled that the immunity provisions conflict with statutes that establish an inspection authority and require the enforcement of codes and regulations. In other words, a community cannot be required to do something and at the same time be immune from liability if it or its officers (inspectors) do a job inadequately or negligently.

Malfeasance — Commission of an unlawful act; committed by a public official.

Negligence — Failure to exercise the same care that a reasonable, prudent, and careful person would under the same or similar circumstances.

Misfeasance — Improper performance of a legal or lawful act.

Nonfeasance — Failing to perform a required duty.

Public-Duty Doctrine — States that a government entity (such as a state or municipality) cannot be held liable for an individual plaintiff's injury resulting from a governmental officer's or employee's breach of a duty owed to the general public rather than to the individual plaintiff.

Outside Technical Assistance

Occasionally, a jurisdiction may determine that a code issue or request to build a unique structure or industrial process is beyond its capabilities to assess. For example, a new building may require use of alternative building materials or specialized fire protection systems to meet the intent of the building and fire code. These situations generally require the involvement of a more highly trained individual such as a fire protection engineer (FPE). Some large fire departments and industrial operations have FPEs on staff; jurisdictions that do not may contract with an engineering or fire protection firm to provide qualified third-party assistance.

Before entering into an agreement with a third party, local code officials must verify that such an agreement is legal and that decisions made by the outside firm can be made binding by the local enforcement official. These legal questions can normally be legitimized through the enabling legislation of the fire inspection division and the fire department. These legal questions can usually be answered by referring to the enabling legislation of the fire inspection division and the fire department. The agreement must also avoid any conflict of interest between the jurisdiction and the individuals or firm providing the technical report or opinion.

Recommend Policies and Modifications

Code enforcement departments need written policies and procedures to operate efficiently and effectively. These policies and procedures describe the organization's expectations and are based on the organizational model and strategic and operational plans. Depending on the AHJ, these policies and procedures may be called standard operating procedures (SOPs), standard operating guidelines (SOGs), or policy and procedures manual (PPM). Regardless of the specific title, any documents must contain information that is current and appropriate. Therefore, the organization should have a process for evaluating and revising existing policies and procedures and for adopting new ones.

Developing Policies and Procedures

Generally, procedures should contain a purpose and scope statement. Each page should contain basic information at the top that assists the reader in navigating the document:

1. **Subject** — What the policy/procedure is about
2. **Procedure number** — Assigned to the specific procedure for tracking purposes
3. **Dates** — Original date of implementation plus any revision dates
4. **Supersedes** — Procedure number that is replaced by the current page
5. **Approvals** — Initials of the authority approving the policy/procedure
6. **Distribution** — List of persons or groups to whom the policy/procedure is issued
7. **Applicability** — Persons or groups to whom the policy/procedure applies
8. **Revision** — Indicates whether the current page is original or a revision
9. **Forms used** — Indicates the appropriate form if applicable used to fulfill the policy/procedure

The mechanics for writing a policy or procedure may be found in a text on basic communication skills. The steps for determining the need for a new policy or procedure are as follows:

- **Identify the problem or requirement for a policy or procedure** — Determine if a policy or procedure is actually necessary to address the problem. Some problems may be best addressed on an individual basis that does not require a formal policy.
- **Collect data to evaluate the need** — Use data that may be quantitative or qualitative and may come from personnel interviews, product literature, or activity reports.
- **Select the evaluation model** — Use one of the evaluation models: goals-based, process-based, or outcome-based.
- **Establish a timetable for making the needs evaluation** — Consider the length of time required to evaluate the problem, which depends on the complexity of the problem and the amount of information that must be evaluated.

- **Conduct the evaluation** — Follow the recommended steps for the model that is most appropriate for the problem.
- **Select the best response to the need** — Determine the best policy or procedure to solve the problem. Remember that this selection may include no policy or procedure at all.
- **Select alternative responses** — Select a second-best choice if a contingency is indicated. External influences may make it necessary to select a policy or procedure other than the first choice. Personal safety, however, should not be compromised.
- **Establish a revision process or schedule** — Create a revision process as part of the policy or procedure. This process may be general for all policies and procedures or one that is specific to the policy that has been selected.
- **Recommend the policy or procedure that best meets the need** — Consider that because policies and procedures may have the effect of law, they may need to be formally adopted by the jurisdiction. Formal approval requires that the policy or procedure be supported by documentation.

Maintaining and Revising Policies

Policies and procedures are most effective if they are regularly reviewed and revised. Responsibility for monitoring policies and procedures rests with the organization's supervisory level staff. They should be familiar with the policies' content, application, and effects. Infractions and unauthorized alterations of policies and procedures should be noted and reported in order to reinforce existing policies and procedures or revise them to meet changing conditions.

A process for revising policies and procedures will give organizations the flexibility to adapt to changes in the operating environment and organizational requirements. Document the revision process and include answers to the following questions:

- When does the policy or procedure need to be revised? Is there a specific timetable?
- What conditions or circumstances would cause the policy or procedure to need a revision?
- How should the policy or procedure be revised: completely, partially, or not at all?

Indications that a policy or procedure needs to be revised may include the following:

- Internal/external customer complaints
- Increase in policy infractions
- Injuries or property loss due to a procedure's failure
- Change in the resources used to accomplish the task
- Change in the problem that the policy or procedure was intended to solve

When it becomes apparent that a policy or procedure must be revised, replaced, or abandoned, the actual process steps are generally the same as those used earlier for the creation of a new policy or procedure.

Creating Forms and Job Aids

Inspection forms, like any other administrative tool, can help the inspector or department create consistency and uniformity, improve the inspection and permitting process, and help provide legal documentation **(Figure 1.12)**. If copies of forms are provided to owners/occupants, they can assist them in maintaining compliance in between inspections. This type of reminder and documentation becomes a valuable tool in assisting an agency in promoting an overall fire and life safety inspection program.

Inspection forms can be developed in-house for a specific operation or set of requirements. Many of these forms are also available from third-party vendors to address specific codes or standards. As with all other procedures, forms and aids need to be reviewed and updated to reflect changes in the law and in inspection requirements.

NOTE: Examples of inspection forms can be found in **Chapter 16,** Inspection Procedures, and in **Appendix D.**

Figure 1.12 Entering inspection notes into the department's forms will help the inspector document important information immediately.

Chapter Summary

Inspectors are responsible for ensuring that the jurisdiction's fire and life safety program is successful. By performing their assigned duties, inspectors review plans, issue permits, inspect new and existing facilities, and respond to complaints. Inspectors may be private or uniformed or nonuniformed public employees. The inspector's authority is based on national, state/provincial, or local laws or the regulations established by their employer. As certified inspectors, they enforce the codes and standards that have been legally adopted by the authority having jurisdiction.

When enforcing or applying code requirements, an inspector should be aware that the owner/occupant has the right to appeal the requirement. The inspector must interpret and apply the code correctly to reduce the chance that it will be appealed. Correct interpretation and application will also ensure that the requirements provide the level of safety and protection intended by the code and the jurisdiction.

An inspector must be familiar with the enforcement process, including the potential for prosecution. Although the situation may be rare, an inspector may be called upon to testify as a witness or expert in the prosecution of a code violation.

Review Questions

1. What are the categories of fire and life safety inspections that inspectors may be authorized to perform?
2. What are possible actions to ensure code compliance?

1. Describe the differences between negligence, malfeasance, misfeasance, and nonfeasance.
2. Describe some indications that a policy or procedure needs to be revised.

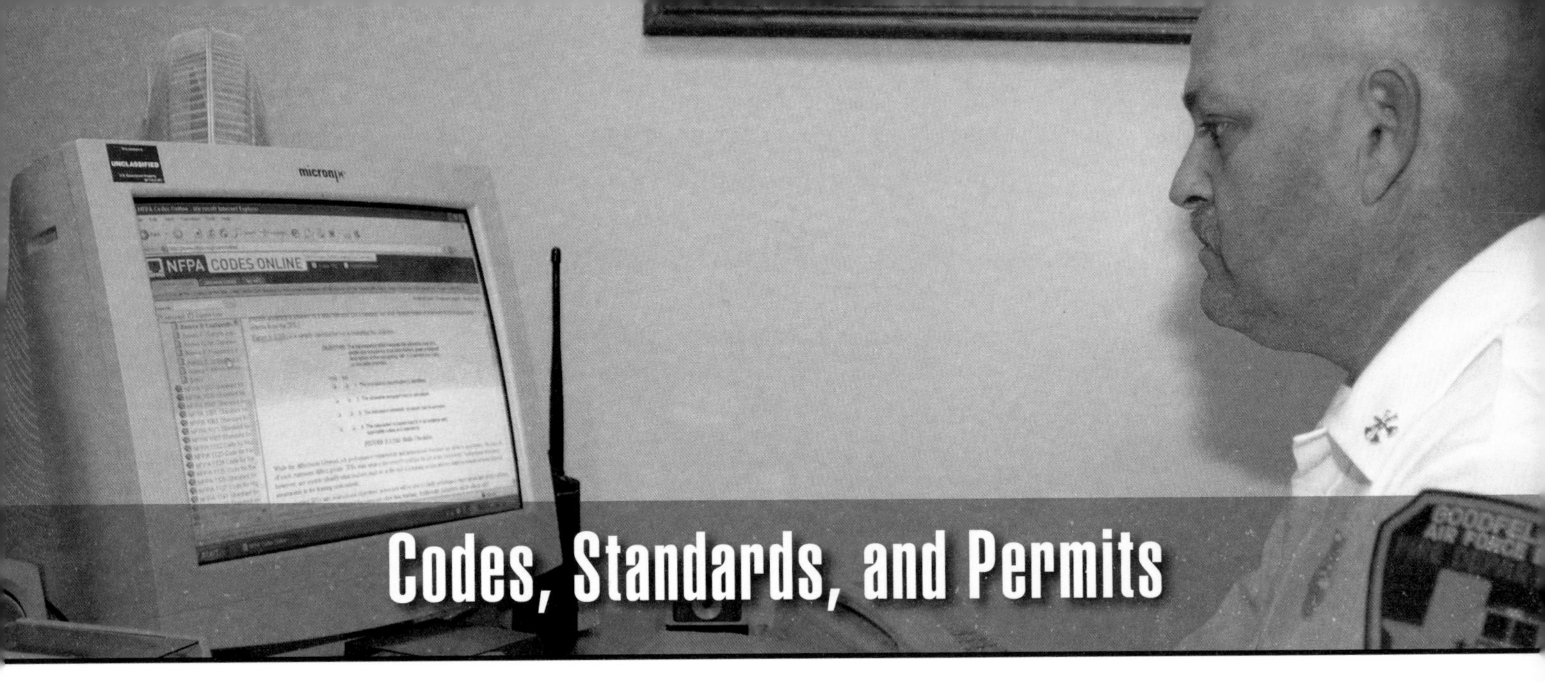

Codes, Standards, and Permits

Chapter Contents

I Case History 41
Codes and Standards 41
Codes 42
Standards 47
Complaint Procedures 51
Permits 52
Operational Permits 52
Construction Permits 55
II Local Code Development Process 56
Identify the Problem 56
Identify Affected Stakeholders 57
Form a Task Force 57
Draft the Proposed Code 58
Submit Code for Legal Review 59
Formally Adopt the Code 51
Code Modification and Appeals Procedures 62
Code Modification 63
Appeals Procedures 65
Permit Process 65
Application 66
Research and Review 66
Issue Permit 66
Evaluate Status 66
Permit Issued in Error 67
Chapter Summary 67
Review Questions 68

Key Terms

Code..42
Consensus Standard 48
Industry standard......................................48
Standard..42
Standardization 49
Sunset Provision 61
Work Session..61

NFPA® Job Performance Requirements

This chapter provides information that addresses the following job performance requirements of NFPA® 1031, *Standard for Professional Qualifications for Fire Inspector and Plan Examiner* (2014).

Fire Inspector I

4.2.2

4.2.4

4.2.5

Fire Inspector II

5.2.1

5.2.4

Codes, Standards, and Permits

Learning Objectives

After reading this chapter, students will be able to:

Inspector I

1. Identify appropriate resources for finding current and applicable codes and standards. (4.2.5)
2. Explain complaint procedures. (4.2.4)
3. Describe the role of an Inspector I in the permitting process. (4.2.2)

Inspector II

1. Explain the role of an Inspector II in the local code development process. (5.2.4)
2. Explain the ways an Inspector II will participate in code modification and appeals procedures. (5.2.4)
3. Describe the role of the Inspector II in the permitting process. (5.2.1)

Chapter 2
Codes, Standards, and Permits

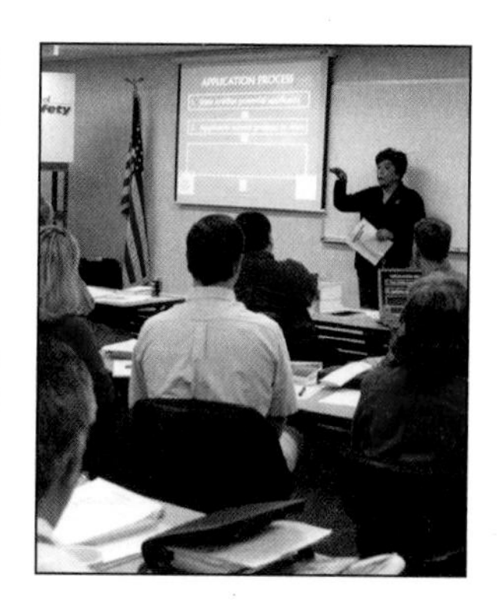

Case History

Due to a change in the recently adopted model code, churches and businesses in a state were concerned that they would not be permitted to display natural-cut Christmas trees during the holiday season. Rather than issue a broad ban, the Department of Commerce, who administers building and fire codes, sought an interpretation in the model code that would provide an exception. The exception would avoid the necessity of changing the model code through the legislature.

The state found that the model code contained a provision that would allow limited quantities of combustible vegetation in any occupancy where the AHJ determined that adequate safeguards were provided. This interpretation allowed jurisdictions to determine whether natural-cut Christmas trees would be allowed. The state included an example of a guideline outlining conditions to be met before obtaining permission that still emphasized proper fire and life safety for each occupancy.

Codes and standards are developed to establish minimum requirements for building design, construction, and use as well as equipment installation, operation, and maintenance. Codes and standards are adopted by a governmental entity and enforced by an AHJ in the building department, fire department, or both.

Fire and life safety codes establish the minimum level of safety that should be present in a structure. Fire and life safety codes regulate construction materials and designs, fire detection and suppression systems, and occupant behavior and processes. These codes must be current and consistent to provide the desired level of protection established by the authority.

This chapter describes codes and standards: how they are developed and their importance to an inspector. Also included are complaint procedures, the processes for developing local codes, amending model codes, and appealing code requirements. Finally, the permit process is outlined.

Codes and Standards

Codes and standards are used to create a minimum level of safety within a jurisdiction. These terms are often used together and sometimes interchangeably. Even the National Fire Protection Association® (NFPA®) tends to assign

Code — A collection of rules and regulations that has been enacted by law in a particular jurisdiction. Codes typically address a single subject area; examples include a mechanical, electrical, building, or fire code.

Standard — A set of principles, protocols, or procedures that explain how to do something or provide a set of minimum standards to be followed. Adhering to a standard is not required by law, although standards may be incorporated in codes, which are legally enforceable.

similar definitions to the terms. Both the NFPA® standards and International Code Council® (ICC®) codes also follow a similar layout in the opening sections. The first sections introduce the reader to the scope or administrative sections of the book and generally provide definitions of vocabulary within the chapters. Nonetheless, an inspector must know the difference between a code and a standard:

- A **code** is a collection of rules and regulations enacted by a legislative body to become law in a particular jurisdiction. A code may be based on a standard or may incorporate an entire standard.
- A **standard** is a developed set of principles, protocols, or procedures. Standards normally explain how to do something or they provide a set of minimum guidelines that are expected to be followed to achieve compliance with a code.

Simply stated, a code is a law that may be based on or may incorporate a standard. A standard, however, only becomes a law when it is legally adopted by a jurisdiction or included as part of a code **(Figure 2.1)**.

Codes

As mentioned previously, codes are legal documents that govern activities at various levels of government. Before the creation of standardized model codes, jurisdictions developed their own sets of codes. This practice led to a wide variety of acceptable minimum levels of safety and some confusion among manufacturers and contractors who sold materials in multiple jurisdictions. As a result, organizations were formed to write consensus or model codes that could be applied universally.

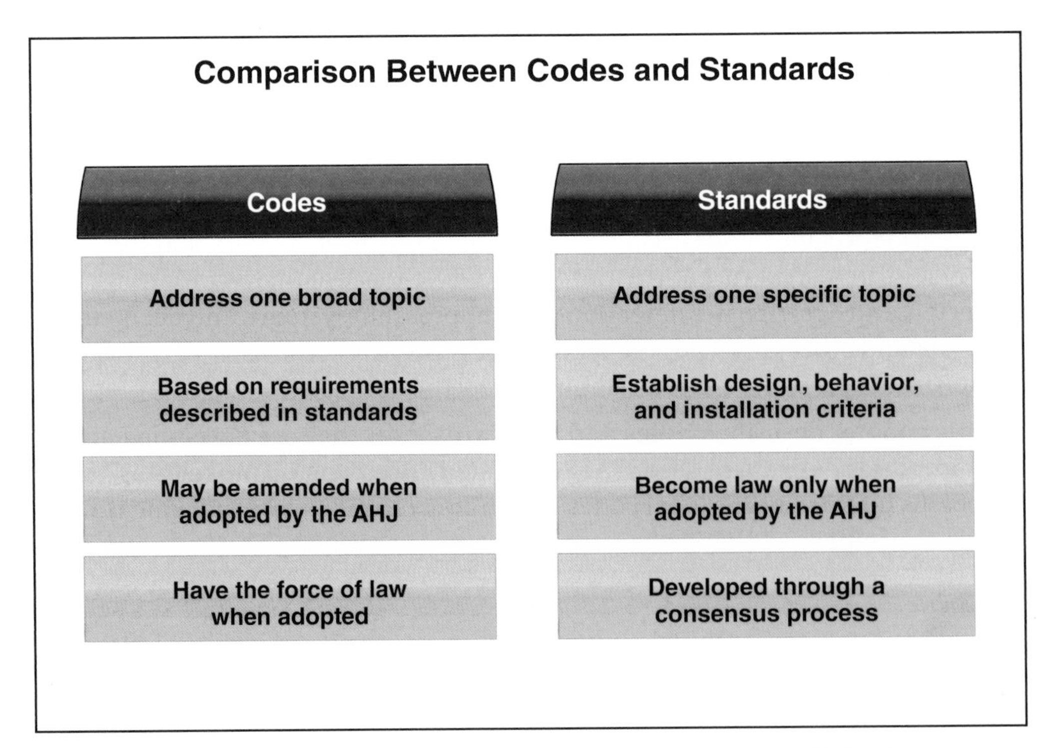

Figure 2.1 Although the terms *codes* and *standards* are sometimes used interchangeably, they have different meanings.

Building and fire codes or standards may be classified as either prescriptive- or performance-based:

- A prescriptive-based (also known as *specification-based*) code/standard describes in detail the types of materials that can be used and how they must be assembled.
- A performance-based code/standard describes an acceptable level of performance that an assembly, material, or system must meet without stating how the item is assembled. A performance-based code/standard gives the designer greater freedom than a prescriptive-based code/standard does.

Prescriptive/Model Codes

The term *model code* describes a set of requirements that are similar to a standard, and these requirements are generally called *prescriptive codes*. A consensus organization such as NFPA® or the ICC® develops model codes that contain agreed-upon requirements for fire protection, building construction, structural safety, building sanitation, and life safety.

Like standards, model codes are only enforceable when the AHJ adopts them. In some communities, the AHJ may adopt a model code without changes. In other instances, the code may be amended to address specific local conditions and needs. An inspector must be thoroughly familiar with the locally adopted code and its amendments. To assist an inspector in interpreting the model codes, explanatory commentaries and handbooks are available for some of the codes. Consensus organizations also provide training sessions and workshops about model codes.

Currently, there are two model code organizations in the United States and one in Canada. Each code organization has a series of codes that, when adopted, can be used to regulate building components — structural, mechanical, electrical, and plumbing — as well as fire and life safety:

- Canadian Commission on Building and Fire Codes (CCBFC)
  - *National Fire Code of Canada (NFC)*
  - *National Building Code of Canada (NBC)*
- International Code Council® (ICC®)
  - *International Fire Code® (IFC®)*
  - *International Building Code® (IBC®)*
- National Fire Protection Association (NFPA®)
  - NFPA 1®, *Fire Code*™
  - NFPA® 101, *Life Safety Code®*
  - NFPA® 5000, *Building Construction and Safety Code®*

Department of Defense (DoD)

Inspectors who are responsible for facilities that are under the jurisdiction of the United States Department of Defense (DoD) enforce a different set of codes. Those codes are known as the Unified Facilities Criteria (UFC). Elements of other model codes and standards may be part of the UFC.

Performance-Based Design

Performance-based designs are employed to address unique buildings or processes when prescriptive requirements do not adequately address their design, construction, operation, and maintenance. The resulting solution must provide a safety and dependability level that is equivalent or superior to the model code requirements.

One benefit of performance-based design (PBD) is that the fire protection solution can be achieved in a more flexible manner than with prescriptive-based design. For example, in a high-rise, total evacuation of occupants may be unrealistic and unnecessary during the initial phase of fire growth. A performance-based design of the means of egress may permit elevators to be used as a component, thus enabling occupants to move from the floor of fire origin to an area of safety but not exiting the structure.

NOTE: Means of Egress is described in Chapter 7.

Although PBD provides additional flexibility, there are difficulties associated with enforcing performance-based codes:

- The AHJ must establish acceptance-testing criteria, methods of evaluating test data, and procedures for acceptance or denial of test results.
- Significant technical expertise is required to review and evaluate performance-based designs. In most cases, professional engineers or third-party inspection agencies are often required to handle these matters.
- Inspectors evaluating PBD options must evaluate how the assembly functions as it is actually used and should be maintained, not only how it functions under its intended use. A PBD building requires that the inspector review the historical design documentation prior to inspection.

NOTE: The inspector may consult the Society of Fire Professional Engineers in addition to the *SFPE Code Official's Guide to Performance-Based Design Review.*

Performance-Based Design During Remodeling

A common application of PBD criteria occurs when historic buildings undergo renovations. As an example, it is difficult to make major improvements to a State Capitol or similar iconic building that is over one hundred years old that complies with modern fire and building codes. Analyses that consider how construction features contribute to fire growth, models of Means of Egress (MOE) travel times, and automatic sprinkler protection and early fire detection systems have been used to ensure adequate fire and life safety coverage.

Application of Codes

Although some jurisdictions use locally developed codes, most jurisdictions adopt model codes. An inspector should be familiar with those model codes and standards that are referenced in the adopted code. For instance, if the AHJ has adopted the ICC® *International Building Code*, it adopts by reference NFPA® 13, *Standard for the Installation of Sprinkler Systems,* thereby making it a legally binding requirement of the building code.

Elements of adopted electrical, plumbing, and mechanical codes that reference the fire code or are referenced by the fire code are also important. For instance, the mechanical code includes installation details for smoke dampers, duct smoke detection, and fire protection for commercial kitchen hoods. Because these requirements are also found in the fire code, the fire inspector must be familiar with them when reviewing plans or making field inspections **(Figure 2.2)**.

NOTE: Installed fire protection systems are covered in Chapter 12, Water-Based Fire Suppression Systems, and Chapter 13, Special-Agent Fire Extinguishing Systems.

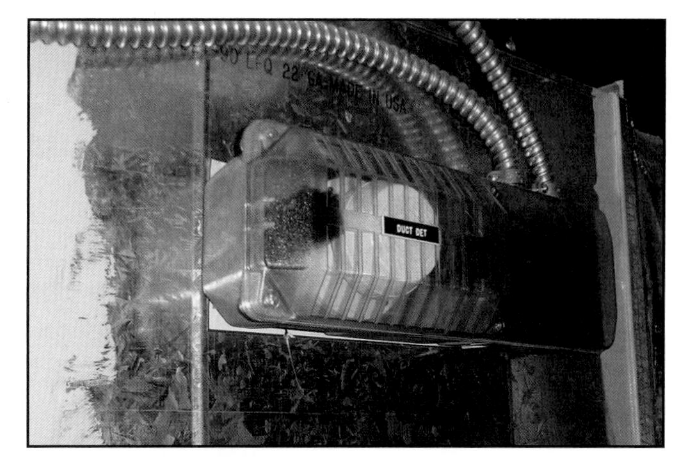

Figure 2.2 Building, mechanical, electrical, and fire codes may be cross-referenced. For example, each of these codes requires ventilation systems to contain smoke detectors like the one pictured.

An inspector must be aware that some facilities may have to meet the requirements of multiple codes. Another important issue in code enforcement is the way codes are applied to new and existing structures.

Multiple codes. A hospital is a common example of a facility that may need to meet several codes. First, the hospital may have to meet the locally adopted building and fire codes. A portion of the same facility may have to meet NFPA® 101, *Life Safety Code®* because of health care licensing requirements in that area. In these instances, an inspector may have to enforce the more restrictive code requirement or coordinate enforcement efforts.

New and existing structures. The current adopted edition of the building code applies to all structures that are built while that code is in effect. Any additions or alterations to existing structures will also be regulated by the current code. An alteration to an existing building or change of use that meets certain criteria will require that the entire structure be brought up to the current and stricter building code requirements.

For example, an existing high-rise structure that was not required to have sprinklers when it was built may be required to have them installed when renovations are made to more than 50 percent of the building. Similarly, when there is a change of use, such as when a factory building is converted into a shopping area, the fire code that governs the new use must be enforced **(Figure 2.3)**.

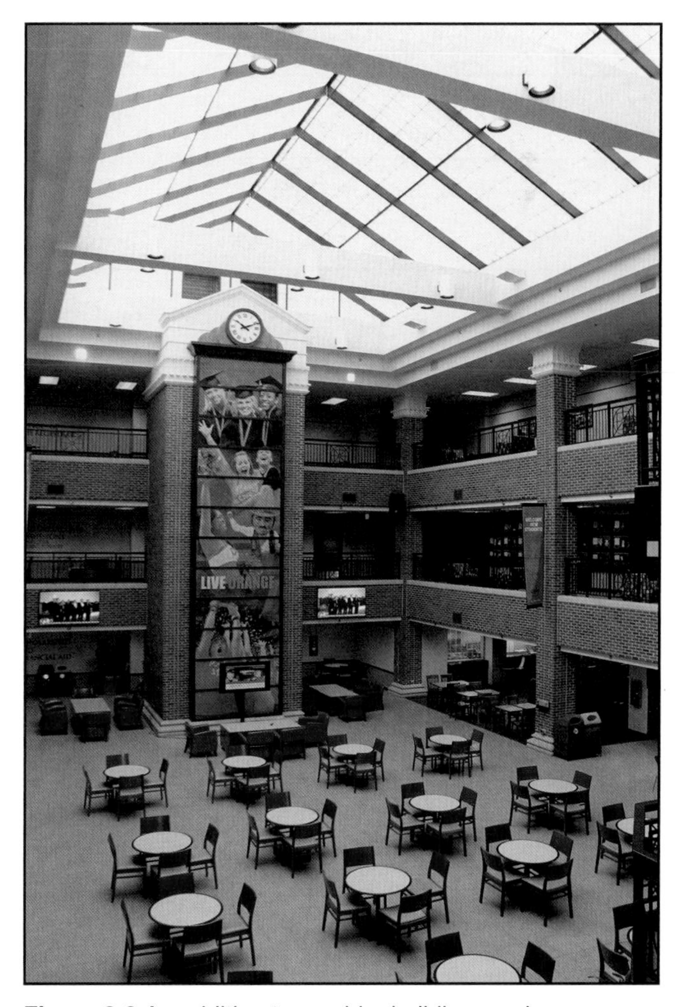

Figure 2.3 An addition to an older building, such as the large new courtyard in this building, will require that builders follow the currently adopted code.

Generally, requirements for existing structures will remain the same as those applied during the original construction. For example, an inspector cannot require the installation of sprinklers in an existing structure unless changes have been made to the building and there is a retroactivity clause in another adopted code. Some codes include a phrase such as, "All new and existing structures shall meet this requirement," which permits the inspector to require stricter code requirements for existing structures.

NOTE: Unless inspectors are specifically given the authority, they cannot apply current building code requirements to existing structures.

Current Codes and Standards

Fire and life safety inspectors must know the adopted editions of codes and standards in their jurisdiction. Some codes may be created locally, such as a zoning code that regulates the size of advertising signs along streets or highways. Others, such as building and fire codes, may be model codes adopted by the jurisdiction as they were published or with amendments.

Most model codes are revised on a regular basis, typically every three to five years. A revised edition of the model code does not take effect unless the AHJ legally adopts the new edition. It is not unusual for an older edition of a model code to continue to be enforced for a number of years after a new edition has been released **(Figure 2.4)**.

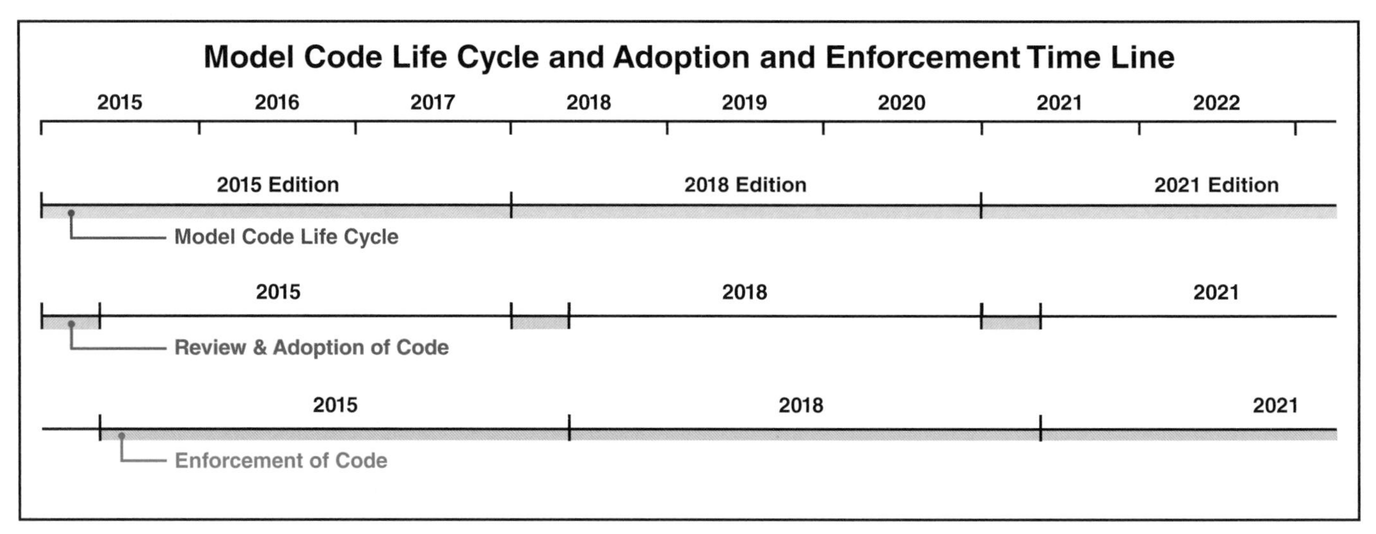

Figure 2.4 Nationally adopted model code editions are adopted on 3- to 5-year cycles. During this time, the older code will be in effect.

Automatic Notification

If fire inspections are an integral part of the procedure for obtaining business licenses and permits, inspectors may automatically receive notification when business and commercial changes occur in the community.

Consistent Codes and Standards

An inspector should continually monitor applicable code provisions enacted at all levels of government. Doing so prevents conflicts between the various codes. Fire and life safety codes, for example, must be consistent with other codes to avoid duplication of effort or the problem of attempting to enforce contradictory regulations. In addition, the interpretation of fire and life safety codes must be clear and consistent within the purpose and intent of the actual fire codes.

An inspector must be familiar with the types of structures in the jurisdiction and develop a good working relationship with other code enforcement and inspection departments. Additional methods for keeping current with building changes are:

- Monitoring business license applications
- Monitoring the issuance of business occupancy and activity permits
- Conducting an annual occupancy inventory survey
- Monitoring new electrical service applications

If a conflict occurs between codes, a cooperative manner and a common sense approach are the best options for finding solutions. A code war is destructive to the code enforcement process as well as to the professional image of all involved departments.

Most codes allow the AHJ some degree of latitude when interpreting code provisions. Inspectors from different departments must maintain open lines of communication and work to produce equivalent, alternate solutions **(Figure 2.5)**. The inspector should inspect for code compliance; if compliance cannot be achieved, an alternative method should be requested from the AHJ.

NOTE: An Inspector I should understand the concept of codes while an Inspector II should know how to evaluate the performance of materials and practices designed to meet the intent of the code requirement.

Figure 2.5 Multiagency meetings like this one are important in maintaining cooperation among various code enforcement personnel. *Courtesy of Federal Management Agency, George Armstrong, Photographer.*

Standards

Standards are an attempt to obtain consistency in design, practice, and materials. Standards offer inspectors a guide to practices and designs that have been proven successful. For example, the ASME/ANSI Standard A17.1 *Safety Code for Elevators and Escalators* is the result of standards developed over the past century. This section outlines the most commonly accepted standard-making organizations in North America.

Industry Standard — Set of published procedures and criteria that peer, professional, or accrediting organizations recognize as acceptable practice.

Consensus Standard — Rules, principles, or measures that are established though agreement of the members of the standards-setting organization.

A number of organizations develop and publish consensus standards that relate to building construction, fire and life safety, and hazardous processes. While these **industry standards** do not have the force of law unless adopted by the jurisdiction's governing body, they are recognized as authoritative documents.

The term **consensus standard** refers to a document that a committee of experts has developed and agreed upon before publication. Consensus standard committees are usually represented by trade associations, scientific and professional societies, special interest groups or persons, government agencies, and standards-developing organizations. Several of the most prominent organizations are as follows:

- National Fire Protection Association® (NFPA®)
- ASTM International (originally known as the *American Society for Testing and Materials [ASTM]*)
- Underwriters Laboratories (UL)
- American National Standards Institute (ANSI)

In Canada, each province adopts approved standards independently from the Canadian federal government. Approvals are given by the Standards Council of Canada (SCC) and ANSI, which facilitate standards development through the consensus process.

National Fire Protection Association® (NFPA®)

The NFPA® develops and publishes the majority of the consensus standards concerning fire protection, electrical systems, and life-safety systems used in the U.S. and Canada. When the need for a standard is recognized, the NFPA® invites participants with expertise in that field to form a committee to develop a draft. The completed draft is then made public for review and comment. The committee reviews the public comments and may or may not incorporate them into the finished document.

The final version of the standard is then submitted to the NFPA® general membership for publication. Although there are hundreds of NFPA® standards, some of the ones that an inspector should be familiar with include:

- NFPA® 1, *Fire Code*™
- NFPA® 13, *Standard for the Installation of Sprinkler Systems*
- NFPA® 14, *Standard for the Installation of Standpipe and Hose Systems*
- NFPA® 25, *Standard for the Inspection, Testing, and Maintenance of Water-Based Fire Protection Systems*
- NFPA® 70, *National Electrical Code®*
- NFPA® 72, *National Fire Alarm and Signaling Code®*
- NFPA® 101, *Life Safety Code®*
- NFPA® 241, *Standard for Safeguarding Construction, Alteration, and Demolition Operations*
- NFPA® 704, *Standard System for the Identification of the Hazards of Materials for Emergency Response*

- NFPA® 1031, *Standard for Professional Qualifications for Fire Inspector and Plan Examiner*
- NFPA® 5000, *Building Construction and Safety Code®*

In addition to standards, the NFPA® publishes handbooks to help inspectors interpret the standards. These handbooks may also be introduced in legal proceedings to demonstrate accepted industry practices. Some of the handbooks that inspectors may use include:

- NFPA® 1: *Fire Code™*
- NFPA® 101: Life Safety Code® Handbook
- NFPA® 72: National Fire Alarm and Signaling Code
- NFPA® 70: *National Electrical Code® Handbook*
- NFPA® 13: *Standard for the Installation of Sprinkler Systems*
- *Fire Protection Handbook®*

ASTM International

ASTM International is a consensus-based standards writing and testing organization. ASTM International develops testing processes that other testing organizations use in the development of safety products. Some of the standards that affect building construction are:

- E84 *Standard Test Method for Surface Burning Characteristics of Building Materials*
- E108 *Standard Test Methods for Fire Tests of Roof Coverings*
- E119 *Standard Test Methods for Fire Tests of Building Construction*

Underwriters Laboratories (UL)

Underwriters Laboratories (UL) is an independent, not-for-profit product safety testing and certification organization. Products that bear the UL label have been tested for their intended use and are certified as safe when properly used and maintained. UL also provides third-party testing and certification.

UL has developed more than 800 standards for safety. Some that directly relate to fire and life safety include:

- UL 260 *Standard for Dry Pipe and Deluge Valves for Fire-Protection Service*
- UL 268 *Smoke Detectors for Fire Alarm Systems*
- UL 299 *Dry Chemical Fire Extinguishers*
- UL 300 *Standard for Fire Testing of Fire Extinguishing Systems for Protection of Commercial Cooking Equipment*
- UL 1626 *Standard for Residential Sprinklers for Fire-Protection Service*

American National Standards Institute

The American National Standards Institute (ANSI) is a private, nonprofit organization that administers and coordinates the voluntary **standardization** and conformity assessment system **(Figure 2.6, p. 50)**. The consensus process is guided by ANSI's principles of consensus, due process, and openness.

Standardization — Process of making or creating things that meet an established criteria.

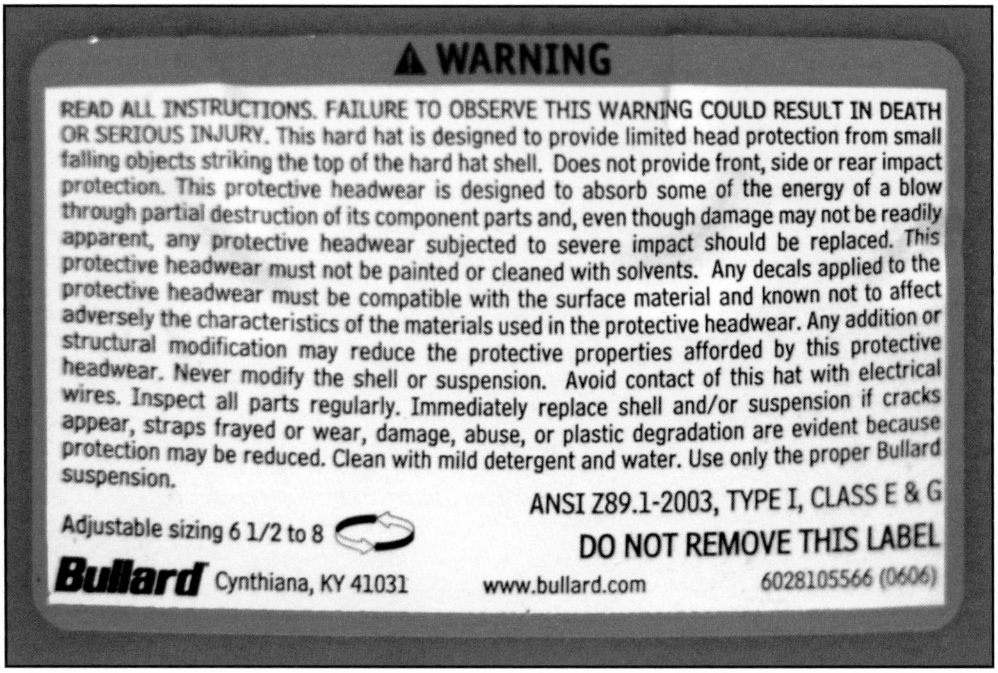

Figure 2.6 ANSI labels appear in personal protective equipment indicating that the equipment meets the established design standards.

Besides accrediting organizations that develop consensus standards, ANSI includes an appeals process for manufacturers who wish to contest test results of their products or materials. ANSI ensures that access to the standards process is made available to anyone directly or materially affected by a standard that is under development. Thousands of individuals, companies, and government agencies voluntarily contribute their efforts to standards development. Many ANSI standards are cross-referenced between NFPA® and Occupational Safety and Health Administration (OSHA) documents.

Standards Council of Canada (SCC)

The Standards Council of Canada (SCC) is a federal Crown corporation with the mandate to promote efficient and effective standardization. The SCC also represents Canada's interests in standards-related matters in foreign and international forums. In general, standards approved by the SCC are the basis for regulations that influence fire and life safety. These standards may be developed by a number of organizations:

- Canadian Standards Association (CSA)
- Underwriters Laboratories of Canada (ULC)
- Canadian General Standards Board (CGSB)
- Bureau de normalisation du Quebec (BNQ)

While many Canadian fire and emergency service agencies use the NFPA® professional qualifications standards for training and certifying personnel, none of those standards are law in Canada. ANSI standards are recognized in Canada but normally are used as references in Canadian codes. For instance, Transport Canada (TC) is responsible for a number of key acts and regulations that govern Canada's transportation system. This organization references ANSI Z26.1-1996, *Safety Code for Safety Glazing Materials for Glazing Motor Vehicles Operating on Land Highways,* in the Motor Vehicle Safety Act.

Complaint Procedures

The inspection organization's standard operating procedures (SOPs) should outline a procedure for receiving and processing complaints. An inspector should process and act upon each complaint in a consistent manner. An inspector taking a complaint should record all pertinent information **(Figure 2.7)**.

Complaints that do not require immediate attention can be routinely assigned to inspection staff or qualified company officers for research and investigation. Complaints that involve a serious life-safety threat require immediate action and rapid correction.

The type of occupancy, location within the occupancy, and severity of the nature of the complaint often determine whether an inspector needs to give advance notice or to obtain an administrative warrant to enter the location. A common situation requiring the inspector's judgment would be a citizen calling to complain about chained and padlocked emergency exits at a nightclub. If the nightclub management knew that an inspector was on the way, they would have the exits unlocked before the inspector arrived. Although voluntary compliance is always a goal, occasionally the issuance of a citation and subsequent penalty sends a strong message to the community regarding the severity of these violations.

An inspector must carry and display appropriate identification and explain the purpose of an unannounced inspection. Upon finding a code violation, an inspector must initiate the process that leads to corrective action. Inspectors must be prepared to deal with negative attitudes when they act on complaints. An inspector must maintain a professional demeanor even if the other person does not.

When a complaint has been resolved, the person who initiated it must be formally notified and thanked. This acknowledgement provides confirmation that the person has contributed to the community's safety.

All complaints must be documented and maintained in the organization's record-keeping system **(Figure 2.8)**. The complaint form may be filed by location, date, or type. When records are kept electronically, they can be cross-referenced for easier retrieval.

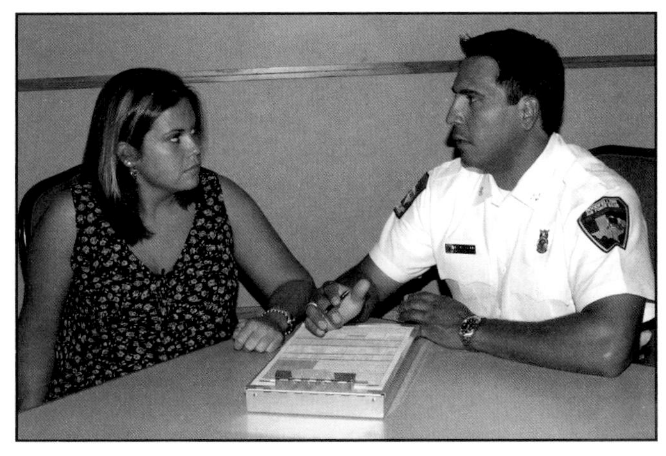

Figure 2.7 A standardized complaint form is useful to see that all information is complete and correct.

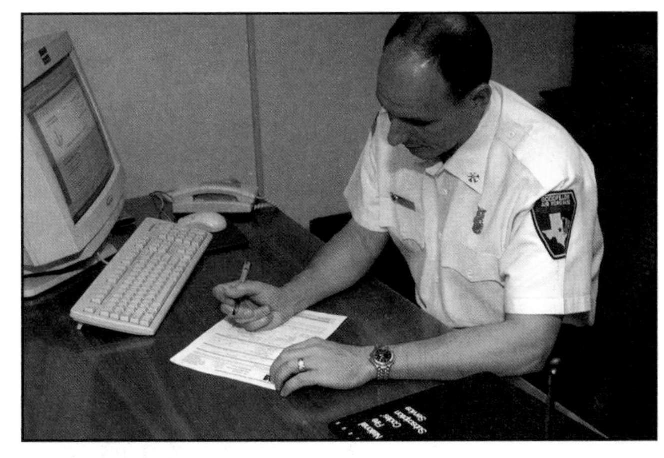

Figure 2.8 Complaints must be documented carefully for accurate record-keeping and use by other members of the department.

Accurate record keeping is essential. Records can confirm when a pattern of violations begins to appear. An inspector should always refer to the complaint records before making any inspection to determine whether violations have been reported about the property previously. It is also possible that the owner of several properties may have a record of code violations involving more than one address. Cross-referencing code violations, property ownership, and property addresses are important means for tracking repeat offenders.

Permits

A permit is an official document that grants a property owner or other party permission to perform a specific activity. Permits may be obtained for a single event, such as a fireworks display, or for a continuing operation or process, such as the manufacture of fireworks. The building, fire, and/or code enforcement departments may be assigned the authority to issue and monitor permits.

An inspector must be aware of all types of permits or licenses issued by the jurisdiction as well as the steps in the application and issuance process. The inspector should also be aware of previous permits that have been issued and of the types of activities the permits cover.

Permits are issued for two reasons:

- To ensure that hazardous situations or conditions are not allowed to develop within the jurisdiction without the knowledge and approval of the AHJ.
- To enable inspection personnel to verify that the conditions meet the applicable code requirements.

Permits are usually issued for a specific condition, location, and time. They are not transferable beyond the conditions stated on the permit. The permit authorizes, by law, the right of entry for the fire inspector at any time to verify compliance with code requirements and the conditions of the permit. In general, permits are not issued to allow a party to avoid having to meet minimum code requirements.

Each model code defines the types of situations that require a permit or license. When adopting codes, local governments may add to or delete permit requirements based on local needs.

Within the codes, the permit requirement is usually contained in the section relating to the activity requiring the permit. The *International Fire Code®* (IFC®) has two types of permits: operational and construction. Operational permits allow a person or group to conduct an operation or business for a specified time or until the permit is renewed or revoked. A construction permit is issued for the installation or alteration of a system or equipment. NFPA® 1 does not make this distinction. **Table 2.1** compares the required operational permits listed in the NFPA® and ICC® fire codes.

Operational Permits

Typically, operational permits that relate to fire and life safety are used to regulate certain activities, including:

- Storage, use, dispensing, and handling of hazardous materials
- Hazardous operations or processes

Table 2.1
Model Code Activities Requiring Permits

International Code	National Fire Protection Association®
Aerosol Products	Aerosol Products
Amusement Buildings	Amusement Parks
Aviation Facilities	
	Aircraft Fuel Servicing
	Airport Terminal Buildings
	Ammonium Nitrate
	Asbestos Removal
	Automatic Fire-Suppression Systems
Battery Systems	Battery Systems
	Automobile Wrecking Yards
	Automotive Fuel Servicing
	Candles, Open Flames, and Portable Cooking
Carnivals and Fairs	Carnivals and Fairs
Cellulose Nitrate Film	Cellulose Nitrate Film
	Cellulose Nitrate Plastic
	Clean Rooms
Combustible Dust-Producing Operations	Dust-Producing Operations
	Combustible Fibers
	Combustible Material Storage
Compressed Gases	Compressed Gases
	Consumer Fireworks (1.4 G)
Covered Mall Buildings	Covered Mall Buildings
Cryogenic Fluids	Cryogens
Cutting and Welding Operations	Cutting and Welding Operations
	Display Fireworks (1.3 G)
Dry Cleaning Plants	Dry Cleaning Plants
Exhibits and Trade Shows	Exhibit and Trade Shows
Explosives	Explosives
	Fire Alarm and Detection Systems and Related Equipment
Fire Hydrants and Valves	Fire Hydrants and Water-Control Valves
	Flame Effects
Flammable and Combustible Liquids	Flammable and Combustible Liquids
Floor Finishing	
Fruit and Crop Ripening	
Fumigation and Thermal Insecticidal Fogging	
	Grandstands and Bleachers, Folding and Telescopic Seating

Continued

Table 2.1 (Continued)

International Code	National Fire Protection Association®
Hazardous materials	Hazardous Materials
Hazardous Materials Production Facilities	
	High-Piled Combustible Storage
	High-Powered Rocketry
Hot Work Operations	Hot Work Operations
Industrial Ovens	Industrial Ovens and Furnaces
	Laboratories
Lumber Yards and Woodworking Plants	Lumberyards and Woodworking Plants
Liquid- or Gas-Fueled Vehicles or Equipment in Assembly Buildings	Liquid- or Gas-Fueled Vehicles
Liquefied Petroleum Gas	Liquefied Petroleum Gases
Magnesium	
	Marine Craft Fuel Servicing
	Membrane Structures, Tents, and Canopies — Permanent
	Membrane Structures, Tents, and Canopies — Temporary
Miscellaneous Combustible Storage	
	Oil- and Gas-Fueled Heating Appliances
Open Burning	Open Burning
Open Flames and Torches	Open Fires
Open Flames and Candles	
Organic Coatings	Organic Coatings
	Organic Peroxide Formulations
	Oxidizers
	Parade Floats
Places of Assembly	Places of Assembly
Private Fire Hydrants	Private Fire Hydrants
Pyrotechnic Special-Effects Material	Pyrotechnic Articles
	Pyrotechnics Before a Proximate Audience
Pyroxylin Plastics	Pyroxylin Plastics
Refrigeration Equipment	Refrigeration Equipment
Repair Garages and Motor Fuel-Dispensing Facilities	Repair Garages and Service Stations
	Rocketry Manufacturing
Rooftop Heliports	Rooftop Heliports
	Solvent Extraction
Spraying or Dipping	Spraying or Dipping of Flammable Finish
	Standpipe Systems

Continued

Table 2.1 (Concluded)

International Code	National Fire Protection Association®
	Special Outdoor Events, Carnivals, and Fairs
Storage of Scrap Tire and Tire Byproducts	
	Tar Kettles
Temporary Membrane Structures, Tents, and Canopies	
Tire-Rebuilding Plants	Tire-Rebuilding Plants
	Tire Storage
Waste Handling	
	Wildland Fire-Prone Areas
	Wood Products

- Installation/operation of equipment in connection with hazardous operations, maintenance, and storage
- Open burning
- Temporary membrane structures or large-area tents

Construction Permits

The IFC® requires the issuance of construction permits for the installation of a fire protection system or the repair, abandonment, removal, storage, or use of particular item, such as:

- Automatic fire extinguishing systems
- Medical gas systems
- Standpipe systems
- Fire alarm and detection systems and related equipment
- Spraying or dipping processes
- Private fire hydrants
- Fire pumps and related equipment
- Industrial ovens
- Flammable and combustible liquids
- Hazardous materials
- Compressed gases and cryogenic fluids
- Liquefied petroleum gas
- Temporary tents and membrane structures

NOTE: The permitting process is covered in the Inspector II portion of this chapter.

Local Code Development Process

The use of model building and fire codes has greatly simplified the process of developing a fire and life safety code for a jurisdiction. Because model codes represent a consensus of expert opinions, local code developers are relieved of the task of justifying each code section before it is formally adopted by the local legislative body.

While jurisdictions take full advantage of the benefits of model codes, special fire and life safety code provisions may need to be developed to address a local condition or fire department requirement. Code amendments are developed locally or regionally (when similar conditions exist in nearby communities). For example, the lack of an adequate water supply might justify a code amendment requiring facilities to provide an impounded water source on site **(Figure 2.9)**. When amendments are developed, officials must take care to avoid conflicts with other portions of the fire and life safety code or other adopted codes of the community.

Most jurisdictions have a process for developing or amending local codes. This process generally includes the following activities:

- Identify the problem.
- Identify affected stakeholders.
- Form a task force.
- Draft the proposed code.
- Submit the code for legal review.
- Formally adopt the code.

Figure 2.9 Code amendments can mandate special features at remote sites, such as an impounded water supply, if there is insufficient access to the utility water supply at the site.

Identify the Problem

The first step in developing a code or amendment involves identifying the reason why it is needed. Ideally, most communities update all their codes at the same time for code consistency and agreement. Code amendments presented to address the community's fire problem must be justifiable to the governing body

Often a need for a change is recognized through observations made during the inspection process. A new condition or situation in a community usually necessitates such a change. Occasionally, a technological change in fire protection is deemed important enough to warrant a change. For example, when carbon monoxide detectors became available, many communities and states adopted fire and life safety code provisions requiring them. Conversely, some jurisdictions removed the requirements because their climate does not require the extensive use of heating systems.

Identify Affected Stakeholders

When a new code or amendment is proposed, the individuals and groups it affects should be included in its development. Collectively these individuals or groups are called *stakeholders*, who may include members of the following:

- Building industry
- Chambers of Commerce
- Insurance companies
- Local citizen groups
- Others who may have a financial or community interest in the laws governing the regulated industry

Local Officials as Stakeholders

Although they are not true stakeholders, elected officials serve at the discretion of the voters. They represent all stakeholders and are very receptive to their collective voice. Therefore, it is of utmost importance that fire code officials fully inform elected officials of the nature and effect of proposed changes or amendments to codes or adoption of new codes.

The most effective way to adopt a code change or amendment is to explain accurately and realistically the benefits that the new code will bring to the community. Equally important is to disclose the costs needed to make the changes. It is better to be forthright with stakeholders who may oppose the changes than it is to take a hard position that may be defeated by the local legislative body. Effective code change should be an attempt to find common ground.

Form a Task Force

A task force may be formed to evaluate or develop a new code or amendment. The task force should be representative of the community's demographics. Representatives from the following groups may be included:

- Building construction trades
- Insurance industries
- Fire protection industries
- Allied health service organizations
- Local human services agencies

- Fire department bargaining units
- Real estate agencies
- Local governing bodies
- Other stakeholders

Duties of the task force may include:

- Schedule meetings and work sessions that are open to the public (**Figure 2.10**). Work sessions are spent reviewing broad fire and life safety code proposals.
- Review model codes to select the one that best meets local needs.
- Discuss the type and number of possible amendments to the model code and make recommendations for their inclusion.

Depending upon local and state/provincial laws governing open public meetings, the task force always should prepare agendas and post meeting information. The local municipal authority, usually the municipal clerk, should coordinate these efforts with the fire code official's assistance. The meeting minutes should be recorded and a transcript prepared for public record.

The task force's final report should include fire and life safety code recommendations that can be used when drafting the proposed fire code. The report should be presented to the mayor, local legislative leader, and/or the municipal manager. They will usually request or assign the fire and life safety code official the task of preparing the new code legislation.

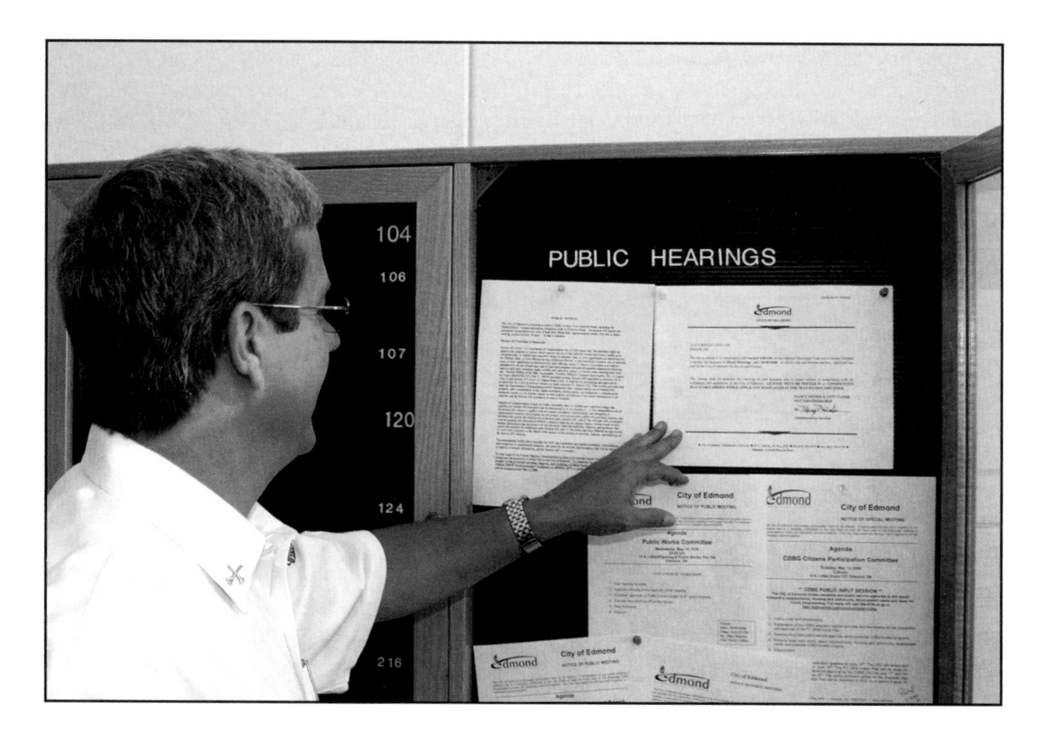

Figure 2.10 Code review meetings will be open to the public.

Draft the Proposed Code

Once the task force has made its recommendations, usually the fire and life safety code official, assisted by other department members, begins to write the new fire and life safety code. If a model code is being adopted, this process is simplified because the code may require little or no modification. After the model code has been selected, the preparation of amendments, additions to sections, and removal of sections occurs. When a model code is altered, the fire and life safety code official must take care to avoid conflicts or inconsistencies between sections of the code.

The process of drafting the code should begin with the most general, broad base of code language and proceed to more specific language. This approach will reduce confusion and misinterpretations. Any local additions or amendments should be inserted alongside the amended code section number. If the proposed code is a model code, number any amendments using the same general system used in the code **(Figure 2.11, p. 60)**.

Submit Code for Legal Review

Once the new code is written, the jurisdiction's legal counsel should review it. This review ensures that the code meets the legal standards of the community, state, and federal government. The code must protect fundamental rights guaranteed under the federal and state constitutions. In essence, the code may not single out or discriminate against individuals by differentiating the way in which provisions of the code are enforced or interpreted.

The fire and life safety code official is responsible for explaining the purpose of the code to the legal counsel. The fire and life safety code official must also clearly describe the provisions of the proposed changes so that the legal language that the counsel prepares completely meets the requirements of the community.

Once the legal review has been completed and the code is in a final draft form, the code drafting committee should review it again to guarantee that all of the desired provisions have been included and the meaning of each section is correct. The final draft of the proposed code is then returned to the fire and life safety code official who prepares the proposal for adoption by the local legislative board.

Formally Adopt the Code

Once the fire and life safety code official has received the final draft of the proposed code, the process of preparing a legislative resolution for its formal adoption begins. Each jurisdiction's code has a legislative procedure for presenting documents for approval and inclusion into the local municipal code. Generally, the process includes preparation of the proposed legislation, study by the legislative body, formal presentation, discussion, and legal action.

Prepare Formal Resolution

The formal resolution to adopt the new fire and life safety code language should include the following elements:

- Nature of the legislation
- Location of the legislation within the framework of the municipal code group
- Authority of the community to adopt such code language
- Need of the community to adopt a new edition of the code to reflect local changes

Amendments to Fire Code

Adopted References

The International Fire Code®, 2003 Edition, as published and copyrighted by the International Code Council including all appendices and added amendments.

The International Building Code®, 2003 Edition, as published and copyrighted by the International Code Council including appendices.

Amendments to IFC:

Key Boxes

506.1 The Fire Chief is authorized to require a key box installed in an approved location.

506.1.2 Key Boxes. The lock box must be an approved model utilized by the Fire Department and shall be installed 60" above finish grade. Authorized lock box order forms are available at the Fire District Administrative Office, Monday through Friday 8:00 am to 4:00 pm. Lock box forms can be transmitted by e-mail or facsimile.

Horn & Strobe

901.4.3.1 Horn & Strobes. Additional Horn & Strobe (weather proof) shall be installed in an approved location on the exterior of the building in addition to a water gong or electric bell. Horn/Strobe shall be installed on the front, or address side of the building as approved by the Fire Chief.

Fire Sprinkler Systems

903.1.1 Commercial development. All new commercial occupancies for which a building or construction permit is obtained shall be protected by a fully automatic Sprinkler System. Installation of the sprinkler system shall be in accordance with the requirements of NFPA 13, unless otherwise approved by the Fire Chief. Existing buildings, structures and occupancies will not require retrofitting with fire sprinkler systems to meet current code standards unless:

- ☑ Building fire resistance has decreased; or
- ☑ Building area has increased to more than 2,500 Sq/feet; or
- ☑ Building occupant load has increased; or
- ☑ Building occupancy classification has changed; or
- ☑ Fire or other structural damage in buildings exceeding 50% of the Sq/footage; or
- ☑ As determined by the Fire Chief.

Standpipes

905.3.1 Building height. Class I standpipe systems shall be installed throughout buildings where the floor level of the highest story is one floor above the lowest level of the fire department vehicle access, or where the floor level of the lowest story is one floor below the highest level of fire department vehicle access.

Fire Alarm Systems

907.2 Where required - New building and structures. All new commercial occupancies for which a building or construction permit is obtained where the occupancy load exceeds 50 or more shall be protected by a fully automatic fire alarm system. Installation of the fire alarm system shall be accordance with the requirements of NFPA® 72, unless otherwise approved by the Fire Chief. All fire alarms systems shall be addressable systems with Class-A wiring. An approved fully automatic fire detection system shall be installed in accordance with the provisions of this code and NFPA® 72. Monitoring shall be by a central station as defined by NFPA® 72 section 3.3.193.1. Devices, combinations of devices, appliances and equipment shall comply with Section 907.1.2. The automatic fire detectors shall be smoke detectors, except that an approved alternative type of detector shall be installed in spaces such as boiler rooms, utilities rooms, and janitor's closet with water heater and sink, where, during normal operation, products of combustion are present in sufficient quantity to actuate a smoke detector.

Figure 2.11 Amendments are usually numbered to indicate the section of the model code they may alter or amend.

- Authority of the community to enforce the legislation
- Term or **sunset provisions**, if any, regarding the proposed code
- Proposed implementation date for the code once adopted

A cover letter is prepared by the fire chief or fire and life safety code official and sent to the municipal manager for review. Some municipalities require an additional document, often called a *council communication* or *executive summary*, to be included with the manager's letter. The manager then forwards copies to all municipal board members and the clerk for inclusion on the Board's agenda.

Sunset Provision — Clause in a law or ordinance that stipulates the periodic review of government agencies and programs in order to continue their existence.

Discuss Legislation and Adopt Code

Once the proposed code has been scheduled, it is common for the legislative body to include discussion of the proposed code during a **work session**. This session is an opportunity for both proponents and opponents of the legislation to express their views before the more formal board meeting.

Work Session — Formal, open meeting by a legislative body to study the merits of the proposed legislation and ask questions of the fire and life safety code official and the public regarding the provisions of the proposed code. At a work session, there are no actions taken.

After a work session, members of the governing body may request that additional provisions be developed and that others be deleted or modified. The AHJ should make the requested changes and resubmit the proposed code for additional consideration.

After the code has progressed through the Board study process, the governing body schedules it for formal consideration. The actual presentation of the resolution is usually made to the legislative body by one of the following people:

- Municipal manager (for example, city manager or county manager)
- Member of a legislative body (for example, council member or board member)
- Manager of a municipal department that has oversight on the code
- Member of the citizen advisory committee appointed to develop new legislation
- Fire and life safety code official acting on behalf of the fire chief

Once the proposed code legislation is presented, the AHJ is typically asked to provide a detailed explanation of the proposed code changes. Then, the Board president or mayor opens a public comment period during which citizens can speak in favor of or in opposition to the proposal. Following discussion, members of the Board vote on the proposal. In most communities, a simple majority is required for passage of the enabling legislation. In some communities, a supermajority may be required for passage of certain types of legislation.

Introduce New Code

After the enabling legislation is passed, the clerk posts the new legislation and maintains a copy for public review. The effective enforcement date is contained within the enabling legislation.

When the new code language is adopted, the fire and life safety code official should immediately begin a formal notification and information process within the community. For example, public service announcements in local newspapers, radio and television stations, electronic media outlets, and letters to all who have an interest in the provisions of the code should be prepared and delivered. It is important to continue to present and promote the new code even after its formal adoption. Promoting the new code helps provide a smoother transition from the old fire code to the new one **(Figure 2.12)**.

Code Adoption and Amendment Process

1. Code official begins code adoption process by setting up the code adoption committee and verifying process through legal counsel.

↓

2. Public comments solicited on new code. Time for this can vary from a few weeks to several months.

↓

3. Code adoption committee begins deliberations on new code language and on public comments. This arduous process will take varying amounts of time.

↓

4. Code adoption committee provides written document with recommendations for any changes or updates to the code official.

↓

5. Code official develops official document with changes and deletions to be presented to the City Council or other legally organized governing body of the jurisdiction.

↓

6. City Council provides documents to and hears testimony from the public according to the legally adopted rules for the jurisdiction.

↓

7. City Council adopts the new rules with modifications made during the process. In addition, sets date for final adoption to be effective.

Figure 2.12 The code adoption process varies by jurisdiction. The one shown is a general example.

Retention of Old Fire Codes

The fire prevention program should include the requirement for maintaining records of fire and life safety inspections and complaint resolution. This archive should include copies of each edition of the fire and building codes, the date of adoption, a list of amendments, and copies of any revisions that were made. This historical data will assist the fire and life safety inspector when variances are requested, occupancy categories change, and when responding to complaints.

Code Modification and Appeals Procedures

While enforcing fire and life safety codes, it may become evident that the code needs to be modified to be more effective or fair. Additionally, citizens may appeal a code requirement they perceive to be unfair or overly burdensome. Fire and life safety inspectors should be aware of the process for modifying the code.

Code Modification

Code modifications may be called *variances, appeals,* or *modifications* depending on the AHJ. The modification must meet the intent of the code as the fire official interprets it. The details of any modification must be recorded as per the local jurisdiction's standards. The goal in all cases is to maintain the required safety level specified by the code.

A number of situations may involve requests to modify the code:

- Desire to use new building materials or technologies that have not yet been recognized by the model codes and standards
- Desire to use a material or technology in place of approved materials and methods
- Confusion or conflicts that occasionally occur between the different model codes and standards
- Desire to implement performance-based design (PBD) concepts
- Need to construct special occupancies due to unique circumstances
- Desire by the owner/contractor to reduce the cost of construction
- Disagreement between the owner/contractor and the fire prevention division regarding the local interpretation and implementation of a code or standard
- Belief that the fire code does not apply to the owner/occupant or contractor

To receive consideration, an applicant must present a written request to the AHJ, usually the fire prevention division. The AHJ then reviews the request to verify that the fire code's intent is being observed and that public safety is maintained.

As previously stated, in most jurisdictions, fire and life safety inspectors acting alone have little authority to approve requests for modification of code requirements. The inspector usually processes the requests and receives a formal interpretation from one of the following authorities:

- Superior inspection bureau officer
- Staff fire protection engineer
- Fire marshal
- Code official
- Contract fire protection consultant
- Local board of appeals
- Code or standards-writing organization

After review and an official's decision, the applicant receives an official reply from the senior fire code official having jurisdiction. The reply includes a detailed explanation of the decision, which the applicant can use for an appeal if necessary. Detailed records of all decisions are generally kept in the department because they are official local interpretations of the code as applied by the jurisdiction **(Figure 2.13, p. 64)**.

PHOENIX FIRE DEPARTMENT
Fire Prevention Section
150 South 12th Street
Phoenix Arizona 85034-2301
(602) 262-6771 FAX: (602) 271-9243

Petition of Appeal to the Fire Marshal

All appeals shall be detailed on this form. Supporting data may be attached and submitted if desired however, all entries and statements on this form shall be complete. Incomplete forms will not be accepted.

INTERNAL USE:		
Log Number:	Date Logged Out:	Date Logged In:
Case/KIVA Number:	Hearing Date:	Hearing Time:
Engineer/FPS Familiar with Project:	Occupancy Type:	Compliance Date:

Business/Occupancy Name:	Address:
Business Owner's or Corporate Agent's Name:	Mailing Address:
Tenant's Name:	Mailing Address:
Appellant's Name:	Mailing Address:

This appeal applies to (Check one):

☒ A project in the plans review stage. Building Safety Log No. ________________

☒ An alleged Fire Code violation.

An appeal is hereby made to the Fire Marshal for a deviation from Section ______________________ of the Phoenix Fire Code.
Briefly state the requirements being appealed.

State in detail what is proposed in lieu of literal compliance with the Fire Code:

Appellant's Signature:	Title:	Phone Number:

Building Owner's Signature:	Building Owner's Phone Number:

INTERNAL USE:

Decision of the Fire Marshal

☒ Approved ☒ Approved with Stipulations

☒ Denied ☒ See Attachment

Fire Department Official:	Date:

DISTRIBUTION: WHITE – Appeals File YELLOW – Appellant BLUE – Fire Prevention *WEB – 3 Completed & Signed Forms Required for Submission

91-48D Rev. 3/05
61582253060-CP

Figure 2.13 Jurisdictions provide a means for appealing or petitioning a decision made by an inspector. Forms are available from the appropriate authority or online. *Courtesy of Phoenix (AZ) Fire Department.*

Appeals Procedures

As stated earlier, usually a Board of Appeals has the authority to interpret the fire code, approve an equivalent method of protection or safety, and issue a ruling. Even if a Board of Appeals has broad, interpretive powers regarding the fire code, it is prohibited from changing the direct intent of the code or waiving any portion of it. The inspector must be familiar with certain issues involving the appeals process and the Board of Appeals:

- Whether the inspector can continue to enforce codes on the property during the appeal process
- Whether the ruling on the question will affect the enforcement of that section of the code or ordinance for other properties in the jurisdiction
- How modification of the code will affect the enforcement of other code sections

Adopted code regulations usually specify a time limit for submitting an appeal. The Board then accepts or rejects the appeal and files an appropriate notice. Rules and regulations used by the Board during its hearings are established by the local municipal code or state law **(Figure 2.14)**.

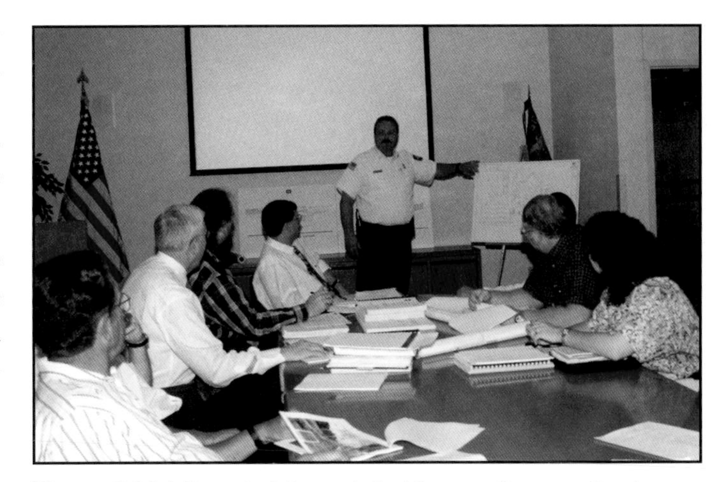

Figure 2.14 A Board of Appeals holds a routine meeting to decide whether to accept or reject modifications to the locally adopted fire code.

Most interpretations and decisions made by the Board are general in nature and apply to similar code circumstances within a jurisdiction. When the Board issues a ruling of this type, fire inspectors must implement it the same way in similar situations.

Occasionally, a Board of Appeals uses a one-time modification, which may be referred to as a *variance*. This decision is binding only for a particular circumstance and is not automatically applied to other situations. A variance is usually approved when a situation arises that is beyond the control of the applicant and cannot be changed in a manner that meets the requirements of the code. The widening of a roadway that blocks the normal emergency access to a structure would be an example. An alternate, less efficient option may be proposed and installed. A decision to grant a variance may be cause for concern by the fire inspector because other applicants may expect to receive similar treatment. Each circumstance will need to be evaluated separately.

Permit Process

The permit process begins when the property owner/occupant recognizes the need to obtain a permit for an operation or condition that will exist on the property. Citizens are not usually familiar with this process; therefore, inspectors are frequently contacted for guidance. The inspector should explain the permit process and emphasize that a conversation is not the same as granting a permit. The permit process is detailed in the model codes and includes steps for processing applications, research and review, issuing permits, evaluating permits, and permits issued in error.

Application

In most jurisdictions, a property owner who wishes to obtain a permit must complete a permit application. Typically, a permit processing fee, which is usually nonrefundable, is required when the application is submitted.

Depending on the type of permit being sought and local requirements, the applicant may be required to submit additional documentation, such as:

- Shop drawings
- Construction documents
- Plot diagrams
- Safety data sheets (SDSs or other chemical documentation)

Research and Review

Once the application for a permit is received, the inspector reviews it for completeness, accuracy, and supporting documents as needed. Then, the inspector needs to research the applicable standards. In some jurisdictions, a permit can be issued immediately upon review and approval. For example, over-the-counter permits may be issued for minor modifications to fire protection systems.

If the inspector's review and research reveal deficiencies, the inspector must advise the applicant. The applicant will be responsible for correcting deficiencies and returning requested documents for further review. Once the inspector is satisfied that all requirements have been met and that codes are followed, the permit can be issued.

Issue Permit

Once a permit is issued, the applicant is generally required to post the permit where the work or activity is being performed. If it is a construction permit, the permit may stipulate that additional fire and safety inspections will occur before construction can proceed.

A permit must clearly explain the conditions under which it has been issued. This explanation includes the actions that are allowed, the guidelines that must be followed, and the time frame for which the permit is applicable. Most enabling legislation that grants the authority to issue permits also specifies time limits for the activity or use. Each jurisdiction should also have procedures for granting extensions if they are needed.

Evaluate Status

The permit is enforced for the time specified. It may be necessary, however, for the permit to be reissued, renewed, or revoked depending on the situation. In all cases, the inspector should review the current activity status and determine if the conditions of the initial permit are still being met. If changes in the permit language are necessary, the owner/occupant may be required to submit additional documents before the permit can be renewed.

The inspector has the authority to revoke a permit if an inspection reveals that the stipulations in the permit are not being followed **(Figure 2.15)**. Another cause for revocation is an inspection revealing that false

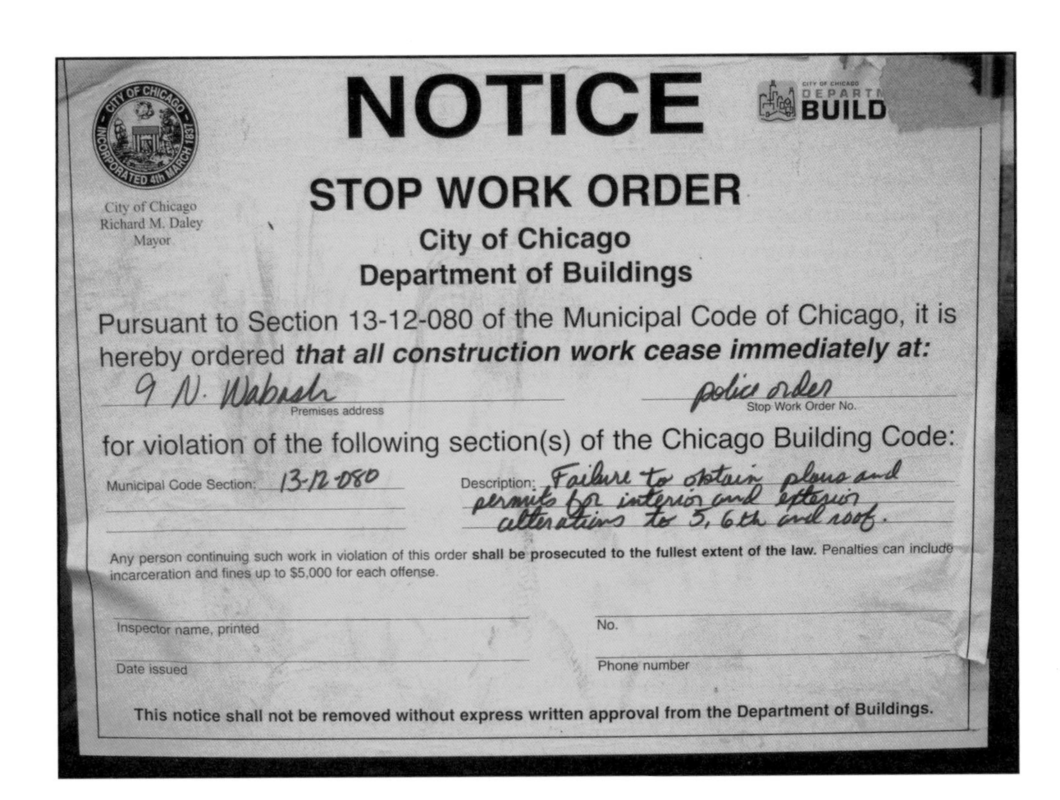

NOTICE

CITY OF CHICAGO DEPARTM[ENT OF] BUILD[INGS]

CITY OF CHICAGO INCORPORATED 4th MARCH 1837

City of Chicago
Richard M. Daley
Mayor

STOP WORK ORDER

City of Chicago
Department of Buildings

Pursuant to Section 13-12-080 of the Municipal Code of Chicago, it is hereby ordered ***that all construction work cease immediately at:***

9 N. Wabash — Premises address

police order — Stop Work Order No.

for violation of the following section(s) of the Chicago Building Code:

Municipal Code Section: 13-12-080

Description: Failure to obtain plans and permits for interior and exterior alterations to 5, 6th and roof.

Any person continuing such work in violation of this order **shall be prosecuted to the fullest extent of the law.** Penalties can include incarceration and fines up to $5,000 for each offense.

Inspector name, printed

No.

Date issued

Phone number

This notice shall not be removed without express written approval from the Department of Buildings.

Figure 2.15 Failing to secure required permits may result in an order to stop work.

statements or misrepresentations of the actual conditions on the property were made. Depending on the circumstances, a permit may be revoked temporarily or permanently pending correction of violations **(Figure 2.15)**.

Permit Issued in Error

Occasionally, a jurisdiction may issue a permit in error. The procedure for dealing with this situation and the legal basis for revocation typically are addressed in either the code or the enabling legislation that adopts the code. This issue tends to be a challenging one due to the sensitive nature of revoking a privilege. The inspector needs to work closely with the AHJ on the proper procedure to handle this situation because it varies greatly from jurisdiction to jurisdiction.

Essentially, if the jurisdiction issues a permit in error, and the error is not due to gross negligence, the applicant/owner will still be responsible for correcting the deficiency. The inspector's job is to verify code compliance. If the design was deficient and the problem was missed by the inspector, the owner is still responsible for ensuring compliance. To avoid issuing permits in error, the AHJ can require special inspections for complex processes or construction and certification of compliance from the special inspector.

NOTE: Complex complaint procedures are described in Chapter 16.

Chapter Summary

Locally adopted codes and standards provide the legal basis for the tasks that an inspector performs. Originally developed through a consensus process, codes and standards are not mandatory until they are adopted by the AHJ. Once adopted, the codes and standards can be modified to meet local needs. An inspector should be familiar with the code adoption and modification process, the appeals process, and the original consensus process. Developing and adopting fire and life safety codes at the local level is a process that requires the involvement of interested and affected citizens and business representatives. Inspectors must be aware of the necessity for including citizens, for permitting comments and discussion, and the value of working to achieve a common goal for the safety of the community.

The inspector must understand the steps needed within a jurisdiction for filing complaints so that concerns are addressed and procedures are followed fairly. The inspector must also be familiar with the permitting process.

Review Questions

1. What are the differences between a code and a standard?
2. What steps are included in the complaint process?
3. What must an Inspector I know about permits?

1. What steps are included in the process most jurisdictions use to develop or amend local codes?
2. What situations might involve a request to modify a code?
3. What is the role of an Inspector II in the permitting process?

Photo courtesy of National Institute of Standards and Technology (NIST).

Chapter Contents

Case History **73**

Science of Fire **74**

Physical and Chemical Changes 74

Modes of Combustion 76

Temperature, Energy, and Heat 77

Sources of Thermal Energy 80

Heat Transfer 83

Passive Agents 87

Fuel 87

Oxidizer 92

Self-Sustained Chemical Reaction 95

Classification of Fires 96

Stages of Fire Development **98**

Factors that Affect Fire Development 99

Incipient Stage 103

Growth Stage 104

Fully Developed Stage 107

Decay Stage 108

Fire Control **108**

Temperature Reduction 109

Fuel Removal 110

Oxygen Exclusion 110

Chemical Flame Inhibition 111

Chapter Summary **111**

Review Questions **111**

Key Terms

Term	Page
Autoignition	79
Autoignition Temperature	80
Buoyant	84
Ceiling Jet	104
Chemical Flame Inhibition	96
Combustion	75
Conduction	83
Convection	83
Endothermic Reaction	75
Energy	74
Exothermic Reaction	75
Fire	75
Fire Point	90
Fire Tetrahedron	76
Fire Triangle	76
Flame	76
Flammable Liquid	90
Flammable Range	94
Flash Point	90
Flashover	106
Flow Path	104
Free Radicals	9
Fuel	74
Fuel Load	102
Heat Flux	83
Heat of Combustion	87
Heat Release Rate (HRR)	87
Ignition	79
Incipient Stage	103
Kinetic Energy	75
Lower Flammable Limit (LFL)	94
Matter	74
Miscible	91
Neutral Plane	105
Oxidation	75
Oxidizing Agent	74
Passive Agents	87
Piloted Ignition	79
Plume	103
Potential Energy	75
Products of Combustion	84
Pyrolysis	79
Radiation	83
Reducing Agent	87
Saponification	98
Self-Heating	80
Solubility	90
Specific Gravity	90
Spontaneous Ignition	80
Temperature	78
Thermal Energy	78
Thermal Layering	104
Upper Flammable Limit (UFL)	94
Upper Layer	87
Vapor Density	89
Vapor Pressure	90
Vaporization	79
Watt	78

NFPA® Job Performance Requirements

This chapter provides information that addresses the following job performance requirements of NFPA® 1031, *Standard for Professional Qualifications for Fire Inspector and Plan Examiner* (2014).

Fire Inspector I	Fire Inspector II
4.3.8	5.3.6
4.3.14	5.3.8
	5.3.9
	5.3.10

Learning Objectives

After reading this chapter, students will be able to:

Inspector I

1. Describe the various components of fire behavior. (4.3.8)
2. Describe fire development. (4.3.14)

Inspector II

1. Describe how the fire tetrahedron model can be used to control or extinguish a fire. (5.3.6, 5.3.8, 5.3.9, 5.3.10)

Chapter 3
Fire Behavior

Case History

A fire on June 18, 2007, in the Sofa Super Store in Charleston, SC resulted in the death of nine firefighters. The fire began outside the loading dock, spread into the enclosed loading dock, and from there into both the retail showroom and warehouse spaces.

The fire was oxygen-controlled, meaning that there were large amounts of heated combustibles in the ceiling. Firefighters at the front of the store were unable to see the progress of the fire at the rear of the store. When the store's front windows were broken or vented, the fire progressed very rapidly through the store, trapping the firefighters.

A number of factors contributed to the growth of the fire:

- Large open spaces and doorways that allowed the fire to move from the exterior of the store to the interior.
- High fuel loads provided by foam-filled furniture.
- Lack of automatic sprinklers to suppress the fire in its early stages
- Metal walls that allowed heat to move from the loading dock to the showrooms and ignite items in these adjacent spaces
- The venting of smoke by breaking of the store's front windows, which provided additional air to the fire.

A NIST review of the circumstances surrounding the fire produced a number of recommendations:

— Adopt model building and fire codes that specifically address high fuel load commercial spaces.

— Implement aggressive fire inspection and enforcement programs and see that inspectors are professionally qualified to a national standard.

— Adopt and enforce model codes requiring automatic sprinkler systems for all new retail furniture stores and all existing retail furniture stores with display areas greater than 2,000 square feet.

A fire inspector's primary duty is to help protect the life safety of the citizens and fire and emergency services responders in a community. To accomplish this duty, a fire inspector must thoroughly examine building plans, inspect structures within the jurisdiction, and apply the adopted fire and life safety codes and standards.

During field inspections, the inspector must evaluate a building's physical characteristics, construction materials, and contents to determine the appropriate level of protection for the occupancy. Even if buildings are constructed of fire-resistant materials, combustible contents, hazardous processes, and fallible humans can compromise their fire safety.

This chapter describes the science of fire, stages of fire development, and other fire behavior characteristics. In order to enhance fire and life safety, the inspector needs to understand the combustion process, as well as know what fire is and how it behaves in different materials and environments. The inspector also needs to know the classifications of fire in order to identify the most appropriate extinguishing agent or systems.

Fuel — A material that will maintain combustion under specified environmental conditions. (NFPA® 921)

Oxidizing Agent — Substance that oxidizes another substance; can cause other materials to combust more readily or make fires burn more strongly. *Also known as* Oxidizer.

Matter — Anything that occupies space and has mass.

Energy — Capacity to perform work; occurs when a force is applied to an object over a distance, or when a chemical, biological, or physical transformation is made in a substance.

Science of Fire

Fire can take a variety of forms, but all fires involve a heat-producing chemical reaction between some type of **fuel** and an **oxidizing agent**, most commonly oxygen in the air. When a material burns, heat is generated faster than it can be dissipated, causing a significant increase in temperature. This process is best explained through the study of physical science.

Physical science is the study of the physical world around us and includes the sciences of chemistry and physics. Such a study necessarily includes the laws related to **matter** and **energy**. This theoretical foundation must be translated into a practical knowledge of fire behavior.

The science of fire can be very complex with many different factors to consider, including the following:

- Physical and chemical changes
- Modes of combustion
- Temperature, energy, and heat
- Sources of thermal energy
- Heat transfer
- Passive agents
- Fuel
- Oxidizer
- Self-sustained chemical reaction
- Classification of fires

The fire inspector must understand the concepts of the science of fire. With this knowledge, he or she can better identify and mitigate significant fire hazards, such as excessive fuel loading, improper storage, and unclean commercial cooking equipment. This knowledge will also better prepare the fire inspector to determine requirements for fire separation, interior finish, and sprinkler coverage.

Physical and Chemical Changes

While matter can undergo many types of physical and chemical changes, this chapter will concentrate on those changes related to fire. A physical change occurs when a substance remains chemically the same but changes in size, shape, or appearance. Examples of physical change are water freezing (liquid to solid) and boiling (liquid to gas).

When a substance changes from one type of matter into another, a chemical reaction occurs. This often involves the reaction of two or more substances combining to form compounds. **Oxidation** is a chemical reaction involving the combination of an oxidizer, such as oxygen in the air, with other materials. Oxidation can be slow, such as when a substance combines with oxygen to form rust, or rapid, as in **combustion** or an explosion **(Figure 3.1)**.

Physical and chemical changes involve an exchange of energy. A fuel's **potential energy** is released during combustion and converted to **kinetic energy**. Reactions that give off energy as they occur are called **exothermic**. **Fire** is an exothermic chemical reaction involving combustion. Reactions that absorb energy as they occur are called **endothermic** **(Figure 3.2)**. For example, converting water from a liquid to a gas (steam) requires the input of energy and is

Timeline of Oxidation

Rust
Smoldering Fire
Methane Ignites
BLEVE
Slow
Ultra Rapid

Figure 3.1 The timeline illustrates the speed difference in types of oxidation from very slow (rust) to explosive (BLEVE).

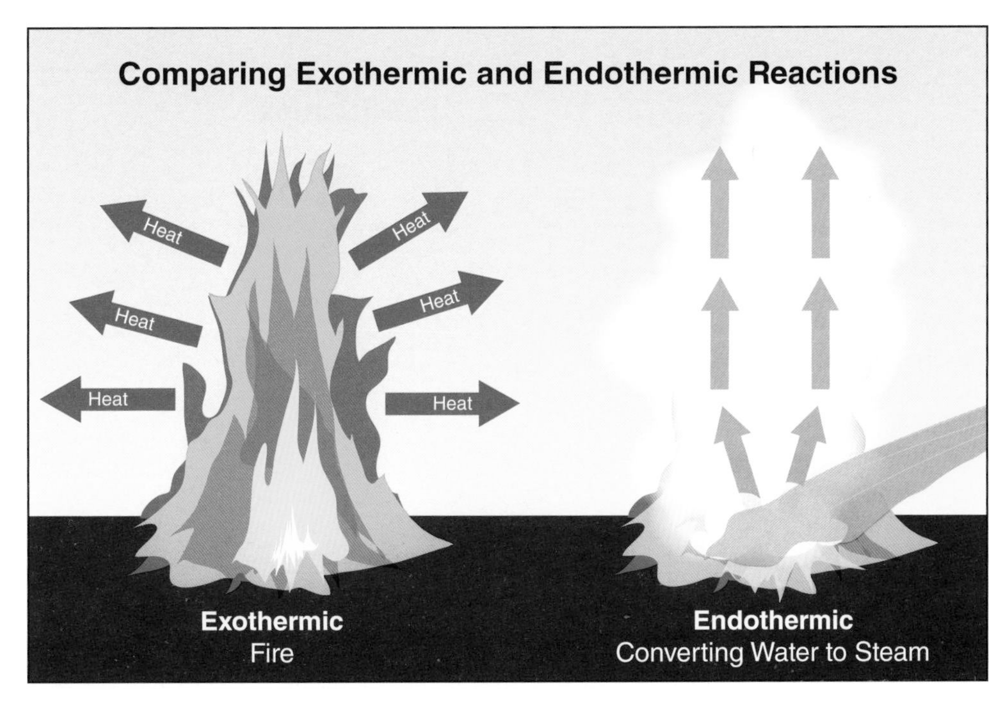

Figure 3.2 Exothermic reactions actively release energy as opposed to endothermic reactions, which absorb energy.

Oxidation — Chemical process that occurs when a substance combines with an oxidizer such as oxygen in the air; a common example is the formation of rust on metal.

Combustion — A chemical process of oxidation that occurs at a rate fast enough to produce heat and usually light in the form of either a glow or flame. (NFPA® 921)

Potential Energy — Stored energy possessed by an object that can be released in the future to perform work once released.

Kinetic Energy — The energy possessed by a body because of its motion.

Exothermic Reaction — Chemical reaction that releases thermal energy or heat.

Fire — A rapid oxidation process, which is a chemical reaction resulting in the evolution of light and heat in varying intensities. (NFPA® 921)

Endothermic Reaction — Chemical reaction that absorbs thermal energy or heat.

an endothermic physical reaction. Converting water to steam is an important part of controlling and extinguishing some types of fires. The inspector uses this knowledge to recognize the ways in which installed and available fire protection systems and equipment will act to help extinguish a fire.

Modes of Combustion

Flame — Visible, luminous body of a burning gas emitting radiant energy, including light of various colors produced by burning gases or vapors during the combustion process.

Fire Triangle — Model used to explain the elements/ conditions necessary for combustion. The sides of the triangle represent heat, oxygen, and fuel.

Fire Tetrahedron — Model of the four elements/ conditions required to have a fire. The four sides of the tetrahedron represent fuel, heat, oxygen, and self-sustaining chemical chain reaction.

The terms *fire* and *combustion* are commonly used to mean the same thing. However, when the two modes of combustion — nonflaming and flaming — are considered, a difference becomes apparent. Nonflaming combustion occurs more slowly at a lower temperature, producing a smoldering glow on the material's surface. Flaming combustion is commonly referred to as fire because it produces a visible **flame** above the material's surface.

Models of Fire

Two models, the **fire triangle** and **fire tetrahedron**, have been developed to explain both the elements of fire and how fires can be extinguished. The fire triangle, which is the oldest and simplest model, illustrates the three elements necessary for fire to occur: fuel, oxygen, and heat **(Figure 3.3)**. Remove any one of these elements and the fire will be extinguished.

Researchers created the fire tetrahedron after determining that an uninhibited chemical reaction is needed for a fire to occur **(Figure 3.4)**. This model explains fires involving certain types of substances and the types of agents necessary to extinguish them.

Nonflaming Combustion

Nonflaming combustion occurs when burning is localized on or near the fuel's surface, where it is in contact with the oxygen. Examples of nonflaming combustion include burning charcoal or smoldering wood or fabric **(Figure 3.5)**.

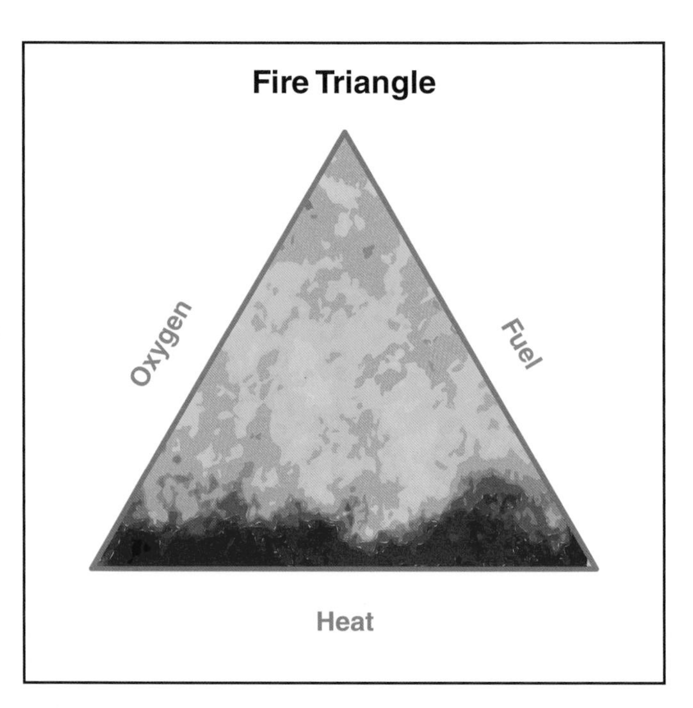

Figure 3.3 The fire triangle illustrates the three components necessary for the existence of fire.

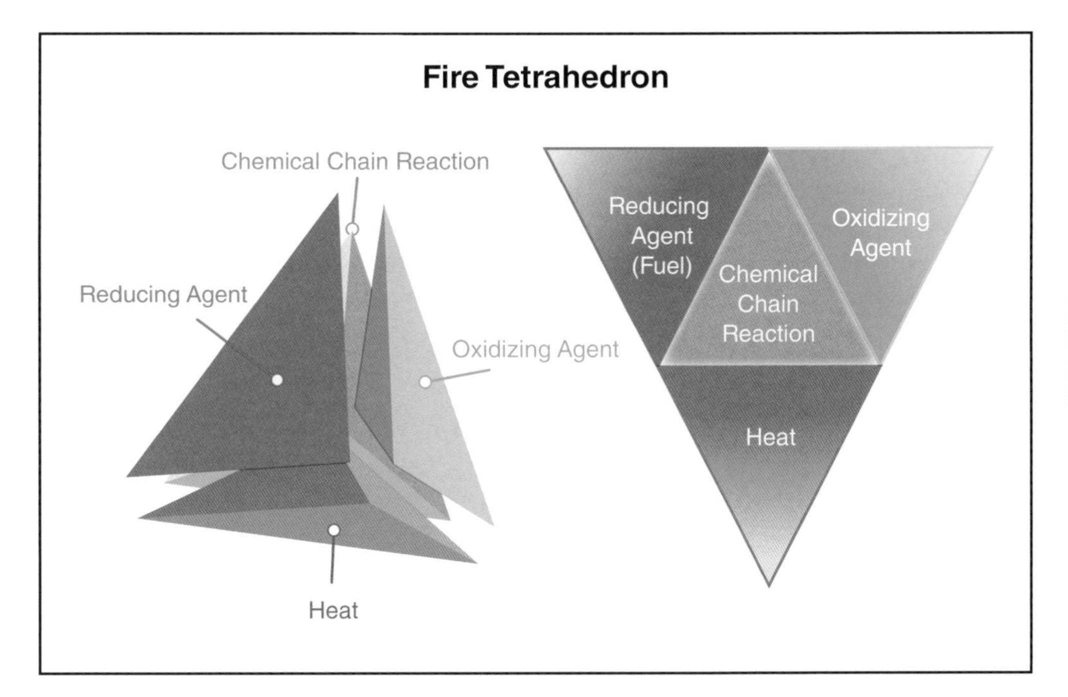

Figure 3.4 The fire tetrahedron accurately reflects the conditions required for flaming combustion.

Figure 3.5 Flaming combustion is characterized by the presence of a visible flame above the fuel, as opposed to nonflaming combustion, which features a lower temperature and smoldering burn.

Flaming Combustion

In flaming combustion, oxidation involves fuel in the gas phase. This chemical process requires liquid or solid fuels to be vaporized or converted to the gas phase through the addition of heat. When heated, both liquid and solid fuels will give off vapors that mix with oxygen, producing flames above the material's surface.

Each element of the tetrahedron must be in place for flaming combustion to occur. Flaming combustion may be stopped by interrupting the chemical chain reaction. However, the fire may continue to smolder (nonflaming combustion) depending on the characteristics of the fuel.

Temperature, Energy, and Heat

Having a knowledge of fire behavior requires an understanding of temperature, energy, and heat. Confusion often results when people use the terms interchangeably without understanding the differences between them.

Temperature — Measure of a material's ability to transfer heat energy to other objects; the greater the energy, the higher the temperature. Measure of the average kinetic energy of the particles in a sample of matter, expressed in terms of units or degrees designated on a standard scale.

Temperature

Temperature is a measure of the average kinetic energy of the particles in a sample of matter, expressed in terms of units or degrees. Several scales are used to measure temperature. The most common are the Celsius scale, which is used in the metric system, and the Fahrenheit scale, which is used in the customary system. The freezing and boiling points of water provide a simple way to compare these two scales **(Figure 3.6)**.

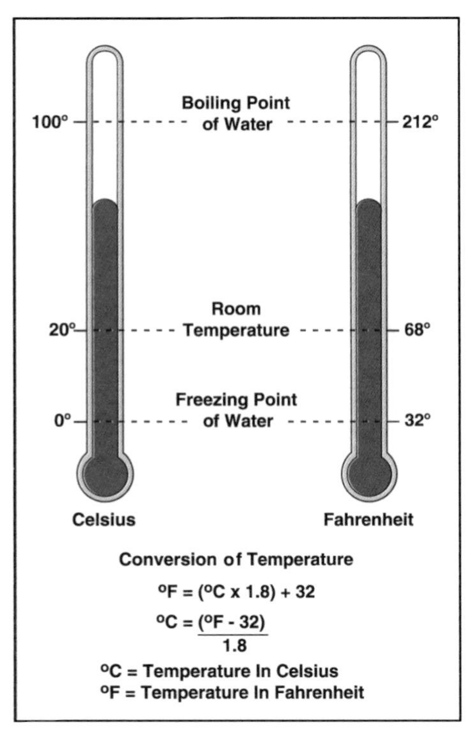

Figure 3.6 Both Celsius and Fahrenheit are commonly used temperature scales, and knowing how to convert between them may save time and effort during an incident response.

Energy

Energy is the capacity to perform work. Work occurs when a force is applied to an object over a distance or when a substance experiences a chemical, biological, or physical transformation. Although it is not possible to measure energy directly, it is necessary to measure the work that it does. In the case of heat, work means increasing the temperature of a substance.

Energy exists in many forms, including chemical, mechanical, electrical, light, nuclear, and sound. These forms are classified as potential and kinetic. Potential energy is the energy possessed by an object that may be released in the future. Kinetic energy is the energy possessed by a moving object. Energy can change from one form to another.

Watt — A unit of measure of power or rate of work equal to one joule per second (J/s).

The concept of power is important to understanding the rate at which energy is released during combustion and the rate of heat transfer. While energy is the ability to perform work, power is the rate at which work is performed or energy is converted from one form to another (such as chemical potential energy to kinetic thermal energy). In the case of combustion, the size of the fire relates to the rate at which energy is released. The standard international (SI) unit for power is the **watt** (W), which is one joule per second (J/s).

Thermal Energy — The kinetic energy associated with the random motions of the molecules of a material or object; often used interchangeably with the terms heat and heat energy. Measured in joules or Btu.

Heat

Heat is energy in transit. **Thermal energy** is a more appropriate description for this type of energy. Heat is the transfer of energy from a high-temperature substance to a low-temperature substance. Because energy is most often in transit, the term heat is commonly used to describe thermal energy.

Heat is kinetic energy associated with the movement of the atoms and molecules that comprise matter. As it relates to fire behavior, a fuel has potential chemical energy before it is ignited. When that fuel burns, the potential chemical energy is converted to thermal kinetic energy **(Figure 3.7)**. The conversion of chemical potential energy in fuel to thermal energy (heat) is particularly important because heat is the energy element of both the fire triangle and fire tetrahedron.

Forms of Ignition

As fuel is heated, its temperature increases. Applying sufficient heat causes **pyrolysis** in solid fuels and **vaporization** in liquid fuels **(Figure 3.8)**. A spark or other external source can provide the energy necessary for **ignition**, or the fuel can be heated until it ignites without a spark or other source. Once ignited, the ignition process continues the production and ignition of fuel vapors or gases so that the combustion reaction is sustained.

Pyrolysis — The chemical decomposition of a solid material by heating. Pyrolysis often precedes combustion.

Vaporization — Physical process that changes a liquid into a gaseous state; the rate of vaporization depends on the substance involved, heat, pressure, and exposed surface area.

Ignition — The process of initiating self-sustained combustion. (NFPA® 921)

Figure 3.7 Potential energy waits in a stored form, as opposed to kinetic energy, which actively releases energy. *Courtesy of Dan Madrzykowski/NIST.*

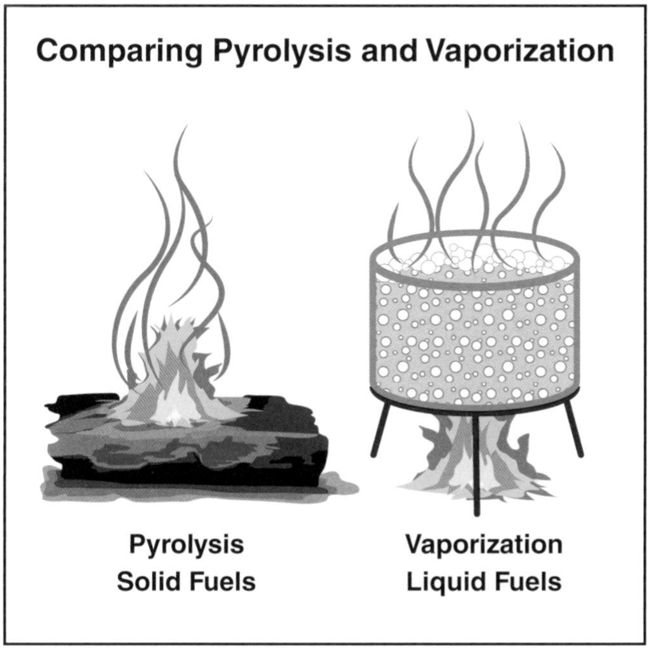

Comparing Pyrolysis and Vaporization

Pyrolysis
Solid Fuels

Vaporization
Liquid Fuels

Figure 3.8 Pyrolysis is the conversion of a solid fuel item into a gas fuel that is capable of supporting combustion, whereas vaporization is the conversion of a liquid to a vapor using the heat energy from combustion.

There are two forms of ignition: **piloted ignition** and **autoignition** (also called nonpiloted) **(Figure 3.9, p. 80)**. Piloted ignition is the most common form of ignition. It occurs when a mixture of fuel and oxygen encounter an external heat source with sufficient heat or thermal energy to start the combustion process. Autoignition occurs without any external flame or spark to ignite the

Piloted Ignition — Moment when a mixture of fuel and oxygen encounters an external heat (ignition) source with sufficient heat or thermal energy to start the combustion reaction.

Autoignition — Initiation of combustion by heat but without a spark or flame. (NFPA® 921). *Also called* nonpiloted ignition.

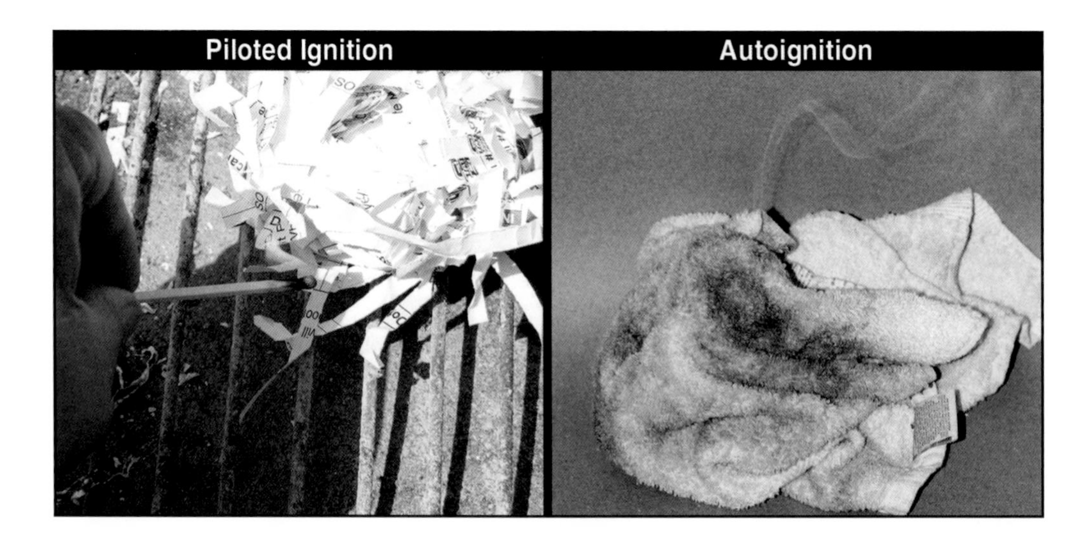

Figure 3.9 Piloted ignition requires the participation of an outside ignition source, whereas autoignition occurs from the confluence of specific conditions.

Autoignition Temperature — The lowest temperature at which a combustible material ignites in air without a spark or flame. (NFPA® 921)

fuel gases or vapors. In this case, the fuel surface is chemically heated to the point at which the combustion reaction occurs. **Autoignition temperature (AIT)** is the minimum temperature to which a fuel in the air must be heated in order to start self-sustained combustion. The autoignition temperature of a substance is always higher than its piloted ignition temperature.

Sources of Thermal Energy

Chemical, mechanical, electrical, light, nuclear, and sound energy can be transformed into thermal energy by causing the temperature of a substance to increase by speeding up the movement of its molecules. Chemical, electrical, and mechanical energy are common sources of heat that result in the ignition of a fuel.

Chemical Energy

Chemical energy is the most common source of heat in combustion reactions. When any combustible is in contact with oxygen, the potential for oxidation exists. The oxidation process almost always results in the production of thermal energy.

Self-Heating — The result of exothermic reactions, occurring spontaneously in some materials under certain conditions, whereby heat is generated at a rate sufficient to raise the temperature of the material. (NFPA® 921)

Spontaneous Ignition — Initiation of combustion of a material by an internal chemical or biological reaction that has produced sufficient heat to ignite the material. (NFPA® 921)

Self-heating may result in combustion and is both a form of oxidation and an exothermic chemical reaction that increases the temperature of a material without the addition of external heat. Normally, thermal energy is produced slowly by oxidation and is lost to the surroundings almost as fast as it is generated. An external heat source, such as sunshine, can initiate or accelerate the process.

In order for self-heating to progress to **spontaneous ignition**, the material must be heated to its autoignition temperature as follows:

- The insulation properties of the material immediately surrounding the fuel must be such that the heat cannot dissipate as quickly as it is being generated.
- The rate of heat production must be great enough to raise the temperature of the material to its autoignition temperature.
- The available air supply in and around the material being heated must be adequate to support combustion.

An example of a situation that could lead to spontaneous ignition would be rags soaked in linseed oil, rolled into a ball, and thrown into a corner. If the heat generated by the natural oxidation of the oil and cloth is not allowed to dissipate, either by movement of air around the rags or some other method of heat transfer, the temperature of the cloth could eventually increase enough to cause ignition.

The oxidation rate, and therefore the heat production, increases as more heat is generated and held by the materials insulating the fuel. The rate of most chemical reactions increases as the temperature of the reacting materials increases. The more heat generated and absorbed by the fuel, the faster the reaction causing the heat generation. When the heat generated by a self-heating reaction exceeds the heat being lost, the material may reach its autoignition temperature and ignite spontaneously. **Table 3.1** lists some common materials that are subject to self-heating.

Table 3.1
Spontaneous Heating Materials and Locations

Type of Material	Possible Locations
Charcoal	Convenience Stores Hardware Stores Industrial Plants Restaurants
Linseed Oil-Soaked Rags	Woodworking Shops Lumber Yards Furniture Repair Shops Picture Frame Shops
Straw and Manure	Farms Feed Stores Arenas Feedlots

Electrical Energy

Electrical energy can generate temperatures high enough to ignite any combustible materials near the heated area. Electrical heating can occur in several ways, including the following:

- ***Resistance heating*** — When electric current flows through a conductor, heat is produced. Some electrical appliances, such as incandescent lamps, ranges, ovens, or portable heaters, are designed to make use of resistance heating. Other electrical equipment is designed to limit resistance heating under normal operating conditions **(Figure 3.10, p. 82)**.
- ***Overcurrent or overload*** — When the current flowing through a conductor exceeds its design limits, it may overheat and present an ignition hazard. Overcurrent or overload is unintended resistance heating **(Figure 3.11, p. 82)**.
- ***Arcing*** — In general, an arc is a high-temperature luminous electric discharge across a gap or through a medium such as charred insulation. Arcs may be generated when a conductor is separated (such as in an electric motor or switch) or by high voltage, static electricity, or lightning.

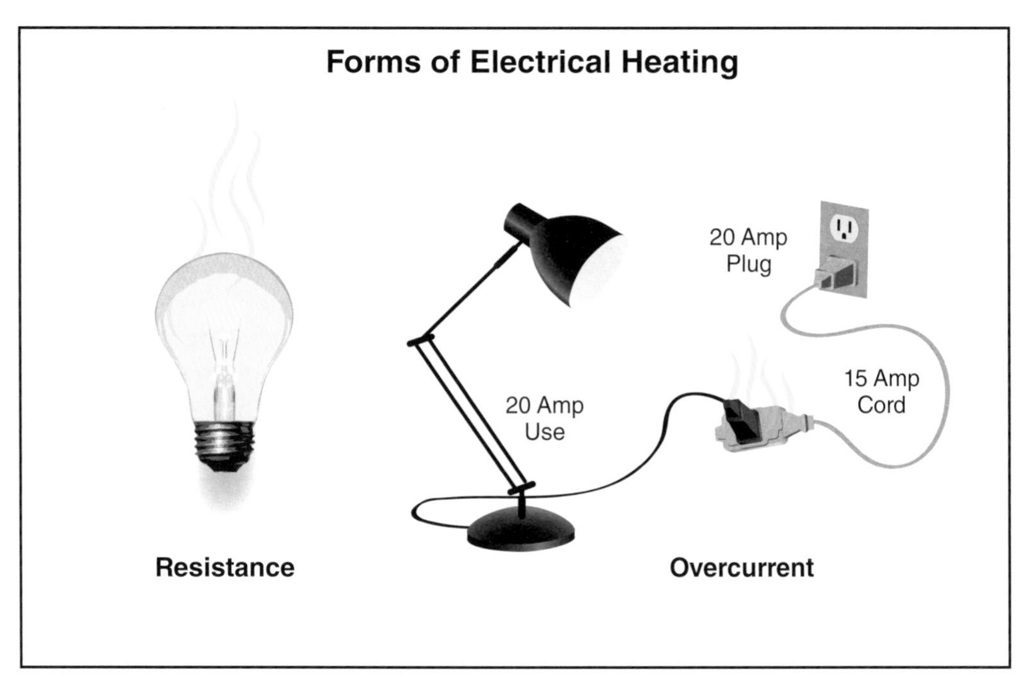

Figure 3.10 Forms of electrical heating may serve a domestic purpose or they may be controlled.

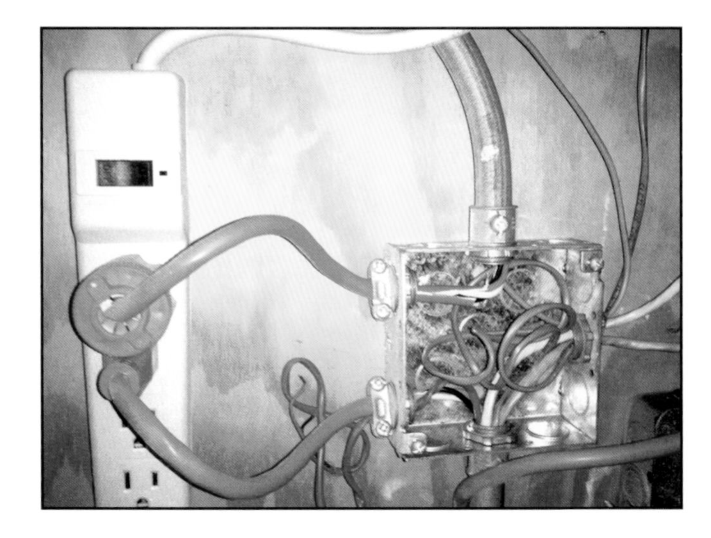

Figure 3.11 Misuse of electrical appliances is a common safety hazard that inspectors will see. *Courtesy of Scott Strassburg.*

- ***Sparking*** — When an electric arc occurs, luminous (glowing) particles can be formed and spatter away from the point of arcing. In electrical terms, sparking refers to this spatter.

Mechanical Energy

Mechanical energy is generated by friction or compression. Heat of friction occurs when two surfaces move against each other. This movement results in heat and/or sparks being generated. Heat of compression is generated when a gas is compressed using mechanical means. Diesel engines use this principle to ignite fuel vapor without a spark plug. The principle is also the reason that self-contained breathing apparatus (SCBA) cylinders feel warm to the touch after they have been filled.

Heat Transfer

Heat transfer from one point or object to another is basic to the study of fire behavior. Heat transfer from the initial fuel package (burning object) to other fuels in and beyond the area of fire origin affects the growth of any fire. The inspector uses knowledge of heat transfer to evaluate how passive and active fire safety features, such as fire-rated walls and sprinkler systems, will contain a fire, allowing sufficient time for building evacuation and emergency operations.

Structural components and interior finishes, such as drywall, masonry, fire doors, as well as fuel loads and combustible waste and storage, have a significant effect on heat transfer. Knowledge of heat transfer is also important when considering the potential for fire to spread beyond the area of origin to adjacent rooms or structures.

For heat to be transferred from one body to another, the two objects must be at different temperatures. Heat moves from warmer objects to cooler objects. The rate at which heat is transferred is related to the temperature differential of the objects and the thermal conductivity of the solid material involved. For any given substance, the greater the temperature differences between the objects, the greater the transfer rate.

Transfer of thermal energy is described as **heat flux** (energy transfer over time per unit of surface area) and is typically measured in kilowatts per meter squared (kW/m^2). Heat can be transferred from one body to another by three mechanisms: **conduction**, **convection**, and **radiation**.

Heat Flux — The measure of the rate of heat transfer to a surface, expressed in kilowatts/m^2, kilojoules/$m^2 \cdot$ sec, or Btu/$ft^2 \cdot$ sec. (NFPA® 921)

Conduction — Transfer of heat through or between solids that are in direct contact.

Convection — Heat transfer by circulation within a medium such as a gas or a liquid. (NFPA® 921)

Radiation — Heat transfer by way of electromagnetic energy. (NFPA® 921)

Conduction

Conduction is the transfer of heat through and between solids **(Figure 3.12)**. Conduction results from increased molecular motion and collisions between the molecules of a substance, resulting in the transfer of energy through the substance. The more closely packed the molecules of a substance are, the more readily it will conduct heat. Conduction occurs when a material is heated as a result of direct contact with a heat source. For example, if a fire heats a metal

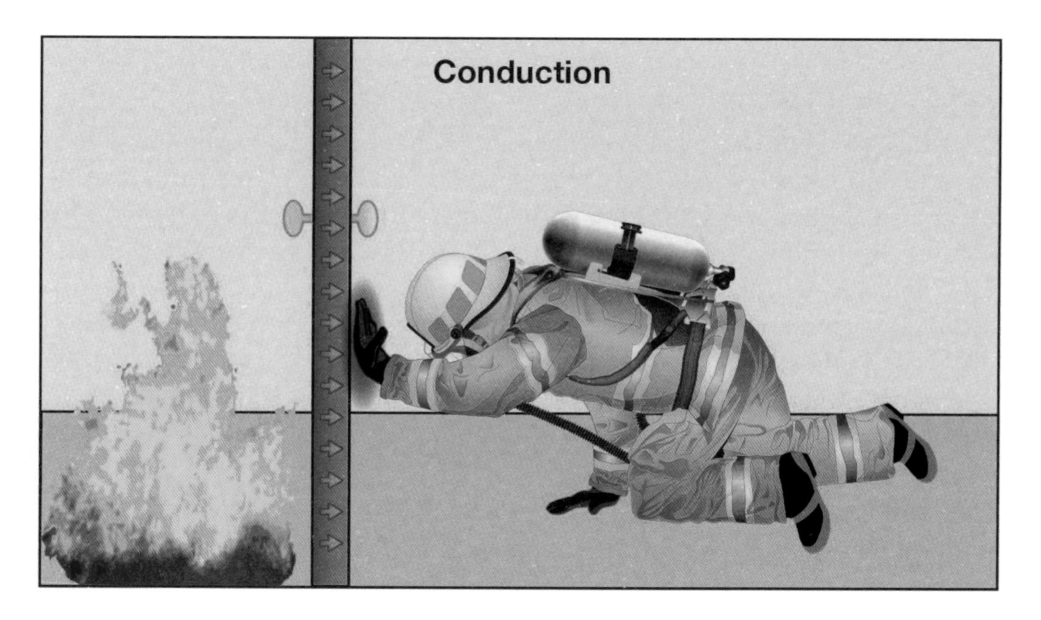

Figure 3.12 Conduction occurs when heat is transferred along a solid object.

pipe on one side of a wall, heat conducted through the pipe can ignite wooden framing components in the wall or nearby combustibles on the other side of the wall.

Heat flow due to conduction is dependent on the area being heated, the temperature difference between the heat source and the material being heated, and the thermal conductivity of the material. **Table 3.2** shows the thermal conductivity of various common materials at the same ambient temperature (68°F/20°C). As shown in the table, copper will conduct heat more than seven times more readily than steel. Likewise, steel is nearly forty times as thermally conductive as concrete. Wood is a poor conductor of heat.

Insulating materials delay the transfer of heat primarily by slowing conduction. Good insulators are materials that do not conduct heat well. Because of their physical makeup, they disrupt the point-to-point transfer of heat. The best commercial insulation materials used in building construction are those made of fine particles or fibers with void spaces between them filled with a gas such as air. Gases do not conduct heat well because their molecules are relatively far apart.

Table 3.2
Thermal Conductivity of Common Substances

Substance	Temperature	Thermal Conductivity (W/mK)
Copper	68°F (20°C)	386.00
Steel	68°F (20°C)	36.00 – 54.00
Concrete	68°F (20°C)	0.8 – 1.28
Wood (pine)	68°F (20°C)	0.13

Convection

Convection is the transfer of thermal energy by the circulation or movement of a fluid (liquid or gas) **(Figure 3.13)**. In the fire environment, this process usually involves the transfer of heat through the movement of hot smoke and fire gases. As with all heat transfer, the flow of heat is from the hot fire gases to the cooler structural surfaces, building contents, and air.

Buoyant — The tendency or capacity to remain afloat in a liquid or rise in air or gas.

Products of Combustion — Materials produced and released during burning.

Convection may occur in any direction. Generally, the movement will be upward because the smoke and fire gases are **buoyant**. As a fire begins to grow, more air is pulled into the fire and heated, adding to the fire and the rising column containing **products of combustion**. Convection currents can also move laterally within a structure. These lateral currents can be caused by the products of combustion moving from areas of higher pressure to those of lower pressure. Lateral movement can be from the fire area or from the openings on the windward side (where the pressure is higher) of the building to those on the leeward (the direction opposite from which the wind is blowing) side.

Radiation

Radiation is the transmission of energy as an electromagnetic wave, such as light waves, radio waves, or X-rays, without an intervening medium **(Figure 3.14)**. Radiant heat can become the dominant mode of heat transfer when the

Figure 3.13 Convection is the heating of surfaces and gases through the circulation molecules within fluids.

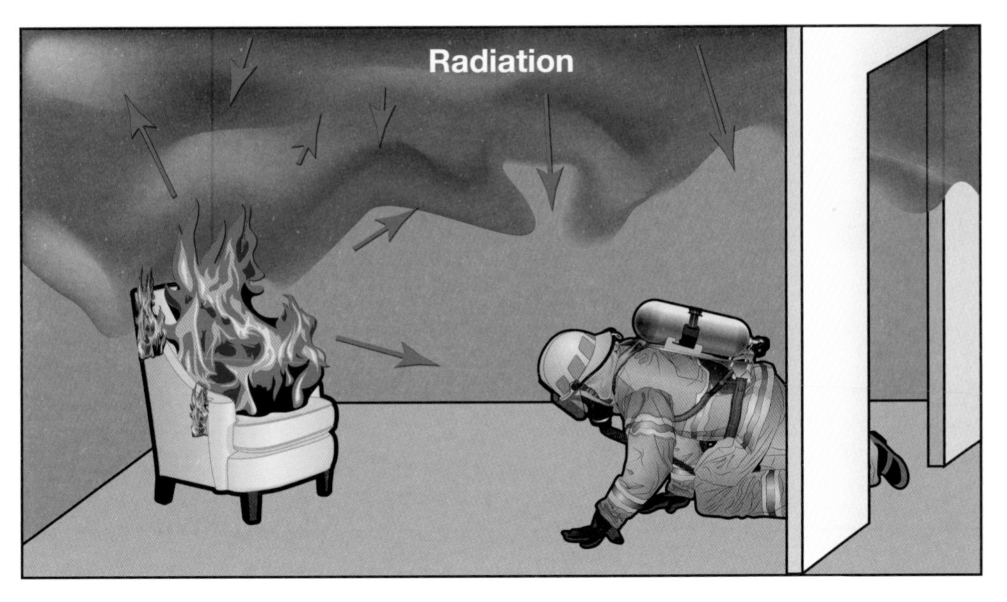

Figure 3.14 Unlike conduction and convection, heat transfer via radiation does not require an additional medium to transmit heat energy.

fire grows in size and can have a significant effect on the ignition of objects located some distance from the fire.

Radiant heat transfer is also a significant factor in fire development and spread in compartments. Inspectors should be aware that radiation is a common method of heat transfer in buildings with radiant heat systems **(Figure 3.15, p. 86)**.

Many factors influence radiant heat transfer, including:

- **The nature of the exposed surfaces** — Dark materials emit and absorb heat more effectively than lighter-colored materials; smooth or highly polished surfaces reflect more radiant heat than rough surfaces.
- **The distance between the heat source and the exposed surfaces** — Increasing distance reduces the effect of radiant heat **(Figure 3.16, p. 86)**.
- **The temperature difference between the heat source and exposed surfaces** — Temperature difference has a major effect on heat transfer through radiation. As the temperature of the heat source increases, the radiant energy increases by a factor to the fourth power **(Figure 3.17, p. 86)**.

Figure 3.15 An example of a radiant heat system. *Courtesy of Greg Havel.*

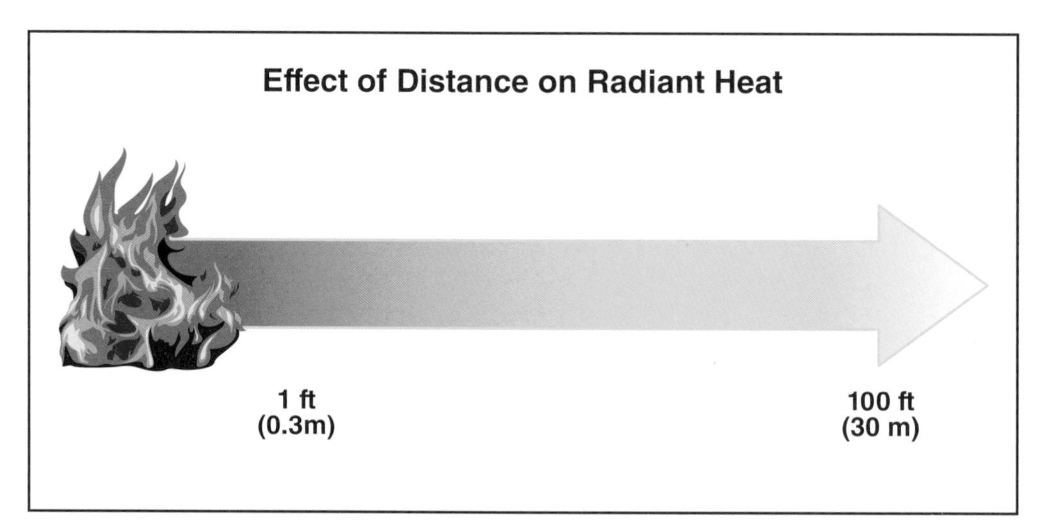

Figure 3.16 The effects of radiant heat decrease as distance between the fire and the exposure increases.

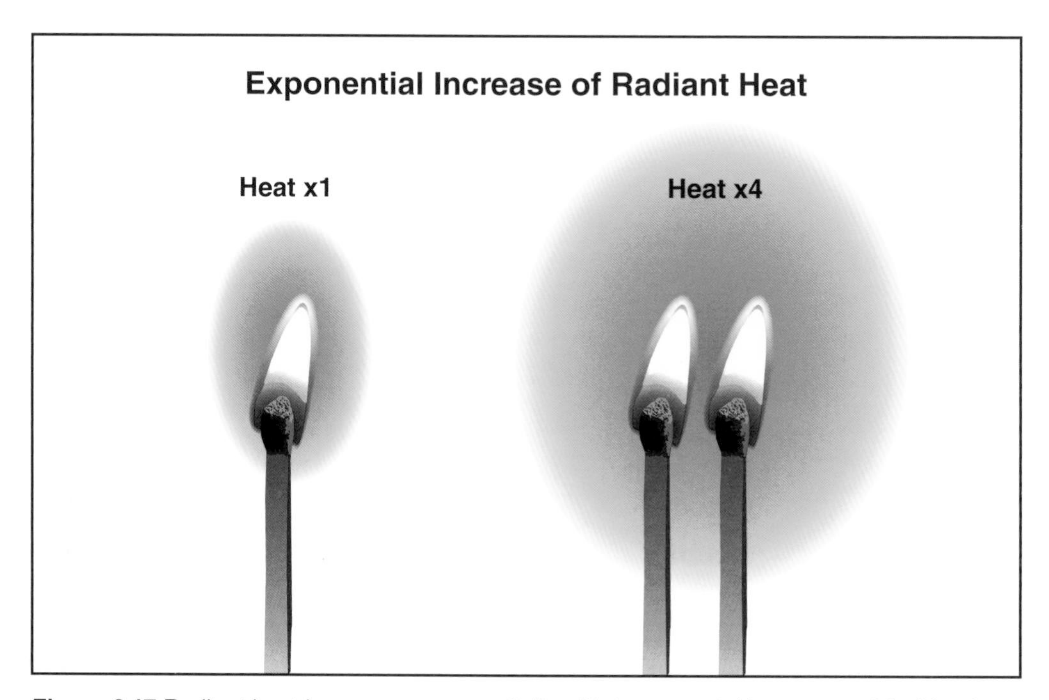

Figure 3.17 Radiant heat increases exponentially with incremental increases of fuel load.

Because energy is an electromagnetic wave, it travels in a straight line at the speed of light. The heat of the sun is an excellent example of heat transfer by radiation. The energy travels at the speed of light from the sun through space (a vacuum) until it collides with and warms the surface of the earth.

Radiation is a common cause of exposure fires (fires ignited in fuel packages or buildings that are remote from the area of origin). As a fire grows, it radiates more energy. An object absorbs the energy, converting it to heat. In large fires, radiated heat can ignite buildings or other fuel packages that are a considerable distance away.

Radiated heat travels through vacuums and air spaces that would disrupt heat transfer by conduction and convection. However, materials that reflect radiated energy will disrupt the transmission of heat, such as fire-rated doors/walls. While flames have high temperature, resulting in the emission of significant radiant energy, hot smoke in the **upper layer** also transmits significant energy.

Upper Layer — Buoyant layer of hot gases and smoke produced by a fire in a compartment.

Passive Agents

Besides fuel, heat, and oxygen, other materials can have a significant effect on both ignition and fire development. **Passive agents** are materials that absorb heat but do not contribute fuel in the combustion reaction. In building construction, one of the most common passive agents is drywall or gypsum board that contains moisture in the form of hydrates. As the drywall heats, the moisture vaporizes, which slows the increase in the temperature of the gypsum board **(Figure 3.18)**.

Passive Agents — Materials that absorb heat but do not participate actively in the combustion process.

Fuel

Fuel is the substance that is oxidized or burned in the combustion process. In scientific terms, the fuel in a combustion reaction is known as the **reducing agent**. Fuels may be inorganic or organic. Inorganic fuels, such as hydrogen or magnesium, do not contain carbon. Most common fuels are organic, containing carbon along with other elements. These fuels can be further divided into hydrocarbon-based fuels, such as gasoline, fuel oil, and plastics, and cellulose-based materials, such as wood and paper.

Reducing Agent — The fuel that is being oxidized or burned during combustion.

The chemical content of any fuel influences both its **heat of combustion** and **heat release rate (HRR)**. The heat of combustion of a fuel is the total amount of thermal energy released when a specific amount of that fuel is oxidized (burned). In other words, different materials release more or less heat or thermal energy than others depending on their chemical makeup. Heat of combus-

Heat of Combustion — Total amount of thermal energy (heat) that could be generated by the combustion (oxidation) reaction if a fuel were completely burned. The heat of combustion is measured in British Thermal Units (Btu) per pound or Kj/g not mJ/Kg.

Heat Release Rate (HRR) — Total amount of heat released per unit time. The HRR is measured in kilowatts (kW) or megawatts (MW) of output.

Figure 3.18 Gypsum panels have a high water content and are used to shield structural elements. *Courtesy of Greg Havel.*

tion is usually expressed in kilojoules/gram (kJ/g). Many plastics, flammable liquids, and flammable gases contain more potential heat or thermal energy than wood **(Table 3.3)**. This knowledge is particularly significant to inspectors given the widespread use of synthetics in building construction materials.

Heat release rate is the energy released per unit of time as a fuel burns and is usually expressed in kilowatts (kW) or megawatts (MW) **(Figure 3.19)**. HRR is dependent on the type, quantity, and orientation of the fuel. The thermal energy released by combustion is directly related to the oxygen consumed. This process will significantly influence the heat release rate in compartment fires where ventilation is limited.

Prefixes for Units of Measure: Kilo and Mega

The standard international system of units (SI) specifies a set of prefixes that precede units of measurement to indicate a multiple or fraction of that unit. Kilo and mega are two common prefixes encountered when discussing energy (Joules) and heat release rate (watts). The prefix kilo indicates a multiple of 1,000 (such as a kilowatt is 1,000 watts) and mega indicates a multiple of 1,000,000 (such as a megawatt is 1,000,000 watts).

Table 3.3
Representative Peak Heat Release Rates (HRR) During Unconfined Burning

Fuel Material	Peak HRR in kilowatts	Common Locations for Material
Small Wastebasket	4-18	Homes, Businesses, Shops
Cotton Mattress	140-350	Homes, Furniture Stores, Motels
Cotton Easy Chair	290-370	Homes, Furniture Stores, Office Buildings
Small Pool of Gasoline	400	Traffic Crash, Fuel Stations
Dry Christmas Tree	500-650	Homes, Trash Facilities, Dumpsters, Recycling Sites
Polyurethane Mattress	810-2630	Homes, Furniture Stores, Motels, Dormitories, Jails
Polyurethane Easy Chair	1350-1990	Homes, Furniture Stores, Motels
Polyurethane Sofa	3120	Homes, Furniture Stores, Motels, Dormitories, Office Buildings

Adapted from NFPA® 921, 2004 edition

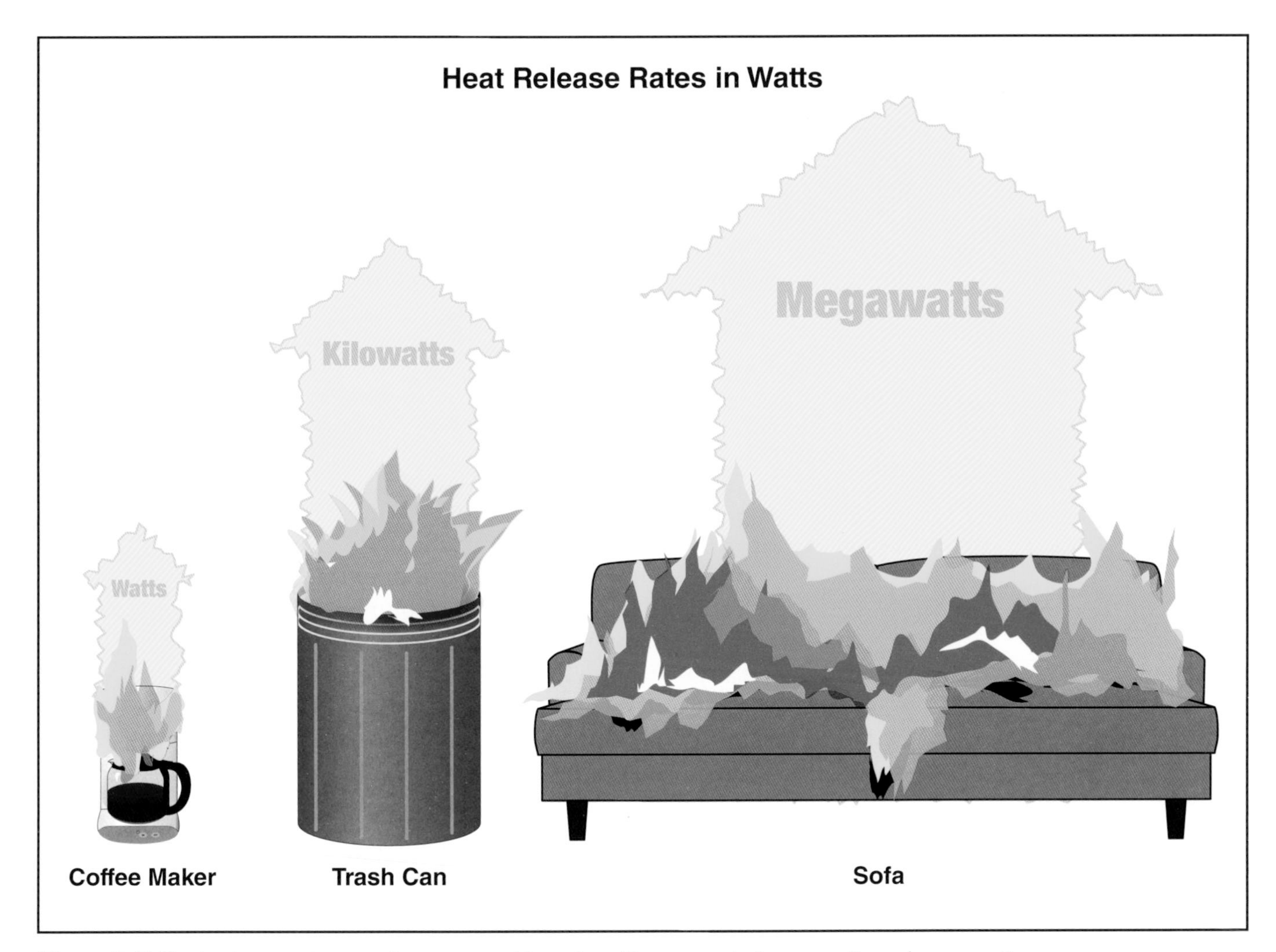

Figure 3.19 Heat release rates may be measured in watts with a range between watts and megawatts.

A fuel may be found in any of three physical states of matter: gas, solid, or liquid. For flaming combustion to occur, fuels must be in the gaseous state. Thermal energy is required to change solids and liquids into the gaseous state.

Gaseous Fuel

Gaseous fuels, such as methane, hydrogen, and acetylene, can be the most dangerous of all fuel types because they are already in the physical state required for ignition. **Table 3.4, p. 90** contains characteristics of common flammable gases. Gases have mass but no definite shape or volume:

- A gas placed in a container will diffuse and fill the available space, assuming the shape and volume of the container.
- When a gas is released from a container, it will rise or sink, depending on its **vapor density**.

Vapor density describes the density of gases in relation to air. Air has been assigned a vapor density of 1. Gases with a vapor density of less than 1, such as methane, will rise while those gases having a vapor density of greater than 1, such as propane, will sink. These densities presume that the gas and air are at the same temperature (generally specified as 68°F (20°C). Heated gases expand and become less dense; when cooled, they contract and become more dense.

Vapor Density — Weight of a given volume of pure vapor or gas compared to the weight of an equal volume of dry air at the same temperature and pressure. A vapor density less than 1 indicates a vapor lighter than air; a vapor density greater than 1 indicates a vapor heavier than air.

Table 3.4
Characteristics of Common Flammable Gases

Material	Vapor Density	Ignition Temperature
Methane (Natural Gas)	0.55	(1004°F) 540°C
Propane (Liquefied Petroleum Gas)	1.52	(842°F) 450°C
Carbon Monoxide	0.96	(1,128°F) 620°C

Source: *Computer Aided Management of Emergency Operations* (CAMEO)

Specific Gravity — Mass (weight) of a substance compared to the mass of an equal volume of water at a given temperature. A specific gravity less than 1 indicates a substance lighter than water; a specific gravity greater than 1 indicates a substance heavier than water.

Flammable Liquid — Any liquid having a flash point below 100°F (37.8°C) and a vapor pressure not exceeding 40 psi absolute (276 kPa) {2.76 bar}.

Vapor Pressure — (1) Measure of the tendency of a substance to evaporate. (2) The pressure at which a vapor is in equilibrium with its liquid phase for a given temperature; liquids that have a greater tendency to evaporate have higher vapor pressures for a given temperature.

Flash Point — Minimum temperature at which a liquid gives off enough vapors to form an ignitable mixture with air near the liquid's surface.

Fire Point — Temperature at which a liquid fuel produces sufficient vapors to support combustion once the fuel is ignited. Fire point must exceed 5 seconds of burning duration during the test. The fire point is usually a few degrees above the flash point.

Solubility — Degree to which a solid, liquid, or gas dissolves in a solvent (usually water).

Liquid Fuel

Liquids have mass and volume but no definite shape, except for a flat surface or when they assume the shape of their container. When released on the ground, liquids will flow downhill and can pool in low areas. Just as gases are compared to air, the density of liquids is compared with that of water. **Specific gravity** is the ratio of the mass of a given volume of a liquid compared with the mass of an equal volume of water at the same temperature. Water has been assigned a specific gravity of 1. Liquids with a specific gravity less than 1, such as gasoline and most **flammable liquids**, are lighter than water and will float on its surface **(Figure 3.20)**. Liquids with a specific gravity greater than 1, such as epichlorohydrin (used in making plastics), are heavier than water, causing them to sink.

Liquids must be vaporized in order to burn. Vaporization is the transformation of a liquid to vapor or gaseous state. In order to vaporize, liquids must overcome the pressure exerted by the atmosphere. At sea level, the atmosphere exerts a pressure of 14.7 psi (102.9 kPa). **Vapor pressure** is the pressure produced or exerted by vapors released by a liquid.

As a liquid is heated, vapor pressure increases along with the rate of vaporization. For example, a puddle of water eventually evaporates. When the same amount of water is heated on a stove, however, it vaporizes much more rapidly because more thermal energy is being applied. The rate of vaporization is determined by the vapor pressure of the substance and the amount of thermal energy applied to it. The volatility or ease with which a liquid gives off vapor influences how easily it can be ignited.

The **flash point** is the minimum temperature at which a liquid gives off sufficient vapors to ignite but not sustain combustion. **Fire point** is the temperature at which sufficient vapors are generated to sustain the combustion reaction. Flash point is commonly used to indicate the flammability hazard of liquid fuels **(Figure 3.21)**. Liquid fuels that vaporize sufficiently to burn at temperatures under 100°F (38°C) present a significant flammability hazard.

The extent to which a liquid will give off vapor is also influenced by how much surface area is exposed to the atmosphere. In many open containers, the surface area of liquid exposed to the atmosphere is limited.

A number of other characteristics of liquid fuels are important to inspectors. Principal among these are density in comparison to water and ability to mix with water (**solubility**). Liquids such as hydrocarbon fuels (including gasoline,

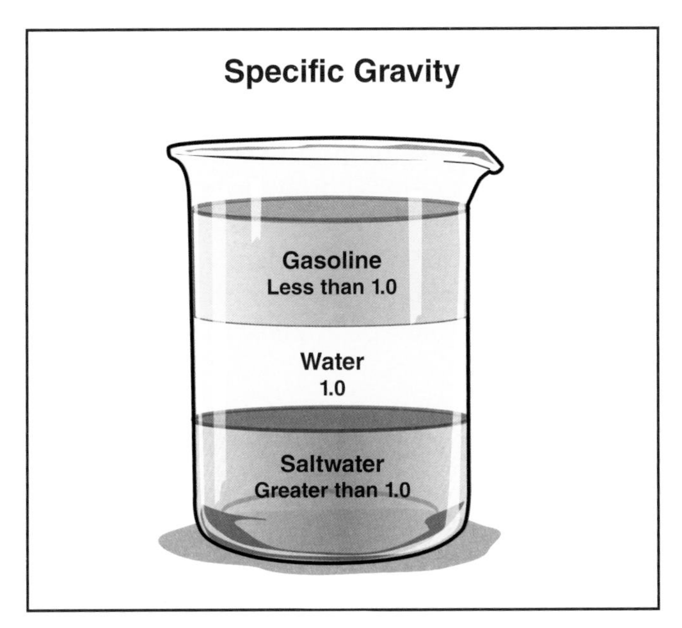

Figure 3.20 The specific gravity of a flammable liquid will have an impact on the choice of an extinguishing agent.

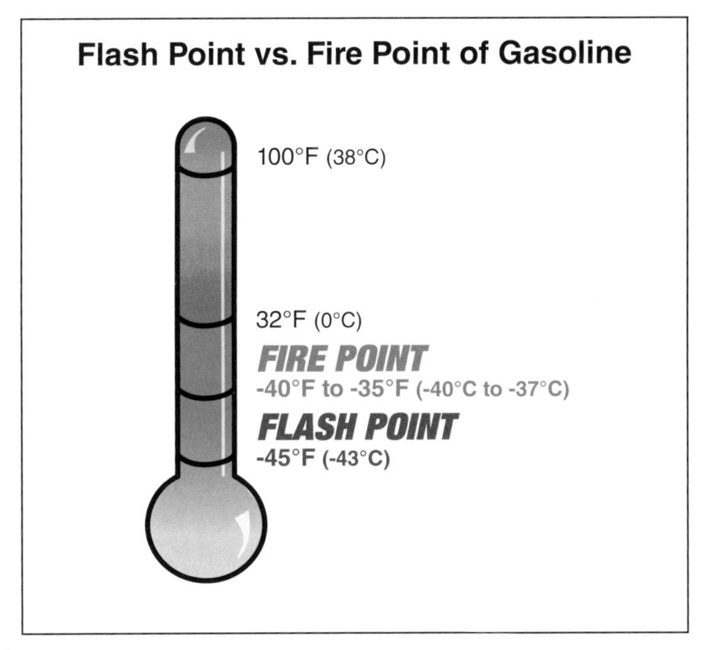

Figure 3.21 The flash point allows the creation of a quick fire, whereas the fire point sustains a combustion reaction.

diesel, and fuel oil) are lighter than water and do not mix with water. Other liquids (called polar solvents) such as alcohols (including methanol, ethanol) will mix readily with water.

Solubility describes the extent to which a substance (in this case a liquid) will mix with water. This may be expressed in qualitative terms (such as slightly, completely) or as a percentage. Materials that are **miscible** in water will mix in any proportion.

Miscible — Materials that are capable of being mixed in all proportions.

Solid Fuel

Solids have definite size and shape. Solids may also react differently when exposed to heat. Some solids (wax, thermoplastics, and metals) will readily change their state and melt, while others (wood and thermosetting plastics) will not.

Pyrolysis occurs as solid fuels are heated and begin to decompose and give off combustible vapors. If there is sufficient fuel and heat, this process generates enough burnable vapors to ignite in the presence of sufficient oxygen or another oxidizer.

Solids, such as wood, paper, fabric, and plastic, are the primary fuels commonly found in room or compartment fires. Pyrolysis must occur to generate the flammable vapors required for combustion. When wood is first heated, water vapor is driven off as the wood dries. As heating continues, the wood begins to pyrolize and decompose into its volatile components and carbon. Pyrolysis of wood begins at temperatures below 400°F (204°C). These temperatures are lower than required for ignition of the vapors being given off, which ranges from approximately 1,000°F to 1,300°F (538°C to 704°C).

The pyrolysis process is similar with synthetic fuels, such as plastics and some fabrics. Unlike wood, though, plastics do not generally contain moisture that heat input must drive off before pyrolysis can occur.

Solid fuels have a defined shape and size. This property significantly affects how easy or difficult they are to ignite. The primary consideration is the surface area of the fuel in proportion to the mass, called the surface-to-mass ratio **(Figure 3.22)**. As an example, the surface area of a log is very low compared to its mass; therefore, the surface-to-mass ratio is low. If the log is sawn into planks, the resulting surface area is increased, thus increasing the surface-to-mass ratio. The chips and sawdust produced as the planks are sawn into boards have an even higher surface-to-mass ratio. If the boards are milled or sanded, the resulting shavings or sawdust have the highest surface-to-mass ratio of any of the examples.

As the surface-to-mass ratio increases, the fuel particles become smaller (more finely divided). Therefore, their ability to be ignited increases tremendously. As the surface area increases, more of the material is exposed to the heat and generates combustible pyrolysis products more quickly, making the fuel easier to ignite as surface-to-mass ratio increases.

The proximity and orientation of a solid fuel relative to the source of heat also affects the way it burns **(Figure 3.23)**. For example, if you ignite one corner of a sheet of ⅛-inch (3 mm) plywood paneling that is lying horizontally (flat), the fire will consume the fuel at a relatively slow rate. The same type of paneling in a vertical position (standing on edge) burns much more rapidly. In this case, more heat is transferred to the solid fuel, speeding fire development. Products such as paneling are listed and tested for vertical or horizontal spread and should be installed according to their listing.

NOTE: Refer to Chapter 6 for more information on testing for flame spread and smoke development.

The Station Nightclub Fire

Several factors contributed to the fast-moving fire that killed 100 people and injured an additional 200 at The Station nightclub in Rhode Island in 2003. Factors that played a major role in the fire were the interior finish materials and the lack of an installed sprinkler system.

Polyurethane foam covered the walls and part of the ceiling in the vicinity of the bandstand, and a portion of the interior walls in the area of the bandstand and dance floor consisted of wood paneling. When pyrotechnics used by the band ignited the polyurethane foam, flames quickly spread to and along the ceiling of the nightclub. The thick, black smoke resulting from the burning foam made it very difficult for attendees to find the emergency exits.

One result of the fire was greater awareness of the need for occupants of any type of assembly occupancy to be aware of and to use the closest exit. The NFPA® and ICC also made changes in fire sprinkler and crowd management requirements for nightclubs and other assembly occupancies.

Oxidizer

After a fuel has been converted into a gaseous state, it must be mixed with an oxidizer in the proper ratio in order for combustion to occur. Oxygen in the air is the primary oxidizing agent in most fires. Air consists of about 21 per-

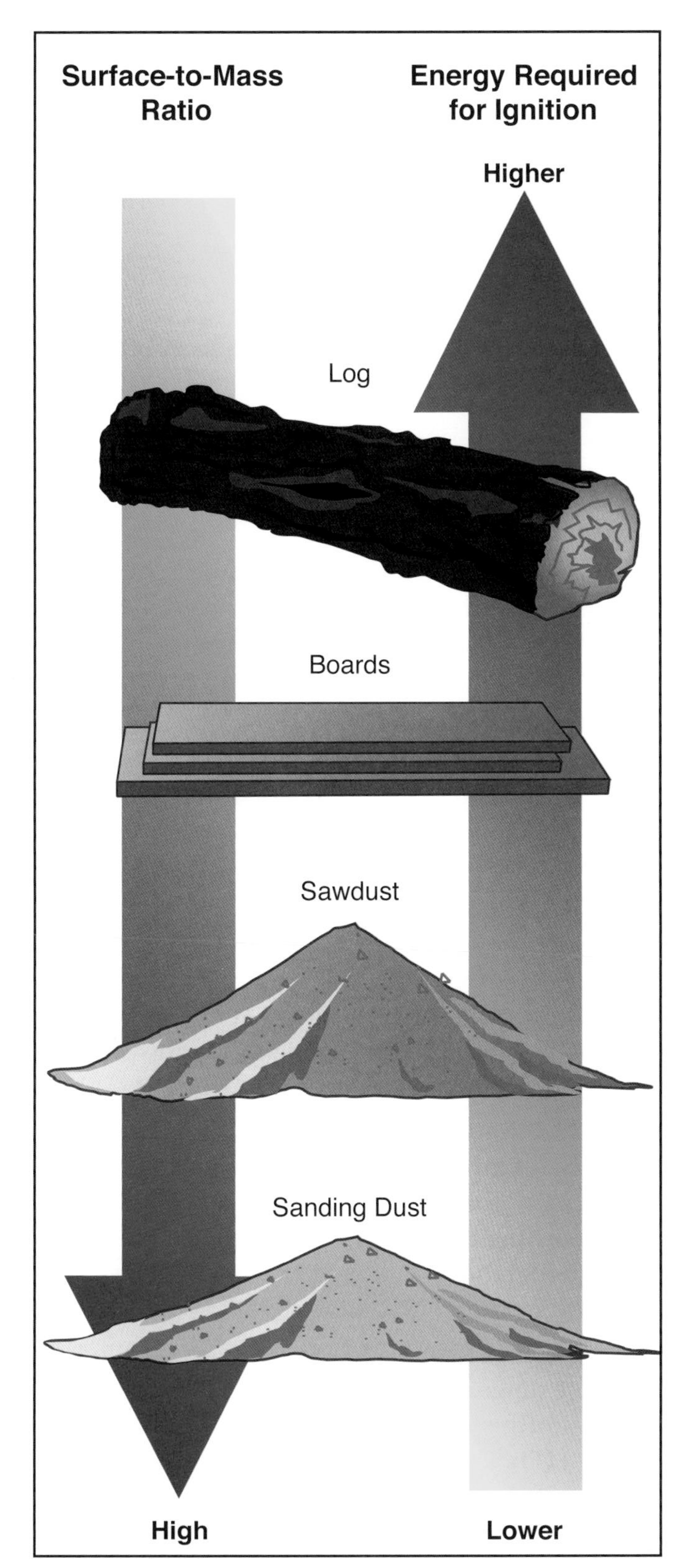

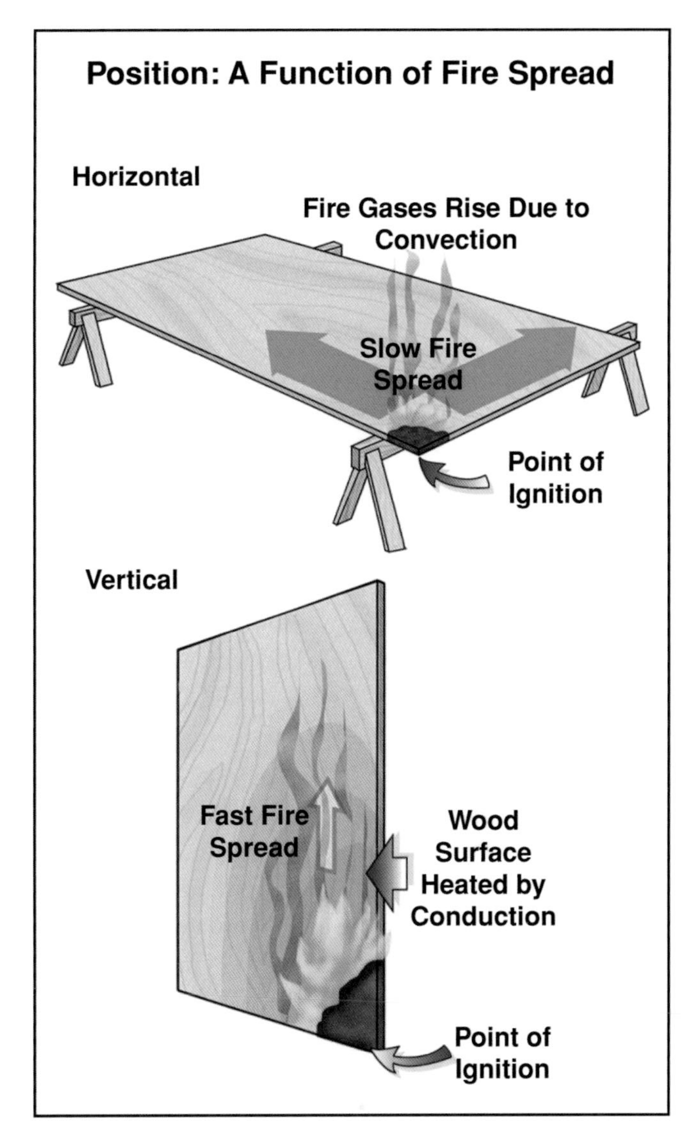

Figure 3.23 A board aligned vertically presents much more surface area to pyrolysis from rising gases as compared to a board in a horizontal position.

Figure 3.22 Fine-grained fuel particles ignite much more readily than large dimension lumber.

cent oxygen. Oxidizers are not combustible, but they will support or enhance combustion. **Table 3.5, p. 94** lists some common oxidizers.

At normal ambient temperatures (68°F/20°C), materials can ignite and burn at oxygen concentrations as low as 14 percent. When oxygen concentration is limited, the flaming combustion will diminish, causing combustion to continue

Table 3.5
Common Oxidizers

Substance	Common Use
Calcium Hypochlorite (granular solid)	Chlorination of water in swimming pools
Chlorine (gas)	Water purification
Ammonium Nitrate (granular solid)	Fertilizer
Hydrogen Peroxide (liquid)	Industrial bleaching (pulp and paper and chemical manufacturing)
Methyl Ethyl Ketone Peroxide	Catalyst in plastics manufacturing

Courtesy of Ed Hartin.

in the nonflaming mode. Nonflaming combustion can continue at extremely low oxygen concentrations, even when the surrounding environment is at a relatively low temperature. However, at high ambient temperatures, flaming combustion may continue at considerably lower oxygen concentrations.

When the oxygen concentration is higher than normal, materials exhibit different burning characteristics. Materials that burn at normal oxygen levels will burn more intensely and may ignite more readily in oxygen-enriched atmospheres. Some petroleum-based materials will autoignite (ignite spontaneously without an external heat source) in oxygen-enriched atmospheres.

Flammable Range — The range between the upper flammable limit and lower flammable limit in which a substance can be ignited.

Lower Flammable Limit (LFL) — Lower limit at which a flammable gas or vapor will ignite and support combustion; below this limit the gas or vapor is too *lean* or *thin* to burn (lacks the proper quantity of fuel). *Also known as* Lower Explosive Limit (LEL).

Upper Flammable (Explosive) Limit (UFL) — Upper limit at which a flammable gas or vapor will ignite; above this limit the gas or vapor is too *rich* to burn (lacks the proper quantity of oxygen). *Also known as* Upper Explosive Limit (UEL).

Many materials that do not burn at normal oxygen levels burn readily in oxygen-enriched atmospheres. One such material is Nomex® fire-resistant fabric, which is used in firefighters' protective clothing. At normal oxygen levels, Nomex® does not burn, but in an oxygen-enriched atmosphere of approximately 31 percent oxygen, it ignites and burns vigorously.

Fires in oxygen-enriched atmospheres are more difficult to extinguish. These conditions are commonly found in hospitals and other health care facilities, some industrial and mercantile occupancies, and even private homes where occupants use breathing equipment containing pure oxygen. The inspector should understand that these occupancies may require special considerations in their automatic sprinkler system design.

The range of concentrations of the fuel vapor and air (oxidizer) is called the **flammable range**. The flammable range of a fuel is reported using the percent by volume of gas or vapor in air for the lower flammable limit (LFL) and for the upper flammable limit (UFL) **(Figure 3.24)**. The **lower flammable limit** is the minimum concentration of fuel vapor and air that supports combustion. Concentrations that are below the LFL are said to be *too lean* to burn. The **upper flammable limit** is the concentration above which combustion cannot take place. Concentrations that are above the UFL are said to be *too rich* to burn. Within the flammable range, there is an ideal concentration at which the exact amount of fuel and oxygen required for combustion is present.

Table 3.6 presents the flammable ranges for some common materials. The flammable limits for combustible gases are presented in chemical handbooks and documents, such as the National Fire Protection Association® (NFPA®)

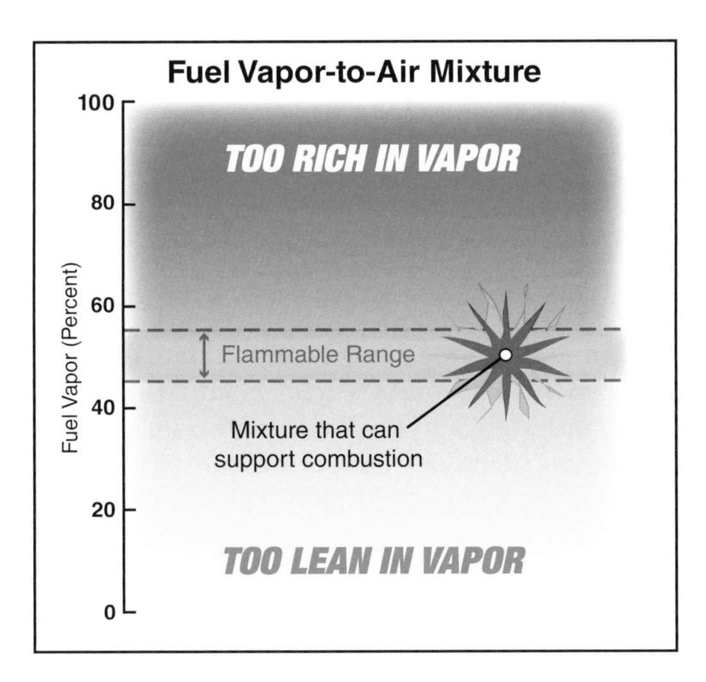

Figure 3.24 The flammable range is a relatively narrow band of conditions between the upper and lower flammable limits.

Table 3.6
Flammable Ranges of Common Flammable Gases and Liquids (Vapor)

Substance	Flammable Range
Methane	5%–15%
Propane	2.1%–9.5%
Carbon Monoxide	12%–75%
Gasoline	1.4%–7.4%
Diesel	1.3%–6%
Ethanol	3.3%–19%
Methanol	6%–35.5%

Source: *Computer Aided Management of Emergency Operations* (CAMEO)

Fire Protection Guide to Hazardous Materials. The limits are normally reported at ambient temperatures and atmospheric pressures. Variations in temperature and pressure can cause the flammable range to vary considerably.

Self-Sustained Chemical Reaction

The self-sustained chemical reaction involved in flaming combustion is complex. Combustion of a simple fuel, such as methane and oxygen, provides a good example of this complexity. Complete oxidation of methane results in production of carbon dioxide and water as well as the release of energy in the form of heat and light.

As flaming combustion occurs, the molecules of methane and oxygen break apart to form **free radicals** (electrically charged, highly reactive parts of molecules). Free radicals combine with oxygen or with the elements that form the fuel material (in the case of methane, carbon and hydrogen). This

Free Radicals — Molecular fragments that are highly reactive.

combination produces intermediate combustion products (new substances), even more free radicals, and increases the speed of the oxidation reaction. At various points in the combustion of methane, this process results in production of carbon monoxide and formaldehyde, which are both flammable and toxic. When more chemically complex fuels burn, this process involves many types of free radicals and intermediate combustion products, many of which are also flammable and toxic.

Chemical Flame Inhibition — Extinguishment of a fire by interruption of the chemical chain reaction.

Flaming combustion is an example of a chemical chain reaction. Sufficient heat will cause fuel and oxygen to form free radicals and initiate the self-sustained chemical reaction. The fire will continue to burn until the fuel or oxygen is exhausted or an extinguishing agent is applied in sufficient quantity to interfere with the ongoing reaction. **Chemical flame inhibition** occurs when an extinguishing agent, such as dry chemical or Halon-replacement agent, interferes with this chemical reaction, forms a stable product, and terminates the combustion reaction.

NOTE: Extinguishing agents are described in more detail in Chapter 13, Special-Agent Fire Extinguishing Systems and Extinguishers.

While the heat from a fire is a danger to anyone directly exposed to it, toxic smoke causes the largest percentage of fire deaths. Smoke is an aerosol comprised of gases, vapor, and solid particulates. Fire gases, such as carbon monoxide, are generally colorless, while vapor and particulates give smoke its varied colors.

The materials that make up smoke vary from fuel to fuel, but generally all smoke is toxic. Remember that the toxic effects of smoke inhalation are not the result of any one gas; it is the interrelated effect of all the toxic products present. **Table 3.7** lists some of the more common products of combustion and their toxic effects.

Classification of Fires

Fires are classified by the type of fuel involved and the type of extinguishing agent or activity that will be required to control and extinguish the fire. There are five classifications of fire.

Class A Fires

Class A fires involve ordinary combustible materials, such as wood, cloth, paper, rubber, grass, and many plastics. The primary method of extinguishment for Class A fires is cooling with water to reduce the temperature of the fuel to slow or stop the release of flammable vapors.

Class B Fires

Class B fires involve such flammable and combustible liquids and gases as gasoline, oil, lacquer, methane, mineral spirits, and alcohol. Structures containing Class B liquids require extinguishing agents such as foam and/or dry chemical. When a facility poses a risk of Class B fires involving gases, the inspector may verify procedures for shutting off the gas supply.

Table 3.7
Common Products of Combustion and Their Toxic Effects

Acetaldehyde	Colorless liquid with a pungent choking odor, which is irritating to the mucous membranes and especially the eyes. Breathing vapors will cause nausea, vomiting, headache, and unconsciousness.
Acrolein	Colorless-to-yellow volatile liquid with a disagreeable choking odor, this material is irritating to the eyes and mucous membranes. This substance is extremely toxic; inhalation of concentrations as little as 10 ppm may be fatal within a few minutes.
Asbestos	A magnesium silicate mineral that occurs as slender, strong, flexible fibers. Breathing of asbestos dust causes asbestosis and lung cancer.
Benzene	Colorless liquid with a petroleum-like odor. Acute exposure to benzene can result in dizziness, excitation, headache, difficulty breathing, nausea, and vomiting. Benzene is also a carcinogen.
Benzaldehyde	Colorless-to-clear yellow liquid with a bitter almond odor. Inhalation of concentrated vapor is irritating to the eyes, nose, and throat.
Carbon Monoxide	Colorless, odorless gas. Inhalation of carbon monoxide causes headache, dizziness, weakness, confusion, nausea, unconsciousness, and death. Exposure to as little as 0.2% carbon monoxide can result in unconsciousness within 30 minutes. Inhalation of a high concentration can result in immediate collapse and unconsciousness.
Formaldehyde	Colorless gas with a pungent, irritating odor that is highly irritating to the nose; 50–100 ppm can cause severe irritation to the respiratory track and serious injury. Exposure to high concentrations can cause injury to the skin. Formaldehyde is a suspected carcinogen.
Glutaraldehyde	Light-yellow liquid that causes severe irritation of the eyes and irritation of the skin.
Hydrogen Chloride	Colorless gas with a sharp, pungent odor. Mixes with water to form hydrochloric acid. Hydrogen chloride is corrosive to human tissue. Exposure to hydrogen chloride can result in irritation of skin and respiratory distress.
Isovaleraldehyde	Colorless liquid with a weak, suffocating odor. Inhalation causes respiratory distress, nausea, vomiting and headache.
Nitrogen Dioxide	Reddish-brown gas or yellowish-brown liquid, which is highly toxic and corrosive.
Particulates	Small particles that can be inhaled and deposited in the mouth, trachea, or the lungs. Exposure to particulates can cause eye irritation and respiratory distress (in addition to health hazards specifically related to the particular substances involved).
Polycyclic Aromatic Hydrocarbons (PAHs)	PAHs are a group of over 100 different chemicals that generally occur as complex mixtures as part of the combustion process. These materials are generally colorless, white, or pale yellow-green solids with a pleasant odor. Some of these materials are human carcinogens.
Sulfur Dioxide	Colorless gas with a choking or suffocating odor. Sulfur dioxide is toxic and corrosive and can irritate the eyes and mucous membranes.

Source: *Computer Aided Management of Emergency Operations (CAMEO) and Toxicological Profile for Polycyclic Aromatic Hydrocarbons.*

Class C Fires

Unlike the other classes, which are based on the fuel type, Class C fires involve energized electrical equipment. Class C fires typically involve household appliances, computers, transformers, electric motors, and overhead transmission lines. However, electricity does not burn, so the actual fuel in a Class C fire is usually insulation on wiring (Class A material) or lubricants (Class B materials). Any extinguishing agent used before de-energizing the equipment must not conduct electricity. The electrical equipment must be de-energized or shut off before extinguishing efforts can begin.

Class D Fires

Class D fires involve combustible metals, such as aluminum, magnesium, potassium, sodium, titanium, and zirconium. These materials are particularly hazardous in their powdered form. In the right concentrations, airborne metal dusts can cause powerful explosions when exposed to an ignition source. The extremely high temperature of some burning metals makes water reactive and other common extinguishing agents ineffective. No single agent effectively controls fires in all combustible metals.

Class D materials may be found in a variety of industrial or storage facilities. Some Class D materials react violently with water or other extinguishing agents, and may produce highly toxic smoke or vapors. Verifying compatibility of extinguishing agents for Class D materials is vital to the fire safety of the structures containing these hazards.

Class K Fires

Saponification — A phenomenon that occurs when mixtures of alkaline-based chemicals and certain cooking oils come into contact resulting in the formation of a soapy film.

Class K fires involve vegetable-based oils and greases normally found in commercial kitchens and food preparation facilities using deep fryers. Inspectors must remember that vegetable cooking oils are heated to higher temperatures and therefore pose greater hazards than other oils.

Class K fires require an extinguishing agent specifically formulated for the involved materials. Through a process known as **saponification**, these agents turn fats and oils into a soapy foam that extinguishes the fire.

Stages of Fire Development

The stages of fire include incipient, growth, fully developed, and decay. These stages of fire development have been developed as a result of laboratory studies examining fire progression in a single compartment **(Figure 3.25)**. Actual conditions within a building made up of multiple compartments can vary widely and fires may not progress in this order. For instance, the compartment of origin may be in the fully developed stage while adjacent compartments may be in the growth stage. In another example, an attic or void space may be in an underventilated decay stage while adjacent compartments are in the growth or

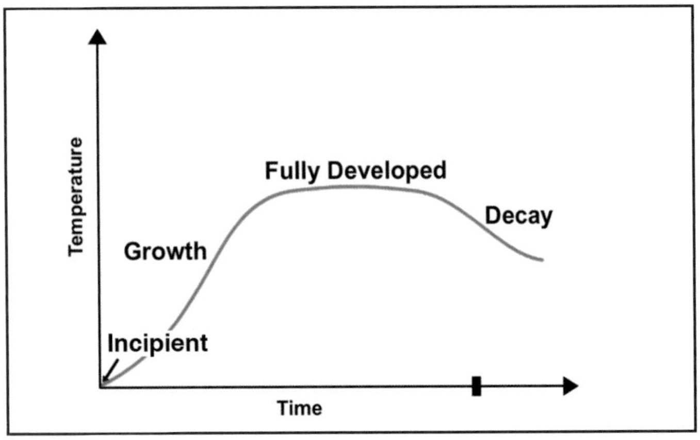

Figure 3.25 This line graph shows the progression of a fire in a lab or controlled environment. *Courtesy of Dan Madrzykowski/NIST.*

fully developed stage. Knowledge of fire development is important as the inspector identifies and evaluates fire extinguishing equipment and installed fire protection systems.

Factors that Affect Fire Development

A number of factors influence fire development within a compartment, including:

- Fuel type
- Availability and location of additional fuels
- Compartment volume and ceiling height
- Ventilation
- Thermal properties of the compartment
- Fuel load

Fuel Type

The type of fuel involved in combustion affects the heat release rate (HRR). Fires involving Class B and C fuels will eventually spread to the building contents and structure, resulting in a primarily Class A fueled fire.

In a compartment fire, surface-to-mass ratio is one of the most fundamental Class A fuel characteristics influencing fire development. Combustible materials with high surface-to-mass ratios are much more easily ignited and will burn more quickly than the same substance with less surface area.

Fires involving Class B flammable/combustible liquids are also influenced by the surface area and type of fuel involved. A liquid fuel spill will increase that liquid's surface-to-volume ratio, generating more flammable vapors than the same liquid in an open container. The increase of vapor due to the spill will also allow more of the fuel to ignite, resulting in greater heat over a shorter period of time.

Structure fires involving single types of fuels are rare. The inspector also needs to remember that today's homes and businesses are largely filled with contents made from petroleum-based materials. These synthetic fuels have a higher heat of combustion and produce higher HRRs than natural materials such as wood. Burning synthetic fuels produce products of combustion that contain large quantities of solid and liquid particulates and unburned gases.

A compartment fire that results from a flammable/combustible gas leak may begin with a rapid ignition of the gas and an explosion. Inspectors should verify that procedures are in place to shut off the fuel source, which will reduce or eliminate the Class B fuel. They should also understand the resulting Class A fire will continue to burn unless a sprinkler system is in place to control it.

Availability and Location of Additional Fuels

Factors that influence the availability and location of additional fuels include the building configuration, contents, construction materials, interior finish materials, fuel proximity, and fire location.

Building configuration refers to the layout of the structure, including:

- Number of stories above or below grade
- Compartmentation

- Floor plan
- Openings between floors
- Continuous voids or concealed spaces
- Barriers to fire spread

Each of these elements may contribute to fire spread or containment. For instance, an open floor plan space may contain furnishings that provide fuel sources on all sides of a point of ignition. On the other hand, a compartmentalized configuration may have fire-rated barriers, such as walls, ceilings, and doors, separating fuel sources and limiting fire development to an individual compartment **(Figures 3.26 a and b)**.

Contents of the building (nonstructural fire load). The contents of a structure are often the most readily available fuel source, significantly influencing fire development in a compartment fire. When contents release a large amount

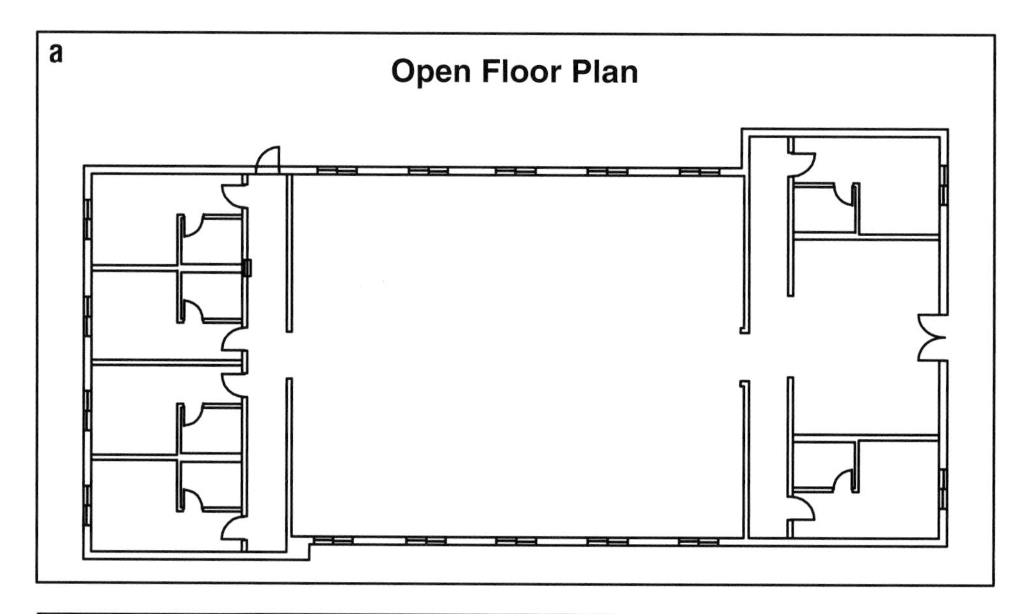

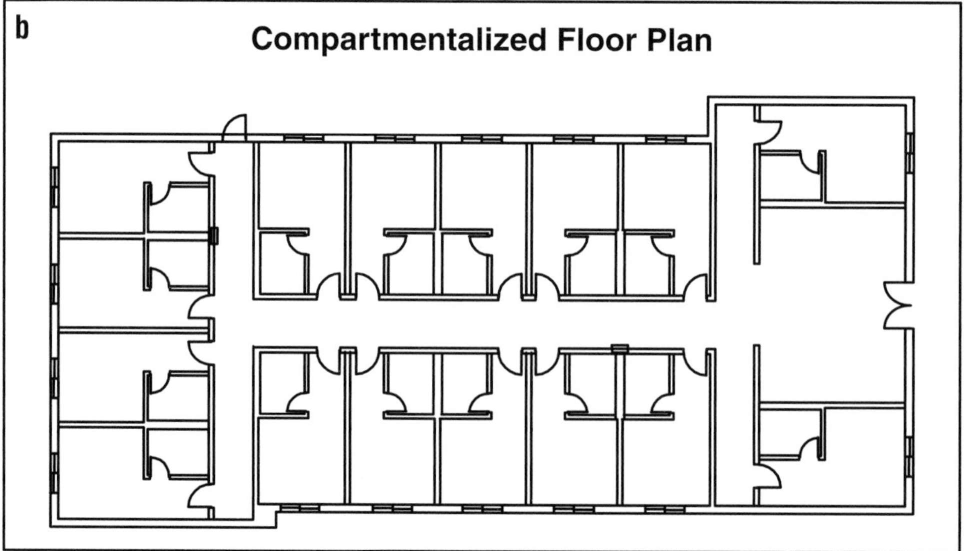

Figures 3.26 a and b (a) An open floor plan features some small compartments and a large open area, often located in the center of the structure. (b) A compartmentalized floor plan limits the amount of space and fuel a fire may access.

of heat rapidly, both the intensity of the fire and speed of development will be increased. For example, synthetic furnishings, such as polyurethane foam, will begin to pyrolize rapidly under fire conditions (even when located some distance from the origin of the fire) due to the chemical makeup of the foam and its high surface-to-mass ratio. This pyrolysis will speed the process of fire development. Polyurethane foam will liquefy and continue to burn.

Construction of the building (structural fire load). The materials used to construct the building, such as wall studs, floor and ceiling joists, and roof supports as well as sheathing, contribute to the fuel load. Each type of construction is symbolized by a different quantity of fuel.

NOTE: See Chapter 4, Construction Types and Occupancy Classifications, for details on construction types.

Construction/interior finish materials. The types of construction materials used in a structure influence fuel load as well. In some types of buildings, the construction materials can be affected by the fire, adding to the fuel load. For example, in wood-frame buildings, the structure itself is a source of fuel (**Figure 3.27**). In addition to structural members, combustible interior finishes, such as wood paneling and window coverings, can be a significant factor influencing fire spread. As described earlier in this chapter, the orientation of these fuels as well as their surface-to-mass ratio will also influence the rate and intensity of fire spread.

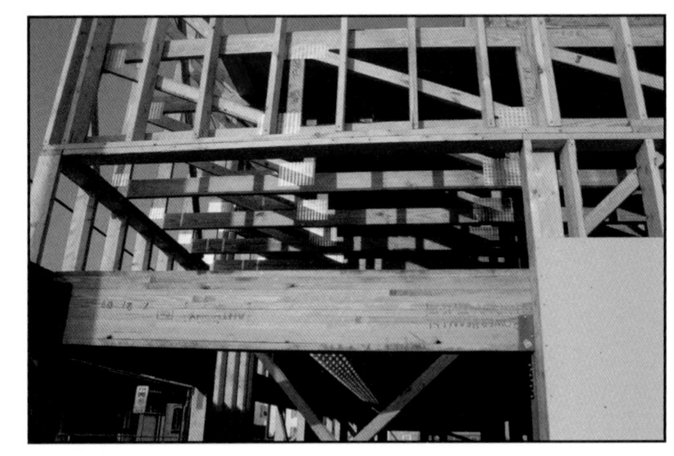

Figure 3.27 Wood-frame structures represent an additional source of fuel. *Courtesy of Walt Schneider.*

Fuel proximity and continuity. The proximity and continuity of contents and structural fuels also influence fire development. Fuels located in the upper level of adjacent compartments will pyrolize more quickly from the effect of the hot gas layer. Continuous fuels (such as combustible interior finishes) will rapidly spread the fire from compartment to compartment.

Fire location. Similarly, the location of the fire within the building will influence fire development. When the fire is located low in the building, such as in the basement or on the first floor, convected heat currents will cause vertical extension through atriums, unprotected stairways, vertical shafts, and concealed spaces. Fires originating on upper levels generally extend downward much more slowly, following the fuel path or as a result of structural collapse.

Compartment Volume and Ceiling Height

All other factors being equal, a fire in a large compartment will develop more slowly than one in a small compartment. Slower fire development is due to the greater volume of air and the increased distance radiated heat must travel from the fire to the contents that must be heated. Remember, though, that this large volume of air will support the development of a larger fire before the lack of ventilation becomes the limiting factor. A high ceiling may also hide the extent of fire development by allowing a large volume of hot smoke and fire gases to accumulate at the ceiling level while conditions at floor level remain relatively unchanged.

Figure 3.28 Fire doors are frequently found to be blocked open. The inspector must work to eliminate this fire hazard. *Courtesy of Scott Strassburg.*

Ventilation

Ventilation in a compartment significantly influences how fire develops and spreads. For the most part, all buildings exchange air inside the structure with the air outside the structure. In some cases, this air exchange is due to constructed openings, such as windows, doors, and passive ventilation devices as well as leakage through cracks and other gaps in construction. In other cases, air exchange occurs through the heating, ventilating, and air conditioning (HVAC) system.

In a compartment fire that involves the contents of the room, fire development is limited by the available air supply and is said to be ventilation controlled. When the available oxygen in the air is depleted, the fire's burning rate will decrease.

When considering fire development, inspectors should be aware that potential openings can change ventilation. Under fire conditions, windows can fail or doors can be left open, increasing ventilation into the compartment. When a fire develops to the point where it becomes ventilation controlled, the available air supply will determine the speed and extent of fire development and the direction of fire travel. Fire will have a tendency to grow in the direction of ventilation openings as the fire seeks fresh air.

Open fire doors can facilitate the movement of fire and smoke. Because inspectors frequently encounter fire doors that have been blocked open, this condition must be noted and corrected **(Figure 3.28)**.

Thermal Properties of the Compartment

The thermal properties of the compartment can contribute to rapid fire development as well as making extinguishment more difficult and reignition possible. Thermal properties of a compartment include:

- **Insulation** — Contains heat within the compartment, causing a localized increase in the temperature and fire growth
- **Heat reflectivity** — Increases fire spread through the transfer of radiant heat from wall surfaces to adjacent fuel sources
- **Retention** — Maintains temperature by absorbing and releasing large amounts of heat slowly

Fuel Load — The total quantity of combustible contents of a building, space, or fire area, including interior finish and trim, expressed in heat units of the equivalent weight in wood.

Fuel Load

The total quantity of combustible contents of a building, space, or fire area is referred to as the **fuel load** (some documents may use the term *fire load*). The fuel load includes all the structure's furnishings, merchandise, interior finishes, and structural components **(Figure 3.29)**. Using a set of mathematical equations, a qualified professional, such as a fire protection engineer, can generate a fairly accurate estimate of a structure's fuel load. However, the inspector may only be able to generate an estimate based on knowledge and experience. For instance, a concrete block structure containing stored steel pipe will have a much smaller fuel load than a steel structure containing

Figure 3.29 The fuel load in this store looks relatively light, but includes many paints and flammable materials.

flammable and combustible liquids. Knowledge of building construction and occupancy types will be essential to determining fuel loads.

Incipient Stage

All fires occur as a result of piloted ignition or autoignition. The **incipient stage** starts with ignition when the three elements of the fire triangle come together and combustion occurs. At this point, the fire is small and confined to the material (fuel) first ignited **(Figure 3.30)**.

Once combustion begins, development of an incipient fire is largely dependent on the characteristics and configuration of the fuel involved (fuel-controlled fire). Air in the compartment provides adequate oxygen to continue fire development. During this initial phase of fire development, radiant heat warms the adjacent fuel and continues the process of pyrolysis.

A **plume** of hot gases and flame rises from the fire and mixes with the cooler air in the room **(Figure 3.31)**. As this plume reaches the ceiling, hot gases begin to spread horizontally across the ceiling in what firefighters have historically

Incipient Stage — First stage of the burning process in a compartment in which the substance being oxidized is producing some heat, but the heat has not spread to other substances nearby. During this phase, the oxygen content of the air has not been significantly reduced and the temperature within the compartment is not significantly higher than ambient temperature.

Plume — The column of hot gases, flames, and smoke rising above a fire; also called *convection column, thermal updraft, or thermal column.* (NFPA® 921)

Figure 3.30 During the incipient stage of a fire, oxygen and fuel are plentiful. *Courtesy of Dan Madrzykowski/NIST.*

Figure 3.31 A fire emits a plume of hot gases and flame that begins the process of pyrolysis on other materials nearby. *Courtesy of Dan Madrzykowski/NIST.*

Ceiling Jet — A relatively thin layer of flowing hot gases that develops under a horizontal surface (e.g., ceiling) as a result of plume impingement and the flowing gas being forced to move horizontally. (NFPA® 921)

Thermal Layering — Outcome of combustion in a confined space in which gases tend to form into layers, according to temperature, with the hottest gases found at the ceiling and the coolest gases at floor level.

Flow Path — Composed of at least one inlet opening, one exhaust opening, and the connecting volume between the openings. The direction of the flow is determined by difference in pressure. Heat and smoke in a high pressure area will flow toward areas of lower pressure (NIST)

called mushrooming; however, in scientific or engineering terms, it is referred to as forming a **ceiling jet**. Hot gases in contact with the surfaces of the compartment and its contents transfer heat to other materials. This complex process of heat transfer begins to increase the overall temperature in the room. **Table 3.8** lists the factors that influence the development of fuel-controlled fires.

In this early stage of fire development, the fire has not yet influenced the environment within the compartment to a significant extent. The temperature, while increasing, is only slightly above ambient and the concentration of products of combustion is low.

During the incipient phase, occupants can more likely escape from the compartment and the fire could be safely extinguished with a portable extinguisher or small hoseline. It is essential to recognize that the transition from incipient to growth stage can occur quite quickly (in seconds in some cases), depending on the type and configuration of fuel involved.

Growth Stage

As the fire transitions from incipient to growth stage, it begins to influence the environment within the compartment **(Figure 3.32)**. Likewise, the fire is influenced by the configuration of the compartment and the amount of ventilation. In a compartment fire, the ceiling and walls affect the plume of hot gases rising from the fire.

The first effect is the amount of air that is entrained into the plume. In a compartment fire, the location of the fuel package in relation to the compartment walls affects the amount of air that is entrained and therefore the amount of cooling that occurs. Unconfined fires draw air from all sides and the entrainment of air cools the plume of hot gases, reducing flame length and vertical extension. This process can occur in the following ways:

- Fuel packages in the middle of the room can entrain air from all sides.
- Fires in fuel packages near walls can only entrain air from three sides.
- Fires in fuel packages in corners can only entrain air from two sides.

Therefore, when the fuel package is not in the middle of the room, the combustion zone (area where sufficient air is available to feed the fire) expands vertically and a higher plume results. This result stretches the flame even further, affecting the temperatures in the developing hot-gas layer at ceiling level and increasing the speed of fire development. In addition, as wall surfaces become hot, burning fuel receives more reflected radiant heat, further increasing the speed of fire development.

Thermal Layering

The **thermal layering** of gases, sometimes referred to as heat stratification and thermal balance, is the tendency of gases to form into layers according to temperature. Generally, the hottest gases tend to be in the upper layer, while the cooler gases form the lower layers. In addition to the effects of heat transfer through radiation and convection described earlier, radiation from the hot gas layer acts to heat the interior surfaces of the compartment and its contents. Changes in ventilation and **flow path** can significantly alter the thermal layering.

Table 3.8
Factors Influencing Development of a Fuel-Controlled Fire

Mass and Surface Area	The greater the surface area for a given mass of fuel, the easier it is for that fuel to be heated to its ignition temperature.
Chemical Content	The chemical makeup of the fuel has a significant impact on the heat released during combustion. Many hydrocarbon-based synthetic materials have a heat of combustion that is more than twice that of cellulose materials such as wood.
Fuel Load	The total amount of fuel available for combustion influences total potential heat release.
Fuel Moisture	While not a factor with all types of fuel, water acts as thermal ballast, slowing the process of heating the fuel to its ignition temperature.
Orientation	Orientation in relation to the fire influences how heat is transferred. For example, a wood wall surface is heated by both convection and radiation, where the floor is more likely to be heated by radiant heat alone.
Continuity	Continuity is the proximity of various fuel elements to one another. The closer (or more continuous) the fuel is, the easier and more rapidly fire will extend. Continuity may be either horizontal (i.e. ceiling surface) or vertical (i.e. wall or rack storage).

Courtesy of Ed Hartin.

Figure 3.32 During the growth stage of a fire, the fire influences the environment and begins to be influenced by the ventilation conditions of the compartment. *Courtesy of Dan Madrzykowski/NIST.*

As the volume and temperature of the hot gas layer increases, so does the pressure. Higher pressure causes the hot gas layer to spread downward within the compartment and out through any openings, such as doors or windows. The pressure of the cool gas layer is lower, resulting in inward movement of air from outside the compartment at the bottom as the hot gases exit through the top of the opening. The level where these two layers meet the pressure is neutral. The interface of the hot and cooler gas layers at the opening is commonly referred to as the **neutral plane**. The neutral plane only exists at openings where hot gases are exiting and cooler air is entering the compartment **(Figure 3.33, p. 106)**. One way an inspector can limit the spread of gases and fire is by verifying that all fire doors are equipped with functioning door closers.

Neutral Plane — The level at a compartment opening where the difference in pressure exerted by expansion and buoyancy of hot smoke flowing out of the opening and the inward pressure of cooler, ambient temperature air flowing in through the opening is equal.

Isolated Flames

As the fire moves through the growth stage and becomes ventilation controlled, isolated flames may be observed moving through the hot gas layer **(Figure 3.34)**. Combustion of these hot gases indicates that portions of the hot gas layer are within their flammable range and that there is sufficient temperature to result in ignition. As these hot gases circulate to the outer edges of the plume, they find sufficient oxygen to ignite. This phenomenon is frequently observed prior to more substantial involvement of flammable products of combustion in the hot gas layer.

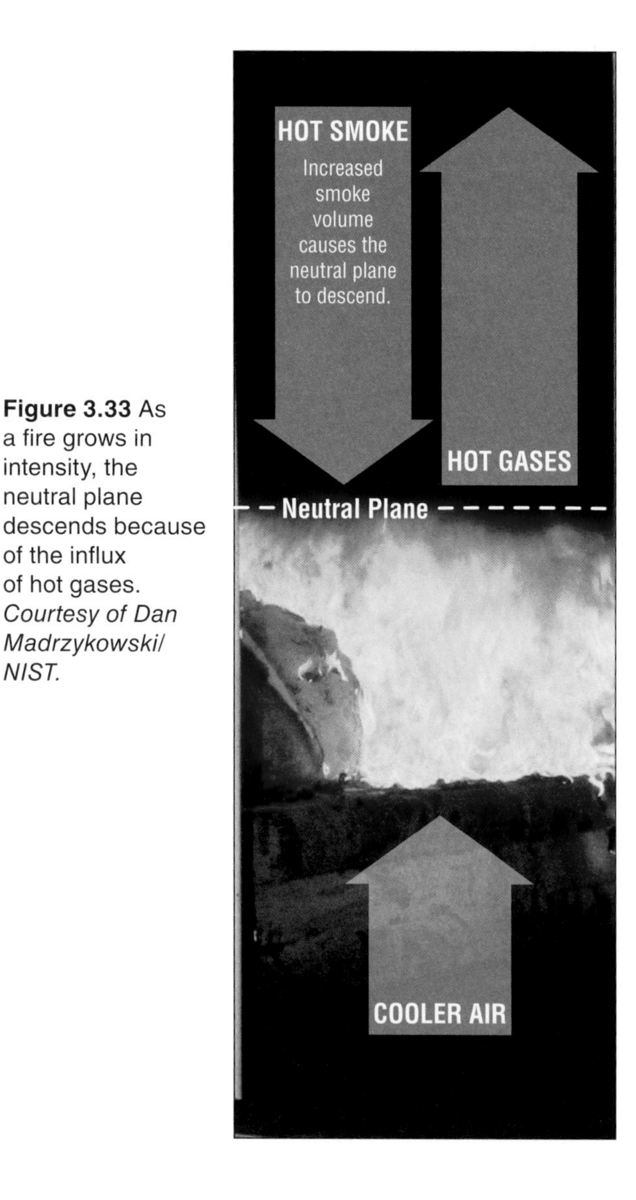

Figure 3.33 As a fire grows in intensity, the neutral plane descends because of the influx of hot gases. *Courtesy of Dan Madrzykowski/NIST.*

Figure 3.34 Isolated flames exhibit ghosting, an indication that the layers of combustible gases above the fire have reached their ignition temperature. *Courtesy of National Institute of Standards and Technology (NIST).*

Rapid Transition

Rapid transition from the growth stage to the fully developed stage is known as **flashover**. The environment of the room is changing from a two-layer condition (hot on top, cooler on the bottom) to a single-layer, well-mixed, hot gas condition from floor to ceiling. This atmosphere will not be survivable even for fully protected firefighters. Depending on the ventilation, combustion may be flaming or nonflaming.

Flashover — A rapid transition from the growth stage to the fully developed stage.

The transition period between pre-flashover fire conditions (growth stage in a fuel-limited case) to post-flashover (fully developed stage) can occur rapidly. During flashover, the fire's volume can increase from approximately ¼ to ½ of the room's upper volume to filling the entire volume of the room and potentially extending out of any openings in the room.

Flashover does not occur in every compartment fire. If the fire becomes ventilation controlled, it may not progress to flashover. This situation will limit the heat release rate and cause the fire to enter the decay stage while continuing the process of pyrolysis and increasing the fuel content of the smoke.

Most fires that develop beyond the incipient stage become ventilation controlled. Even when doors and windows are open, there is often insufficient air to allow the fire to continue to develop based on the available fuel. When windows are intact and doors are closed, the fire may move into a ventilation-controlled state even more quickly. While this change reduces the heat release rate, fuel will continue to pyrolize, creating extremely fuel-rich smoke.

Because fuel load and interior finish materials contribute to fire growth, the inspector should be aware of the type and amount of combustible furnishings and fuel load during an inspection. These include decorations, furniture (such as a couch in a bar), and wall coverings.

Fully Developed Stage

In the fully developed stage, all combustible materials in the compartment are burning, releasing the maximum amount of heat possible for the available fuel and oxygen, and producing large volumes of fire gases **(Figure 3.35)**. The fire is ventilation controlled because the heat release is dependent on the compartment openings that are providing oxygen to support the ongoing combustion reaction. An increase in the available air supply will result in higher heat release.

Figure 3.35 A fully developed fire involves everything within a compartment and can be severely increased by a rapid influx of outside air because fully developed fires tend to be ventilation controlled.

Figure 3.36 During the decay stage, the fire begins to cool and die as fuel and/or ventilation run out of supply. *Courtesy of Dan Madrzykowski/NIST.*

During this stage, flammable products of combustion are likely to flow from the compartment of origin into adjacent compartments or out through exterior openings. Because there is insufficient oxygen for complete combustion in the compartment, flames will extend out of the compartment openings. If there are no openings, it is unlikely that the fire will reach a fully developed stage due to limited ventilation.

Decay Stage

A compartment fire will decay as the fuel is consumed or if the oxygen concentration falls to the point that flaming combustion is diminished **(Figure 3.36)**. Both of these situations can result in the combustion reaction coming to a stop. However, decay due to reduced oxygen concentration can follow a considerably different path if the ventilation of the compartment changes before combustion ceases and temperature in the compartment lowers.

Fuel Consumption

The fire enters the decay stage as it consumes the available fuel in the compartment and the heat release rate begins to decline. If there is adequate ventilation, the fire becomes fuel controlled. The heat release rate will drop, but the temperature in the compartment may remain high for some time. During this stage, the flammable products of combustion can accumulate within the compartment or adjacent spaces.

Limited Ventilation

When a compartment fire enters the decay stage due to a lack of oxygen, the rate of heat release also declines. However, the continuing combustion reaction (based on available fuel and the limited oxygen available to the fire) may maintain an extremely high temperature within the compartment. Temperature drops, but pyrolysis can continue.

Fire Control

Fire is controlled and extinguished by limiting or interrupting one or more of the essential elements in the combustion process depicted by the fire tetrahedron model. Firefighters influence fire behavior in the following ways:

- Temperature reduction
- Fuel removal
- Oxygen exclusion
- Chemical flame inhibition

Using fire prevention efforts, fire inspectors influence fire behavior through temperature reduction, fuel removal, oxygen exclusion, and chemical flame inhibition. Inspectors are responsible for the following tasks, among others:

- Verifying that extinguishing systems are installed and functioning correctly so they can act to cool a fire, interrupt its chemical chain reaction, and, in some cases, exclude oxygen.
- Checking the installation and maintenance of standpipes and hose systems, which provide emergency fire crews with the ability to disrupt combustion on elevated stories or in large facilities.
- Verifying proper placement and maintenance of fire extinguishers, enabling building occupants the opportunity to suppress incipient fires.
- Monitoring and regulating hazardous conditions, including fire load and interior finishes, thereby helping to reduce the potential for fire growth.

Temperature Reduction

Cooling with water is one of the most common methods of fire control and extinguishment. This process reduces the temperature of a fuel to a point where it does not produce sufficient vapor to burn. Solid fuels and liquid fuels with high flash points can be extinguished by cooling. The use of water is also the most effective method available for extinguishing smoldering fires.

Cooling with water is accomplished through installed sprinkler systems, which are designed to extinguish a fire by reducing its temperature **(Figure 3.37)**. Enough water must be applied to the burning fuel to absorb the heat being generated by combustion. Nonetheless, cooling with water cannot sufficiently reduce vapor production to extinguish fires involving low flash point flammable liquids and gases.

In addition to extinguishment by cooling, water can also be used to control burning gases and reduce the temperature of hot products of combustion in the upper layer. This approach slows the pyrolysis process of combustible materials, reducing both radiant heat flux from the upper layer and the potential for flashover.

Figure 3.37 Sprinkler systems are designed to control fires by cooling them.

Water absorbs significant heat as its temperature is raised, but it has its greatest effect when it is vaporized into steam **(Figure 3.38)**. When water is converted to steam at 212°F (100°C), it expands approximately 1,700 times.

Figure 3.38 An indirect attack involves directing a stream toward the ceiling of a room.

Heat Capacity of Water

One pound of water has the ability to absorb about 970 BTUs (970 kJ) of heat. One gallon (4 L) of water is capable of absorbing about 8,050 BTUs (8 050 kJ) of energy, given that it has a density of 8.3 pounds/gallon (1 kg/L).

Fuel Removal

Removing the fuel is one way to avoid or end an unwanted fire. Unfortunately, this method is not possible in buildings constructed from combustible materials or in occupancies where combustibles are stored or displayed. Building codes commonly require the use of fire-resistive construction to isolate certain fire hazards or materials that have a higher heat of combustion or burning rate. Isolating the fuel using fire-resistive construction reduces the likelihood of ignition. If ignition occurs, the fire-resistive separation will compartmentalize the combustibles to limit its spread beyond the room or area of origin.

Oxygen Exclusion

Reducing the oxygen available to the combustion process reduces a fire's growth and may extinguish it over time. Flooding an area with an inert gas, such as carbon dioxide, displaces the oxygen and disrupts the combustion process. Oxygen can also be separated from some fuels by blanketing them with foam. Of course, neither of these methods works on oxygen-enhanced fires such as those involving oxidizers.

Chemical Flame Inhibition

Extinguishing agents, such as some dry chemicals, halogenated agents (Halons), and Halon-replacement agents, interrupt the combustion reaction and stop flame production. This method of extinguishment is effective on gas and liquid fuels because they must flame to burn.

These agents do not easily extinguish nonflaming combustion because there is no chemical chain reaction to inhibit. The high agent concentrations and extended periods necessary to extinguish smoldering fires make these agents impractical in these cases.

Chapter Summary

The inspector's job is to reduce risk. To this end, the inspector needs to understand the combustion process and how fire behaves in different materials and environments. Understanding fire behavior will help the inspector evaluate fire protection systems that are designed to prevent fire or fire spread. The inspector must also be able to recognize how building design, construction materials, fuel loads, and combustible furnishings affect fire behavior and fire spread. The inspector needs to understand that a solid working knowledge of fire behavior will help reduce the loss of life and property from fire and help businesses to operate more safely and provide for business continuance.

Review Questions

1. Describe and name an example of chemical, electrical, and mechanical energy.
2. What are the ways heat can be transferred to an object?
3. Describe the five classifications of fire.
4. List some factors that influence fire development within a compartment.
5. What are the stages of fire?

1. What are the four components of the fire tetrahedron model?

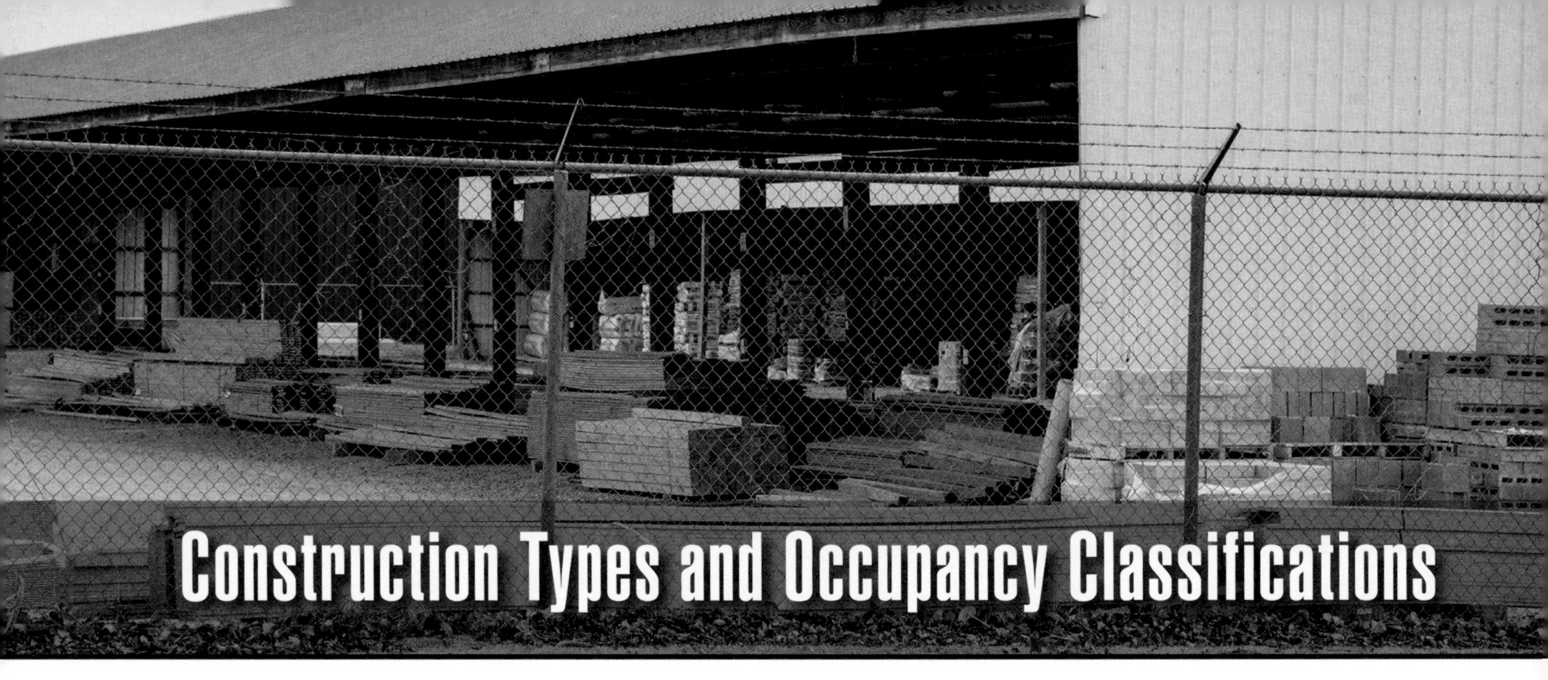

Construction Types and Occupancy Classifications

Chapter Contents

Case History 115

Construction Types 116

- United States Construction 117
- Canadian Construction 125

Occupancy Classifications 125

- Assembly Occupancies 132
- Business Occupancies..133
- Educational Occupancies...133
- Factory/Industrial Occupancies.................................134
- Institutional Occupancies..134
- Mercantile Occupancies 138
- Residential Occupancies 139
- Storage Occupancies...144
- Utility/Miscellaneous Occupancies 145

Multiple-Use Occupancies 146

- Incidental Use...147
- Mixed-Use..147

Chapter Summary 148

Review Questions 148

Key Terms

Assembly...132
Fire Load...125
Fire-Stop...122
Glued-Laminated Beam..........................122
Load-Bearing Wall...................................117
Noncombustible.....................................118
Nonload-Bearing Wall.............................117
Oriented Strand Board (OSB).................124
Protected Steel..118

NFPA® Job Performance Requirements

This chapter provides information that addresses the following job performance requirements of NFPA® 1031, *Standard for Professional Qualifications for Fire Inspector and Plan Examiner* (2014).

Fire Inspector I

4.3.1

4.3.4

Fire Inspector II

5.3.2

Construction Types and Occupancy Classifications

Learning Objectives

After reading this chapter, students will be able to:

Inspector I

1. Identify construction types. (4.3.4)
2. Identify single-use occupancy classifications. (4.3.1)

Inspector II

1. Identify multiple-use occupancies. (5.3.2)

Chapter 4
Construction Types and Occupancy Classifications

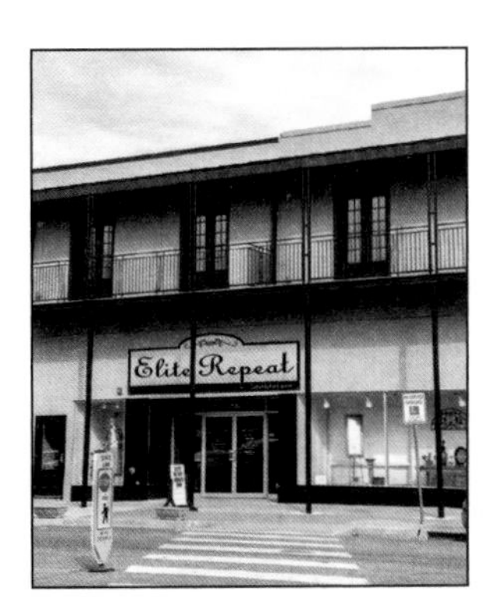

Case History

October, 2007, Las Vegas NV: County building officials ordered two floors of the Rio Ipanema hotel, containing 140 guestrooms, to be closed for inspection. An inspection conducted earlier that year after allegations of remodeling without proper permits or safety inspections found no problems. After an investigation by the Review-Journal newspaper, however, Clark County Building officials reopened the investigation. Of particular concern were allegations from former workers that holes for piping or wiring lacked proper sealing and that post-tensioned cables had been severed without any county review.

The inspections were conducted with additional oversight, and the county announced that it planned to hire an outside consultant to review the process by which the Building Division handles complaints. The hotel rooms were scheduled to reopen after inspections were completed.

Construction types and occupancy classifications are two critical elements an inspector must understand when performing plan reviews, issuing permits, or making field inspections of new, renovated, or existing structures. A building's construction type is based on the fire resistance of the materials and design of the structure, while the occupancy classification is based on the use of the structure.

Although the type of construction used in a building may remain the same for the life of the structure, it is common for an inspector to discover that the building's occupancy has changed since the previous inspection. Occupancy changes may result in a higher level of risk than the building was constructed to handle **(Figure 4.1, p. 116)**.

This chapter provides an overview of the construction types and occupancy classifications established by the model building codes. Each construction type and occupancy group is described and comparisons are made between the model codes.

NOTE: Chapter 5, Building Construction: Materials and Structural Systems, provides an introduction to building construction, including the various materials and types of structural systems commonly used. Chapter 15, Plans Review, provides more information on the plans review process.

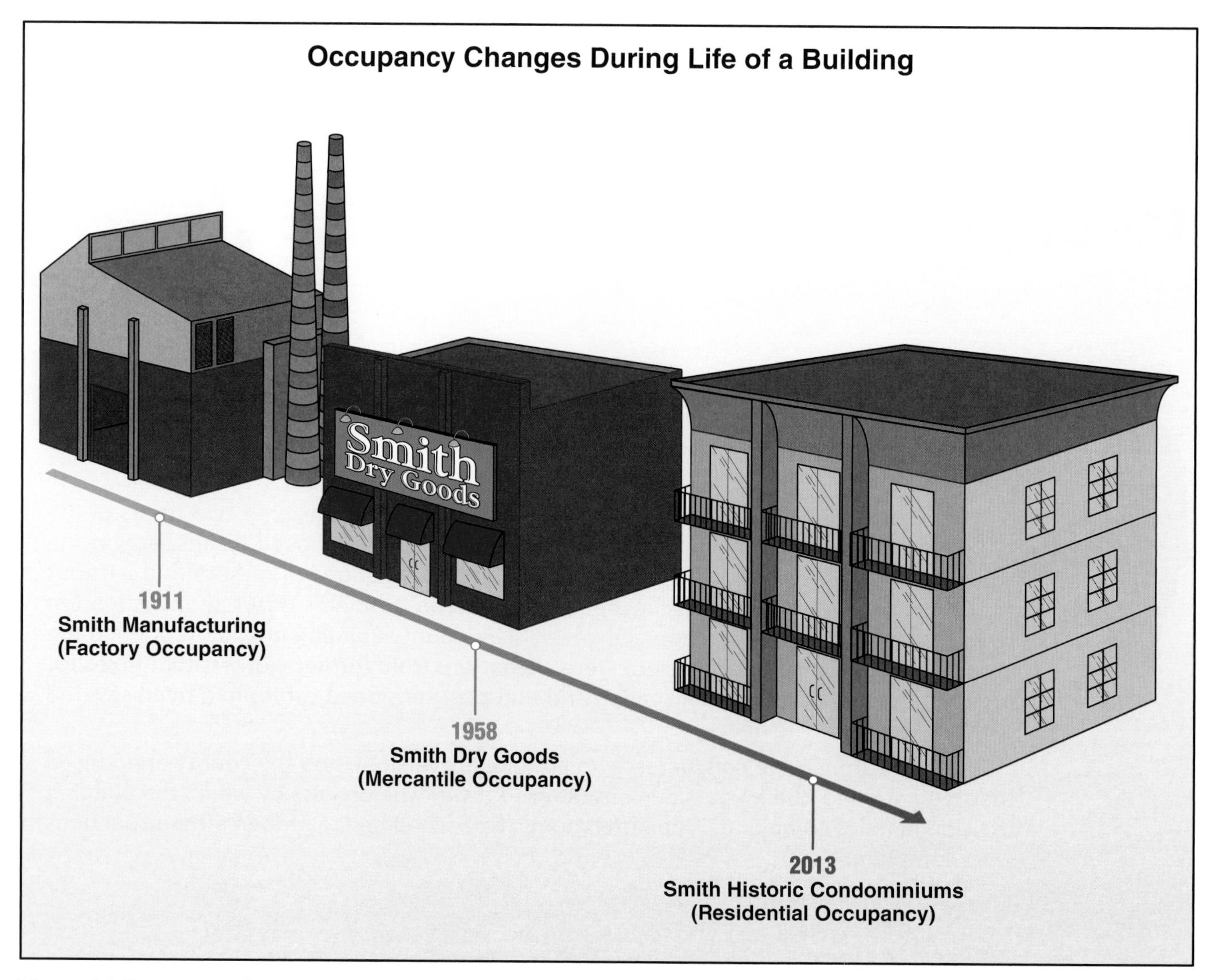

Figure 4.1 Occupancy changes are common during the life of a building.

Construction Types

The AHJ's adopted building codes establish construction types and occupancy classifications. The majority of these building codes are based on model codes developed by such third-party organizations as the National Fire Protection Association® (NFPA®), International Code Council® (ICC®), or the Canadian Commission on Building and Fire Codes®. Inspectors must have a thorough understanding of these occupancy classifications and construction types in order to review plans or conduct inspections.

The architect or design professional of record typically determines the type of construction used in building or remodeling based on the following factors:

- Occupancy type
- Building size
- Presence or lack of an automatic fire suppression system
- Construction materials used and their fire resistance

- Cost (from the owner's standpoint)
- Other applicable factors

The inspector must be familiar with the specific descriptions of these construction types as they are defined in the locally adopted codes. This section describes the various types of construction recognized in the United States and Canada.

NOTE: For more information about building construction, refer to the IFSTA **Building Construction for the Fire Service** manual.

United States Construction

Both the *International Building Code®* and the NFPA® recognize five types of construction: Type I through Type V. Each type is further divided into subcategories, depending on the code and construction type. Although the building codes have subtle differences regarding construction types, there are significant common areas.

Load-Bearing Wall — Walls of a building that by design carry at least some part of the structural load of the building in the direction of the ground or base.

Nonload-Bearing Wall — Wall, usually interior, that supports only its own weight. These walls can be breached or removed without compromising the structural integrity of the building.

Every structure is composed of basic building elements:

- Structural frame
- **Load-bearing walls**, both interior and exterior **(Figure 4.2)**
- **Nonload-bearing walls** and partitions, both interior and exterior

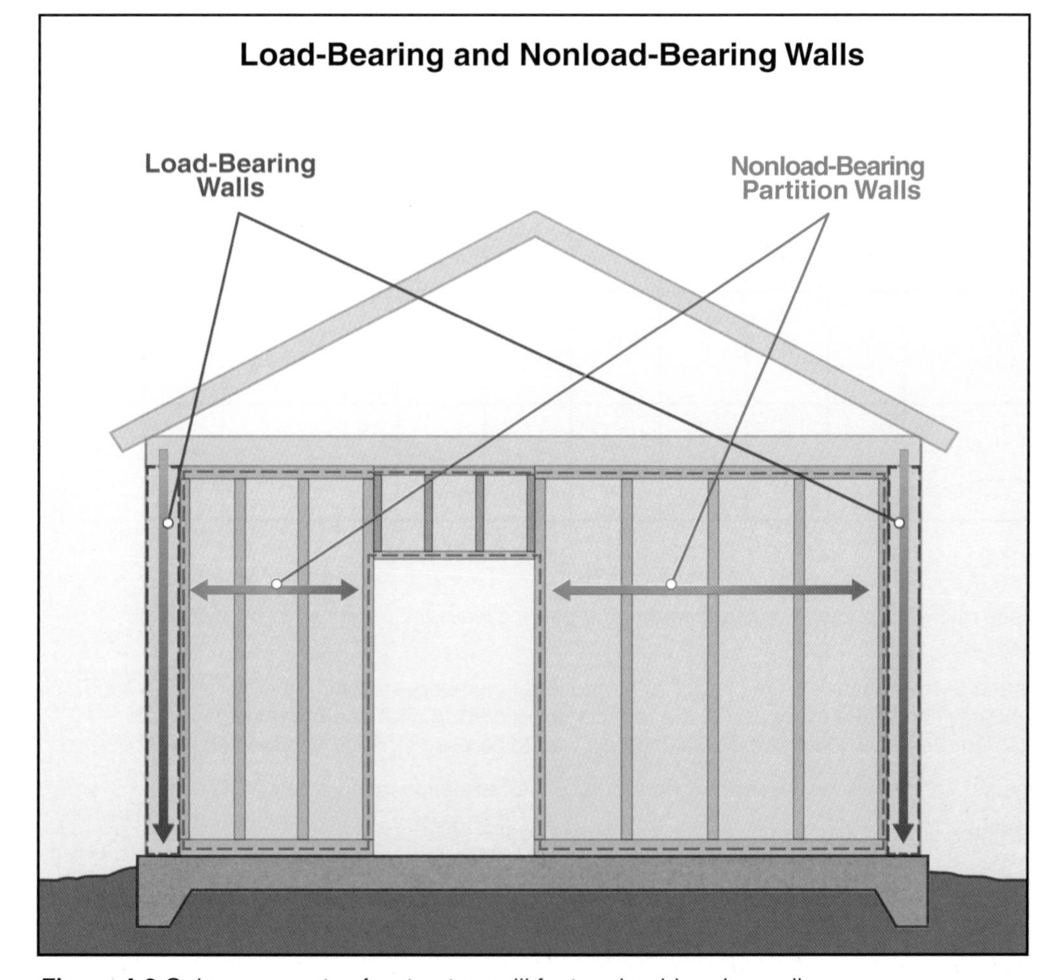

Figure 4.2 Only some parts of a structure will feature load-bearing walls.

- Floor construction
- Roof construction

The construction materials and their performance under fire conditions determine the construction type of the structure. Refer to **Tables 4.1** and **4.2** from the *International Building Code®* for use as references for this section.

Noncombustible — Incapable of supporting combustion under normal circumstances.

Protected Steel — Steel beams that are covered with spray-on fireproofing (an insulating barrier) or are fully encased in an Underwriters Laboratories Inc. (UL) designed system.

Type I

In Type I construction, all structural members are composed of only **noncombustible** materials that possess a high fire-resistance rating. Type I construction can be expected to remain structurally stable during a fire for the duration of the structural members' fire resistance rating. Reinforced and precast concrete, masonry, and **protected steel**-frame construction meet the criteria for Type I construction **(Figure 4.3)**.

Table 4.1
Fire-Resistance Rating Requirements for Building Elements (Hours)

Building Element	Type I		Type II		Type III		Type IV	Type V	
	A	B	A	B	A	B	HT	A	B
Primary Structural Frame[f] (see Section 202)	3[a]	2[a]	1	0	1	0	HT	1	0
Bearing Walls Exterior[e,f] Interior	 3 3[a]	 2 2[a]	 1 1	 0 0	 2 1	 2 0	 2 1/HT	 1 1	 0 0
Nonbearing Walls and Partitions Exterior	See Table 4.2								
Nonbearing Walls and Partitions Interior[d]	0	0	0	0	0	0	See Section 602.4.6*	0	0
Floor Construction and associated secondary members (see Section 202)	2	2	1	0	1	0	HT	1	0
Roof Construction and associated secondary members (see Section 202)	1½[b]	1[b, c]	1[b, c]	0[c]	1[b, c]	0	HT	1[b, c]	0

For SI: 1 foot = 304.8 mm
HT = Heavy Timber

a. Roof supports: Fire-resistance ratings of primary structural frame and bearing walls are permitted to be reduced by 1 hour where supporting a roof only.

b. Except in Group F-1, H, M, and S-1 occupancies, fire protection of structural members shall not be required, including protection of roof framing and decking where every part of the roof construction is 20 feet or more above any floor immediately below. Fire-retardant-treated wood members shall be allowed to be used for such unprotected members.

c. In all occupancies, heavy timber shall be allowed where a 1-hour or less fire-resistance rating is required.

d. Not less than the fire-resistance rating required by other sections* of this code.

e. Not less than the fire-resistance rating based on fire separation distance (see Table 4.2)

f. Not less than the fire-resistance rating as referenced in Section 704.10.*

* Section numbers refer to sections in the *2015 International Building Code®*.
Courtesy of the International Code Council®, 2015 International Building Code®, Table 601.

Table 4.2
Fire-Resistance Rating Requirements for Exterior Walls Based On Fire Separation Distance[a, d, g]

Fire Separation Distance = X (feet)	Types of Construction	Occupancy Group H[e]	Occupancy Group F-1, M, S-1[f]	Occupancy Group A, B, E, F-2, I, R, S-2, U
X < 5[b]	All	3	2	1
5 ≤ X <10	IA Others	3 2	2 1	1 1
10 ≤ X < 30	IA, IB IIB, VB Others	2 1 1	1 0 1	1[c] 0 1[c]
X ≥ 30	All	0	0	0

For SI:1 foot = 304.8 mm

a. Load-bearing exterior walls shall also comply with the fire-resistance rating requirements of Table 4.1.

b. See Section 706.1.1* for party walls.

c. Open parking garages complying with Section 406* shall not be required to have a fire-resistance rating.

d. The fire-resistance rating of an exterior wall is determined based upon the fire separation distance of the exterior wall and the story in which the wall is located.

e. For special requirements for Group H occupancies, see Section 415.6.*

f. For special requirements for Group S aircraft hangars, see Section 412.4.1.*

g. Where Table 705.8* permits nonbearing exterior walls with unlimited area of unprotected openings, the required fire-resistance rating for the exterior walls in 0 hours.

* Section numbers refer to sections in the 2015 *International Building Code®*.
Courtesy of the International Code Council®, 2015 International Building Code®, Table 602.

Figure 4.3 This Type I, concrete structure has a high fire-resistance rating. *Courtesy of Ron Moore and McKinney (TX) Fire Department.*

Figure 4.4 The amount and type of combustibles in a Type I structure will have a significant impact on its fire resistance.

Specific combustible materials in small quantities are occasionally allowed for use in Type I construction. These materials include some types of roof coverings, wood trim, finished flooring, and wall coverings. Examples of these exceptions are described in detail in each of the model building codes.

Owner/occupants may install greater amounts of these combustible materials without the knowledge of the building or fire department inspectors. The inspector must be aware that the unregulated use of certain materials in Type I construction may be prohibited because the materials contribute to an unacceptable increase in risk.

Structures that are Type I construction are often referred to incorrectly as being *fireproof*. This perceived characteristic is frequently cited as justification for reducing automatic sprinklers or other fire suppression provisions. Although the use of Type I construction provides structural stability during a fire and limits fire spread by virtue of fire barriers, it may not, by itself, offer greater life safety or loss reduction. The amount of combustible materials in the structure, such as furniture, wall and window coverings, and merchandise, may compromise the structure's fire resistance and structural integrity **(Figure 4.4)**.

Type II

Buildings classified as Type II construction are composed of materials that will not contribute to fire development or spread. This construction type consists of noncombustible materials that do not meet the stricter requirements of materials used in the Type I building classification. Structures with metal framing members, metal cladding, or concrete block walls with metal deck roofs supported by unprotected open-web steel joists are the most common form of Type II construction **(Figures 4.5 a and b)**. Type II construction is normally used when fire risk is expected to be low or when fire suppression and detection systems are designed to meet the hazard load.

The inspector must keep in mind that the term *noncombustible* does not always reflect the true nature of the structure. Noncombustible buildings often incorporate combustible materials into their construction. This practice is most notable with some combustible roof systems, flooring, and display areas. Additionally, combustible features can be included on the exterior of Type II structures, including balconies or wall coverings added for aesthetic purposes.

Type III

Type III construction is common in churches, schools, apartment buildings, and mercantile structures **(Figure 4.6)**. This construction type requires that exterior walls be constructed of noncombustible materials. Interior elements can be constructed of any material permitted by the adopted code. Brick, concrete, and reinforced concrete are typical materials used in exterior walls and interior nonbearing walls. Floors, roofs, and interior nonbearing framing and partitions are constructed of small-dimension wood or metal stud systems.

Figures 4.5 a and b Steel-framed structures are typical of Type II construction. Note that B features fire-resistive treated whole logs used as structural support of the roof. This is an example of mixed construction type used in a Type II building. A *Courtesy of Ron Moore and McKinney (TX) Fire Department. B Courtesy of Scott Strassburg.*

Figure 4.6 The outer walls of this structure will be constructed of noncombustible materials so that the building will meet specifications for Type III construction.

Unprotected steel and aluminum nonbearing wall framing members are also found in Type III construction. It is common to find older buildings of Type III construction with wood or steel trusses, while new buildings tend to have more wood trusses and floor joist systems.

Fire-Stop — Materials used to prevent or limit the spread of fire in hollow walls or floors, above false ceilings, in penetrations for plumbing or electrical installations, or in cocklofts and crawl spaces.

The inspector should be aware of several factors with Type III construction:

- Voids exist inside the wooden channels created by roof and truss systems that will allow fire spread unless proper **fire-stopping** is applied.

NOTE: Fire-stopping is required for all construction with vertical or horizontal penetrations.

- Renovations in older Type III structures may have resulted in greater fire risk due to the creation of large voids above ceilings and below floors.
- New construction materials used during renovations. This substitution may result in reducing the load-carrying capacity of the supporting structural member.
- The original use of the structure may have changed to one that requires a greater load-carrying capacity than that of the original design.

Type IV

Type IV construction is often referred to as heavy-timber construction. This construction type is characterized by the use of large-dimension timber (greater than 4 inches [100 mm]) for all structural elements **(Figure 4.7)**. The dimensions of all structural elements, including columns, beams, joists, girders, and roof sheathing (planks), must adhere to minimum dimension sizing.

In Type IV construction, exterior walls are constructed of noncombustible materials. Interior building elements are solid or laminated wood with no concealed spaces. Fire-retardant-treated wood framing is permitted within interior wall assemblies. Floors and roofs are constructed of wood and generally have no void or concealed spaces that could provide a means for fire to travel. Any other materials used in construction and not composed of wood must have a fire-resistance rating of at least one hour.

Type IV structures are stable and more resistant to collapse due to the effects of fire than other construction types that are not protected by a fire suppression system. When involved in a fire, the heavy-timber structural elements form an insulating effect derived from the timbers' own char that reduces heat penetration to the inside of the beam.

Glued-Laminated Beam — Term used to describe wood members produced by joining small, flat strips of wood with glue. *Also known as* Glulam.

Modern Type IV construction materials may include smaller-dimension lumber that is glued together to form a **glued-laminated beam**. Laminated beams are strong and exhibit the same or improved fire performance characteristics as older solid-sawn beams. They are used for many types of buildings, most often in churches, auditoriums, and other large, vaulted facilities **(Figure 4.8)**.

Type V

Type V construction is commonly known as *wood frame* or *frame* construction **(Figure 4.9)**. The exterior bearing walls may be composed of wood and other combustible materials. A brick or stone veneer may be constructed over the wood framing. The veneer offers the appearance of masonry-type construction but provides little additional fire protection to the structure. A single-family dwelling or residence is perhaps the most common example of this type of construction.

Figure 4.7 Large-dimension lumber comprises all of the structural elements in Type IV construction. *Courtesy of Dave Coombs.*

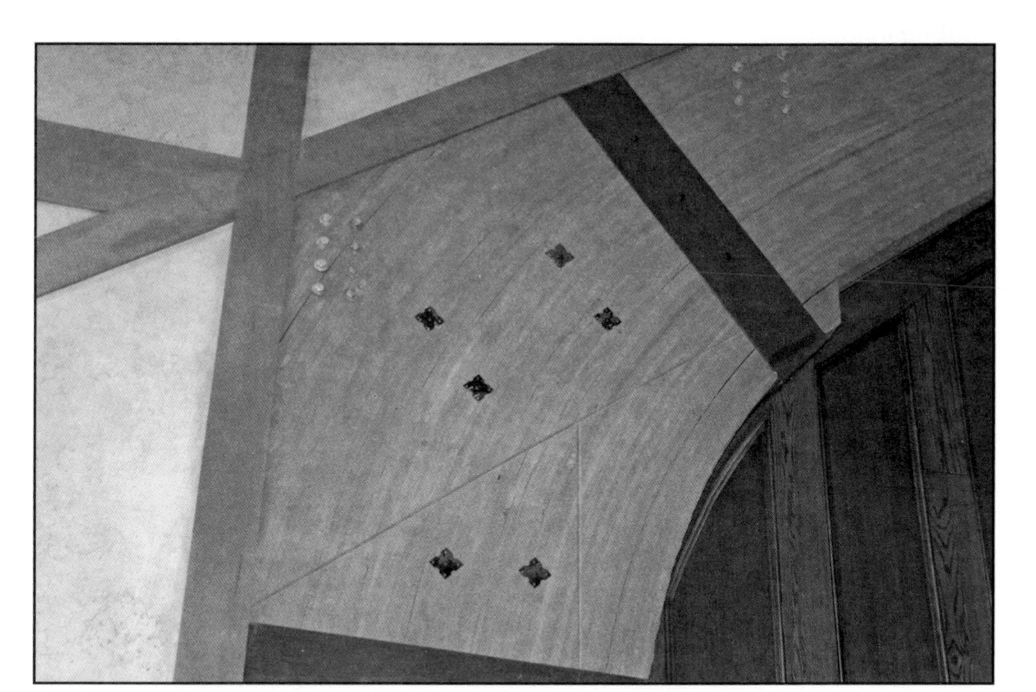

Figure 4.8 Curved glulam beams used to support a church roof.

Figure 4.9 In Type V construction, the exterior bearing walls may be composed of wood and other combustible materials. *Courtesy of Walt Schneider.*

Most often, Type V construction consists of framing materials that include wood 2 × 4-inch (50 mm by 100 mm) studs or wood sill plates. Sheathing, which is applied directly to the stud face, may consist of the following:

Oriented Strand Board (OSB) — A wooden structural panel formed by gluing and compressing wood strands together under pressure.

- Wooden boards
- Plywood
- **Oriented strand board (OSB)**
- Rigid insulation boards
- Any combination of these

The exterior surface is then covered with a covering material, such as shingles, shakes, wood clapboards, sheet metal, plastic siding, and stucco. Exterior siding is attached by nails, screws, or glue. Stucco is spread over a screen lattice that is attached to the framing studs.

Over the years, Type V construction has evolved to include the use of more lightweight engineered lumber systems to support both roof and floor loads. The lightweight wood truss can be assembled in many configurations to support large loads over relatively long spans **(Figure 4.10)**. Trusses typically use a triangular configuration to provide stability and efficiently transfer the applied loads.

These trusses are an efficient use of lumber materials and provide many open areas to install mechanical, plumbing, and electrical systems throughout a structure. The disadvantage with truss construction is that structural redundancy is reduced. All truss members are critical to maintain the strength and stability of the truss, and the loss of any member can be catastrophic.

NOTE: See Chapter 5 for additional information about wood truss systems.

Like the lightweight wood truss, wood I-joists are efficient at spanning large distances with little material. This building element is an assembly of two wood flanges connected by a wooden web. The web originally was ⅜-inch or $^{7}/_{16}$-inch plywood; in recent years OSB of a similar thickness has become

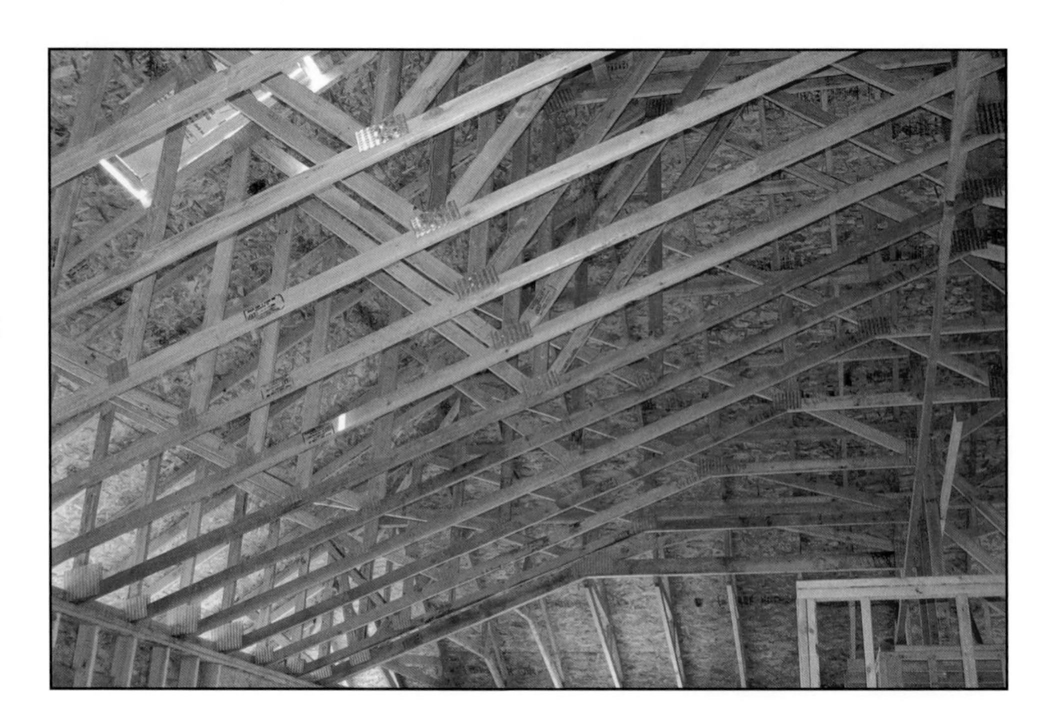

Figure 4.10 A typical lightweight wooden truss. *Courtesy of Dave Coombs.*

more common. The flanges can be made of either solid-sawn dimension lumber joined using finger joints or may be an engineered wood product such as laminated veneer lumber.

The efficiency of wood I-joists also means that the structural member is light and susceptible to fire damage. In many cases, the web will burn prior to the floor sheathing, resulting in the floor system collapsing with little or no warning.

Canadian Construction

The *National Building Code of Canada (NBC)* defines three types of building construction:

1. **Combustible construction** — Construction that does not meet the acceptance criteria for noncombustible construction.
2. **Noncombustible construction** — Construction in which the degree of fire safety is attained by the use of noncombustible materials for structural members and other building assemblies.
3. **Heavy-timber construction** — Combustible construction in which a degree of fire safety is attained by placing limitations on the sizes of wood structural methods and the thickness and composition of wood floors and roofs; it also avoids concealed spaces under floors and roofs.

To help Canadian code users understand these definitions, the NBC specifies requirements and limitations on materials used for each type of construction within the code. These requirements are listed in table formats that are easy to read and understand based on the occupancy classification and construction type.

Occupancy Classifications

An occupancy classification can be defined as the use of all or a portion of a building or structure **(Figures 4.11 a–c, p. 126)**. Occupancy classifications are established because certain occupancies have higher **fire loads** (quantity of combustible materials) and greater numbers of occupants within them than other buildings. For example, an elementary school (educational classification) would be expected to have fewer combustibles than a warehouse (storage classification). Using occupancy classifications, building officials and code enforcement personnel can gain a reasonable expectation of the hazard level in a particular building.

Fire Load — The amount of fuel within a compartment expressed in pounds per square foot obtained by dividing the amount of fuel present by the floor area. Fire load is used as a measure of the potential heat release of a fire within a compartment.

Model code organizations have developed classifications that separate each occupancy into risk categories based upon the use of the structure or space **(Table 4.3, p. 127)**. Although there are many similarities among the various model codes, each one has differences that inspectors must understand.

For the purpose of comparison, Table 4.3 uses general occupancy classifications. The three model codes are then grouped according to general descriptions:

- Assembly
- Business
- Educational
- Factory/Industrial

Figures 4.11 a-c A warehouse is likely to have higher fire loads than a shopping mall, but a healthcare occupancy has special hazards related to healthcare equipment. Shopping malls have special concerns for temporary occupants, hospitals for patients.

Table 4.3
Comparative Overview of Occupancy Categories

This table is a general comparative overview of the occupancy categories for three major model code systems. Readers must consult the locally adopted code and amendments for complete information regarding each of these occupancies.

Occupancy	ICC	NFPA®	NBC
Assembly	**A-1** — Occupancies with fixed seating that are intended for the production and viewing of performing arts or motion picture films. **A-2** — Those that include the serving of food and beverages; occupancies have nonfixed seating. Nonfixed seating is not attached to the structure and can be rearranged as needed. **A-3** — Occupancies used for worship, recreation, or amusement, such as churches, art galleries, bowling alleys, amusement arcades, as well as those that are not classified elsewhere in this section. **A-4** — Occupancies used for viewing of indoor sporting events and other activities that have spectator seating. **A-5** — Outdoor viewing areas; these are typically open air venues but may also contain covered canopy areas as well as interior concourses that provide locations for vendors and other commercial kiosks.	**Assembly Occupancy** — An occupancy (1) used for a gathering of 50 or more persons for deliberation, worship, entertainment, eating, drinking, amusement, awaiting transportation, or similar uses; or (2) used as a special amusement building, regardless of occupant load.	**Group A Division 1** — Occupancies intended for the production and viewing of the performing arts. **Group A Division 2** — Occupancies not classified elsewhere in Group A. **Group A Division 3** — Occupancies of the arena type. **Group A Division 4** — Occupancies in which occupants are gathered in open air.
Business	***Business Group B*** — Buildings used as offices to deliver service-type or professional transactions, including the storage of records and accounts. Characterized by office configurations to include: desks, conference rooms, cubicles, laboratory benches, computer/data terminals, filing cabinets, and educational occupancies above the 12th grade.	***Business*** — Occupancy used for the transaction of business other than mercantile.	**Group D** — Business and personal services occupancies

Table 4.3 (Continued)

Occupancy	ICC	NFPA®	NBC
Educational	***Educational Group E*** — Buildings providing facilities for six or more persons at one time for educational purposes in grades kindergarten through twelfth grade. Religious educational rooms and auditoriums that are part of a place of worship, which have occupant loads of less than 100 persons, retain a classification of Group A-3.	**Educational Occupancy** — Occupancy used for educational purposes through the twelfth grade by six or more persons for 4 or more hours per day or more than 12 hours per week.	Covered under Group A
Factory Industrial	***Factory Industrial Group F*** — Occupancies used for assembling, disassembling, fabrication, finishing, manufacturing, packaging, repair, or processing operations. - ***Factory Industrial F-1 Moderate Hazard*** (examples include but not limited to: aircraft, furniture, metals, and millwork) - ***Factory Industrial F-2 Low Hazard*** (examples include but not limited to: brick and masonry, foundries, glass products, and gypsum) ***High Hazard Group H*** — Buildings used in manufacturing or storage of materials that constitute a physical or health hazard. - ***High-Hazard Group H-1*** — Detonation hazard - ***High-Hazard Group H-2*** — Deflagration or accelerated burning hazard - ***High-Hazard Group H-3*** — Materials that readily support combustion or pose a physical hazard - ***High-Hazard Group H-4*** — Health hazards - ***High-Hazard Group H-5*** — Hazardous production	**Industrial Occupancy** — Occupancy in which products are manufactured or in which processing, assembling, mixing, packaging, finishing, decorating, or repair operations are conducted.	**Group F Division 1** — High-hazard industrial occupancies **Group F Division 2** — Medium-hazard occupancies **Group F Division 3** — Low-hazard industrial occupancies

Table 4.3 (Continued)

Occupancy	ICC	NFPA®	NBC
Occupancy Institutional (Care and Detention)	***Institutional Group I*** **Group I-1** — Assisted living facilities holding more than 16 persons on a 24 hour basis. These persons are capable of self rescue. **Group I-2** — Medical, surgical, psychiatric, or nursing care facilities for more than 5 people who are not capable of self-preservation or need assistance to evacuate. **Group I-3** — Prisons and detention facilities for more than 5 people under restraint. **Group I-4** — Child and adult day care facilities.	**Ambulatory Health Care** — Building (or portion thereof) used to provide outpatient services or treatment simultaneously to four or more patients that renders the patients incapable of taking action for self-preservation under emergency conditions without the assistance of others. **Health Care** — An occupancy used for purposes of medical or other treatment, or care of four or more persons where such occupants are mostly incapable of self-preservation due to age, physical or mental disability, or because of security measures not under the occupants' control. **Residential Board and Care** — Building or portion thereof that is used for lodging and boarding of four or more residents, not related by blood or marriage to the owners or operators, for the purpose of providing personal care services. **Detention and Correctional** — An occupancy used to house one or more persons under varied degrees of restraint or security where such occupants are mostly incapable of self-preservation because of security measures not under the occupants' control.	**Group B Division 1** — Care or detention occupancies in which persons are under restraint or are incapable of self-preservation because of security measures not under their control. **Group B Division 2** — Care or detention occupancies in which persons having cognitive or physical limitations require special care or treatment.
Mercantile	***Mercantile Group M*** — Occupancies open to the public that are used to store, display, and sell merchandise with incidental inventory storage.	**Mercantile** - An occupancy used for the display and sale of merchandise.	**Group E** — Mercantile occupancies
Residential	***Residential Group R*** **R-1** — Residential occupancies containing sleeping units where the occupants are primarily transient in nature (boarding houses, hotels, and motels). **R-2** — Residential occupancies containing sleeping units or more than 2 dwelling units where the occupants are primarily permanent in nature (apartments, convents, non-transient hotels, etc…).	**Residential Occupancy** — Provides sleeping accommodations for purposes other than health care or detention and correctional. **One- and Two-Family Dwelling Unit** — Building that contains not more than two dwelling units with independent cooking and bathroom facilities.	**Group C** — Residential occupancies

Table 4.3 (Continued)

Occupancy	ICC	NFPA®	NBC
Residential (continued)	**R-3** — Residential occupancies where the occupants are primarily permanent in nature and not classified as Group R-1, R-2, R-4, or I. **R-4** — Residential occupancies shall include occupancies buildings arranged for occupancy as residential care/assisted living facilities for more than 5 but less than 16 occupants (excluding staff).	**Lodging or Rooming House** — Building (or portion thereof) that does not qualify as a one- or two-family dwelling, that provides sleeping accommodations for a total of 16 or fewer people on a transient or permanent basis, without personal care services, with or without meals, but without separate cooking facilities for individual occupants. **Hotel** — Building or groups of buildings under the same management in which there are sleeping accommodations for more than 16 persons and primarily used by transients for lodging with or without meals. **Dormitory** — A building or a space in a building in which group sleeping accommodations are provided for more than 16 persons who are not members of the same family in one room, or a series of closely associated rooms, under joint occupancy and single management, with or without meals, but without individual cooking facilities. **Apartment Building** — Building (or portion thereof) containing three or more dwelling units with independent cooking and bathroom facilities.	**Group C** — Residential occupancies
Storage	***Storage Group S*** — Structures or portions of structures that are used for storage and are not classified as hazardous occupancies. - ***Moderate-Hazard Storage, Group S-1*** (examples include but not limited to: bags, books, linoleum, and lumber) - ***Low-Hazard Storage, Group S-2*** (examples include but not limited to: asbestos, bagged cement, electric motors, glass, and metal parts)	**Storage Occupancy** — An occupancy used primarily for the storage or sheltering of goods, merchandise, products, vehicles, or animals.	Covered under Group F
Utility/ Miscellaneous	***Utility/Miscellaneous Group U*** — These are accessory buildings and other miscellaneous structures that are not classified in any specific occupancy (agricultural facilities such as barns, sheds, and fences over 6ft [2m]).	—	—

- Institutional
- Mercantile
- Residential
- Storage
- Utility/Miscellaneous

Some of the codes contain the term *mixed occupancy.* It is not an official occupancy category but instead describes situations in which a variety of occupancies may be included in the same structure.

The Inspector II portion of this chapter covers multiple-use occupancies. Mixed/Mixed Use/Multiple Use are terms used in varying codes. Sometimes more than one term is used in one code (NFPA® 101 for example). The information in the sections that follows is general in nature and most closely follows NFPA® 1, *Fire Code*™, and NFPA® 101, *Life Safety Code*®. When ICC® codes and Canadian codes differ, this fact is noted. More detailed information is given in the respective model building code or standard.

In addition to being familiar with general occupancy classifications, inspectors should be aware that building conditions rarely remain the same. These changes often result in uses for which a building was not intended **(Figure 4.12)**. For instance, a structure that was originally a warehouse and classified as a storage occupancy may be converted into loft apartments or condominiums. Any fire protection systems originally installed in the warehouse would be inadequate for the life safety requirements of a residential multifamily occupancy. Exit requirements, including fire-rated corridor walls, signage, and emergency lighting, may need to be added to meet higher requirements. Some owners or developers may neglect to make these changes, believing that the original structure was approved for occupancy.

Figure 4.12 This older department store structure now contains apartments.

A storage occupancy that has more combustible contents than originally used in the building is another example of an occupancy change that affects fire safety. In this situation, the design of the fire sprinkler system and other protection features may also be inadequate for protection of the new contents. The inspector should be aware that this situation is an all-too-frequent cause of a substantial loss.

If an inspector is reviewing plans for a renovation, issuing a new occupancy permit, or making a field inspection, two questions must be answered:

- Have the building's construction materials or the building's purpose changed since the last inspection?
- What effect will those changes have on the fire and life safety provisions for the structure?

Assembly — Occupancy classification of buildings, structures, or compartments (rooms) that are used for the gathering of 50 or more persons.

Assembly Occupancies

An **assembly** occupancy is any structure or compartment (room) that is used for the gathering of 50 or more persons **(Figure 4.13)**. A wide variety of possible uses fall under this general classification, including houses of worship, theaters, restaurants, and arenas.

The assembly classification is divided into subclassifications based upon the type of activities that take place and the perceived hazard associated with those activities. For example, a theater with fixed seats will have different requirements than a dinner club with movable tables and chairs or a gymnasium with bleachers. The exception to this requirement is the classification used by NFPA® that subdivides the assembly occupancies by occupant load.

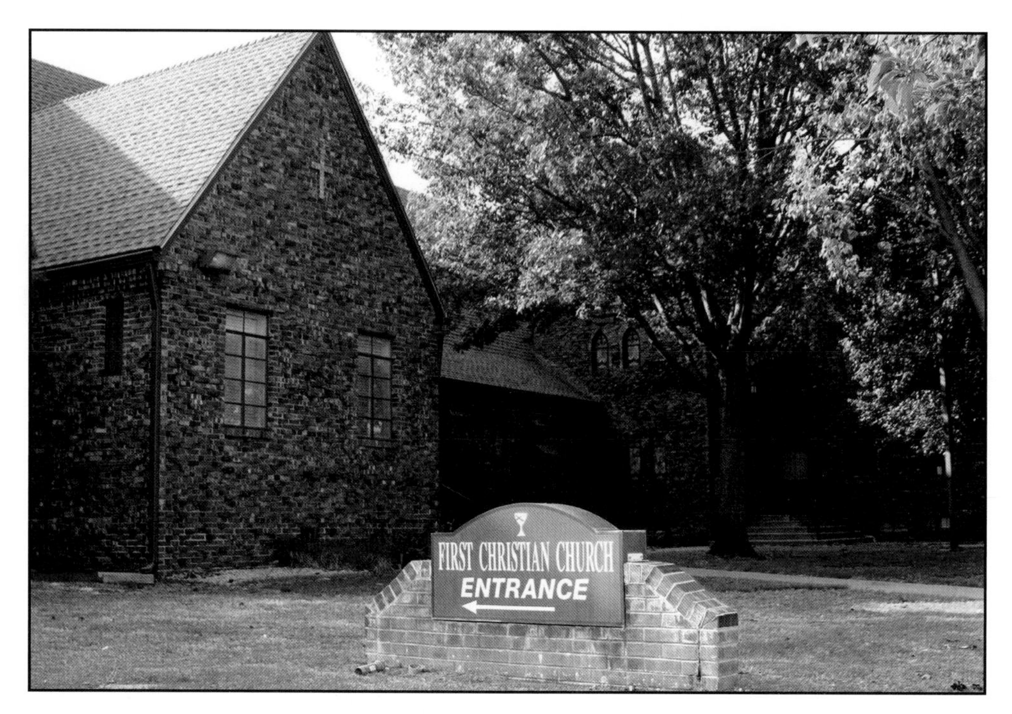

Figure 4.13 A church is an example of an assembly occupancy.

Assembly Occupancies Under Different Standards

NFPA® divides assembly occupancies into the following classes:

- Class A for structures that have occupant loads over 1,000 persons
- Class B for occupant loads of 301 to 1,000
- Class C for occupant loads of 50 to 300

The ICC uses five classification groups based upon specific use, A-1 through A-5. The Canadian codes employ four classification divisions that are grouped by their general use. As with other occupancies, inspectors should address specific code requirements by consulting locally adopted fire and building codes.

Business Occupancies

Business occupancies provide a working place for large numbers of occupants in an office environment, including:

- General offices
- Medical offices
- Air traffic control towers
- Municipal buildings
- College and university instructional buildings
- Dry cleaning and laundry facilities
- Barber and beauty shops

The ICC® also includes buildings that house outpatient clinics where patients are ambulatory and not incapacitated by anesthetic. NFPA® classifies outpatient clinics as ambulatory health care occupancies.

Business occupancies are normally divided into group areas and individual working spaces. This division has the indirect positive effect of compartmentalizing the large office space and separating it into fire and smoke zones. This separation was lost with the advent of the open-plan office space concept. Fire and smoke can move unimpeded throughout these large spaces.

Educational Occupancies

NFPA® 101, NFPA® 5000, *Building Construction and Safety Code*®, and ICC describe an educational occupancy as a structure that is used by six or more persons through the 12th grade that meets for four or more hours in a day or more than 12 hours in a week. Inspecting educational occupancies often presents significant challenges. The wide variety of uses in these facilities can include large spectator events, food preparation, laboratory experimentation areas, and industrial machining areas **(Figure 4.14)**. The age and capabilities of the occupants of an educational facility can vary from pre-kindergarten through adulthood.

Figure 4.14 Educational occupancies like this school offer significant challenges to inspectors because they house services ranging from classrooms and assembly occupancies to machine shops, science labs, and industrial-size kitchens.

These codes allow one person for every 20 square feet (1.85 m^2) of classroom space. Laboratories and vocational shops are permitted one person for every 50 square feet (4.65 m^2). Gymnasiums, lecture halls, and dining halls follow the requirements for assembly occupancies. Day care facilities providing supervision or personal care service for six or more children older than 2½ years old are also classified within the educational category.

NOTE: Religious educational rooms and auditoriums associated with a church that have occupant loads of less than 100 are classified with assembly occupancies.

Although educational facilities are not intended for 24-hour operations or boarding, maintenance staff and other building personnel may be in the buildings at any hour. The risk of fire in these occupancies is usually considered low to moderate; however, these facilities pose hazards due to their high occupant loads and wide variety of uses.

Because each code describes educational facilities in a different manner, inspectors must be familiar with their jurisdiction's adopted building and fire codes in order to correctly classify these occupancies. This knowledge is crucial when determining the occupancy load permitted for these structures.

The Canadian codes do not classify educational facilities separately. Instead, these buildings are grouped with Group A Assembly occupancies. Although schools or other educational facilities are not specifically described in the code, the requirements limiting occupant loads are similar to those found in the United States. Vocational training facilities, however, are far more stringent, requiring almost double the square meter space per person than in the ICC® or NFPA® codes (9.3 m^2 versus 4.6 m^2 [100 ft^2 versus 49.5 ft^2]).

Factory/Industrial Occupancies

Each model code classifies manufacturing and processing facilities differently. The codes further separate this category into several subdivisions based upon the relative hazard or risk to life created by the process or activity.

NFPA® classifies manufacturing and processing facilities as industrial occupancies. Within this broad classification are three subdivisions:

- General Purpose
- Special Purpose
- High Hazard

The ICC® codes describe factory/industrial occupancies as buildings that are used in manufacturing, packaging, finishing, assembling, or disassembling products that are not classified as hazardous. Hazardous materials and processes are not included in the ICC® factory/industrial occupancy classification. Occupancies that contain hazardous materials, involve processes that generate hazardous materials, or constitute a physical or health hazard may need to be classified as high-hazard occupancies by the ICC® codes.

The Canadian codes divide factory and manufacturing facilities into three subdivisions listed as Group F Factory/Industrial Occupancies **(Table 4.4)**.

An inspector should evaluate the structure based upon those requirements that pertain to the structure's primary use. For example, offices and administrative areas should be inspected as business occupancies and cafeterias and dining areas as assemblies **(Figure 4.15)**.

Institutional Occupancies

Buildings classified as Group I Institutional by the ICC® codes are occupancies in which people with physical limitations due to health or age are cared for or provided medical treatment. Additionally, this broad classification includes facilities where individuals are detained for penal or correctional purposes. Equivalent occupancy classifications contained in NFPA® 1™ and NFPA® 101® include the following:

- Health Care
- Ambulatory Health Care
- Detention and Correctional
- Residential Board and Care
- Day Care (portions of the occupancy requirements)

Table 4.4 Canadian Codes: Factory/Industrial Occupancies		
Group F: Division 1 High-Hazard Industrial Occupancies	**Group F: Division 2 Medium-Hazard Occupancies**	**Group F: Division 3 Low-Hazard Occupancies**
• Bulk plants for flammable liquids • Flour mills • Bulk storage warehouses • Grain elevators for hazardous substances • Lacquer factories • Cereal mills • Mattress factories • Chemicals manufacturing • Distilleries • Spray painting operations • Dry cleaning plants	• Aircraft hangars • Mattress factories • Box factories • Planning mills • Candy plants • Printing plants • Cold storage plants • Repair garages • Television studios admitting a viewing audience • Factories • Warehouses	• Creameries • Storage garages including open air • Factories • Parking garages • Laboratories • Storage rooms • Power plants • Workshops • Samples display rooms

Figure 4.15 A restaurant like this would be inspected as an assembly occupancy.

Group B: Division 1 Occupancies	Group B: Division 2 Occupancies
• Jails • Psychiatric hospitals with detention quarters • Reformatories with detention quarters • Penitentiaries • Police stations with detention quarters • Prisons	• Children's custodial homes • Psychiatric hospitals without detention quarters • Convalescent homes • Reformatories without detention quarters • Hospitals or sanatoria without detention quarters • Infirmaries • Nursing homes • Orphanages

Table 4.5
Canadian Codes: Industrial Occupancies

The Canadian codes classify these occupancies as Group B Care or Detention Occupancies, dividing the group into Divisions 1 and 2 **(Table 4.5)**. These divisions broadly encompass all the provisions found in the ICC® code for this occupancy classification.

Health Care and Ambulatory Health Care Occupancies

Health care occupancies provide health or medical services to four or more individuals who cannot evacuate themselves during an emergency without assistance from staff or emergency responders. A broad range of conditions may render patients unable to escape, including age and mental, physical, or security provisions. This occupancy classification is identified only in NFPA® 1™ and NFPA® 101®.

The ICC® codes include similar requirements that are included as part of Group I Institutional I-2, a subclassification. The Canadian codes include similar provisions within their Group B Care of Detention Occupancies classification. As is the case with several other occupancy classifications, there are differences among the various codes.

In institutional/health care facilities, an inspector may find that more than one fire and life safety code applies. The adopted local code may be applied to the design of the building, while NFPA® 101® may pertain to areas such as long-term care in order for the institution to obtain Medicare certification. When this situation occurs, both code documents are used to ensure that all requirements are met.

Ambulatory health care occupancies are buildings or portions of buildings that provide medical services to four or more patients on an outpatient basis. Patients in these facilities may be incapable of self-preservation without assistance due to anesthesia or medication. Urgent care facilities are included in this category due to the nature of treatment and inability of patients to care for themselves **(Figure 4.16)**. The ICC® codes classify outpatient care facilities as Group B Business Occupancies.

Increasing numbers of one-and two-family dwellings are being used as residential care facilities for nonambulatory occupants. States may regulate only those dwellings with four or more such residents, leaving residences with three or fewer nonambulatory residents unregulated. Inspectors should familiarize themselves with state/provincial legislation for authority regarding life safety hazards.

Detention and Correctional Occupancies

As defined in NFPA® 1™ and NFPA® 101®, detention and correctional facilities are locations where the occupants are held under restraint or security. In emergency situations, the occupants are prevented from taking anything but limited life-preservation actions without direct assistance from staff personnel **(Figure 4.17)**. The ICC® codes include these facilities within the classification of Group I Institutional Subdivision I-3.

Residential Board and Care Occupancies

NFPA® 1™ and NFPA® 101® describe residential board and care occupancies as locations where lodging, boarding, and personal care are provided to four or more residents who are unrelated to the owner or operator. NFPA® differentiates these occupancies through the delivery of personal care services to the residents. The ICC® codes classify similar occupancies within Group I Institutional or Group R Residential, while the Canadian codes employ either the Group B Care or Detention Occupancies or Group C Residential Occupancies.

Staff members in these facilities are responsible for the safety and welfare of the residents, but they do not provide medical treatment or nursing care. Staff members typically provide meals for residents, ensure that they have taken their medications, and monitor their well-being.

The evacuation capabilities of a residential board and care occupancy must be properly classified as *slow, prompt,* or *impractical.* The occupancy as a whole must be classified based on the resident posing the most significant

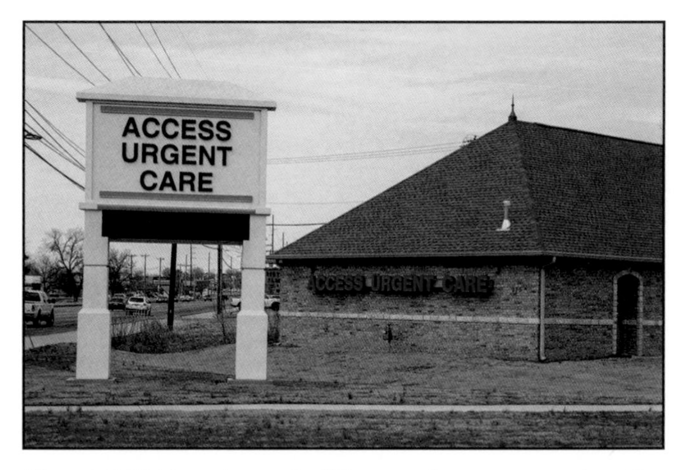

Figure 4.16 Urgent care facilities can have occupants with a wide variety of ailments and mobility issues at any time.

Figure 4.17 Detention occupancies have controlled ingress and egress systems, so prisoners are unable to self-evacuate during an emergency.

risk. For example, if one resident has slow evacuation capabilities and the rest have prompt capabilities, the entire occupancy must be classified as *slow.* The requirements for determining which classification each occupancy falls under can be found in NFPA® 101® and NFPA® 101A, *Guide on Alternative Approaches to Life Safety.* During each inspection of the facility, the inspector should be familiar with any changes that may have occurred regarding the proper classification of occupants.

Day Care Occupancies

Day care occupancies provide care, maintenance, and supervision of persons of any age for periods of less than twenty-four hours per day **(Figure 4.18)**. Someone other than a relative or legal guardian provides client care in these facilities.

An inspector must be aware of the locally adopted code because the numbers of clients that establish these occupancies differ in each code. The inspector also needs to be aware that other regulating agencies, such as state health departments, may have requirements that differ from the locally adopted code.

Although the model codes provide requirements for these facilities, each code differs in how it classifies them. For example, NFPA® 101® places day care facilities into a separate classification category. The ICC® model codes include provisions for day care facilities within the classifications of Group E Educational and Group I Institutional Subdivision I-4.

The inspection of a day care facility must be tailored to meet the specific needs of the residents. By understanding the capabilities of the occupant groups being served, whether by age group or physical condition, the inspector can best observe and mitigate potential fire and life safety hazards. For example, age groups that are capable of ambulatory self-rescue and can recognize dangerous situations pose a lower risk than nonambulatory, developmentally disabled individuals.

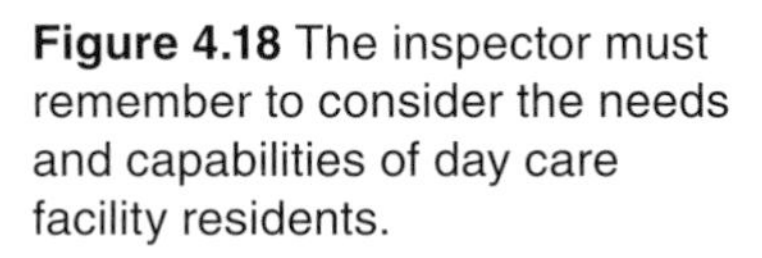

Figure 4.18 The inspector must remember to consider the needs and capabilities of day care facility residents.

Mercantile Occupancies

Each of the model codes defines a mercantile occupancy as any building that is used to display or sell merchandise. Some examples:

- Department stores
- Pharmacies **(Figure 4.19)**
- Supermarkets

Figure 4.19 Drug stores and pharmacies are mercantile occupancies that present many hazards because of the wide variety of products in the store and in the stockrooms.

- Shopping centers
- Malls
- Other retail locations

Mercantile occupancies contain both large quantities of combustible materials and the potential for high life loss. The arrangement of the merchandise, both on display and in storage, can result in high fire loads of combustible materials and at the same time restrict exit access for customers. Product displays are rarely fixed to the floor and can be moved to create new access patterns in the showrooms. Changes in displays can alter the original automatic sprinkler discharge patterns, thereby decreasing the level of protection in the structure.

Residential Occupancies

Generally, residential occupancies provide sleeping accommodations under conditions other than those defined for health care or detention and correctional occupancies. All model codes require that residential occupancies meet minimum fire and life safety requirements. Some of those requirements are included in codes separate from the building code, such as one- and two-family dwelling codes.

The NFPA® divides residential occupancies into five categories:

- One- and two-family dwelling
- Lodging or rooming house
- Hotel
- Dormitory
- Apartment building

One- and Two-Family Dwellings

NFPA® 1™ and NFPA® 101® define one- and two-family dwellings as those structures that have no more than two dwelling units, including detached units, semidetached units, and duplexes **(Figure 4.20, p. 140)**. The degree of fire separation between units determines their occupancy classifications. For example, a complex of dwelling units separated by a fire-rated wall allows the dwellings to be classified as individual units.

Figure 4.20 Although single-family dwellings must meet all the standards of the locally adopted codes, they do not have to be inspected on a regular basis.

One- and two-family dwellings are not exempted in the model codes. However, they are not subject to periodic inspections in most jurisdictions. A homeowner, however, may request municipal inspectors to conduct a voluntary fire prevention inspection.

Lodging (Boarding) or Rooming Houses

Lodging (boarding) and rooming houses are facilities that provide sleeping accommodations for rent. The management may provide meals, but separate cooking facilities are not included for individual occupants. NFPA® 1™ and NFPA® 101® and IFC® (International Fire Code®) (R-1 boarding houses) use this classification to describe occupancies that include guest houses, foster homes, bed and breakfasts, and motels that provide twenty-four hour accommodations for sixteen or fewer individuals without cooking facilities.

The ICC® and Canadian codes include provisions that are similar to these within the broader occupancy classification of Group R Residential. Examples are as follows:

- Guest houses
- Foster homes
- Bed and breakfasts
- Motels

A boarding or rooming house is usually a separate, distinct occupancy. When a building shares space with a boarding or rooming house with services and operation intermingled, the most restrictive code provisions for these occupancies are applied. A boarding or rooming house cannot be located above a mercantile occupancy unless it is separated by a 1-hour fire-separation barrier or the mercantile occupancy is equipped with an approved automatic fire sprinkler system.

In order to increase rentable space, new rooms are often created in areas that were not designed to serve in that capacity. Often, the inspector will find that the exit halls have been converted into sleeping or living areas. These

Figure 4.21 Clearly, a door has been eliminated from this facility during a remodel. The inspector needs to make sure that exit capacity is adequate.

modifications may block or eliminate exit passageways **(Figure 4.21)**. The inspector must verify that the maximum number of residents does not exceed sixteen individuals.

With any occupancy designed for temporary use, the inspector must remember that occupants are unfamiliar with their surroundings and thus at greater risk during an emergency. To help increase safety, protection levels may need to be increased in the form of additional signage, maps showing exit paths, and additional fire protection features.

NOTE: Refer to Chapter 7, Means of Egress, for additional information on exiting signs.

Hotels

The model building codes define a hotel as any building or group of buildings that provides sleeping rooms for transient guests. Hotels present a wide range of fire and life safety challenges for the inspector.

NOTE: Transients are often defined as persons staying in an occupancy for less than thirty days.

In addition to sleeping accommodations, hotels often include many other occupancies, such as:

- Meeting rooms
- Convention areas
- Casinos
- Ballrooms
- Theaters
- Restaurants and kitchens
- Bars and lounges

- Storage areas
- Swimming pools
- Physical fitness centers
- Boutique retail shops
- Business offices
- Parking garages
- Mechanical spaces

Unless these various occupancies are separated from other building elements, the inspector must consider the requirements and challenges that each occupancy classification or use poses to the rest of the facility.

The ICC® and Canadian codes do not have occupancy categories that differentiate hotel uses. Each code describes these occupancies within the broader residential group classification. The ICC® codes include hotels within Group R Residential Occupancies and the Canadian codes include them as Group C Residential Occupancies.

Dormitories

A dormitory is any building or portion of a building that provides sleeping accommodations to sixteen or more persons who are not related **(Figure 4.22)**. The sleeping accommodations may be in one room or a series of smaller rooms. Individual cooking facilities are not provided, but a cafeteria or dining hall may be part of the facility. Residential fraternity houses may fall into this category according to the ICC®. Fires in college, university, and boarding school dormitories, as well as fraternity and sorority houses, have caused an increased awareness for the need for improved fire and life safety in these types of structures.

Figure 4.22 Inspectors should take special care in inspecting college dormitories. Fires in such occupancies have raised awareness about fire and life safety hazards in these residential occupancies.

Apartment Buildings

Apartment buildings may be single or multistory structures containing three or more independent dwelling units equipped with cooking and bathroom facilities. Units may have direct access to the exterior or be designed with interior corridors that lead to protected or unprotected stairways. Apartment buildings that are greater than seven stories may be considered high-rise structures that require higher levels of fire and life safety protection.

Each apartment building presents a unique set of problems to a fire inspector because of its location and structure age as well as the age and economic status of the occupants **(Figures 4.23 a and b)**. Fire protection features and requirements must be evaluated based upon the risk levels in each build-

Figure 4.23 a Old-style apartments generally have interior hallways that provide access to individual units, high ceilings, and heating systems located in basements. *Courtesy of Ron Jeffers.*

Figure 4.23 b Apartments constructed in the past 50 years tend to include exterior access, individual heating and cooling systems, and lightweight construction materials. Fire walls may be required between specific numbers of units to prevent fire spread through common attic spaces.

ing. All design features of the apartment must be maintained throughout its inhabited life. The inspection division should keep official records as part of the building's permanent file and use them as a reference for all inspection activities.

An addition to a building's occupancy is of special concern to the inspector. For example, stores or other business-type occupancies may have been added to an apartment building. Some of these changes introduce a higher hazard level than when the structure was simply an apartment building. An inspector must verify that proper fire separations, as directed by the code, have been included between the apartments and the new occupancies.

Another concern is structures that were originally intended for one use, such as warehouses, factories, and office buildings, that have been converted into another use, such as residential occupancies. These conversions usually result in multiple apartments or individually owned condominiums in structures that were not intended for residential use. Increased fire and life safety requirements are usually established during the plan review stage of the conversion project.

Storage Occupancies

Occupancies that are used to store goods, merchandise, vehicles, or animals are generally referred to as storage facilities. Each model code approaches this classification in somewhat different ways. NFPA® uses a broad approach that includes the following facilities **(Figures 4.24 a and b)**:

- Warehouses
- Storage units
- Freight terminals
- Parking garages
- Aircraft hangars
- Grain elevators
- Barns
- Stables

The ICC® codes, however, describe storage within several occupancy classifications. The primary classification is Group S Storage, while the storage of hazardous goods in excess of maximum allowable quantities falls into Group H Hazardous classification. A fire inspector must be familiar with the locally adopted code and amendments to adequately conduct an inspection of these structures.

NOTE: Refer to Chapter 10, Hazardous Materials, for additional descriptions of maximum allowable quantities.

The combustibility or flammability of the contents inside a storage facility usually determines the classification of the occupancy. Additional consideration is given to the method of storage employed in the facility. The inspector should carefully evaluate a change to more combustible contents or to the storage method that was used in the original occupancy. The design of the fire sprinkler system and other protection features may be inadequate for protection of the new contents. This situation has led to many substantial losses.

Figure 4.24 a Individual storage units may contain a wide variety of contents. Inspectors will rarely have the authority to inspect these units unless a complaint is filed.

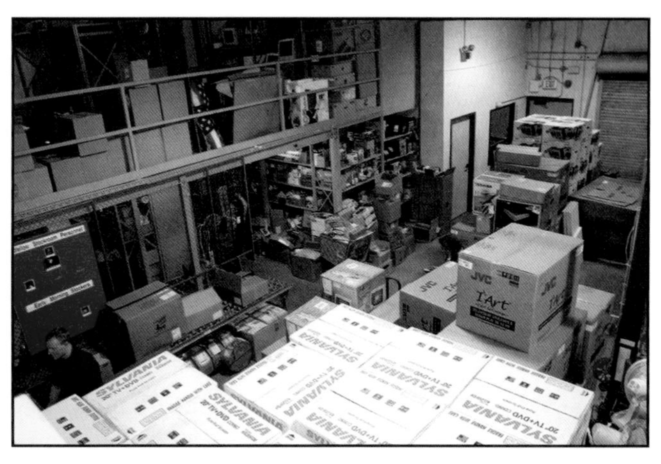

Figure 4.24 b Local jurisdictions may authorize periodic inspections of warehouses that may be part of an industrial complex or individual facilities used for storing materials during transit.

Utility/Miscellaneous Occupancies

The utility/miscellaneous classification, which is only recognized by the ICC®, is used for buildings or structures that do not fit any other classification **(Figures 4.25 a and b)**. Generally, the buildings or structures are incidental or accessory to the primary occupancy and may not pose a hazard to it. Examples of utility and miscellaneous occupancies are:

- Barns
- Livestock shelters
- Carports
- Towers
- Sheds
- Fences over 6 feet (1.8 m) in height
- Retaining walls

Figure 4.25 a and b Barns are separate structures not connected to houses or stables; therefore, they pose less risk to other structures. Tall fences are classified as miscellaneous occupancies.

Multiple-Use Occupancies

Determining occupancy classification for a structure that has a single function or use is a relatively simple task. When structures have two or more types of uses, determining occupancy classification is considerably more complicated. Multiple-use occupancies usually require inspectors to determine which parts of the facility fall under a particular occupancy classification.

Inspectors making a routine inspection should examine occupancy certificates or inspection reports to verify which occupancy classifications have been assigned. In most cases, inspection personnel will only need to verify that the structure still falls under the same classification. The inspector should consult with the building official if a change is found or suspected.

In general, NFPA® 101® does not require the separation of occupancies in the same structure. Local codes may require separations, but this requirement varies from jurisdiction to jurisdiction. The following two situations concerning multiple-use occupancies are of particular importance to inspection personnel:

1. Buildings that have different uses in distinctly different portions of the building
2. Buildings that have different uses in an intermingled manner that make it impractical to separate portions of the building for different classifications

A school with a large auditorium is an example of the first situation **(Figure 4.26)**. The majority of the building would be classified as an educational occupancy, while the auditorium would be considered a place of assembly. Each portion of the building would be required to meet the code requirements for that occupancy.

Figure 4.26 An auditorium in a school is classified as an assembly occupancy.

Occupancies that feature intermingled uses represent a significant challenge in determining occupancy classification. An example of intermingled uses would be a large commercial printing operation (industrial occupancy) that has bulk paper storage (storage occupancy) throughout the facility. NFPA® 101® recommends that the most restrictive fire and life safety requirements be enforced in this situation. Following this recommendation could mean that either all the requirements for one type of occupancy will take precedence or that various requirements from each of the two occupancies will be used.

The ICC® does not have a separate occupancy classification for multiple-use occupancies. However, like NFPA®, the ICC® recognizes that many structures may have two or more occupancy types that constitute large portions of a building.

The Canadian codes do not address the issue of multiple use, accessory use, or incidental use in the same way that codes in the US do. The Canadian codes require separations between these occupancies in most instances, and restrict combinations of occupancies when a high-hazard occupancy is involved.

Incidental Use

According to the ICC®, any area that is designated as incidental use must conform to the occupancy requirements of the building or portion of the building it occupies. Therefore, fire separation walls, assemblies, or protection must conform to the requirements for the primary use group, such as Institutional or Mercantile. If this condition is possible, the building must conform to the requirements of a mixed-use occupancy as defined in the following section. Examples of incidental-use areas, as listed in the ICC® codes, include the following:

- Furnace rooms
- Parking garages
- Incinerator rooms
- Laboratories and vocational shops, located in Group E Educational and Group I Institutional, Subdivision I-2 (medical), Occupancies that are not classified as Group H High Hazard under Factory/Industrial Occupancies
- Waste and linen collection rooms that are over 100 square feet (9.3 m^2)

Model codes usually require fire-resistance-rated separations regarding incidental use. Where permitted, an automatic fire extinguishing system without a fire barrier can be installed as long as construction features separate the rest of the building from the incidental-use area. These construction features must be capable of restricting the passage of fire from the incidental-use area to other portions of the structure.

When they are used, partitions must extend from the floor of the use area to the bottom of a fire-resistance-rated floor or ceiling above. Doors leading into these areas must be self-closing or the automatic-closing type that closes when the fire alarm is activated. Doors in these areas shall not have air-transfer openings and must meet the provisions outlined in NFPA® 80, *Standard for Fire Doors and Other Opening Protectives*.

Mixed-Use

Structures containing multiple occupancy types are considered to be mixed-use occupancies. Each structure is individually classified by its primary occupancy classification and separated from the other occupancies by the appropriate fire wall separation. When an automatic fire suppression system protects the entire structure, the fire-resistance rating requirements may be reduced.

NOTE: Fire walls are described in Chapter 5.

Some occupancy classifications are not permitted to share the same building regardless of fire separation features or fire suppression system. For instance, assembly and educational occupancies are not permitted in the same building with a Factory/Industrial High Hazard Group H-1 Occupancy classification. The inspector must remember this fact when old buildings are altered for multiple uses. The ICC® further divides mixed occupancies into accessory, nonseparated, and separated occupancies.

Accessory

Accessory occupancies are subsidiary to the main occupancy of a structure and limited to no more than 10 percent of the area of the story on which they are located. This limitation does not include accessory assembly occupancies that are less than 750 square feet (70 m^2) that remain as part of the main occupancy classification.

Similarly, assembly areas that are accessory to ICC® Group E Educational Occupancies are not considered separate, except when applying the code's assembly occupancy requirements. Examples are cafeterias and theaters attached to schools.

Group H High-Hazard Groups H-2, H-3, H-4, or H-5 Occupancies must be separated from all other occupancies. Other occupancy types do not require separation, except for particular requirements of the individual classification. Examples of accessory uses could include a storage area accessory to an office building, an office area accessory to a manufacturing area, or a sales area accessory to a storage area.

Nonseparated

Nonseparated occupancies are buildings or portions of buildings that have two or more occupancy uses, with each classified according to its own use. However, the restrictive provisions outlined in the ICC® codes regarding high-rises and those of fire protection systems are applied to the entire structure.

Separated

Entire structures or portions of structures that have several occupancy types contained within them are referred to as separated occupancies. Adopted fire code restrictions are applied separately, not to the structure as a whole. Each occupancy complies with adopted fire code requirements for its occupancy type. Separation between each of these occupancies is defined in the ICC® codes and in **Table 4.6**.

Chapter Summary

An inspector's ability to determine the construction type and occupancy classification of a structure is a critical part of an inspection. The architect or construction engineer evaluates the construction materials and determines the construction type. The appropriate occupancy classification must be assigned during the plans review process and verified when the certificate of occupancy is issued before the owner/occupant takes possession of the structure.

During periodic inspections, the inspector should note any changes in use and contents. When alterations to the structure are made, an inspector must verify that the fire and life safety requirements are not compromised. Model building and fire codes provide the inspector with the guidelines for

Table 4.6
Required Separation of Occupancies (Hours)

Occupancy	A, E		I-1[a],I-3, I-4		I-2		R[a]		F-2, S-2[b], U		B[e], F-1, M, S-1		H-1		H-2		H-3, H-4		H-5	
	S	NS	S	NS	S	NS	S	NS	S	NS	S	NS	S	NS	S	NS	S	NS	S	NS
A, E	N	N	1	2	2	NP	1	2	N	1	1	2	NP	NP	3	4	2	3	2	NP
I-1[a], I-3, I-4	—	—	N	N	2	NP	1	NP	1	2	1	2	NP	NP	3	NP	2	NP	2	NP
I-2	—	—	—	—	N	N	2	NP	2	NP	2	NP	NP	NP	3	NP	2	NP	2	NP
R[a]	—	—	—	—	—	—	N	N	1[c]	2[c]	1	2	NP	NP	3	NP	2	NP	2	NP
F-2, S-2[b], U	—	—	—	—	—	—	—	—	N	N	1	2	NP	NP	3	4	2	3	2	NP
B[e], F-1, M, S-1	—	—	—	—	—	—	—	—	—	—	N	N	NP	NP	2	3	1	2	1	NP
H-1	—	—	—	—	—	—	—	—	—	—	—	—	N	NP	NP	NP	NP	NP	NP	NP
H-2	—	—	—	—	—	—	—	—	—	—	—	—	—	—	N	NP	1	NP	1	NP
H-3, H-4	—	—	—	—	—	—	—	—	—	—	—	—	—	—	—	—	1[d]	NP	1	NP
H-5	—	—	—	—	—	—	—	—	—	—	—	—	—	—	—	—	—	—	N	NP

S = Buildings equipped throughout with an automatic sprinkler system installed in accordance with Section 903.3.1.1.*
NS = Buildings not equipped throughout with an automatic sprinkler system installed in accordance with Section 903.3.1.1.*
N = No separation requirement.
NP = Not permitted

a. See Section 420.*

b. The required separation from areas used only for private or pleasure vehicles shall be reduced by 1 hour but not to less than 1 hour.

c. See Section 406.3.4.*

d. Separation is not required between occupancies of the same classification.

e. See Section 422.2* for ambulatory care facilities.

* Section numbers refer to sections in the 2015 *International Building Code®*.
Courtesy of the International Code Council®, 2015 International Building Code®, Table 508.4.

documenting that life safety requirements are met. Because the information in this chapter is general in nature, inspectors must become familiar with their locally adopted codes.

Review Questions

1. How many types of construction are recognized by both the International Building Code® and the NFPA®?
2. What are the single-use occupancy classifications?

3. What are the different types of multiple-use occupancies?

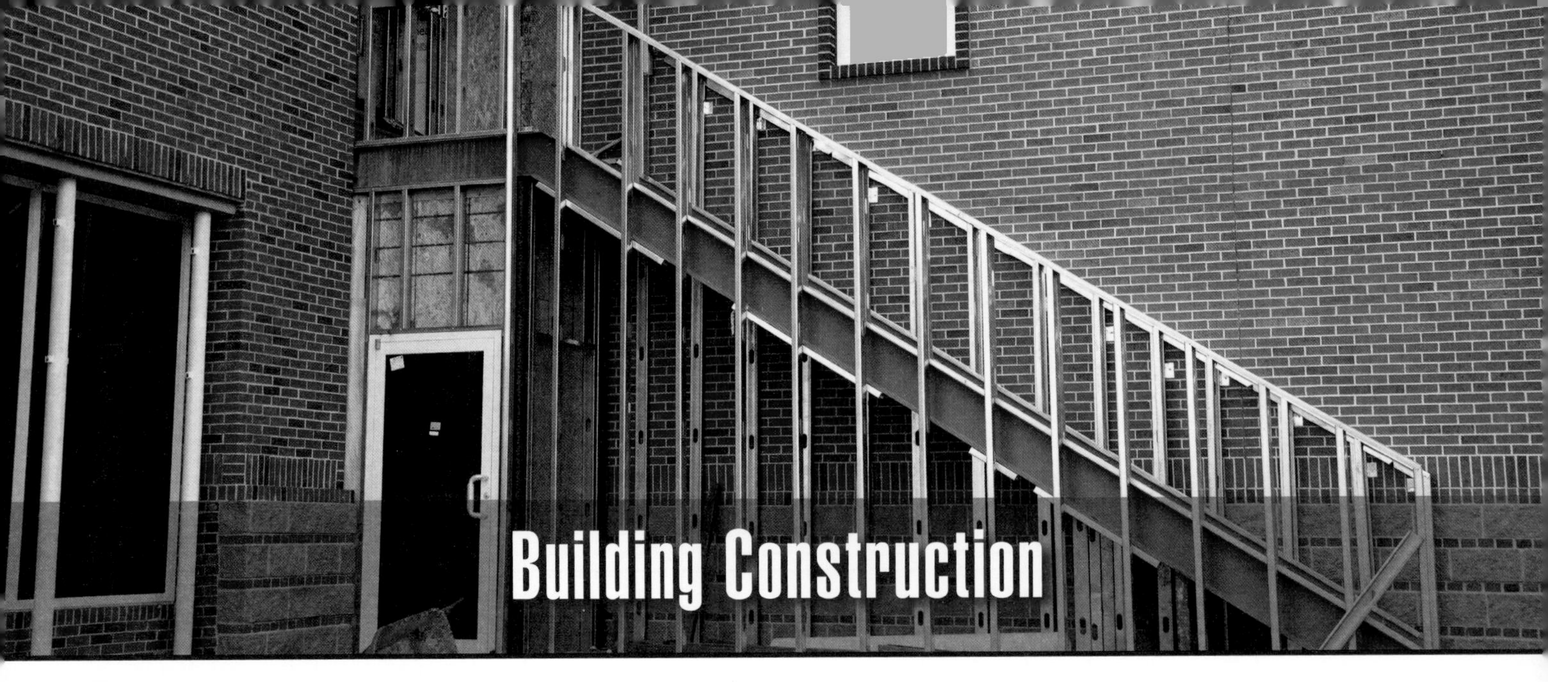

Chapter Contents

Case History **153**
I Construction Materials **153**
Fire Protection and Resistance Terminology 154
Wood 154
Masonry 160
Concrete 163
Steel 163
Glazing 166
Gypsum Board 167
Plastics 168
II Structural Systems **171**
Bearing Wall Structures 172
Loads 172
Shell and Membrane Structures 174
Frame Structural Systems 176
Wood Structural Systems 178
Masonry Structural Systems 181
Steel Structural Systems 184
Concrete Structural Systems 186
Chapter Summary **188**
Review Questions **189**

Key Terms

Balloon-Frame Construction 180
Bearing Wall Structure 172
Capital 176
Compressive Loads 172
Concrete Block 161
Course 181
Dead Load 172
Dimensional Lumber 155
Drop Panel...176
Exterior Insulation and Finish Systems (EIFS) 170
Finger Joint 156
Fire Damper 165
Fire Retardant 157
Frame 176
Glazing...166
Glued-Laminated Beam 155
Header Course 181
Intumescent Coating...............................157
Live Load 172
Membrane Structure 174
Post and Beam Construction 176
Post-Tensioned Reinforcement 188
Pretensioned Reinforcement 187
Rebar 163
Rigid Frame 178
Roof Deck 186
Scarf Joint 156
Stud..176
Tensile 163
Tension 163
Thrust Plate 183
Wythe 181

NFPA® Job Performance Requirements

This chapter provides information that addresses the following job performance requirements of NFPA® 1031, *Standard for Professional Qualifications for Fire Inspector and Plan Examiner* (2014).

Fire Inspector I	Fire Inspector II
4.3.4	5.3.3
	5.3.10
	5.4.6

Building Construction

Learning Objectives

After reading this chapter, students will be able to:

Inspector I

1. Identify accepted types of construction building materials and the fire risks associated with them. (4.3.4)

Inspector II

1. Identify the different types of structural systems used in building construction and the fire risks associated with each. (5.3.3, 5.3.10, 5.4.6)

Chapter 5
Building Construction

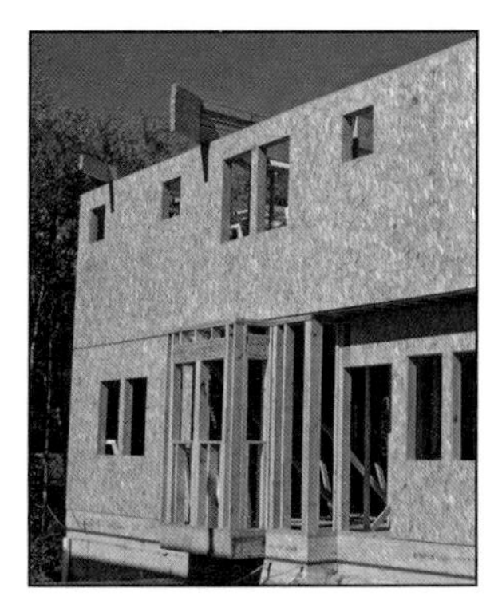

Case History

The Hubert H. Humphrey Metrodome is a domed sports stadium in Minneapolis, Minnesota. At least twenty fans provide the positive-pressure air needed to support the roof, which is composed of two layers of fabric. Dead air space between the two layers serves to provide insulation.

To keep snow from accumulating on the roof, workers would climb to the roof and use steam and high-powered hot-water hoses to melt the snow. In December of 2010, a heavy snowstorm of 17 inches and strong winds made snow removal activities unsafe. The next morning the roof collapsed as three panels of the fabric roof failed. Fortunately, no one was injured. Three days later, a fourth panel ripped open. Home sports events were played at other facilities until the entire roof was replaced and reinflated.

During plans reviews or field inspections, the inspector must be able to recognize building materials and structural systems and compare them to the building code requirements for a particular occupancy classification. This chapter provides an overview of commonly used building materials and their use in structural building systems.

Construction Materials

In modern practice, the materials most commonly used for construction are:

- Wood
- Masonry
- Concrete
- Steel

In addition, other materials are used in a nonload-carrying capacity, including:

- Metals, such as aluminum, cast iron, copper, zinc
- Glass
- Gypsum board
- Plastics
- Fabric

The chemical and mechanical properties of characteristics of different building materials determine their usefulness in architectural applications. Some of the physical properties that the fire inspector should be familiar with are:

- Combustibility **(Figure 5.1)**
- Thermal conductivity
- Rate of thermal expansion
- Variation of strength with temperature. As temperature increases from fire involvement or exposure, materials weaken. When materials are weakened, their ability to carry a dead or live load is compromised.

  NOTE: The Structural Systems portion of this chapter describes loads.

During a fire, the behavior of building materials is affected by their chemical and physical properties as well as the method of fabrication. Therefore, the inspector should have a basic knowledge of the properties of common building materials that are seen during construction and field inspections.

Fire Protection and Resistance Terminology

The following terms sound similar but have unique applications. They are defined here for the sake of clarity.

- **Fire protection rating** – Rating assigned to an opening in a fire wall to indicate the length of time a protective assembly (door, window) can withstand fire conditions. This application may use a lower rating than the surrounding fire-resistant barriers because of the assumption that the fire load will not be close to these assemblies.
- **Fire-resistant materials** – Wall assemblies that are rated through laboratory testing to determine their ability to withstand fire conditions over a set amount of time.
- **Fire-retardant coatings** – Applied to the surface of combustible materials to suppress, reduce, or delay the flame-spread rating of a material. Fire retardant coatings behave in unique ways depending on their chemical make-up and application.
- **Ignition-resistant construction** – Materials used to decrease the vulnerability of structures to hazards, such as wildland fires. Features include the use of fire-resistant roof coverings.

Wood

Wood structural members provide a large amount of fuel for combustion and a building constructed of wood can be destroyed by fire. The voids created in floor, attic, and wall cavities result in many square feet (square meters) of combustible surface area surrounded by large volumes of air for combustion. The different properties of wood also affect its use in construction. Because the strength of wood varies significantly, the International Building Code® (IBC) requires that lumber be graded (labeled) to denote its strength. Wood is primarily used in construction as solid lumber and as engineered wood products.

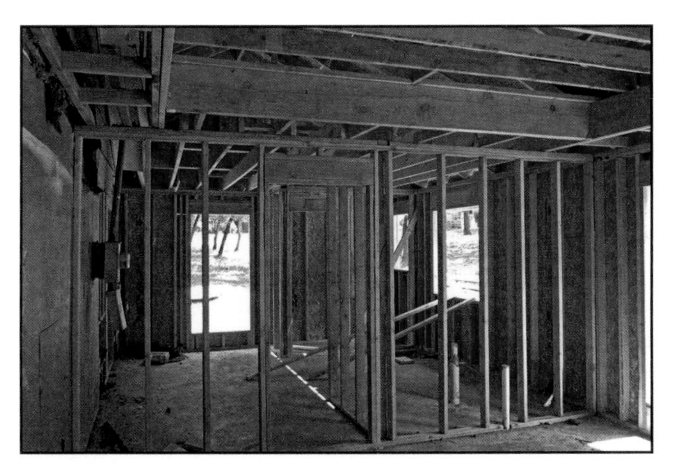

Figure 5.1 In a wood structure like this one, the actual building components can contribute heavily to a fire. *Courtesy of Joe Elam.*

Figure 5.2 Dimensional lumber matches its stated measurements. *Courtesy of NIST.*

Solid Lumber

Solid lumber includes **dimensional lumber**, boards, and timbers **(Figure 5.2)**. With dimensional lumber, the standard measurements of dimensional lumber match the stated measurements. Dimensional lumber is available in lengths from 8 to 24 feet (2.5 to 7 m) in 2-foot (600 mm) increments.

Dimensional Lumber — Lumber with standard, nominal measurements for use in building construction. Dimensional lumber is also available in rough, green components with actual dimensions that match the nominal dimensions.

Engineered Wood Products

Engineered wood products are prefabricated from such components as dimensional lumber, panels, adhesives, and metal fasteners and then shipped to the construction site. Manufactured members include trusses, box beams, I-beams, and panel components **(Figures 5.3 a and b)**. These products can include laminated members and wood panel products.

Laminated members. Laminated members are produced by joining smaller pieces of wood with glue. The beams produced by this method are known as **glued-laminated beams**. The advantage to manufacturing laminated members is that sizes and shapes can be produced that are not available from solid wood. These include curves and varying cross-sections.

Glued-Laminated Beam — Term used to describe wood members produced by joining small, flat strips of wood with glue. *Also known as* Glulam.

Figure 5.3 a and b This box beam and center beam are manufactured wood. *A courtesy of Dave Coombs; B Courtesy of Ron Moore and McKinney (TX) Fire Department.*

Scarf Joint — Connection between two parts made by the cutting of overlapping mating parts and securing them by glue or fasteners so that the joint is not enlarged and the patterns are complementary.

Finger Joint — Connection between two parts made by cutting complementary mating parts, and then securing the joint with glue.

To obtain the necessary length from shorter pieces, **scarf joints** or **finger joints** are used in constructing laminated members **(Figure 5.4)**. Scarf joints and finger joints provide additional surface area for the glue to allow the transmission of a greater amount of force than a typical butt joint. The inspector should know that these joints hold laminated members together.

Wood panel products. Wood panel products include plywood, oriented strand board (OSB), and nonveneered panels. Panel products possess several advantages from a construction standpoint and are widely used for roofs, subflooring, and siding **(Figures 5.5 a and b)**. Although solid wood pieces may be stronger in the direction parallel to the grain, panel products are more equal in strength along their two major axes.

Wood panel products are graded for their structural use and their exposure durability. A grade stamp appears on the back of a structural panel that indicates its intended structural application and its suitability for exposure to water.

Stressed skin panels consist of an interior frame or plastic foam core to which a skin of plywood or OSB is attached. A common use of manufactured panels is in modular buildings. OSB has replaced plywood and planking in many construction applications, including roof decks, walls, and subfloors.

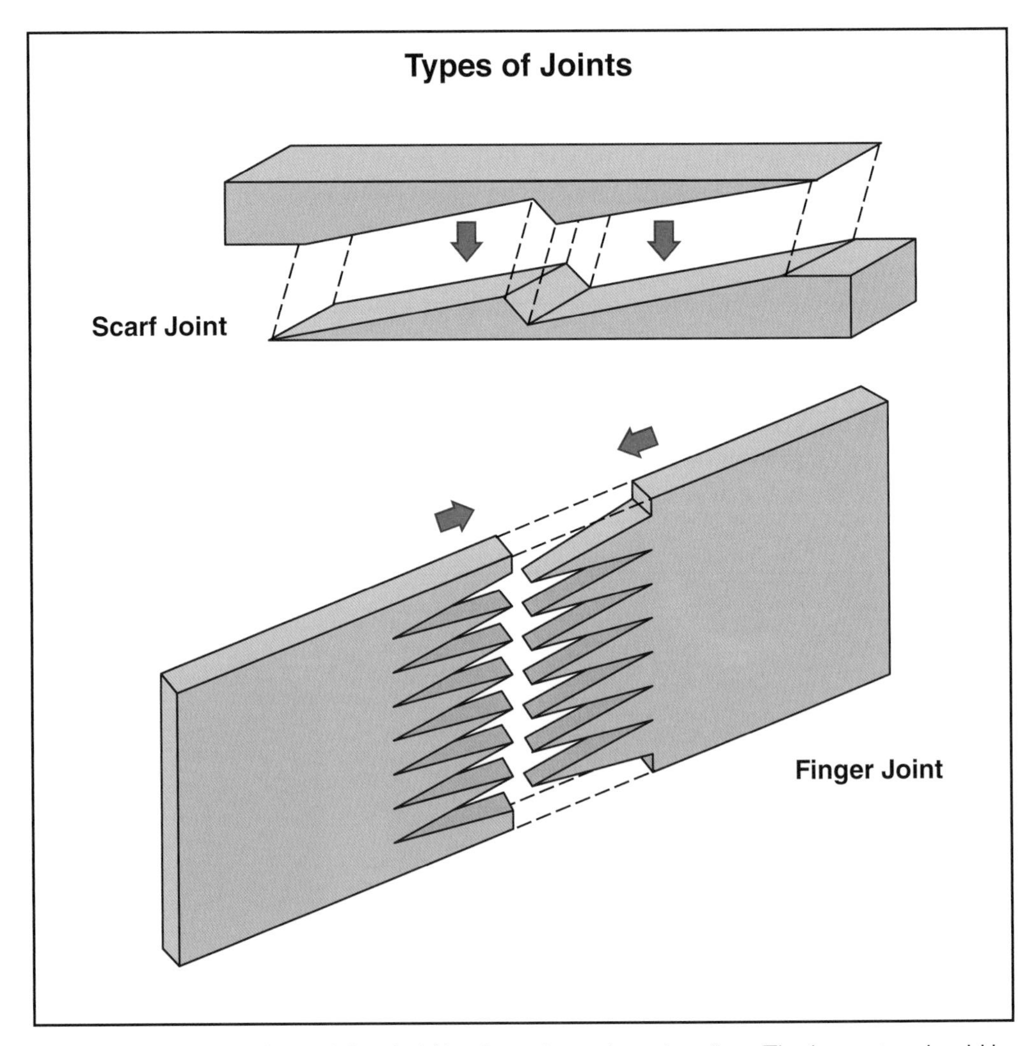

Figure 5.4 Scarf and finger joints hold laminated members together. The inspector should be aware that weakness can occur if these joints become delaminated.

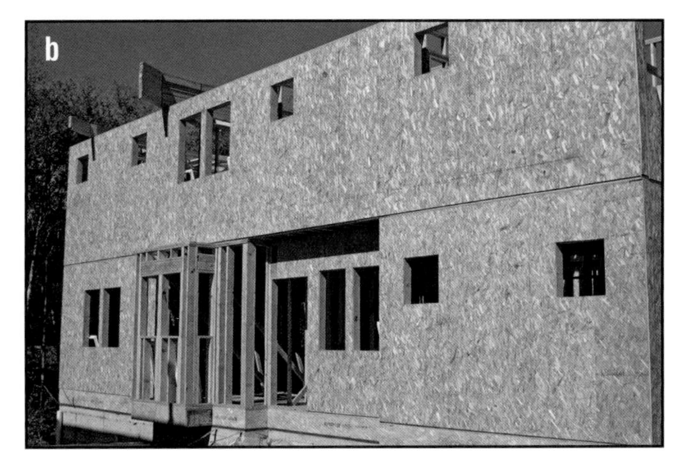

Figures 5.5 a and b Foam core is often used in building panels, and OSB is a common siding material. *Photo a courtesy of Greg Havel.*

Fire-Retardant Treated Wood

Wood can be chemically altered with a **fire retardant** to reduce its susceptibility to ignition. Building codes permit the use of fire-retardant-treated wood for certain applications. However, wood that is listed as fire-retardant treated should not be confused with materials that are noncombustible or fire resistive.

Fire Retardant — Chemical applied to material or another substance that is designed to slow ignition or the spread of fire.

NOTE: Chapter 6, Building Components, contains descriptions of surface burning and flame spread tests.

The two main methods of fire-retardant treatment of wood are pressure impregnation and surface coating. Only these treatments can produce wood that is listed as fire-retardant-treated wood.

The most commonly used fire-retardant treatments are combinations of inorganic or organic salts. Surface coating is used primarily to reduce the surface burning and flame spread characteristics of wood. Fire-retardant chemicals used in the pressurization process are proprietary products so their exact formulations generally are not available.

Fire-retardant treatment of wood has some disadvantages and limitations, especially with regard to water solubility and structural strength. Some treatments use chemicals that are water soluble, prohibiting their use for exterior applications. Others can be used in interior applications where high humidity does not exist. Some fire-retardant chemicals may adversely affect the structural strength of wood as a result of elevated temperature and humidity.

Intumescent coatings can be applied to wood to make its surface fire-resistant. The challenge for inspectors is that these coatings are often painted over or not maintained. In particular, the inspector should look for missing coating.

Intumescent Coating — Coating or paint like product that expands when exposed to the heat of a fire to create an insulating barrier.

NOTE: Intumescent coatings are also commonly applied to steel to provide fire protection.

Exterior Wall Materials

In addition to the structural framing, the exterior walls of a wood-frame building include materials that provide resistance to environmental and pest infiltration. Sidings are often chosen for aesthetic reasons, and underlayer components may be chosen for compatibility with sidings.

Sheathing. Sheathing is a layer of material installed outside the studs to provide structural stability, insulation, and an underlayer for the siding. The most common sheathings are plywood, OSB, particle board, or exterior gypsum sheathing.

Building wrap. A layer of building wrap between the sheathing and the siding acts as a vapor barrier. The wrap reduces the infiltration of moisture and air. Modern synthetic building wraps are much more fire-resistant than the felt or tar paper used in older installations **(Figure 5.6)**.

Foam insulation. The use of a combustible insulation does somewhat increase the possibility of a fire starting within the wall. An example is from an electrical malfunction igniting the insulation. The extent to which foam insulation will increase fire spread within a wood-framed wall depends on the existence of air space. Air space between the foam and the wall surface will contribute to rapid fire development within the wall space. If the space is completely filled with the foam, however, the fire has to burn upward through the material and will progress much more slowly.

Noncombustible materials used for insulation include:

- Glass wool and rock wool in the form of batts or blankets with combustible paper or foil coverings
- Fiberglass **(Figure 5.7)**

NOTE: Some older vermiculite and batt insulation also contains asbestos.

Figure 5.7 Fiber glass insulation is used for sound and temperature control.

Figure 5.6 Synthetic building wraps reduce moisture and are more fire resistant than older wrap materials.

Insulation can also take the form of loose-fill material, including:

- Granulated rock wool
- Mineral wool and glass wool, either blown into stud spaces or manually packed
- Cellulose fiber and shredded wood, treated with water-soluble salts to reduce combustibility (fire in this material will slowly smolder)

Solid-fill foam insulations are applied as soft foam that hardens after application. The foam is treated with flame retardants before application. This style of insulation is gaining popularity and is referred to as "building icing" because of its appearance. Two types of solid-fill foam insulations are polyurethane foam and urea formaldehyde foam.

Siding. Siding provides the exterior cladding of a wood-frame building. Siding provides weather protection and can contribute to the appearance of a building. Some siding materials are noncombustible; others are combustible **(Figures 5.8 a-c)**. The combustibility of a siding material can affect fire behavior by allowing exterior fire travel or ignition due to an exposure fire.

When a building is remodeled, new siding material is frequently applied over the existing siding. Older buildings, therefore, may have multiple layers of siding. Materials used for siding are not limited to wood and include:

- Aluminum
- Asphalt siding/shingles
- Cement board

Figures 5.8 a-c Wood shakes, stucco, and vinyl siding are common siding materials.

- Plywood
- Stone
- Stucco
- Wood boards
- Wood shingles
- Vinyl

Asbestos Components

Asbestos can be found in a wide variety of building products, including:

- Siding
- Insulation
- Construction adhesives
- Drywall compound
- Electrical equipment
- Heating equipment
- Floor tile
- Ceiling tile

Asbestos is no longer used as a building material because it is known to cause respiratory ailments. However, it was commonly used in the U.S. from the 1930s until the 1970s and is still found in many older buildings. Asbestos siding may be covered with another siding material, or, in some cases, still be exposed.

The inspector needs to be aware of the health hazards associated with asbestos, and take care not to disturb asbestos materials when conducting inspections. Exercise caution while in proximity to components that may potentially contain asbestos. Any activity near or affecting asbestos products must include proper PPE and decontamination procedures.

In addition to respiratory hazards, asbestos fibers are also difficult to remove from contaminated clothing and other resources. The AHJ may require contaminated resources to be discarded after use. For additional information, refer to NFPA® 1851, *Standard of Selection, Care and Maintenance of Protective Ensembles, Structural Firefighting and Proximity Firefighting.*

Masonry

Masonry is one of the oldest and simplest building materials, dating back thousands of years. Although masonry can be used for fencing or stone work trim, an inspector's chief interest is masonry's use for wall construction. The fundamental construction technique consists of stacking individual masonry units on top of one another and bonding them into a solid mass through the use of a bonding agent. Masonry units can be made of several materials:

- Brick
- Concrete block

- Stone
- Clay tile block
- Gypsum block

Although masonry units are inherently fire-resistive, the mortar joints may deteriorate over time or when exposed to fire and result in weakening of the structure.

Brick

Bricks are produced from a variety of locally available clay and shale. Bricks are manufactured by placing clay in molds, removing the molded clay, and drying the bricks. The bricks are fired in a kiln during which they are subjected to temperatures as high as 2,400°F (1 300°C). This intense heat converts them to a ceramic material **(Figure 5.9)**.

Figure 5.9 Bricks have long been used to construct walls or to provide an attractive veneer.

Concrete Block

Concrete blocks are also known as concrete masonry units (CMUs). The most commonly used concrete block is the hollow concrete block, sometimes referred to as cinder block. Hollow concrete blocks are produced in a number of sizes and shapes, but the most commonly used is the nominal 8- × 8- × 16-inch (200 mm by 200 mm by 400 mm) block. In addition to the hollow block, concrete masonry units can be produced as either bricks or as solid blocks **(Figure 5.10, p. 162)**.

Concrete Block — Large rectangular brick used in construction; the most common type is the hollow concrete block. *Also known as* Concrete Masonry Units (CMU).

Stone

Stone masonry consists of pieces of rock that have been removed from a quarry and cut to the size and shape desired. The principal types of stone used in construction are:

- Granite
- Limestone

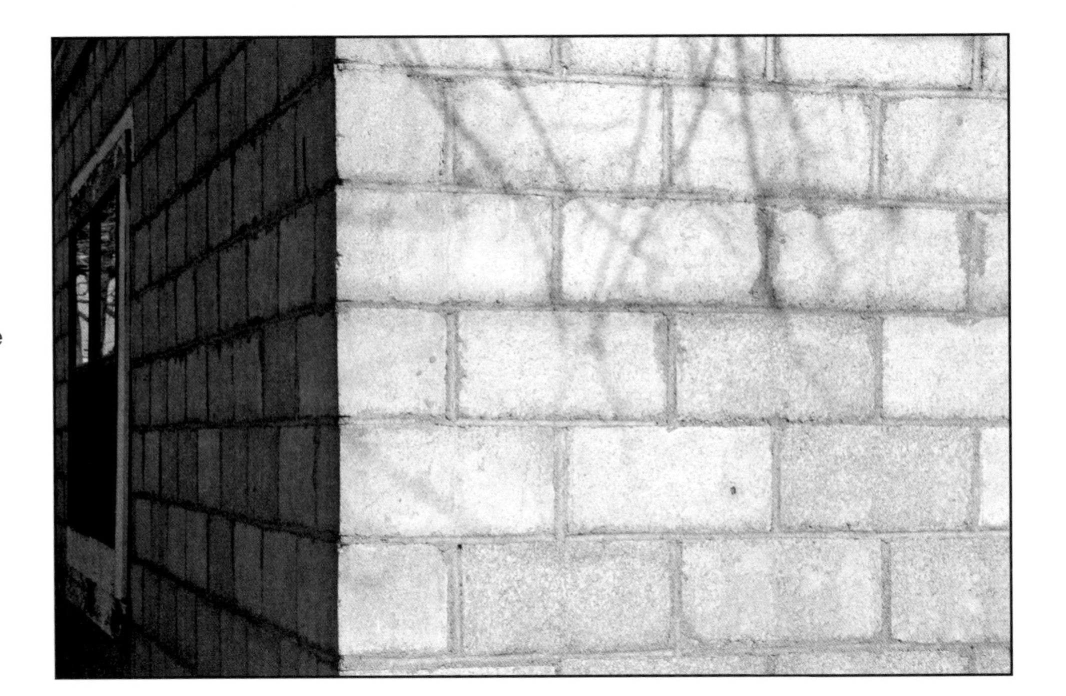

Figure 5.10 Concrete blocks are larger than ordinary bricks.

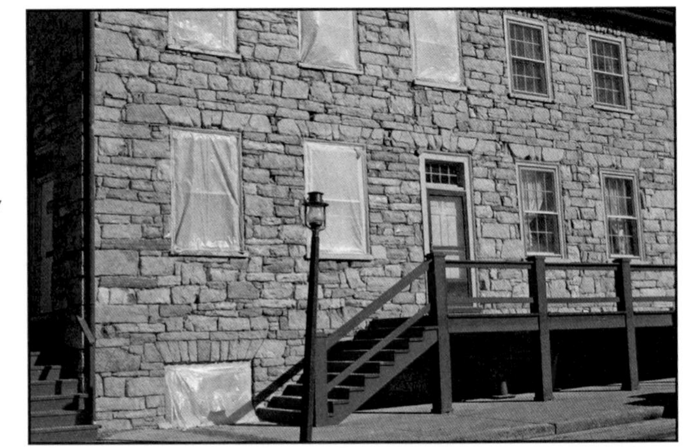

Figure 5.11 An older masonry building. *Courtesy of Ed Prendergast.*

- Sandstone
- Slate
- Marble

Stone can be used in two ways in construction. It can be laid with mortar to form walls similar to brick or concrete block, or it can be used as an exterior veneer attached by supports to the structural frame of a building **(Figure 5.11)**.

Clay Tile Block and Gypsum Block

Clay tile blocks are used for foundations and walls in areas where clay is available as a building material. The blocks are hollow, providing a space for fill material. However, the blocks do deteriorate over time and are susceptible to damage from water and freezing temperatures.

Gypsum block use has diminished in modern practice, although blocks can still be found in many existing buildings. They are not applicable for exterior use due to their ability to absorb water.

Clay tile blocks and gypsum blocks were once widely used for construction of interior partitions. Structural glazed tile is still frequently used where a smooth surface is desired, such as in a shower room.

Concrete

Concrete is used for the following:

- Pavement
- Foundations
- Columns
- Floors
- Walls
- Concrete masonry units

Like masonry, concrete does not burn, and it resists insects and the effects of contact with soil. It can be placed in forms to create a variety of architectural shapes. Concrete types include ordinary and lightweight.

NOTE: Concrete is poured and then allowed to harden; masonry is not.

Because concrete is weak in **tension**, it cannot be used alone where significant **tensile** forces occur in a structure. To resist these forces, concrete is reinforced through the use of steel bars (**rebar**) or cables placed within the concrete **(Figure 5.12)**. The Inspector II portion of this chapter contains more information about reinforcing concrete.

Tension — Vertical or horizontal forces that tend to pull things apart; for example, the force exerted on the bottom chord of a truss.

Tensile — Force of pulling apart or stretching.

Rebar — Short for reinforcing bar. These steel bars are placed in concrete forms before the cement is poured. When the concrete sets (hardens), the rebar within it adds considerable strength and reinforcement.

Steel

Steel is basically an alloy of iron and carbon. Common structural steel has less than three tenths of one percent carbon. Cast iron, by contrast, has a carbon content of three to four percent. The higher carbon content of cast iron produces a material that is hard but brittle, while the lower carbon content of steel makes it less likely to fracture or break.

Figure 5.12 Vertical reinforcing bars in a concrete column. *Courtesy of Ed Prendergast.*

Characteristics

Steel is the strongest of the structural materials. It is nonrotting, resistant to aging, and dimensionally stable **(Figure 5.13)**. Steel is also flame resistant. Steel is a relatively expensive material, but its strength and the variety of forms in which it is produced allow it to be used in smaller quantities than other materials. Steel is used for applications varying from heavy beams and columns to door frames and nails.

The inspector should be concerned with the disadvantages of steel as a structural building material and methods employed to overcome these disadvantages. Steel possesses two inherent disadvantages:

- Steel tends to rust when exposed to air and moisture.
- Steel loses strength and elongates when exposed to the heat of a fire.

Inspectors should be alert to the presence of rust because it is an indication of deterioration of steel and a weakening of the structure. In particular, they should notice exposed steel fire escapes that can become unsafe due to the rusting of supports and attachment points. Methods to protect steel from rusting include painting the surface with a rust-inhibiting paint or coating it with zinc or aluminum. Steel can also be produced using ingredients that resist rust, as in the case of stainless steel.

Steel's loss of strength and elongation due to fire can cause exterior walls to fail and eventually lead to building collapse. Steel can also melt when exposed to the extreme levels of heat common in compartment fires **(Figure 5.14)**. The rate at which unprotected steel fails when exposed to fire depends on several factors:

- Mass of the steel members and surface area exposed
- Intensity of the exposing fire
- Load supported by the steel
- Type of structural connections used to join the steel members

Fire Protection

Steel must be protected from the heat of a fire so that builders can use it in fire-resistive construction. The inspector must be able to recognize the ma-

Figure 5.13 When completed, steel-framed buildings may be covered with brick or stone veneers.

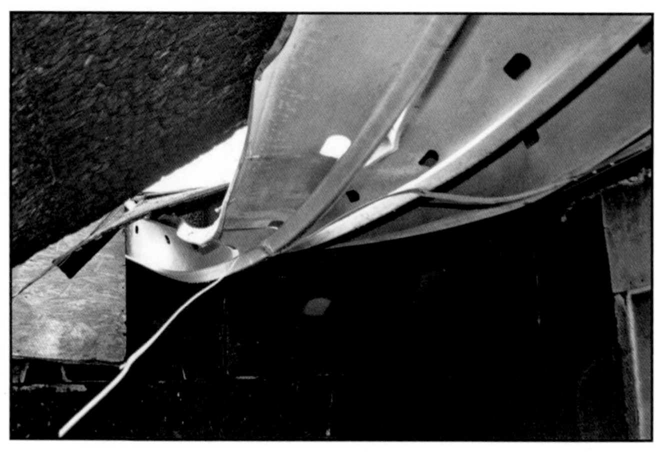

Figure 5.14 Steel can elongate and lose its strength at very high temperatures. *Courtesy of NIST.*

terials and methods that are commonly used to provide fire protection for steel. The most common method of providing fire protection is through the use of an insulating material **(Figure 5.15)**. In newer buildings, steel can be protected in a number of ways:

- Metal lath and plaster
- Gypsum board
- Sprayed-on cement coating
- Mineral and fiberboard

Figure 5.15 Spray-on coatings are a common way to provide steel with fire protection.

The lighter-weight insulating materials are usually more fragile than heavier materials and are more susceptible to being damaged or dislodged. This damage may reduce or negate the fire rating of the member. Damage needs repair whether it happens during construction or over time.

Another method of providing fire protection for steel is the use of intumescent coatings. These paint-like coatings can be applied in thicknesses of a fraction of an inch (millimeter) and provide fire-resistance ratings of up to three hours when applied as required by its listing.

Membrane Ceilings

Steel floor support systems are frequently protected using a noncombustible ceiling system or a special insulating ceiling tile known as a membrane ceiling **(Figure 5.16)**. The membrane ceiling is popular with designers because it allows ductwork and electrical conduit to be hidden above the ceiling while providing an attractive finished surface. The rated ceiling acts as a thermal barrier to the heat of a fire below it.

Workers often penetrate ceilings to install light fixtures or ventilation ducts. In these cases, such special provisions as increased insulation or **fire dampers** may be required to maintain the integrity of the ceiling **(Figure 5.17)**.

NOTE: Chapters 6 and 14 contains additional descriptions of fire dampers.

Fire Damper — Device installed in air ducts that penetrate fire-resistance-rated vertical or horizontal assemblies; limits the transfer of heat and passage of flames through the ducts at the point where the duct passes through the assembly.

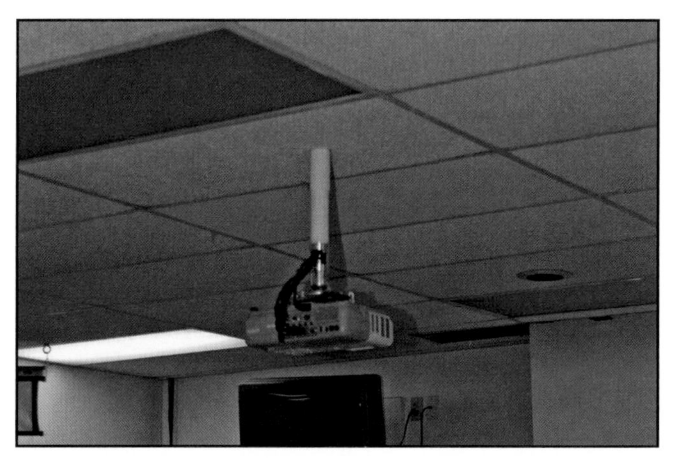

Figure 5.16 Membrane ceilings are a popular way to cover ductwork and electrical conduit. The inspector should remember that these spaces can conceal smoke and fire.

Figure 5.17 Fire dampers are designed to stop fire extension through a rated wall to maintain the rating of the wall.

Glazing — Glass or thermoplastic panel in a window that allows light to pass.

Glazing

Glazing (glass) is present in most buildings. Its obvious use is for windows, skylights, storefronts, and anywhere light transmission is desirable. The architectural applications of glass extend to church windows, partition walls, and some structural applications for the exterior curtain walls of buildings **(Figure 5.18)**. The inspector must be aware that when glazing is broken, it is often replaced with the wrong (inferior) glass, such as refurbishing an old structure and failing to install fire glazing in exit doors.

The most commonly encountered types of glazing are as follows:

- **Single-strength annealed** — Glass produced by slowly cooling the hot glass during its production, which permits the release of thermal stresses that would form if the glass were cooled rapidly.
- **Heat-strengthened** — Glass with a residual surface compression stronger than annealed glass of the same size and thickness. As a result, the material is more resistant to thermally induced stresses, cyclic wind pressures, and impacts by windborne objects.
- **Fully tempered** — Glass having a residual surface compression even stronger than heat-strengthened glass. Fully tempered glass breaks into small granules rather than large, sharp-edged chunks. It is used in windows that might be subject to high wind forces and exterior doors that people might walk into accidentally.
- **Laminated** — Glass that consists of two layers of glass with a transparent layer of vinyl bonded into the center. When laminated glass is broken, the inner core of vinyl holds the broken pieces of glass in place. This glass is used in security windows and to reduce noise transmission.
- **Glass block** — Glass produced either as solid or hollow non-load-bearing units with different surface patterns. Glass blocks may be used for the protection of limited-size openings in fire-rated walls when permitted by the local building code.

When glass is heated, internal thermal stresses cause it to shatter and fall out of its frame; however, fire-rated glass is available that can be used in fire-rated assemblies and as a structural component. The most common of these is wired glass, although other proprietary unobstructed products are available.

- **Wired glass** — Wired glass is used in both interior and exterior applications, in fire doors, in windows adjacent to fire escapes, in corridor separations, and to protect against exterior exposures. When the glass breaks, the wires hold the glass in place, permitting it to act as a barrier to a fire.

  NOTE: Not all wired glass is fire rated.

- **Fire-rated glass** — Glass that does not use interior wires. These fire-rated glass panels are made from a combination of glass and plastic. Fire-rated glazing must be installed in a fire-rated framing that is rated as an assembly.

  NOTE: All fire-rated glass is marked, whether it is wired glass or not, and the inspector should check for the appropriate markings **(Figure 5.19)**.

Figure 5.18 Inspectors must look for situations where glazing has been replaced with inferior-quality glazing.

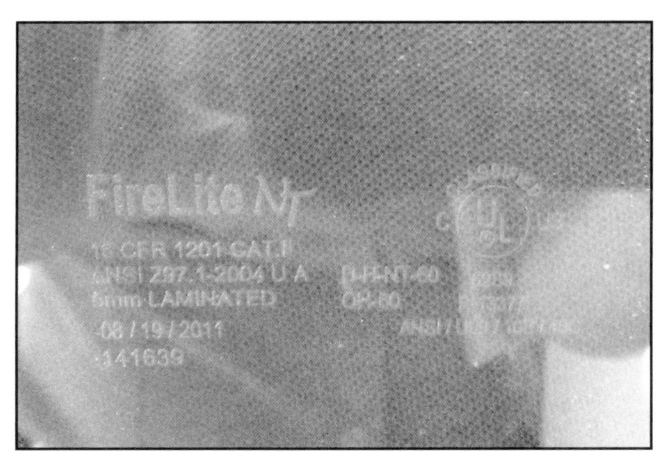

Figure 5.19 Markings on fire-rated glass. *Courtesy of Scott Strassburg.*

Gypsum Board

Gypsum, also known as drywall or Sheetrock®, is a mineral-based product from which plaster and wallboards are constructed **(Figure 5.20)**. Gypsum's high water content gives it excellent heat-resistant and fire-retardant properties. Because gypsum breaks down gradually under fire conditions, it is commonly used in the fire protection of individual beams and girders as well as in fire-resistive assemblies, such as:

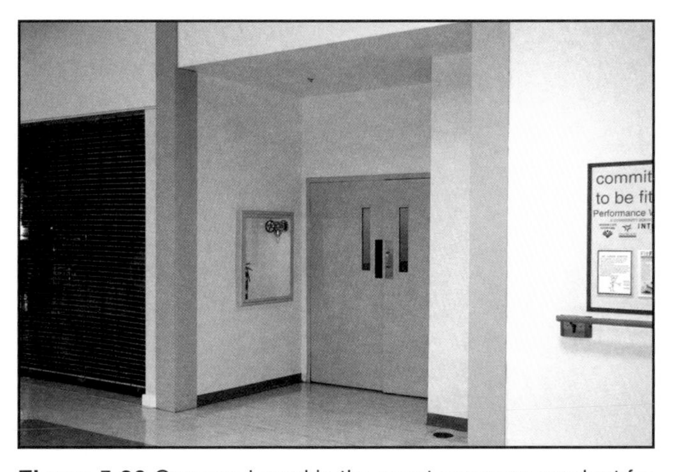

Figure 5.20 Gypsum board is the most common product for applying a smooth interior finish.

- Corridor partitions
- Stair enclosures
- Shaft walls
- Column protection
- Membrane ceilings

Type X gypsum board used in fire-rated assemblies is produced with glass fibers that act as reinforcement. The glass fibers provide tensile strength for the inner gypsum core and prevent its deterioration when exposed to fire. Type C gypsum board contains vermiculite that expands as it is exposed to heat. This expansion allows the gypsum to maintain its integrity and remain dimensionally stable for longer periods of exposure.

NOTE: As with other types of innovations, the inspector must remember that they can improve efficiency at the cost of safe fire fighting operations. For example, a new type of acoustic board adds a steel sheet between two drywall panels. This addition is intended to improve the speech intelligibility or other sound qualities of a space. Unintended consequences in fire conditions include increased difficulty of forcible entry, extinguishment, and overhaul.

NOTE: Chapter 6 describes the ASTM E-119 test, which evaluates the ability of structural members to carry loads and to act as a fire barrier.

Several types of gypsum board are produced for different purposes. The inspector may need to research the original building plans to verify that the correct product is used. Types of gypsum board include:

- **Regular gypsum board** — Used for most applications

- **Water-resistant gypsum board** — Produced with a water-repellent paper facing for use where it may be exposed to moisture
- **Type X gypsum board** — Used in fire-rated assemblies
- **Type C gypsum board** — Used in fire-rated assemblies
- **Foil-backed gypsum board** — Used to replace the vapor barrier in outside walls
- **Gypsum backing board** — Used as a backing layer in multilayer assemblies
- **Coreboard** — Used for shaft walls and solid partitions

Lath and Plaster Wall Construction

Before the introduction of gypsum board, the most common interior finish was lath and plaster **(Figure 5.21)**. The lath, which could be wood or metal, was nailed horizontally to the wall studs. A ¼-inch (6 mm) space was left between each strip. A ¼-inch (6 mm) thick layer of plaster was applied over the lath and pressed into the spaces to hold the plaster in place. A top coat of plaster was applied as a finish coat that would be painted or covered with wall paper or canvas.

Although lath and plaster construction is seldom used today, an inspector will encounter structures built before 1960 that contain lath and plaster walls. Adopted building codes may require that damaged lath and plaster be repaired or replaced to meet newer standards.

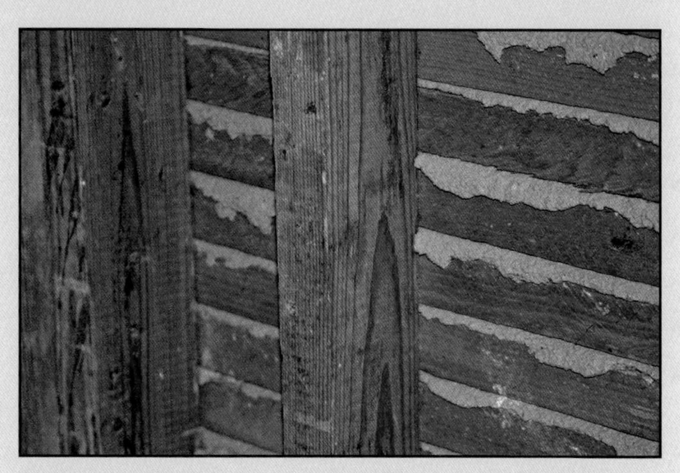

Figure 5.21 Before the introduction of gypsum board, building interiors were finished with lath and plaster. *Courtesy of Bob Billen, Edmond (OK) Fire Department.*

Plastics

The large variety of plastics available permits their use in many interior and exterior applications. In building construction, plastics are used for such components as the following:

- Siding
- Floor covering
- Insulation
- Tub and shower enclosures
- Vapor barriers
- Pipe and pipe fittings **(Figures 5.22 a and b)**

- Lighting fixtures
- Skylights and roof domes
- Sprinkler piping
- Moldings
- Wall coverings
- Gratings
- Mantel pieces

The strength of plastic materials is close to that of wood, although glass-fiber-reinforced plastic may be nearly as strong as steel. Plastics are not usually used for structural applications in buildings because of their generally lower strength and their greater tendency to bend.

Table 5.1, p. 170, lists the major groups of plastics that are used in building construction and their typical applications. It must be emphasized that the materials listed in this table are major plastic *groups*. Within these groups are a number of specific subgroups.

The inspector must recognize that the use of plastics increases the amount of combustible materials — and therefore the fire hazards — both inside and outside a structure. Some plastics used as thermal barriers may be invisible to the inspector once the structure is completed while other plastics, used for external veneer, may be difficult to recognize. The sections that follow address these issues.

Figures 5.22 a and b Plastic piping may be left exposed or concealed in interior wall construction.

Flammability

The flammability of plastics varies widely. Some plastics, such as cellulose nitrate, burn so rapidly that they constitute a unique fire hazard. Other plastics may burn slowly and stop burning when the ignition source is removed. Fire retardants can be added to some plastics to reduce their flammability and ignition sensitivity. However, even plastics with low flammability are subject to deterioration and may give off toxic gases at temperatures above 500°F (260°C).

Plastic materials frequently exhibit burning properties different from other materials. For example, nylon usually melts and drips when it burns. Some plastics generate enormous quantities of heavy smoke. The products of combustion of some plastics are more toxic than nonplastic materials. The combustion of vinyl chloride, for example, produces hydrogen chloride, the gaseous form of hydrochloric acid. Hydrogen chloride is corrosive as well as toxic and increases the damage done to sensitive electrical equipment.

Fire Hazards

Plastics used in building construction increase the fire hazard because they add to the amount of fuel in a building and increase the toxicity of products of combustion. Where large amounts of plastic are used, such as foam plastic

Table 5.1
Plastics Commonly Used in Construction

Plastic Group	Construction Application
Acrylonitrile-Butadiene Styrene (ABS)	Water and gas supply lines, drain and waste systems
Acrylics	Skylights, translucent ceiling panels, light diffuser, and glazing
Cellulosics	Piping and pipe fittings, outdoor lighting fixtures
Epoxies	Adhesive for sandwich panels, floor tiles, and patching concrete
Fluorocarbons	Chemical piping
Nylons	Carpeting, fabric in air-supported structures
Phenolics	Electrical parts, foamed insulation, and sandwich panels
Polycarbonates	Safety glazing, lighting fixtures, and lighted signs
Polyesters	Translucent sheeting, molded bathtubs, shower stalls, and sinks
Polyethylene	Vapor barrier in wall assemblies, wire and cable insulation
Polystyrene (foamed)	Duct and pipe insulation, insulation in freezers, refrigerators, walls, and ceilings
Polyurethanes	Wall and ceiling insulation, pipe and duct insulation, upholstery
Vinyl	Tile flooring, gutters, molding, window frames, siding, and exterior finish

in wall insulation, the fire hazard is greatly increased. Special treatment, such as covering the surface of the insulation with a noncombustible material or installing automatic sprinklers, is required in these instances.

Of special importance to the inspector are plastics used in buildings that are classified as either noncombustible or fire-resistive under the provisions of a building code. For example, the use of plastic shower enclosures in a fire-resistive hotel may provide a path for fire to communicate vertically through a plumbing chase.

NOTE: See Chapter 6 for more information about chases.

Thermal Barriers

A thermal barrier is a material installed to reduce the heat rise of the material it is protecting.

Adopted building codes may require that a thermal barrier be used to separate plastics and other combustible materials from the interior of the building.

Exterior Insulation and Finish Systems (EIFS)

One application for plastic in an exterior veneer system is known as **Exterior Insulation and Finish Systems (EIFS)**. EIFS consists of fiberglass insulation, gypsum board, and expanded polystyrene beadboard or extruded foam with a noncombustible hand-troweled finish. The finished product closely resembles a stucco or etched-concrete finish.

EIFS is a way to improve the appearance of an existing masonry wall that has become deteriorated. Because it is flammable, however, EIFS can be ignited from an exterior source or from the radiant heat of an exposed fire. For these reasons, an inspector needs to recognize the use of EIFS in a building's construction **(Figure 5.23)**.

Exterior Insulation and Finish Systems (EIFS) — Exterior cladding or covering systems composed of an adhesively or mechanically fastened foam insulation board, reinforcing mesh, a base coat, and an outer finish coat. *Also known as* Synthetic Stucco.

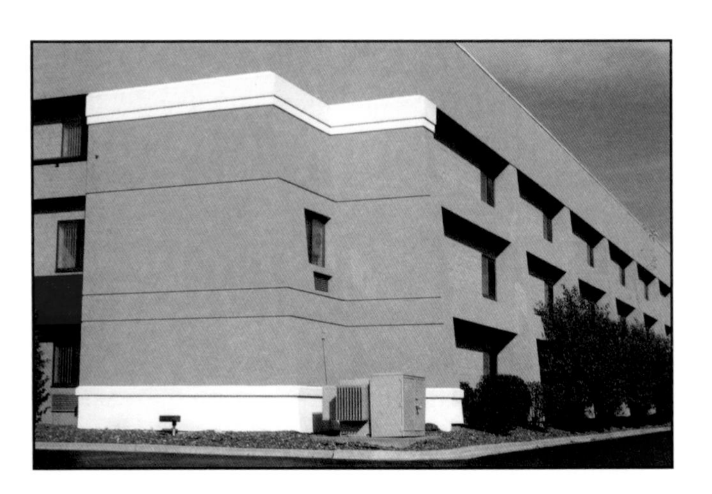

Figure 5.23 Even though this finish looks like concrete, it is made from flammable materials that can ignite and burn. *Courtesy of Ed Prendergast.*

EIFS are used to increase the insulative properties of a building. Building codes impose stringent regulations on the use of EIFS because foam insulation is combustible and because flame spreads rapidly over its surface. Typically, a code will require that foam insulation be faced with a thermal barrier, such as gypsum wallboard, to prevent or slow surface ignition of the foam.

Structural Systems

Basic structural components are of little value unless they can be assembled into a composite system that will support a building. From an inspector's viewpoint, construction types and basic structural systems are related. The construction classification types result from the materials and structural systems used in buildings. Some examples:

- Reinforced concrete or protected steel framing is typical of Type I, fire-resistive buildings.
- The use of unprotected steel in a Type I building results in a building being classified as a Type II, noncombustible building.
- Wood and masonry together are found in ordinary construction (Type III), heavy-timber construction (Type IV), and wood-frame construction (Type V).

An inspector must keep in mind that these relationships are not absolute. For example, masonry curtain walls can be found in a building with a steel frame, and unprotected steel beams can be used to support a concrete slab. Furthermore, basic structural systems can be used in combination **(Figure 5.24)**. It is possible for a building to use both reinforced concrete and a steel frame, or steel and masonry, for different parts of the building. The design of a structural system involves the blending of the properties of one or more of the construction materials into a unit that will withstand the forces applied to it in a reliable, economical, functional, and attractive manner .

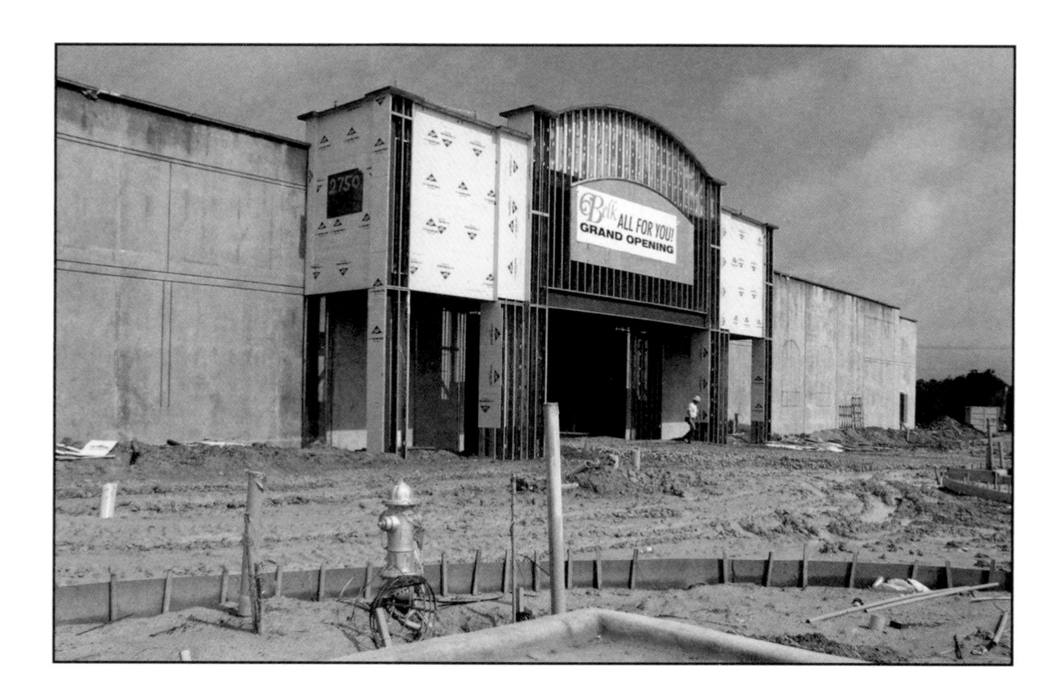

Figure 5.24 Structural systems are frequently used in combination, as this concrete and steel structure. *Courtesy of Ron Moore and McKinney FD.*

When evaluating a particular structure, the inspector may need to refer to the building department or permit office files for the construction documents that were submitted with the original building permit application. The fire department's inspection file may also contain a description of the construction system used. Frequently, all that an inspector can do is to become familiar with prevailing construction methods in the jurisdiction.

This section examines general types of structural systems, including the following:

- Bearing wall structures
- Loads
- Shell and membrane structures
- Frame structural systems
- Wood structural systems
- Masonry structural systems
- Steel structural systems
- Concrete structural systems

Bearing Wall Structure — Common type of structure that uses the walls of a building to support spanning elements such as beams, trusses, and pre-cast concrete slabs.

Compressive Loads — Vertical and/or horizontal forces that tend to push the mass of a material together; for example, the force exerted on the top chord of a truss.

Dead Load — Weight of the structure, structural members, building components, and any other features permanently attached to the building that are constant and immobile.

Live Load — (1) Items within a building that are movable but are not included as a permanent part of the structure. (2) Force placed upon a structure by the addition of people, objects, or weather.

Bearing Wall Structures

Bearing wall structures use the walls of a building to support spanning elements, such as beams, trusses, and precast concrete slabs **(Figure 5.25)**. Bearing walls provide lateral support to the structure along the direction of the wall. Usually, the exterior walls serve as the bearing walls, with the interior support system consisting of columns and beams. However, it is possible to use interior walls for structural support.

In a bearing wall structure, the walls are subjected to **compressive loads**. The walls may be continuous or they may be interrupted for door and window openings. Materials used for bearing walls include concrete blocks, bricks, stone, concrete panels, and wood. A log cabin is an example of the use of solid wood for a bearing wall.

Loads

The forces on a building resulting from gravity are classified into two types: **dead loads** and **live loads**. A dead load has the characteristic of being fixed in location and accurately quantifiable. A dead load is the weight of any permanent part of a building, including the weight of the building components, such as:

- Roofs
- Floor slabs or decks **(5.26 a)**
- Interior walls
- Stair systems
- Exterior walls
- Columns

Although dead loads usually remain the same, they can change, such as when an air conditioning unit is installed on the roof of a building. The air conditioning unit is considered a dead load, but its new position on the roof will substantially increase the load on the roof's supports. The dead load represented by the weight of the roof can also increase over time when additional layers of roofing material are added in the course of resurfacing.

A live load is any load that is not fixed or permanent, including:

- Wind and seismic loads **(Figure 5.26 b)**
- Building contents
- Building occupants
- Weight of snow or rain on the roof

Usually the actual weight and distribution of a building's contents are not known exactly. Because live loads vary by occupancy, building codes specify minimum live loads to be used in the design process for different occupancies. **Table 5.2, p. 174,** shows some of the uniformly distributed live loads required by the IBC. Building codes specify that when the live load for a given occupancy is known and exceeds the values contained in the code, the actual load must be used in the design calculations.

It must be emphasized that the loads shown in Table 5.2 are uniformly distributed loads applied over a large area. A concentrated load is one that is applied at one point or over a small area. Concentrated loads produce high localized

Figure 5.25 Interior support systems. *Courtesy of McKinney (TX) Fire Department.*

Figure 5.26 a The beam is part of the dead weight of a structure.

Figure 5.26 b Winds are an example of live load forces. This parapet was toppled by high winds. *Courtesy of Ed Prendergast.*

Table 5.2
Minimum Uniformly Distributed Live Loads

Occupancy	Pounds per Square Foot (psf)	Kilograms per Square Meter (kg/m²)
Assembly Areas and Theaters		
Fixed Seats	60	290
Lobbies	100	490
Movable Seats	100	490
Stages	150	730
Catwalks	40	195
Bowling Alleys	75	365
Dining Rooms and Restaurants	100	490
Fire Escapes	100	490
Gymnasiums	100	490
Manufacturing		
Light	125	610
Heavy	250	1200
Stores		
Retail, First Floor	100	490
Retail, Upper Floors	75	365
Wholesale, all Floors	125	610

Source: 2015 International Building Code®

forces and non-uniform loads in the supporting structural members. Building codes require that a specified minimum concentrated load be used in the structural analysis when it creates greater load effects than the uniform load.

Shell and Membrane Structures

Membrane Structure — Weather-resistant, flexible, or semiflexible covering consisting of layers of materials over a supporting framework.

Although fabric has long been used for tents or other temporary shelters, newer fabrics can be used as part of the walls and roofs of permanent structures **(Figures 5.27 a and b)**. Shell and **membrane structures** consist primarily of an enclosing, waterproof surface; the stresses resulting from the applied loads occur within the surface bearing wall structures.

A membrane structure can be distinguished from a simple tent by its permanence. Tents are used for short periods, while membrane structures are permanent. Building codes typically address membrane structures with a life of 180 days or more, while fire codes address those used for fewer than 180 days.

Membrane structures possess several advantages from a design standpoint:

- They can usually be erected in less time than a rigid structural system.
- The fabric of the membrane can flex and absorb some of the stresses from seismic and wind forces.
- The use of fabric also permits the development of innovative architectural shapes.
- The fabrics weigh less than other roof systems.

Figures 5.27 a and b (a) A membrane structure supported by a frame. (b) Air-supported structures are supported by the pressurization of the entire surface inside the membrane.

There are disadvantages to the use of membrane structures. Because fabrics cannot support compressive forces, they must be supported by cables and masts or a tubular framework. The support system must provide sufficient rigidity to avoid shaking or flapping.

Similarly, the fabrics used in membrane structures cannot be used to support building appliances, such as lighting or heating equipment. This equipment must be supported from the framework that supports the membrane. When a membrane structure requires an automatic sprinkler system, the sprinkler piping must also be supported from the framework used to support the surface material.

Fabrics are considerably thinner than other assemblies used for roof systems. To overcome the problem of providing adequate thermal insulation, two layers of material separated by a distance of several inches (millimeters) can be used. The intervening air space acts as a thermal barrier.

The materials used in membrane structures must be noncombustible or must meet the fire propagation performance requirements in the locally adopted code. Membrane structures can also be used in fire-resistive construction when a building code only requires a noncombustible roof for a fire-resistive building.

Frame — Internal system of structural supports within a building.

Stud — Vertical structural member within a wall in frame buildings; most are made of wood, but some are made of light-gauge metal.

Post and Beam Construction — Construction style using vertical elements to support horizontal elements. Historically associated with heavy wood beams and columns; modern construction may use materials including heavy timber, steel, and precast concrete.

Frame Structural Systems

In a **frame** structure, structural support is provided in a manner similar to the way the skeleton supports the human body. The walls act as the 'skin' to enclose the frame. The walls may also provide lateral stiffness but provide no structural support. In the fire service, the term frame construction often refers to a wood-frame building, but frame structural systems are also built using other materials. In addition to the framing associated with wood construction, the following sections describe other types of structural frame systems.

Steel Stud Wall Framing

Steel **stud** wall construction uses relatively closely spaced vertical steel studs connected by top and bottom horizontal members. Historically, stud-wall frame construction has been associated with the use of 2 x 4-inch (50 mm x 100 mm) wood studs, although the use of steel studs has become more common in recent years. A steel stud wall is frequently provided with diagonal bracing for stability **(Figure 5.28)**. Both sides of a stud wall may be covered with paneling and sheathing.

Post and Beam Framing

Post and beam construction uses a series of vertical elements (the posts) to support horizontal elements (the beams) that are subject to transverse loads **(Figure 5.29)**. Historically, this system evolved from the use of tree trunks for framing and is still commonly associated with wood beams and columns. Other materials can be used, however, including masonry for the posts and steel and precast concrete for the posts and beams.

The distinctive characteristic of post and beam framing is the spacing of the vertical posts and the cross-sectional dimension of the members. The vertical posts may be spaced up to 24 inches (600 mm) apart, unlike stud wall construction where the studs are 12 to 16 inches (300 to 400 mm) apart. The minimum dimensions used for the wood posts and beams are larger than the studs in stud wall construction. Typical dimensions for the posts are 6 x 8 inches (150 mm x 200 mm) when supporting roofs only. Post and beam construction requires the addition of other members, such as diagonal braces to withstand lateral loads.

Truss Frames

Trusses can be adapted to a variety of applications. It is possible to build components of a frame using a series of trusses, as with the arch or the rigid frame shown in **(Figure 5.30)**.

Capital — Broad top surface of a column or pilaster, designed to spread the load held by a column.

Drop Panel — Type of concrete floor construction in which the portion of the floor above each column is dropped below the bottom level of the rest of the slab, increasing the floor thickness at the column.

Slab and Column Frames

Slab and column frames are most frequently encountered in concrete structures. The floors of a multistory, reinforced-concrete building can be designed by several methods, depending on the loads to be supported. These floor slabs are supported by concrete columns. Because of the high stress load in the connection, the intersection between the slab and column is usually reinforced by additional material in the form of a **capital** or a **drop panel (Figure 5.31, p. 178)**. In addition to concrete slabs, horizontal systems that can be used to support floor loads include wood decks and metal decks supported by beams and columns.

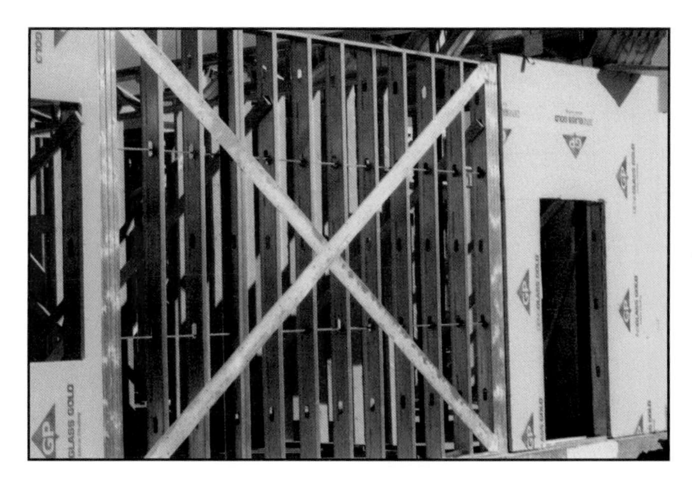

Figure 5.28 Steel stud wall with diagonal bracing. *Courtesy of Ed Prendergast.*

Figure 5.29 In post and beam construction, the vertical posts support the horizontal members.

Figure 5.30 A steel truss frame used in a new structure.

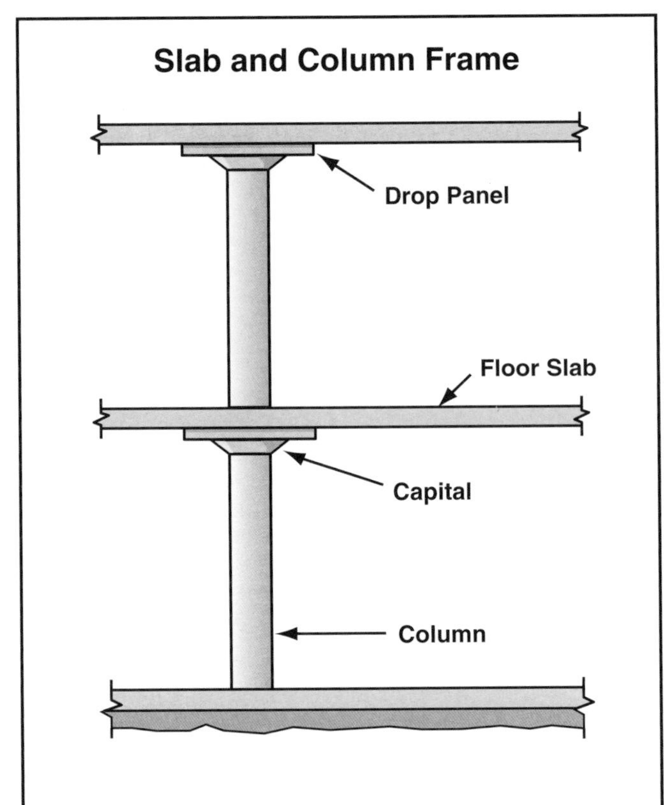

Figure 5.31 Reinforcement of concrete floor.

Rigid Frame — Load-bearing system constructed with a skeletal frame and reinforcement between a column and beam.

Rigid Frames

A **rigid frame** structural system is used when the joints between a column and a beam are reinforced so bending stresses can be transmitted through the joints **(Figure 5.32)**. The most easily recognized rigid-frame structure is the single-story, gabled-roof and rigid-frame building. The peak of the roof is usually provided with a hinged connection to allow for slight movement between the two halves of the frame. This type of rigid frame can be constructed of steel, laminated wood, or reinforced concrete. Rigid frames are used in other types of structures, including multistory and multispan designs. The joints will be the last portion of the assembly to fail under fire conditions.

Figure 5.32 This home is typical of a rigid frame building. *Courtesy of McKinney (TX) Fire Department.*

Wood Structural Systems

Wood is almost always used in Type IV and Type V construction as interior and exterior framing. This use includes:

- One- and two-family dwellings

- Townhouses
- Apartment buildings
- Hotels of five stories or less

The wood framing systems most frequently encountered are classified into three basic types:

- Wood framing (Type V)
- Post and beam framing
- Heavy-timber framing (Type IV)

NOTE: The previous section on framing systems included a description of post and beam framing.

Wood Framing

The most popular form of wood framing is known as wood-frame construction. The walls are formed from vertical members known as studs. The floors are supported by joists or trusses, and inclined roofs are supported by rafters or light trusses. The two most common types of wood framing are balloon framing and platform framing.

Wood framing is usually covered with an interior finish of plaster or drywall. The interior finish will act to retard the spread of fire into the stud spaces. However, a fire that originates in the stud space or spreads into it can readily spread from the vertical cavity into the horizontal joists and into the attic space. Fire-stopping must be provided in addition to the structural members **(Figure 5.33)**.

Figures 5.33 a and b In a wood-frame structure, firestopping can be accomplished by using gypsum board or by placing wood horizontally to help stop vertical fire spread. *Both photos courtesy of McKinney (TX) Fire Department.*

Platform framing. In platform framing, the exterior wall vertical studs are not continuous **(Figure 5.34 a, p. 180)**. The first floor is constructed as a platform upon which the exterior vertical studs are erected. The second-story framing is erected on the platform formed by the story joists and flooring. In a platform frame, the plate installed on the top of the studs provides a fire-stop that tends to block the spread of fire from floor to floor within the walls.

Figure 5.34 a Platform frame building under construction. Notice that the studs are not continuous from floor to floor. *Courtesy of Ed Prendergast.*

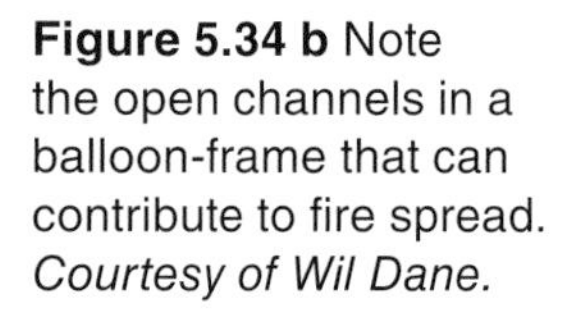

Figure 5.34 b Note the open channels in a balloon-frame that can contribute to fire spread. *Courtesy of Wil Dane.*

Balloon-Frame Construction — Type of structural framing used in some single-story and multistory wood frame buildings wherein the studs are continuous from the foundation to the roof. There may be no fire stops between the studs.

Balloon framing. **Balloon framing** uses closely spaced members (studs) that are continuous from the sill to the top plate of the roof line **(Figure 5.34b)**. The joists that support the second floor are nailed directly to the studs. The vertical combustible spaces between the studs in balloon-frame construction provide a channel for the rapid communication of fire from floor to floor.

Seasonal Attractions

Seasonal attractions such as haunted houses are a challenge for inspectors and fire marshals. When Inspector IIs evaluate these structures for safety, they must consider many factors, such as structural supports, combustible finishes, limited lighting, and various types of props. Egress components need to be considered and are detailed elsewhere in this manual. The Inspector II needs to evaluate proper compatibility and structural stability during inspections and approval of these facilities. Familiarity with the combustibility of various materials, the ability to separate and compartmentalize fires that can occur, and establishing performance-based evaluations are critical.

Heavy-Timber Framing

In a heavy-timber design, the basic structural support is provided by a framework of beams and columns that are made of wooden timbers **(Figure 5.35)**. The exterior walls may be non-load-bearing panels with an exterior siding that may be any of several materials. Ordinary corrugated sheet metal is sometimes used for the exterior walls of small storage or industrial buildings with heavy-timber frames.

In both heavy-timber framing and post and beam framing, the interior wood surface is usually left exposed. The exposed wood framing eliminates combustible voids. Architecturally, the exposed surface of the wood creates an attractive, rustic finish.

Figure 5.35 The distinguishing feature of heavy-timber framing is the size of the wooden structural members. *Courtesy of Ed Prendergast.*

Masonry Structural Systems

Masonry can be used to construct bearing walls that provide the basic structural support for a building, or it can be used for non-load-bearing curtain walls or partition walls **(Figure 5.36)**. Masonry exterior walls are used with a variety of framing systems including unprotected steel, protected steel, and wood. Therefore, both fire-resistive and non-fire-resistive buildings may feature masonry walls.

NOTE: Refer to Chapter 4 for an illustration of a load-bearing wall. Curtain walls and partition walls are described in Chapter 6.

Figure 5.36 Masonry is being used as a veneer for this structure. *Courtesy of Ron Moore, McKinney (TX) Fire Department*

Masonry Walls

When a masonry wall is constructed, the masonry units are laid side by side in a horizontal layer known as a **course**. The horizontal courses of brick are laid on top of each other in a vertical layer known as a **wythe** **(Figure 5.37, p. 182)**. The simplest brick wall consists of a wall with a single wythe. Multiple wythes are commonly used to supply the necessary strength and stability in a masonry wall. Other placements are made for appearance or strength:

- A *stretcher course* has bricks placed end-to-end
- A *soldier course* has bricks placed vertically on end
- A ***header course*** has bricks placed with the end facing out

Course — Horizontal layer of individual masonry units.

Wythe — Single vertical row of multiple rows of masonry units in a wall, usually brick.

Header Course — Course of bricks with the ends of the bricks facing outward.

In a plain (nonreinforced) masonry wall, the strength and stability of the wall are derived from the weight of the masonry and horizontal bonding between adjacent wythes and the vertical compressive strength of the masonry units. Different types of corrosion-resistant metal ties can be used to provide horizontal bonding. The metal ties are commonly used when the wall is constructed of brick and concrete block.

An exterior brick wall usually is constructed with a vertical cavity between the exterior wythe and interior wythes. The cavity prevents the seepage of water through the mortar joints to the interior of the building and increases the

Figure 5.37 Parallel wythes of bricks can be bonded using a header course every sixth course. A header course has the ends of the bricks facing outward.

thermal insulating value of the wall. In the case of a cavity wall, the placement of metal ties is important because the use of a brick header course usually is not practical.

Structural Stability

Placing vertical steel rods (similar to those used with reinforced concrete) in a cavity between two adjacent wythes of a brick wall effectively reinforces masonry walls **(Figure 5.38)**. The cavity is then filled with grout (a mixture of portland cement, aggregate, and water).

NOTE: Placing the steel rods in the openings in the individual blocks and filling the opening with grout reinforces concrete block walls in a similar manner.

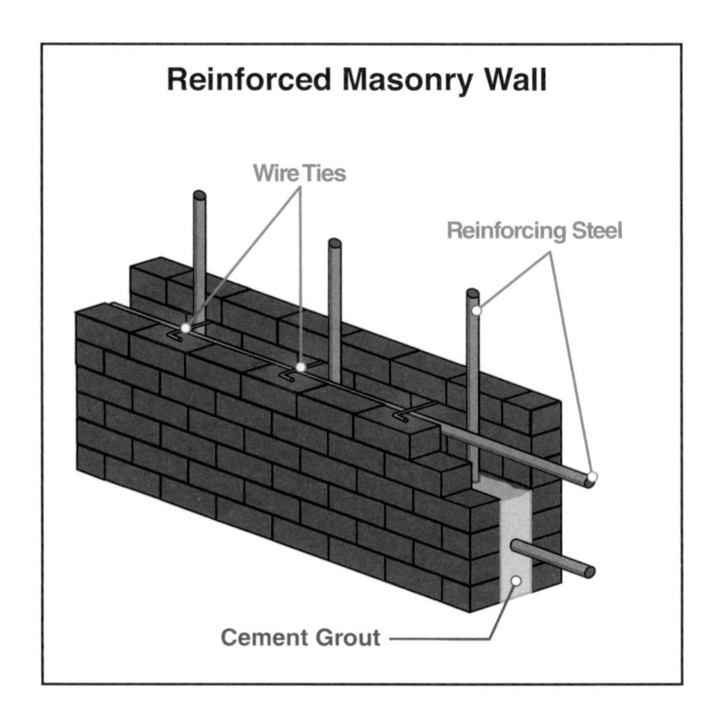

Figure 5.38 Details of reinforcement in a reinforced masonry wall.

As the height of a wall is increased, the thickness of the wall at the base must be increased as the load-bearing area of the wall increases **(Figure 5.39)**. The increasing weight of a load-bearing wall with increased height makes very tall masonry structures impractical or more costly than alternative designs unless the masonry is reinforced with steel.

One method of reinforcing a masonry structure is by extending tension rods through the masonry walls and attaching them to **thrust plates** on the outside. The thrust plates are usually visible on the outside and may indicate that the building has undergone repairs **(Figure 5.40).**

Thrust Plate — Steel plate located on the exterior of a masonry building to which a tension rod is anchored.

Interior Structural Framing

The interior structural framing in masonry buildings can include a combination of unprotected steel, protected steel, and wood. For example, it is not uncommon to have a wood or concrete floor deck supported by steel beams, which in turn are supported by a masonry wall. Joists, beams, or trusses are used to transfer the gravity load of the building contents. Masonry buildings with wood interior framing are classified as ordinary (Type III) or heavy-timber (Type IV) depending on the size of the wooden structural members **(Figure 5.41)**.

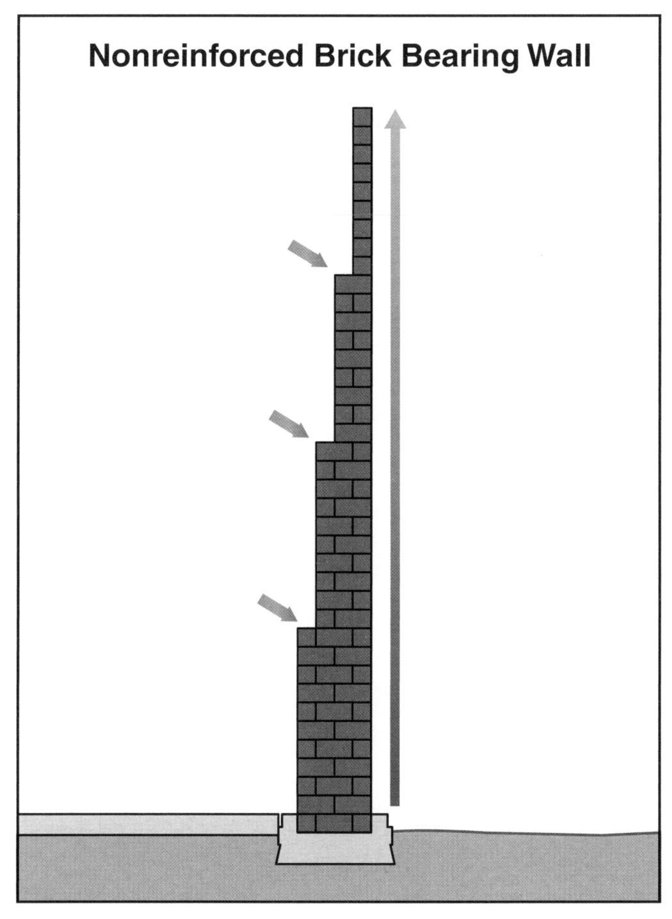

Figure 5.39 In older brick masonry buildings, the wall thickness increases at the base to handle the increasing load and to provide stability.

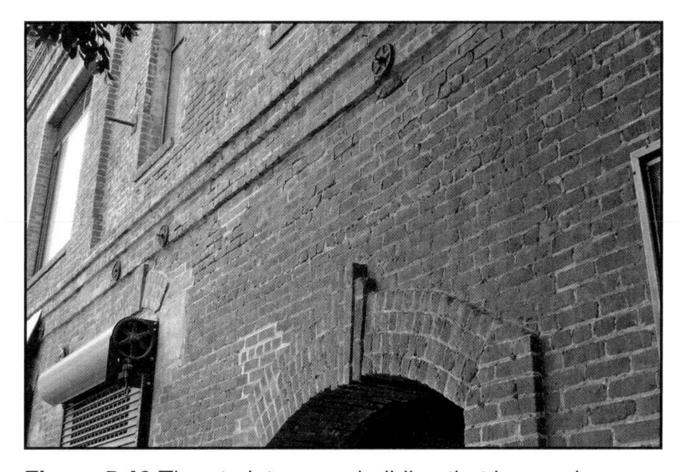

Figure 5.40 Thrust plates on a building that has undergone repairs. *Courtesy of Dave Coombs.*

Figure 5.41 Wood ceiling trusses used in a concrete structure. *Courtesy of Dave Coombs.*

Fire Resistance

The fire resistance of a masonry wall depends on the type of masonry units used and the thickness of the wall. As the thickness of a masonry wall increases, generally the fire resistance increases. Building codes usually permit less clearance (separation) between buildings with masonry or other fire-resistive exterior walls because they are less likely to communicate fire from structure to structure.

Steel Structural Systems

Steel is a primary material used for structural support in the construction of large modern buildings **(Figures 5.42 a and b)**. Steel structural shapes can be used to construct a frame of columns, beams, and girders. Steel also can be used in heavy or lightweight trusses to support roofs and floors. Rigid frames and arches can be constructed from steel. Steel cables or rods can be used to support roofs. Cold-rolled steel studs are being used to construct exterior walls.

Because steel is a strong but very dense material, it is not efficient to use it in the form of solid slabs or panels as is done with other materials such as wood or concrete. Steel in sheet form, however, is used for applications such

Figures 5.42 a and b Steel is used to frame buildings and to form roof trusses. *Both figures Ron Moore, McKinney (TX) Fire Department.*

as floor decking and exterior curtain walls. The exterior envelope of a steel-frame building can consist of concrete, masonry, or glass **(Figure 5.43)**. Steel trusses can be fabricated in a variety of shapes and are frequently used in three-dimensional space frames. Trusses can also be constructed of wood or a combination of steel and wood.

Figure 5.43 A curtain wall in a commercial building. *Courtesy of Donny Howard.*

Steel Trusses

Steel trusses provide a structural member that can carry loads across greater spans more economically than beams can. Two commonly encountered applications of the basic steel truss are the open web joist and the joist girder **(Figures 5.44 a and b)**.

Open web joists are mass-produced and are available in varying depths and spans. The top and bottom chords of an open web joist can be made from two angles, two bars, or a T-shaped member. The diagonal members can be made from flat bars welded to the top and bottom chords, or they can be a continuous round bar bent back and forth and welded to the chords.

When round bars are used for the diagonal members, the open web truss is known as a bar joist. Bar joists are often used in closely spaced configurations for the support of floors and roof decks.

Joist girders are heavy steel trusses that are used to take the place of steel beams as part of the primary structural frame. Steel joist girders are open web, primary load-carrying members used for the support of floors and roofs.

Figure 5.44 a Open web bar joist. *Courtesy of Dave Coombs.*

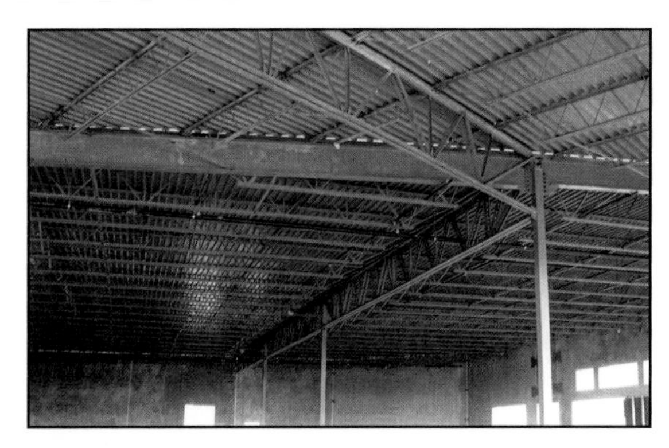

Figure 5.44 b Open web bar Joist. *Courtesy of McKinney (TX) Fire Department.*

Figure 5.44 c Joist girder. *Courtesy of Ed Prendergast.*

Concrete Structural Systems

Roof Deck — Bottom components of the roof assembly that support the roof covering; may be constructed of plywood, wood studs (2 x 4 inches or larger), lath strips, or other materials.

Concrete is used most commonly to form foundation stem walls, floor slabs, driveways, and walks. It may also be used for interior and exterior walls, **roof decks**, and the floors of upper stories as well as stairways and elevated walkways. Concrete can also be used to create roof tiles and decorative interior finish. Concrete may be precast or cast-in-place and may also be the construction system for entire structures.

Cast-in-Place Concrete

Cast-in-place concrete permits the designer to cast the concrete in a wider variety of shapes **(Figure 5.45)**. Cast-in-place concrete does not develop its design strength until after it has been placed where it will be cured and used.

Precast Concrete

Precast concrete is placed in forms and cured at a precasting plant away from the job site. Precast structural shapes, including slabs, wall panels, and columns, are transported to the job site and hoisted into position **(Figure 5.46)**.

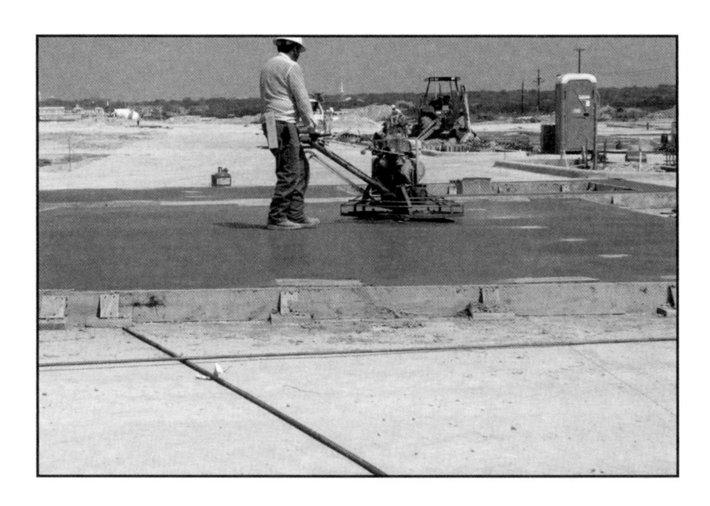

Figure 5.45 Cast in place concrete is smoothed to make sure it fills all the spaces in the formwork. *Courtesy of McKinney (TX) Fire Department.*

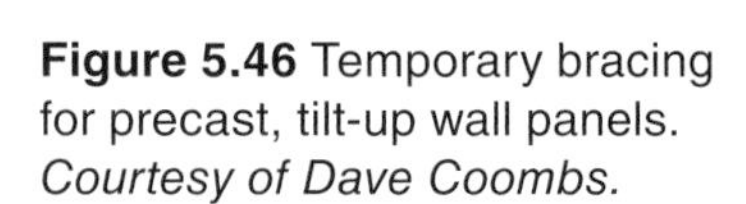

Figure 5.46 Temporary bracing for precast, tilt-up wall panels. *Courtesy of Dave Coombs.*

Precast concrete buildings can be built using whole precast modular units; however, precast parts assembled into a framework for the building are more common. Therefore, from a construction standpoint, precast concrete structures have more in common with steel-framed buildings than with cast-in-place concrete buildings.

Ordinary Reinforcing

With ordinary reinforcing, steel bars are placed in the formwork and the wet concrete is placed in the formwork around the bars. The concrete must be properly compacted as it is placed in the forms to completely surround the rebars (reinforcing bars) and to avoid cavities in the hardened concrete **(Figure 5.47)**. When the concrete has hardened, it adheres to the reinforcing bars because of textural shaping on the surface of the bars.

Pretensioning and Posttensioning

When a concrete beam or floor slab supports a load, more efficient use of concrete is made by either pretensioning or posttensioning the concrete. Both processes use the same basic materials.

Pretensioned Reinforcement — Concrete reinforcement method. Steel strands are stretched, producing a tensile force in the steel. Concrete is then placed around the steel strands and allowed to harden.

Pretensioned reinforcement uses steel strands stretched between anchors producing a tensile force in the steel. Concrete is then placed around the steel strands and allowed to harden. After the concrete has hardened sufficiently, the force applied to the steel strands is released and the strands exert a compressive force in the concrete **(Figure 5.48)**.

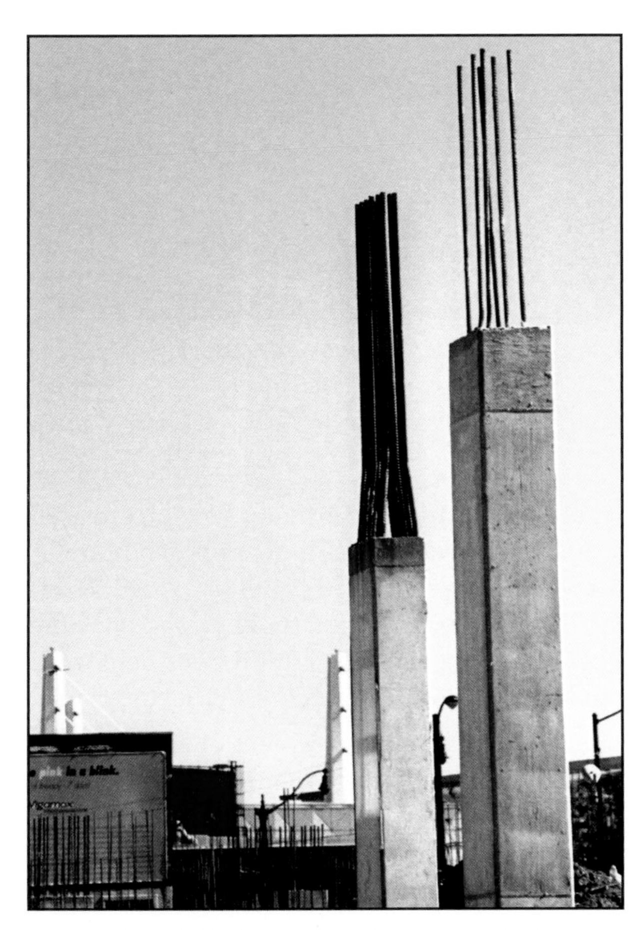

Figure 5.47 Vertical reinforcing bars in a concrete column. *Courtesy of Ed Prendergast.*

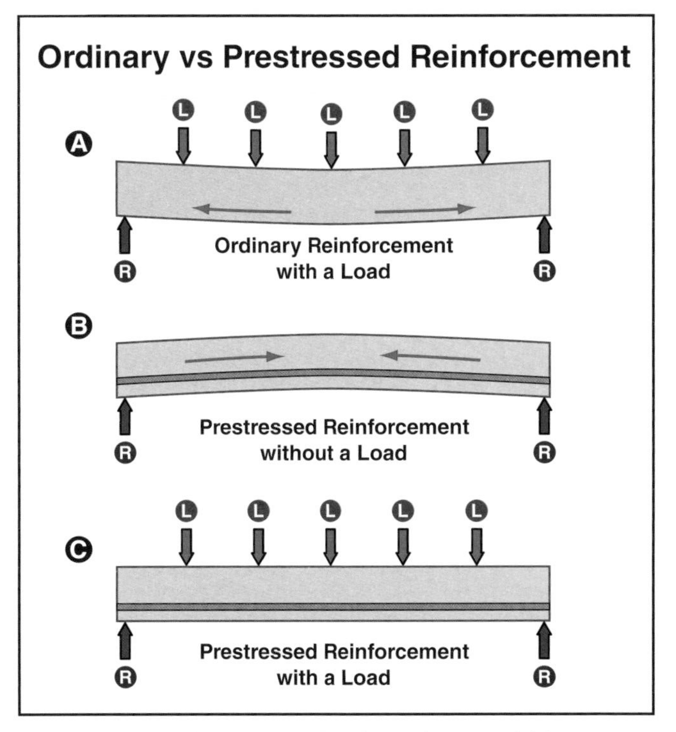

Figure 5.48 “A” illustrates a load on a beam, which causes downward deflection and tensile stresses in the lower part of the beam. “B” shows how prestressing the beam with reinforcing steel causes an upward deformation and creates compression forces in the beam. “C” illustrates that when a load is applied to the beam, the tensile and compressive forces equalize and the slight upward deflection disappears.

Post-Tensioned Reinforcement — Concrete reinforcement method. Reinforcing steel strands in the concrete are tensioned after the concrete has hardened.

Post-tensioned reinforcement uses reinforcing steel that is not tensioned until after the concrete has hardened. The reinforcing strands are placed in formwork and covered with grease or a plastic tubing to prevent binding with the concrete **(Figure 5.49)**. When the concrete has hardened, the strands are anchored against one end of the concrete member and a jack is positioned at the other end. The jack is used to apply a large tensile force to the steel that stretches the steel and results in a compressive force in the concrete. The pulled end of the reinforcing strand is anchored to the concrete and the reinforcing cables are trimmed to the edge of the concrete and grouted.

It is difficult to determine which type of reinforcing has been used in a concrete frame building. It may also be impossible to distinguish between cast-in-place concrete and precast concrete after a building is completed. In addition, some systems, such as stucco and EIFS, may appear to be concrete.

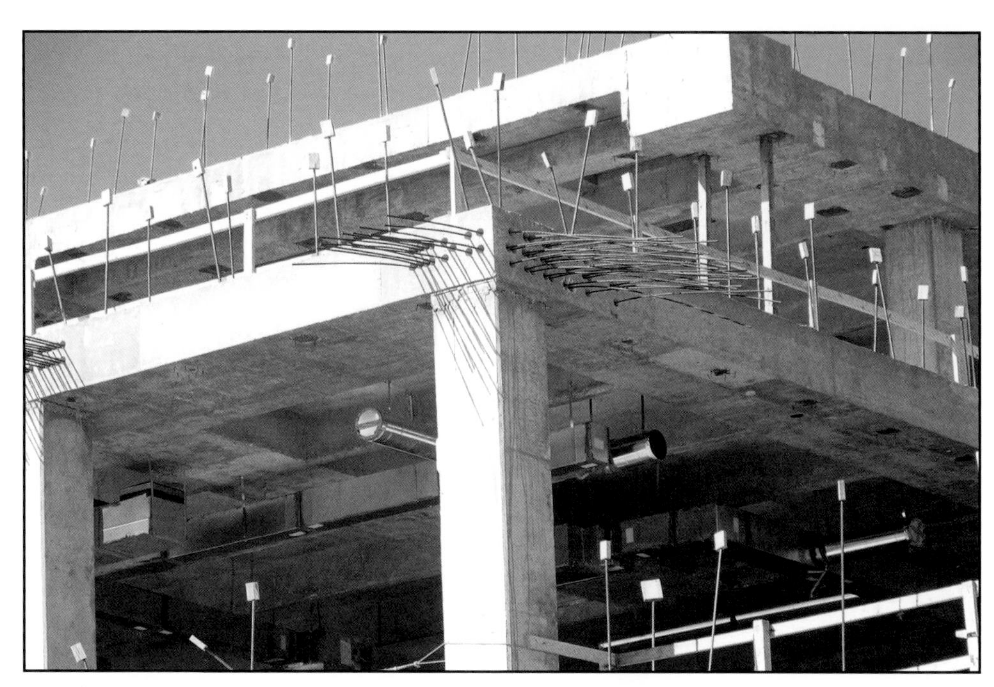

Figure 5.49 Posttensioning strands in a concrete slab. *Courtesy of McKinney (TX) Fire Department.*

Chapter Summary

The inspector must have a clear understanding of the materials that are used to construct a building and the common structural components of buildings. This knowledge is necessary to evaluate the structure's ability to resist the effects of fire. Combined with knowledge of construction types presented in the previous chapter, the inspector can readily categorize a building and determine the necessary code enforcement approaches and applications that apply to it.

The type of construction determines the needed protection systems and structural separations that some occupancy classifications require. The building materials and the structural systems are components of the specific building type and will add to or subtract from the fire-resistive nature of the building. The inspector must have a strong working knowledge of these building elements and how they reduce risk in case a fire does occur.

Review Questions

1. What are the materials most commonly used for construction?
3. What are the types of wood structural members?
4. Why is it important to know where asbestos can be found?
5. What are the different types of masonry?
6. What is concrete used for in building construction?
7. What are some of the characteristics of steel?
8. What are the most commonly encountered types of glazing (glass)?
9. In a structure, where are you most likely to find gypsum board?
10. Why is it important for an inspector to recognize the use of plastics in building construction?

1. What are the forces on a building as a result of gravity called?
2. What are some of the advantages of membrance structures?
3. What are the different types of frame structural systems?
4. Where are you most likely to find wood structural systems as interior and exterior framing?
5. What is one method of reinforcing a masonry structure?
6. In what type of structure are you most likely to find steel structural systems?
7. What can concrete structural systems be used for in building construction?

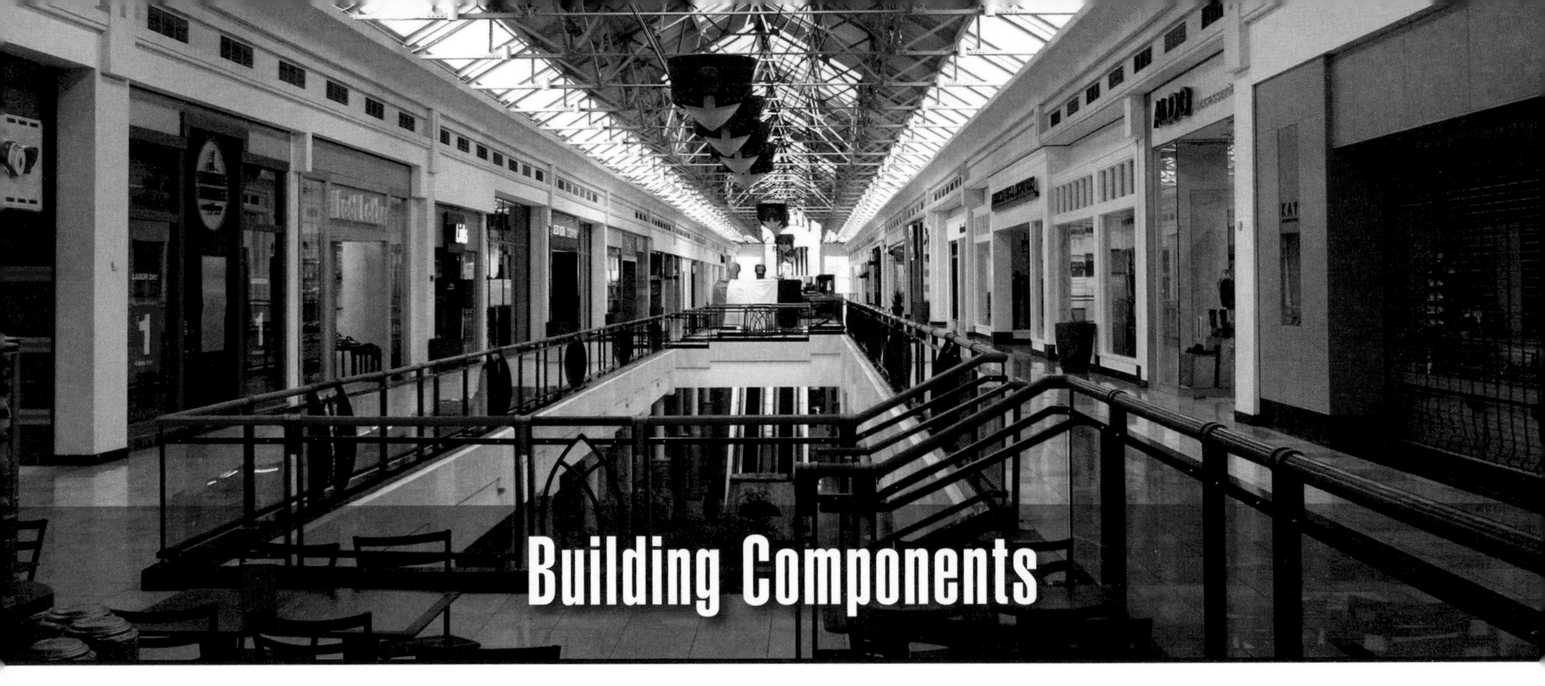

Chapter Contents

Case History 193
Walls 194
Fire Walls 194
Party Walls 195
Fire Partitions and Fire Barriers 196
Enclosure and Shaft Walls 196
Curtain Walls 197
Roofs 199
Floors 200
Construction Materials 200
Supports and Coverings 200
Penetrations and Openings 202
Ceilings 203
Stairs 203
Doors 204
Types 204
Styles and Construction Materials 206
Fire Doors 207
Classifications 209
Frames and Hardware 210
Construction and Operational Types 211
Windows 213
Components 213
Types 214
Security 215
Interior Finishes 216
Building Services 217
Elevator Hoistways and Doors 217
Utility Chases and Vertical Shafts 218
Evaluating Fire Walls 220
Evaluating Rooftop Photovoltaic Systems 221
Firefighter Safety 221
Access Pathways and Smoke Ventilation 221
Evaluating Structural Stability 222
Evaluating Interior Components 222
Floor Finishes 222
Ceilings 223
Stairs 223
Evaluating Fire Doors 223
Evaluating Interior Finishes 225
Flame Spread Ratings 225
Smoke-Developed Index 226
Fire-Retardant Coatings 228
Evaluating Building Services 228
Heating, Ventilating, and Air Conditioning Systems 229
Refrigeration Systems 231
Electrical Systems 232
Elevator Hoistways and Doors 234
Utility Chases and Vertical Shafts 235
Chapter Summary 236
Review Questions 237

Key Terms

Compartmentation System 220
Curtain Wall 197
Fire Wall..194
Flame Spread Rating 225
Interior Finish 216
Interstitial Space 223
Opening Protective.................................194
Parapet 194
Return-Air Plenum 223
Smoke-Developed Index.........................225
Steiner Tunnel Test 225
Toxicity 226

NFPA® Job Performance Requirements

This chapter provides information that addresses the following job performance requirements of NFPA® 1031, *Standard for Professional Qualifications for Fire Inspector and Plan Examiner* (2014).

Fire Inspector I	Fire Inspector II
4.3.1	5.3.12
4.3.4	5.4.6

Building Components

Learning Objectives

After reading this chapter, students will be able to:

Inspector I

1. Describe different types of walls found in structures and the fire hazards they present. (4.3.1, 4.3.4)
2. Identify roof types and coverings and the fire hazards they present. (4.3.1, 4.3.4)
3. Identify floor characteristics. (4.3.4)
4. Describe ceiling characteristics. (4.3.4)
5. Describe stair characteristics important to inspectors. (4.3.4)
6. Describe the fire risks posed by how doors operate. (4.3.1, 4.3.4)
7. Differentiate among fire doors based on construction and operation. (4.3.4)
8. Describe different types of windows and how they operate. (4.3.4)
9. Describe how interior finishes can contribute to fire spread. (4.3.4)
10. Explain the fire and life safety aspects of building services. (4.3.4)

Inspector II

1. Describe the characteristics of fire walls. (5.4.6)
2. Describe the hazards solar panels pose for firefighters. (5.4.6)
3. Identify the standardized testing method currently accepted by building codes. (5.4.6)
4. Describe the characteristics of interior components to be evaluated during fire inspections. (5.4.6)
5. Explain the inconsistencies in fire door classifications. (5.4.6)
6. Identify the testing methods used to evaluate interior finishes. (5.4.6)
7. Describe building service characteristics that require inspector evaluations. (5.3.12, 5.4.6)

Chapter 6
Building Components

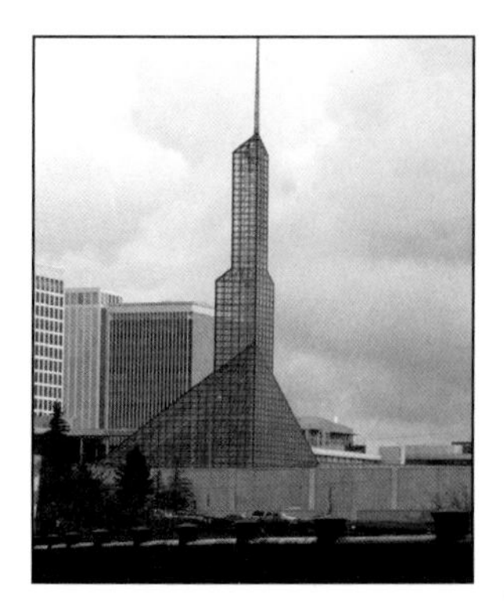

Case History

Cincinnati, Ohio, March 25, 2015: As 30-year veteran firefighter Daryl Gordon searched for trapped residents in an apartment fire, he opened what appeared to be an apartment door. It was actually a nonfunctioning elevator door. Gordon stepped into an open elevator shaft and fell several stories. He died from his injuries.

City building inspection reports describe a litany of problems at the five-story structure over the past decade, from broken pipes and windows to animals gnawing their way into bathrooms. Apartment residents there said the building's only elevator was unreliable and many feared using it.

One national elevator safety expert suggested that something may have been wrong with the elevator door's locking system, allowing it to be opened even though the elevator car was stopped on a different floor.

Building codes define the design and construction of each building component and its relationship to the occupancy classification. The presence or lack of fire protection systems, including automatic sprinklers, standpipes, special-agent systems, and fire detection and alarm systems, also affect requirements for construction components.

NOTE: Chapters 12-14 address fire protection systems.

In addition to a building's structural components, an inspector must be familiar with other components that affect the fire and life safety of the occupants, including the following:

- Walls
- Roofs
- Floors
- Ceilings
- Stairs
- Doors
- Fire doors
- Windows
- Interior finishes
- Building services

This chapter describes the identification of these components, their fire-resistive qualities, and hazards.

Walls

Within a building, interior walls subdivide areas for such purposes as security, privacy, and separation of occupancies. Walls can also be used for fire protection purposes with fire walls by enclosing stairwells and elevator hoistways.

All types of walls affect the development of fire in a building. Walls with flammable finishes will contribute to the spread of fire. Fire-rated partition walls compartmentalize the space and act as barriers to fire spread. Therefore, all building codes address the design of wall assemblies, whether they are load-bearing or not. Inspectors will encounter a number of types of walls while performing plans review and field inspections:

- Fire walls
- Party walls
- Fire partitions and fire barriers
- Enclosure and shaft walls
- Curtain walls

Fire Wall — Wall with a specified degree of fire resistance that is designed to prevent the spread of fire within a structure or between adjacent structures.

Fire Walls

A **fire wall** is a fire-resistance-rated wall with protected openings that restricts the spread of fire. A fire wall extends continuously from the foundation to or through the roof. Under fire conditions, a fire wall must have sufficient structural stability to allow the collapse of construction on either side without the wall itself collapsing.

Construction Types

Fire walls are typically constructed of masonry, although other fire-resistive materials like concrete or gypsum board can be used in a fire-rated assembly. Fire walls can be constructed with fire-resistance ratings of 2 or more hours **(Figure 6.1)**. The rating used depends on the occupancies being separated and the reason for the separation.

Parapet — (1) Portion of the exterior walls of a building that extends above the roof. A low wall at the edge of a roof. (2) Any required fire walls surrounding or dividing a roof or surrounding roof openings such as light/ventilation shafts.

Building codes typically allow a reduction in a fire wall's fire-resistance rating if the building is equipped with an automatic sprinkler system. A complete automatic sprinkler system will also allow building areas to be increased and may reduce or eliminate the need for fire walls.

Fire walls may extend beyond walls and roofs to prevent the radiant heat of flames from igniting adjacent surfaces **(Figure 6.2)**. This extension is accomplished by continuing the fire wall through the roof in the form of a **parapet**. Building codes determine the parapet height above a combustible roof.

NOTE: A building code may require a parapet to provide a barrier to the communication of fire between closely spaced buildings.

Opening Protective — A device installed over an opening or within a wall assembly to protect it against smoke, flame, and heated gases.

Openings

All openings in fire walls must be protected by approved **opening protectives**, such as fire shutters or fire-rated doors. The openings must have a fire-protection rating consistent with the fire resistance rating of the wall.

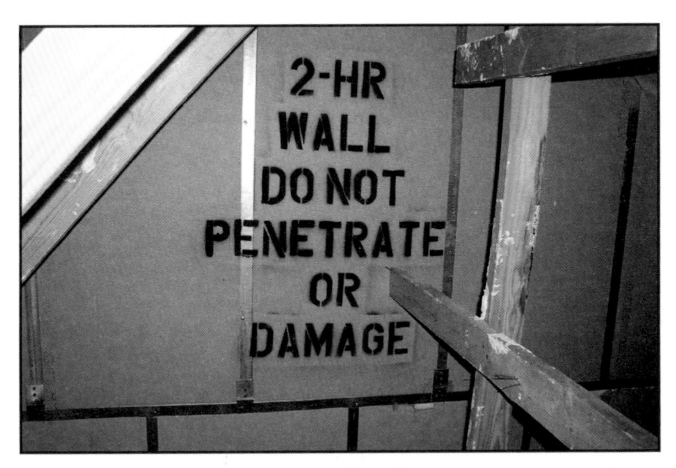

Figure 6.1 A fire wall's rating depends on the occupancies and reasons for separation. *Courtesy of McKinney (TX) Fire Department.*

Figure 6.2 A fire wall with a parapet is designed to provide extra protection against fire extension.

When ducts for heating, ventilating, and air conditioning (HVAC) systems penetrate fire walls that have a fire-resistance rating of 2 hours or greater, the ducts must be equipped with fire and smoke dampers within the duct. Rated activating devices and smoke detectors are installed in the ducts, causing the dampers to operate.

NOTE: Refer to Chapter 5 for additional descriptions of fire and smoke dampers.

Party Walls

A party wall is a wall that is located on a lot line between two buildings and is common to both buildings **(Figure 6.3)**. Party walls are almost always load-bearing walls.

Fire inspectors should be aware that it is against code regulations to breach a party wall for any reason. Such openings make it possible for fires to communicate from building to building. Attic areas and cocklofts are common locations for breaches in party walls. An inspector who discovers a breach in a party wall must take steps to have the building owner correct the condition. In addition, the fire department operations section should be informed of the existence of these breaches.

Figure 6.3 These row houses are separated by party walls. *Courtesy of Ron Jeffers.*

Fire Partitions and Fire Barriers

Fire partitions and fire barriers are interior walls that subdivide a floor or area of a building but do not qualify as fire walls. They are frequently required for such applications as corridor walls and occupancy separations. The material chosen depends on the required fire resistance and the construction type of the building.

Fire partitions and fire barriers are constructed from a variety of materials, including:

- Gypsum wallboard
- Concrete block
- Lath and plaster
- Combinations of these and other materials

Fire partitions may not extend continuously through a building. These walls are erected from a floor to the underside of the floor or roof above or to the bottom of a fire-rated floor/ceiling assembly. Fire partitions are typically 1-hour-rated structures. For example, the partition walls separating adjacent units in an apartment building may be required to have a 1-hour fire-resistance rating **(Figure 6.4)**.

The construction of fire barriers is identical to fire partitions, except that fire barriers are required to terminate at the floor or roof deck. Fire barriers have greater fire-resistance requirements and can be 1- to 4-hour-rated structures. Fire barriers are used for the following:

- Area separation
- Hazard protection
- Stairwells
- Shafts
- Areas of refuge in health care facilities

For example, fire-rated corridor walls in apartment buildings are an important means of protecting an exit corridor from a fire.

Figure 6.4 The units in this apartment building are separated by fire partitions.

Enclosure and Shaft Walls

Enclosure walls are designed to block the vertical spread of fire through a building. They are placed around such vertical openings as stairwells, elevator shafts, and pipe chases that extend from floor-to-floor in a building. Enclosure walls are required to have a fire-resistance rating and are usually non-load-bearing. In stairwells, enclosure walls are designed to protect a means of egress.

NOTE: See Chapter 7, Means of Egress, for more information about means of egress requirements.

Most common construction materials can be used for enclosure walls, including gypsum board with steel or wood studs, lath and plaster, or concrete

Figure 6.5 Gypsum board is a common material used in enclosure walls.

block **(Figure 6.5)**. Hollow, clay-tile enclosure walls exist in old, fire-resistive structures. Load-bearing masonry stair enclosures are found in some old buildings.

Fire-rated glazing (glass) can also be used in conjunction with stair enclosures. As in the case of fire barriers, the use of fire-rated glazing provides a fire barrier while permitting observation of the stair enclosure, which can enhance security.

Curtain Walls

When a building is constructed using a structural frame, the exterior wall functions only to enclose the building and is known as a **curtain wall** (or sometimes as cladding). The design function of a curtain wall is to separate the interior environment from the exterior environment. A curtain wall tends to conceal the structural details of a building, making it difficult to accurately identify the structural system by observation alone.

Curtain Wall — Nonload-bearing exterior wall attached to the outside of a building.

Installing a curtain wall commonly results in a gap between the edge of the floor and the curtain wall **(Figure 6.6, p. 198)**. This gap, which may be several inches (millimeters) wide, provides a path for fire to communicate up the inside of the curtain wall. Suitable firestopping material is needed to maintain the continuity of the floor as a fire-resistive barrier.

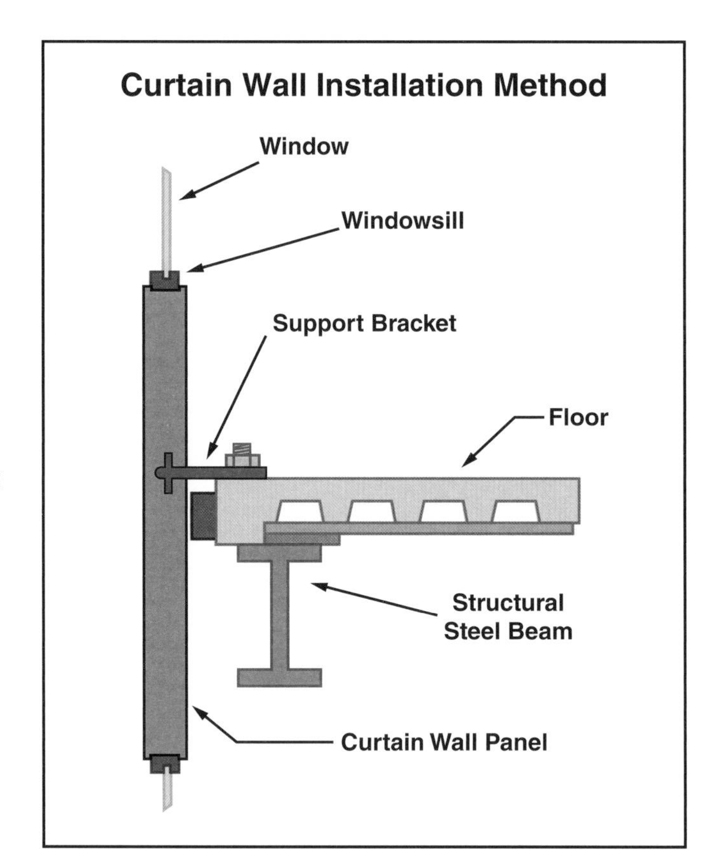

Figure 6.6 This cross section of a curtain wall shows the small gap created between a curtain wall and structural members of a building.

Figure 6.7 Glass curtain walls help protect against weather but may have little fire resistance.

Some curtain wall assemblies, such as those made of aluminum and glass, have little fire resistance **(Figure 6.7)**. However, building codes may require that exterior walls, including curtain walls, have some degree of fire resistance to reduce the communication of fire between buildings. The required fire resistance depends on the separation distance between buildings and the building's occupancy classification.

Roofs

Roof coverings provide the water-resistant barrier for the roof system. The type of roof covering used depends on the form of the roof structure, slope of the roof, climate, and appearance desired. Common coverings include wood shingles, asphalt shingles, clay tiles, and metal **(Figures 6.8 a-d)**. Their resistance to fire varies.

Wood roof shingles and shakes can be easily ignited if burning embers land on them, and wood shingles have contributed to fires involving entire neighborhoods. Wood shingles and shakes can be pressure-impregnated with a fire-retardant chemical to reduce their combustibility and meet building code requirements. Experience has indicated that the treatment remains effective after exposure to the elements. Fire-retardant shingles and shakes are shipped to the job site with a paper label identifying them. Once they are in place, however, identification can be difficult.

Figure 6.8 a Wood shingles present a considerable fire danger because they communicate fire very easily from rooftop to rooftop.

Figure 6.8 b Asphalt shingles are a very popular roof covering because they provide good resistance to the weather and protection from burning embers from external fires.

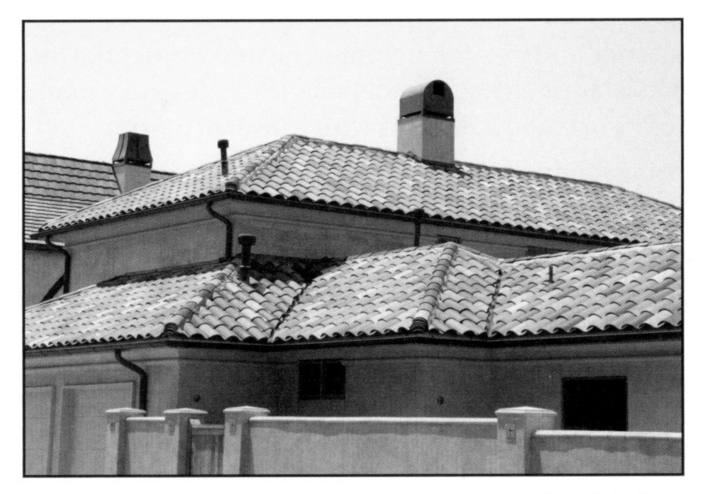

Figure 6.8 c Clay, slate, or cement tiles like the Spanish tiles shown add a greater weight load to roof components than asphalt but also prevent flying embers from nearby fires from igniting a structure.

Figure 6.8 d Metal roof coverings are noncombustible and very hardy; however, under extreme fire conditions, metal roofs will weaken and melt.

Asphalt shingles are fundamentally combustible. They tend to drip and run under fire conditions and produce a characteristically heavy black smoke. Asphalt shingles used for roofs are typically produced with a grit surface that reduces their ease of ignition and permits their use under the provisions of building codes.

Clay, slate, and cement-based tiles are noncombustible and produce fire-resistant roof coverings that have excellent resistance to flying embers. Flying embers, however, can be blown under tiles, such as Spanish tiles that do not lie flat, and could ignite the roof deck. Metal roof coverings are noncombustible and will protect the structure from flying embers.

Roof coverings also have installation requirements that dictate the use of appropriate fastening methods as well as the use of underlayment materials. Other important issues include the insulation of the roof structure and the use of many different types of materials from plastics to polyurethane foam. The codes also provide information about rework and replacement. For example, no more than two layers of roofing material can be installed on a roof.

Floors

Floors are designed to provide a safe structure upon which people can stand and work, and to provide a support system for furniture and machinery. Generally, some type of floor covering is placed over a subfloor, which is designed to support the building's live load. The inspector should be familiar with the materials used in floor construction, floor support systems, and the effects of penetrations and openings in floor assemblies.

Construction Materials

Construction materials for floors vary depending on the design and use of the building and the weight of the equipment or building materials that the floor will support **(Figures 6.9 a-d)**:

- **Concrete** — Commonly used in fire-resistive construction because it is inherently fire resistive. The fire resistance of the floor depends on the fireproofing of the supporting steel.
- **Clay tiles** — Used in interior and exterior commercial and residential applications. The flame resistance of clay is excellent. Clay tiles are usually attached to rough concrete or wood subfloors with a layer of mortar.
- **Bricks** — Used for flooring for many years. Bricks can be laid directly onto the soil or secured to a subfloor with mortar. They may be full height or thin brick veneer.
- **Wood** — Consists of a subfloor made of planks or panel products covered with a hardwood floor that is finished and sealed. The inspector should consider that the wood floor's finish can be questionable regarding flame resistance.

Supports and Coverings

Floors that are installed over a crawl space or above grade level must have some form of support beneath them, such as masonry, wood, or steel **(Figure 6.10)**. In old buildings, many types of materials or structures support the floors. The exact type of floor system in a building may not be readily apparent to an inspector.

Figure 6.9 a Concrete flooring is fire-resistant but may show the effects of spalling under fire conditions.

Figure 6.9 b This terrazzo floor comprised of marble and stone chips held together with an epoxy resin is an excellent, flame-resistant floor covering.

Figure 6.9 c A masonry or brick floor provides excellent flame resistance. It may be laid on soil subfloor or mortared directly to a concrete slab.

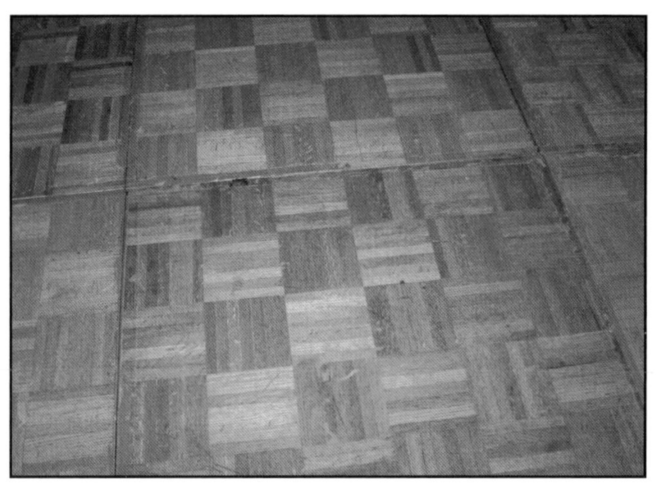

Figure 6.9 d Wood floors are extremely popular. They are manufactured in a wide variety of styles and types.

A structural floor system usually is covered with a finish material for appearance and sound deadening. A variety of finishing materials may cover floor construction:

- Carpet
- Paint
- Laminated wood
- Vinyl
- Ceramic tile

At one time, the flammability of floor coverings was considered insignificant in overall building fire safety. Because the heat from a fire rises, it was assumed that anything located at floor level would be at the coolest part of the room. However, fires that spread over the surface of some thick carpeting can increase the fire hazard within a structure.

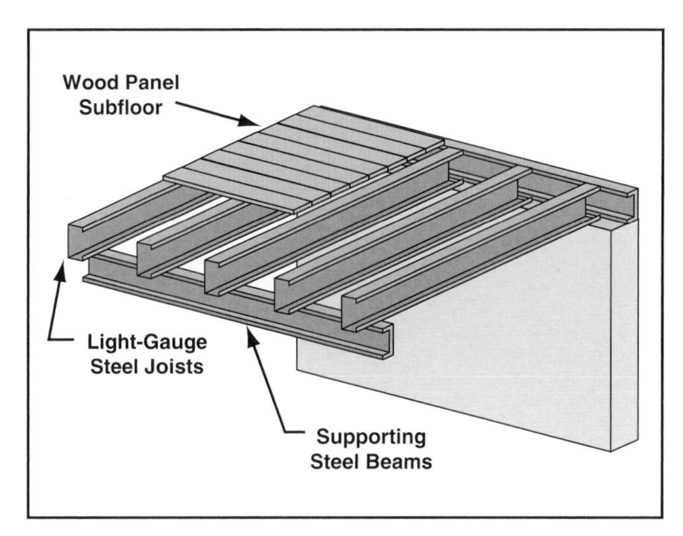

Figure 6.10 Light-gauge steel joist system.

Penetrations and Openings

The fire resistance of a floor assembly plays a primary role in preventing the vertical spread of fire and products of combustion through a building. Building codes, therefore, contain requirements that vertical penetrations of floors for such purposes as elevator shafts, stairwells, and service shafts be provided with a fire-resistive enclosure. However, it is frequently desirable from an architectural or operational perspective to have communicating openings between floors. Examples of floor penetrations and openings include the following:

- Atriums
- Convenience stairs
- Escalator openings
- Covered malls

Ideally, all vertical openings should be enclosed. In many cases, an architect can specify that doors be held open by automatic devices that release upon activation of a smoke detector. Fire-rated glazing can also be used to enclose a vertical opening. In open areas such as atriums and malls, building codes require automatic sprinkler and smoke management systems **(Figure 6.11)**.

An unprotected opening is one without any provisions to stop the passage of fire and the products of combustion. Small vertical openings can allow fire and smoke to spread as efficiently as a chimney. Vertical openings that are commonly left unprotected include the annular (or ring-shaped) space around pipes and cables where they pass through floor and ceiling decks **(Figure 6.12)**.

Figure 6.11 Open areas like atriums in shopping malls will need to be sprinklered.

Figure 6.12 Vertical openings can allow fire and smoke to spread.

Ceilings

Ceilings as a distinct building component usually do not play a structural role. They can be installed simply to provide an attractive interior finish. However, a ceiling frequently has a functional role in the design of a building. Ceilings can be designed to control the diffusion of light, control the distribution of air in a room, act as a sound barrier, and act as part of a fire-resistive assembly to separate one floor from another. Ceiling materials, such as gypsum board or mineral tiles, are often a required fire-resistance component for the floor/ceiling system **(Figure 6.13)**.

Figure 6.13 Lay-in ceiling tiles are often made of gypsum board. The tiles shown also feature a recessed sprinkler as part of the building's automatic sprinkler system.

Stairs

Stairs are the basic architectural feature of buildings that provide access to different levels of a structure or serve as a means of egress. Inspectors must be able to recognize the fire and life safety requirements for stairs that are designated as part of a means of egress.

All types of stairs have similar components **(Figure 6.14)**. Building codes specify the acceptable dimensions for stairs, typically expressed as minimum and maximum rise and tread measurements, also known as the stair's

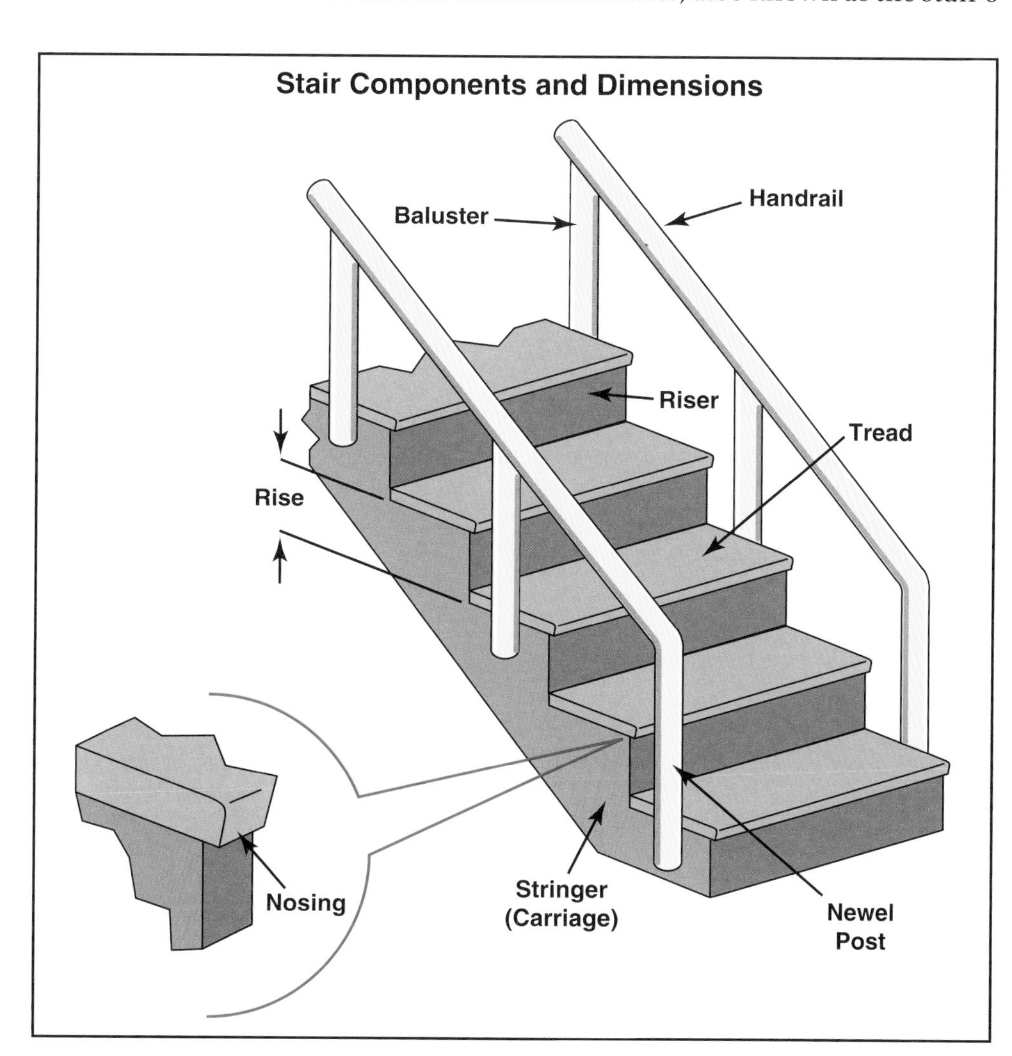

Figure 6.14 This illustration shows the components involved in staircase construction.

rise and run. The design or layout of a set of stairs may take any of several different forms.

Regardless of the type of stair, the building and fire codes specify requirements for construction and use of stairs. Several stair configurations, including spiral stairs, winders (circular stairs), and alternating tread stairs, are allowed by the codes under specific conditions.

NOTE: See Chapter 7, Means of Egress, for further descriptions of stairs and other components of means of egress.

Doors

Inspectors will encounter a variety of doors when evaluating building construction components. The inspector must be able to recognize the basic operational types of doors, design styles, and construction materials used in door construction.

Types

Doors may be classified by the way they operate. Generally, five types of doors are used in modern construction **(Figure 6.15)**:

- Swinging
- Sliding

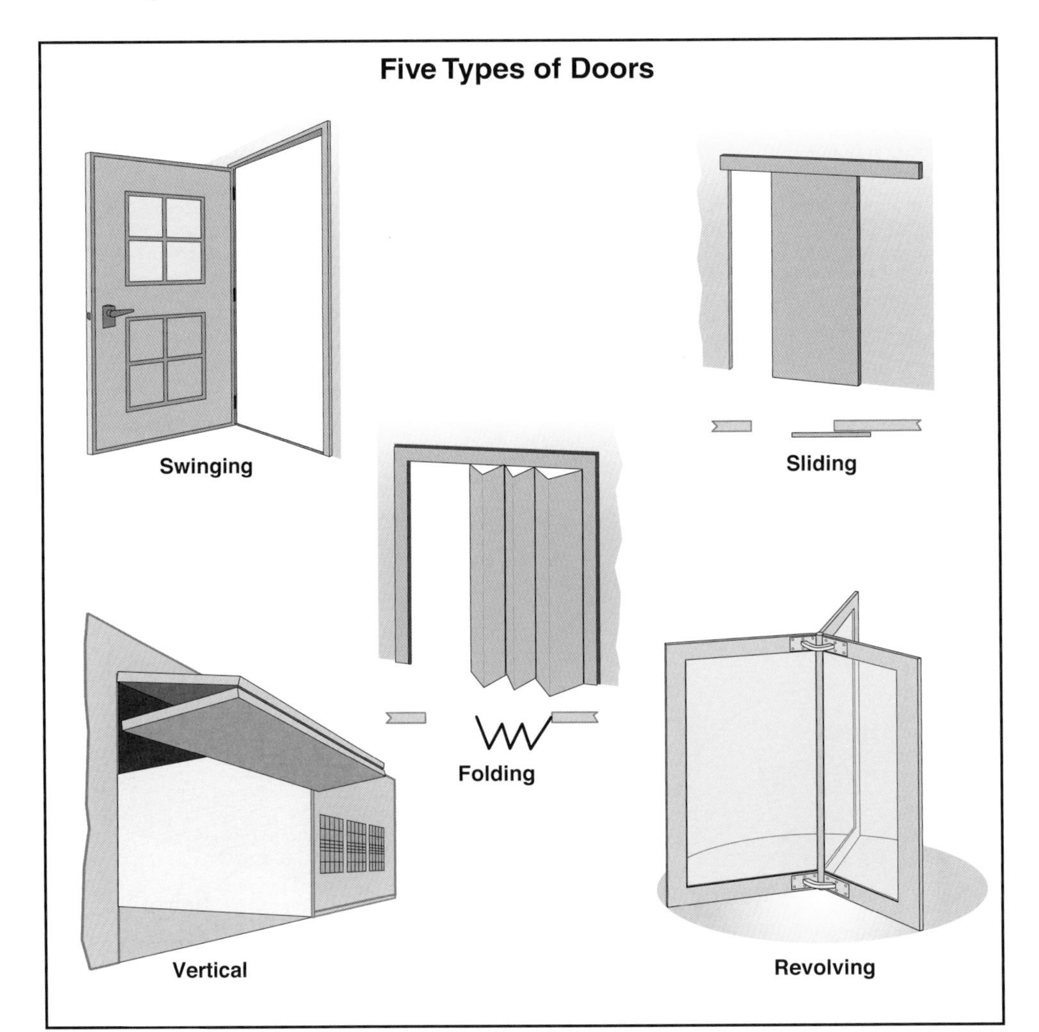

Figure 6.15 These doors are used in modern construction and are classified according to how they operate.

- Folding
- Vertical
- Revolving

Swinging Doors

A swinging door rotates around a vertical axis by means of hinges secured to the side jambs of the doorway framing. It may also operate on pivot posts supported at the top and bottom. A swinging door can be either single or double leaf. It may also be single acting (swinging in one direction) or double acting (swinging in two directions). Generally, exit doors in a means of egress must be swinging doors, although other types of doors can be used under certain conditions.

Sliding Doors

A sliding door is suspended from an overhead track and may use steel or nylon rollers. Floor guides or tracks prevent the door from swinging laterally. A sliding door can be designed as surface sliding, pocket sliding, or bypass sliding.

A sliding door's main advantage is that it eliminates a door swing that might interfere with the use of interior space. For example, a pocket sliding door slides into the wall assembly and is frequently used within residential units because it is out of sight when open. Sliding doors are also used for elevators, power-operated doors in storefront entrances, and fire doors to protect openings that are not a part of the means of egress.

Folding Doors

A folding door is hung from an overhead track with rollers or glides similar to those used by a sliding door. A folding door can be either bifolding or multifolding. Folding doors are used in places of assembly to divide large conference areas into smaller rooms, in residential occupancies, and as horizontal fire doors. Horizontal fire door assemblies must meet specific requirements and be tested and listed for use in a means of egress.

Vertical Doors

An overhead door opens in a vertical plane. These doors are often used in industrial occupancies as loading dock doors, garage doors, freight elevator doors, or fire doors protecting openings that are not part of the required means of egress. A vertical operating door can be a single leaf that is raised in vertical guides along the edge of the doorway, or it can consist of two or more horizontal panels. Many vertical rolling doors consist of interlocking metal slats.

A door that operates vertically usually has a counterbalance mechanism — either actual weights or springs — to help overcome the weight of the door. A vertical door can be raised manually, mechanically via chain hoist, or power-operated. A swinging door can be installed in large overhead doors to act as a means of egress when required to meet the maximum travel distance for the space. The swinging door in the vertical rolling door must meet the requirements of the locally adopted code if it is used as a means of egress.

Figure 6.16 Revolving doors must have swinging doors near them.

Revolving Doors

A revolving door is constructed with three or four sections or wings that rotate in a circular frame. A revolving door is designed to reduce heat or cooling losses by minimizing the flow of air through a door opening.

A revolving door prevents the movement of hose or equipment into a building, which can present a problem for firefighters. Furthermore, a crowd of people attempting to flee during an emergency cannot move through a revolving door as quickly as they can through a comparable swinging door. To overcome these restrictions under emergency conditions, the wings of the door must be collapsible to provide an unobstructed opening. Codes also limit the amount of force required to collapse the wings.

Several types of mechanisms hold the wings of revolving doors in place. Older models use simple chain keepers or stretcher bars between the wings. Newer models use spring-loaded, cam-in-groove, or bullet-detent hardware. Most models employ a collapsing mechanism that allows the wings to open to a book-fold position when they are pushed in opposite directions. The use of revolving doors in a means of egress is restricted; a swinging door must be located within 10 feet (3 m) of the revolving door **(Figure 6.16)**.

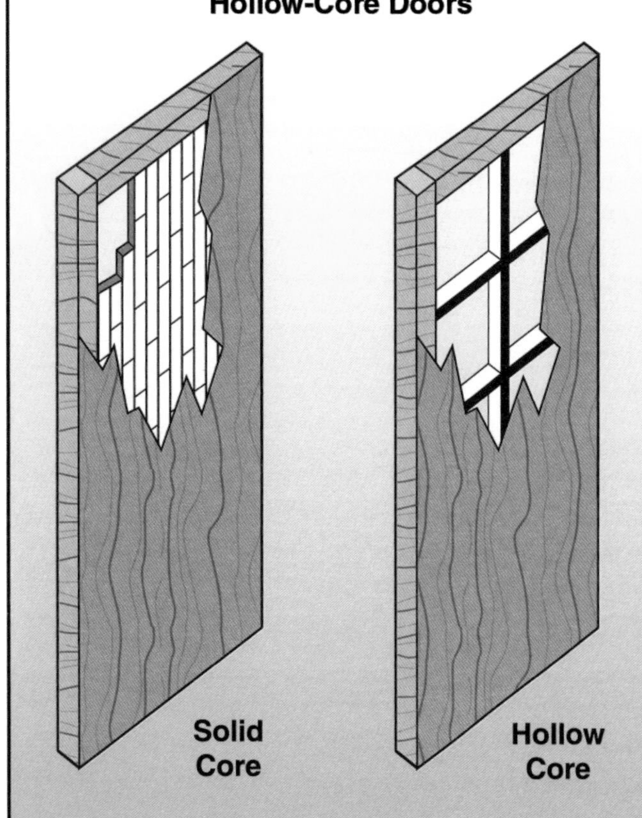

Figure 6.17 Doors may be either solid or hollow core. Hollow-core doors are less expensive than solid-core doors; however, they are not as fire-resistant.

Styles and Construction Materials

Fire-rated door assemblies are manufactured, assembled, and installed in accordance with specific requirements that have been tested and listed by an independent testing laboratory, such as Underwriters Laboratories (UL). Common types of doors are wood panel and flush doors, glass doors, and metal doors.

Wood Panel and Flush Doors

The wood panel door is a common type of swinging door that consists of vertical and horizontal members that frame a rectangular area. Thin panels of wood, glass, or louvers are placed within the framed rectangular area.

A flush door (sometimes referred to as a slab door) consists of flat face panels that are the full height and width of the door. The panels are attached to a solid or hollow core **(Figure 6.17)**.

Glass Doors

Glass doors are used for both exterior and interior applications and are most commonly used in office and mercantile buildings. Glass doors can be either framed or frameless. In a frameless glass door, the door consists of a single sheet of glass to which door hardware, such as handles, are attached. In a framed door, the glass is surrounded by a metal or wood frame with the required door hardware attached to the frame.

Codes require glass doors to be made of tempered glass that resists breakage. Various nonglass components are often used in the framed door to provide additional security.

Metal Doors

A common type of metal door is a hollow metal door made from steel or aluminum. A hollow metal door can be either panel or flush. A flush door consists of smooth sheet metal face panels. Vertical sheet metal ribs within the door separate the face panels of a steel door from one another.

Fire Doors

Fire doors protect the egress pathways leading through fire-rated walls. When properly maintained and operated, they are effective at limiting the spread of fire and total fire damage **(Figure 6.18)**. Fire doors differ from ordinary (nonfire) doors in their construction, their hardware, and the extent to which they may be required to automatically close.

During normal daily operations, many fire doors may be held open or are easy to open to allow for ordinary movement of building occupants. When a fire occurs, fire doors must be closed in order to properly perform their function. Fire doors can be either automatic or self-closing. An automatic door is normally held open and closes when an operating device is activated. In order to activate the closing device and shut a fire door that is normally held open, a detection device must be installed to sense a fire or detect smoke.

NOTE: The inspector will commonly find fire doors propped open or equipped with kick-down door stops (**Figures 6.19 a and b, p. 208**). The inspector must require the removal of these devices and educate the building owner or occupant about the importance of these doors remaining closed. Only approved devices, such as magnetic hold-open devices that release on fire alarm activation, are permitted.

The oldest and simplest detection device is a fusible link that detects the heat from an approaching fire and melts, allowing the door to close. A fusible link has the advantage of being inexpensive, relatively rugged, and easy to maintain. However, because it depends on heat from a fire, a fusible link is slower to operate than detection devices that react to smoke or temperature rise. A significant amount of smoke may flow through a door opening before a fusible link can release a fire door.

In order to close the fire door more quickly than a fusible link, a smoke detector may be used to activate the door release. A smoke detector may cost more and

Figure 6.18 Fire doors must remain closed to serve as a barrier to heat and smoke.

Figures 6.19 a and b Fire doors are frequently found to be propped or wedged open. The inspector must help educate the building occupants that this practice defeats the purpose of the doors. *B Courtesy of Rich Mahaney.*

requires periodic cleaning, but it also permits easy testing of the fire door. As is the case with all smoke detectors, they must be properly positioned with respect to dead-air spaces or ventilation ducts.

Different types of fire door closures consist of the following devices:

- **Fire door closers** — Used for overhead rolling, sliding, or swinging fire doors. They can incorporate a hold-open device or be self-closing **(Figure 6.20)**. Self-closing fire door closers are commonly used for applications such as stairwell doors and doors from hotel rooms to corridors. One commonly used self-closer uses a spring hinge to close the door when released.
- **Electromagnetic door holders** — Can be used with swinging, sliding, or rolling fire doors. They are intended to be used with a suitable door closer and a smoke detector that releases the holder **(Figure 6.21)**. The smoke detectors are sufficiently sensitive so that the doors quickly close under fire conditions. Having the fire doors held open prevents the practice of blocking the doors in an open position.

Figure 6.20 A typical self-closing fire door closer.

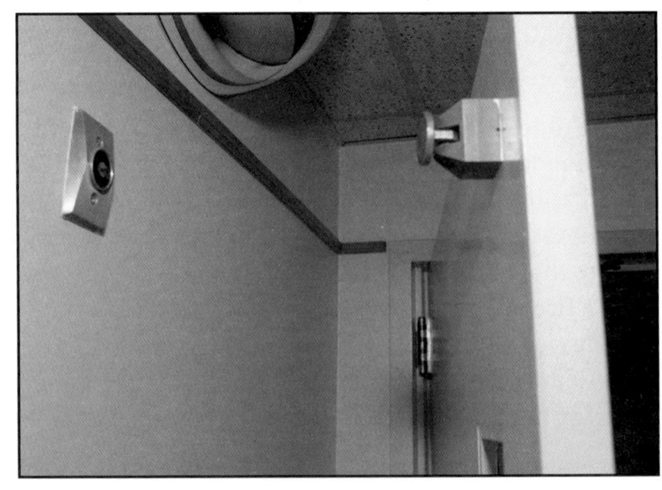

Figure 6.21 A typical arrangement of an electromagnetic door holder.

This arrangement is useful in areas with a large volume of traffic, such as school stair enclosures and corridors. Electromagnetic door holders are also often used for corridor doors in health care occupancies where they can be released by operation of a fire alarm system.

- **Door operating devices** — Intended for use with sliding fire doors that are mounted on either a level or inclined track **(Figure 6.22)**. This device consists of an electric motor that opens and closes the door for normal usage. Under fire conditions, a fusible link disconnects the door from the operating device and allows it to close by means of a spring-powered door closer or a system of suspended weights.

Figure 6.22 One type of horizontal fire door closer that incorporates a fusible-link device that acts to release the door under fire conditions.

Classifications

Three methods have been used for classifying fire doors:

1. **Hourly fire protection rating** — Indicates the length of time the door will resist fire before it is breached.
2. **Alphabetical letter designation** — Indicates the type of opening that is to be protected based on the location and required level of protection. Although this form of classification is no longer used, inspectors will encounter fire doors that use this designation during field inspections and building alterations.

   NOTE: See the information box that follows for examples of this designation.
3. **Combination of hour and letter** — Indicates the fire protection rating and type of opening for which the door assembly has been tested and certified. This classification may be found on existing fire doors, although the system is no longer used.

Alphabetical Letter Designations

Although letter designations are no longer used to describe fire doors, inspectors may encounter them in older buildings. They are described as follows:

- **Class A** — Openings in fire walls and walls that divide a structure into separate fire areas
- **Class B** — Openings in vertical shafts, such as stairwells and openings, in 2-hour rated partitions
- **Class C** — Openings between rooms and corridors having a fire resistance of 1-hour or less
- **Class D** — Openings in exterior walls subject to severe fire exposure from the outside of a building
- **Class E** — Openings in exterior walls subject to moderate or light exposure from the outside

Some examples of fire door classifications that an inspector may find include the following:

- A fire door intended to protect an opening into an exit stairwell classified as a *Class B* door
- A fire door with a combination classification, such as a *Class B 1½-hour rating,* meaning that the door is intended to protect an opening in a vertical shaft and has a 1½-hour rating

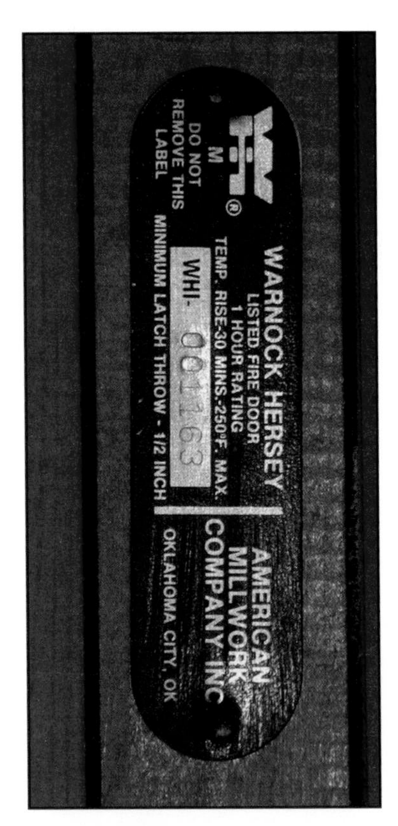

Figure 6.23 Inspectors should look for labels such as this one when attempting to determine a door's fire rating.

Rated fire doors are identified with a label indicating the door type, hourly rating, and identifying logo of the testing laboratory **(Figure 6.23)**. Building and fire inspectors can use fire door labels to identify fire doors in the field. However, the labels are commonly painted over in the course of building maintenance, a practice that voids the labeling system. In these cases, the door will need to be recertified.

NOTE: There has also been at least one documented case of counterfeit laboratory labels.

Frames and Hardware

For a fire door to effectively block the spread of fire, the door must remain closed and attached to the fire-rated wall under fire conditions **(Figures 6.24 a and b)**. Therefore, the door hardware and frame must withstand the stresses and pressures of fire exposure. The testing of fire doors includes frames and hardware, which the testing laboratories also list for use with fire doors.

The fire door is part of an assembly that must be rated, and this assembly is certified as a single unit for a specified time. To qualify as a rated fire door, the entire assembly, including the door, hardware (hinges, latches, locks), door seal, and frame, must pass a test by a third-party testing agency and be listed. Glass panels in fire doors must achieve the same test results as the rest of the assembly. The sections that follow describe fire resistance classifications, testing, frames and hardware, and construction types.

NOTE: Fire resistance testing is covered later in this chapter.

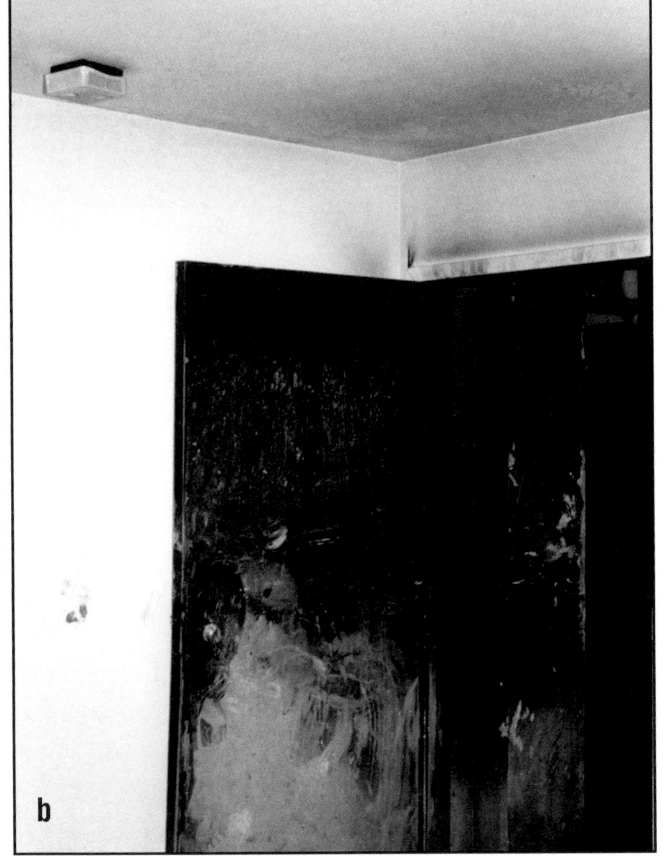

Figures 6.24 a and b A- Apartment side of a corridor door in a nonsprinklered building closed at the time of a fire. B- When the door is opened, note the extent of fire exposure from the corridor side. *Courtesy of Ed Prendergast.*

The hardware used on fire doors is referred to as either builder's hardware or fire door hardware **(Figure 6.25)**. Fire door hardware is used on both sliding and swinging fire doors and is normally shipped with the fire doors. The hardware includes the following:

- Hinges
- Locks
- Latches
- Bolts
- Closers

Modifying the door or using any hardware other than that allowed by the manufacturer can result in a loss of listing for that door and may require a field inspection from the listing organization.

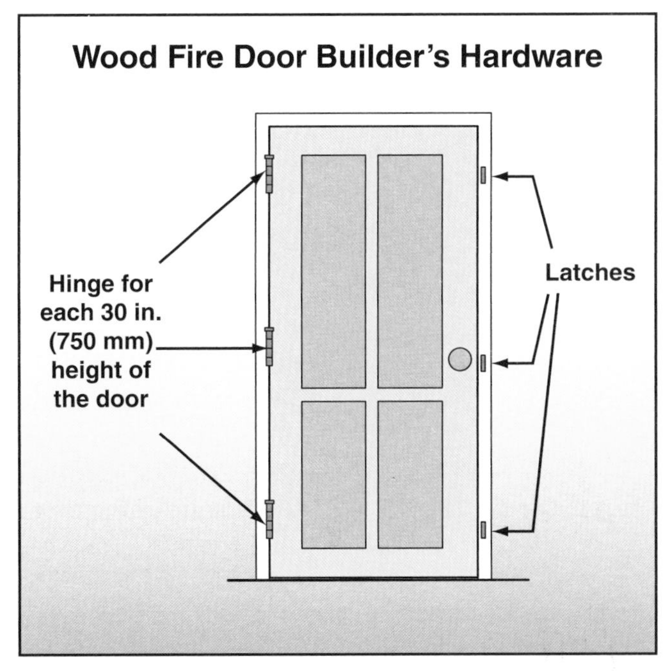

Figure 6.25 Builder's hardware examples.

Construction and Operational Types

The construction and operation of fire doors depends on the type of occupancy, the amount of space around the door opening, and the required fire protection rating for the door. Fire doors may roll, slide, or swing into place when released. The sections that follow describe the following:

- Rolling steel fire doors
- Horizontal sliding fire doors
- Swinging fire doors
- Special types
- Glass panels and louvers

Rolling Steel Fire Doors

An overhead rolling steel fire door is commonly used to protect an opening in a fire wall in an industrial occupancy or an opening in a wall separating buildings into fire areas. An overhead rolling steel fire door may be used on one or both sides of a wall opening. This type of door cannot be used on any opening that is required to be part of the means of egress.

An overhead rolling fire door is constructed of interlocking steel slats. Other operating components include:

- Releasing devices
- Governors
- Counterbalance mechanisms
- Wall guides

This type of door ordinarily closes under the force of gravity when a fusible link melts, but motor-driven doors are available that are activated by smoke detection **(Figure 6.26)**. Inspectors should be aware that without a conventional swinging door at the location, the overhead rolling fire door may create a dangerous dead-end corridor when it closes.

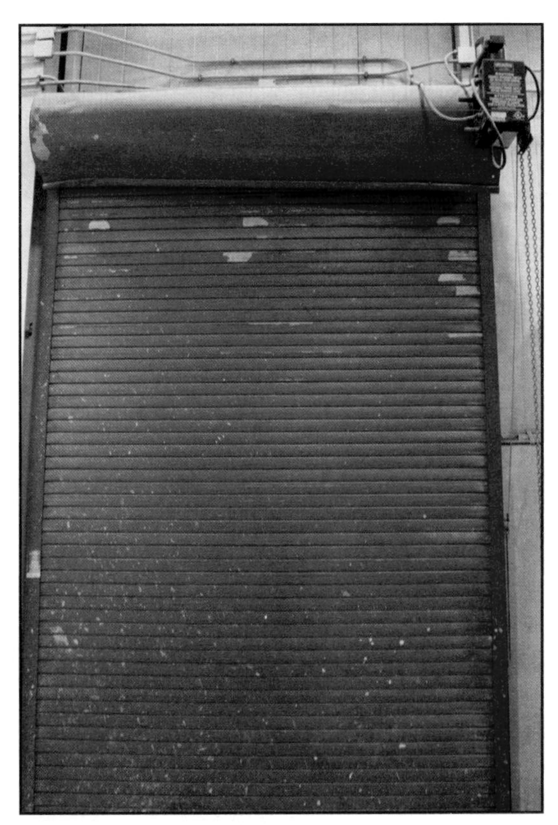
Figure 6.26 Overhead rolling fire doors close under the force of gravity when the fusible link that holds them in place melts due to the heat from a fire.

NOTE: In some schools, the inspector will also encounter security rolling doors that are tied to the detection/alarm systems. These doors roll up on activation of the system and need to be tested during an inspection.

Horizontal Sliding Fire Doors

Horizontal sliding fire doors are often found in old industrial buildings and are usually held open by a fusible link. They slide into position along a track either by gravity or by the force of a counterweight **(Figure 6.27)**. The metals used to cover the wood core include steel and galvanized sheet metal, and are often referred to as tin-clad doors.

If they are certified as such, some horizontal sliding doors can be used to protect openings in walls that are required parts of a means of egress **(Figure 6.28)**. These doors are used to separate large areas and are connected to a fire alarm system that closes automatically upon activation of a fire alarm.

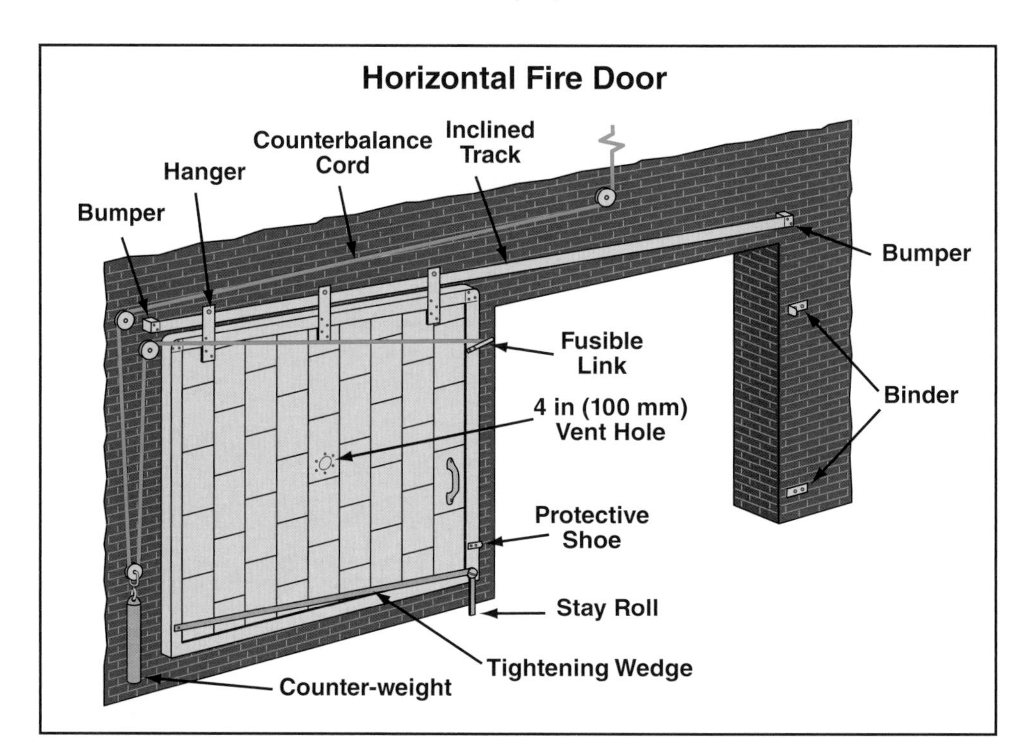

Figure 6.27 Old industrial buildings often contain horizontal sliding fire doors like the one illustrated.

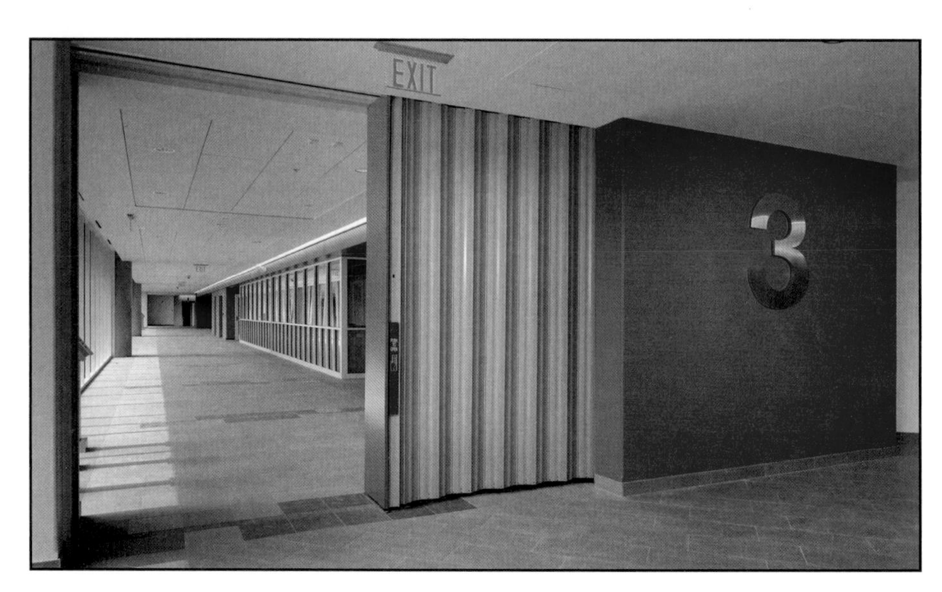

Figure 6.28 Horizontal sliding fire doors can be used to protect areas in a means of egress. *Courtesy of Won-Door Corporation.*

Swinging Fire Doors

When fire doors are needed on either side of a wall, the swing of the doors may impede an exit path through the doors **(Figure 6.29)**. When this situation arises, both doors are required to swing in the direction of egress travel. A fire-resistive-constructed vestibule between the doors provides the space needed for the doors to swing in the same direction.

Figure 6.29 Swinging fire doors are commonly used in stairwell enclosures or corridors that require a fire door.

Special Types

Special types of fire doors are available for the following:

- Freight and passenger elevators
- Service counter openings
- Security (bullet-resisting) doors
- Dumbwaiters
- Chute openings

If a fire-rated partition is required and the designer does not wish to provide a fixed wall to create an unobstructed floor plan, a horizontal-folding fire door is frequently used. A signal from a smoke detector or fire alarm system initiates the door's closing. A battery powers the motor if the regular power supply is interrupted.

Glass Panels and Louvers

A fire door often contains glass vision panels, which enhance safety and security by permitting observation through a closed door. Any glazing material used in fire doors must be fire-rated. There are restrictions on the allowable area of glass in fire doors.

Occasionally, it is desirable to install louvers in a fire door to permit ventilation while the door is closed, such as in the case of a furnace room enclosure. The louvers in a fire door must close during a fire to protect the opening. Usually, this closing is accomplished by means of a fusible link. Testing laboratories list the louvers separately. Only those fire doors that are listed for the installation of louvers can have louvers installed.

Windows

Windows have long been relied upon as a means of light, ventilation, access, and rescue. Although older buildings used openable windows for light and ventilation, modern buildings often rely on heating and air conditioning systems and artificial lighting. Installing fixed windows enhances energy efficiency because it reduces air infiltration around windows.

Components

A window consists of a frame, one or more sashes, and the necessary hardware to make a complete unit. A window frame includes the members that form the perimeter of a window and is fixed to the surrounding wall or other supports **(Figure 6.30, p. 214)**.

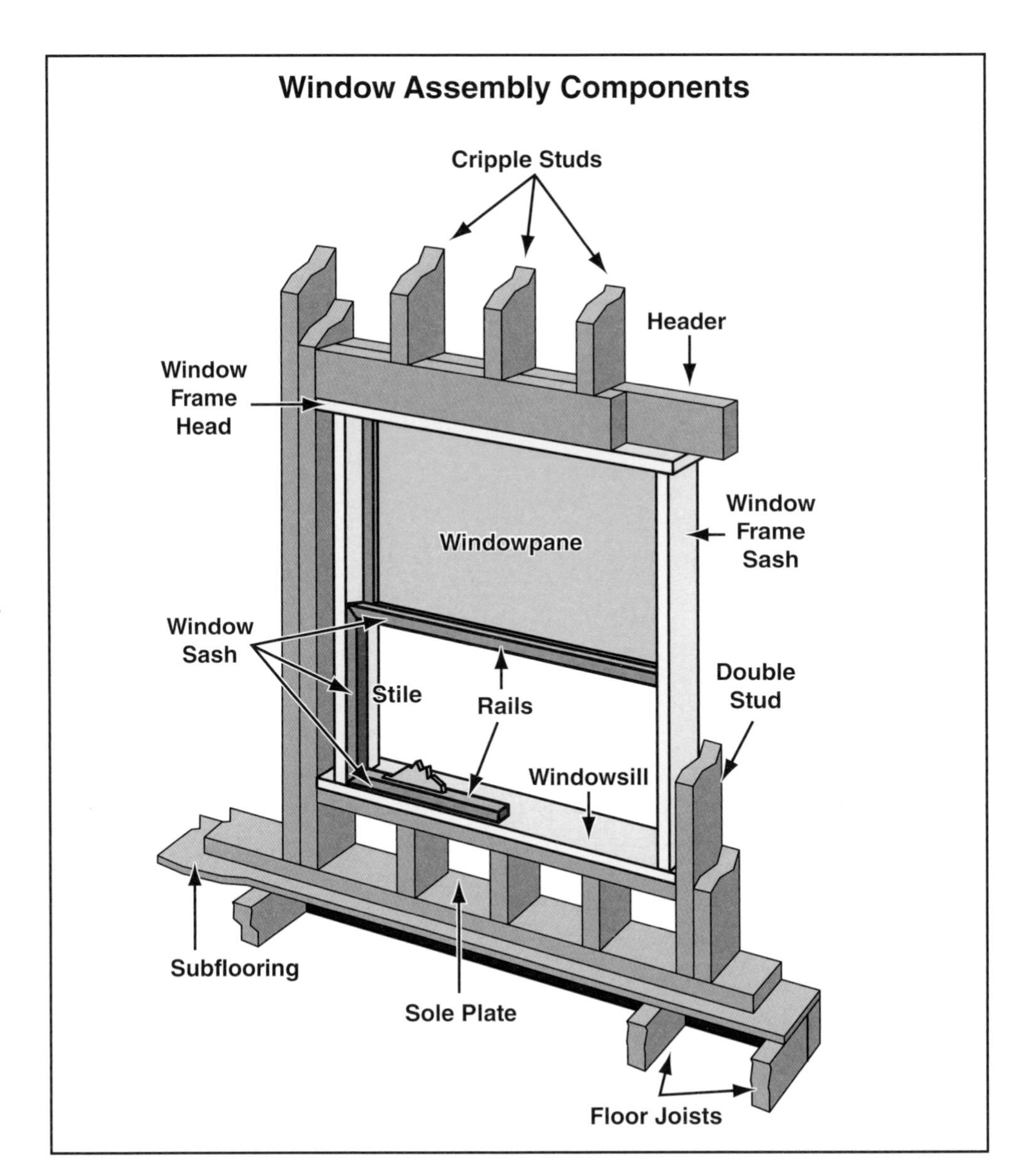

Figure 6.30 This illustration identifies and locates the various parts of a typical window assembly.

Types

Windows are broadly classified as fixed or movable. Windows that contain both fixed and movable parts are generally included in the movable classification.

A fixed window consists only of a frame and a glazed stationary sash, and it can be used alone or in combination with movable windows **(Figure 6.31)**. The large windows found in mercantile occupancies and high-rise office buildings are common examples of fixed windows.

A movable window is designed in several common configurations **(Figure 6.32)**:

- **Double-hung** — Has two sashes that can move past each other in a vertical plane.
- **Single-hung** — Has only the lower sash capable of being opened.
- **Casement** — Has a side-hinged sash that is usually installed to swing outward.
- **Horizontal sliding** — Has two or more sashes of which at least one moves horizontally within the window frame.

- **Awning** — Has one or more top-hinged, outward-swinging sashes.
- **Jalousie** — Includes a large number of narrow overlapping glass sections swinging outward (basic concept of the awning window).
- **Projecting** — Swings outward at the top or bottom and slides upward or downward in grooves.
- **Pivoting** — Has a sash that pivots horizontally or vertically about a central axis.

Security

Metal bars and screens are commonly used on windows to provide additional security. The metal bars may be fastened to the building, embedded in masonry, or mounted on hinges and locked with padlocks or other locking devices. Security windows are available with movable sashes and fixed bars so that the windows can be opened for ventilation.

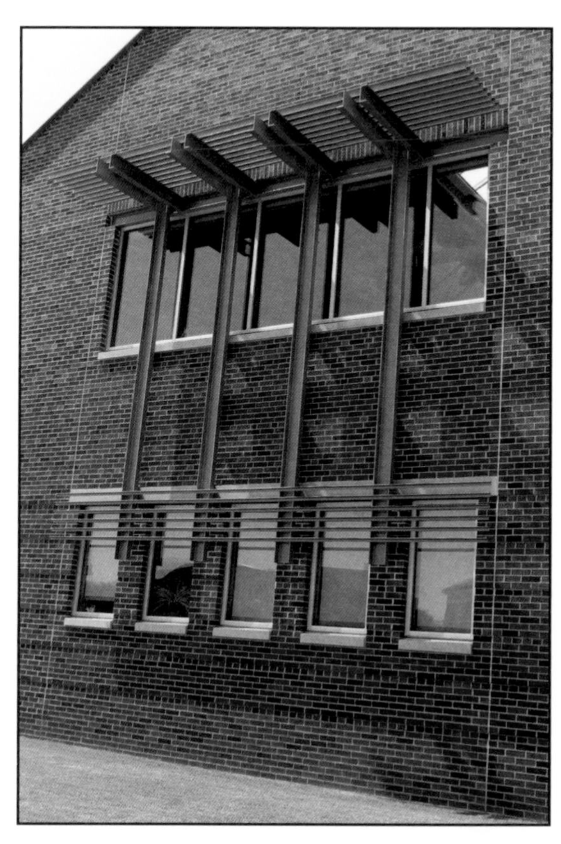

Figure 6.31 Fixed windows are designed to let in light but cannot be opened like conventional windows.

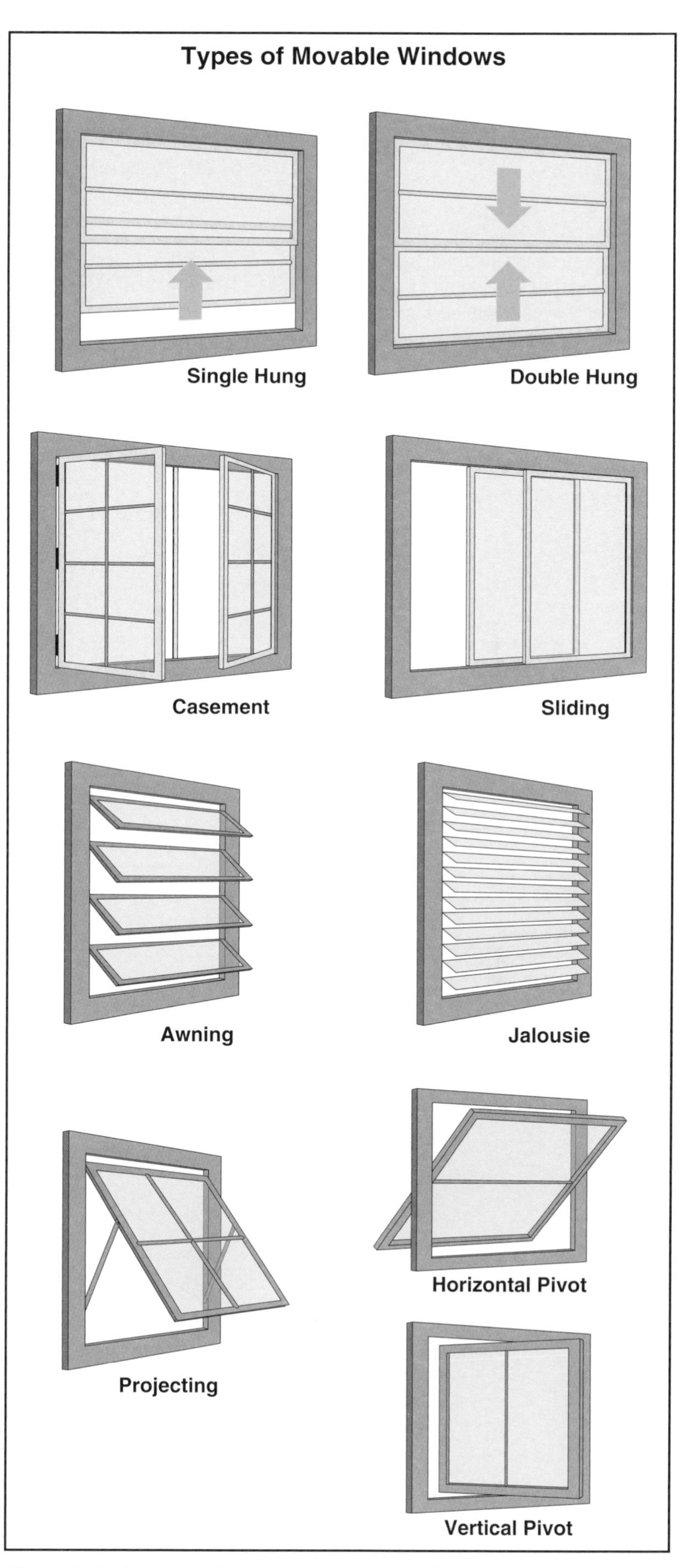

Figure 6.32 A variety of movable windows are illustrated.

A concern with security bars or screens is that they can prevent the escape of trapped occupants and firefighters and slow access time for emergency responders. An inspector should be aware of the restrictions placed on the installation of security bars and grilles as defined by the locally adopted codes **(Figure 6.33)**. If metal bars or grilles are operable, they should be operated during an inspection.

Interior Finish — Exposed interior surfaces of buildings including, but not limited to, fixed or movable walls and partitions, columns, and ceilings. Commonly refers to finish on walls and ceilings and not floor coverings.

Interior Finishes

The term **interior finish** refers to the materials used for the exposed surfaces of the walls and ceilings of a building **(Figure 6.34)**. Inspectors should be alert to materials used as interior finish that have not been approved for that application, such as carpeting on walls. Approved interior finish materials typically include the following:

- Plaster
- Gypsum wallboard
- Wood paneling
- Ceiling tiles
- Plastic
- Fiberboard
- Fabric
- Other wall coverings

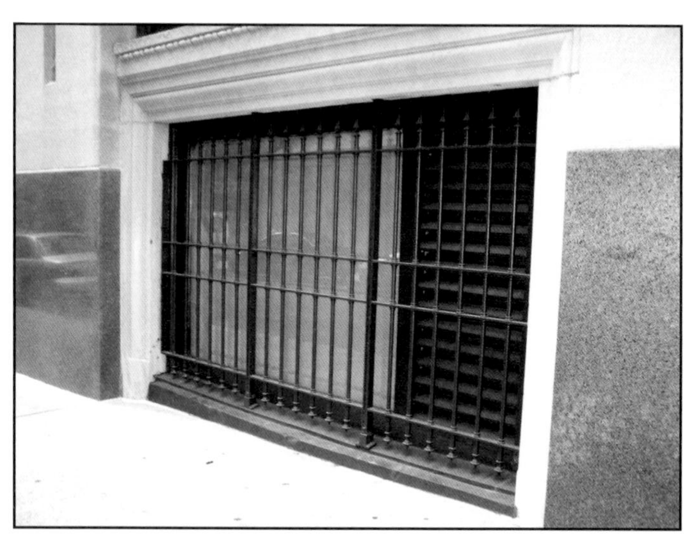

Figure 6.33 Because security bars can impede egress, the inspector must be aware of any restrictions that may be placed on their use.

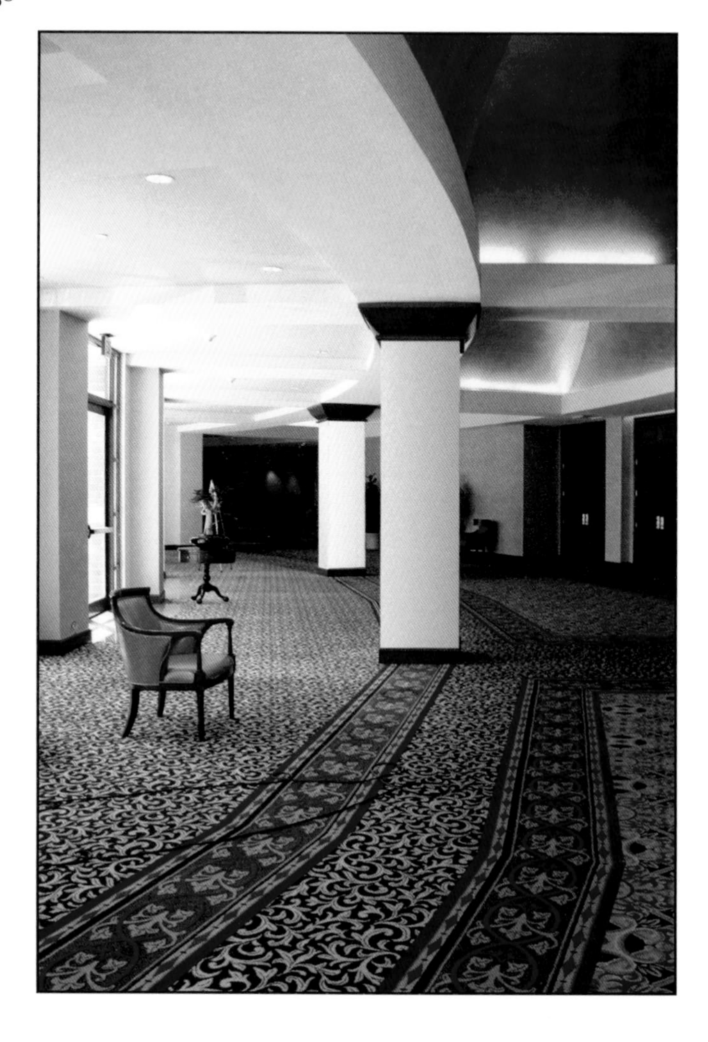

Figure 6.34 The interior finishes in this hotel can all contribute to fire spread. *Courtesy of Tara Gladden.*

A number of fires over the past century have emphasized the effect that interior finishes have had on fire behavior. In recent years, the National Institute for Standards and Technology (NIST) has attempted to study the relationship between various types of finishes and fire behavior. The combustibility of an interior finish has been shown to contribute to fire behavior in at least four ways:

1. Contributes to fire extension by flame spread over its surface
2. Affects the rate of fire growth to flashover
3. Adds to the intensity of a fire because it contributes fuel
4. Produces smoke and toxic gases that can contribute to the life hazard

NOTE: The Inspector II portion of this chapter covers fire-retardant coatings and tests for flame spread ratings.

Building codes may define doors, door and window frames, chair railings, and wainscotings as interior finishes. Hanging fabrics, such as draperies and curtains, are not treated as interior finishes unless they are applied to a ceiling or wall. Building codes usually exclude surface treatments such as paint.

Interior finishes may be part of the building structural materials. For example, if wood paneling is applied over a concrete block wall, the wood paneling would be considered the interior finish. However, a masonry wall left exposed as part of the decor of a restaurant is part of the interior finish and part of the structural components.

Building Services

Buildings contain a variety of services and subsystems designed to provide safety, comfort, and convenience for occupants. A building may also contain services and subsystems provided for a special purpose, depending on the facility's use. Subsystems must be appropriately installed, inspected, tested, and maintained in order to positively affect overall building system safety. Conversely, a defective subsystem will have detrimental effects on the entire system.

The inspector's responsibility may be to verify that the owner/occupant is maintaining building services in compliance with applicable fire and life safety building codes. The sections that follow address the fire and life safety aspects of building services and subsystems, such as elevator hoistways and doors, utility chases, and vertical shafts.

NOTE: The Inspector II section of this chapter includes additional information about building systems.

Elevator Hoistways and Doors

An elevator hoistway is the vertical shaft in which the elevator car travels **(Figure 6.35)**. Hoistways are constructed of fire-resistive materials and equipped with fire-rated door assemblies to stop the spread of fire throughout a building. Any penetrations through

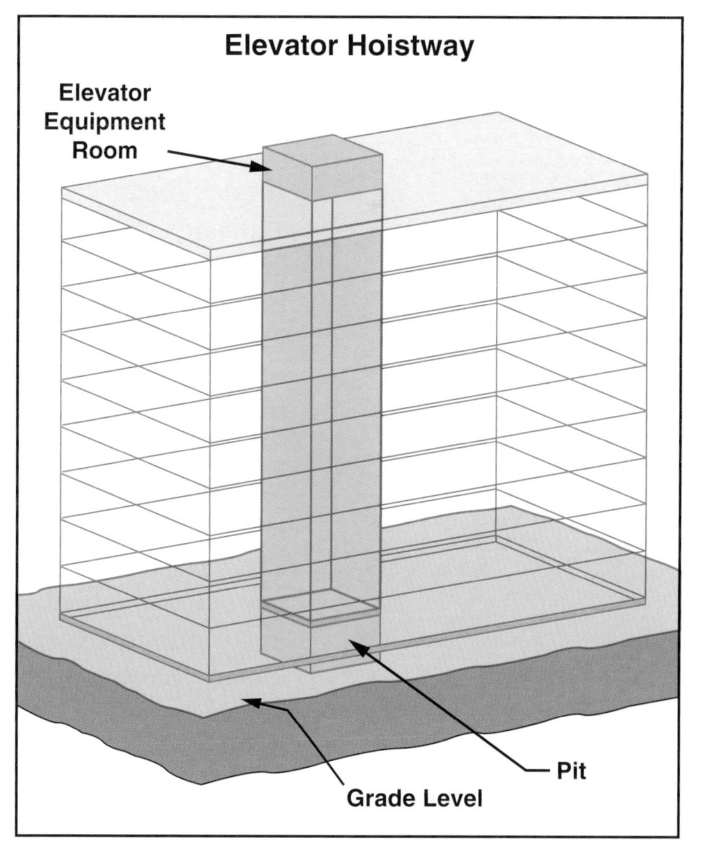

Figure 6.35 Elevator hoistways are constructed of fire-resistive materials and feature a pit beneath them that extends below grade level.

the hoistway walls must be done with an appropriately rated assembly such as a listed and labeled door and frame. No wiring, ductwork, or piping should be run within the hoistway unless it is required for the elevator itself.

Utility Chases and Vertical Shafts

The term *utility chase* generally refers to a vertical pathway in a building that contains building services. Utility chases can be used for a variety of items, such as the following:

- Plumbing
- Electrical raceways
- Telecommunications
- Data cables
- Ductwork for HVAC and grease

A vertical shaft is also a vertical pathway in a building used for other means. Inspectors should note that utility chases are unoccupied, whereas vertical shafts may be occupied. Vertical shafts are primarily used for refuse and linen chutes, as well as light shafts and material lifts.

Refuse and Linen Chutes

A refuse or linen chute is a large vertical chute that extends through a building and has openings on each floor for depositing trash or soiled linens **(Figure 6.36)**. The chute terminates at grade level or in a basement where the material is collected. Both refuse and linen chutes are required to be constructed of fire-resistive construction and may require sprinkler protection.

Grease Ducts

Typically, grease ducts are installed over cooking appliances that produce grease-laden vapors. The ducts carry grease vapors to the outside of a building. A proper installation has no areas, such as dips or horizontal runs, where grease may become trapped. The application, design, and protection required for grease ducts are specified in codes. The inspector should examine the duct to see that it is clean and free from grease buildup **(Figure 6.37)**.

A single-exhaust duct is one that only serves one exhaust hood. It typically has one fire suppression system to include the appliance, hood, and all ductwork. As with refuse and linen chutes, a grease duct is designed to withstand a fire inside the duct without causing the surrounding structure to ignite. The use of noncombustible duct materials, liquid-tight welds and connections, and minimum clearances from any combustible building components help ensure that grease ducts do not communicate fires. A commercial cooking venting system may require automatic fire suppression equipment (**Figure 6.38**).

Kitchen grease exhaust ducts always require 1-hour fire-resistive separation. This requirement can be accomplished with conventional duct construction, but usually this is accomplished with the use of a fire-resistive-rated duct wrap system.

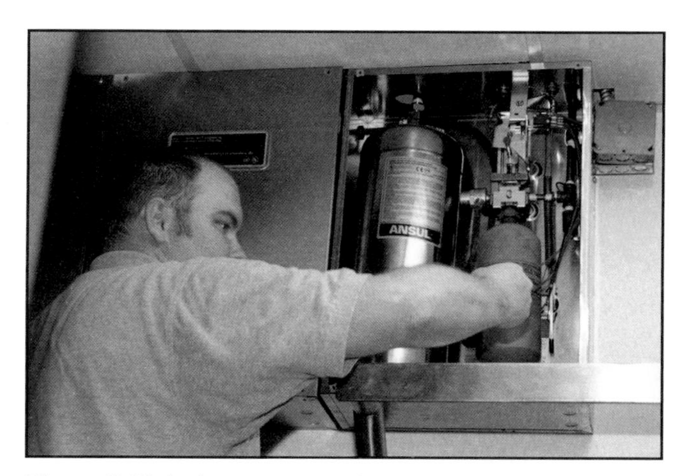

Figure 6.38 An inspector examines a restaurant extinguishing system.

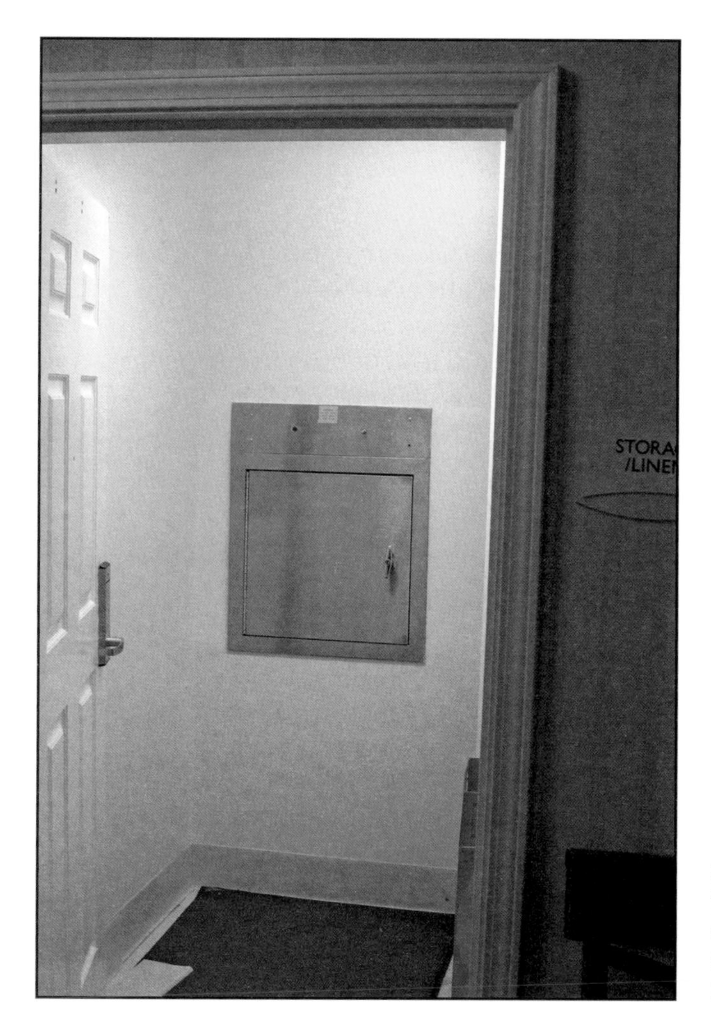

Figure 6.36 Refuse and linen chutes generally extend the entire height of a building and therefore can communicate smoke and fire throughout a structure if they are not constructed with proper fire-resistant materials.

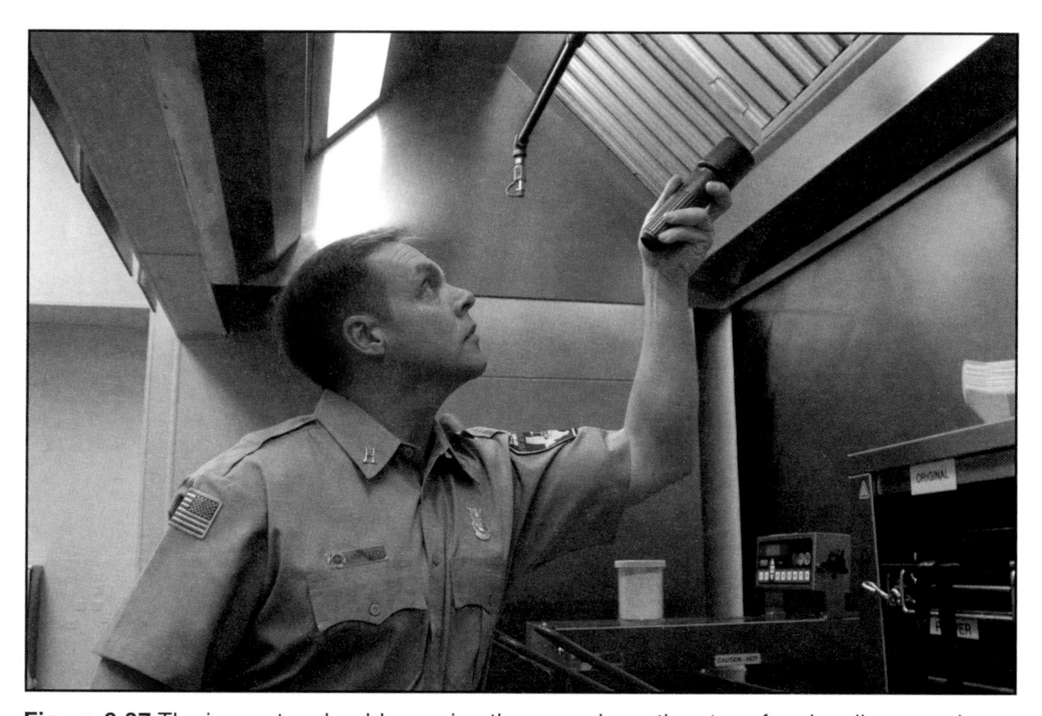

Figure 6.37 The inspector should examine the area above the stove for cleanliness and accumulated grease.

II

Evaluating Fire Walls

The evaluation of fire walls is a critical task for the Inspector II. Fire walls subdivide a building into smaller areas so that a fire in one portion of a building is limited to that area. For example, fire walls could divide a 100,000-square foot (9 000 m^2) factory into four 25,000-square-foot (2 250 m^2) areas. Fire walls also can separate various functions within a structure, such as separating hazardous processes and storage from the remainder of the facility.

Fire walls can be constructed as freestanding walls or tied walls. Freestanding fire walls are self-supporting and independent of the building frame; a tied wall requires support from the surrounding structure. The tied wall often is lighter, but it often may be a double-wall system to meet the code requirements for fire walls.

Any structural members incorporated into a fire wall must have the same degree of fire resistance as that required for the fire wall. The structural framework must have sufficient strength to resist the lateral pull of the collapse of framework on either side. No combustible structural members may penetrate a fire wall. Where combustible members are framed into a fire wall, they must be designed to fall away freely from the wall under fire conditions.

Compartmentation System — Series of barriers designed to keep flames, smoke, and heat from spreading from one room or floor to another; barriers may be doors, extra walls or partitions, fire-stopping materials inside walls or other concealed spaces, or floors.

When the floor area of a building is subdivided with fire-resistive partitions or walls, it is said to be compartmentalized or part of a **compartmentation system (Figure 6.39)**. Fire-rated partitions or fire barriers are designed to contain fire to one area. These structures are not self-supporting and freestanding like fire walls. Naturally, any openings in these fire-rated partitions or barriers, such as doors, windows, and access panels, nullify the value of the partitions unless these openings are protected by fire doors, shutters, or other rated building components.

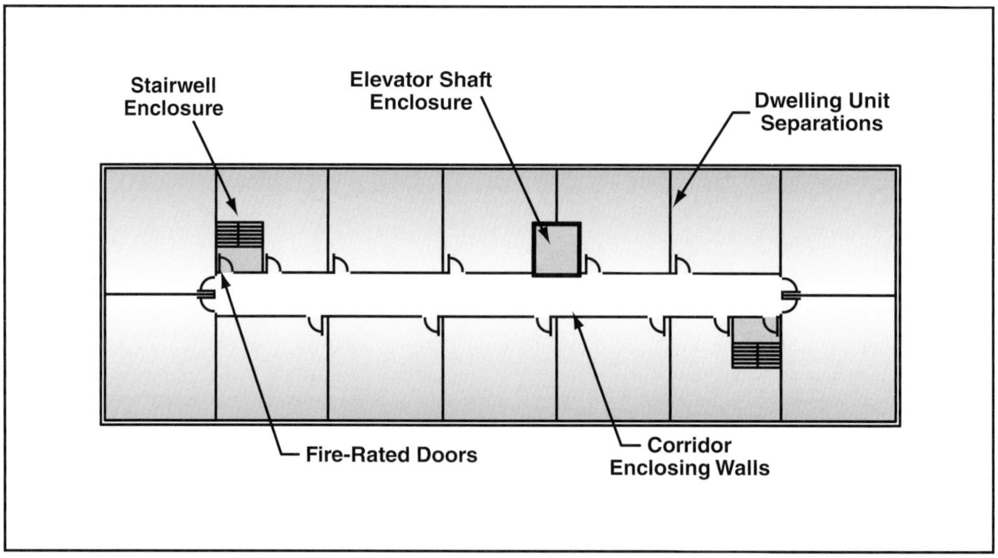

Figure 6.39 Typical compartment arrangement in the upper floor of a residential occupancy.

Evaluating Rooftop Photovoltaic Systems

Recent technological advancements and increased energy costs have created a market for photovoltaic (PV) systems, commonly referred to as solar panels. PV systems convert energy from the sun into usable electricity. These systems can be found on both residential and commercial buildings. Often these systems are designed to provide electrical power to the building and, during peak times, send electricity into the public utilities grid, providing a rebate to the system owner.

Firefighter Safety

There are hazards associated with PV systems. Depending on the system design and available sunlight, PV systems can generate up to 8 amps and 600 volts of electricity. During a daytime structure fire, the PV system may continue to produce electricity that could potentially cause harm to firefighters. If light is available--even light from apparatus headlights or lighting equipment--the panels will continue to produce power. A disconnect switch will only prohibit the electricity from the disconnect switch to the structure's main electrical system. It will not discontinue the electricity between the PV panels and the disconnect switch. All conduit and system disconnect switches should be marked according to model codes to warn firefighters of the possible danger. Fire inspectors should work closely with local electrical inspectors to verify that the systems are designed and installed with the safety of both occupants and firefighters in mind **(Figure 6.40)**.

Figure 6.40 Because photovoltaic panels post an electrical hazard, inspectors need to make sure the disconnect switches are marked according to the locally adopted code. *Courtesy of McKinney (TX) Fire Department.*

Access Pathways and Smoke Ventilation

PV systems are designed to produce the maximum amount of electricity they can. The more PV panels that are placed on a roof, the more electricity that can be potentially produced. This design becomes a problem for firefighters to access/egress pathways to and on a roof. Once on the roof, areas must be provided for smoke ventilation operations. Model building codes have criteria for these paths and ventilation points. Conversations with the responding fire fighting support companies could be beneficial for the design and the familiarization for the firefighters before a fire occurs.

Evaluating Structural Stability

Despite the limitations of laboratory-controlled testing, the ASTM E-119 test is the only standardized test method currently universally accepted by building codes. The use of the fire ratings developed over the years has contributed significantly to the safety of individual buildings and collectively to the fire safety of communities.

The E-119 standard test evaluates the ability of structural assemblies to carry a structural load and to act as a fire barrier. The test does not provide the following information:

- Information about performance of assemblies constructed with components or lengths other than those tested.
- Evaluation of the extent to which the assembly may generate smoke, toxic gases, or other products of combustion.
- Measurement of the degree of control or limitation of the passage of smoke or products of combustion.
- Fire behavior of joints between building elements, such as floor-to-wall or wall-to-wall connections.
- Measurement of flame spread over the surface of the tested material.
- The effect on fire endurance of openings in an assembly, such as electrical outlets and plumbing openings unless specifically provided for in the construction tested.

Although all of the above limitations are important, the last is of particular interest to building and fire prevention inspectors. When the continuity of an assembly is destroyed, it cannot function as a fire barrier. Over time, and particularly during renovation, fire-resistive assemblies may be penetrated. Penetrations of fire-resistive assemblies may be made for ductwork, plumbing, electrical, and communication purposes and not be adequately firestopped.

NOTE: Joint systems for floor-to-wall and wall-to-wall connections are tested in accordance with UL Standard 2079, "Standard for Fire Tests of Joint Systems."

Evaluating Interior Components

Substandard or unsafe interior finishes have contributed to many loss-of-life fires. The job of an Inspector II is to evaluate whether interior components conform to the locally adopted fire and safety codes. These include floor finishes, ceilings, and stairs.

Floor Finishes

Floor finishes are classified by the amount of heat that can be applied to the material before it ignites and fire spreads. There are two classes of floor finishes: Class I and Class II. Class I can withstand higher temperatures before igniting than Class II can withstand.

The more critical the fire exposure, the higher the classification required. Generally, the exits, exit passageways, and corridors require Class I interior floor finishes. However, the use of an automatic sprinkler system will decrease the requirement to Class II.

Ceilings

In a multistory building, ceiling materials can be attached directly to the underside of floor joists or trusses or installed at a distance below the floor supports, creating a large concealed space. It is common for old buildings to have a new ceiling installed below an existing ceiling as a means of creating new interior decor. As in the case of finished flooring, ceiling materials can conceal the type of floor or roof structure above **(Figure 6.41)**.

The space above the ceiling can be used to conceal air-conditioning ducts, electrical and communications wiring, and plumbing and sprinkler piping. This space is known as an **interstitial space**. The interstitial space can be used as a **return-air plenum** and when that is the case, the fire inspector must verify that materials, wires, ducts, pipes, etc. are noncombustible or rated for plenum use.

Interstitial Space — In building construction, refers to generally inaccessible spaces between layers of building materials. May be large enough to provide a potential space for fire to spread unseen to other parts of the building.

Return-Air Plenum — Unoccupied space within a building through which air flows back to the heating, ventilating, and air-conditioning (HVAC) system; normally is immediately above a ceiling and below an insulated roof or the floor above.

Stairs

Access or convenience stairs are not required to be a part of the means of egress system and typically connect no more than two levels. Stairs can be classified as either interior or exterior, depending on their location.

Stairs that are a part of the required means of egress must be evaluated so that they provide protection for the occupants as they travel to safety. Stairs meeting these requirements are called protected or enclosed because they are built to resist the spread of fire and smoke.

NOTE: Chapter 7, Means of Egress, contains additional information about stairs.

Although fire escapes, escalators, and fixed ladders have been used as a means of egress in the past, they are no longer allowed as components in the required means of egress from normally occupied spaces in new construction. The slide escape was another device used for some time that is no longer permitted for newly constructed buildings **(Figures 6.42 a and b, p. 224)**.

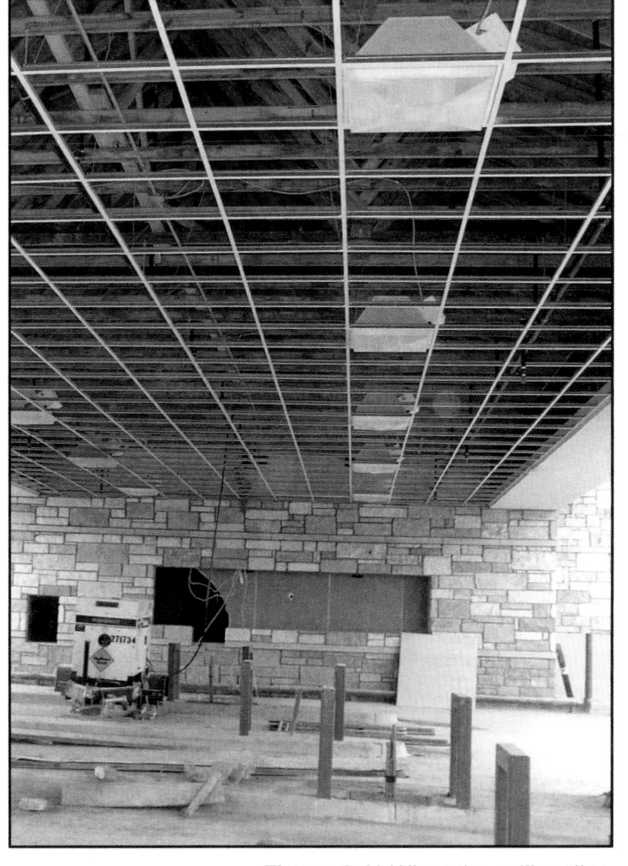

Figure 6.41 When the ceiling tiles are placed in this new structure, the space above and the building materials will be concealed. *Courtesy of Ron Moore, McKinney (TX) Fire Deparatment.*

Evaluating Fire Doors

Based on the building's age, the Inspector II may have to evaluate different classifications of doors. There are a few apparent inconsistencies in regard to fire door classifications, and an Inspector II should be able to evaluate them based on the locally adopted code. For example:

- A code may permit an opening in a 2-hour-rated stairwell enclosure to be protected with a 1½-hour-rated fire door rather than a 2-hour-rated door.
- A code may also require two 3-hour-rated fire doors to protect an opening in a 4-hour-rated wall and may not permit a 3-hour-rated door to be used in combination with a 1½-hour-rated door to satisfy the requirement.
- A ⅓-hour-rated door may be found being used as a smoke barrier and an opening to a corridor.

NOTE: The reason for a difference between the rating of a fire door and a fire-rated wall is that the test criteria for the two are different.

Figure 6.42 Fire escape ladders (a) and fire escape slides (b) are not authorized in new construction. Where they are permitted, however, they must meet all applicable codes enacted by the AHJ.

Testing of Fire Doors and Fire Door Assemblies

Fire doors are tested in accordance with the procedures contained in NFPA® 252, *Standard Methods of Fire Tests of Door Assemblies*, which is also designated ASTM E-152 (from ASTM International, originally known as the American Society for Testing and Materials). The test procedure uses a furnace to expose the fire doors to the same time and temperature curve used to establish the fire resistance rating of structural assemblies. However, the conditions for passing the test for door assemblies are not as rigid as those required for fire-rated walls.

For fire doors, the primary criterion for acceptability is that the fire door remains in place during the test. Some warping of the door is permitted, and intermittent passage of flames is permitted after the first 30 minutes of the test **(Figure 6.43)**. There is also no maximum surface temperature rise permitted on the unexposed side of the door.

In addition to the fire exposure test, a fire door assembly must remain in place when subjected to a hose stream immediately following the fire test. The use of a hose stream subjects the door assembly to cooling and impact effects that might accompany fire suppression operations. A door with a ⅓-hour rating may not be subjected to the hose test.

Figure 6.43 During testing, fire doors must remain in place.

Evaluating Interior Finishes

Because flammable interior finishes can intensify fires, UL and other organizations developed testing methods to establish flame-spread ratings for interior finish materials. Ratings based on smoke development were also developed. In an attempt to provide an acceptable level of protection, fire-retardant coatings have also been developed, tested, and approved. Each of these topics is described in the sections that follow.

Inspectors should be aware that modifications to building construction components, such as interior finishes, can affect the component's fire rating. Modifications or repairs to a component should conform to the same level that was originally approved. Inspectors may need to research original plans to determine these fire ratings.

Flame Spread Ratings

Because interior finishes typically cover a large area and are relatively thin compared to structural components, they pose a great potential fire danger to occupants. For example, if wood paneling is used over a concrete block wall, fire can spread rapidly over the surface of the paneling, even though the quantity of wood is small compared to the concrete block. As a result, the early investigations of the role of interior finish materials concentrated on evaluating the speed with which flame can spread over the surface of a material (surface-burning characteristic). The **Steiner Tunnel Test** is the most commonly used method for evaluating surface-burning characteristics of materials **(Figure 6.44)**. The tunnel test produces a numerical evaluation of the flammability of interior materials, which is known as the **flame spread rating** and the **smoke-developed index**.

NOTE: The Steiner Tunnel Test is referred to as the tunnel test, but it is formally identified as ASTM E-84 and UL-753. It is also found in NFPA® 255, *Standard Method of Test of Surface Burning Characteristics of Building Materials.*

Steiner Tunnel Test — Unofficial name for the test used to determine the flame spread ratings of various materials.

Flame Spread Rating — Numerical rating assigned to a material based on the speed and extent to which flame travels over its surface.

Smoke-Developed Index — A measure of the concentration of smoke a material emits as it burns.

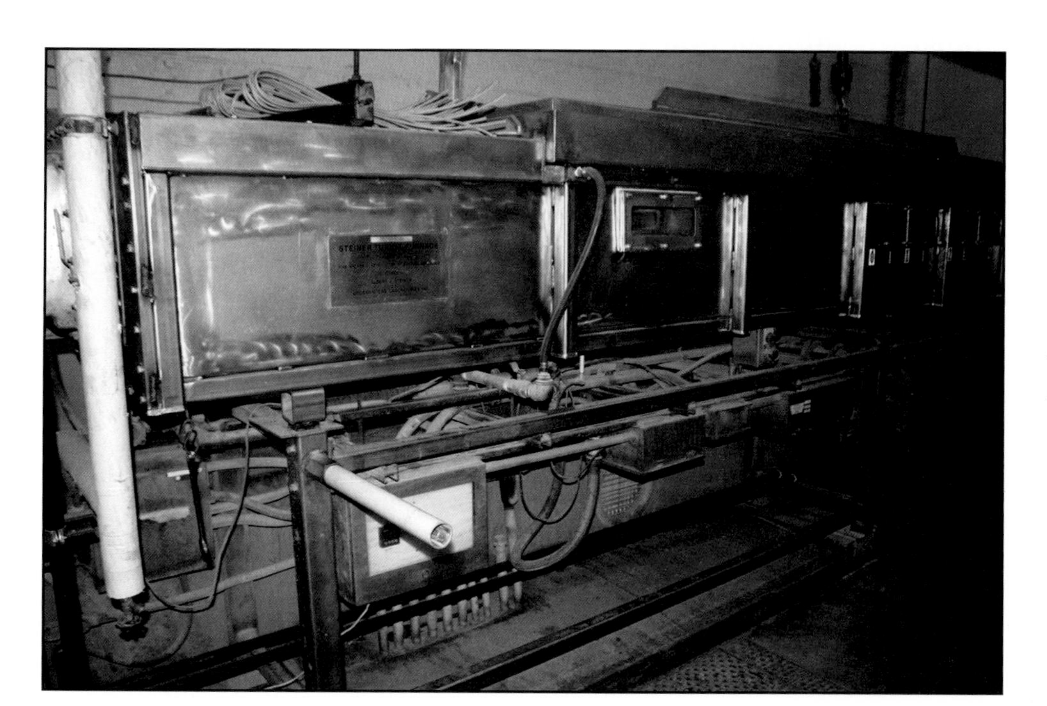

Figure 6.44 The Steiner Tunnel Test is used to develop flame-spread ratings.

The flame spread rating developed in the tunnel test is a means of comparing the surface flammability of a material to standard materials under controlled test conditions. For comparison, red oak flooring was assigned a flame-spread and smoke developed index of 100 during a 10-minute tunnel test. All materials are tested against this rating. Building codes use the flame spread ratings of materials to establish some control over interior finishes. The codes establish three classifications of interior finishes using letter designations as shown in **Table 6.1**. Typical use of the classifications of materials is to restrict the materials in vertical exits and exit corridors to those with low flame spreads and to permit materials with a higher flame-spread rating in other areas.

The flame spread rating, while useful, is NOT an absolute measure of the spread of fire travel. Such factors as room shape and dimensions and the fuel load within the compartment have a significant effect on surface burning. The inspector should be aware that differences between field applications and test conditions create a differing behavior in the field. For example, when interior finish materials are intended to be used in varying thicknesses, they must be tested at those thicknesses.

The flame spread rating developed in the tunnel test does not apply to floor coverings. However, if a floor covering such as carpeting is used for a wall or ceiling finish, it must meet the same flame spread criteria as other wall and ceiling finishes.

Unfortunately for inspectors, there is no way that flame spread ratings can be positively determined in the field. The relative surface burning can be assumed for some interior surfaces, such as concrete block or plaster. Other materials, such as corrugated paper, can usually be assumed to have a higher flame spread rating. However, the flame spread rating of many materials, especially composite materials, simply cannot be determined in the field unless the manufacturer can be identified and contacted.

Smoke-Developed Index

In addition to the flame spread rating, the tunnel test provides another measure of flammability: the smoke-developed index, which is a measure of the relative visual obscurity created by the smoke from a tested material. It is measured by means of a photoelectric cell and a light source located at the end of the tunnel furnace.

As with the flame spread rating, red oak is used as a standard and has the same rating of 100 for smoke. Therefore, under test conditions, a material that has a smoke-developed index of 200 produces smoke that is twice as visually obscuring as red oak. Flame-retardant coatings can have an effect on both the smoke-developed rating and on flame spread ratings.

Toxicity — Ability of a substance to do harm within the body.

The smoke-developed index is not an indication of the **toxicity** of products of combustion of interior finish materials. The tunnel test will not detect or measure a completely transparent product of combustion such as carbon monoxide. Furthermore, the tunnel test does not measure the combined effects of heat, irritation, and toxicity.

Table 6.1
Interior Wall and Ceiling Finish Requirements by Occupancy[k]

Group	Sprinklered[l]			Nonsprinklered		
	Interior exit stairways, interior exit ramps and exit passageways[a,b]	Corridors and enclosure for exit access stairways and exit access ramps	Rooms and enclosed spaces[c]	Interior exit stairways, interior exit ramps and exit passageways[a,b]	Corridors and enclosure for exit access stairways and exit access ramps	Rooms and enclosed spaces[c]
A-1 & A-2	B	B	C	A	A[d]	B[e]
A-3[f], A-4, A-5	B	B	C	A	A[d]	C
B, E, M, R-1,	B	C	C	A	B	C
R-4	B	C	C	A	B	B
F	C	C	C	B	C	C
H	B	B	C[g]	A	A	B
I-1	B	C	C	A	B	B
I-2	B	B	B[h,i]	A	A	B
I-3	A	A[j]	C	A	A	B
I-4	B	B	B[h,i]	A	A	B
R-2	C	C	C	B	B	C
R-3	C	C	C	C	C	C
S	C	C	C	B	B	C
U	No Restrictions			No Restrictions		

For SI: 1 inch = 25.4 mm, 1 square foot = 0.0929 m^2

a. Class C interior finish materials shall be permitted for wainscoting or paneling of not more than 1,000 square feet of applied surface area in the grade lobby where applied directly to a noncombustible base or over furring strips applied to a noncombustible base and fireblocked as required by Section 803.13.1.*

b. In other than Group I-3 occupancies in buildings less than three stories above grade plane, Class B interior finish for nonsprinklered buildings and Class C interior finish for sprinklered buildings shall be permitted in interior exit stairways and ramps.

c. Requirements for rooms and enclosed spaces shall be based upon spaces enclosed by partitions. Where a fire-resistance rating is required for structural elements, the enclosing partitions shall extend from the floor to the ceiling. Partitions that do not comply with this shall be considered enclosing spaces and the rooms or spaces on both sides shall be considered one. In determining the applicable requirements for rooms and enclosed spaces, the specific occupancy thereof shall be the governing factor regardless of the group classification of the building or structure.

d. Lobby areas in Group A-1, A-2 and A-3 occupancies shall not be less than Class B materials.

e. Class C interior finish materials shall be permitted in places of assembly with an occupant load of 300 persons or less.

f. For places of religious worship, wood used for ornamental purposes, trusses, paneling or chancel furnishing shall be permitted.

g. Class B material is required where the building exceeds two stories.

h. Class C interior finish materials shall be permitted in administrative spaces.

i. Class C interior finish materials shall be permitted in rooms with a capacity of four persons or less.

j. Class B materials shall be permitted as wainscoting extending not more than 48 inches above the finished floor in corridors and exit access stairways and ramps.

k. Finish materials as provided for in other sections* of this code.

l. Applies when protected by an automatic sprinkler system installed in accordance with Section 903.3.1.1* or 903.3.1.2.*

* Section numbers refer to sections in the 2015 *International Building Code®*.

Courtesy of the International Code Council®, 2015 International Building Code®, Table 803.11.

Fire-Retardant Coatings

The flame spread rating of some interior finishes, most notably wood materials, can be reduced through the use of retardant coatings, such as:

- Intumescent paints
- Mastics
- Cementitous (cement-based) and mineral-fiber coatings
- Topical fire-retardant coatings

These types of coatings behave in different ways **(Figure 6.45)**. For example, intumescent paints expand upon exposure to heat to create a thick, puffy coating that insulates the surface beneath the wood. The mastic coatings form a thick, noncombustible membrane over the surface of the wood.

Listed and approved fire-retardant coatings are acceptable treatments for reducing surface burning when they are applied as directed; however, they are susceptible to misuse. They must be applied at a specified rate of square feet (meters) per gallon (liter) and may require more than one coat.

Figure 6.45 Sprayed-on fire retardant coatings need to remain intact if they are to be effective.

Limitations of Fire-Retardant Coatings

Fire-retardant coatings only affect the material's surface. In addition, a material that is listed as a fire-retardant coating does not increase the fire resistance of structural components or assemblies unless it has also been tested and listed for use in a fire-resistive assembly. However, because fire-retardant coatings can be field-applied, contractors and design professionals may attempt to substitute them as an inexpensive remedy for other fire protection shortcomings.

Evaluating Building Services

Buildings contain a variety of services and subsystems designed to provide safety, convenience, and comfort for occupants. A building may also have specific services and subsystems, depending on the particular use of the facility.

All services and subsystems can have a potential effect on fire and life safety, either positively or negatively. Subsystems must be appropriately designed, installed, inspected, tested, and maintained in order to contribute positively to the overall building system safety. As with any complex machine, a defective subsystem can have detrimental effects on the entire system.

Many building services and subsystems penetrate the built-in structural elements that create compartmentalization. An inspector's responsibility is to verify that the designer and owner have installed and maintained building services in compliance with the locally adopted fire and life safety building codes. The sections that follow address the fire and life safety aspects of building services and subsystems, such as the following:

- HVAC systems
- Electrical systems
- Utility chases and vertical shafts
- Elevator hoistways and shafts
- Grease ducts

Heating, Ventilating, and Air Conditioning Systems

HVAC systems are designed primarily to maintain a comfortable environment for occupants. They also regulate the intake of outdoor air and the recirculation of indoor air. The components of an HVAC system vary based on the size and use of the structure **(Figure 6.46)**. As with all building systems, HVAC systems have the potential to significantly affect any fire event.

The inspector should verify that HVAC equipment is being properly maintained by reviewing maintenance records and conducting field observations. The inspector should verify documentation of regular maintenance of the following systems:

- Outside air intakes
- Fans
- Air filtration devices
- Exhaust properly sealed into chimneys
- Air heating and cooling equipment
- Air ducts
- Smoke/fire dampers

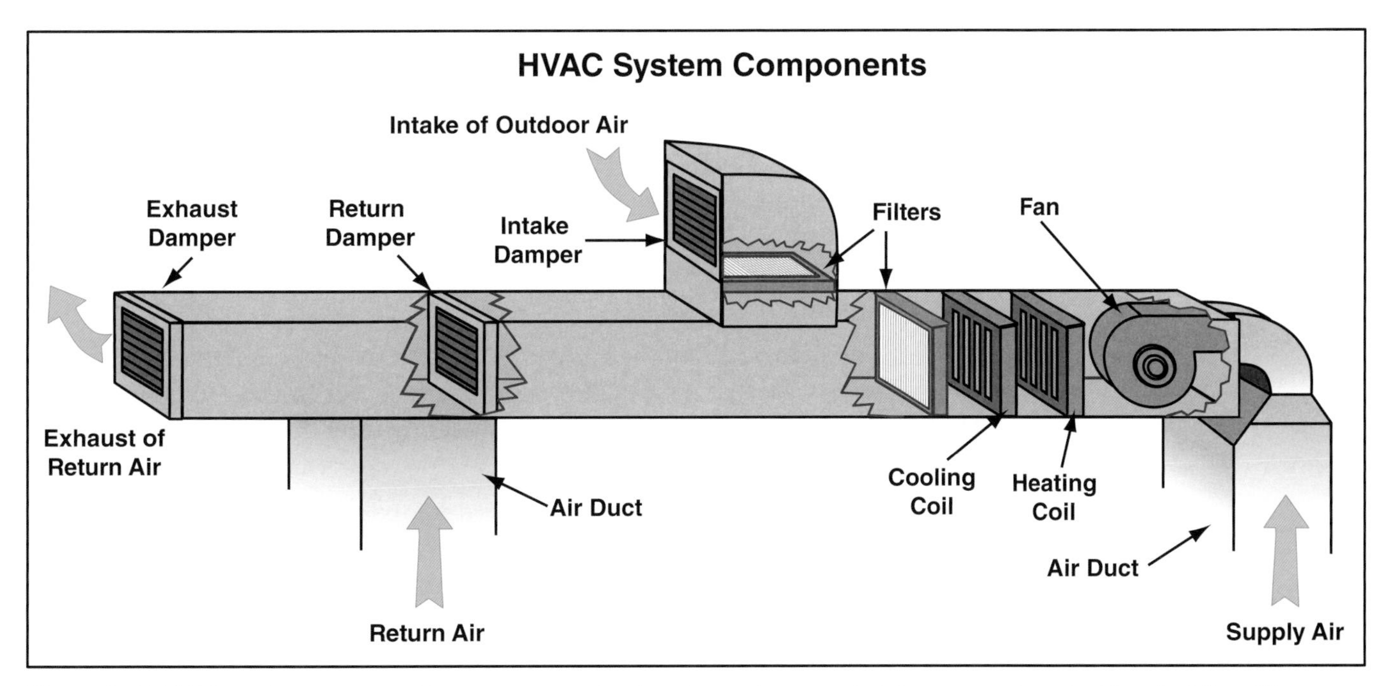

Figure 6.46 HVAC systems are used in modern construction to create a comfortable atmosphere inside a building. The various components of HVAC systems are illustrated.

Figure 6.47 Smoke dampers in ductwork protect penetrations in fire-rated assemblies and are usually activated in conjunction with smoke or fire alarm systems.

Codes may require the installation of smoke or fire dampers to protect duct penetrations through fire-rated assemblies **(Figure 6.47)**. A smoke damper usually is actuated by an associated smoke detector and closes by active mechanical action. Smoke dampers may also be actuated by an automatic alarm signal from the building fire alarm system.

A fire damper is usually a spring-loaded shutter that is held open by a fusible link. The shutter closes when the fusible link melts in response to heat and releases the spring-loaded mechanism. Combination smoke and fire dampers close in response to either heat or smoke.

Most codes require that an HVAC system above a certain capacity have an internal duct smoke detection device to automatically turn off the system when smoke is detected. The intent is to prevent the system from spreading and recirculating smoke from a fire either outside or inside the HVAC unit. Duct smoke detection devices should never be considered a substitute for area smoke detection devices.

Ductwork for HVAC systems can provide a ready means for the products of combustion to travel throughout the entire area served by the system. Provisions must be made to control the flow of smoke within the building through HVAC systems. Although some HVAC systems are actively used to exhaust smoke from a building during fire conditions, most are not. Most systems are designed to turn off immediately upon detection of fire conditions within the system.

Instead of using the HVAC system, most buildings use a system of passive smoke control that prevents smoke from traveling through the system into other areas of the building. Once fire conditions are detected, the HVAC system fan deactivates and smoke, fire, or combination dampers within the system and activate at points where the HVAC ducts pass through smoke or fire-barrier walls within the building. This action helps to compartmentalize the fire, impeding its progress throughout the rest of the HVAC system. Dampers used within a smoke control system should be inspected by a qualified person and also exercised.

Smoke detectors within the HVAC system may be located in the main supply ducts, downstream of the air filters and cleaners, or in the return air ducts. Their locations depend on the mechanical building code under which they were installed.

When the HVAC system is used for active smoke control, the system operates in a special mode that no longer delivers supply air to the fire area. Air is instead drawn from the fire area and discharged to the outside without recirculating or contaminating the air intake. At the same time, the return air supply to all or part of the building ceases. In this type of active smoke-control system, fire dampers are usually omitted. These types of systems are common in covered malls, underground buildings, and high-rise structures **(Figure 6.48)**.

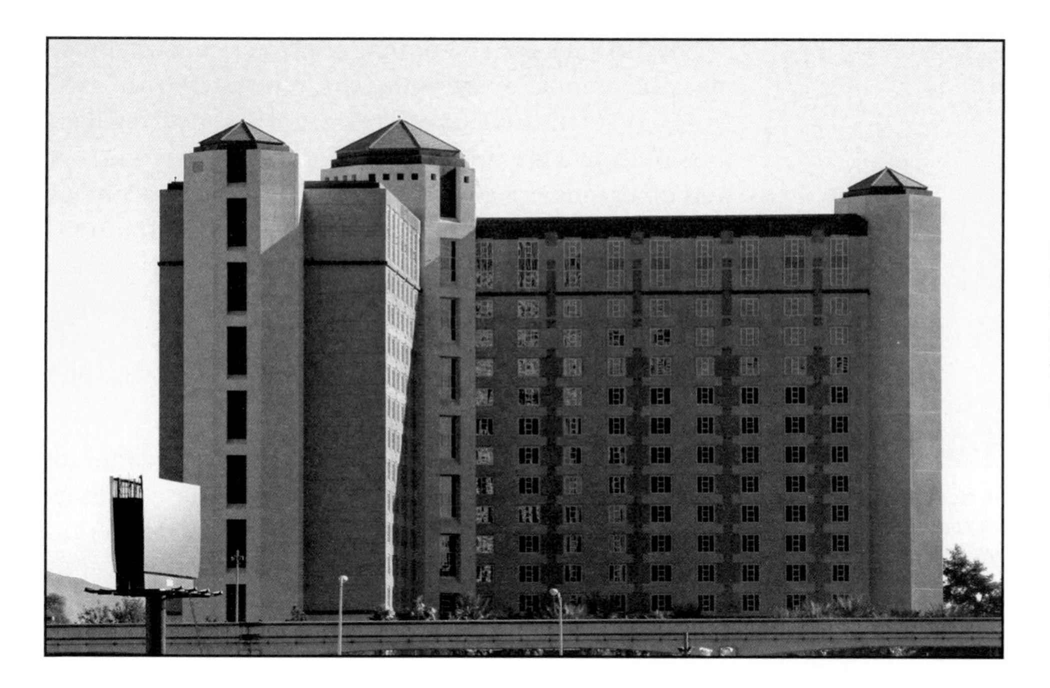

Figure 6.48 Large areas like this high-rise require active smoke control systems because of the time it takes to evacuate the structure and gain access to a fire.

Active smoke-control systems are especially applicable in high-rise structures because a lengthy time is needed for evacuation, aerial ladders cannot reach upper floors, and strong air-flow patterns are found in vertical shafts that carry smoke into upper floors of the structure. They are also important in underground structures where exit travel is into upper floors. The system must be capable of maintaining smoke- and heat-free exit routes for occupants. It must also be designed to allow for sufficient evacuation time for occupants to either leave the structure or move to areas of refuge.

The *National Building Code of Canada* (NBC) identifies 14 different methods of smoke control in buildings (Measures A through N) that have specific inspection and testing requirements that an inspector must verify. The 14 measures do not include smoke-control measures that have been implemented as equivalent methods or alternatives permitted by objective-based codes. Inspectors need to be familiar with the testing requirements and the frequency for testing these systems. Always request documentation that the required testing and maintenance has been performed by a qualified person.

Refrigeration Systems

Refrigeration systems are installed in buildings to add or remove heat. Refrigeration controls the air temperatures for heating and cooling of building occupants or processes/items. For an inspector of commercial buildings, these systems are also used to preserve foods, maintain the temperature of time-sensitive pharmaceuticals, and to ensure that certain hazardous materials do not become unstable.

Refrigerants are classified by the *International Mechanical Code* (IMC) and ASHRAE (American Society of Heating, Refrigeration and Air Conditioning Engineers) for flammability and toxicity. ASHRAE Standard 34, *Designation and Safety Classifications of Refrigerants,* is used to assign their classification. Inspectors require a basic understanding of the ASHRAE standard because the amount of refrigerant in the system and its classification dictate when the jurisdiction's fire code requirements for these systems are applicable.

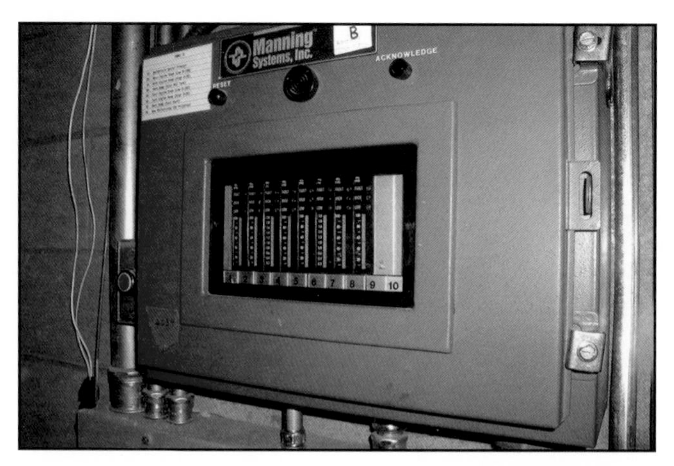

Figure 6.49 Gas detection systems are designed to alert plant personnel of refrigerant leaks. *Courtesy of Scott Stookey, International Code Council, Washington DC.*

Figure 6.50 A manual emergency switch enables employees to activate the mechanical ventilation system. *Courtesy of Scott Stookey, International Code Council, Washington DC.*

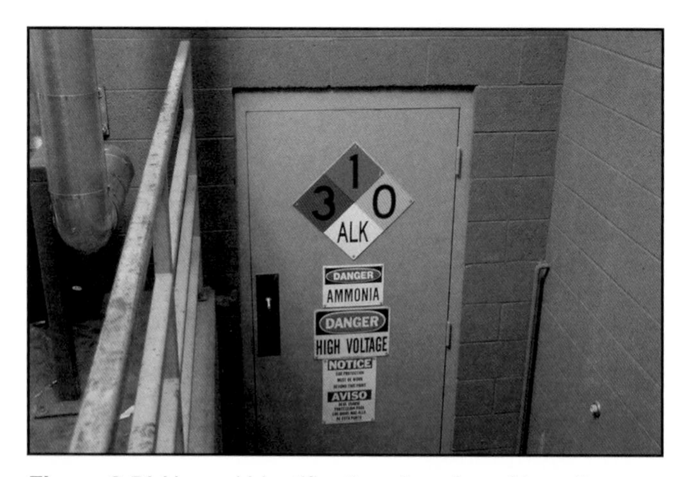

Figure 6.51 Hazard identification signs for refrigeration mechanical room. *Courtesy of Scott Stookey, International Code Council, Washington DC.*

ASHRAE 34 uses an alphabet-numerical system for classification of refrigerants. The letters A or B are used to designate the toxicity of a refrigerant. Type B refrigerants present a greater health hazard to facility workers and emergency responders when compared to Type A refrigerants. Refrigerants are also assigned a flammability rating on a scale of 1 to 3, with Class 1 refrigerants being noncombustible and Class 3 refrigerants being flammable. For example, anhydrous ammonia, which is classified as refrigerant 717 by ASHRAE 34, is classified as a B2 refrigerant.

Model codes require refrigeration machinery rooms to be equipped with gas detection systems. These systems are used to alert facility personnel of possible refrigerant leaks. Activation of the gas detection system must activate occupant notification devices and the alarm must be transmitted to an approved location **(Figure 6.49)**.

Another safety feature required in refrigeration machinery rooms is a mechanical ventilation system. Mechanical codes require this system to provide a means of exhausting the room in the event of a refrigerant leak. The mechanical ventilation system is activated by the gas detection system.

In the event the gas detection or mechanical ventilation system fails to operate, model mechanical and fire codes require a manual means of controlling the systems in an emergency. Model fire code require a break-glass type switch near the machinery room in an approved location that is designed to shut down compressors, pumps, and automatic valves that are required to remain in the closed position. A second emergency switch is also required to manually activate the machinery room's mechanical ventilation system **(Figure 6.50)**.

The health, flammability, and instability hazard of the refrigerant being used is required to be clearly identified to emergency responders. The model fire codes require hazard identification that signs comply with NFPA® 704, *Standard System for the Identification of the Hazards of Materials for Emergency Response* **(Figure 6.51)**.

Electrical Systems

Electrical systems have equipment that may be installed in separate rooms or vaults or in buildings separate from the main structure. The installations may be above or below grade level or on individual floors. Codes may require electrical service panels, switch gear, generators, and transformers to be separated from the rest of the building by fire-rated construction. The sections that follow present each of these electrical system components.

Electrical Service Panels

All structures that have electrical power systems have electrical service panels, which distribute the electrical power that arrives at the panel into individual circuits. Circuits are designed to evenly distribute the electricity, preventing wiring from becoming overloaded and ensuring adequate power for equipment connected to the circuit. Circuit breakers also turn off power to the circuit if there is a short.

Although circuit breakers have replaced fuses in electrical systems, fuse boxes are still found in structures built before 1950. Fuses may still be found on electrical equipment such as air-conditioning units. Master control switches are present within either the circuit breaker or fuse box service panel. These master switches control all the electricity that enters the panel. When it is necessary to work on the panel or turn off power to the entire building, these switches or fuses may be used **(Figures 6.52 a and b)**.

Switch Gear

Switch gear is a term used to describe electrical equipment that is used to isolate circuits and energized equipment. It may be located in electrical power stations, industrial complexes, or electrical equipment rooms. The switch gear contains multiple circuit breakers or switches that will prevent a short in the system or reset it after a short has occurred.

Figures 6.52 a and b Master switches control electricity entering and exiting a circuit breaker or fuse box panel. They should be turned off if any electrical work is done inside a building.

Generators

Many buildings rely on generators to supply backup power during an emergency. These generators may be limited to operating the fire protection systems and emergency lighting systems or they may have the capacity to provide power to the entire building or operation.

Generators are generally located outside a building but may also be found in basement areas **(Figure 6.53)**. Natural gas and diesel are the main types of fuel for the generators. A power transfer switch starts the generator when there is a loss of power and turns it off when the primary power supply reengages.

Transformers

Transformers convert high-voltage electricity supplied by the electric utility service to an appropriate voltage for use in a building **(Figure 6.54)**. Some dedicated transformers supply special systems and equipment in industrial and commercial buildings.

Transformers generate heat; air cooling and oil cooling are the two most common cooling methods:

1. **Air-cooled transformers** — Use the surrounding air to cool the transformer through fins and heat sinks installed on the body of the transformer (also called dry transformers).
2. **Oil-cooled or oil-filled transformers** — Contain oil to conduct heat away from the core and to electrically insulate internal components. In addition to the hazard of being energized electrical equipment, they also have the potential to be the source of a combustible liquid leak.

CAUTION

Be cautious when examining any transformers. Transformer hazards are the same as those with any other energized electrical equipment, especially electrocution or fires caused by shorts, arcs, and sparks.

Elevator Hoistways and Doors

Hoistways have the potential to act as a vertical chimney and spread fire and smoke throughout a building. If the hoistway is not vented at the top, the hot gases and smoke may accumulate and spread horizontally to the upper floors. To prevent this accumulation, the model building codes require venting at the top of nearly every hoistway built today.

Elevator hoistway enclosures are usually required to be fire-rated assemblies. Any penetrations through the hoistway walls must be done with the installation of an appropriately rated assembly, such as a listed and labeled door and frame. No wiring, ductwork, or piping should be run within the hoistway unless it is required for the elevator itself.

Hoistway doors are rated assemblies that work in conjunction with the car doors and, with the exception of freight elevators, depend upon the car doors for their power. Hoistway doors are of the same types as car doors with one addition: a swinging door is installed on some hoistways.

Swinging elevator doors are like regular doors in that they are hinged at one side and swing outward from the hoistway. Like most other hoistway doors, swinging doors are not powered, so they are equipped with a handle for manual operation. With this type of hoistway door, the elevator car door will usually be of the single-slide type.

Elevator hoistway doors may not completely prevent the passage of smoke from the hoistway into the building because some door clearance is required for operation. The model codes typically require a lobby arrangement with rated doors separating the lobby from the rest of the building. This separation provides an additional barrier to prevent the passage of smoke and fire gases into the rest of the building. The swinging door is typically provided to meet the requirement of an additional barrier from smoke travel out of the hoistway.

Hoistway door assemblies are listed and labeled with a fire-resistance rating. The fire-resistance rating should be clearly visible on the label, which is on the hoistway side of the door **(Figure 6.55)**. The hoistway door assembly must be installed in accordance with the manufacturer's instructions in order to satisfy the listing criteria. Any hardware, such as floor sill, header, and closure equipment, must also be labeled.

Utility Chases and Vertical Shafts

The inspector should have knowledge of chases and shafts because they can provide a vertical path for smoke and fire as well as serve as the area of origin for fires. Vertical shaft enclosures are built with fire-rated construction methods but contain combustible materials, such as pipe and electrical wiring within the shaft. These areas are critical for the inspector to verify that the required firestopping has been installed and maintained.

As described earlier in the chapter, both refuse and linen chutes are examples of vertical shafts. These types of shafts must be constructed of noncombustible material with rated doors. A fire-rated enclosure must surround the chute. Automatic sprinklers may be required at the top of the chute and in its termination room. A fire in a properly designed, installed, and maintained refuse chute should be contained within the chute.

Figure 6.53 Emergency backup generators like this one are required systems in malls and buildings with large atriums.

Figure 6.54 Electrical transformers located outside buildings generate heat and must have a cooling solution in place to function safely and efficiently.

Figure 6.55 An inspector checks for a fire-resistance label on an elevator hoistway door.

Because of poor maintenance or loss of operational integrity, it is common for some smoke to leak out of refuse chutes through the doors. Smoke is transferred throughout floors, and there may be heavy smoke in upper floors. Inspectors should be aware of these weak points and evaluate that the components of these vertical shafts are intact.

In evaluating grease ducts, the Inspector II may need to consider manifold exhaust ducts, which serve more than one exhaust hood. The duct is smallest as it leaves each single exhaust hood and increases in cross-section as it joins with another branch duct. Manifold ducts typically have separate fire suppression systems for each hood. Each branch duct, however, will operate simultaneously for the protection of the common duct. These ducts are normally found in such areas as food courts in shopping malls. Because of the common ductwork, fire may spread over large areas within the building if the system has not been properly maintained.

Solid Fuel Appliances

A solid fuel appliance is required to have its own hood and its own ducts.

Some exhaust systems have additional grease-removal devices in the duct system known as extractors. They may be located in false ceiling spaces, in a mezzanine, or on the roof. These systems may present additional fire hazards due to the accumulation of grease on filters and the fusible links that activate the fire suppression system. Some design applications include horizontal ducts, which must slope to a vertical section and cannot have a trap or dam.

Grease extraction and control equipment may be installed to remove odors from the exhaust stream. Most of this equipment is installed on the roof but may also be located in ceiling spaces. Some systems are designed with water wash or odor-control chemical spray systems. The extractor may contain the following:

- Filters
- Electrostatic precipitators
- Catalysts
- Odor absorbers
- Gas-fired afterburners

Accessibility of Grease System

Inspectors may need to evaluate that the grease system is fully accessible to verify that systems are being cleaned throughout their length to the fan.

Chapter Summary

Components that make a building habitable are incorporated into a building's structural system. Exterior walls, roofs, and floors enclose the structural components to define a building's limits. Within these components, interior walls, floors, and ceilings further divide the space to create individual work and living compartments or rooms.

Stairs, doors, and windows provide access to a structure and between the individual spaces. Walls, floors, and ceilings are finished with interior finishes that may or may not contribute to fire spread by increasing or limiting the fuel load of the compartment. Building services include elevators, HVAC systems, and others that may also increase or decrease the inherent fire hazards within a building. An inspector must be able to evaluate these building components and determine the level of fire protection provided by them.

Review Questions

1. What are the different types of walls found in structures and the fire-related hazards they present?
2. How does fire resistance vary between the different roof types and roof coverings?
3. List flooring characteristics and components an inspector may encounter.
4. What functional roles do ceilings play?
5. What is the best resource for determining appropriate stair dimensions or configurations?
6. How do door operations impact fire risk?
7. What types of door closure components are used on fire doors?
8. Describe the common window configurations and operations.
9. What are the four ways in which interior coverings contribute to fire behavior?
10. What are some of the fire risks associated with building services?

1. What are some of the common characteristics of fire walls?
2. What are the hazards solar panels pose for firefighters?
3. List the information that cannot be obtained from e-911 standard testing.
4. What are some of the characteristics of interior components an Inspector II should be prepared to evaluate?
5. Provide an example of an inconsistent door classification.
6. What are the common testing methods used to evaluate interior finishes?
7. What are some of the building service hazards inspectors should be aware of during an inspection?

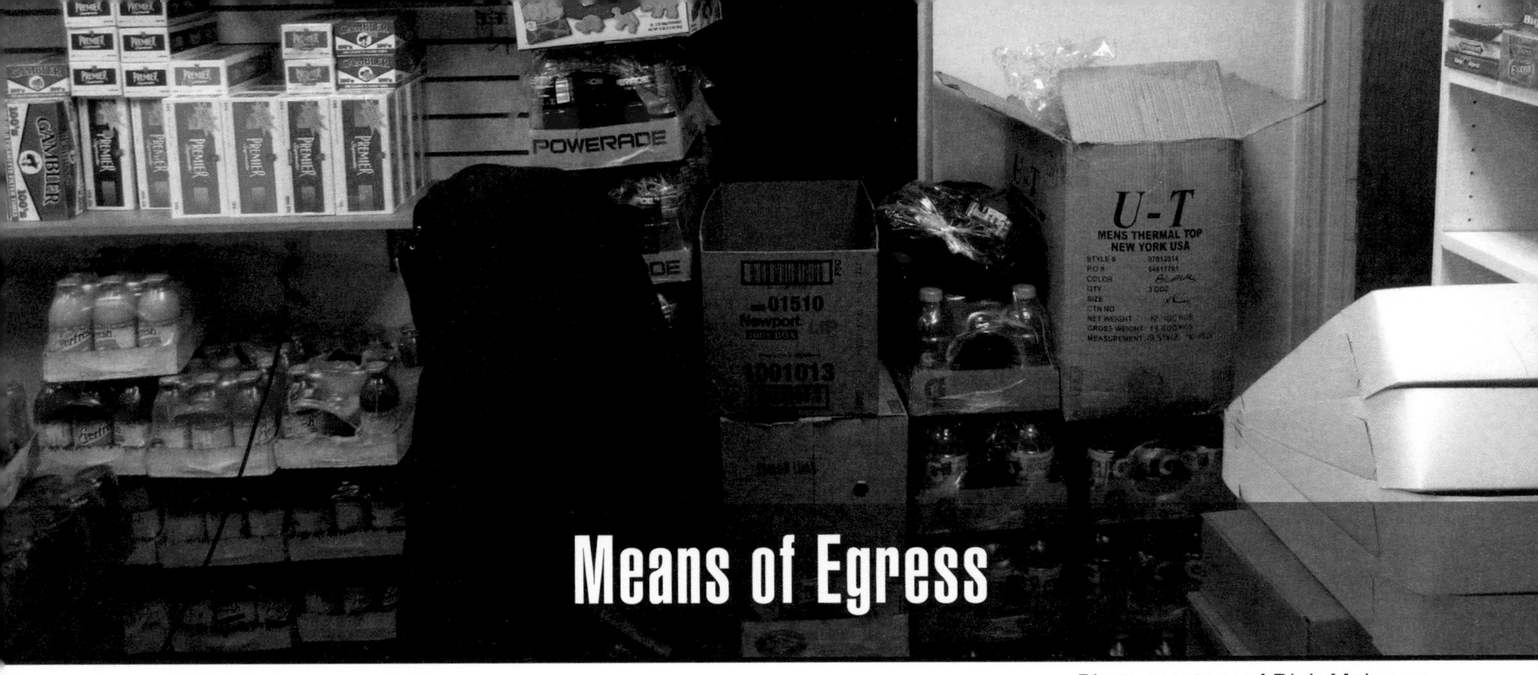

Means of Egress

Photo courtesy of Rich Mahaney.

Chapter Contents

Case History 241
Means of Egress Systems 241
Elements 246
Building Components 251
Exit Illumination and Markings 262
Occupant Loads 265
Multiuse Occupant Loads 268
Means of Egress Determinations 268
Capacity 268
Arrangement.. 270
Effectiveness 273
Chapter Summary 275
Calculations 276
Review Questions 280

chapter 7

Key Terms

Americans with Disabilities Act (ADA) of 1990 - Public Law 101-336 245
Area of Refuge 244
Clear Width 244
Common Path of Travel 246
Exit 246
Exit Access 246
Exit Discharge 250
Horizontal Exit 248
Means of Egress 242
Occupant Load 265
Panic Hardware 254
Public Way 244
Smokeproof Enclosure 249
Travel Distance 272

NFPA® Job Performance Requirements

This chapter provides information that addresses the following job performance requirements of NFPA® 1031, *Standard for Professional Qualifications for Fire Inspector and Plan Examiner* (2014).

Fire Inspector I	Fire Inspector II	
4.3.2	5.2.2	5.3.11
4.3.3	5.3.1	5.4.1
	5.3.2	5.4.2
	5.3.3	5.4.5
	5.3.5	

Means of Egress

Learning Objectives

After reading this chapter, students will be able to:

Inspector I

1. Describe means of egress systems. (4.3.3)
2. Explain the way to calculate occupant load for single-use occupancies (4.3.2)

Inspector II

1. Explain the way to calculate occupant loads for multiuse occupancy. (5.2.2, 5.3.1, 5.3.2, 5.3.3, 5.4.1, 5.4.2)
2. Explain the steps in determining means of egress. (5.3.5, 5.3.11, 5.4.5)

Chapter 7
Means of Egress

Case History

San Antonio, Texas, August 1012: Fire in the Amistad Residential Facility, a group home for persons with mental disabilities, claimed the lives of four of the thirteen residents. The fire, which started in a closet, spread up the stairs to block the only functioning exit. Another exit was blocked by a half-completed addition that had been halted by the city because the owner lacked a building permit. The home had no sprinkler system and no round-the-clock staff.

In previous years, the state had responded to numerous complaints at the facility and had found evidence of abuse, neglect, and exploitation of clients. In 2010, the Texas legislature stopped requiring facilities such as Amistad to carry state licenses. Before then, they were categorized as assisted living facilities and required to undergo annual safety inspections. Their new classification termed them as boarding homes. Inspections fell to city or county governments. Another fire during the same month in a similar facility prompted the city to expedite regulation of such homes.

An inspector performs a vital role in fire and life safety by evaluating exits for safety and capability. Beginning with the plans review process for new or remodeled buildings, an inspector verifies the occupancy classification and notes that the means of egress meets the adopted local fire and building code requirements. During inspections, an inspector monitors the condition of existing means of egress and verifies that conditions have not changed regarding occupancy, use, or exit requirements for the structure. This chapter discusses means of egress systems, occupant loads, and means of egress determinations. Examples of calculations for means of egress are also included.

Means of Egress Systems

The ability of occupants to evacuate a structure rapidly and safely during a fire or other emergency is a crucial aspect of building design. Unfortunately, many lives have been lost during emergencies because exits were locked, blocked, or poorly marked **(Table 7.1, p. 242)**. While there have been many improvements in fire protection and code revisions and updates as a result of large loss-of-life fires, common factors continue to be cited even today:

- Locked or blocked exit doors **(Figures 7.1 a - c, p. 243)**
- Improperly designed or marked exits
- Inaccessible exits

Table 7.1 Factors in Loss-of-Life Fires				
Year	**Fire Event**	**Occupancy**	**Deaths**	**Contributing Factors**
1911	Triangle Shirtwaist Co, New York, NY	Industrial, Manufacturing	146	Locked exits, heavy fuel load, Inward-opening exit doors
1944	Winecoff Hotel, Atlanta, GA	Residential, Hotel	119	Lack of installed fire protection, combustible interior finishes, blocked exits, faulty wiring
1977	Beverly Hills Supper Club, Southgate, KY	Assembly, Nightclub	164	Overcrowding, inadequate exits, faulty wiring
1980	MGM Grand Hotel, Las Vegas, NV	Residential, Hotel	84	Faulty installation, lack of fire protection, flammable interior materials
1991	Imperial Foods Processing, Hamlet, NC	Industrial, Plant	25	Locked exits, lack of safety inspections, lack of fire protection systems
2000	The Station, West Warwick, RI	Assembly, Nightclub	100	Combustible interior materials, pyrotechnics, heavy smoke, over-capacity for exits
2003	Cook County, Chicago, IL	Business, Office	6	Locked stairwells, smoke
2003	Greenwood Health Center, Hartford, CT	Institutional (Nursing Home)	16	Lack of sprinklers, inability to self-evacuate
2014	L'Isle Verte, Quebec	Institutional (Senior Residence)	32	Lack of fire protection, inability of residents to self-evacuate, limited staff

Table 7.1: Despite improvements in building codes and a greater awareness of the importance of installed fire protection systems and adequate exiting capacities, people continue to die in fires because they are unable to quickly exit the structure or are overcome by toxic products of smoke.

- Inadequate or nonexistent fire protection systems
- Combustible interior materials or contents
- Overcrowded occupancies

Means of Egress — Safe, continuous path of travel from any point in a structure to a public way; the means of egress is composed of three parts: the exit access, the exit, and the exit discharge.

Because of these potentially life-threatening conditions, the **means of egress** is one of the most important factors inspectors must consider in determining whether a structure can be considered safe. A means of egress system is composed of three basic elements **(Figure 7.2)**:

- The exit access
- The exit
- The exit discharge

Figure 7.1 These locked exits were found during fire inspections. Inspectors must work immediately to correct these life safety hazards. *a and b courtesy of Rich Mahaney; c courtesy of Scott Strassburg.*

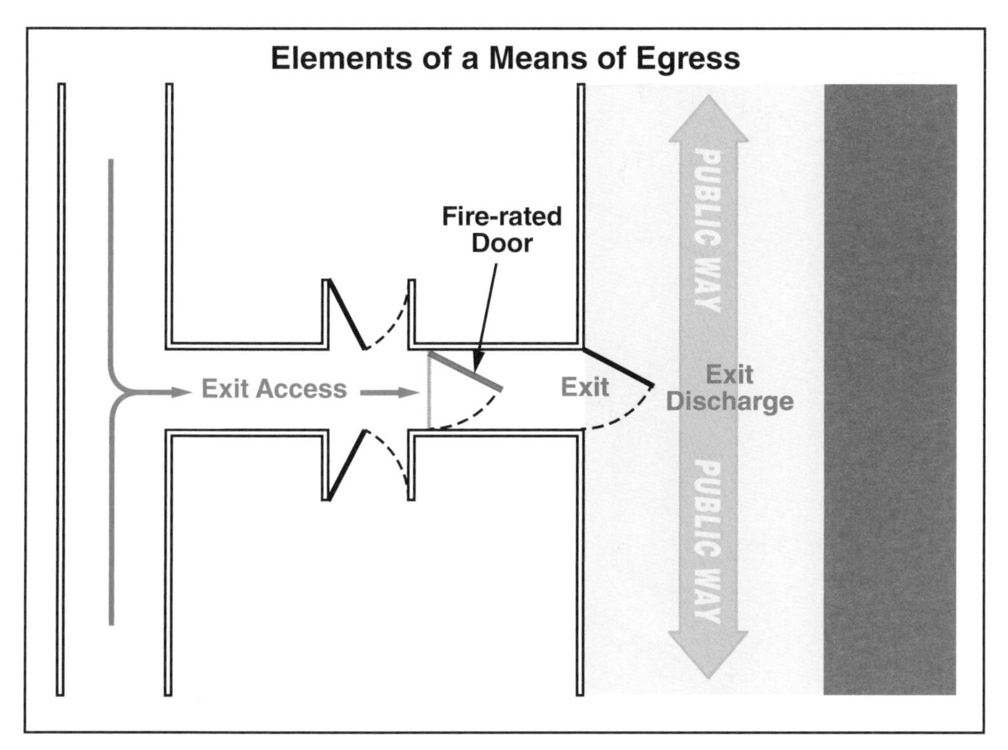

Figure 7.2 The elements of a means of egress are the exit access, the exit, and the exit discharge.

These elements must work together to allow occupants to safely exit from anywhere in a structure during an emergency.

A means of egress system relies on a number of subsystems to protect and guide occupants during their escape. These subsystems may include such active fire protection systems as automatic sprinklers, or passive fire protection, such as fire-resistance-rated construction of doors and walls.

Other means of egress system components include:

- Exit signs
- Exit illumination
- Door hardware
- Handrails

According to NFPA® 101, *Life Safety Code®*, the means of egress system may pass through the following:

- Intervening room spaces
- Doorways
- Corridors
- Passageways
- Balconies
- Ramps
- Stairs
- Courts
- Yards
- Horizontal exits

Public Way — Parcel of land such as a street or sidewalk that is essentially open to the outside and is used by the public to move from one location to another.

Area of Refuge — An area where persons unable to use stairways can remain temporarily to await instructions or assistance during emergency evacuation.

Clear Width — Actual unobstructed opening size of an exit.

The means of egress must terminate in a **public way** or an **area of refuge**. A public way is a street, alley, or similar parcel of land essentially open to the outside that is used by the public. A public way must also have a **clear width** and height of 10 feet (3 m) **(Figure 7.3)**.

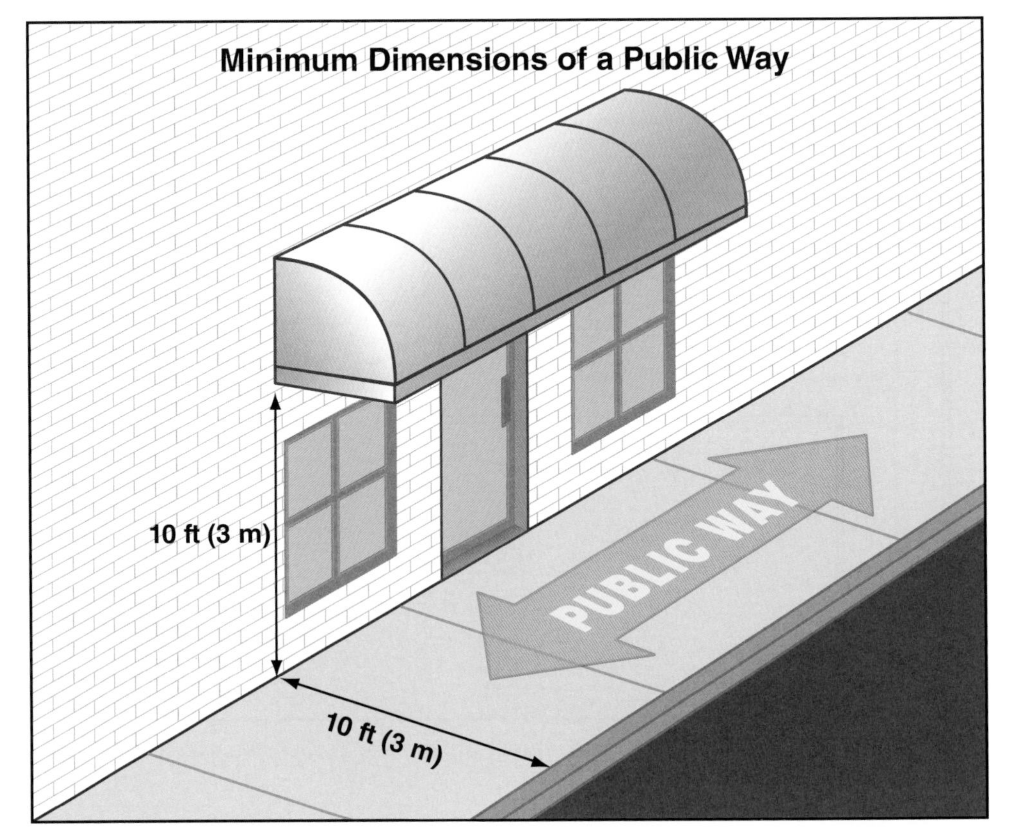

Figure 7.3 Model building codes establish the minimum dimensions for a public way.

An area of refuge should provide protection to those occupants who cannot evacuate a building. Areas of refuge can be located in an approved elevator lobby or a rated stairwell **(Figures 7.4 a and b)**. These areas should be separated and resist the passage of smoke. If two-way communication is required in an area of refuge, the inspector should verify that communication devices are present and in working order.

With the enactment of the **Americans with Disabilities Act (ADA)** in 1990, areas of refuge were required in some occupancies as a way to protect people who are mobility impaired. The design of the area of refuge must meet the requirements of the ICC/ANSI 117.1, Standard on Accessible and Usable Buildings and Facilities, from the International Code Council®/American National Standards Institute. An inspector should be aware of the ways the ADA requirements may affect the configuration of these areas.

Americans with Disabilities Act (ADA) of 1990 - Public Law 101-336 — Federal statute intended to remove barriers, physical and otherwise, that limit access by individuals with disabilities.

Figures 7.4 a and b – *a* The fire doors that can close off an elevator or stairwell enable the spaces inside to serve as an area of refuge. *b Courtesy of Rich Mahaney.*

Elements

All of the elements of a means of egress must be free from obstructions and combustible materials if they are to be useful during an emergency. For example, no furnishings or decorations may be allowed to obstruct or conceal an exit or exit access at any time. The following sections describe each of the elements of a means of egress.

Exit Access — Portion of a means of egress that leads to the exit; for example, hallways, corridors, and aisles.

Common Path of Travel — The route of travel used to determine measured egress distances in code enforcement. The common path of travel is considered to be down the center of a straight corridor and a 1-foot radius around each corner. Also called the normal path of travel.

Exit Access

The **exit access** leads from an occupied portion of a building or structure to the exit **(Figures 7.5 a and b)**. Examples of an exit access include:

- Corridor leading to the exit opening
- Aisle within an assembly occupancy that is designed to accommodate and guide people to an exit
- **Common path of travel** leading from inside a space to an exit
- Unenclosed ramp or stairs
- Occupied room or space

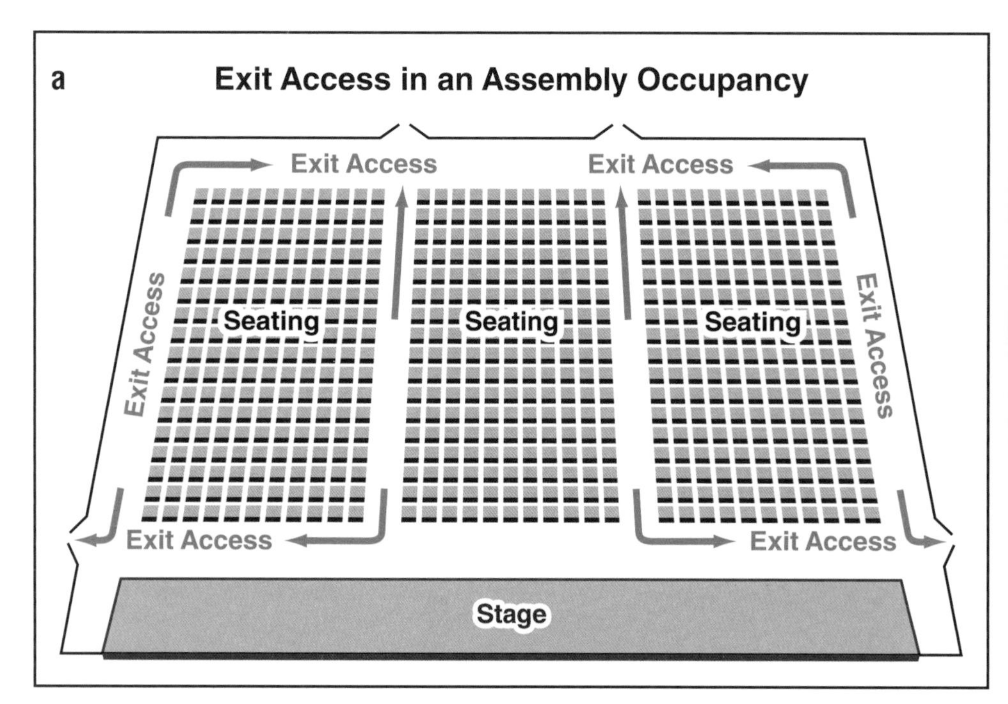

Figures 7.5 a and b These two examples illustrate an occupancy's exit access, which is the route occupants travel to leave occupied areas and locate an exit.

Exit — Portion of a means of egress that is separated from all other spaces of the building structure by construction or equipment, and provides a protected way of travel to the exit discharge.

Exit

An **exit** is a protected path consisting of exit components constructed of approved fire-resistance-rated assemblies, such as exit passageways and enclosures. The exit is separated from the area of the building from which escape is to be made **(Figures 7.6 a and b)**. Some examples of exits include the following:

- Doors at ground level that lead directly to the outside of the building (See Components section for more information on doors.)
- Exit passageway to the outside
- Horizontal exit
- Smokeproof enclosure (a stairway that is enclosed by fire-resistance-rated walls and self-closing rated doors)

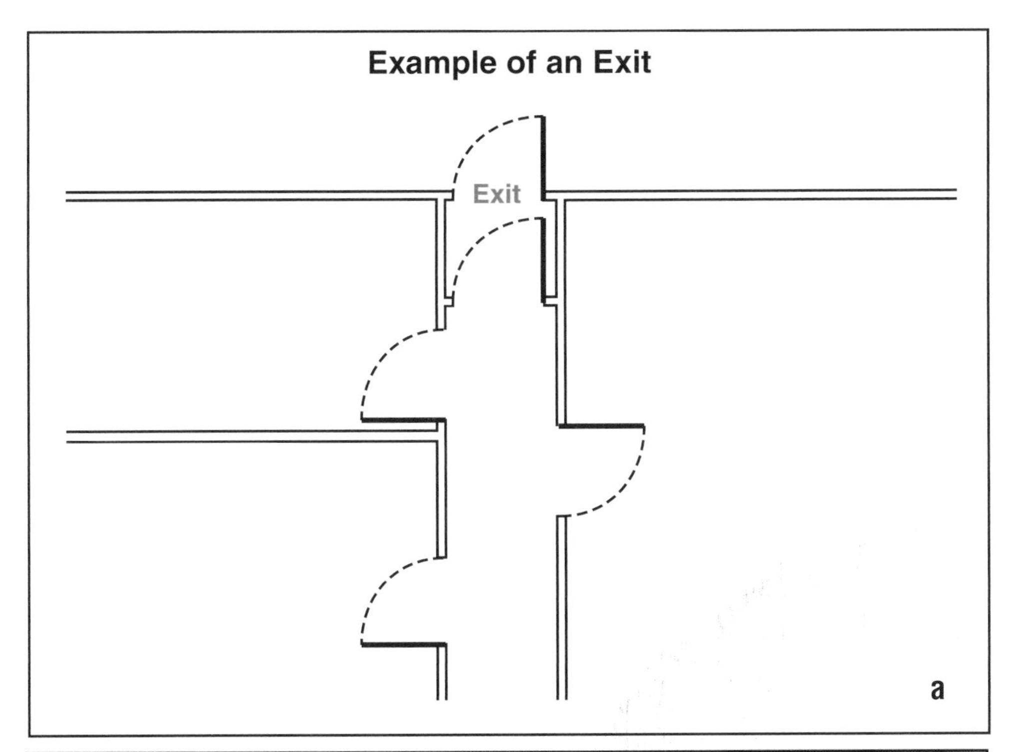

Figures 7.6 a and b Exits should be clearly marked and constructed with design features that make them safe means of egress even under emergency conditions.

Exit passageways. An exit passageway is similar to an exit stair enclosure, except that it is constructed in a horizontal plane. Exit passageways are designed to connect an interior exit stair or a hallway without stairs on the same level to an exit door on the structure's exterior. Exit passageways must be constructed of the same fire-resistant-rated material as the exit stairs. In large-area buildings, such as shopping malls, factories, or industrial complexes, the exit passageways can be used to shorten the travel distance to an exit.

NOTE: Travel distances are described later in this chapter.

Horizontal Exit — A path of egress travel from one building to an area in another building on approximately the same level, or a path of egress travel through or around a wall or partition to an area on approximately the same level in the same building.

An exit passageway must be wide enough to accommodate the total capacity of all exits that discharge through it. Therefore, its design must be calculated using the expected occupant load for that portion of the structure and not just the immediate occupant load.

NOTE: Occupant loads are described more fully later in this chapter.

Horizontal exits. **Horizontal exits** are commonly used in (but not limited to) high-rise buildings and hospitals as a means of passing through a fire-barrier wall that separates two fire compartments in a structure **(Figures 7.7 a and b)**. Horizontal exits may be substituted for other exits if they do not compose more than 50 percent of the total exit capacity of the building. Two basic types of horizontal exits are:

- A means of egress from one building to an area of refuge in another building on approximately the same level.
- A means of egress through a fire barrier or fire wall to an area of refuge at approximately the same level in the building that provides protection from smoke and fire.

These exits require fire walls or fire-barrier walls with at least a 2-hour fire-resistance rating. A 1½-hour fire-rated door assembly would be installed in the fire barrier to permit movement between the two compartments. Because doors must swing in the direction of exit travel, double doors are required if the area on both sides of the wall is used as part of the mean of egress.

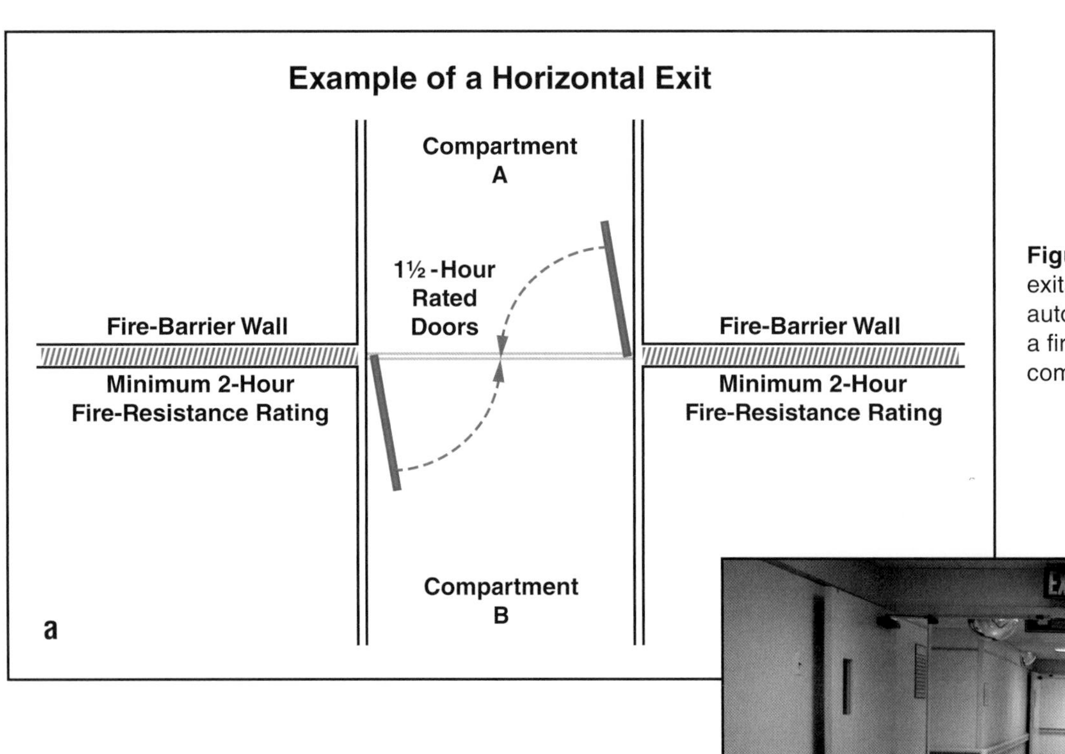

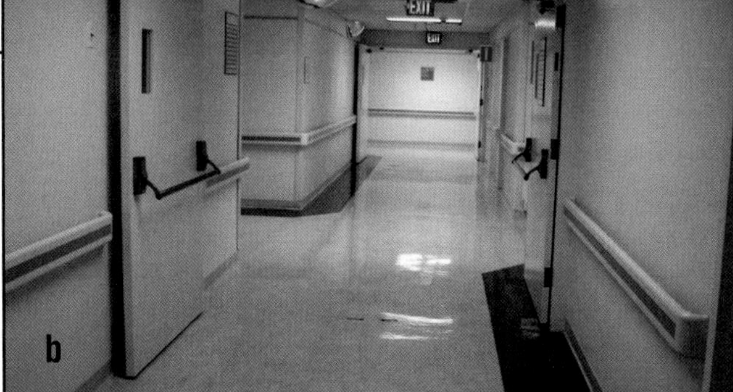

Figures 7.7 a and b Horizontal exits include doors that close automatically to complete a fire barrier between two compartments.

Smokeproof enclosures. **Smokeproof enclosures** are stairways or stair enclosures (sometimes referred to as smoke towers) that are designed to limit the penetration of smoke, heat, and toxic gases into the stairway. Smokeproof enclosures provide the highest degree of fire protection for stair enclosures that the model codes require **(Figures 7.8 a and b)**. Two ways to make the stair enclosure smokeproof are to pressurize it or to use natural or mechanical ventilation.

Smokeproof Enclosures — Stairways that are designed to limit the penetration of smoke, heat, and toxic gases from a fire on a floor of a building into the stairway and that serve as part of a means of egress.

Occupants enter a smokeproof enclosure through a vestibule or outside balcony. This arrangement prevents smoke from entering the stairwell when corridor doors are opened.

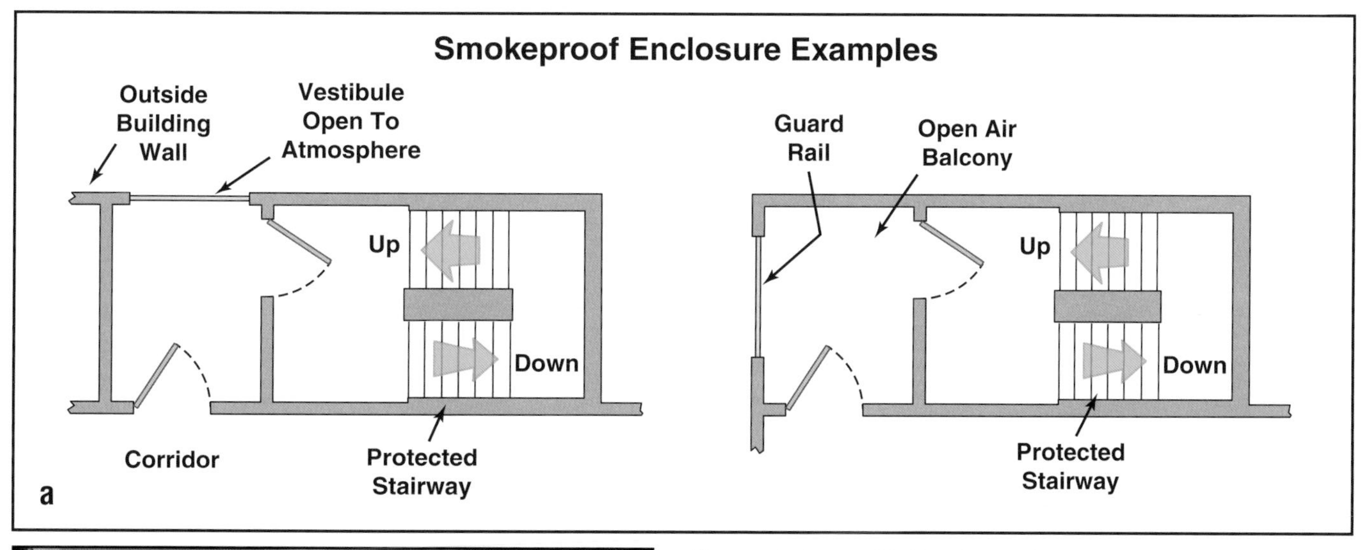

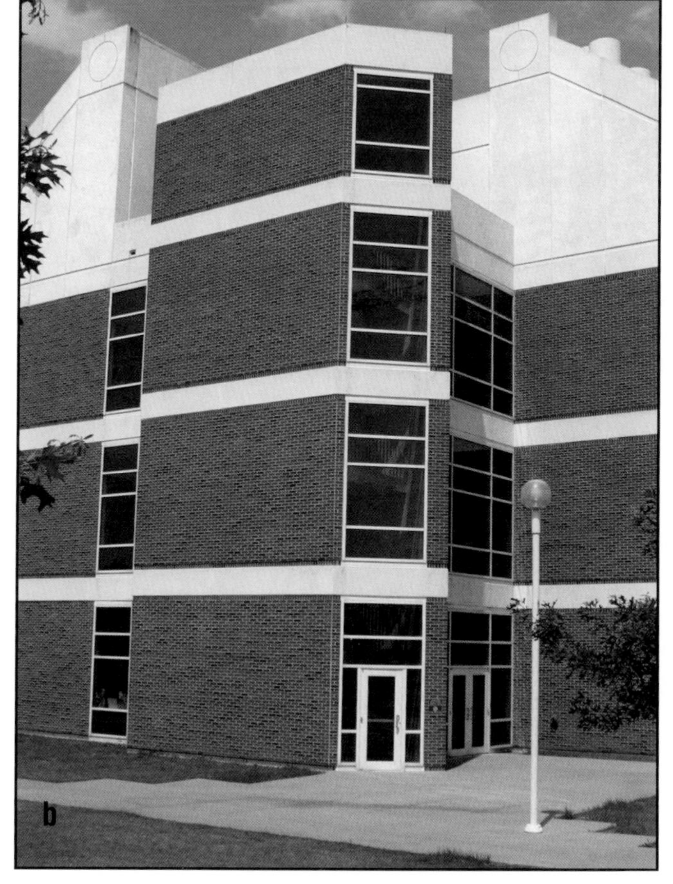

Figures 7.8 a and b Smokeproof enclosures protect stairways on the exterior of a structure from the communication of smoke and fire and also provide the highest level of means-of-egress protection that model codes require.

Exit Discharge — That portion of the exit that is between the exit and a public way.

Exit Discharge

The **exit discharge** is that portion of a means of egress that is between the exit and a public way **(Figures 7.9 a and b)**. Two examples of exit discharges are an exterior walkway along the side of a structure from the exit to a public way and a privately owned drive or alley that connects the exit to a public way. Inspectors frequently discover that exit discharges are blocked or cluttered to such a degree that occupant safety is compromised.

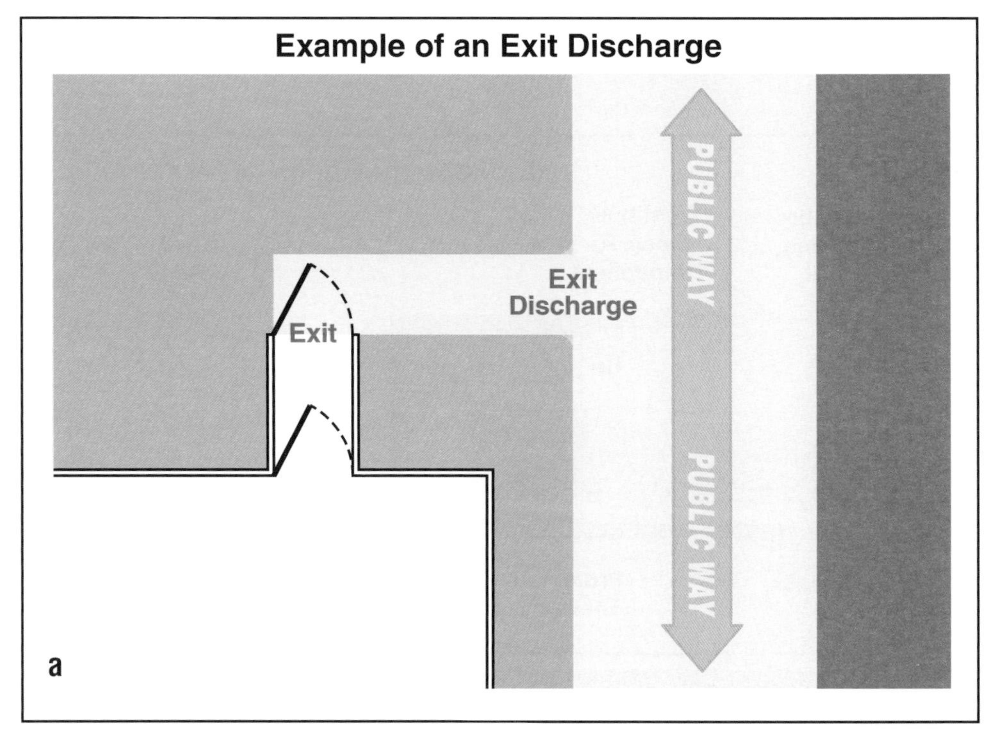

Figure 7.9 a and b An exit discharge between an exit and a public way must remain free of clutter and obstructions.

Escalators, Elevators, and Moving Walkways

None of the model codes permit escalators, elevators, or moving walkways to be considered as part of a means of egress in new occupancies unless designed accordingly. Some jurisdictions may allow escalators to be counted as part of the means of egress in certain older occupancies. These situations are limited and should not be assumed to exist in all municipalities.

Building Components

As mentioned previously, means of egress components may be active, such as an automatic sprinkler system, or passive, such as doors, walls, and floors. Part of the inspector's job is to:

- Verify that doors, stairs, and walls conform to the adopted building and safety codes.
- Recognize and evaluate those building components that serve as components of a means of egress.
- Be aware of any exemptions in the means of egress requirements that are permitted in structures that are protected by automatic sprinkler systems. NFPA® 101, *Life Safety Code®*, contains specific examples of automatic sprinkler exemptions.

Building components that make up fire-resistance-rated assemblies provide occupants with separation from products of combustion. These assemblies include but are not limited to doors, walls, ceilings, floors, stairs, and ramps. In addition, panic hardware, handrails, and guardrails may assist occupants in opening doors and protect them on stairs and ramps.

NOTE: This chapter section describes fire protection components. Chapters 12 and 13 describe active components, such as sprinkler systems and fire extinguishers.

Doors

Two of the most important life safety functions of doors are to:

- Separate occupants from the movement of fire, smoke, and other toxic gases
- Serve as components of a means of egress system

NOTE: Exit doors also provide a means of access for fire and emergency services personnel.

Doors that serve as components of a means of egress must be constructed so that the way of exit travel is obvious. Exit doors may be required to open in the direction of travel toward the primary exit, depending on occupancy and occupant load **(Figure 7.10, p. 252)**. Each door opening must be wide enough to accommodate the number of people expected to travel through the door in an emergency.

In new buildings, model codes require that swinging or hinged doors serving as a component of a means of egress provide a minimum of 32 inches (800 mm) of clear unobstructed width **(Figure 7.11, p. 252)**. There are exceptions to this rule for existing buildings and other special situations. The inspector should refer to the NFPA® *Life Safety Code®* or the applicable building code for information about exceptions to widths of exits.

The floor or landing on each side of the door must be level and at the same elevation on both sides of the doorframe. The door must not open over a set

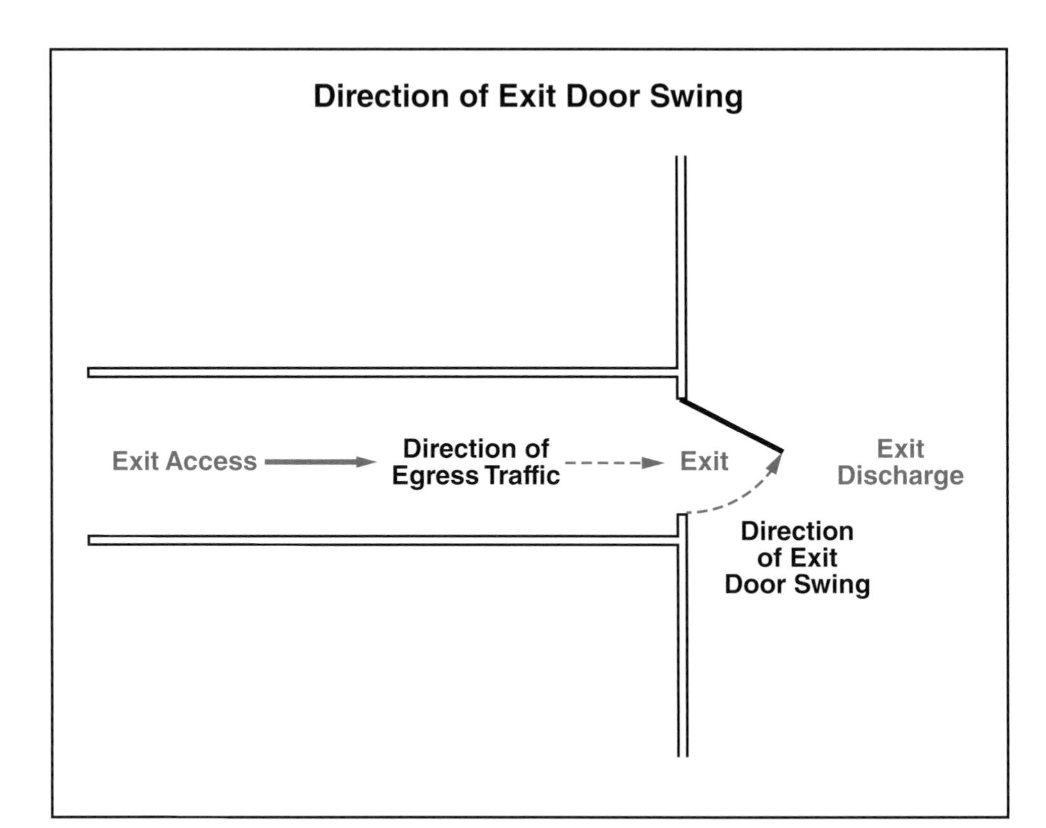

Figure 7.10 Doors that are part of a means of egress system must open in the direction of occupant travel.

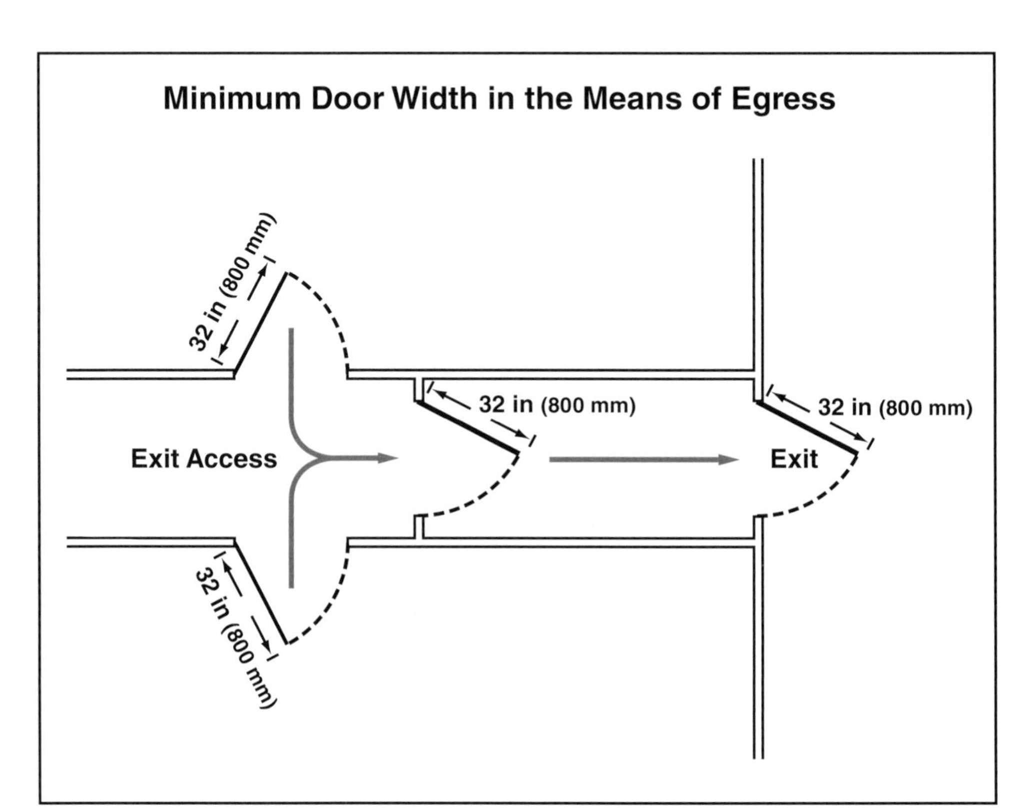

Figure 7.11 Exit doors are required to have specific widths like those illustrated. Inspectors should be familiar with the regulations and exceptions.

of stairs. While the building is occupied, the exit doors must open easily. If 50 or more people are in a room or occupancy, the exit requirements for the building take on a higher level of significance and additional exiting system requirements must be met. Doors that open into corridors or hallways must be placed so that the door does not obstruct more than half of the required exit width during any point in its swing **(Figures 7.12 a and b)**.

Building codes allow a variety of security options for entrance and egress doors. Inspectors should become familiar with the operation and acceptance of different types of locking hardware. Locking hardware allows exit doors to remain locked from the outside but still permits slightly delayed or unrestricted egress at all times. Other security devices cause exit doors to remain locked until activated by fire detection, waterflow, or other alarm devices.

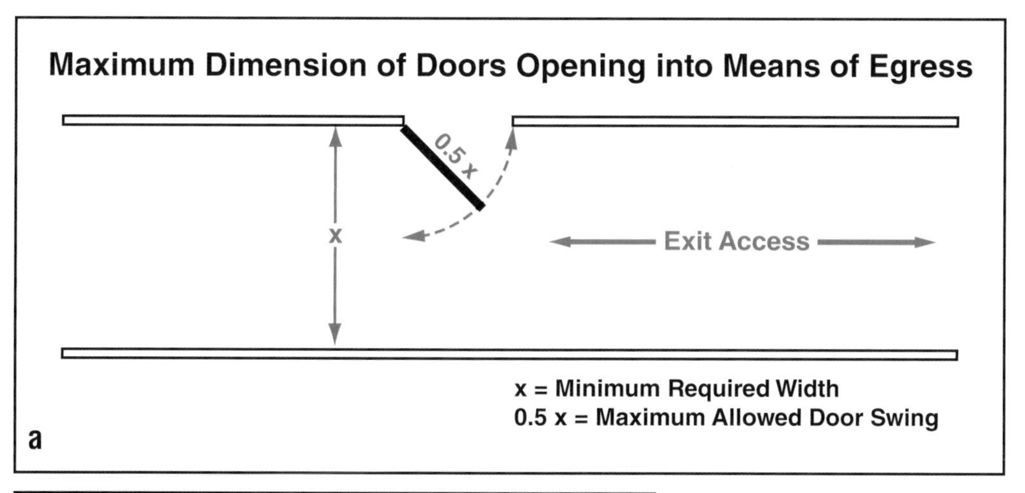

Figures 7.12 a and b Doors opening into means-of-egress corridors must not impede the ability of occupants to navigate the corridor to available exits in an emergency.

Several requirements have to be met to use the different types of locking and security hardware. The inspector should be familiar with these requirements, both for the safety of the public as well as any responding firefighters. Some of the most common locking hardware includes the following:

- Panic hardware
- Bolt locks

- Access-controlled egress
- Delayed egress
- Electromagnetically locked
- Special locking

Panic Hardware — Hardware mounted on exit doors in public buildings that unlocks from the inside and enables doors to be opened when pressure is applied to the release mechanism. Also called Exit Device.

Panic hardware. Panic hardware may be required based upon occupancy classification and occupant load. Occupants should be able to operate panic hardware as follows:

— Causing the latch to release by applying a force of no more than 15 pounds (7.5 N).

— Setting the door in motion by applying a force of no more than 30 pounds (15 N) **(Figure 7.13)**.

In some occupancies, operating panic hardware to open exit doors activates an alarm. This alarm is intended to notify the employees or management or to help prevent theft.

Bolt locks. In buildings that do not require panic hardware on exit doors, the model building codes define the types of door locks that may be installed:

— In most cases, exit doors can feature deadbolt locks that operate with the action of the latch and do not require keys, special tools, or knowledge to operate.

— Thumb-turn bolts that operate independently of the door latch are examples of unacceptable locking devices on the egress (inside) side of exit doors (**Figure 7.14**).

— Manually operated flush bolts or surface bolts are not permitted.

Access-controlled egress. This type of hardware is designed to allow occupants to egress freely while enabling the door to be secured from the outside **(Figure 7.15)**. The occupant must use a security card or enter a code to gain entrance. Several things are required for this type of hardware:

Figure 7.13 Typical panic hardware on an exit door should be easy for any occupant to operate. Here, an inspector verifies that an exit door opens easily.

Figure 7.14 A tubular deadbolt lock uses a thumb-turn knob on the inside and a keyway on the outside of the door. This type of locking device is unacceptable.

— A sensor on the egress side detects an occupant approaching.

— Loss of power to the hardware automatically unlocks the doors.

— The doors must be arranged to unlock from a manual unlocking device that delays relocking for a minimum of 30 seconds and has a sign stating, "PUSH TO EXIT."

— Activation of the building fire alarm, fire detection, or automatic sprinkler system will unlock the doors.

— Some entrance doors in certain occupancy classes may not be allowed to be secured during periods of occupancies - or at all.

Delayed egress. Doors unlock upon operation of automatic sprinkler system or automatic fire detection system. Doors also unlock when power controlling the locking mechanism is lost. Additional requirements:

— Initiating the release mechanism by applying a force of no more than 15 lbs (67 N) must permit the door to open within 15 seconds, or in some cases, 30 seconds.

— Opening the door will activate an audible signal.

— Doors must be labeled with a sign stating, "PUSH UNTIL ALARM SOUNDS. DOOR CAN BE REOPENED IN 15/30 SECONDS."

— Occupants must not have to pass through more than one door equipped with delayed egress before reaching an exit.

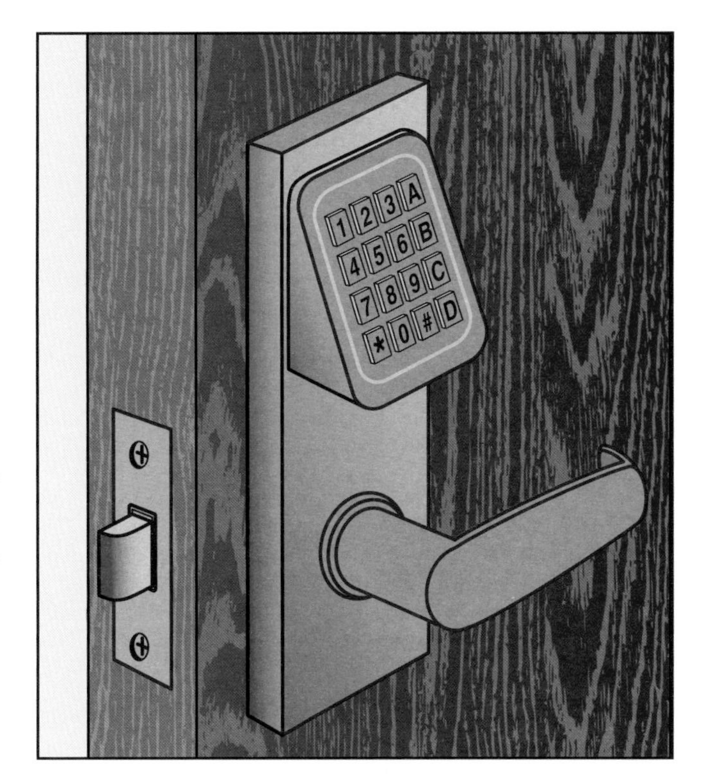

Figure 7.15 Electronic keyless locks provide higher security by maintaining continuous security and controlled access.

Figure 7.16 Electromagnetic locks hold a door securely as long as an electric current passes through the electromagnet and the armature plate.

Electromagnetically locked. This type of hardware may be used on doors in the means of egress that are not otherwise required to have panic hardware. These doors must have listed hardware that incorporates a built-in switch. This switch must directly supply power to the electromagnetic lock, not through a computer system or other device as would be used in access-controlled egress systems. This hardware is a basic system that detects a push on the hardware and supplies power to the mag-lock in the doorframe **(Figure 7.16)**. Card readers or punched-in codes are not needed to gain access. System requirements:

— Listed hardware that is affixed to the door and has an obvious method of operation.

— Must be capable of being operated with one hand.

— Operation of the hardware releases to the electromagnetic lock and immediately unlocks the door.

— Loss of power to the hardware automatically unlocks the door.

Special locking. Health care facilities may also have controlled egress security devices on some doors to prevent unauthorized exits. Examples of these types of occupancies are drug and alcohol rehabilitation and mental health or memory care units. In these situations, alarm devices, key pads, or similar staff-activated devices control exit door locks. A loss of power in the facility allows the doors to open.

Inspectors must also be familiar with self-closing doors and latching doors that are activated by fire detection, waterflow, or other alarm devices. Self-closing doors — usually held open by magnetic devices — close automatically to provide a smoke or fire barrier in a fire-barrier wall. Self-closing doors may also be found in the following locations:

- Openings into mechanical or electrical spaces
- Walls separating connected buildings
- Walls that separate two different occupancy use groups **(Figure 7.17)**.

In occupancies such as business, factory, mercantile and storage, owner/occupants may install security bars or grilles over exit doors to prevent unauthorized entry. Security bars slow access by fire and emergency services personnel and have been responsible for many fatalities. Requirements include:

- Operation of locks on the egress side of the security bars must conform to the requirements for exit doors **(Figure 7.18)**.
- Horizontal sliding or vertical security grilles at the main exit must be operable from the inside without the use of a key or special knowledge or effort during periods of occupation.
- During periods of occupancy by the general public, the grilles must remain secured in the full-open position.
- Where two or more exits are required, no more than half of the exits can be equipped with security grilles.

Walls

Fire-resistant-rated walls are used to separate designated exits from other parts of the building. The fire-resistance rating is based on the type of occupancy, building height, type of building construction, and whether the building has an automatic sprinkler system **(Figure 7.19)**. Because not all older buildings have been retrofitted to meet current requirements, an inspector should become familiar with the building codes that were in effect when the structure was originally built. Modifications to the structure or changes in the occupancy classification may require changes in the exit-wall requirements. An inspector should also look for any concealed penetrations in the walls above decorative-type ceilings.

The type of occupancy group also dictates the wall finish (paint, wallpaper, fabric, or other materials). Generally, NFPA® permits only the use of Class A (0–25 flame-spread rating) or Class B (26–75 flame-spread rating) interior finishes in exits or exit-access corridors.

NOTE: Refer to Chapter 6, Building Components, for more information on interior finishes and flame-spread ratings.

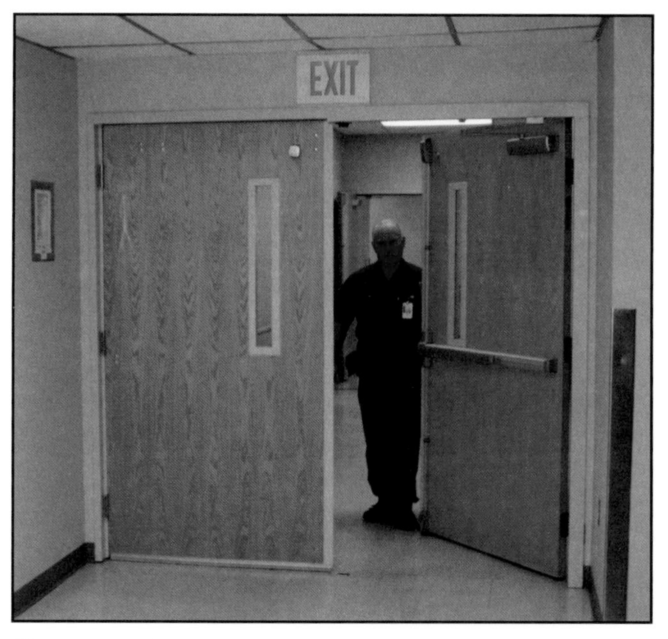

Figure 7.17 Self-closing doors provide an automatic impediment to the communication of fire and smoke.

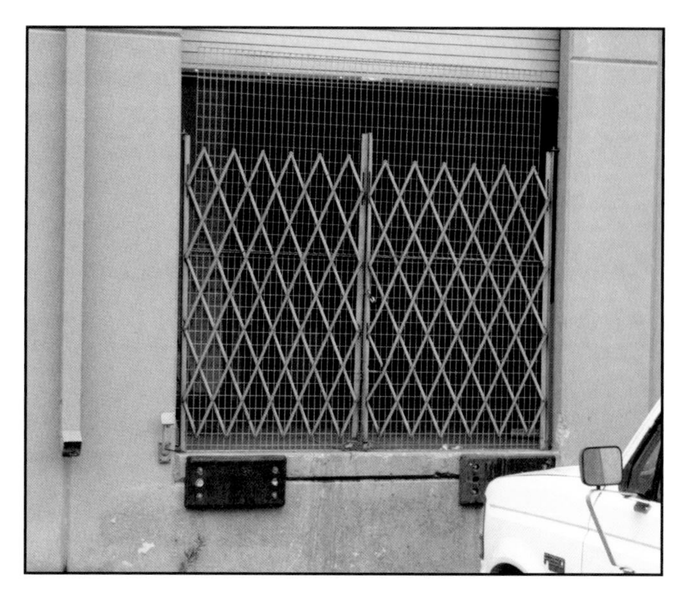

Figure 7.18 Security grilles like the gate on this loading dock's overhead door may prevent unauthorized entry but can be a barrier to a means-of-egress system.

Figure 7.19 The 1-hour fire-resistive walls in this hospital corridor are designed to protect the means of egress from fires originating in adjacent spaces.

Ceilings

Ceilings complete the enclosure for the exit or exit-access corridor. Depending on the design requirements, walls may extend to the bottom of the roof deck or floor above, or to the ceiling of the exit. In the first case, the ceiling may conceal wiring and HVAC ductwork, or act as a return-air plenum. A decorative ceiling may not be required to have any fire-resistance rating, but a ceiling that is part of a means of egress must have the same fire-resistance rating as the walls.

Floors

The type of floor covering is a frequent source of concern due to the flammability of the materials used. Floor coverings, including subflooring, must meet flame- and smoke-generation tests for the given means of egress and be installed according to the locally adopted code. Because unapproved materials are often used during remodeling, the inspector must pay particular attention to these areas.

Stairs

Stairs that serve as part of the required means of egress must meet strict requirements in regard to the fire-resistance rating of the enclosure, or separation when exterior stairs are utilized. Other requirements also increase safety during both emergency and nonemergency use.

In multistoried buildings, exit stairs are critical components of the means of egress. Stairways must be at least 44 inches (1 100 mm) wide unless the total occupant load of all floors served by the stairway is less than 50 people. In this case, stairways must be at least 36 inches (900 mm) wide. Other requirements include the following:

- Stair treads must be solid and slip resistant.
- Landings must be provided so that no flights of stairs are greater than 12 feet (3.5 m) high **(Figure 7.20)**.
- Handrails may be required for both sides of the stairs.
- Stairs that are exceptionally wide may be required to have intermediate handrails in the middle of the stairs.
- Stair treads and risers must be in good condition to prevent tripping.

To provide a protected path of travel and qualify as an exit, fire-resistant construction must separate interior stairs from other parts of the building. The construction that encloses the exit must have at least a 1-hour minimum fire-resistance rating when the exit connects three stories or fewer. This minimum applies whether the stories connected are above or below the story at which the exit discharge begins.

When the exit connects four or more stories, the separating construction must have a fire-resistance rating of at least two hours. Again, there are some exceptions to this general rule and an inspector must consult the locally adopted code.

A self-closing and latching fire-resistance-rated exit door must protect any opening in the exit stairs **(Figure 7.21)**. The stairway door must be rated properly, depending on the height of the building. The only permissible openings in exit stairs are those that allow people to enter the stairs from inside the building and those that empty to the exit discharge.

Door Openings in Exit Stairways

Exit doors must have a 1-hour rating when used in a 1-hour-rated enclosure and at least a 1½-hour rating when used in a 2-hour-rated enclosure.

Exterior stairs may serve as exit stairs if they meet the adopted code requirements for outside stairs **(Figure 7.22)**. In certain occupancies, open stairs used as part of the means of egress are only allowed in buildings equipped with an automatic sprinkler system. People often use stairways to store trash, furniture, or other items. However, these areas must remain clear so that egress is not impaired. Emergency responders also need cleared stairwells for building access. Inspectors should verify that exit stairways are not used for any purposes other than as a means of egress.

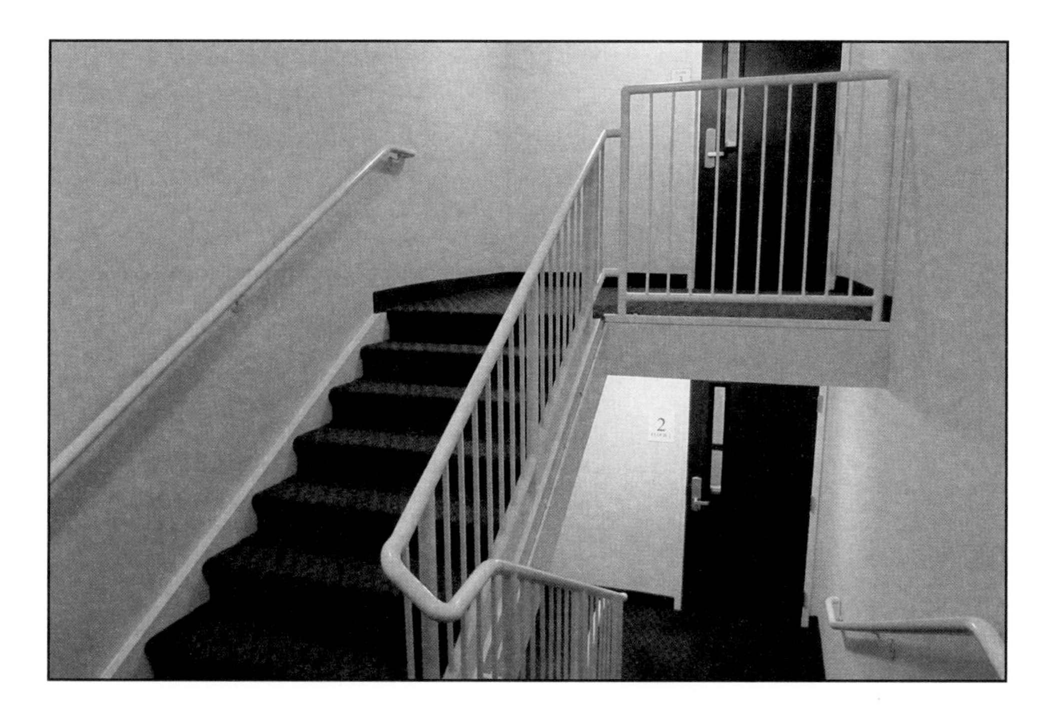

Figure 7.20 All exit stairways must have handrails, and each flight of stairs between landings must be no longer than 12 feet (4 m).

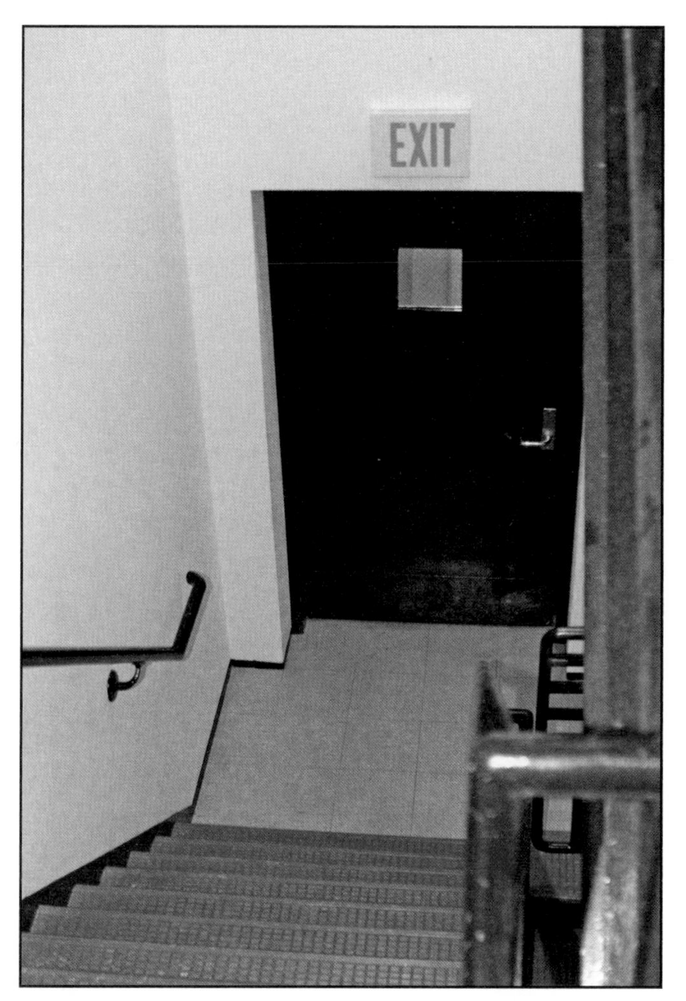

Figure 7.21 Self-closing exit doors in stairways are similar to those in corridors but are required to have fire ratings applicable to the stairways that they protect.

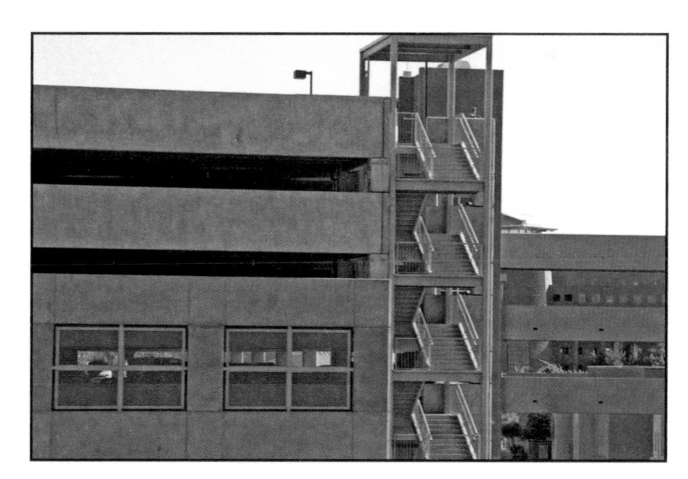

Figure 7.22 These exterior stairs serve as an additional means of egress from this parking garage.

NOTE: Additional information on exit stairs is included in the model codes, including occupancy-specific requirements.

Ramps

Ramps became common in many occupancies in the U.S. with the adoption of the Americans with Disabilities Act (ADA) of 1990. Each model code addresses the requirements for ramps in a similar manner. Fire-resistance-rated construction similar to that used for exit stairs must enclose and protect exterior ramps. Exterior ramps must offer the same degree of protection as exterior stairs, with some exceptions for certain occupancies. An inspector should consult the provisions of the locally adopted code and amendments to adequately address this issue.

Specifications for Ramps

New ramps must be at least 44 inches (1 100 mm) wide with a maximum slope of 1 to 12 (1 foot [300 mm] of rise for every 12 feet [3.5 m] of horizontal distance). The maximum length for a single ramp is 30 feet (9 m) without a landing (**Figure 7.23**).

Fire Escape Stairs, Ladders, and Slides

Because they are unsafe and unreliable, external fire escape stairs may not be used as any part of a means of egress in new construction **(Figure 7.24)**. Where these stairs are still allowed on existing buildings, they may constitute only half of the required means of egress. Inspectors should be wary when inspecting buildings that still have fire escape stairs because they may be:

- Poorly maintained and rusted through
- Not properly secured to the wall
- Snow-covered or icy
- Exposed to high winds
- Cluttered with garbage and other debris
- Inaccessible due to locked or secured access windows and doors

Fire escape stairs have also lost favor because many occupants are unaccustomed to using them on a regular basis. People who have a fear of heights may find them uncomfortable to use, slowing the progress of other people. In addition, people with physical impairments cannot safely access or use them.

To avoid trapping occupants, fire escape stairs must be exposed to the smallest number of door or window openings as possible. Each of the model codes provides detailed guidelines for the protection of openings that may expose fire escape stairs. Windows may be used as access to fire escapes in existing buildings if they meet certain criteria concerning the size of the opening. Windows used for this purpose must open with a minimum of effort, or, in some cases, open outward in a manner that does not obstruct the exit path. Access to fire escape stairs must be directly from a balcony, landing, or platform.

Fire escape ladders are only allowed for limited purposes if approved by the AHJ **(Figure 7.25)**. Fire escape slides, also called slidescapes, may be used as a means of egress where they are authorized. They must be of an approved type and rated at one exit unit per slide with a rated capacity of 60 persons.

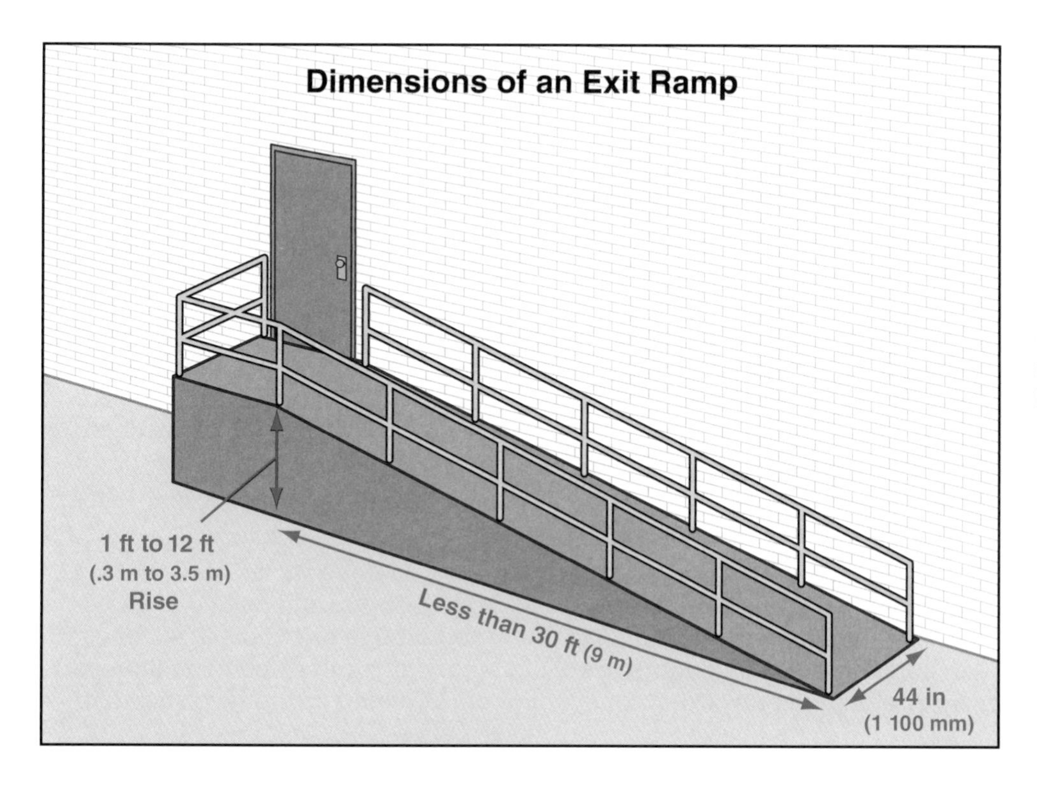

Figure 7.23 Exit ramps must meet specific rise, width, and length requirements.

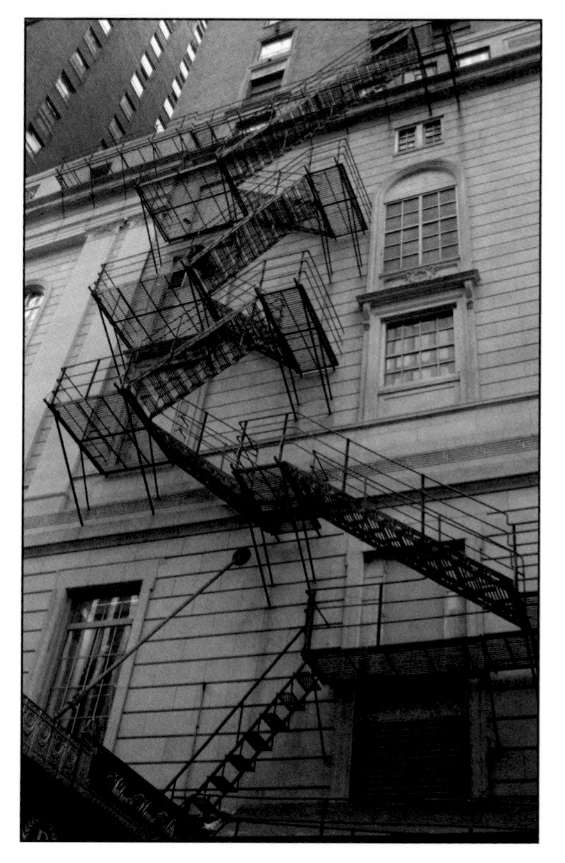

Figure 7.24 Exterior fire escape stairs may be poorly maintained, obstructed, or badly deteriorated and should not be considered a sole means of egress in buildings where they are installed.

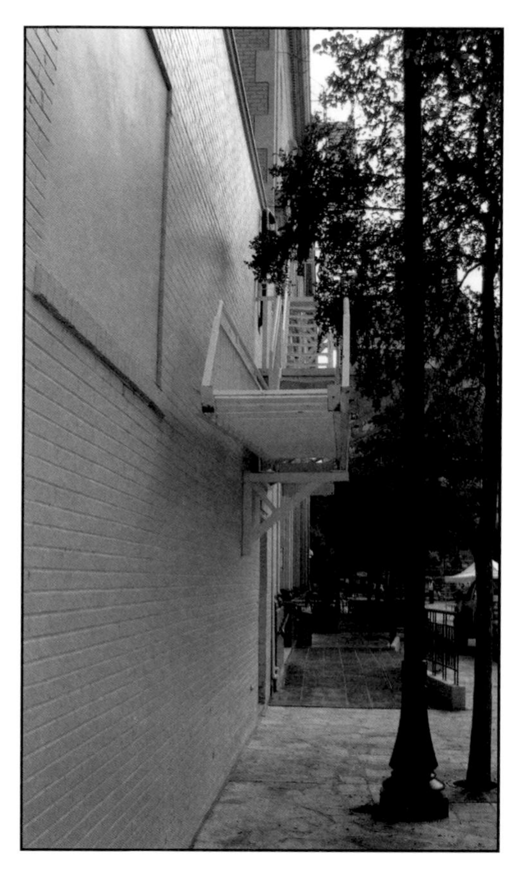

Figure 7.25 Fire escape ladders and fire escape slides are two additional means of egress that are not universally authorized throughout all jurisdictions. Where they are allowed, they must meet all applicable codes enacted by the AHJ.

Exit Illumination and Markings

NFPA® studies have cited lack of adequate illumination as a contributing factor to hundreds of fatalities in public places over the past century. Two of these incidents – the Cocoanut Grove fire and the Winecoff Hotel fire — resulted in the development of NFPA® standards for exits. Today, exit illumination, emergency lighting, and markings vary with each type of occupancy classification but are consistent among model code organizations.

Cocoanut Grove and Winecoff Hotel Tragedies

Cocoanut Grove, Boston, MA: On November 28, 1942, a fire in the Cocoanut Grove nightclub resulted in the death of 492 people and injuries to hundreds more. A fire that began in a downstairs area quickly spread upstairs through flammable decorations, where it engulfed the entire nightclub. The occupancy was at least double the approved capacity, and the only exit was a single revolving door, which quickly became blocked as panicked occupants tried to flee. Other exits were obscured by decorations, chained closed, or blocked to prevent patrons from entering or leaving without paying. The few unlocked side exits opened inward and became blocked as patrons tried to leave.

Following the fire, Massachusetts and other states enacted laws banning flammable decorations, inward-swinging exit doors, and bolting or chaining of emergency exits. Exit signs were required to be visible at all times. In addition, revolving doors used in a means of egress had to be flanked by at least one outward-swinging door or retrofitted so the revolving door leaves could be folded flat.

Winecoff Hotel, Atlanta, GA: On December 7, 1946, a fire ignited in the 15-story Winecoff Hotel. Even though the structure had no fire alarms, fire escapes, or sprinkler system, the hotel had been advertised as fireproof because it was built with masonry walls. The hotel was constructed with centrally located elevators and wooden stairs that facilitated the vertical travel of smoke and flames, which blocked the only means of egress. In all, 119 people died; 80 percent of the fatalities were guests who stayed above the 8th floor. The fire still represents the largest loss of life in a United States hotel.

Within days of the fire, local officials across the United States met to upgrade fire and life safety codes. For example, the Georgia state legislature adopted the *Building Exits Code* in 1948 to ensure that people could escape buildings in the event of a fire. The enormous loss of life in a "fireproof" structure led to the establishment of the national fire safety codes that are in effect today.

Exit Illumination

When codes require exit illumination, the illumination must be continuous during periods of occupancy. Furthermore, if a single lighting unit such as a bulb fails, the other lights must continue to function so the area is not left in darkness. Battery-operated units may not be used for primary exit illumination but instead as backup supplies.

Emergency Lighting

Figure 7.26 Emergency lights like these must have an auxiliary power source to provide adequate illumination in the case of a primary power failure during an emergency.

Emergency lighting is designed to provide illumination when normal power is lost. Emergency lighting is required in certain occupancies, such as places of assembly, educational facilities, health care facilities, and high-rise structures. Depending on the locally adopted code, emergency lighting may also be required in underground or limited-access structures **(Figure 7.26)**.

Emergency lights are powered by batteries or an auxiliary power system. The emergency lighting system must provide the proper amount of illumination (1 foot-candle [10 lux]) for 90 minutes. Floors must be illuminated at no less than 1 foot-candle (10 lux) at floor level. A reduction of 0.2 foot-candle (2 lux) is permitted in auditoriums, theaters, concert or opera halls, and other places of assembly during performances as long as the lighting automatically adjusts to 1 foot-candle (10 lux) when a fire alarm is activated.

Markings

Markings are signs that direct occupants to the nearest exit. They are required in most occupancies **(Figures 7.27 a and b)**. These illuminated exit signs must be positioned so that no point in the exit access is more than 100 feet (30 m) from the nearest visible sign. The letters on exit signs must be at least 6 inches (150 mm) high and the principal strokes of the letters at least ¾-inch (19 mm) wide.

Exit signs may be wired into the building electrical system or be self-illuminating. Wired exit signs must have an auxiliary power source — usually an internal battery - or a separate electrical circuit powered by an auxiliary power supply such as a generator.

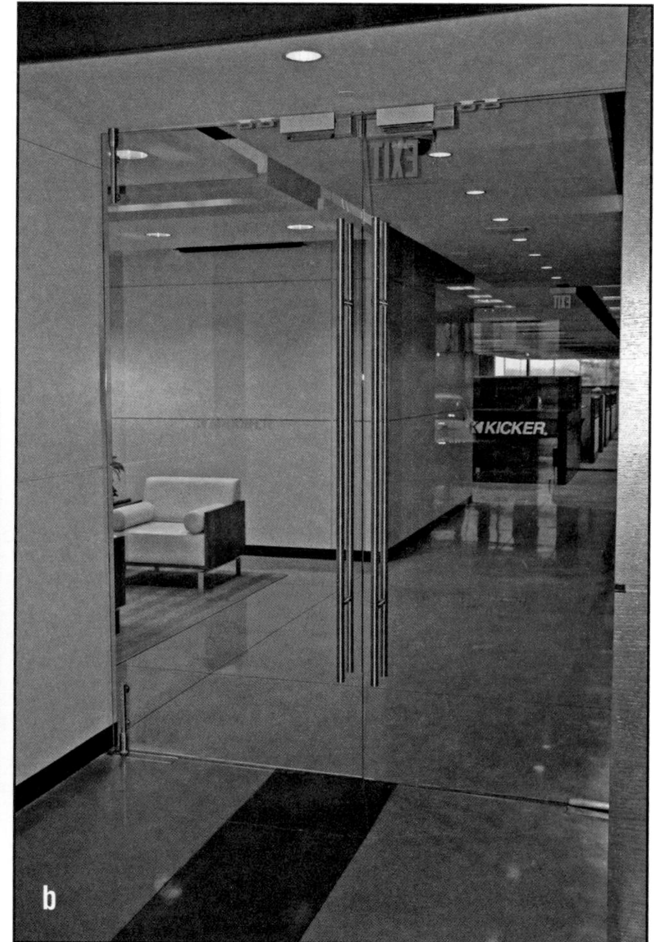

Figures 7.27 a and b Exit markings should be illuminated at all times and provided with secondary powered systems to ensure that they are also illuminated during emergency situations. *a Courtesy of Dennis Marx.*

Floor proximity exit signs or floor-level exit signs, along with floor proximity egress path markings, have become popular in some jurisdictions and are required in some occupancy classifications. These signs are placed to allow occupants to identify the location of an exit as they crawl along a floor **(Figures 7.28 a and b)**. Floor-level signs are used in addition to standard ceiling-level signs; they are not a substitute.

The NFPA *Life Safety Code®* and other model codes specify that the bottom of a floor proximity exit sign must be between 6 and 8 inches (150 and 200 mm) above the floor surface. The requirements for illumination and character size for floor-level signs are the same as for other exit signs.

Figures 7.28 a and b Floor-level signs are used in addition to conventional exit signs and are placed at a height lower than the anticipated level of smoke in an emergency.

Auxiliary Power

Auxiliary power can originate from batteries or a generator. When a generator is used to power emergency lighting and exit signs, it is connected to the building's electrical system by a switchover device that activates the generator when the building power supply is terminated. Depending on code requirements, natural gas, liquid petroleum gas (LPG), or diesel fuel may be used to fuel generators **(Figure 7.29)**.

An inspector should be familiar with the test procedures for both the auxiliary power supply and emergency lighting system. Codes require a monthly and annual test of the system. The inspector should review the documentation to verify that these tests have been conducted as specified.

NOTE: IFC Section 604.5 requires the building owner to document the test results and make the records available to the inspector.

Figure 7.29 Auxiliary power systems are necessary to ensure that emergency power activates when the primary power in a structure fails or is deactivated due to an emergency.

Occupant Loads

One of the functions of the model codes is to help the building official or plans examiner in determining the number of people who may safely occupy a building. Even more important, the codes provide an accepted way to determine how many people can safely exit a structure during an emergency. Even a properly designed and maintained means of egress cannot function effectively when the total occupant load has been exceeded and too many people are trying to move through an exit at the same time.

Occupant Load — Total number of people for which the means of egress of a building or portion thereof is designed.

The term **occupant load** is defined as the total number of persons for which the means of egress of a building or portion thereof is designed. The occupant load for a building or room is established during the plans review process. A design professional typically calculates the proposed occupant load, and the plans examiner verifies the calculations. For example, in assembly occupancies, the approved maximum occupant load for a structure or room must be visible, legible, and posted on a sign near the entry to the structure or area used for assembly (**Figure 7.30**).

NOTE: See Chapter 15 for more information on plans review.

For purposes of fire and life safety, inspectors must be able to verify the occupant load of existing occupancies during field inspections. All inspectors should be familiar with the methods for completing these calculations. In particular, they must be able to verify a new occupant load when a structure changes occupancy classification. An inspector must verify that the occupant load for a given location is still correct and that a sign, if required, is posted in an appropriate place. Any increase in occupant load must be supported by building design and verification that the means of egress is sufficient. Because each occupancy classification is different, any change in occupancy may also change exit requirements.

Figure 7.30 Inspectors are responsible for calculating occupancy loads and verifying that the maximum occupancy is clearly posted and adhered to by the owner/occupant.

NOTE: A change in occupancy classification should initiate a plans review and a new certificate of occupancy.

Determining occupant load requires more information than counting square footage and number of exits. The width of each means of egress and the distance occupants will need to travel to reach an exit are just some of the factors inspectors need to take into account. The following information is needed to determine occupant load:

- Minimum width of each means of egress
- Capacity of each individual means of egress
- Total capacity of all means of egress
- Number of exits required
- Maximum travel distance to an exit

One part of calculating occupant load is similar for each of the model codes. Each code assigns an occupant load factor per person based on the function of the space, meaning the maximum floor area allowed per person stated in square feet (m^2). The formula for determining the occupant load of a structure, room, or area is as follows:

Occupant Load = Net Floor Area ÷ Area per Person (Factor)

For example, a stage or platform occupant load is calculated using 15 square feet (1.3935 m^2) per person while the audience area with nonfixed seating would use 7 square feet (0.6503 m^2) per person in movable chairs. The occupant load factor is divided into the square footage (m^2) of the room to determine the maximum occupant load. A 150-square foot (13.94 m^2) stage would be capable of holding 10 persons, while a seating area of 2,100 square feet (190 m^2) could accommodate 300 occupants.

Table 7.2 provides the International Building Code® (IBC®) and IFC maximum floor area allowances per occupant. The table contains the words *gross* and *net* to describe the allowable square feet (m^2) per occupant in various types of occupancies:

- **Gross** — Refers to the entire square footage (m^2) of a space measured wall to wall with no deductions for desks, files, movable partitions, or other items.
- **Net** — Refers to the gross square footage (m^2) minus any space taken up by equipment, furniture, corridors, or other space that is not used for the occupancy. For example, in an assembly occupancy such as a pool hall, only the clear floor space available for use is calculated — the gross area of the space minus the area taken up by the pool tables, counters, restroom spaces, office areas, and storage spaces.

NOTE: There are other factors that affect the occupant load calculations beyond the locally adopted fire code. For example, bathroom facilities affect the number of people who can reasonably occupy a space. The inspector should check with the local building official before posting a new occupancy limit.

Examples 1 and 2 at the end of the chapter illustrate how occupant loads can be calculated for single-use occupancies. Example calculations are in both Customary System units and International System of Units (SI):

- **Example 1: Occupant Load Calculation for Single-Use Occupancy**
- **Example 2: Occupant load calculation for single-use occupancy – Single story Business office**

Table 7.2
Maximum Floor Area Allowances per Occupant

Function of Space	Floor Area In Square Feet Per Occupant
Accessory Storage Areas, Mechanical Equipment Room	300 gross
Agricultural Building	300 gross
Aircraft Hangars	500 gross
Airport Terminal	
Baggage claim	20 gross
Baggage handling	300 gross
Concourse	100 gross
Waiting areas	15 gross
Assembly	
Gaming floors (keno, slots, etc,)	11 gross
Assembly with Fixed Seats	See Section 1004.4*
Assembly without Fixed Seats	
Concentrated (chairs only — not fixed)	7 net
Standing space	5 net
Unconcentrated (tables and chairs)	15 net
Bowling Centers, allow 5 persons for each lane, including 15 feet of runway, and for additional areas	7 net
Business Areas	100 gross
Courtrooms — other than fixed seating areas	40 net
Day Care	35 net
Dormitories	50 gross
Educational	
Classroom area	20 net
Shops and other vocational room areas	50 net
Exercise Rooms	50 gross
H-5 Fabrication and Manufacturing areas	200 gross
Industrial Areas	100 gross
Institutional Areas	
Inpatient treatment areas	240 gross
Outpatient areas	100 gross
Sleeping areas	120 gross
Kitchens, Commercial	200 gross
Library	
Reading rooms	50 net
Stack area	100 gross
Locker Rooms	50 gross
Mercantile	
Areas on other floors	60 gross
Storage, stock, shipping areas	300 gross
Parking Garages	200 gross
Residential	200 gross
Skating Rinks, Swimming Pools	
Rink and pool	50 gross
Decks	15 gross
Stages and Platforms	15 net
Warehouses	500 gross

For SI: 1 square foot = 0.0929 m^2

* Section number refers to sections in the *2006 International Building Code*®.

Multiuse Occupant Loads

Inspectors frequently encounter buildings that are used for more than one purpose or that contain more than one occupancy classification. Inspection personnel must know how to calculate the occupancy load for multiuse occupancies and multiple occupancies in one structure. In general, the following procedure for making the calculations is the same for each of the model codes:

1. ***Multiuse occupancy*** — For a building or a portion of a building that has more than one use, the occupant load is determined by the use that allows for the largest number of persons to occupy the building.
2. ***Multiple occupancies in one building*** — For a building or a portion of a building that contains two or more distinct occupancies, the occupant load is determined by calculating the occupant load for each of the occupancies separately and then adding them. **Examples 3 and 4** highlight each of these two possibilities:

- **Example 3: Occupant Load Calculation for Multiuse Occupancy**
- **Example 4:** Occupant Load Calculation for Multiple Occupancies in One Structure

Means of Egress Determinations

In addition to determining how many people may safely occupy a particular room or building, inspectors must determine the number of exits a room or building must have to enable people to leave quickly and safely. The steps in determining means of egress are:

- Determine the capacity of the means of egress, which is the number of people who can move along the means of egress.
- Determine the arrangement of the exits from the room or building.
- Evaluate the effectiveness of the means of egress.

Capacity

The means of egress capacity must be at least equal to the occupant load, determined by the floor area and based upon the occupancy type. If a building or room has a means of egress capacity that is lower than the floor area occupant load, additional exits must be constructed to handle the entire floor area occupant load. In short, a room or building must have enough exits, which are located remotely from each other and wide enough to allow people to safely exit.

To determine the required width of the means of egress, an inspector must apply a numerical factor, expressed in terms of inches (millimeters) per person. The most common numerical factors that are expressed in the codes are 0.3 inches (7.62 mm) per person for stairways and 0.2 inches (5.08 mm) per person for ramps or level exit components, such as doors and corridors **(Table 7.3)**.

Table 7.3
Egress Width per Occupant Served

Occupancy	Without Sprinkler System		With Sprinkler System[a]	
	Stairways (inches per occupant)	Other Egress Components (inches per occupant)	Stairways (inches per occupant)	Other Egress Components (inches per occupant)
Occupancies other than those listed below	0.3	0.2	0.2	0.15
Hazardous: H-1, H-2, H-3, and H-4	0.7	0.4	0.3	0.2
Institutional: I-2	NA	NA	0.3	0.2

For SI: 1 inch = 25.4 mm
NA = Not applicable

a. Buildings equipped throughout with an automatic sprinkler system in accordance with Section 903.3.1.1 or 903.1.2.*

* Section numbers refer to sections in the *2006 International Building Code®*.

Courtesy of the International Code Council®, International Building Code®, 2006, Table 1005.1.

Each of the model codes contains other factors for selected occupancies. For example, buildings that are considered hazardous occupancies use 0.7 inches (17.78 mm) per person for stairways and 0.4 inches (10.16 mm) per person for ramps or level exit components. Each of the model codes also makes allowances based on whether the occupancy has an automatic sprinkler system.

To determine the exit capacity, the inspector must use mathematical calculations that are based on the established minimum exit widths outlined in the model building codes. The capacities of each individual means of egress are added to determine the total exit capacity for the room or building. Total exit capacity is used to determine the number of required exits.

Exit Capacity

To determine the capacity of an exit, the inspector must calculate the capacity for each of the three means of egress elements: exit access, exit, and exit discharge. Measure each element in clear (unobstructed) width at the narrowest point. **The element with the smallest capacity will determine the total capacity for the means of egress**. See **Example 5: Calculating Egress Capacities**, to illustrate this point.

Total Exit Capacity

In order to calculate the total exit capacity of a space or building, determine the capacity of each exit and then add the resulting figures. The procedures for determining total exit capacity differ depending on whether the exit is at grade level. See **Examples 6 and 7:**

- **Example 6: Calculating Total Exit Capacity (Business Occupancy)**
- **Example 7: Calculating Total Exit Capacity (Residential Occupancy)**

Stairways must be able to accommodate the greatest number of people expected to occupy any single floor above. During an evacuation, not everyone from all floors will be in the same area of the stairwell at the same time **(Figure 7.31)**. For example, by the time the people from the third floor reach the second floor, the people from the second floor should have exited the building.

The International Code Council® (ICC®) model codes also state the following:

- The means of egress cannot become narrower in the direction of egress travel or beyond any point where two or more exits converge.
- The total of the exit capacities cannot be less than the sum of all the capacities.

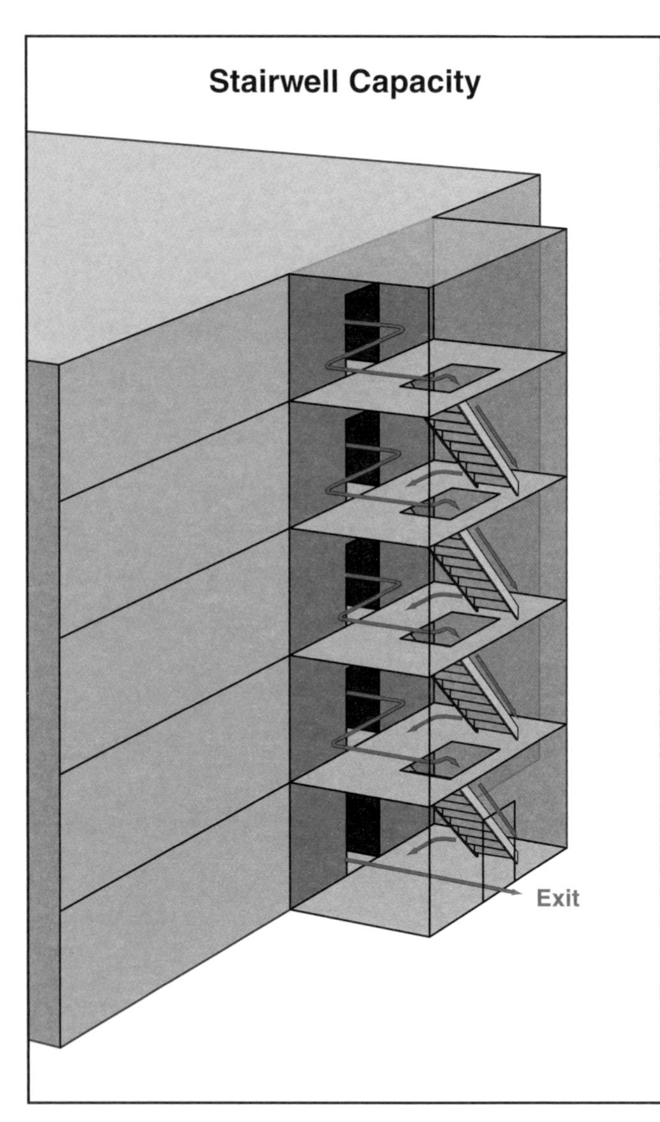

Figure 7.31 Each arrow in this illustration represents the occupants' escape into the stairway enclosure. The stairway must be able to accommodate the greatest number of people on any floor above that portion of the stairway.

In multistory structures, personnel can use the emergency notification system to alert occupants to exit in order of proximity to the emergency (or fire) floor. Therefore, the three floors closest to the fire floor exit first, then the next three, and so on until the structure is evacuated. In some jurisdictions, high-rise structures are required to have designated fire wardens on each floor to perform this control function.

Required Number of Exits

Inspectors and building officials must be familiar with the requirements for the minimum number of exits in the buildings they inspect. The means of egress chapters of the model codes provide general information on minimum number of exits. The model codes also detail special requirements and exceptions for specific occupancy types.

In most cases, the codes require that there be at least two exits from any balcony, mezzanine, story, or portion thereof that has an occupant load of 500 persons or less. For occupant loads greater than 500 persons, the codes generally require a minimum of three separate exits, with four or more exits required for 1,000 or more people.

When inspectors encounter situations where the occupant load is consistently exceeded, they should explore all possible alternatives for correction within the framework of their local codes. Ultimately, enforcement through citation may become the only remedy.

Arrangement

Multiple means of egress enable occupants to quickly locate an exit and leave quickly during an emergency. An inspector should be able to determine the location of multiple exits for new construction buildings and decide if the maximum travel distance to existing exits is acceptable. If exits are insufficient, an inspector should be able to recommend alternatives for meeting the means of egress requirements.

The codes also specify that the exits should be as remote from each other as possible. This spacing helps minimize travel distance to an exit. It also increases the chances of finding an alternative exit if the closest exit is unreachable due to fire conditions. Each of the model codes specify certain situations where a single exit is permitted.

Location of Exits

When more than one exit is required, it would be meaningless to place them close to each other. In this situation, the *one-half diagonal rule* is applied to assist in determining the placement of the exits:

- When two exits are required, they should be located no less than half the length of the overall diagonal dimension of the room or building area. The diagonal is measured as a straight line from the nearest edges of the exit or exit access doors to each other **(Figure 7.32)**.
- In spaces or structures where there is a requirement for more than two exits or exit access doors, at least two doors will meet the minimum separation requirements while the other exits should be located in a manner so that they will be available if the primary exits become blocked or unavailable.

An inspector must be aware that regardless of configuration, each room, space, or building must have adequate and appropriate exiting capabilities. During inspections, the inspector must verify that designed exit plans follow all of the code-required factors for fire and life safety, even if changes have been made to the occupancy. For example, it is common to find modifications to assembly occupancies that dramatically change the fire and life safety aspects of the structure. Blocking an exit door in one location and building a new one in another location may not provide a full measure of safety for occupants.

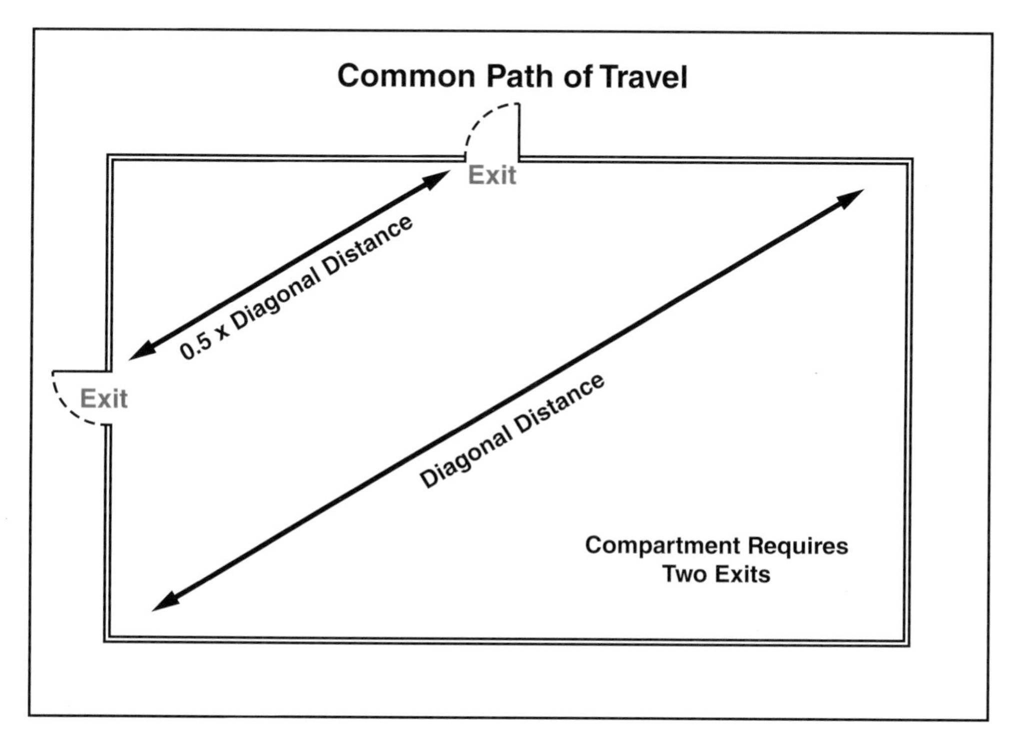

Figure 7.32 The diagonal rule of exit placement, illustrated here, requires that exits be located not less than one-half the length of the overall diagonal dimension of the room or building area.

Travel Distance — Distance from any given area in a structure to the nearest exit or to a fire extinguisher.

Maximum Travel Distance to an Exit

Each of the model codes establishes the maximum allowable **travel distance** to the nearest exit and limits the length of dead-end corridors. The travel distance refers to the total length of travel necessary to reach the protection of an exit **(Figure 7.33)**.

The measurement starts from the most remote portion of the occupancy, curving around any corners or obstructions with a 1-foot (300 mm) clearance, and ends at the center of the exit doorway or other point at which the exit begins. In cases where the layout of the room is unknown or may change, the maximum travel distance is calculated by starting at the most remote point and proceeding by following the walls to the door of the room.

Within each model code, maximum travel distances vary depending on the type of occupancy and whether the building has an automatic sprinkler system. Generally, longer travel distances are allowed in buildings that are protected by automatic sprinkler systems; these distances are specified in the applicable codes.

An inspector should also be aware of the following two means of egress concepts that may be present in existing structures or appear on new or remodel construction plans:

1. **Dead-end corridor** — Condition that exists when a corridor has no outlet to a means of egress and is more than 20 feet (6 m) in length **(Figure 7.34)**.
2. **Common path of travel** — Path that all occupants must travel in one direction before reaching a point where they may choose between two separate and distinct paths leading to two separate exits **(Figure 7.35)**. This distance is used in the calculation of total travel distance to an exit for the purpose of locating exits for a building or area.

Table 7.4, p. 274, provides distances for dead-end, common paths of travel, and total travel distance limits.

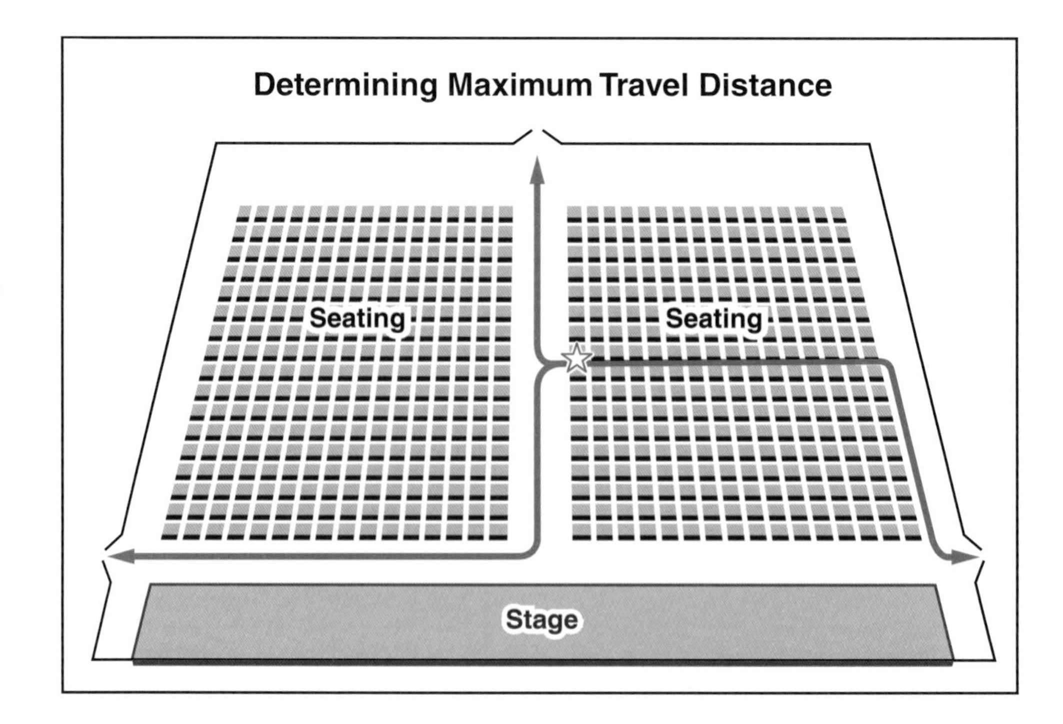

Figure 7.33 The travel distance is measured on the floor or walking surface along the centerline of the natural path of travel.

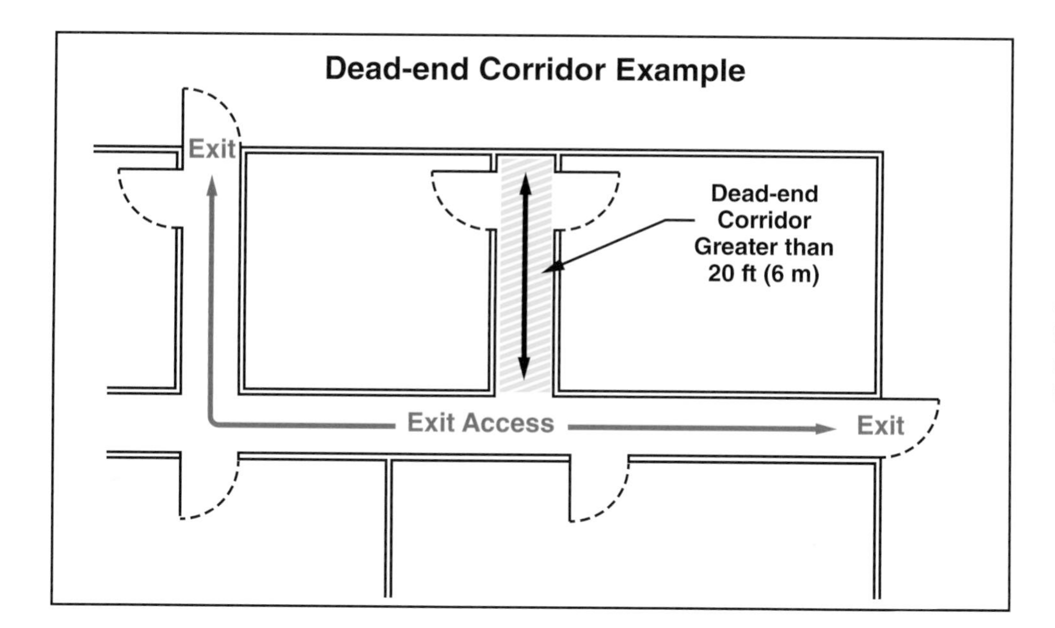

Figure 7.34 The dead-end corridor identified in the illustration has no outlet to a means of egress and is longer than 20 ft (6 m).

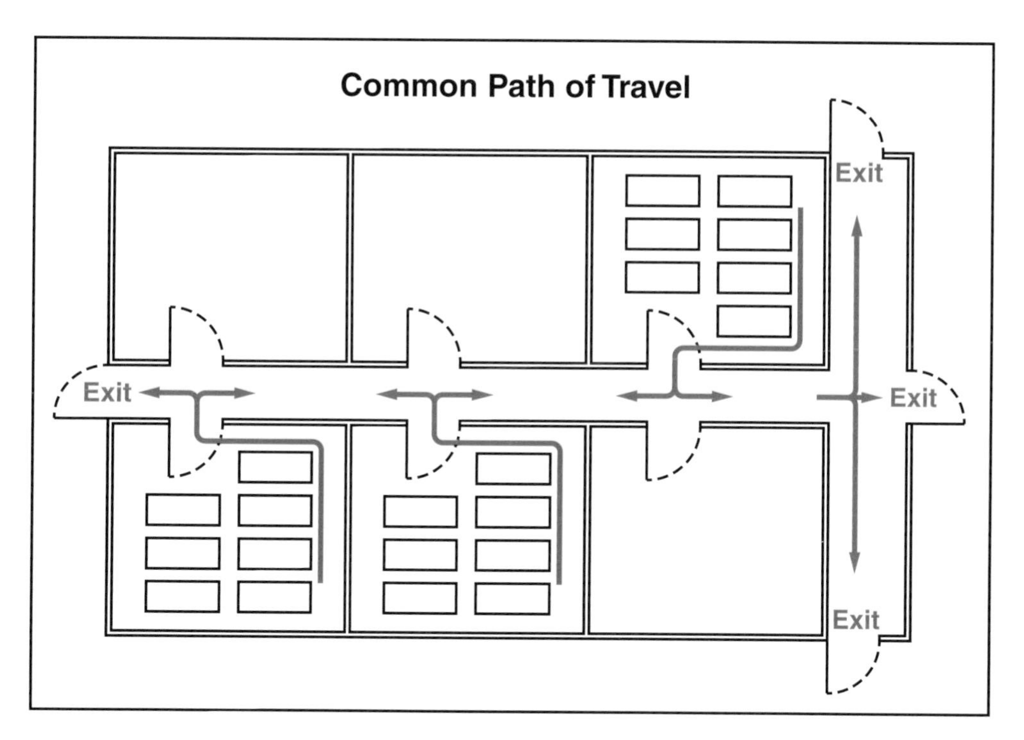

Figure 7.35 Many different compartments in an occupancy may share a means of egress, known as a common path of travel.

Effectiveness

As the inspector determines whether the means of egress in a particular structure meets the requirements of the locally adopted code, the following questions must be asked and answered:

- What is the occupancy classification for the structure or portion of the structure?
- Do the various parts of the means of egress comply with those allowed for the occupancy classification?
- What is the total exit capacity in inches (mm), people, or both?
- Does the travel distance to the nearest exit fall within the maximum distances for the occupancy classification?

Table 7.4
Common Path, Dead-end and Travel Distance Limits (By Occupancy)

Occupancy	Common Path Limit		Dead-end Limit		Travel Distance Limit	
	Unsprinklered (feet)	Sprinklered (feet)	Unsprinklered (feet)	Sprinklered (feet)	Unsprinklered (feet)	Sprinklered (feet)
Group A	20/75[a]	20/75[a]	20[b]	20[b]	200	250
Group B	75	100	50	50	200	250
Group E	75	75	20	20	200	250
Groups F-1, S-1[d]	75	100	50	50	200	250
Groups F-2, S-2[d]	75	100	50	50	300	400
Group H-1	25	25	0	0	75	75
Group H-2	50	100	0	0	75	100
Group H-3	50	100	20	20	100	150
Group H-4	75	75	20	20	150	175
Group H-5	75	75	20	50	150	200
Group I-1	75	75	20	20	200	250
Group I-2 (Health Care)	NR	NR	NR	NR	150	200[c]
Group I-3 (Detention and Correctional — Use Conditions II, III, IV, V)	100	100	NR	NR	150[c]	200[c]
Group I-4 (Day Care Centers)	NR	NR	20	20	200	250
Group M (Covered Mall)	75	100	50	50	200	400
Group M (Mercantile)	75	100	50	50	200	250
Group R-1 (Hotels)	75	75	50	50	200	250
Group R-2 (Apartments)	75	75	50	50	200	250
Group R-3 (One- and Two- Family); Group R-4 (Residential Care/ Assisted Living)	NR	NR	NR	NR	NR	NR
Group U	75	75	20	20	200	250

For SI: 1 foot = 304.8 mm.

NR = No requirements.

a. 20 feet for common path serving 50 or more persons; 75 feet for common path serving less than 50 persons.

b. See Section 1025.9.5* for dead-end aisles in Group A occupancies.

c. This dimension is for the total travel distance, assuming incremental portions have fully utilized their allowable maximums. For travel distance within the room, and from the room exit access door to the exit, see the appropriate occupancy chapter*.

d. See the *International Building Code®* for special requirements on spacing of doors in aircraft hangars.

* Section number and chapter reference refer to the 2006 *International Fire Code®*.

Courtesy of the International Code Council®, International Fire Code®, Table 1027.17.2.

- Are there dead-end corridors or common paths of travel that must be considered?
- Are all of the exits easily identifiable and accessible to different occupancy groups?
- Are the means of egress properly illuminated and marked?
- Do exit doors open easily? Are they equipped with panic hardware where required?
- Are the parts of the means of egress free of obstructions?
- What is the maximum number of occupants allowed for the particular occupancy or structure?
- Are interior finishes, surface trim materials, coverings, and decorations within the adopted code specified as to flame spread and smoke development limits for the occupancy?
- Is the existing means of egress adequate for the occupancy classification or are additional exits needed?
- When a building's occupancy classification changes or alterations are made, are alterations to the means of egress requirements necessary?

An inspector should use a consistent and logical approach to determining the means of egress requirements for both new and existing structures. One approach would be to follow a series of steps when calculating the requirements. Those steps are as follows:

Step 1: Determine occupant load.

Step 2: Determine clear width of each component.

Step 3: Determine egress capacity of each component.

Step 4: Determine most restrictive component of each route.

Step 5: Determine if egress capacity is sufficient.

Chapter Summary

Analysis of fires dating back over a century has shown that inadequate or blocked exits have been a significant factor in the loss of lives. These tragic events have resulted in model code requirements for means of egress for all types of occupancy classifications.

The means of egress routes must conform to the requirements of the locally adopted building and fire code. Equally important, emergency responders must be able to gain access to the structure.

An inspector is responsible for verifying that building occupants have an unimpeded ability to evacuate a structure or move to an area of refuge within the structure. This ability is especially important for persons who are mobility impaired or who are not otherwise capable of seeking safety on their own.

Calculations

Inspector I Calculations

Example 1: Occupant Load Calculation for Single-Use Occupancy

An inspector must calculate a portion of the occupant load of a building that was formerly a warehouse and has recently been turned into a nightclub. The area of this structure measures 100 feet by 150 feet (30.48 m by 45.72 m). As a nightclub, it will contain unconcentrated seating, (tables and chairs). The inspector would use the following steps to determine the occupant load:

Step 1: Determine the total square footage (m^2) of the nightclub by multiplying the facility's length by its width:

$(100 \text{ feet} \times 150 \text{ feet}) = 15{,}000 \text{ ft}^2$

In SI

$(30.48 \text{ m} \times 45.72 \text{ m}) = 1\ 393.54 \text{ m}^2$

Step 2: Consult **Table 7.2, p. 267**, to determine the maximum allowable floor area per person in an assembly occupancy with unconcentrated tables and chairs.

Step 3: Allow 15 square feet (1.3935 m^2) per person based on the requirements found in Table 7.2 for an assembly occupancy unconcentrated with tables and chairs.

Occupant Load = $(15{,}000 \text{ ft}^2 \div 15 \text{ ft}^2) = 1{,}000$ persons

In SI

Occupant Load = $(1\ 393.54 \text{ m}^2 \div 1.3935 \text{ m}^2) = 1{,}000$ persons

NOTE: The 1,000-person occupant load will also depend on an adequate number of exits that are of an acceptable width and correctly spaced. These exits must be arranged in such a manner that they do not exceed maximum allowable travel distance.

Example 2: Occupant load calculation for single-use occupancy – Single story Business office

An inspector must calculate the occupant load for a medical billing office that will replace a store in a strip mall. The office will be an open-plan design with cubicles. The area of the office space measures 100 feet by 75 feet (30.48 m by 22.86 m).

Step 1: Determine the floor area of the facility using the following dimensions: 100 feet by 75 feet (30.48 m by 22.86 m).

$(100 \text{ feet} \times 75 \text{ feet}) = 7{,}500 \text{ ft}^2$

In SI

$(30.48 \text{ m} \times 22.86 \text{ m}) = 696.77 \text{ m}^2$

Step 2: Consult Table 7.2 to determine the maximum allowable floor area per person in an office building.

Step 3: Divide the area per person, 100 square feet (9.29 m^2), into the floor area of the structure to determine occupant load:

Occupant load = $(7{,}500 \text{ ft}^2 \div 100 \text{ ft}^2) = 75$ people

In SI

Occupant load = $(696.77 \text{ m}^2 \div 9.29 \text{ m}^2) = 75$ people

Inspector II Calculations

Example 3: Occupant Load Calculation for Multiuse Occupancy

Inspectors are assigned to determine the occupant load on a new youth-oriented entertainment facility being constructed within their jurisdiction. The building's main room is 100 feet by 250 feet (30.48 m by 76.2 m). Depending on the day of operation, this room is used as either a dance hall or an exercise area. The inspector would use the following steps to determine the occupant load.

Step 1: Determine the floor area in the main room.

(100 feet × 250 feet) = 25,000 ft^2

In SI

(30.48 m × 76.2 m) = 2 322.5 m^2

Step 2: Consult **Table 7.2** for the floor area per person that is required for this occupancy. A dance hall is considered a *concentrated use* and each person is allowed 7 square feet (0.6503 m^2). An exercise area is calculated at 50 square feet (4.645 m^2) per person.

Step 3: Use Formula 1 to determine occupant load.

Occupant Load (when used as a dance hall) = (25,000 ft^2 ÷ 7 ft^2) = 3,571 persons

Occupant Load (when used as an exercise area) = (25,000 ft^2 ÷ 50 ft^2) = 500 persons

In SI

Occupant Load (when used as a dance hall) = (2 322.5 m^2 ÷ 0.6503 m^2) = 3,571 persons

Occupant Load (when used as an exercise area) = (2 322.5 m^2 ÷ 4.645 m^2) = 500 persons

Given these results, and the requirement that the occupant load is based on the maximum number of people that will use the facility at one time, the maximum occupant load for this building should be established at 3,571 persons.

Example 4: Occupant Load Calculation for Multiple Occupancies in One Structure

An inspector has been given plans for a new single-story mercantile store. The inspector must determine the occupant load for this structure. The sales floor portion of the building is 100 feet by 125 feet (30.48 m by 38.1 m) and contains general displays for the merchandise. A storage area is located at the rear of the building for storage of additional stock and shipping and receiving activities. It is 75 feet by 75 feet (22.86 m by 22.86 m). The inspector would use the following steps to determine the occupant load:

Step 1: Determine the floor area of the facility.

Floor area of the Sales Area = (100 feet × 125 feet) = 12,500 ft2

Floor area of the Storage Area = (75 feet × 75 feet) = 5,625 ft2

In SI

Floor area of the Sales Area = (30.48 m × 38.1 m) = 1 161.28 m2

Floor area of the Storage Area = (22.86 m × 22.86 m) = 522.58 m2

Step 2: Consult Table 7.2 to determine the maximum allowable area per person per square foot (m^2). Mercantile occupancies are allowed 30 square feet (2.787 m^2) per person for single story structures. The storage area is allowed 300 square feet (27.87 m^2) per person.

Step 3: Determine the occupant load for each portion of the structure.

Occupant Load for the Sales Area = (12,500 $ft^2 \div 30\ ft^2$) = 416 persons

Occupant Load for the Storage Area = (5,625 $ft^2 \div 300\ ft^2$) = 18 persons

In SI

Occupant Load for the Sales Area = (1 161.28 $m^2 \div 2.787\ m^2$) = 416 persons

Occupant Load for the Storage Area = (522.58 $m^2 \div 27.87\ m^2$) = 18 persons

Step 4: Determine the occupant load for the whole building.

416 Persons (Sales) + 18 persons (Storage) = 434 Persons (Total Occupant Load)

(Please note that these examples are simply for illustration.)

Example 5: Calculating Egress Capacities

A means of egress has the following dimensions:

- **Exit access** — Corridor that is 44 inches (1 117.6 mm) wide
- **Exit** — Stairway that is 44 inches (1 117.6 mm) wide
- **Exit discharge** — Alley that is 12 feet (3.6576 m) wide

A corridor is a level exit component; therefore, it is capable of handling one person for every 0.2 inch (5.08 mm). As a result:

Corridor Capacity = (44 in ÷ 0.2 in) = 220 persons

In SI

Corridor Capacity = (1 117.6 mm ÷ 5.08 mm) = 220 persons

The numerical factor for stairs is 0.3 inches (7.62 mm) per person. As a result:

Stairway Capacity = (44 in ÷ 0.3 in) = 146 persons

In SI

Stairway Capacity = (1 117.6 mm ÷ 7.62 mm) = 146 persons

The alley is considered a level component, so the factor of 0.2 inch (5.08 mm) per person factor is used. In order to calculate correctly, convert the alley's dimensions from feet (meters) to inches (mm).

12 feet × 12 inches/foot = 144 inches

In SI

3.6576 m × 1 000 mm/m = 3 657.6 mm

Alley Capacity = 144 in ÷ 0.2 in = 720 persons

In SI

Alley Capacity = 3 657.6 mm ÷ 5.08 mm = 720 persons

The most restrictive number for the three computations is 146 persons, so the total capacity of the means of egress will be 146.

Example 6: Calculating Total Exit Capacity (Business Occupancy)

An inspector is reviewing plans for a new business occupancy that shows the building has three swinging exit doors that have a clear width of 36 inches (914.4 mm). All these exits are on ground level. Use the following steps to determine the total exit capacity:

Step 1: Determine the numerical factor that should be used for a level egress in a business occupancy classification. From Table 7.2, we can determine that level components may be rated at 0.2 inches (5.08 mm) per person.

Step 2: Determine the capacity of each exit:

Exit Capacity = 36 in ÷ 0.2 in = 180 persons

In SI

Exit Capacity = 914.4 mm ÷ 5.08 mm = 180 persons

Step 3: Determine the total exit capacity. Because three doors are the same size, the calculation is relatively simple.

Total Exit Capacity = 180 persons × 3 = 540 persons

Example 7: Calculating Total Exit Capacity (Residential Occupancy)

An inspector is assigned to determine the total exit capacity for a new four-story apartment building. There are two exits from each floor, and the stairways have a clear (unobstructed) width of 44 inches (1 117.6 mm). The inspector would use the following steps to determine the total exit capacity:

Step 1: Determine the numerical factor that should be used for a stairway in a residential occupancy. Based on Table 7.2, stairways are rated at 0.3 inches (7.62 mm) per person.

Step 2: Determine the capacity of each exit stairway.

Stairway Capacity = 44 in ÷ 0.3 = 146 persons

In SI

Stairway Capacity = 1 117.6 mm ÷ 7.62 mm = 146 persons

Step 3: Determine the total exit capacity. Because there are two stairways of the same size, the calculation is relatively simple.

Total Exit Capacity = 146 persons × 2 = 292 persons

Review Questions

1. What are the three elements of a means of egress?
2. What information is needed to determine the occupant load for a single-use occupancy?

1. How is the occupancy load calculated for multiple occupancies in one building?
2. Explain the steps used to determine the means of egress requirements for new and existing structures.

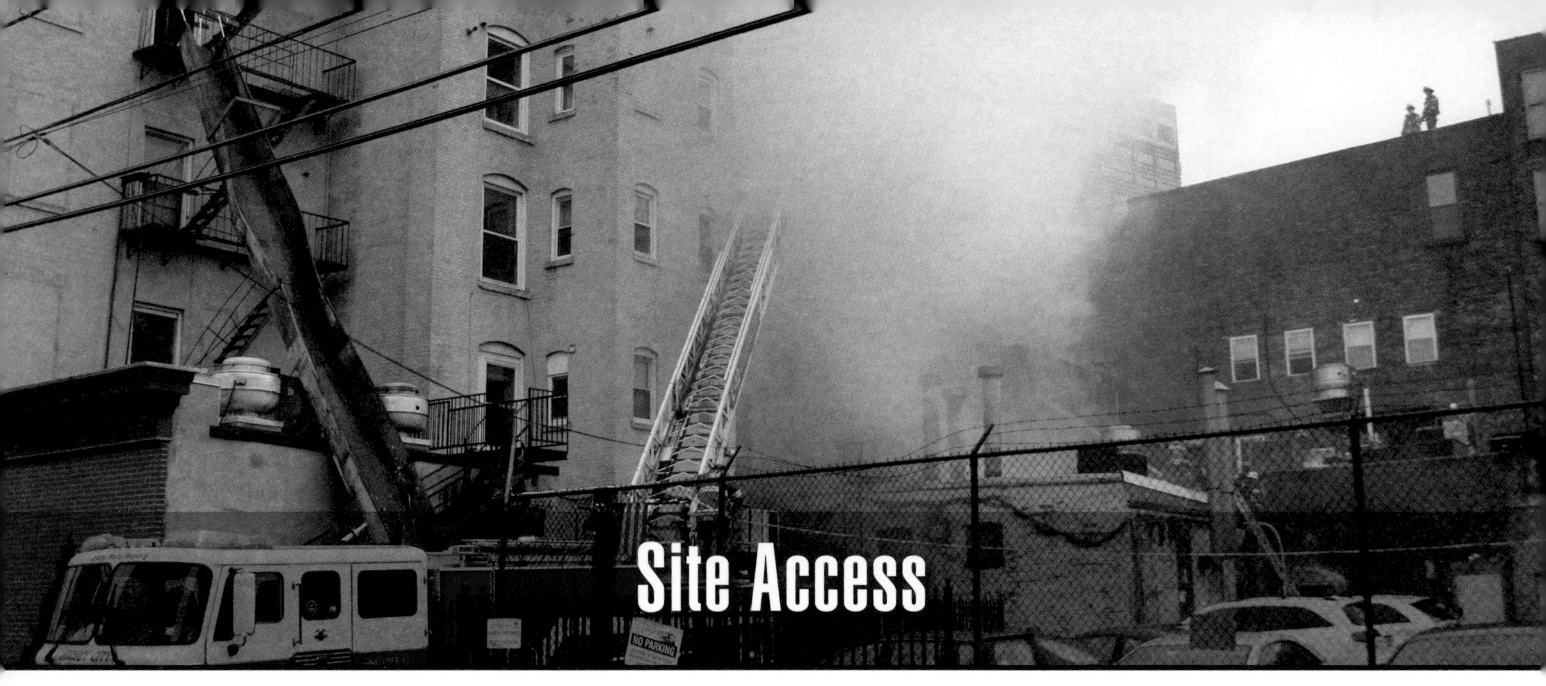

Site Access

Courtesy of Ron Jeffers.

Chapter Contents

Case History **285**

Fire Lanes and Fire Apparatus Access Roads **285**

Requirements 286

Dead-End Access Roads 288

Road Markings and Signs 289

Construction and Demolition Sites **291**

Structure Access Barriers **294**

Exterior Access... 294

Interior Access 299

Chapter Summary **302**

Review Questions **302**

Key Terms

Angle of Approach/Departure 286
False Front 302
Fire Department Connection (FDC) 298
Site Plan 292
Topographical 298

NFPA® Job Performance Requirements

This chapter provides information that addresses the following job performance requirements of NFPA® 1031, *Standard for Professional Qualifications for Fire Inspector and Plan Examiner* (2014).

Fire Inspector I

4.3.4

Site Access

Learning Objectives

After reading this chapter, students will be able to:

Inspector I

1. Describe types of fire lanes and fire apparatus access roads. [NFPA® 1031, 4.3.11]
2. Explain site access considerations for construction and demolition sites. [NFPA® 1031, 4.3.11]
3. Identify structure access barriers. [NFPA® 1031, 4.3.11]

Chapter 8
Site Access

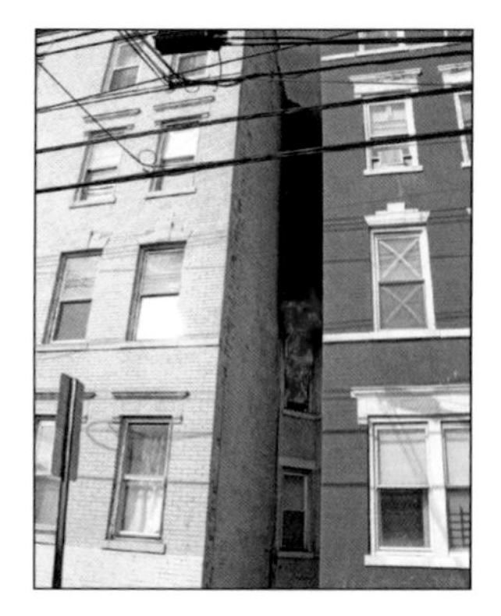

Case History

Units responded to a reported structure fire in a subdivision under development. The initial call stated the structure fire was burning excessively with heavy smoke. As units arrived, they observed a fully involved two-story structure from approximately 200 yards (180 m) away. The development of the area had caused the normal access road to be blocked, and the only access was a wooded trail. It was nighttime and dark. A captain and firefighter left the engine and traveled through the woods to locate an access road. As the captain approached the burning structure, he fell into a swimming pool that was covered with leaves and growing vegetation. The captain, in full structural PPE including SCBA, was immediately submerged. He struggled to swim to the side of the pool, where a firefighter assisted him in climbing out of the pool.

An emergency scene at night, without a proper access road, can be hazardous to emergency responders. Personnel must take all precautions to be aware of their surroundings, especially in limited lighting conditions. This incident reinforced the necessity for working in pairs.

Successful emergency operations depend on the ability to gain access to the emergency scene with apparatus, personnel, and equipment. The Authority Having Jurisdiction (AHJ) has the responsibility to determine whether access is available. Fire codes define the minimum site access requirements and permit the local code official to require additional access.

This chapter describes the basic requirements inspectors should be aware of concerning site access. Topics include fire lanes and fire apparatus access roads, construction and demolition sites, and structure access barriers.

Fire Lanes and Fire Apparatus Access Roads

To effectively respond to an emergency, fire and emergency responders must be able to travel from a public roadway to the incident scene. Utilizing fire lanes and fire apparatus access roads facilitates the approach to structures that are located away from public roadways. These access roads must be capable of supporting the size and weight requirements of fire apparatus. This section covers specific size requirements of fire lanes and access roads, as well as information on dead-end access roads, road markings, and signs.

Firefighters often use fire lanes and fire apparatus access roads when responding to an emergency scene:

- A fire lane is a required road or passageway designed to provide a means of access and parking for fire apparatus. A fire lane is not necessarily designed or intended for any other purpose. The AHJ must designate and identify it as a fire lane, prohibit unauthorized parking, and verify that minimum widths and clearances are maintained.

NOTE: Parking and materials storage on designated fire apparatus access roads and in fire lanes is prohibited.

- A fire apparatus access road can include any street or highway that provides access for emergency vehicles. Requirements for fire lanes and fire apparatus access roads may be found in the following sources:
  - International Fire Code® (IFC®) from the International Code Council® (ICC®)
  - NFPA® 1, *Fire Code*
  - NFPA® 1141, *Standard for Fire Protection Infrastructure for Land Development in Wildland, Rural, and Suburban Areas*
- NFPA® 5000, *Building Construction and Safety Code®*

Requirements

Fire lanes and fire apparatus access roads must be appropriately sized to allow emergency apparatus unimpeded passage to and around a facility. Most nationally developed fire codes contain specifications for the following:

- **Road extension:** Fire lanes and fire apparatus access roads must extend to within 150 feet (45 m) of all portions of a building, although local codes may require otherwise **(Figure 8.1)**.
- **Vertical clearance:** Vertical clearance over a fire lane or fire apparatus access road must be a minimum of 13 feet, 6 inches (4 m).
- **Width:** Fire lanes and fire apparatus access roads must be a minimum width of 20 feet (6 m). Width requirements are intended to allow a vehicle to pass a parked apparatus, including an aerial device with outriggers deployed. This measurement does not include parking widths.
- **Capacity:** Fire lanes and fire apparatus access roads are designed and maintained to support the expected load of emergency vehicles.

NOTE: These specifications can include lanes over underground structures, bridges, or ditches.

When inspecting or designating a fire lane or fire apparatus access road, inspectors must be familiar with the turning radius requirements as well as the weight specifications of their departments' heaviest apparatus. An additional safety margin should be included for future design changes in the weight of new apparatus.

Angle of Approach/Departure — Relationship described in degrees that is created by an incline from or to a road surface.

An inspector should also determine whether the **angle of approach and departure** between the fire lane or fire apparatus access road and the public street meets the fire code requirements. Generally, an angle not exceeding 8 degrees is necessary to allow apparatus to drive onto the lane without the tailboard or front bumper striking the ground **(Figure 8.2)**.

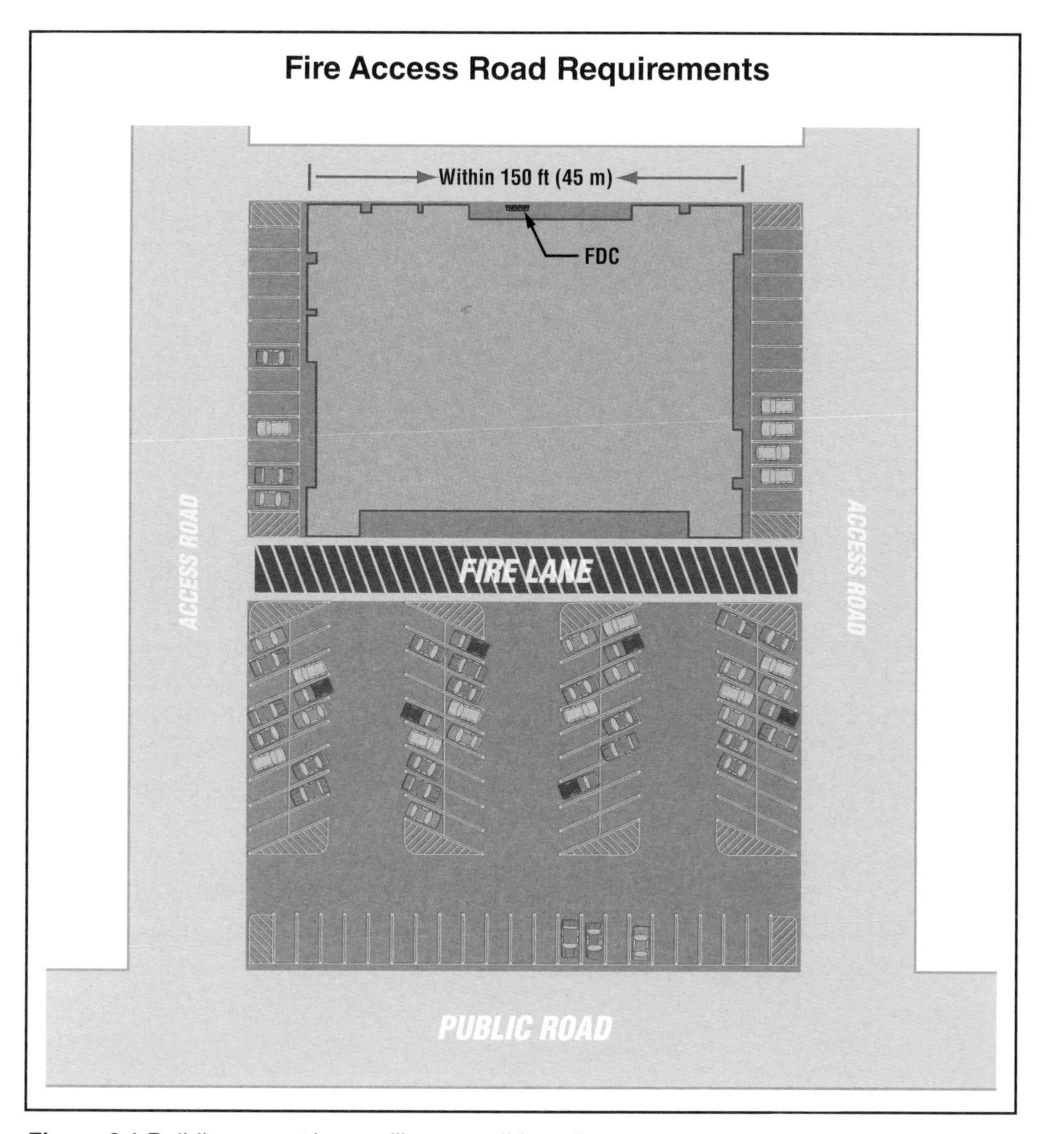

Figure 8.1 Buildings must be readily accessible to fire apparatus for emergency operations. Accessibility is based on the ability to extend a fire hose to a point no greater than 150 feet (45 m) from the apparatus.

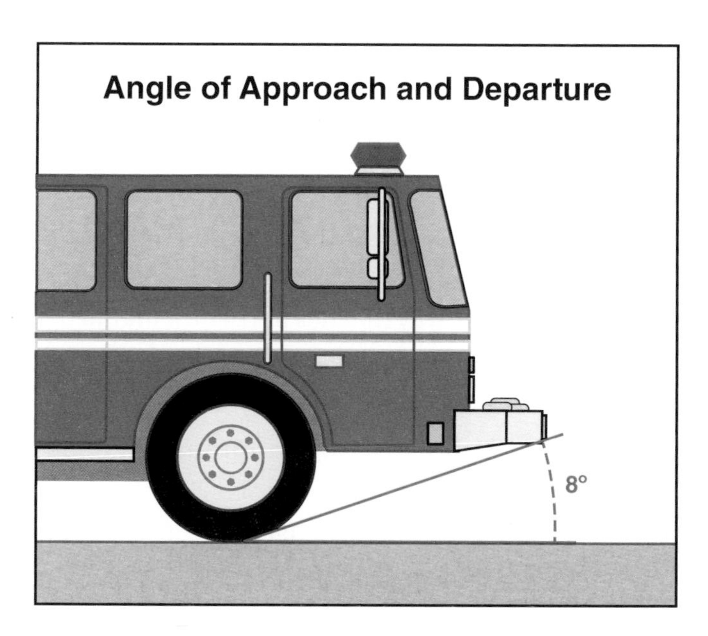

Figure 8.2 An angle not exceeding 8 degrees is necessary to allow apparatus to drive onto the access road without striking the ground with the tailboard or front bumper.

Unified Facility Criteria for Fire Lanes and Fire Apparatus Access Roads

The U.S. Department of Defense (DoD), through the Unified Facility Criteria, UFC 3-600-01, *Fire Protection Engineering for Facilities*, has established requirements for emergency service access roads. Although these requirements resemble those of the other code groups, the DoD fire inspector must be familiar with modifications described in the criteria, including the following:

- Structures that are larger than 5,000 square feet (465 m²) or taller than two stories must have at least one all-weather ground access for emergency vehicles. The AHJ may require that this access be paved and that it must terminate no further than 33 feet (10 m) from the structure.
- All base residential facilities must be provided with all-weather ground access to three sides, with a minimum of two sides having access to sleeping rooms.
- Facilities that are four or more stories tall, as well as all new warehouse structures, must have an all-weather access capable of supporting aerial emergency apparatus on a minimum of two sides of the structure's perimeter.
- Installed force protective devices (base security), including bollards and gates, must be capable of being removed or opened by one person. The fire department or 24-hour security personnel located within the facility must directly control locking devices for these installations.
- Access ways leading to fire department connections (FDCs) must be constructed as an all-weather surface for pumping apparatus. Access ways must extend to within 150 feet (45 m) of the FDC.

Dead-End Access Roads

Wherever possible, fire lanes or fire apparatus access roads should encircle a structure to provide access to all sides. Fire lanes or fire apparatus access roads that do not circle around a structure or return to a public street or road must provide a method for emergency vehicles to turn around. This provision is important so that emergency vehicles do not have to back up excessive distances.

According to the model fire codes, dead-end fire lanes and access roads are those that extend farther than 150 feet (45 m) from a public street or road. A dead-end fire lane or fire apparatus access road must conform to one of the following types of turnarounds: (**Figure 8.3**)

- **T or hammerhead** — Short section of roadway that lies perpendicular to the end of a dead-end street with equal sections on each side of the dead-end.

  NOTE: This turnaround may also be arranged in the form of a Y.
- **Cul-de-sac** — Street closed at one end designed to the minimum dimensions as required by the municipality.
- **Alley dock** — Configuration that allows a vehicle to back into a space and turn around.

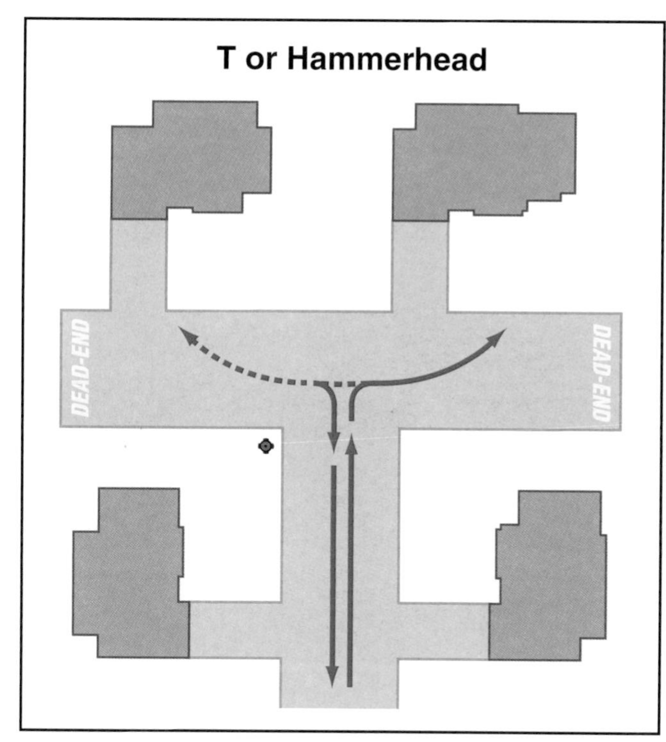

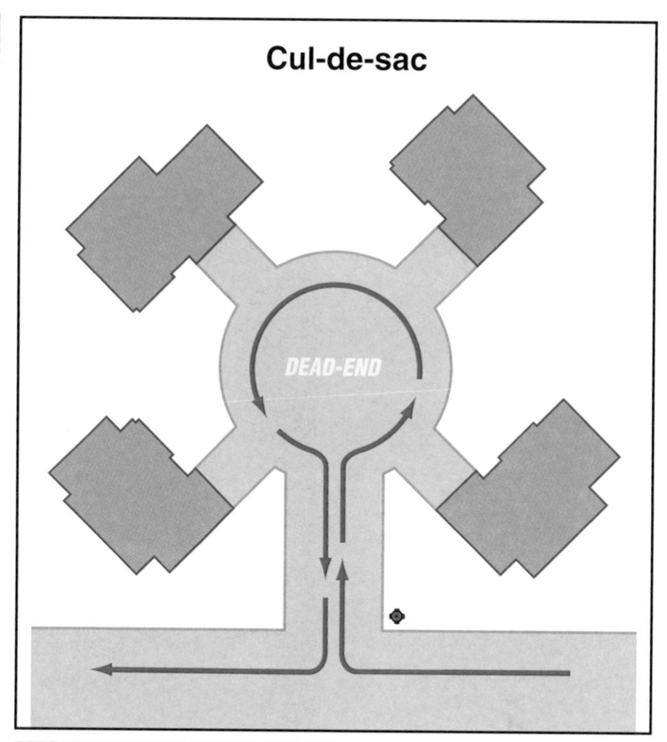

Figure 8.3 Fire and emergency vehicles must be able to turn around on dead-end roads. Commonly accepted turnarounds include the T (hammerhead) or U, cul-de-sac, or alley dock.

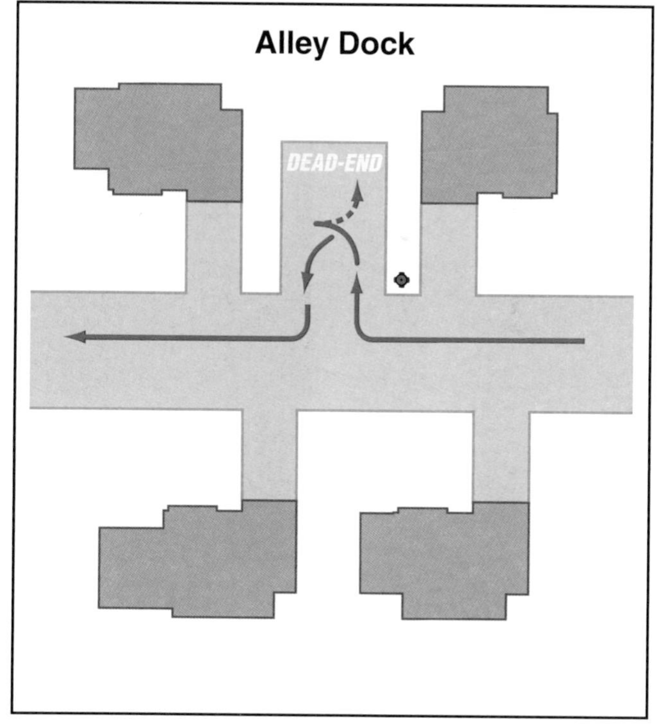

Inspectors will need to review and approve the designs of these dead-end access roads because the size and turning radius of fire apparatus factors into the road's dimensions. Once the AHJ approves the access road, the property owner/occupant is responsible for properly marking and maintaining these turnarounds.

Road Markings and Signs

The local jurisdiction generally establishes requirements for marking fire lanes and fire apparatus access roads. Inspectors may be involved in recommending the location of road markings and signs, as well as offering guidance on parking restrictions based on the need for site access. These markings and signs indicate to emergency responders the areas that are designed to support the weight of the apparatus and provide the most effective access. They also indicate to the public the areas that must remain clear at all times.

Markings take two general forms: painted curbs and posted signs **(Figures 8.4 a and b, p. 290)**. Curbs are generally painted red or red with white lettering. In some locations, fire lane designations are painted in large letters on the road surface. Signs conform to the local or state design for information signs, generally a reflective white background with red lettering, and may feature

Figures 8.4 a and b Fire lanes are designated in a variety of ways. Shown are posted signs and painted road surfaces.
a courtesy of Ron Jeffers.

the words *No Parking Fire Lane* or *No Stopping Fire Lane*. **Appendix D** contains sample fire lane sign requirements developed and used by the Phoenix (AZ) Fire Department.

Construction of Fire Apparatus Access Roads

Fire apparatus access roads may be constructed with subsurface materials that will support the weight of an apparatus, such as precast concrete forms or grasscrete. These forms are covered with soil and sod to give the appearance of a grass lawn **(Figure 8.5)**. Some local jurisdictions require the application of a paving system or another material to make sure the roadway stays visible.

The AHJ should locate signs at the entry and exit points of fire lane or fire apparatus access roads with subsurface materials. Reflective markings or other indicators placed along the edge of the subsurface construction indicate the area designed to support the weight of an apparatus. An inspector should verify that the markings are in the correct location and unobscured by vegetation. Emergency responders should be able to recognize the edge of the roadway to prevent the apparatus from getting stuck in the soft soil next to it.

In some multistory structures, parking garages or plazas surround the grade-level entrance to the building. When this situation occurs, the AHJ usually requires that a portion of the parking garage or plaza is designed to support fire apparatus. Inspectors must verify that the fire lanes are accurately indicated and that these markings are maintained.

Construction and Demolition Sites

Buildings under construction or demolition are subject to rapid fire spread due to the large quantities of unprotected combustible materials on site **(Figure 8.6)**. Fire inspectors must gain and maintain access to these high fire risk sites. NFPA® 241, *Standard for Safeguarding Construction, Alteration, and*

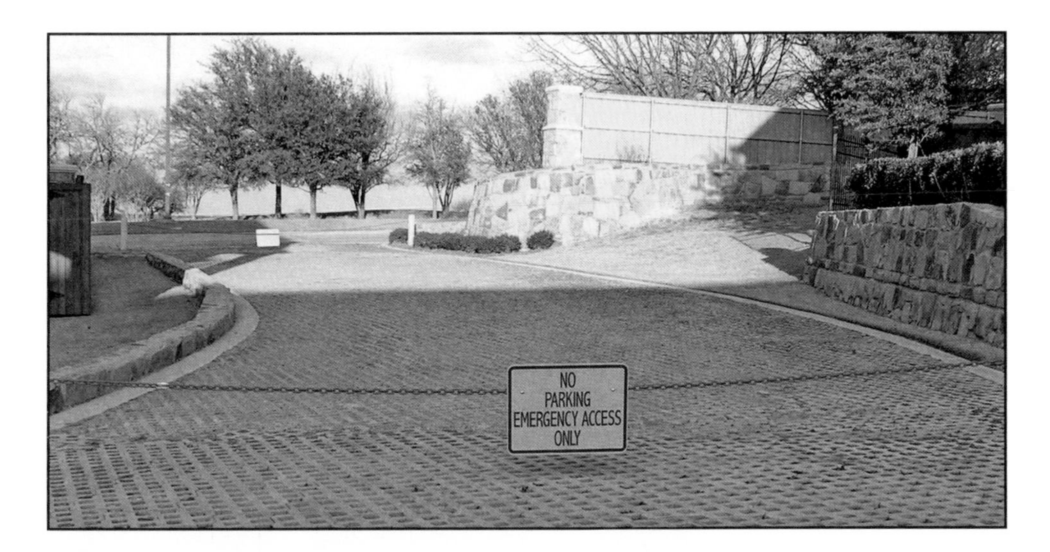

Figure 8.5 Grasscrete is cast on site to provide a road surface. *Courtesy of Ron Moore, McKinney (TX) Fire Department.*

Figure 8.6 At construction sites, inspectors must verify that trash receptacles are not positioned in such a way that they would interfere with emergency vehicle site access. *Courtesy of Ron Moore, McKinney (TX) Fire Department.*

Demolition Operations, provides requirements for designating a Command Post, accessibility of portable fire extinguishing equipment, and lock boxes at construction and demolition sites. The conditions at these sites change frequently due to the nature of the work, and the inspector is responsible for monitoring the sites to verify that the following practices are in place:

Site Plan — Drawing that provides a view of the proposed construction in relation to existing conditions; includes survey information and information on contours and grades; generally the first sheet on a set of drawings.

- A means to call the fire department is available with emergency numbers posted near the phones.
- **Site plans**, access keys, and emergency contact information are located in the construction or security office, commonly known as the Command Post.
- All fire protection systems and hydrants are accessible and operational **(Figure 8.7)**.
- Fire extinguishers are accessible in accordance with the requirements of the adopted fire and building codes.
- All flammable and combustible liquids are stored in compliance with the locally adopted codes.
- Asphalt and tar kettles are equipped with lids that can be fully closed.
- Appropriate regulations for temporary or portable heating devices are followed.
- Trash is disposed of on a daily basis.

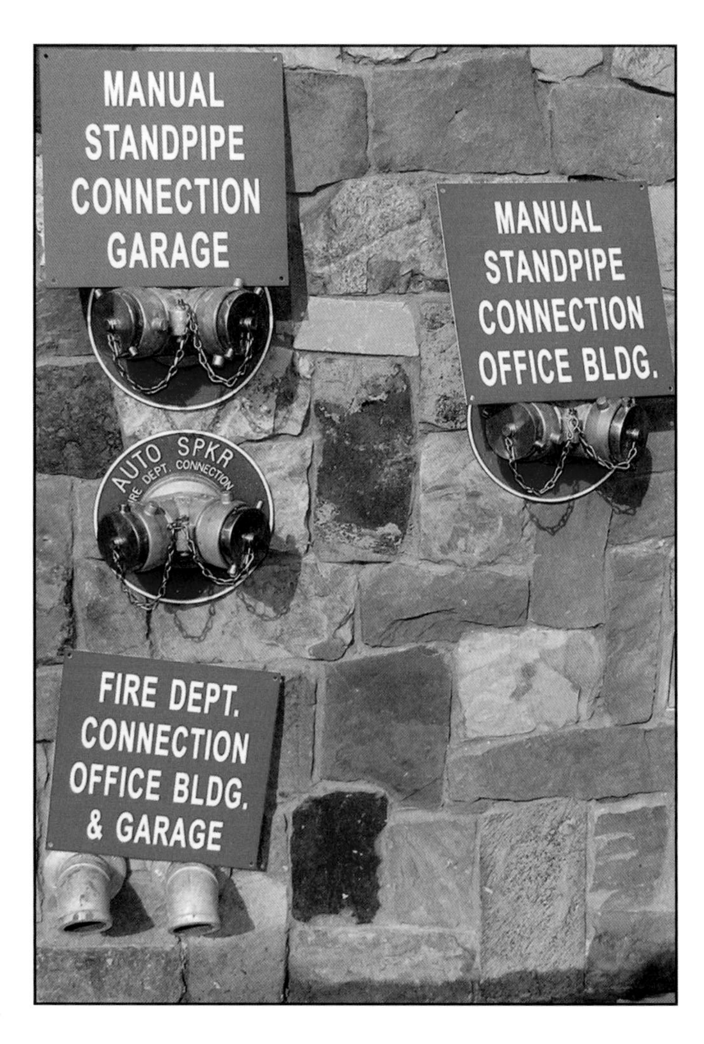

Figure 8.7 Fire protection systems must be labeled for emergency use. *Courtesy of McKinney (TX) Fire Department.*

Fire lane and fire apparatus access road requirements also apply to structures that are under construction or demolition. An inspector must verify that all designated access roads are unobstructed and maintained in serviceable condition. The following are common access problems: **(Figures 8.8 a-c)**

- Realignment of a travel surface that limits access to the structure
- Roadway not maintained in a serviceable condition
- Access to fire department water supplies and fire department connections (FDCs) not maintained
- Roadways narrowed below the minimum width required by the code
- Ditches dug across temporary roadways without alternate routes provided
- Workers using a fire lane as a parking area for personal or construction vehicles
- Temporary gates, fences, or walls that block site access, fire lanes, or fire apparatus access roads
- Demolition debris or large trash receptacles located along or in designated fire lanes or fire apparatus access roads

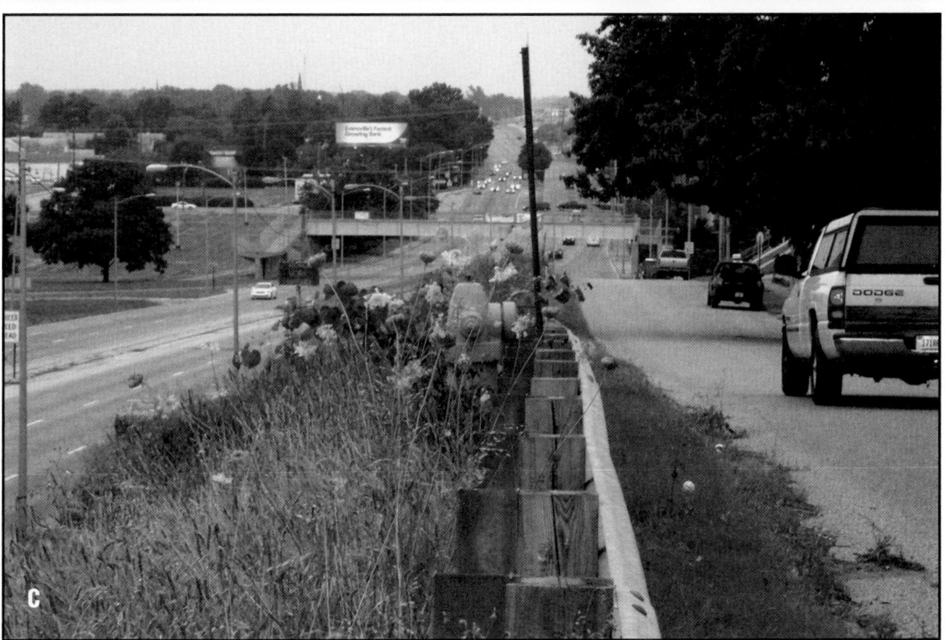

Figures 8.8 a - c Roadways that have become too narrow, are blocked by equipment, or hydrants that have become inaccessible need to be reported to the fire department. *Photo c Courtesy of Dennis Marx.*

An inspector who discovers any of these conditions at a construction or demolition site should immediately notify the general contractor. The inspector and contractor should review the required access road design to make sure that the road is returned to the approved design.

Occasionally, contractors or demolition operators may be reluctant to comply with these requirements. In these cases, an inspector may need to resort to more rigorous enforcement procedures, including formal citations against the general contractor and other individual violators. As a last resort, an inspector may decide to close the site until the contractor complies with the access requirements.

Structure Access Barriers

Upon arrival at an emergency site, emergency responders may find two types of structure access barriers that interfere with the placement of apparatus or equipment. Exterior access barriers, such as security gates and overhead obstructions, may slow, limit, or prohibit emergency responders from accessing the incident scene. Interior access barriers, such as security doors and barred windows, hinder emergency responder entry into the building. An inspector must be aware of a facility's structure access barriers and the possibility of these conditions changing over the life of the facility.

Exterior Access

Inspectors must be aware of site conditions that could hinder or block emergency access to the exterior of a structure. Inspectors are responsible for verifying that buildings are accessible and not impeded. They should document the existence and condition of the following features: **(Figures 8.9 a and b)**.

- Driveways
- Parking lots
- Gates
- Overhangs
- Vegetation
- Setbacks
- Security barriers
- Awnings
- Overhead obstructions
- Temporary fencing around construction sites

NOTE: Gates across fire access lanes must open to provide a minimum clearance of 20 feet (6 m) unless the AHJ approves another distance.

When considering exit access and determining driveway and entrance requirements, the AHJ should take into account the weight, height, length, and width of the largest fire apparatus. A performance-based driveway design, namely, one capable of permitting emergency vehicle passage, is more effective than simply following a minimum radius for the turn stated in the code. The AHJ should specify these requirements to ensure exterior access from a public street or road is not obstructed. Where necessary, the AHJ should

Figures 8.9 a and b Gates and security barriers need to be identified ahead of time so they do not delay emergency response crews. *b courtesy of Dennis Marx.*

designate the driveway as a fire lane or fire apparatus access road and apply those criteria to its design, maintenance, and markings. The following items may also cause issues with exterior access:

- Weight requirements
- Illegal parking
- Overhead obstructions
- Landscape issues
- Topographical conditions
- Season climate conditions

Weight Requirements

The local municipality must develop minimum weight requirements for fire lanes and fire apparatus access roads. Usually these requirements are in a collection of standard designs that the local highway/transportation engineer

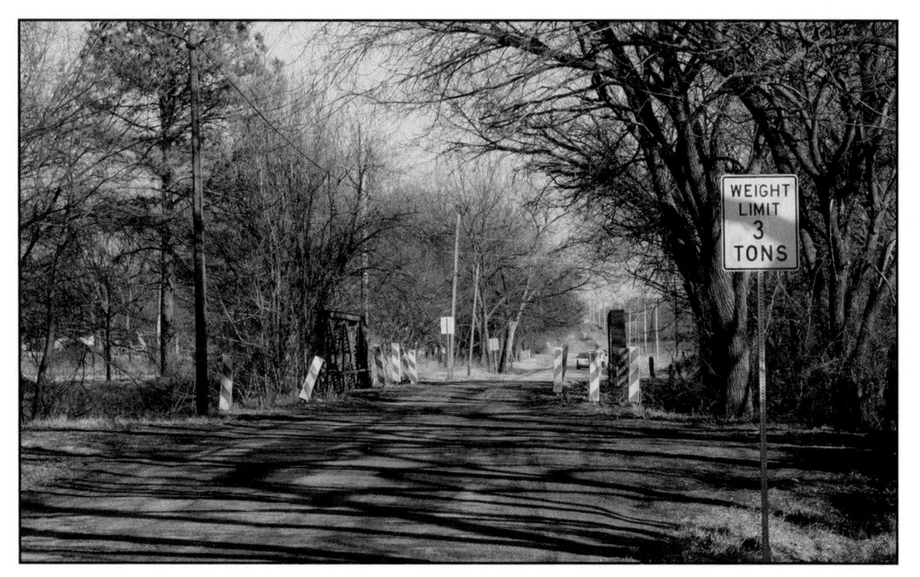

Figure 8.10 It is important to know whether outlying roads or bridges will support the weight of emergency vehicles.

has approved and the locally adopted codes enforce. These roads should be designed to bear the weight of the jurisdiction's heaviest fire apparatus. Private bridges and driveways over culverts must also adhere to the minimum size requirements for fire lanes and fire apparatus access roads.

The AHJ may require labels designating a bridge's capacity to prevent personnel from attempting to cross bridges that will not support the weight of their fire apparatus **(Figure 8.10)**. While these requirements are primarily an issue at the time of construction, inspectors may encounter parking lots, bridges, and other roadways that were constructed before adoption of the fire code or that have been modified since their construction.

Private driveways and parking lots present more concerns than public streets or roads. These surfaces may not be designed or constructed to support the weight of a fully loaded fire apparatus. While this condition may not necessarily prevent fire apparatus from accessing the structure, it could affect the operation of aerial apparatus that need to set stabilizers and deploy an aerial device. Inspectors can rarely mandate changes unless the AHJ requires that existing privately owned driveways and parking lots adhere to current weight requirements.

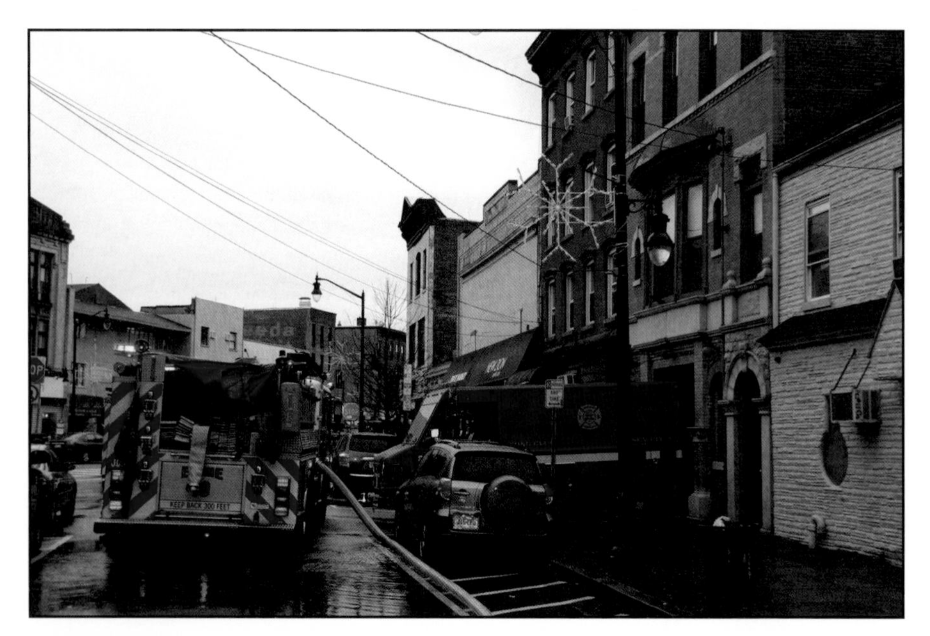

Figure 8.11 As a result of illegal parking in a fire zone, this emergency apparatus has to be positioned out in the street. *Courtesy of Ron Jeffers.*

Illegal Parking

Fire lane areas may be free of obstructions during daytime inspections but subject to illegal parking after hours. An example is an apartment complex where many occupants are away during the day and return during the evening. Illegal parking may also occur around convention centers, sports arenas, and colleges where a large number of attendees may exceed the number of available parking spaces. In addition, drivers may park in designated fire lanes, impeding placement of emergency vehicles **(Figure 8.11)**. An inspector should work with the local law enforcement agency and property owners to correct these conditions and provide the required fire department access at all times.

Overhead Obstructions

When inspecting a facility, inspectors should look for overhead obstructions that block apparatus access, obstruct aerial device operations, or hinder the use of ground ladders. Many potential overhead obstructions can impede the 13-feet, 6-inch (4 m) minimum unobstructed height requirement above fire lanes or fire apparatus access roads, such as the following **(Figures 8.12 a and b)**:

- Electrical lines or utility poles
- Trees, vines, and other vegetation
- Parking lot lights or street lights
- Building canopies, marquees, or overhangs
- Flagpoles
- Signs
- Satellite dishes
- Overhead pipes in industrial complexes

Properly applying locally adopted codes during the plans review, construction, and initial inspection process should control overhead obstructions. However, an inspector may still need to document these conditions during annual inspections so that owners can make corrections.

Figures 8.12 a and b *a* The overhead wires pictured are far too close to nearby structure and could interfere with aerial apparatus or with firefighters' ability to raise ground ladders. *Courtesy of Ron Jeffers. b* Canopies like the one on this building can obstruct access to the building in case of an emergency. Inspectors must verify that such canopies are not an obstruction and that access is readily available.

Landscaping

Landscaping can create exterior access barriers in several ways. Overgrown vegetation may block fire hydrants from view, decorative boulders may hide utility control valves, and fences may obstruct electric shutoff switches **(Figure 8.13)**. Some locally adopted codes require access to these services to be clear and marked. Property owners have been known to purposely design their landscapes to minimize the visibility of necessary services. Inspectors should evaluate and mitigate all landscape situations that can potentially interfere with emergency responders' access. Inspectors should also be aware of any changes in a building's landscape design because these alterations may require another evaluation of potential exterior access barriers.

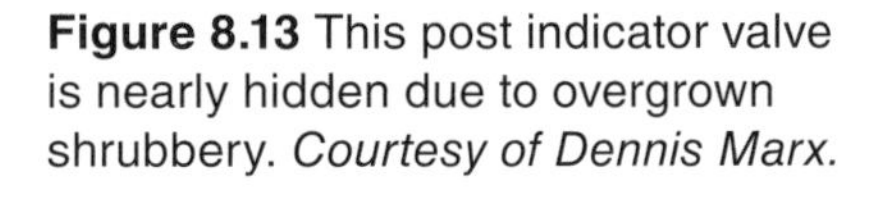

Figure 8.13 This post indicator valve is nearly hidden due to overgrown shrubbery. *Courtesy of Dennis Marx.*

Topographical — Pertaining to configuration of the land or terrain.

Topographical Conditions

Topographical conditions may limit or prohibit access to a building or facility. In these situations, alternative access points should be established if practical. Examples of these types of limitations or barriers include but are not limited to **(Figures 8.14 a and b)**:

- Railroad tracks
- Drainage ditches, streams, rivers, canals, ponds, or lakes
- Steep slopes, cliffs, or variations in topographical grades
- Unstable terrain (sand, loose or soggy soil)
- Narrow alleys or passageways between structures
- Retaining walls

Seasonal Climate Conditions

Seasonal climate conditions, such as snow, rain, and drought, may prevent access to structures or facilities. Responsibility for mitigating climate-related obstructions may belong to the owner/occupant or the AHJ.

Fire Department Connection (FDC) — Point at which the fire department can connect into a sprinkler or standpipe system to boost the water flow in the system.

In areas subject to prolonged and deep accumulations of snow, inspectors should work with outside agencies to implement snow removal procedures. This task may be the responsibility of the jurisdiction's street department or snow removal contractor. The location of fire apparatus access roads must be visible following snowfalls. Snow piles should not block fire apparatus access roads, fire hydrants, or **fire department connections (FDC)**. On private property, the owner/occupant is responsible for keeping exterior exit doors and pathways free of snow and ice.

Figures 8.14 a and b Shown are some topographical challenges for emergency responders: very narrow space between structures and water on one side of a structure. *a Courtesy of Ron Jeffers.*

Periods of heavy rain can overwhelm systems designed to control surface runoff, causing street flooding. Accumulated water on fire lanes and fire apparatus access roads, in parking lots, and in detention ponds can create problems for emergency responders **(Figure 8.15, p. 300)**. In particular, non-hard-surface roads can become saturated and impassible. Fire access lanes that use subsurface materials can cease to support their designed weight limit due to the saturation of the surrounding soil. Inspectors should be aware of potential problems caused by heavy rains and work with owners/occupants and building officials to correct them before they occur.

Droughts can also have an effect on site access. Fire lanes or access roads that are not constructed of asphalt or concrete can become dry, creating a potential source for dust that can obscure buildings, other vehicles, or people. Droughts also create a potential for grass fires around facilities. Building owners/occupants may be required to clear all dead or dry vegetation away from structure and fences.

Interior Access

Inspectors should pay attention to items that may prevent emergency responders from accessing the interior of a structure **(Figure 8.16, p. 300)**. Although inspectors may not be concerned with how firefighters breach a wall or open a locked door, they should share applicable information with emergency responders. Information concerning the type of wall construction, door locks, windows, and other interior access barriers will help firefighters perform their duties more effectively and safely.

Figure 8.15 Floods can severely hamper fire department access. *Courtesy of Dennis Marx.*

Figure 8.16 The lock box shown in this photo holds the keys to the structure behind it. In most jurisdictions, the owner of the property is required to provide a key for the lock box to the fire department.

Firefighters primarily access a building using doors and windows. Where local ordinances require it, lock boxes containing keys to the facility or structure may be located near the doors into the structure. Inspectors should note the location, key, or access codes to lock boxes and provide this information to the fire department's operations section. Security fences and gates may also have lock boxes located near an entry point. Structures that have 24-hour security on site may not be required to have lock boxes.

Rapid Entry Systems

Emergency responders may use additional emergency access systems to gain entry. Some systems, such as Knox and Supra, use keys stored in wall-mounted safes. Emergency responders carry a master key that opens the safe. Other access systems use radio-controlled signals to enable firefighters to open doors and gates **(Figure 8.17)**.

Figure 8.17 This entry system uses radio-controlled system to enable emergency responders to enter. *Courtesy of Click2Enter, Inc.*

When doors and windows fail to provide access to a building, firefighters may utilize building features, such as fire escapes, roof access doors, and other openings. When inspectors examine these interior access features, they must consider the feature's condition, location, and operation. Fire escapes should be securely fastened to the structure, free of rust, operate properly, and meet the construction, testing, and certification requirements of the locally adopted building code. Inspectors should examine roof access doors, hatches, skylights, and other openings as well. They should also evaluate access to below-grade floors through doors, stairs, or windows **(Figures 8.18 a and b)**.

Some buildings that were constructed before national building codes existed may have insufficient access. Inspectors should inform the owner/occupant of these hazards and recommend alternatives. At the very least, inspectors should notify emergency response personnel and recommend that they perform a site survey to list factors that limit access.

Windowless buildings and underground structures pose a special issue because of their significant lack of interior access points. Most building and fire codes

Figures 8.18 a and b *a* This out-of-the way exit door is locked. *b* Access to this roof should be checked to be sure it is functional. *a courtesy of Rich Mahaney.*

require the installation of an automatic sprinkler system in these structures. This requirement should be identified during the plans review process.

Brick buildings have been constructed with sufficient access points, but windows and doors are often closed and sealed as these buildings age. In some cases, it may not be technically or economically possible to install a sprinkler system in an existing building, and the inspector must be able to determine acceptable alternatives and provide them to the owner/occupant.

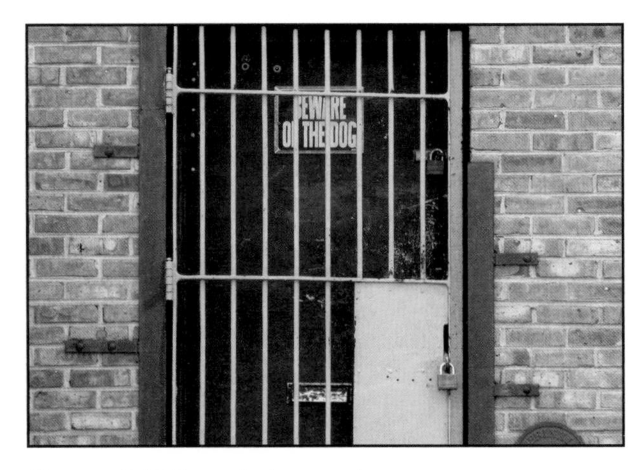

Figures 8.19 Security bars and grilles are intended to prevent illegal access to a structure. At the same time, they can prevent rapid access and egress by firefighters during an emergency incident. *Courtesy of Rich Mahaney.*

False Front — Additional facade on the front of a building applied after the original construction for decoration; often creating a concealed space.

The shell of a building acts as a barrier by preventing access at any point other than a door or window. The following is a list of additional building features that create interior access problems:

- Blank and ornamental walls
- Sunscreens
- Security bars on doors and windows **(Figure 8.19)**
- Grilles
- Hurricane shutters
- **False fronts**

Chapter Summary

Fire and life safety inspections are critical to maintaining site access for fire apparatus and emergency responders. An inspector should note the current access condition and identify any obstructions to fire lanes and fire apparatus access roads. In addition, an inspector should monitor site access to construction and demolition sites and recognize the potential structural access barriers.

Whenever inspectors discover a circumstance that affects fire and emergency responders' ability to access a structure, they should immediately implement appropriate corrective actions. If this situation proves to be a permanent access obstruction, the inspector should inform the fire department operations section of the details.

Review Questions

1. What fire lane and access road characteristics are important for inspectors to verify?
2. What are some of the practices an inspector must verify are in place at construction and demolition sites?
3. What are the two types of structure access barriers emergency responders may encounter?
4. Name some of the exterior access hindrances that can block emergency access to structures.
5. List the interior access hindrances that can block emergency access to structures.

Fire Hazard Recognition

Photo courtesy of Rich Mahaney.

Chapter Contents

Case History 307

I Unsafe Behaviors 308

- Inadequate Housekeeping 308
- Unintentional Ignition Sources 308
- Open Burning 309
- Improper Use of Electrical Equipment 311
- Improper Use and Storage of Flammable and Combustible Liquids 313
- Improper Use and Storage of Compressed/Liquefied Gases 317

Unsafe Conditions 319

- Electrical Hazards 320
- Material Storage Facilities 322
- Recycling Facilities 331

II Building Systems 335

- Heating, Ventilating, and Air Conditioning Equipment/Systems 335
- Cooking Equipment 344
- Fire Suppression.. 347
- Industrial Furnaces and Ovens 348
- Powered Industrial Trucks....................................... 349
- Tents 350

Hazardous Processes 352

- Welding and Thermal Cutting Operations 352
- Flammable Finishing Operations 356
- Quenching Operations... 359
- Dry Cleaning Operations.. 361
- Dust Hazards 362
- Torch-Applied Roofing Materials............................. 372
- Asphalt and Tar Kettles.. 372
- Distilleries.. 373

Chapter Summary 374

Review Questions 375

Key Terms

Arc....320
Commodity....324
Encapsulating....323
Explosion....362
Hot Work....352
Ignition Temperature....320
Intrinsically Safe Equipment....312
Plenum....357
Safety Data Sheet (SDS)....334
Vessel....338

NFPA® Job Performance Requirements

This chapter provides information that addresses the following job performance requirements of NFPA® 1031, *Standard for Professional Qualifications for Fire Inspector and Plan Examiner* (2014).

Fire Inspector I	Fire Inspector II
4.3.8	5.3.6
4.3.12	5.3.8
4.3.13	5.3.9
4.3.14	5.3.11
4.3.15	5.3.12

Fire Hazard Recognition

Learning Objectives

After reading this chapter, students will be able to:

Inspector I

1. Identify unsafe behaviors that may require code enforcement. (4.3.8, 4.3.15)
2. Identify improper use or storage of flammable and combustible liquids. (4.3.8, 4.3.12, 4.3.13, 4.3.15)
3. Recognize unsafe conditions that have hazardous fire growth potential and may require code enforcement. (4.3.8, 4.3.14)

Inspector II

1. Evaluate hazardous conditions involving building systems. (5.3.6, 5.3.8, 5.3.11, 5.3.12)
2. Distinguish among hazardous processes that contribute to increased fire risk. (5.3.8, 5.3.9, 5.3.11)

Chapter 9
Fire Hazard Recognition

Case History

On August 17, 2012, a fire started in a San Antonio warehouse that stored approximately 35,000 plastic pallets. While firefighters were on scene battling the 3-alarm fire, explosions could be heard, possibly from the transformers inside the building and fuel tanks in nearby trucks. Guests at a nearby hotel had to be evacuated due to the enormous quantities of thick, black smoke.

An investigation into the circumstances of the fire revealed several fire code violations. Some plastic pallets were stacked up to 18 feet high – much higher than codes permit. In addition, the facility sprinkler riser valves had been shut off, rendering the sprinkler systems inoperable. Fortunately, no one was injured in the massive fire that resulted in damages estimated at $9 million.

A fire inspector typically encounters three types of fire hazards: unsafe behavior, unsafe conditions, and hazardous processes. It is the inspector's responsibility to recognize these situations and serve as part of the community's risk reduction efforts. Inspectors have the opportunity to observe common hazards as part of the job. In working to correct hazards through education, the inspector increases fire prevention awareness and community safety.

According to data collected through USFA Fire Incident Survey (2003-2012), nonresidential fires in nonresidential structures included the following (in order of frequency):

- Cooking
- Intentional
- Unintentional, carelessness
- Heating
- Electrical malfunction
- Other heat
- Appliances
- Other equipment
- Equipment malfunction
- Smoking

Source: U.S. Department of Homeland Security, U.S. Fire Administration National Fire Data Center/USFA Fire Estimate Summary, 2014.

Some causes of fire, such as lightning strikes, are beyond the ability of an inspector to influence, although their potential to cause harm can be reduced with proper grounding. Incendiary or intentionally set fires are also outside an inspector's influence. The inspector can work to reduce the remaining causes of fire by providing fire and life safety education during inspections, acting to eliminate hazards, and enacting and enforcing fire codes.

This chapter provides an overview of some of the fire and life safety hazards that inspectors may encounter during field inspections. It is not a complete list of every situation by any means. Before beginning an inspection, an inspector must research the specific processes based on the type of occupancy and operation to be inspected. An inspector must be able to identify the process or equipment, be familiar with its purpose and operation, and understand the applicable basic safety principles.

Unsafe Behaviors

An inspector who observes unsafe behaviors that increase fire hazards or other dangerous activities can educate occupants about the hazards of the behaviors or require that the condition be eliminated. Fire company personnel can also perform these functions during facility surveys and preincident inspections. Providing information to help occupants/owners change behaviors and reduce risks is one of the most important ways in which an inspector can make a difference. Some of the most important problem areas are:

- Inadequate housekeeping
- Unintentional ignition sources
- Open burning
- Improper use of electrical equipment
- Improper use and storage of flammable and combustible liquids

Inadequate Housekeeping

In terms of fire and life safety, housekeeping involves creating and maintaining a clean and safe environment. Inadequate housekeeping can contribute to the facility's fire load and prevent occupants from exiting safely during an emergency **(Figures 9.1 a and b)**. An inspector can help increase safety and reduce or eliminate fire hazards by enforcing appropriate sections of the locally adopted fire code. The following are some examples of poor housekeeping:

- Accumulations of trash or litter, particularly near ignition sources
- Oily rags in open waste containers
- Overgrown grass, weeds, and brush
- Obstructions that prevent a safe and orderly exit and/or clear exit access

Unintentional Ignition Sources

An ignition source is any material that can initiate combustion in a fuel. Examples of ignition sources include the following:

- Open flame
- Spark **(Figure 9.2)**
- Heat that friction creates
- Heating device
- Overheated circuit, or damaged cord or electrical equipment
- Any number of other sources

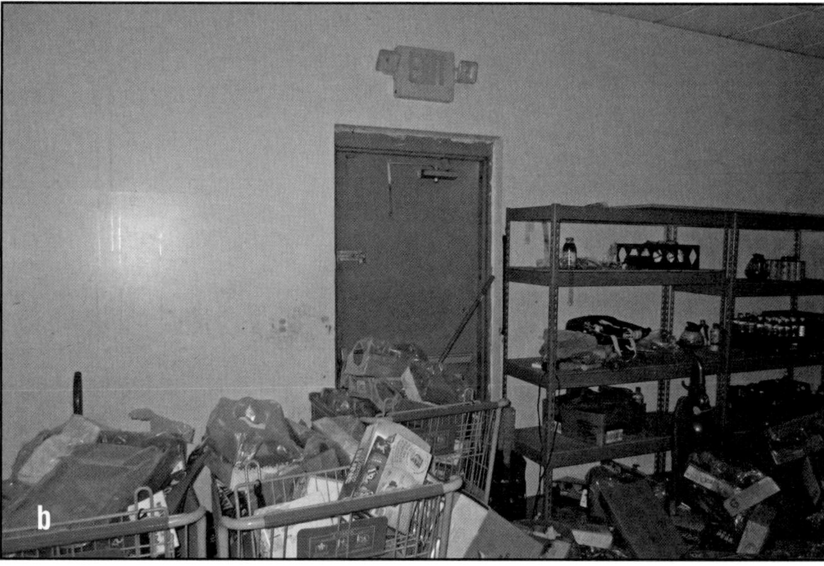

Figures 9.1 a and b Inspectors commonly find evidence of poor housekeeping: trash that is a health hazard and blocked exits that impede egress. *a courtesy of Scott Strassburg; b courtesy of Rich Mahaney.*

People generally take ignition sources for granted when they use them to light smoking materials, cook food, and provide warmth. Becoming complacent in the use of ignition sources can lead to an attitude that the sources are always under control and will never cause a hostile or uncontrollable fire. The following unsafe behaviors are related to ignition sources:

- Disposing of smoking materials or matches improperly
- Igniting a flame in a flammable atmosphere
- Ignoring prohibitions on open flames or smoking
- Leaving cooking utensils and food unattended on a stove
- Placing portable heating devices too close to combustible materials
- Leaving any open fire unattended or failing to verify that fire is completely out
- Allowing children access to matches, lighters, or candles
- Using candles improperly
- Plugging too many electrical appliances into a circuit
- Failing to control or supervise hotwork

  NOTE: Hotwork is described later in this chapter.

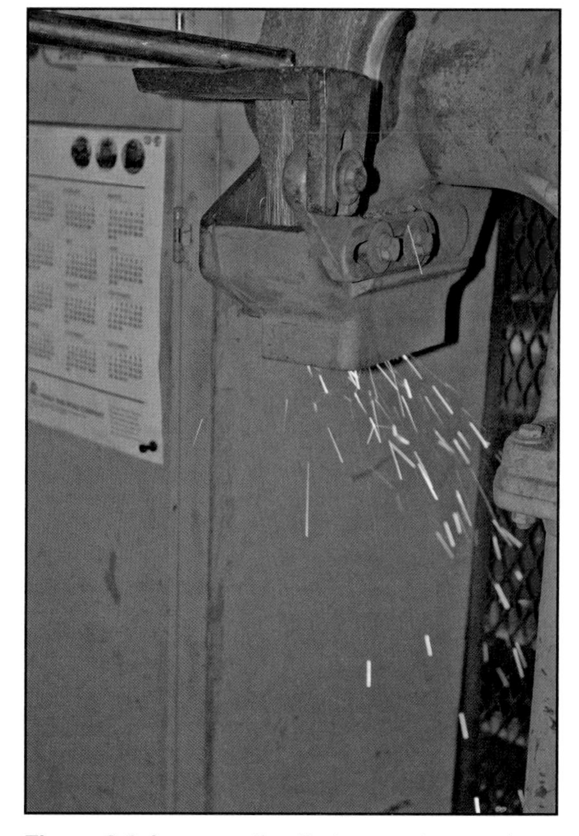

Figure 9.2 Any operation that generates sparks has the potential to start a fire.

Open Burning

Many jurisdictions have regulations that govern open burning. Such agencies as the agricultural commissioner, air quality management district, and/or fire district are responsible for regulating open burning within their jurisdiction. These regulations include the following items as they apply to open burning:

- Permissible (or forbidden) times, air quality, or weather conditions
- Mandatory attendance during the burning period
- Availability of fire extinguishing equipment
- Separation distances from exposure buildings or properties
- Clearance of vegetation around the burn site
- Acceptable fuels used to ignite combustible materials

Most model fire codes require that an individual apply for a permit before the desired burn date. Upon receiving an application, an inspector should visit the site to verify that conditions are in compliance with adopted regulations. With an approved burn site, inspectors may issue a burn permit. The permit should describe the required conditions under which the burn will be held.

NOTE: Inspectors should only approve burn permits when all conditions are in compliance with the jurisdiction's open burning regulations.

According to most model fire codes, open burning includes the following activities:

- Burning of combustible waste (usually organic waste, such as trees, leaves, and brush)
  - In metal drums or excavated pits
  - In debris piles
  - At construction or demolition sites (combustible construction materials)

  NOTE: Certain areas no longer allow the use of barrels for burning of trash and other debris. Check your local open burning regulations for specific requirements.
- Prescribed burning of vegetation
  - To control dead fuels in wildland-urban interface areas **(Figure 9.3)**
  - To control natural areas, such as forests and agricultural fields
- Using torches
  - For paint removal from the exterior of a building
  - For roof membrane installation
- Preparing flaming food or beverages in restaurants **(Figure 9.4)**
- Disposing of a worn or damaged American Flag (as per 1989 Flag Protection Act)
- Recreational fires for cooking, pleasure, religious, or ceremonial reasons

NOTE: Recreational fires do not include burning of rubbish and waste, or fires that are contained in an incinerator, outdoor fireplace, barbeque pit, or similar appliance. Cooking equipment hazards are described later in this chapter. Fires in outdoor fireplaces and barbeque grills/pits that are used to cook food and provide warmth are typically not classified as open burning.

NOTE: The jurisdiction's locally adopted fire codes may limit recreational fire activities because of unfavorable weather conditions or materials that exceed fire area and height restrictions.

Figure 9.3 Open burning of underbrush or other vegetation is individually sanctioned by county or state/provincial regulations and prohibited without attaining proper authority in the form of a permit.

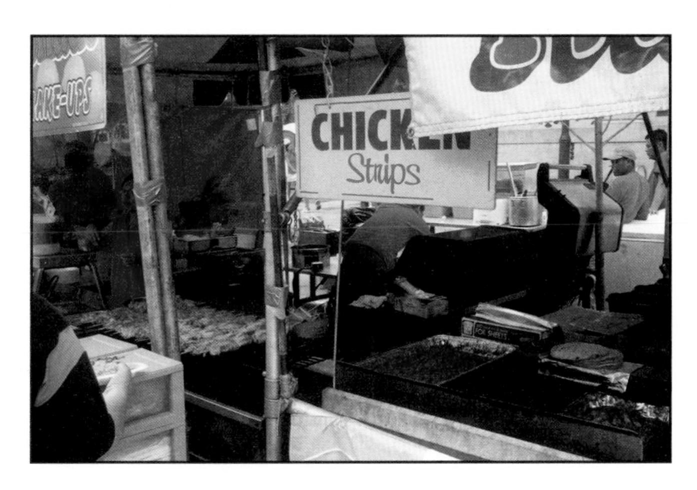

Figure 9.4 Flaming foods are usually considered open burning. *Courtesy of Scott Strassburg.*

Improper Use of Electrical Equipment

Failing to install or maintain electrical equipment or systems properly can create a hazardous condition. Inspectors regularly encounter a number of unsafe uses of electricity during field inspections, including:

- Overloaded wiring or receptacles
- Daisy chains
- Frayed or damaged extension cords
- Extension cords used for permanent wiring, affixed to walls, or run through doors, ceilings, or joists **(Figures 9.5 a and b, p. 312)**
- Extension cords that are undersized in amperage capacity for the equipment plugged into them
- Tools that can cause electrical shorts
- Unprotected temporary lights or wiring **(Figure 9.6, p. 312)**
- Tampering with grounded circuits

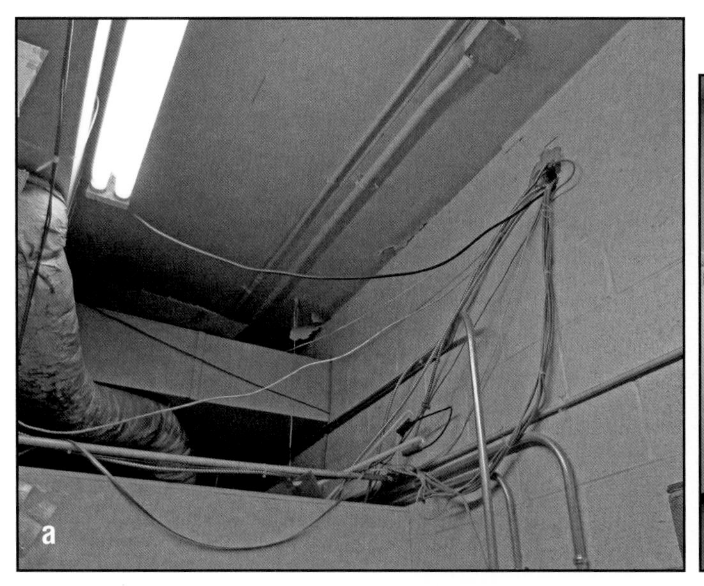

Figures 9.5 a and b Unapproved wiring and extension cords run through ceilings and attached to one another are another typical hazard. The inspector will need to apply codes to evaluate whether such wiring is temporary or considered permanent. Most wiring must be made permanent within 90 days. *Courtesy of Rich Mahaney.*

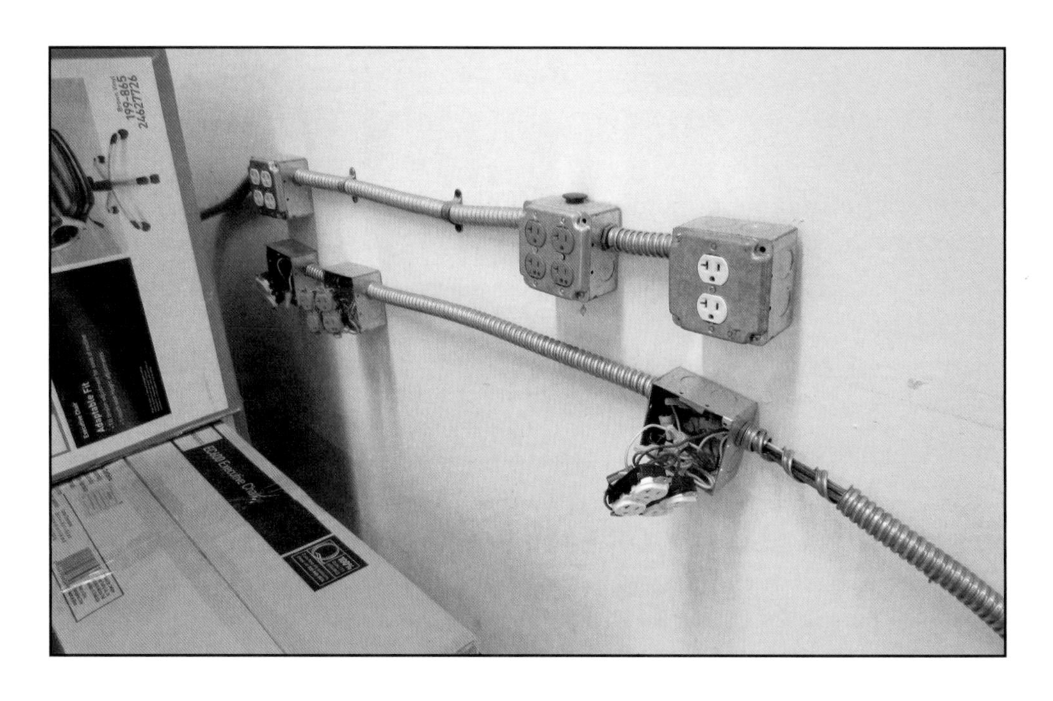

Figures 9.6 Unprotected temporary wiring presents another common hazard. *Courtesy of Rich Mahaney.*

- Unauthorized alterations to electrical systems **(Figure 9.7)**
- Disabled or bypassed shutoff switches, circuit breakers, or other required electrical disconnects

These behaviors not only create a potential ignition source, but also increase the danger of electrocution or shock. If a situation appears to be unsafe, the inspector should notify the authority's electrical inspector.

Intrinsically Safe Equipment — Equipment designed and approved for use in flammable atmospheres that is incapable of releasing sufficient electrical energy to cause the ignition of a flammable atmospheric mixture.

General-purpose electrical equipment can cause explosions in certain atmospheres. Equipment used in areas where explosive concentrations of dusts or vapors may exist must be compliant with NFPA® 70, National Electrical Code®, and listed for this environment and use. Motors in particular may need to be nonsparking and **intrinsically safe equipment**.

NOTE: Combustible dust hazards are described later in this chapter.

NFPA®70, *National Electrical Code®,* classifies electrical hazards according to the hazardous materials in the surrounding area:

- Class I locations — Flammable vapors and gases may be present.
- Class II locations — Combustible dust may be found.
- Class III locations — Ignitable fibers may pose a hazard.

Hazards are further broken down into groups based on materials and divisions based on operating conditions. Analyzing each location for hazardous or explosive atmospheres enables protective measures to be taken that account for both cost and safety factors.

Improper Use and Storage of Flammable and Combustible Liquids

Flammable and combustible liquids are often used and stored improperly. Many people do not recognize the potential for an accident to occur as a result of improper use or storage of flammable and combustible liquids until it is too late. An inspector should be aware that the unsafe use of flammable and combustible liquids can take many forms, including:

- Improperly dispensing, mixing, or transferring flammable and combustible liquids
- Using flammable liquids for cleaning
- Storing flammable or combustible liquids in unapproved containers or locations **(Figure 9.8)**
- Using a flammable liquid such as gasoline to ignite a solid fuel, such as charcoal briquettes

NOTE: Powered industrial trucks can also be ignition sources, as can vapors from distilleries. They are described later in this chapter.

NOTE: Chapter 10, Hazardous Materials, contains additional information about flammable and combustible liquids.

Figure 9.7 Improperly altered electric circuits are a fire hazard. *Courtesy of Rich Mahaney.*

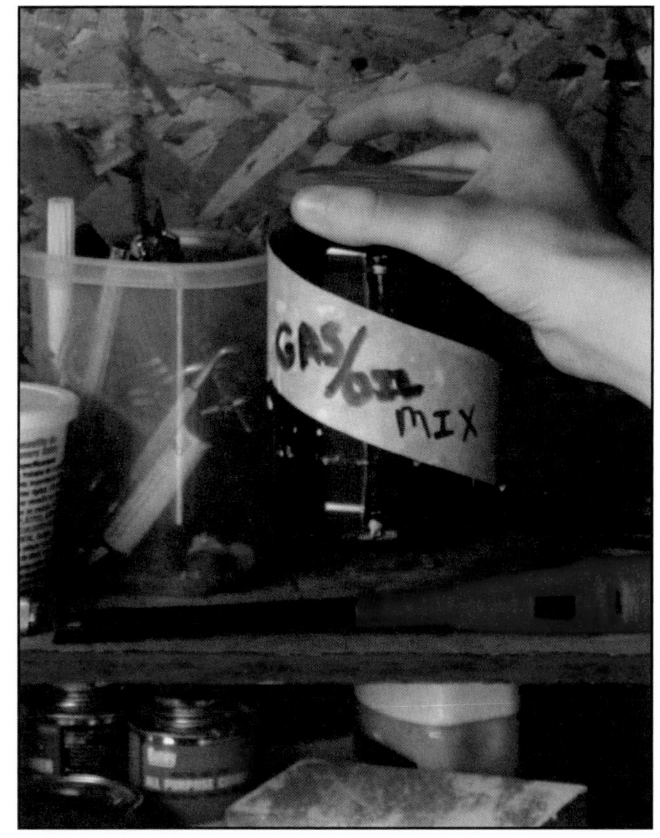

Figure 9.8 Storage of flammable liquids in unapproved containers can result in a fire or personal injury.

Inspectors must be aware that unsafe behaviors involving flammable/combustible liquids can result in dangerous conditions. The following guidelines apply to the safe handling of small amounts of these liquids:

- Classes IA through IC and Class II liquids must be kept in covered safety containers when they are not actually in use.
- Classes IA through IC liquids must not be used in the presence of any possible ignition source, such as open flames, electrical arcs, or heating elements.
- Classes IA through IC liquids cannot be stored in containers that are pressurized with air. In some circumstances, they may be stored in containers that are pressurized with an inert gas.
- Appropriate electrical bonding and grounding procedures must always be followed **(Figure 9.9)**.

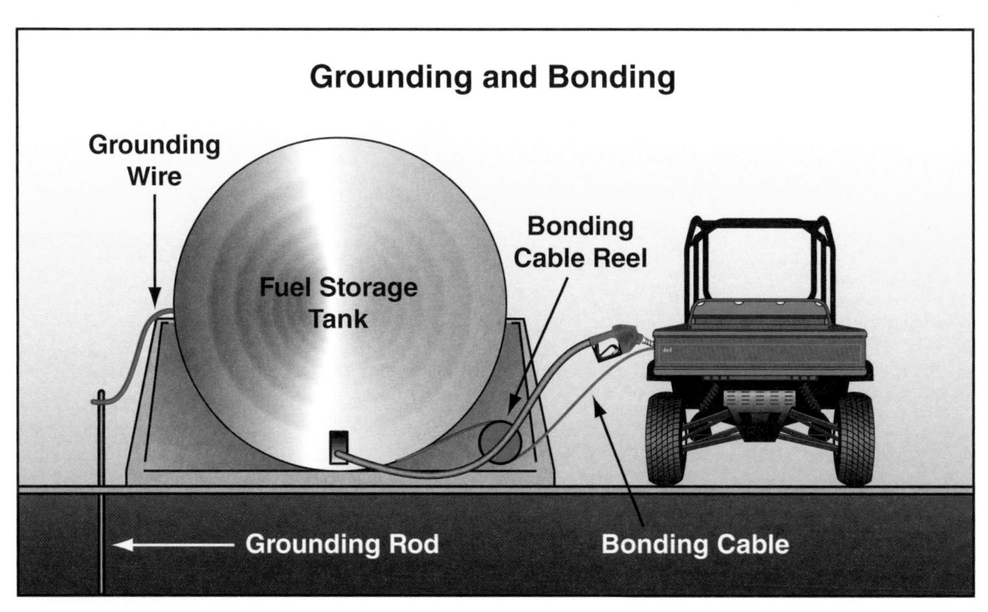

Figure 9.9 The proper use of grounding and bonding reduces the risk of ignition from a static discharge.

Dispensing

Spills, releases, or leaks frequently occur during the dispensing of liquids from one container to another. Although dispensing can be hazardous regardless of the amount of liquid being transferred, loading and unloading procedures at bulk-handling operations are particular concerns. The following rules apply to dispensing hazardous liquids:

- Loading and unloading stations for Classes IA through IC liquids must be located no closer than 25 feet (8 m) from storage tanks, property lines, or adjacent buildings **(Figure 9.10)**.
- Loading and unloading stations for Class II and Classes IIIA and IIIB liquids must be located no closer than 15 feet (5 m) from these same objects **(Figure 9.11)**.
- Loading and unloading stations must be constructed on level ground.
- Curbs, drains, natural ground slope, or other means are required to keep any spills in the original area.

- Adequate ventilation (natural or mechanical) must be maintained for loading and unloading stations.

Some types of liquids, including light fuel oils, toluene, gasoline, and jet fuels, can develop static electrical charges on their surfaces. If a static discharge occurs at the same time that a flammable mixture is near, a fire or explosion can result. To protect against static discharges, tanks must be bonded together with a metal chain or strap **(Figure 9.12, p. 316)**. Tanks also need to be grounded to neutralize static charges when Classes IA through IC liquids are loaded and unloaded.

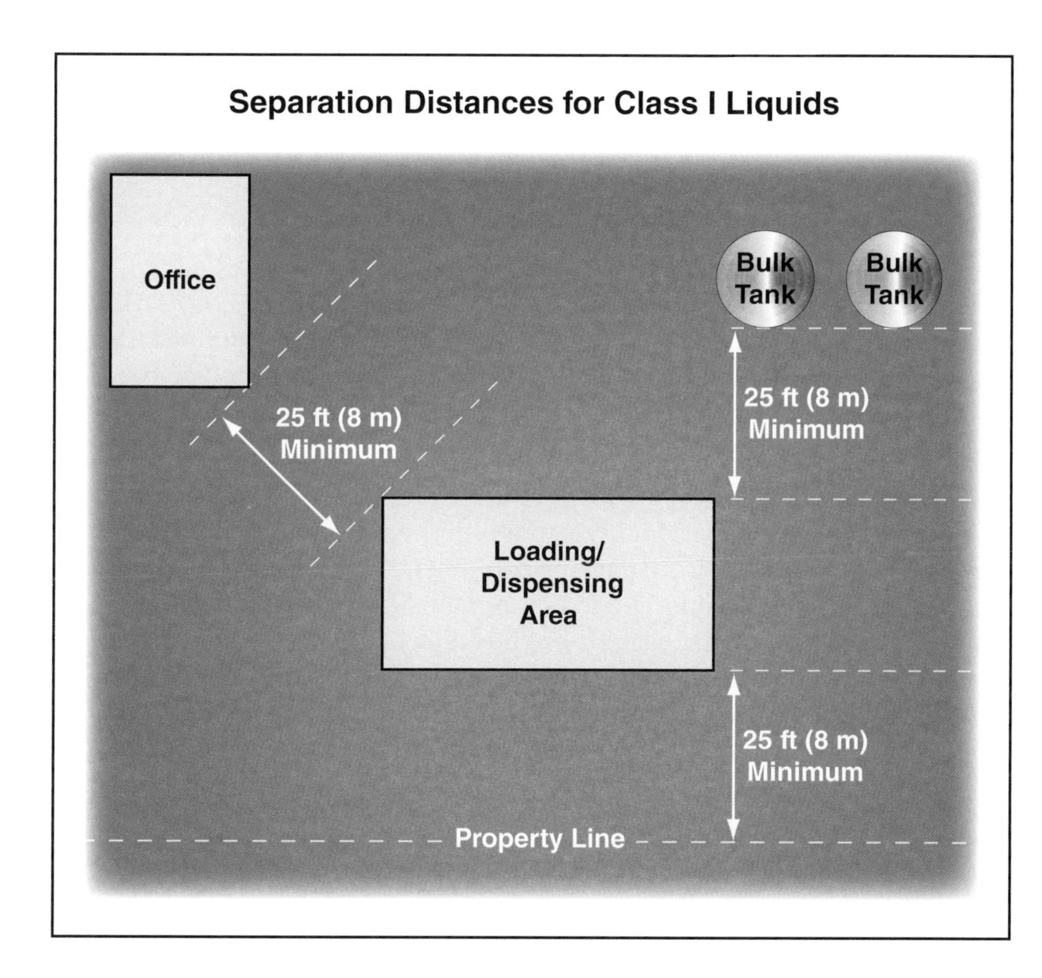

Figure 9.10 Proper distance must be maintained between storage tanks, loading/dispensing areas, and structures.

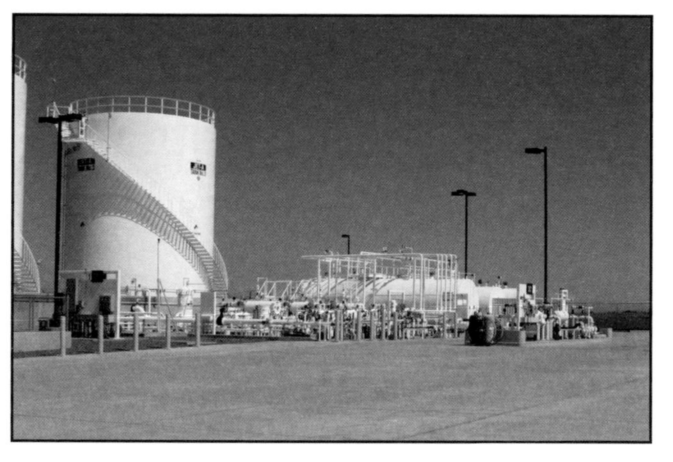

Figure 9.11 Loading/unloading stations must be located at least 15 ft (5 m) from storage tanks, adjacent structures, and property lines.

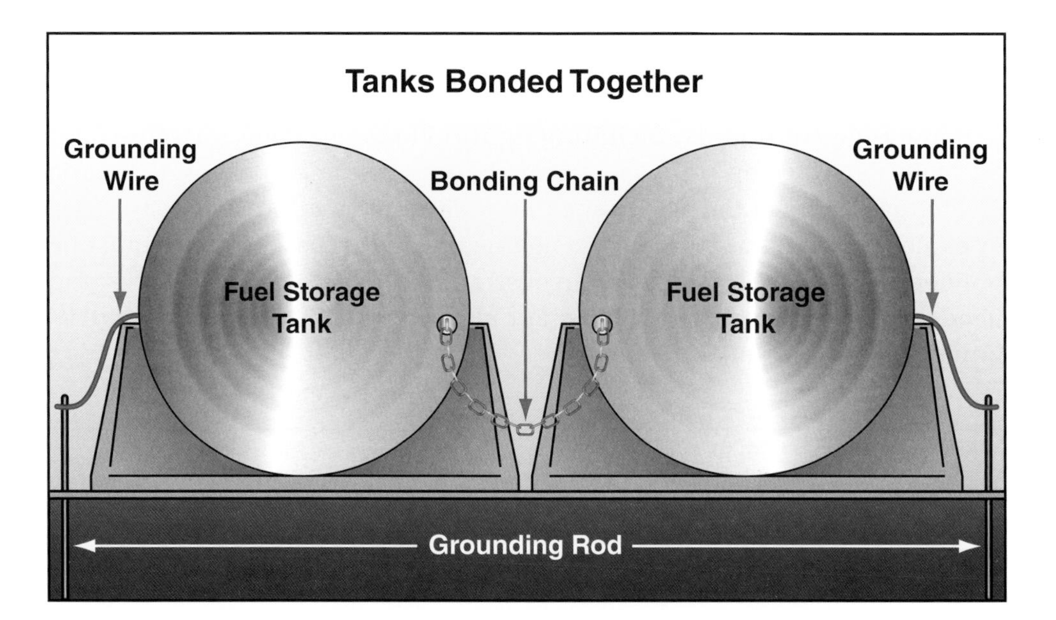

Figure 9.12 A bonding chain between two metal storage tanks is used to protect against static discharges.

Bonding and grounding are also necessary when Class II or Classes IIIA and IIIB liquids are loaded into containers that have previously contained Classes IA through IC liquids. All bonding connections must be completed before dome covers, lids, or caps are removed. Bonding is also needed between the dispensing device and receiving containers.

Perhaps the safest procedure for handling flammable or combustible liquid from a fire protection standpoint is to pump them from underground tanks through a piping system to dispensing equipment. The dispensing equipment must be kept in specially designed rooms or outside the structure. Piping materials must resist any corrosion from the material being transferred. Piping must also be able to withstand the maximum service pressure and temperature as well as thermal shock and physical damage.

All transfer systems must be designed so that liquids do not continue to flow by gravity or by siphoning if a pipe breaks. Control valves must be located at easily accessible points along the piping to control the flow. Inspectors must verify that all piping and connections are inspected on a regular basis for signs of deterioration or leakage and replaced as necessary.

Compressed/liquefied gases are usually transferred by replacing storage cylinders or refilling cylinders or tanks. There must be a standard procedure for all disconnect and transfer operations. When deliveries are made, the receiver must verify that the cylinders are in good condition and labeled properly. A written standard operating procedure (SOP) must be in place to ensure that the correct product is being loaded into the correct vessel. Delivery truck drivers must be warned when unloading operations are in progress so that they do not attempt to drive away while hoses are still connected.

All cylinders on loading stations must be clearly labeled with the name of the gas being used in order to reduce the possibility that someone might connect the wrong containers. Color coding can be useful, but must not be relied upon totally. The loading hose furnished by the receiver must be properly maintained and replaced at least every 5 years. When not in use, hoses must be stored away from direct sunlight. Bonding is generally not required if the hose has vapor-tight connections.

Static Discharge

Static electricity is a potential ignition source where flammable/combustible liquids, compressed/liquefied gases, and oxidizers are transferred. Static electric charge buildup increases when the weather is dry and relative humidity is less than 30 percent. Flowing liquid or gas that is contaminated with metallic oxides, scale, or liquid particles may develop a static electrical charge. If electrically charged gas contacts an ungrounded conductive material, the charge will be transferred to that body.

Bonding the two receptacles with a metal chain or cord and grounding them prevents the static discharge. If the liquid or gas is transferred in a completely closed system, however, no charges can be transferred; and the system need not be electrically grounded or bonded.

Dispensing of flammable solids or explosives and blasting agents may also result in spills or releases. These spills, however, are not as dangerous as the release of liquids or gases. They are also easier to contain and clean up. Disposal of spilled flammable solids must conform to the appropriate federal regulations for hazardous materials.

Transporting

The most obvious forms of transport for flammable/combustible liquids, compressed/liquefied/cryogenic gases, corrosives, and oxidizers are the bulk packaging containers discussed earlier. They may be found on railway lines, highways, and at loading and unloading facilities. Less obvious are the boxcars, cargo trucks, and delivery vehicles that contain loads of nonbulk packages, which can include all types of hazardous materials. Some of these shipments may be contained on pallets or in mixed shipments of other products.

An inspector must be familiar with the local fire and zoning codes that regulate the parking of cargo tank trucks or hazardous materials shipments within the limits of the jurisdiction. Other ordinances may regulate the streets or roads that trucks must take when passing through a jurisdiction.

Finally, an inspector must be aware that small commodities of hazardous materials may be mixed in with other commodities that in total create a much greater hazard than the single package. At the same time, the vehicle containing these small commodities may not be required to have external markings.

Improper Use and Storage of Compressed/Liquefied Gases

When inspectors are evaluating handling procedures for compressed/liquefied gas cylinders, they must verify that workers are properly trained and under competent supervision. It can be useful to review a company's administrative and engineering control policies/guidelines to expose potential unsafe behaviors. The following actions must be taken to control hazards when handling compressed/liquefied gas cylinders:

- Use only cylinders approved for interstate transportation of compressed/liquefied gases (those marked with appropriate DOT approval codes).
- Verify that numbers or marks stamped on cylinders remain in place and unchanged with the exception of those made by hydrostatic testing organizations.

Figure 9.13 Cylinders should be rolled on the cylinder edge to prevent damage to the cylinder and reduce the risk of the cylinder falling.

- Roll cylinders on their bottom edges instead of dragging when moving them into position **(Figure 9.13)**.
- Protect cylinders from cuts or other physical damage.
- Use manufacturer-approved lifting devices and methods for moving cylinders (an electromagnet may not be used).
- Ensure that employees do not drop cylinders or let them strike each other.
- Verify that employees do not use cylinders for rollers, supports, or any purpose other than to contain gas.
- Ensure that employees do not tamper with safety devices on the valves of cylinders.
- Contact the supplier or manufacturer when in doubt about the proper handling of a compressed gas cylinder or its contents.
- Mark empty cylinders with the word *Empty* or the letters *MT* (phonetic equivalent of *Empty*), and secure their valve caps in place **(Figure 9.14)**.

Inspectors must ensure that the following safety precautions are observed in areas containing compressed/liquefied gas cylinders:

- Prohibit smoking in the area where cylinders are stored or used.
- Provide adequate ventilation to prevent accumulation of any flammable gas vapors.
- Enclose electrical elements in glass to prevent direct contact with flammable gases, and keep electric lights in a fixed position.
- Locate electrical switch devices outside the storage room and ensure that they are designed to be explosion-resistant.
- Equip glass light enclosures with a guard to prevent breakage.
- Cover cylinder valves with safety caps when cylinders are not in use in order to protect valves and male threads.
- Anchor stored cylinders to a wall or other secure object with chains around the tank so that they cannot be knocked over accidentally **(Figure 9.15)**.
- Nest bulk cylinders as opposed to anchoring and chaining them.
- Ensure that the owner/occupant provides a written list of procedures for handling and storage of all gases present.
- Store and/or use cylinders at ground (grade) level and in areas that have sufficient ventilation.
- Always disconnect a cutting torch head or instrument from the supply hose before storing it in a tool box or cabinet.
- Ensure that oxygen cylinders, hoses, and connections are free of oil, grease, or other hydrocarbon (petroleum-based) lubricants.
- Secure cylinders properly with caps in place.
- Maintain gas cylinders as prescribed by DOT specifications.
- Store incompatible gases in separate areas.
- Store cylinders in an area that is clean and free of accumulated combustible materials.

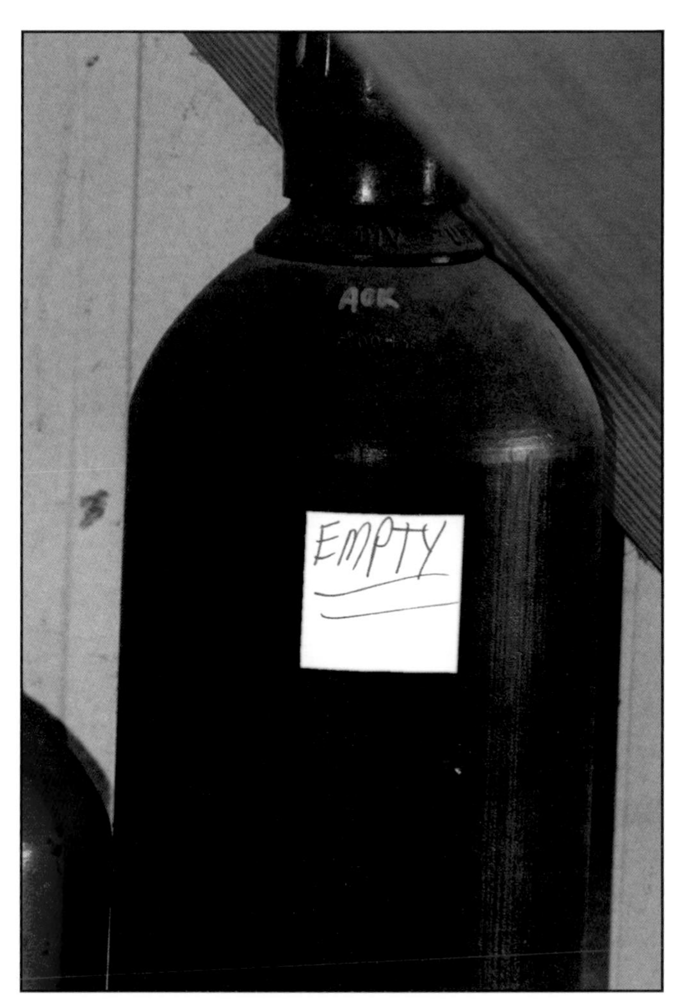

Figure 9.14 Empty cylinders should be clearly marked either with the word as shown or with the letters *MT*.

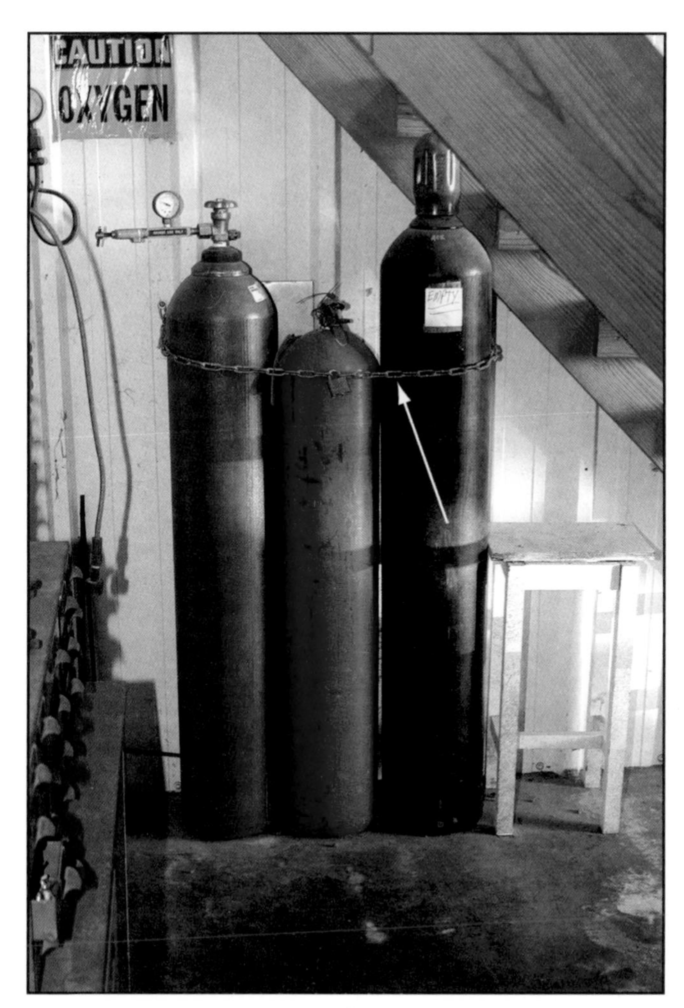

Figure 9.15 Compressed gas cylinders must be secured to a wall or other secure object with a chain to prevent damage to the cylinders.

- Remove cylinders that have been abused, dented, arc-struck, damaged, or are severely corroded from service and replace.
- Remove cylinders that have expired hydrostatic test dates from service.
- Install and maintain pressure-relief devices with appropriate settings and adequate pressure-relief capacities.

Unsafe Conditions

Unsafe conditions that result in fire hazards may arise from the design or use of a building, facility, or equipment that may be related to behaviors or processes. Unsafe conditions may be specific to a certain type of occupancy or be found in all occupancies. Proper application of the locally adopted codes will reduce these potential sources of fire hazards, in particular the following:

- Electrical hazard conditions **(Figures 9.16 a and b, p. 320)**
- Material storage facilities
- Recycling facilities

Figures 9.16 a and b Wrapping cords with duct tape or bunching them together, and facilitating heat buildup are common practices that result in unsafe conditions. *Both photos courtesy of Rich Mahaney.*

Electrical Hazards

The improper use of electricity and the uncontrolled properties of electricity can create fire hazards. When principles of fire safety and code provisions are not followed, electrical energy can generate extremely high temperatures that can result in ignition.

Detailed electrical inspections are typically the responsibility of qualified electrical inspectors. However, fire inspectors should have a basic knowledge of electrical theory so they can understand how electrical energy can be the cause of fires. The inspector needs to be able to recognize potential electrical hazards and report any concerns to electrical inspectors for correction. The next sections discuss the most common causes of electrical hazard fires:

- Worn electrical equipment
- Improper use of electrical equipment
- Defective or improper electrical installations

Arc — A luminous discharge of electricity across a gap. Arcs produce very high temperature.

Ignition Temperature — Minimum temperature to which a fuel, other than a liquid, in air must be heated in order to start self-sustained combustion independent of the heating source.

Worn Electrical Equipment

Worn electrical equipment may be prone to **arcing** and subsequent overheating of surrounding materials to their ignition point. This wear may be due to improper maintenance or equipment that has reached the end of its lifespan and needs replacement. For example, an extension cord placed under a rug can overheat and cause the rug to ignite. In addition, dust, water, or other foreign materials can be drawn into a motor, causing a short circuit, increased load, or jammed armature that overheats.

Arcing is the movement of a current between two electrodes or conductors. Arcing inside electrical equipment may heat the appliance housing or nearby materials to **ignition temperature**. In addition, arcing can ignite combustible materials. This ignition usually occurs when electrical connections have become loose through usage, vibration, or normal wear.

Improper Use of Electrical Equipment

A common electrical hazard occurs when electrical equipment is used improperly. Overloading a circuit beyond its design capabilities is an example **(Figures 9.17 a and b)**. As the circuit is overloaded, resistance increases, resulting in

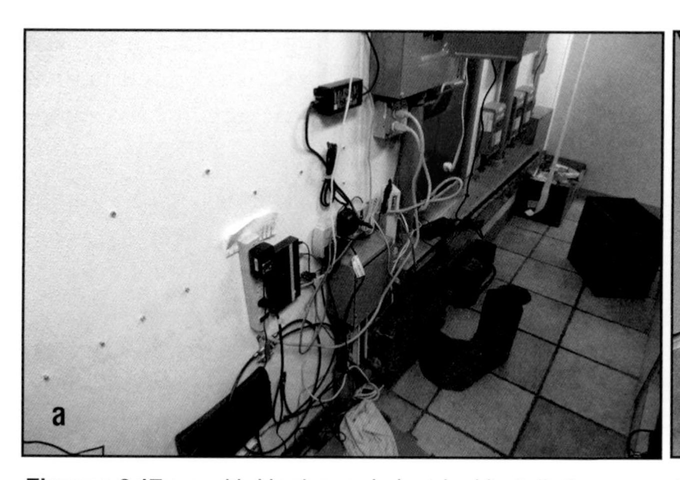

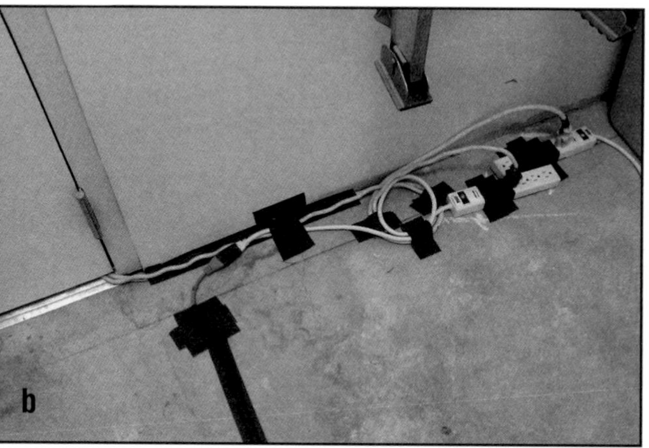

Figures 9.17 a and b Haphazard electrical installations are often dangerous and overload circuits. *a courtesy of Rich Mahaney; b courtesy of Scott Strassburg.*

exponential growth of heat. When this happens, other nearby combustibles will also be exposed to the heat and may ignite. Use of extension cords that are not protected is another example of misuse of electrical equipment.

Defective or Improper Electrical Installations

Fire inspectors may discover electrical equipment that is defective or improperly installed. This issue can develop due to unpermitted electrical work, unskilled technicians, or installation mistakes. If a permit or electrical plan exists, inspectors can consult the document(s) to help determine the cause of the problem. This documentation is created at the time of system design or certificate of occupancy.

Properly installed electrical equipment meets the manufacturer's specifications and requirements as well as those specified by the electrical engineer. The installation must also meet the requirements found in the NFPA® 70, *National Electrical Code®, ICC Electrical Code*™ from the International Code Council® (ICC®), and any other applicable codes and standards.

NOTE: The Consumer Product Safety Commission (CPSC) can recall even properly installed equipment or the equipment can become obsolete or unrepairable.

Improperly installed or maintained electrical equipment is a significant contributing factor in fires caused by electricity. Usually, the affected equipment is not suited for the type of installation, location, or intended use. Examples of improper installations include:

- A recessed light not having the required clearance from combustibles
- Installed appliances using inadequate-sized wiring
- Appliance installed in inappropriate location

Most fires in electrical equipment are caused by arcing or overheating of surrounding materials to their ignition point. For example, in an ordinary household lamp, less than 10 percent of the electricity, or *current,* is converted into light. Over 90 percent of the energy is converted into heat. Uncontrolled, this heat can ignite surrounding combustibles.

As changes are made to an electrical system, a building custodian or responsible party should track the changes. Many of these changes are documented electronically, although older records may still be hard copy or missing altogether.

NOTE: Static electricity, another electrical hazard, is covered in Chapter 10, Hazardous Materials.

Material Storage Facilities

In any type of occupancy, from small grocery stores to big-box stores, material storage can present potential hazards. These hazards include the following:

- Storing products too close to ceilings or electrical panels **(Figure 9.18)**
- Obstructing or blocking fire sprinkler system components
- Improperly storing flammable liquids

Storage facilities frequently change stock inventories and location. Inspectors should be aware of these changes when conducting site inspections and act to confirm compliance with locally adopted codes. Compliance will reduce the risks to the business owner, firefighters, and the surrounding community in the event of an incident. Inspectors should pay special attention to the following:

- Changes in contents may increase the fuel load and can result in a hazard that is under-protected, even in a facility with an automatic sprinkler system.
- Changes in storage arrangement or height may also require changes in the storage systems, in-rack sprinkler arrangements, and width of access aisles.

The growth of big-box stores in recent years has resulted in additional storage concerns. These retail outlets feature thousands of square feet (m^2) of space under a high roof. Within this space, products are stored and sold from high-piled shelving. The products range from lawn and garden supplies to hardware and appliances. Combustible materials such as lumber and tires may be stored inside the store, outside the store, or both **(Figure 9.19)**.

Figure 9.18 Materials stored too close to electrical panels. *Courtesy of Rich Mahaney.*

Figure 9.19 Tires are usually stored vertically in racks for ease of access, although some may be stored in stacks on their sides. Inside storage of tires should be in accordance with the adopted local fire code.

NOTE: ICC *High-Piled Combustible Storage Application Guide* is a resource for the inspector.

Many retail outlets built in recent years have made a commitment to fire protection systems. However, the store owners are usually responsible for the design, installation, and maintenance of these systems. Inspectors should review the documentation to verify that the systems are maintained according to the manufacturer's specifications. Examples of these fire protection systems include:

- Supervised automatic sprinkler systems
- Fire hose valves in areas of high-piled storage
- Smoke removal systems
- Portable fire extinguishers

NOTE: Sprinkler systems are described in Chapter 12.

Sprinklers in Big-Box Stores

Sprinklers in big-box stores may be designed for specific hazards, such as stored lumber. Any movement of those materials may require additional inspection. Changes in amount or material and its location may require that the sprinkler company and the store provide documentation that the system is adequate to protect the new hazard.

The inspector is responsible for accurately identifying and describing conditions within storage facilities and big-box stores and informing fire suppression units of the conditions present. The inspector needs to be able to recognize appropriate distances between materials, note housekeeping issues, and be aware of fire protection measures specified for materials regardless of location. This type of information will help firefighters in preparing accurate preincident plans.

Storage Methods

Each of the model codes has specific requirements on storage methods, including requirements for minimum aisle widths, compatible materials, and fire protection systems. Inspectors should know the code requirements for the three primary storage methods:

- Pallet storage
- Rack storage
- Solid piling

Pallet storage. This method uses pallets to store materials. With multiple stacked pallets, it is more difficult for water from a sprinkler to reach, wet, and control a fire. Pallet storage may include **encapsulating** the pallets with plastic wrap, which creates additional hazards. Other pallet stacking issues include the following:

Encapsulating — Completely enclosed or surrounded as in a capsule.

- Materials are generally stacked 3 to 4 feet (1 m to 1.2 m) high on each pallet, allowing several pallets to be stacked on top of each other and approaching 30 feet (10 m) in height.

- The air space between the top and bottom layers of the pallets (generally about 4 inches [100 mm]) creates a significant fire hazard because it provides a natural path for promoting the spread of fire.
- Materials on pallets are often completely encapsulated (also known as cocooning) in plastic wrap **(Figure 9.20)**.
- Plastic encapsulation, which covers the top of the material, prevents water from a fire suppression system from penetrating the surface of the materials stored.
- Packing/wrapping materials also add to the fuel load and may be regulated for type and amount used.
- Pallets stacked several layers high also prevent water from sprinklers from reaching bottom layers.
- Plastic pallets will change the **commodity** class if the sprinkler system was not designed for Group A plastics **(Figure 9.21)**.

Commodity — Combination of material and packaging that is for sale.

Rack storage. This method uses a structural framework (racks) onto which pallets or other materials are placed **(Figure 9.22)**. Rack height varies depending on the inside height of the building and the equipment available to load and retrieve items. In addition to the typical sprinklers found at the ceiling level, warehouses and retail outlets with rack storage may have sprinklers located within the racks themselves **(Figure 9.23)**. More information on these requirements may be found in NFPA® 13, *Standard for the Installation of Sprinkler Systems*.

Rack storage represents the most challenging fire protection problem of the three primary storage methods because it exposes more surface area to a fire. Rack storage also provides pathways for the spread of fire, such as flue spaces.

Solid piling. This method involves stacking materials directly on top of each other, which can be done manually or with a forklift **(Figure 9.24)**. Compared to the other

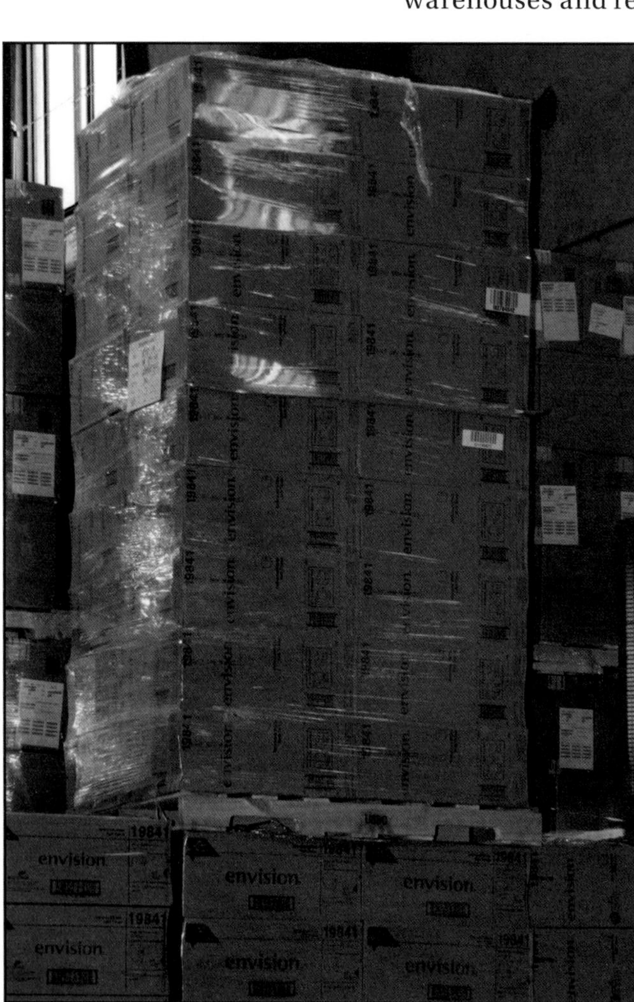

Figure 9.20 These materials show typical plastic encapsulation.

Figure 9.21 Plastic pallets are manufactured from recycled plastic products. Large amounts of plastics add to the fuel load in a storage facility.

Figure 9.22 Rack storage in warehouse facilities presents similar fire hazards to pallet storage.

Figure 9.23 In-rack sprinklers must not be blocked by materials storage.

Figure 9.24 Solid piling like this solid stack storage limits the communication of air between fuel units, reducing the chance of fire development.

two methods of storage, solid piling gives a fire the least chance to develop because it limits the air space between the fuel units.

When certain commodities are stored a specific height above the finished floor, the storage is considered high-piled combustible storage and must meet the requirements of the locally adopted fire code. These commodities may include:

- Vehicle tires
- Plastics
- Flammable liquids
- Unused pallets

Inspection Guidelines

Inspectors should review storage documentation before conducting an inspection at a storage facility or retail outlet. When they are onsite, they should verify information, including:

- Building construction characteristics
- Location of fire walls, fire door assemblies, and fire barriers

- Types and classifications of stored commodities, including plastics
- Type of packing material used
- Methods of storage **(Figures 9.25 a and b)**
- Maximum storage height of commodities
- Type and design of automatic sprinkler system and the fire pump if provided
- Type of water system on site; should meet minimum flow and pressure for the system
- Locations of water sources, hydrants, standpipes, and fire department connections
- Locations and types of smoke and fire detectors
- Method of smoke removal
- Training records for facility personnel
- Disposal method/storage of waste and packing materials

As an example, an inspector must verify that the aisles are kept clear so that materials left on the floor will not help a fire travel to the next storage rack. Aisles and exterior doors must be kept clear to reduce the potential of fire spread across the storage aisle. The inspector should request information regarding the method used to change inventory and the way inventory is located and positioned. If there is a pattern of higher activity in the warehouse when it approaches capacity or when higher-risk materials are stored, the inspector should return to conduct an inspection at those times. An example might be when pesticides or herbicides are inventoried in the late winter or early spring and stored in greater quantities than at other times of the year.

Of particular importance to inspectors are the following issues:

- Introduction of high-risk commodities
- Maintaining the required clearances between automatic sprinklers and the stored materials
- Maintaining specified aisle dimensions to help prevent the spread of fire and to enable emergency responders to gain access during an emergency

The inspector will need to evaluate the following:

- Fire detection and reporting systems
- Security systems **(Figure 9.26)**
- Structural and electrical systems
- Storage methods and practices
- Locator and lot identification systems
- Documents to include household goods descriptive inventories, weight tickets, warehouse receipt, and service orders.

If a more challenging commodity is introduced or the height of storage is increased, the automatic sprinkler system must be evaluated to confirm that the protection is adequate. If it is not, the sprinkler system must be modified or the storage must be eliminated or reduced.

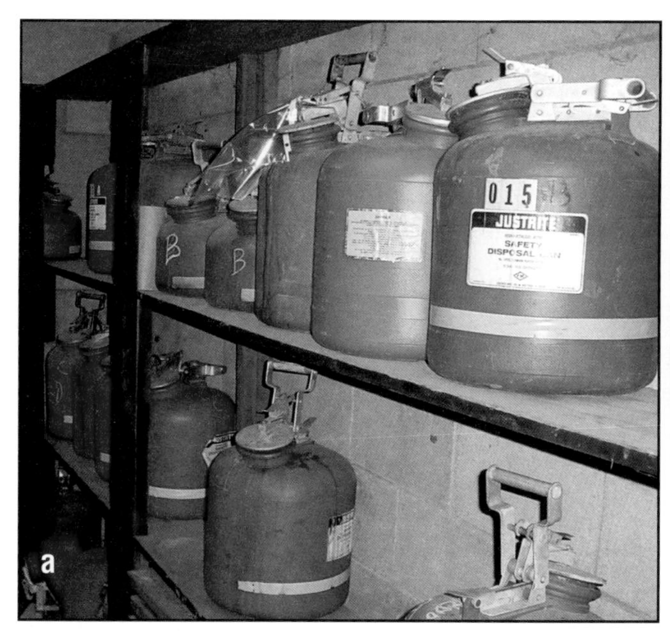

Figures 9.25 a and b Approved nonbulk storage containers include safety cans and polyethylene containers.

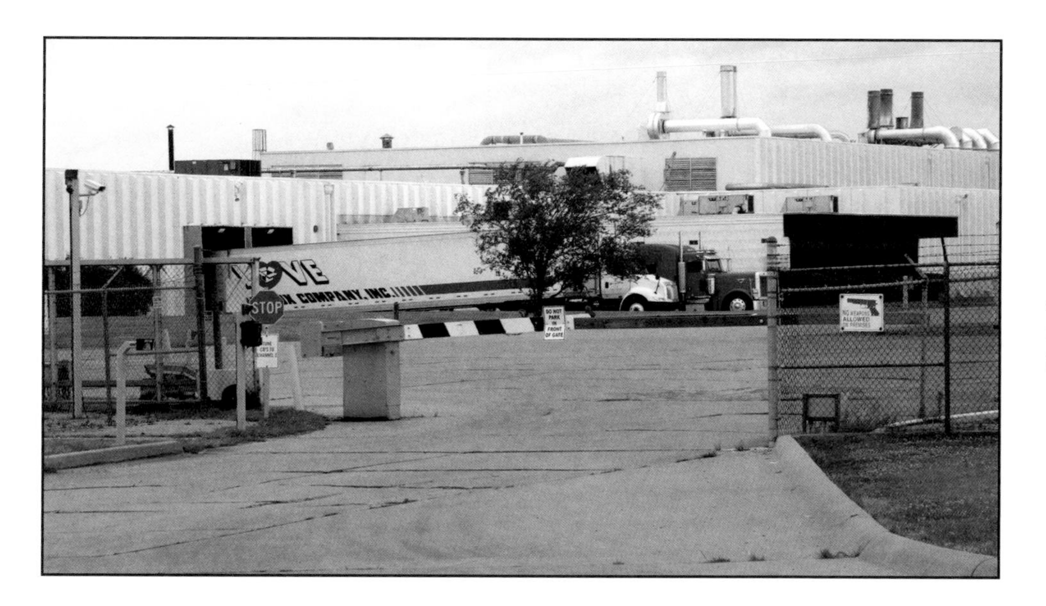

Figure 9.26 Evaluating security systems will help emergency responders know if there are likely to be access problems.

Warehouses

Because warehouses present extreme challenges during fire suppression operations due to their heavy fuel loads, inspectors must be alert for conditions that will contribute to rapid fire growth **(Figure 9.27, p. 328)**. Inspectors must assess the configuration and type of fuels within the structure or storage area and verify that the storage methods do not impede fire detection systems.

It is critical that inspectors diligently inspect a facility's automatic sprinkler system. The inspector should examine the system's records to verify that it is functional and being serviced as required. An inspector should then examine the storage heights of the stock to determine whether materials are too close to the sprinklers, which would render them ineffective. More information is contained in NFPA® 25, *Standard for the Inspection, Testing, and Maintenance of Water-Based Fire Protection Systems.*

For exposure protection, a number of steps are recommended:

- Store or stack noncombustible materials, such as sand or stone, on the perimeter of the yard to act as a barrier between the yard and adjacent properties or buildings.
- Maintain the required separation distances between materials specified in the fire code. If separation is not possible, fire-resistive walls can be used to reduce the separation distances or the height and area of storage can be reduced.
- Control weeds and vegetation away from storage areas. Mow areas prone to vegetative growth or treat them with an effective weed-control herbicide.
- Do not allow waste materials, such as sawdust, dry vegetable matter, and bark, to accumulate.
- Permit smoking and hot-work operations only in designated areas **(Figure 9.28)**. Signs should be posted so that workers know when and where they are allowed to smoke.
- Do not allow portable heating devices or open fires in the lumber storage area. In areas where heating devices are permitted, they should be approved-type equipment installed in an appropriate manner.

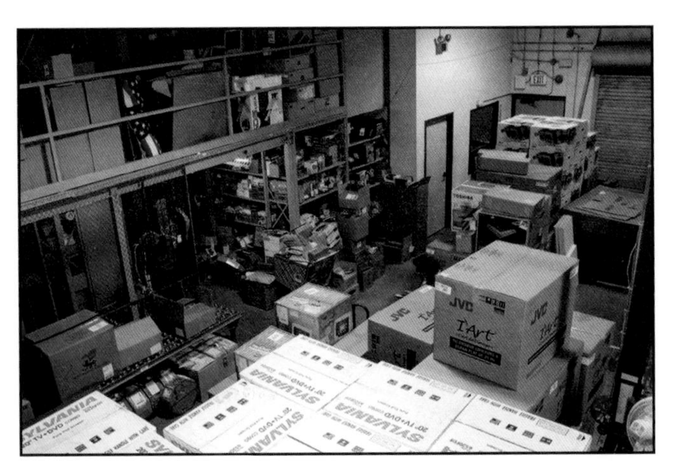

Figure 9.27 Warehouses often have extremely high fuel loads that create the possibility of rapid fire spread.

Figure 9.28 Sparks produced in welding operations present a fire hazard that should be controlled through a work environment that follows applicable codes. Note the No Smoking sign.

Lumberyards

The primary hazard associated with lumberyards is the high fuel load created by the large quantities of combustible materials stored on the site. The quantity and variety of these materials can contribute to a fire that can overwhelm local fire suppression resources. Due to the hazardous nature of lumberyards, they may be required to adhere to different building code requirements.

A number of conditions can influence the growth of lumberyard fires:

- Large, undivided stacks of materials
- Congested storage conditions
- Delayed fire detection
- Inadequate fire protection
- Combinations of Class A and Class B materials in close proximity

For open-yard storage, several important code provisions must be implemented:

- Lumber stacks must be located on solid ground, preferably paved or surfaced with materials such as cinders, fine gravel, or stone **(Figures 9.29 a and b)**.
- The heights of stacks must not exceed 20 feet (6 m) to verify that they remain relatively stable and limit the amount of material available in the event of an unwanted fire

Figures 9.29 a and b Outside lumber storage must be on a level surface and spaced to prevent fire spread. *Both photos courtesy of Rich Mahaney.*

- Gates and driveways must be wide enough to accommodate the fire department's largest vehicle.
- The turning radius of all driveways must meet the requirements of the largest fire apparatus that could respond to that location.
- Fire department-approved padlocks or key boxes must be used to secure the facility and permit emergency access.

Tire Storage Facilities

Fires involving tires produce intense heat, enormous amounts of smoke, and toxic oil. This oil may create pollution and groundwater contamination problems for a community. Because of their configurations, tires readily allow for the spread of heat and flames; in addition, fires involving tires not protected by sprinklers usually burn for long periods. Adequate separation of manageable areas of tires must be planned and enforced.

Inside tire storage may be found in tire or automotive stores, warehouses, big-box retail outlets, and tire manufacturing facilities. Inspectors usually encounter outside tire storage in junkyards or scrap tire storage facilities **(Figure 9.30)**. Problems associated with large-scale scrap tire storage facilities have resulted in numerous serious fires. These facilities are often closely related to scrap tire processing and burning operations.

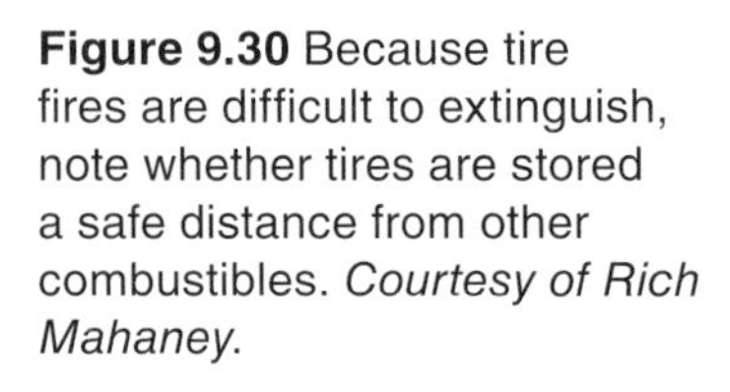

Figure 9.30 Because tire fires are difficult to extinguish, note whether tires are stored a safe distance from other combustibles. *Courtesy of Rich Mahaney.*

Pallet Storage Facilities

Facilities that store, ship, and receive large quantities of materials will have some quantity of pallets on site. Pallets that are loaded with materials will be present on loading docks and in storage areas. Idle pallets can be stored inside or outside the facility. Damaged pallets slated for disposal may be found in piles around the perimeter of the property. Eliminating this type of hazard is similar to controlling waste inside a facility. Pallets may be constructed of wood, plastic, metal, or paper; however, the majority of pallets are wooden **(Figure 9.31)**.

NFPA® 13, *Standard for the Installation of Sprinkler Systems*, sets stricter standards for warehouses using plastic pallets because of their greater heat release rate during fires. Storage facilities with the highest levels of built-in protection can safely use plastic pallets that have no fire retardants, but older facilities with lower levels of fire protection cannot.

Figures 9.31 a and b (a) Pallets stored well away from the facility. (b) These pallets are stored too close to the building. If they should catch fire, it will easily communicate to the building. b *courtesy of Rich Mahaney.*

Recycling Facilities

Two main hazards are associated with recycling plants: bulk storage of combustible materials and hazardous processes conducted on the premises. In particular, facilities that handle wastepaper and cardboard have extremely high fuel loads. These materials are usually contained in large bundles **(Figure 9.32)**.

Figure 9.32 At recycling centers, bulk storage of combustible materials in combination with hazardous processes conducted at the facility can contribute to high fuel loads.

The same general rules already described for outdoor lumber storage apply to bulk storage outside a building. An inspector should also be aware that bundled materials may be found at large retail outlets when cartons are emptied and prepared for shipment to recycling facilities.

Many other types of materials can be recycled in addition to paper and cardboard. Recycling facilities often process different types of materials on the same property. These materials include:

- Most types of metals and plastics
- Batteries

- Motor oil
- Cooking oils from restaurants

Waste recycling operations may also be located inside buildings. The buildings are usually large, with separations for the different materials being processed. Conveyor belts may run through the structure to move recyclable materials between different areas of the building. The structure may also be designed to enable tractor-trailer-type trucks to drive through to unload recycled materials for sorting, creating additional challenges.

Inspectors should check all material-handling equipment within the occupancy to verify that it is properly maintained and operated, including:

- Waste chutes and handling systems
- Shredders
- Extruders
- Conveyor belts
- Compacting machines **(Figure 9.33)**

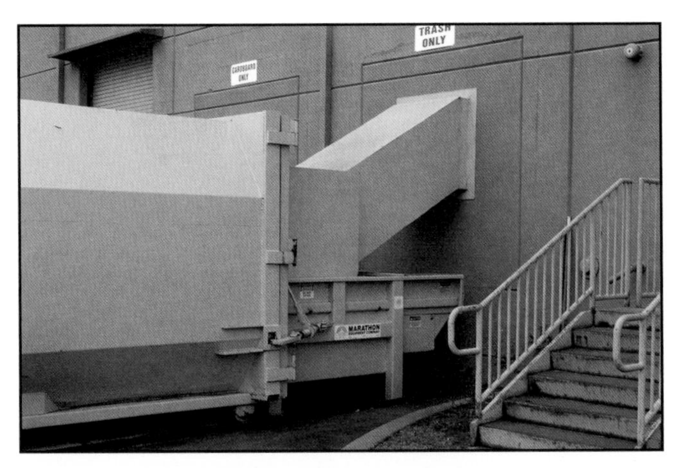

Figure 9.33 Occupancies often include materials handling equipment, such as this paper compactor. Inspectors are also responsible for inspecting this equipment as well as the occupancy itself.

An inspector must verify that safety measures are taken during processing and handling of different types of materials in the same area. For example, piles of combustible materials should not be stored in the same area where cutting torches are used to salvage scrap metal. Hot ashes, coals, cinders, or materials subject to spontaneous heating should be stored in approved metal containers, and these containers should not be located near stored combustible materials.

Waste-Handling Facilities

Waste-handling operations involve the disposal of nonrecyclable materials, including everything from municipal trash to hazardous or biological materials. Operations of this type are extremely complex and require an ever-increasing level of expertise on the part of an inspector. These facilities present unique challenges to inspection and fire suppression personnel due to the quantity of materials processed, the presence of both treated and untreated materials on site, and the process used to treat the materials.

A variety of methods are used to dispose of waste materials besides simply burning them. Steaming, microwaving, and ozone sanitizing are also used to clean, disinfect, and sanitize hazardous and biohazardous waste materials before they are deposited in a landfill. Inspection and code enforcement personnel must work closely with fire suppression personnel to limit the potential risk to the community and emergency response personnel resulting from incidents occurring at waste-handling facilities.

Waste-handling operations, like recycling operations, accumulate large quantities of combustible materials during the time between collection and disposal. Collected materials are retained at a transfer station until they are compacted and removed to a waste-disposal/dump site. Fire protection and access requirements should be similar to those at other outside storage yards, and inspectors should verify that adequate access and fire lane widths are provided and maintained.

Sites have specific design requirements, including the type of soil they are located on (to avoid ground contamination) and safety features (such as plans and placards). In some specialized sites, even inspectors are not permitted on the site unless they have the required training/qualifications and proper personal protective clothing. Rights of entry for inspection do not outweigh safety requirements designed to protect persons from hazards.

An inspector should be aware of different wastes and their hazards on any site. The inspector must follow all safety rules that apply to the facility being inspected, including the use of personal protective equipment (PPE).

Canada Dangerous Goods

In Canada, hazardous materials are known as dangerous goods and include hazardous waste. Both provincial and federal government agencies regulate all waste materials. Federal agencies such as Transport Canada regulate waste while it is being transported through the Transportation of Dangerous Goods (TDG) Regulations. Environment Canada regulates disposal of the waste material through the Canadian Environmental Protection Act (CEPA). Permits to store or transport, licenses for operators, and training for employees are required for hazardous waste operations. The U.S. Environmental Protection Agency (EPA) also has rules for transporting hazardous materials.

Incinerators

Although most waste, including hazardous waste in Canada, is disposed of by compaction and burial, a simpler solution is to burn it **(Figure 9.34, p. 334)**. Incineration of most domestic garbage, once metals and glass have been removed, is acceptable in many areas if the proper environmental controls are followed. Local authorities may require a permit before waste is burned.

Incineration can reduce the bulk of waste by as much as 95 percent. It is also effective in destroying medical, chemical, and biological wastes that might otherwise pose a threat in a conventional landfill or recycling operation. Common hazards associated with incineration include:

- Overheating
- Structural failure
- Corrosion due to scrubber acids
- Failure of scrubber systems

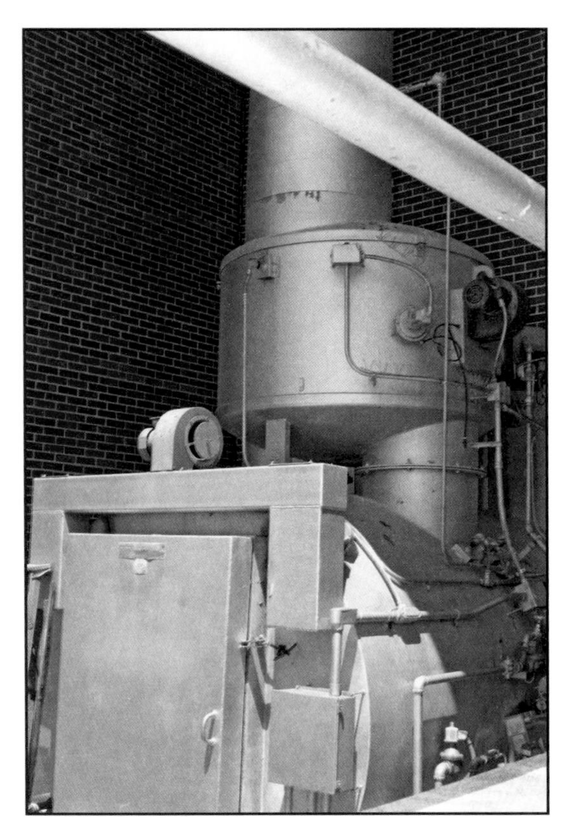

Figure 9.34 As long as incinerators are constructed according to local environmental codes, they can be a very efficient method of disposing of waste.

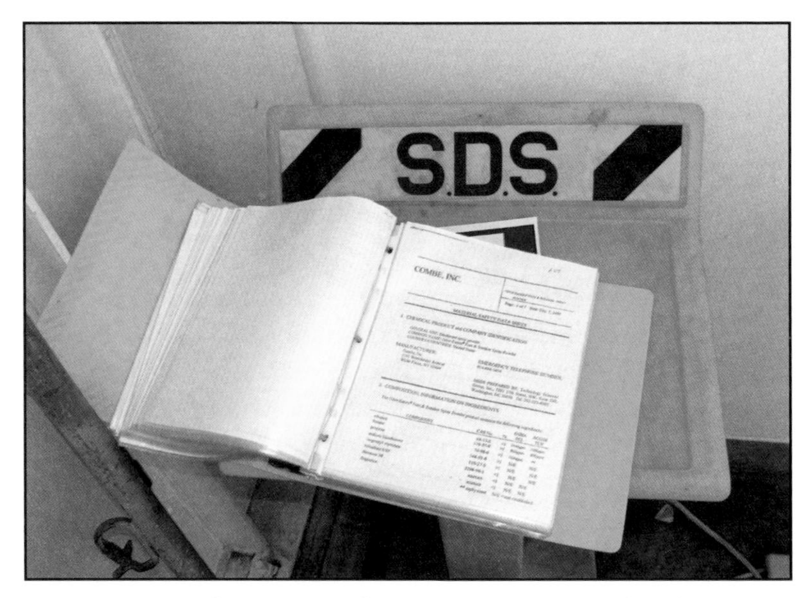

Figure 9.35 The Occupational Safety and Health Administration (OSHA) mandates that safety data sheets (SDS) be maintained on site for all products used or stored at a facility.

An inspector must be familiar with the type of incinerator being used in a community and the safe operating parameters that the facility must employ. The design of the incinerator and its safety systems must comply with the current edition of NFPA® 82, *Standard on Incinerators and Waste and Linen Handling Systems and Equipment.*

As the costs of carbon-based fuels (natural gas, coal, and fuel oil) have risen, burning waste to generate electricity has become a popular option. An inspector must become familiar with the additional hazards associated with both waste incineration and power generation within the same facility.

When inspecting incinerators, inspectors should look for a number of signs that the unit is being operated properly:

- Fuel-fired incinerators are preheated for 30 minutes before use.
- Competent, trained personnel operate the unit at all times. Request documentation of personnel training and credentialing requirements.
- The feed door closes fully after loading for the entire combustion cycle.
- The waste material ash compartment is cleaned on a regular basis, and disposal is accomplished in an approved manner.
- All particulate and emissions scrubbers are properly maintained and operational. No dense smoke or serious odors are emitted during operations.
- **Safety data sheets (SDS)** required by federal environmental and hazardous materials laws are available and in the appropriate locations **(Figure 9.35)**.
- The emergency response plan for the facility is available and placed in the appropriate location for use during an incident.

NOTE: Chapter 10 contains additional information about safety data sheets.

Safety Data Sheet (SDS) — Form provided by the manufacturer and blender of chemicals that contains information about chemical composition, physical and chemical properties, health and safety hazards, emergency response procedures, and waste disposal procedures of the specified material.

Building Systems

An inspector will encounter a number of different building systems. These systems all present unique challenges, and the inspector often needs to do extensive research before conducting an inspection. Inspectors should refer to locally adopted codes regarding these building systems, which include:

- Heating, Ventilating, and Air Conditioning Equipment/Systems
- Cooking Equipment
- Industrial Furnaces and Ovens
- Powered Industrial Trucks
- Tents

Heating, Ventilating, and Air Conditioning Equipment/Systems

Inspectors must be able to recognize the various types and operating conditions of a variety of H VAC systems. HVAC systems provide a variety of services, including:

- Air heating or cooling
- Air humidifying or dehumidifying
- Air filtering or cleaning

The HVAC information in this section is generic in nature. An inspector should consult the adopted codes for jurisdiction-specific information. Requirements for HVAC systems are contained in all model building and fire codes, and several NFPA standards. Some inspection items inspectors are responsible for verifying include the following:

- Qualified personnel are properly inspecting and maintaining the HVAC system.
- Unit clearances are maintained.
- Operating instructions for the equipment are provided.

The major components within an HVAC system include:

- Heaters
- Air conditioners
- Fans
- Ducts **(Figures 9.36 a and b)**
- Heat exchangers
- Thermostatic controls

Although a number of fire conditions can be attributed to HVAC systems, the two primary hazards they pose are the spread of fire and products of combustion through the air handling system and fire hazards from

Figures 9.36 a and b HVAC ducts will be hidden in the building's structure, or left exposed as part of a building design.

heating appliances. Of the two, fire hazards from heating appliances are the greater concern. Common causes of fires related to heating appliances include:

Figure 9.37 Fan belts must be kept in good condition and adjusted properly.

- Improperly adjusted or worn fan drive belts **(Figure 9.37)**
- Clogged filters
- Combustible lint or debris near the burners
- Poorly maintained motors

Inspectors should also take note of the HVAC system room or mechanical equipment room. Only authorized personnel should have access to this room. The type and size of refrigeration or heating equipment will determine the fire separation requirements for the room. If required, fire-resistive construction and the appropriate protective openings should be used to meet these requirements. Inspectors should also look for combustible storage inside the room; generally, this storage is not permitted. Inspectors should consult the appropriate codes for issues with HVAC system rooms.

The sections that follow address various common heating and cooling appliances, such as:

- Furnaces
- Boilers
- Unit heaters
- Room heaters
- Temporary/portable heating equipment
- Air conditioning systems
- Ventilation systems
- Filtering Devices

Furnaces

Warm-air furnaces are the most common type of central heating appliance in use today **(Figure 9.38)**. Most of these furnaces use forced-air principles to move air, but some gravity furnaces are still in use.

Gravity furnaces, which are considered obsolete, operate primarily by the circulation of air without the use of fans. These furnaces are located in the basements of houses. Heated air rises while colder air settles and enters the return-air ducts and becomes makeup air for the furnace.

Forced-air furnaces rely on a fan to move heated air through the system. Because these units contain plenums and ductwork that may become hot enough to ignite adjacent, unprotected woodwork, they require the use of appropriate clearances and insulation. Automatic controls are provided to turn off the furnace when the temperature within the ductwork or plenum reaches 250°F (120°C).

Fire problems with warm-air furnaces are principally due to the following:

- Inadequate clearances
- Lack of proper limit control
- Heat-exchanger burnouts (damage to heat-exchanger piping due to extreme temperatures or prolonged periods of high temperatures)
- Improper installations
- Improper maintenance procedures

Most floor furnaces are designed and approved to be installed underneath combustible floors **(Figure 9.39)**. They also require temperature-limit controls and proper clearances from miscellaneous combustibles. Floor furnaces must have safety controls that turn off the fuel source if the pilot light is extinguished. These furnaces must be properly vented so they cannot contribute to carbon monoxide buildup.

Figure 9.38 Warm-air furnaces like the one shown are the most common heating appliances used today.

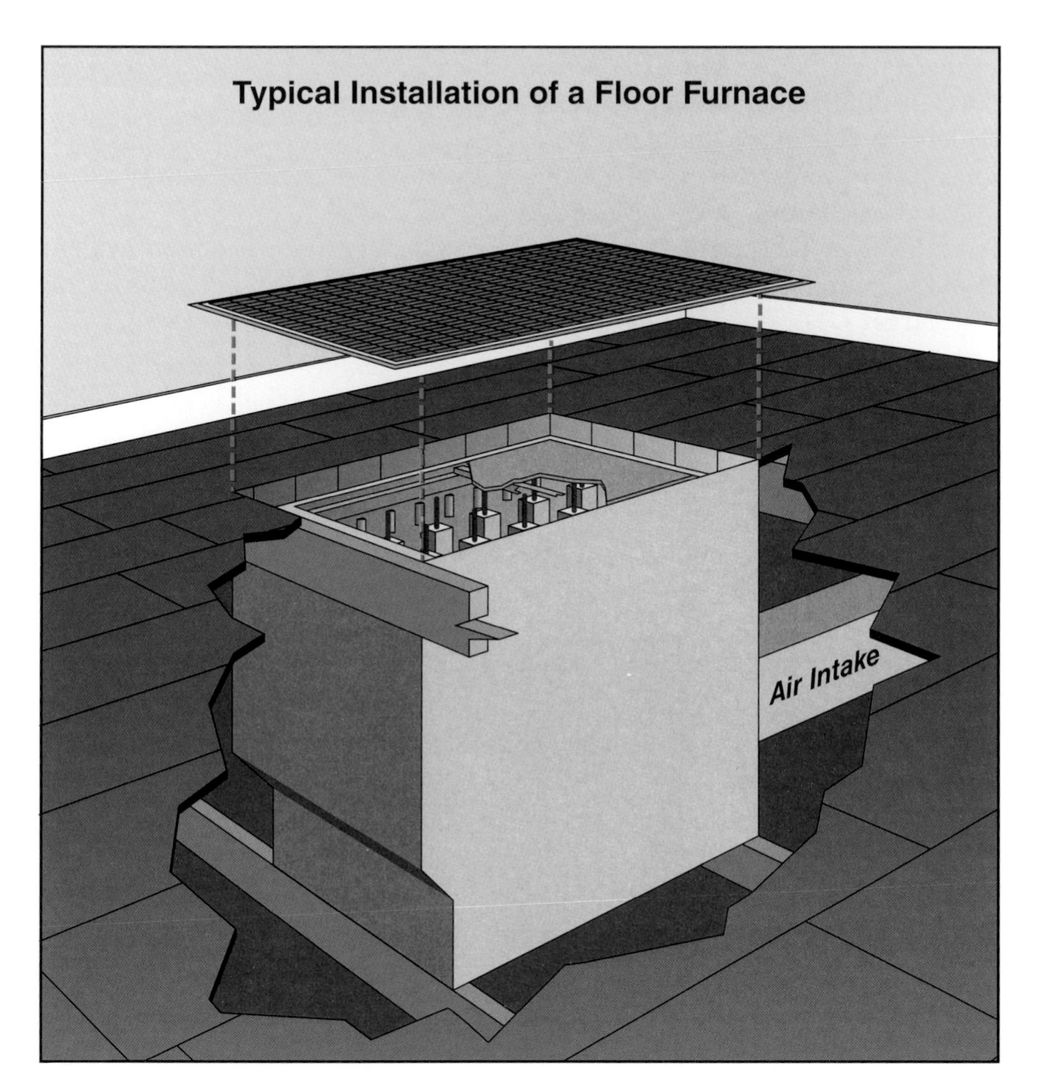

Figure 9.39 Floor furnaces are found primarily in homes built before the 1960s. These furnaces require special venting and adequate distance from all combustibles.

Floor furnaces are mostly found in single-family dwellings built before the 1960s, although some commercial occupancies may use them. Inspectors should warn individuals in occupancies with floor furnaces not to cover the register or attempt to dry wet items, such as clothes or newspapers, over them.

Wall furnaces are self-contained electric, indirect-fired gas, or oil heaters installed within or on a wall **(Figure 9.40)**. Gas or oil heaters may have a direct vent to the exterior or be equipped with an indirect vent or chimney. These units are found in residential occupancies and small office spaces.

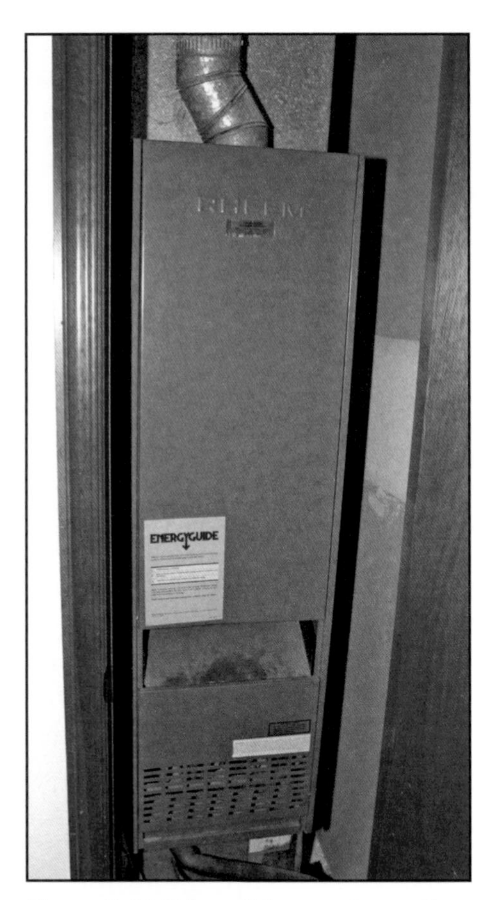

Figure 9.40 This wall furnace is a self-contained unit and must have sufficient ventilation to the exterior of the structure.

Vessel — Tank or container used to store a commodity that may or may not be pressurized.

Boilers

A boiler is a pressure **vessel** that is used for a variety of industrial processes, including generating steam to power machinery and providing heat that the industrial processes require **(Figure 9.41)**. In nonindustrial applications, boilers provide heating and hot water for cooking and bathing through a system of radiators. Boilers are found in both single- and multi-family residences. While they are more prevalent in old structures, newer, more efficient boiler-based heating systems are on the market.

Manufacturers must strictly adhere to all applicable codes when building boilers. In many cases, manufacturers use the American Society of Mechanical Engineers (ASME) Boiler and Pressure Vessel Codes to govern the design, construction, and operation of boilers. All boilers should meet fire and explosion prevention requirements. When finished, the entire system will include interlocks, detectors, alarms, and sensors to turn off fuel sources when a problem occurs.

Figure 9.41 A commercial boiler like the one shown may be used to generate the steam necessary to operate machinery or heat a facility.

When conducting an inspection, inspectors should verify a boiler's code compliance by performing the following:

- Check for ASME nameplate
- Verify inspection history via an up-to-date inspection tag or certificate
- Note the pressure relief device

Two common types of boilers are fire-tube and water-tube. In fire-tube boilers, the combustion gases pass through tubes that are immersed in circulating water, which is converted to steam. A water-tube boiler is a steam-generating unit in which steam and water circulate through a series of small drums and tubes, while the combustion gases pass over the outside of these steam- and water-containing elements.

NOTE: Most large boilers are of the water-tube type.

The combustion process in a boiler results from a continual introduction of fuel and air in a flammable mixture. The following fuels are usually used in boilers:

- **Natural gas** — Either supplied to the burner premixed or externally mixed. In a premix, the air and fuel are mixed before they reach the burner.
- **Coal** — Units deliver air and pulverized coal to the burners for combustion.
- **Oil** — Oil is atomized at the burner, which causes it to mix with air and burn most efficiently.

Hazards associated with boilers vary depending upon a number of fuel and malfunction possibilities:

- Gases can build in the system and ignite when a mechanical failure occurs.
- The main hazard in natural-gas-fired boilers is the leaking of gas and the buildup of fuel-rich mixtures if a burner fails.
- Coal-fired boilers present hazards when debris is mixed with coal.
- Hazards in oil-fired boilers come from the fuels, oil leaks, or interruptions in the oil flow caused by water or sludge in the fuel.
- An additional hazard occurs when coal dust accumulates in areas adjacent to the unit's fire box. An explosive atmosphere can develop if the dust becomes airborne.
- Boilers typically operate at elevated pressures and failure of the vessel can be catastrophic.
- The deterioration or modification of unit exhaust may cause a dangerous buildup of carbon monoxide in the structure. Deterioration or modification requires additional investigation.

Operator training is an important part of any hazard reduction plan. Workers who are educated in the process and function of the equipment around them will be better able to eliminate a potential hazardous situation. In some circumstances, workers are more knowledgeable than inspectors regarding the boilers, and can provide the inspectors with useful information during an inspection. Together, workers and inspectors can identify and report hazardous issues as well as promote solutions.

Unit Heaters

Unit heaters are self-contained devices that are thermostatically controlled. They may be mounted on the floor or suspended from a wall or ceiling **(Figure 9.42)**. However, certain types of heaters are not permitted in various occupancies as per guidelines in NFPA® 1, *Fire Code* and the International Fire Code (IFC).

The model fire codes regulate the type and amount of fuel to operate the heaters. Unit heaters may use the following types of fuel:

- Propane
- Natural gas
- Kerosene
- Electricity

NOTE: The heating element and fan are enclosed in a common operating unit. The inspector should verify that heaters are located with appropriate clearances from combustibles. An inspector should also look for proper storage and handling of the fuel source.

Listing Heaters

All heaters must be listed by an accredited testing agency, such as Underwriters Laboratories (UL), FM Global, or Underwriters Laboratories of Canada (ULC). Heaters that are listed meet safety requirements and are designed for use in specified applications.

Room Heaters

Room heaters are self-contained units designed to heat the immediate surrounding area. They use the circulation of radiant heat as the heating medium. Room heaters usually incorporate manual or thermostatically controlled drafts. These heaters also require that nearby floors and walls be properly protected. Some examples of room heaters include **(Figure 9.43)**:

- Wood and coal stoves
- Electric and gas logs
- Electric baseboard heaters
- Open-front heaters

Hazards associated with room heaters include the following:

- Overfiring the device
- Careless handling of fuel and ashes
- Inadequate clearances from fuel cans or other combustible materials
- Accumulation of creosote in flues and chimneys as applicable
- Use of insufficient extension cords with electric heaters
- Refueling with the wrong fuel (for example, pouring gasoline into a kerosene heater)

Figure 9.42 Self-contained heating devices, like the ceiling-mounted heater shown, enclose the heating element and the fan in a single operating unit.

Figure 9.43 Baseboard heaters are good examples of room heaters that use the circulation of radiant heat as a heating medium.

Solid-fuel room heaters are free-standing appliances that vent to the exterior of the building and are designed to burn coal or wood. These heaters are primarily found in single-family dwellings; however, they may also be found in restaurants, apartment clubhouses, or bars where a living-room atmosphere is desired.

When inspecting room heaters, inspectors should look for the following:

- Solid fuels for wood or coal stoves are stored a safe distance from the stoves.
- Provisions are made for safely removing and storing ashes and cinders.
- All chimneys and flues are kept clear of excessive creosote **(Figure 9.44)**.
- Combustibles are stored a safe distance from heating units.
- All heating appliances have proper industry approvals and are labeled accordingly.
- All heating sources are vented (combustion gases) and provided with adequate ventilation (supply air) in accordance with manufacturer's instructions.

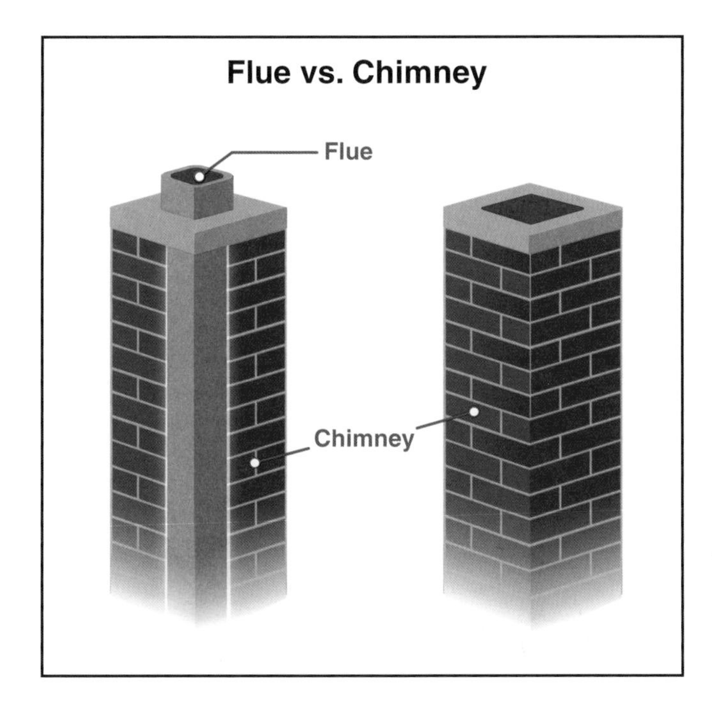

Figure 9.44 The flue is the duct that carries heat, smoke, and fire gases up the chimney. It needs to be kept free of excess creosote to avoid chimney fires.

Temporary/Portable Heating Equipment

Portable heating equipment can be found in nearly every type of occupancy, including those under construction or demolition **(Figures 9.45 a-c)**. Energy sources include:

- Electricity
- Natural gas
- Propane
- Kerosene
- Oil

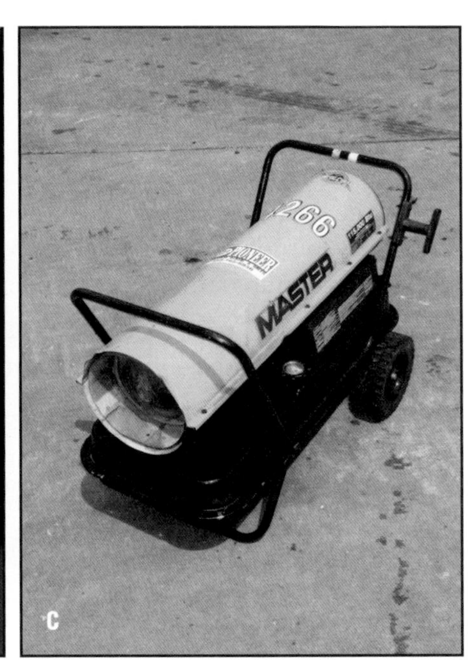

Figures 9.45 a-c Portable heating equipment like the three examples pictured can be powered by (a) kerosene, (b) electricity, or (c) propane.

Portable heaters are hazardous because they can easily be misused, placed in poor locations, or damaged. They are often poorly maintained. Heaters placed close to combustibles such as bedding, clothing, and furniture have resulted in numerous tragic life-loss fires **(Figures 9.46 a and b)**. Another hazard associated with fueled portable heaters is a lack of adequate ventilation, which can cause lethal exposure to carbon monoxide.

Many communities have enacted laws regulating or prohibiting certain types of heaters in certain occupancies. Inspectors need to be familiar with the locally adopted code requirements and with manufacturers' requirements for use. If it is determined that a heater is prohibited, it must be removed from service and taken from the premises as soon as possible.

NOTE: The 2012 edition of the International Fire Code® (IFC) and the International Building Code® (IBC) contain requirements for the installation of CO detection in new and existing Group-R and Group-I occupancies. These would include hotels, dormitories, apartment buildings, hospitals, and nursing homes. These alarms must be installed and maintained in accordance with NFPA® 720, *Standard for the Installation of Carbon Monox-*

ide (CO) Detection and Warning Equipment, as well as the manufacturer's instructions. The inspector must be familiar with these requirements and exceptions.

When the use of a temporary or portable heater is permitted in an occupancy, an inspector should verify the following information:

- The heater is well maintained and in safe working condition.
- The heater is equipped with safety switches that turn the heating element off if the heater tips over.
- The appropriate fuel, if applicable, is used for the heater.
- Electric heaters are **only** plugged into an electrical receptacle/circuit that is designed to carry the current required to operate the heater.
- The heater is located away from the route of egress from the structure.
- Heaters are placed a safe distance from any flammable or combustible materials.
- Fuel for kerosene heaters is stored outside the structure, and the heater is refilled outside the structure.

Figures 9.46 a and b Portable heaters placed too close to combustibles are a common fire hazard. They can also be tripping hazards. *Courtesy of Rich Mahaney.*

Air Conditioning Systems

An air conditioning system is actually a form of refrigeration system where:

- Air is filtered
- Heat is removed through a heat-exchange method
- Air is dehumidified

To accomplish this process, any one of a number of gases, such as chlorinated fluorocarbons (trademarked as Freon® by DuPont Chemical) or anhydrous ammonia, is compressed into a liquid within a coil and then allowed to expand back into a gas. During the expansion process, the refrigerant absorbs the heat in a space and the surrounding air is cooled.

Most commercial refrigerants have a classified level of toxicity and flammability as defined in American National Standards Institute (ANSI) and the American Society of Heating, Refrigeration, and Air-Conditioning Engineers

(ASHRAE) in ANSI/ASHRAE Standard 34, Designation and Safety Classification of Refrigerants. Refrigerants are assigned to one of two toxicity classes (A or B) based on what would be a permissible exposure. Inspectors must remember that they may be dealing with a refrigerant that is not very flammable but is very toxic.

Ventilation Systems

Ventilation systems are mechanical air-moving or air-distribution systems. Inspectors should understand that these systems are tied into both the heating appliance and the air conditioning equipment, but no cooling or heating of the air occurs within the system. The quality and characteristics of the air remain unchanged. The primary fire safety concern associated with ventilation systems is the potential for the transmission of fire, smoke, and products of combustion through the air distribution system **(Figure 9.47)**.

NOTE: An inspector must recognize the difference between an air conditioning system and a ventilation system.

Filtering Devices

To clean the air and remove particulate dust and pollens, HVAC systems contain filtering devices. Three types of filters are used in HVAC systems:

- Fibrous media filters
- Renewable media filters
- Electronic air cleaners **(Figure 9.48)**

Each type of filtering device presents hazards. Fibrous and renewable media filters may become filled with trapped particulates. These particulates can generate large quantities of smoke if they become ignited. Electronic air filters can develop an electrical malfunction that ignites the filters or particulate matter inside the HVAC system.

UL classifies filters into two categories based on flame propagation and smoke development: Class 1 and Class 2. Class 1 filters, when clean, will not contribute fuel to a fire, and they emit very small quantities of smoke when flames strike them. Class 2 filters, when clean, will burn moderately, and they emit moderate quantities of smoke.

Cooking Equipment

Cooking is a common cause of fires in both residential and commercial occupancies. Facilities with cooking equipment represent a large percentage of the occupancies that inspectors will enter. Common cooking equipment includes ranges, ovens, and fryers. Other kitchen equipment may include items used for:

- Mixing
- Cutting
- Exhausting
- Warming
- Washing

In commercial cooking establishments, much of an inspector's attention will be focused on the ventilation hood, exhaust, and fire protection systems installed above the cooking area (**Figure 9.49**). These systems should meet the locally adopted code requirements. Type I hoods are placed over grease-producing equipment, Type II hoods over such equipment as dishwashers and ovens.

An inspector must be able to recognize the hazards associated with cooking equipment. The following are some general conditions that inspectors should look for on all cooking equipment, to include the ventilation hood system:

- Cleanliness and functionality
- No grease accumulation **(Figure 9.50)**
- Proper provisions for grease removal, to include regularly scheduled cleaning
- Clearance of at least 18 inches (450 mm) from any combustible material unless the equipment is specially designed for a lesser clearance
- Fire extinguishers should be accessible, and personnel should have been trained in their proper use. Extinguishers should be compatible with the suppression system.

Figure 9.47 Ventilation systems do not change the temperature of the air but contain ducts that can transmit products of combustion.

Figure 9.48 Electronic air cleaners can become a fire hazard if they malfunction.

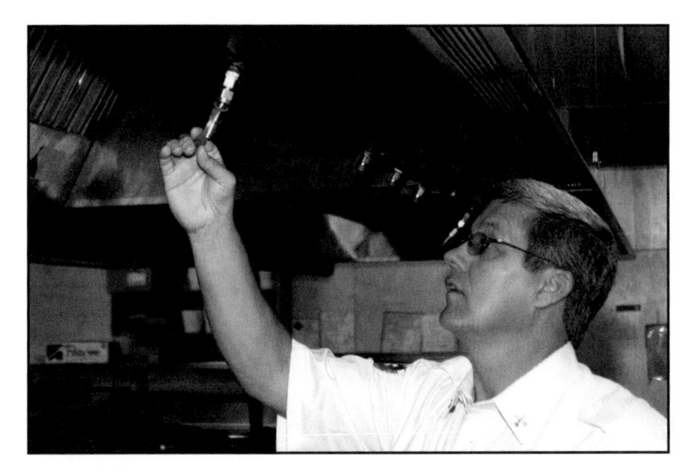

Figure 9.49 The equipment above the cooking area is especially important with regard to cleanliness and fire protection.

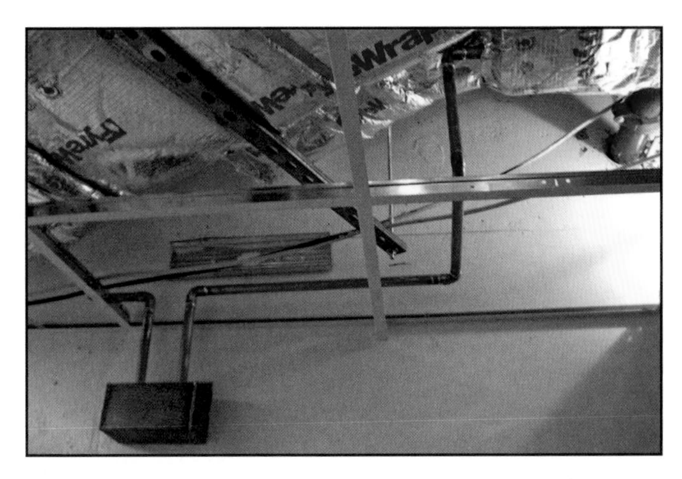

Figure 9.50 This type of grease trap is unacceptable because it can allow grease to accumulate. *Courtesy of Jon Roberts.*

Ventilation-Hood Systems

When inspecting ventilation-hood systems and their associated fire protection equipment, inspectors should look for a number of things, such as:

- Verify general cleanliness and maintenance of the cooking area and ductwork.
- Examine the system for any obvious damage, including mechanical damage.
- Verify that nozzles are properly oriented toward the protected hazard.
- Verify all parts of the system to confirm that they are in their proper locations and connected.
- Inspect manual actuators for obstructions.
- Inspect the tamper indicators and seals to verify that they are intact.
- Check the maintenance tag or certificate for proper placement.
- Verify that a qualified technician has inspected the system within the code specified time parameters (**Figure 9.51**).
- Check the pressure gauges to certify that they read within their operable ranges.
- Verify that a detector or fusible link is provided for each protected cooking appliance located under the hood.
  - — Fusible links that control the activation of the fire extinguishing system should be replaced at least annually.
  - — Fusible links that appear to be distorted from frequent exposure to heat may need to be replaced more often.

Solid-Fuel Cooking Equipment

Solid-fuel cooking equipment is becoming common in many restaurants. This equipment includes wood-fired brick ovens and meat grilling or smoking equipment (**Figure 9.52**). There are many safety requirements pertinent to this type of cooking equipment, including:

- Equipment must be installed on noncombustible floors that extend 3 feet (900 mm) from the outside of the appliance in every direction.
- Equipment requires grease removal devices.
- Equipment requires a ventilation hood system to exhaust smoke and gases.
  - — Spark arrestors must be located inside the hoods ahead of the filter system.
- Only a one-day supply of solid fuel is allowed in the same room as the appliance.
- Ashes must be removed from the firebox regularly so that the buildup of ashes does not interfere with the draft. These ashes must be disposed of in an approved container that is separate from containers containing other combustible waste.

NOTE: Restaurants that operate open-air patios may also have propane or butane heaters or wood-burning fireplaces. These heaters or fireplaces may be in contact with or near awnings or other combustible materials, creating an exposure hazard through radiant heat or burning embers.

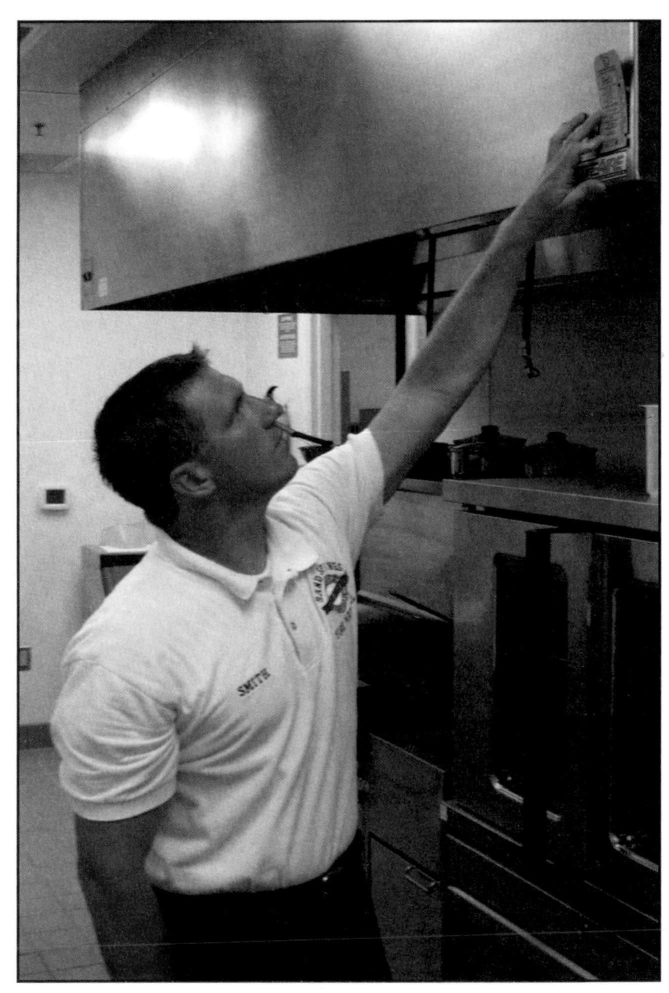

Figure 9.51 Be sure to check that inspection tags are up-to-date.

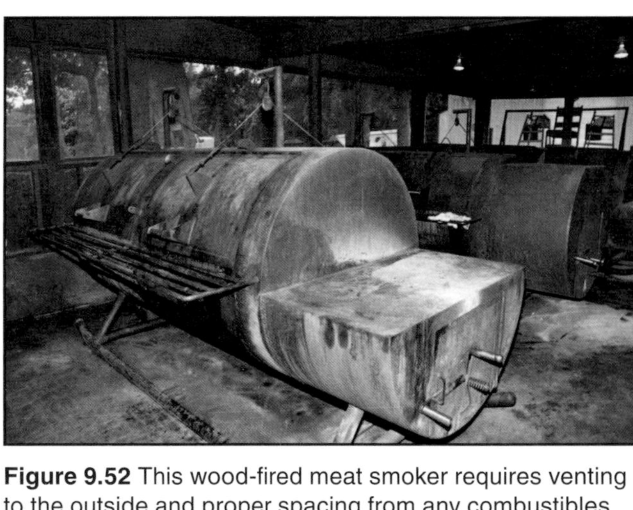

Figure 9.52 This wood-fired meat smoker requires venting to the outside and proper spacing from any combustibles.

Fire Suppression

Fire suppression systems are a vital part of commercial cooking equipment. Inspectors should be familiar with the systems in use in their jurisdiction. Some notable inspection items for inspectors include checking for the following:

- Cooking appliances protected by the system have not been changed or switched out. If the wrong appliances are in place, the systems will not function properly **(Figure 9.53)**.
- An instruction sign for manual system operation is present and employees have been trained to perform manual operations.
- Suppression system has been maintained per NFPA® standards.

Many commercial cooking establishments utilize wet-chemical systems. These systems must meet the requirements in NFPA® 17A, *Standard for Wet Chemical Extinguishing Systems*. Requirements for these systems are also contained in UL Standard 300, *Fire Testing of Fire Extinguishing Systems for Protection of Restaurant Cooking Areas*. See Chapter 13, Special-Agent Fire Extinguishing Systems and Extinguishers, for explanations of these types of fire extinguishing systems.

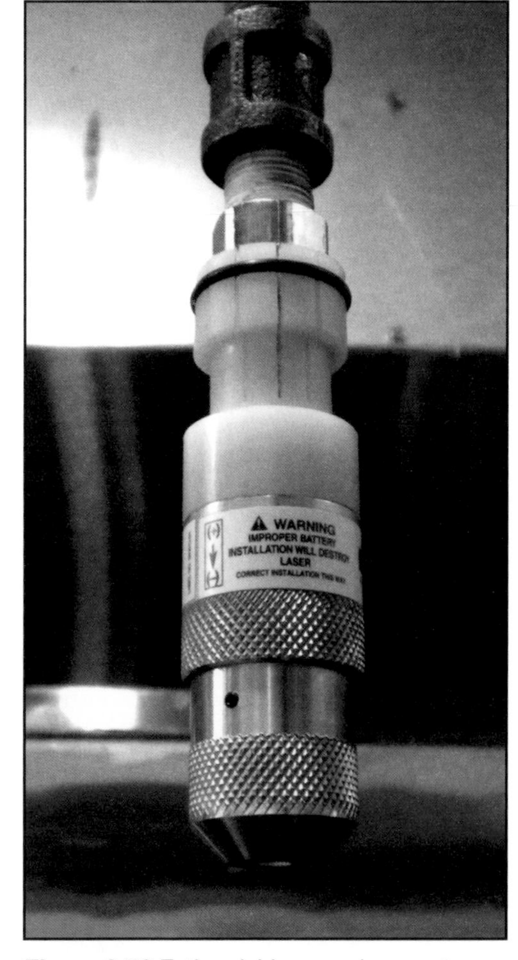

Figure 9.53 Extinguishing nozzles must be checked to verify correct placement. *Courtesy of George Apple, Cosumnes Fire Department.*

NOTE: An inspector should be aware that changes in the type of cooking oil used in fryers can require that a dry-chemical system be replaced by a wet-chemical system designed to control Class K fires. The adopted code may allow the continued use of the dry-chemical system as long as the system can be serviced and maintained.

Industrial Furnaces and Ovens

Furnaces and ovens can be found in a number of industrial applications that require melting, tempering, or drying of materials or products. Explosions and fires in fuel-fired and electric-heat-utilization equipment constitute a potential loss of life, property, and production. The cause of most failures can be traced to inadequately trained operators, lack of proper maintenance, or improper application of equipment. The sections that follow discuss the classes of industrial ovens and furnaces as well as the hazards associated with them.

Classes

NFPA® 86, *Standard for Ovens and Furnaces*, gives information on all types of industrial furnaces, ovens, thermal oxidizers, and dryers. The model codes and standards classify industrial ovens and furnaces into one of the following four classes:

- ***Class A*** — Operates at approximately atmospheric pressure and has a potential for explosion or fire hazard when flammable volatiles or combustible materials are processed or heated.
- ***Class B*** — Operates at approximately atmospheric pressure where no flammable volatiles or combustible materials are being heated.
- ***Class C*** — Has a potential hazard due to a flammable material being used or a special atmosphere is present; can use any type of heating system and includes a special atmosphere supply system; includes integral quench furnaces and molten salt bath furnaces.
- ***Class D*** — Operates at temperatures from above ambient to over 5,000°F (2 750°C); also operates at pressures normally below atmospheric using any type of heating system; can include the use of special processing atmospheres.

Special Considerations

During the design and installation process of industrial furnaces and ovens, certain conditions, such as failing to maintain required clearances, can create fire and explosion hazards. Other areas that can result in hazardous conditions include the following:

- Ventilation systems, including makeup air and exhaust systems
- Heat, gas, and smoke removal systems
- Temperatures required for operation
- Material handling equipment

Automatic sprinklers or water spray systems should be equipped on ovens that contain or process sufficient combustion materials to sustain a fire. Performing proper housekeeping techniques such as the following can reduce other industrial furnace and oven hazards:

- Remove combustible items from the area around the oven or furnace
- Remove general dirt and debris on a regular basis
- Maintain required clearances

Automated controls and alarms control overheating and other malfunctions in industrial furnaces and ovens. These safety controls must constantly evaluate the system and deactivate it if a problem occurs. All safety devices should be approved for their application. Installation and maintenance of these devices should be in accordance with manufacturer's instructions. Components commonly found on the safety controls include:

- Switches
- Valves
- Relays
- Controllers
- Detectors
- Interlocks

NOTE: It is important to check the schedule and the extent of the inspection, testing, and maintenance programs.

It is important to interlock the entire system so that in the event of a malfunction, the equipment will deactivate and remove the remaining fuel from the system. Deactivation of the heating system by any safety feature or safety device requires that an operator manually intervene to reestablish normal system operation.

Powered Industrial Trucks

The most common type of powered industrial truck is the forklift **(Figure 9.54)**. However, there are many other kinds of industrial trucks, including:

- Personnel transportation vehicles
- Special material handling vehicles
- Emergency response vehicles

Industrial trucks generally use the following fuels:

- Diesel
- Gasoline
- Liquefied petroleum gas (LPG)
- Compressed natural gas (CNG)
- Electric batteries
- Dual fuels (gasoline/LPG)

Figure 9.54 Inspectors must be familiar with local codes that address powered industrial vehicles like this forklift.

Electric battery and LPG models are the most common types in use. For a list of truck types and information about the hazards and types of atmospheres in which they can be used, refer to NFPA® 505, *Fire Safety Standard for Powered Industrial Trucks Including Type Designations, Areas of Use, Conversions, Maintenance, and Operations.*

Powered vehicles are found at industrial settings as well as in malls, schools, and even golf courses. Additionally, numerous age-restricted communities are encouraging vehicle use of this type. Power lifts can be found in convention centers, in hotels, and on construction sites **(Figure 9.55)**.

Requirements for powered industrial trucks are contained in most of the model fire codes. In general, an inspector should be alert for the following conditions when inspecting facilities that contain powered industrial trucks:

- Chargers for electric battery-operated units should be kept at least 5 feet (1.5 m) from combustible materials. Charging areas should not be accessible to the public. The charging area should be vented to prevent an accumulation of hydrogen gas.
- Liquid- or gas-fueled vehicles should only be fueled outside buildings. Fuel storage should be in compliance with locally adopted fire codes.
- Workers should perform repairs in areas designated for that function.
- An industrial truck should only be used in the atmosphere for which it is specified. Hazardous atmospheres need special equipment that must be approved for those conditions. Posted signs should warn operators whether a certain type of truck is safe for that area.
- Industrial trucks should be marked with signs indicating the class of truck and any operational restrictions if they are used in a hazardous electrical location.
- Industrial trucks must be maintained in accordance with manufacturer's instructions.

Tents

NFPA® 102, *Standard for Grandstands, Folding and Telescoping Seating, Tents, and Membrane Structures*, and other standards like it have brought about great changes in the guidelines regulating tent materials **(Figure 9.56)**. Tents must be made of flame-resistant materials that are approved and tagged as such. Materials that go inside the tent or on the tent such as banners or decorations must be treated with or made of flame-retardant material. The adopted building code includes permanent tents; temporary tents fall under the jurisdiction of the adopted fire code. See **Appendix E** for sample tent guidelines.

NOTE: While 180 days within a 12-month period is a typical time for a tent to be considered to be temporary, local policy or practice may differ.

NOTE: Membrane structures are covered in Chapter 5.

Tent Safety

Prior to July 6, 1944, tents were not thought of as a major safety concern. However, on that day in Hartford, Connecticut, 168 people died when the circus tent under which they were seated caught fire. This event resulted in the development of NFPA® 102, *Standard for Grandstands, Folding and Telescopic Seating, Tents, and Membrane Structures*.

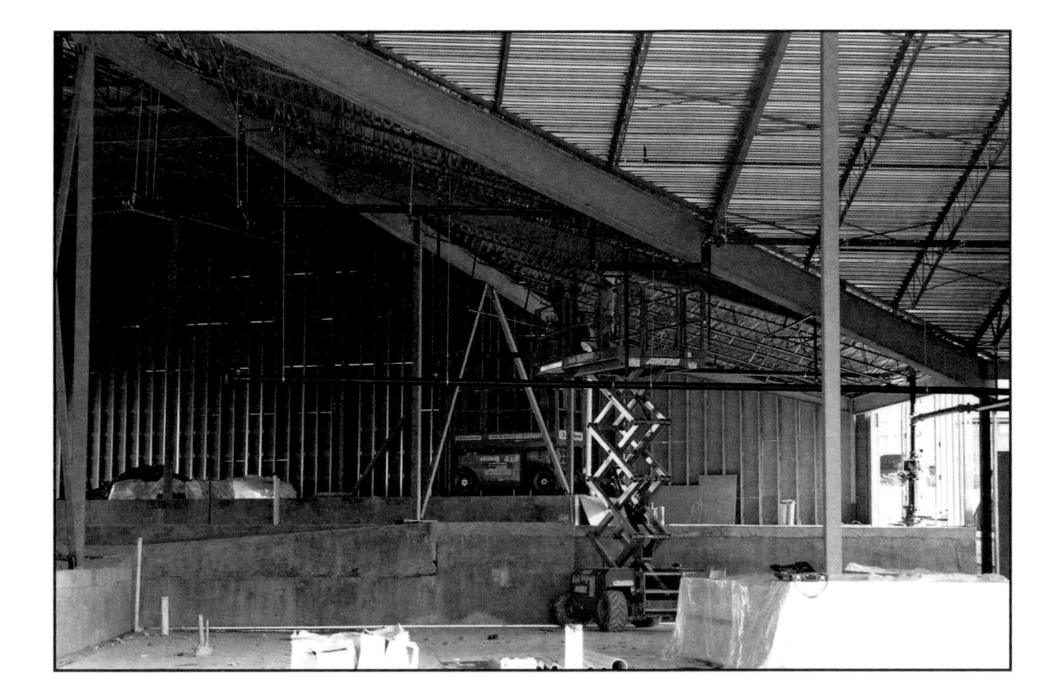

Figure 9.55 Scissor lifts are used in large structures, such as warehouses, arenas, and on construction sites.

Figure 9.56 Inspectors must reference NFPA® standards about the flammability of tent materials and the contents allowed in tents being temporarily used for commercial purposes.

All tents must meet the requirements of the large-scale test contained in NFPA® 701, *Standard Methods of Fire Tests for Flame Propagation of Textiles and Films*. The tents must be tagged or certified that they meet the requirements. Inspectors should not attempt to test the flammability of materials on their own.

Tents must be located according to the requirements of the adopted fire and building codes. Tents must meet other provisions:

- Specified distances from property lines, vehicles, and other tents.
- Means of egress requirements as contained in the locally adopted building and fire codes.
- Storage and handling of flammable or combustible liquids in accordance with fire code requirements.
- Electrical installments in accordance with locally adopted codes (including extension cords).
- Pyrotechnics and open flames are prohibited in any tents or temporary membrane structures.
- Access for emergency vehicles.

NOTE: When multiple tents are put together, the increased area under consideration may involve additional safety requirements and permits.

Combustible vegetation must be cleared both inside the tent and up to 10 feet (3 m) around the outside of the tent. The inspecting authority agrees if vegetation has been removed properly. Hay, straw, and shavings may require treatment with fire-retardant chemicals or limited to amounts sufficient for daily care and feeding of animals. Sawdust may need to be treated with fire-retardant chemicals or kept damp, depending on code requirements.

Hazardous Processes

There are a number of potential hazardous processes an inspector may encounter. These processes may exist temporarily or as a constant function of an industrial site. Examples of hazardous processes include the following:

- Welding, dipping, and quenching operations, as well as the application of flammable finishes in:
  - Manufacturing facilities
  - Auto repair and painting shops
  - Fabrication businesses
  - Furniture repair shops
- Dry cleaning operations
- Dust hazards that can result in both explosions and fires in many types of facilities
- Semiconductors/electronics manufacturing
- Distilleries

Other locations with numerous hazards on site are construction and demolition sites. These sites can be large-scale, commercial operations, or individual residential homes. Some of the operations that can generate ignition sources or hazardous atmospheres include:

- Welding
- Cutting
- Grinding
- Painting
- Presence of asphalt and tar kettles, used for replacing roof coverings

Inspectors should be aware of these processes and their potential locations. Some of these processes will be contained at specific facilities designed for their operations while others may be present during inspections of other types of occupancies. There are locations where these hazardous processes may exist only during the use of the process and will require a permit to perform. An inspector should be familiar with the code requirements for these processes and recognize any code infractions when they are encountered.

Hot Work — Any operation that requires the use of tools or machines that may produce a source of ignition.

Welding and Thermal Cutting Operations

Welding and thermal cutting, also known as **hot work**, are processes used to join or sever materials. The operations may use a variety of energy sources, including a gas flame or an electric arc. For a number of reasons, welding and thermal cutting operations have been a significant cause of fires in commercial and industrial occupancies as well as on construction sites **(Figure 9.57)**.

Welding and thermal cutting operations use two processes. The first process uses an electrical arc as a heat source. The second process uses a combination of oxygen and fuel gas – most commonly acetylene – to produce a flame. Both processes generate temperatures in excess of the melting point of the metals. Inspectors should primarily be concerned with the following hot work fire safety issues:

- Equipment is well maintained.
- Equipment is in working order.
- Equipment is stored properly.
- Combustible materials are kept well away from the areas containing hot work operations.
- Oxygen and fuel gases in cylinders (oxy-fuel gas equipment) are stored properly **(Figure 9.58)**.
- Gas regulators are maintained at their proper settings.

NOTE: Electric welding and thermal cutting equipment should be maintained according to the manufacturer's instructions and adopted electrical codes.

Thermal cutting operations as well as certain arc welding operations produce thousands of ignition sources in the form of sparks and hot slag, a glass-like residue that the welding process creates. Inspectors must verify that sparks or slag cannot come in contact with combustible materials. Combustible materials should not be stored in the welding or cutting areas. Inspectors should check to be sure that shop rags or towels are kept in approved metal containers with lids at all times **(Figure 9.59, p. 354)**.

When welding and thermal cutting work must be performed on an as-needed basis, removing combustibles or protecting them from ignition should make the area fire safe. Welding and cutting should never be allowed in the following situations:

- Inside buildings normally protected with automatic sprinkler systems when the systems are not in operation
- In the presence of explosive atmospheres
- On drums, tanks, or other containers that have previously held flammable liquids
- Near large quantities of exposed combustible materials

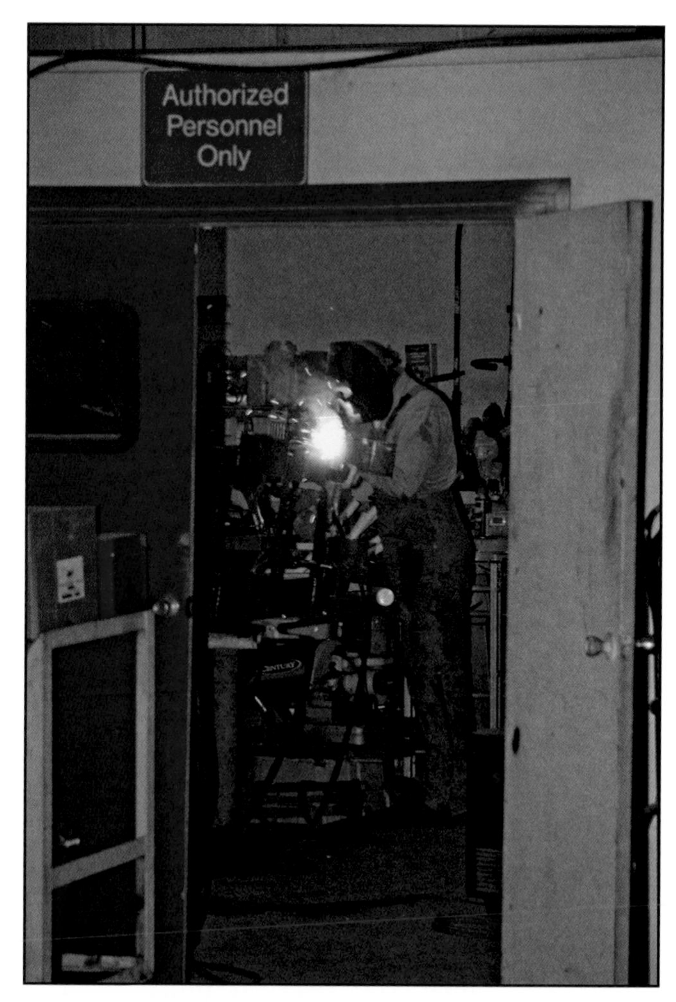

Figure 9.57 Hot work operations must be separated from other areas that have combustible materials.

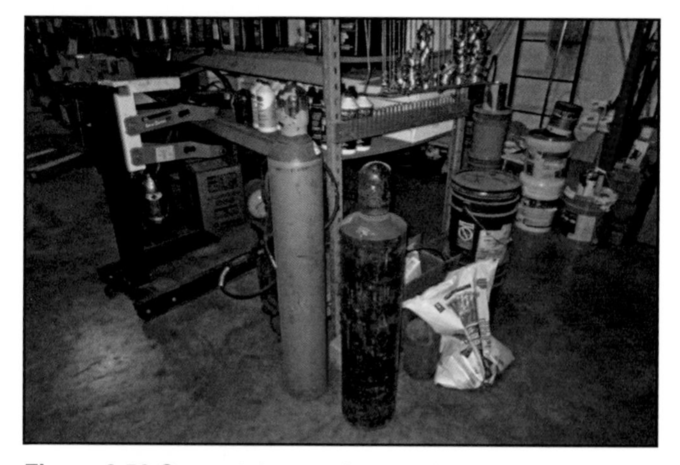

Figure 9.58 Oxyacetylene tanks must be secured when not in use. These loose tanks are not secured properly. *Courtesy of Rich Mahaney.*

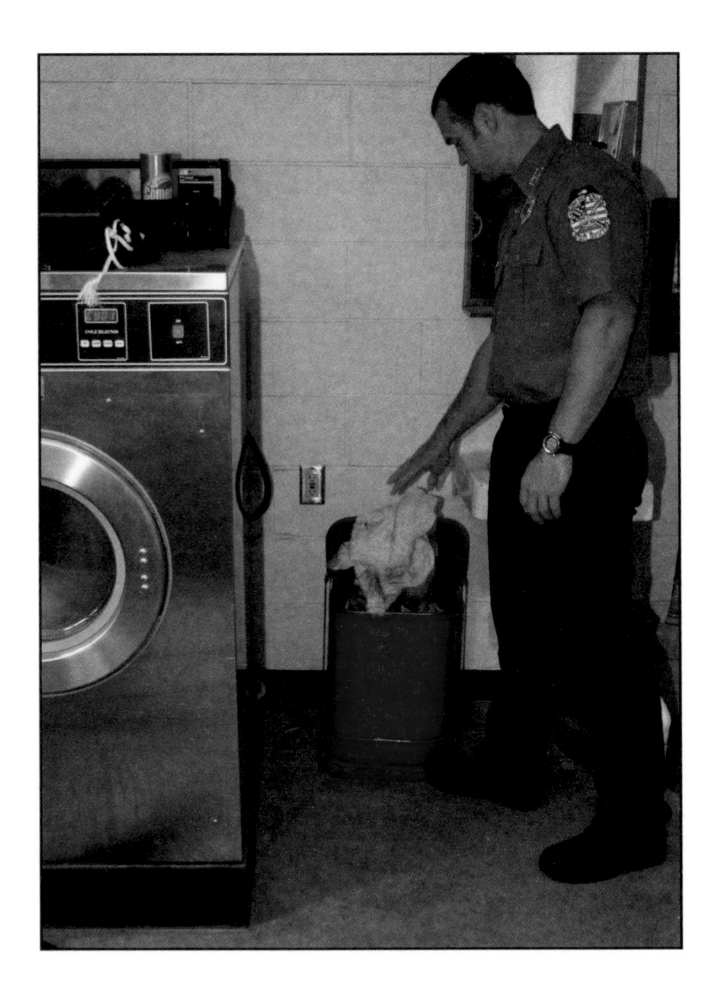

Figure 9.59 Because dirty and oily rags are fire hazards, their proper storage and disposal is vital to a safe work environment.

Where combustible materials, such as paper clippings, wood shavings, or textile fibers, are on the floor, an area with a radius of at least 35 feet (10.7 m) must be cleared before welding or cutting operations begin. When it is impractical to create this size radius, the materials must be covered with approved fire-resistant guards or curtains.

For more information on these types of operations, consult the following sources:

- NFPA® 51, *Standard for the Design and Installation of Oxygen-Fuel Gas Systems for Welding, Cutting, and Allied Processes*
- NFPA® 51B, *Standard for Fire Prevention During Welding, Cutting, and Other Hot Work*
- FM Global Loss Prevention Data Sheets

As a means of reducing hazards associated with welding and thermal cutting operations, most fire codes require the establishment of fire watches, issuance of permits, and development of hot-work programs. Inspectors will need to understand local requirements for each of these means.

Hot-Work Program

A hot work program requires a written plan approved by the AHJ. The plan establishes procedures to prevent fires resulting from temporary operations that involve an open flame or produces heat, sparks, or hot slag. These operations include:

- Welding
- Brazing
- Cutting
- Grinding
- Soldering
- Thawing pipes
- Applying roofing with a torch

Appendix F from FM Global shows an example of a hot work permit for the fire safety supervisor, for the employees performing the hot work operation, and a Warning sign to be posted in the area.

Permits

A work permit is required in some jurisdictions when performing thermal cutting or welding operations. Local municipal jurisdictions may find it prudent to adopt similar requirements regarding the regulation of as-needed welding and cutting operations. Additionally, an annual permit to conduct these operations in fixed locations should be considered for inclusion in the municipal fire code.

The owner/occupant is responsible for developing the hot-work program, establishing a written policy for hot work, and applying for the permit. An inspector will need to evaluate the area where the work is to be performed before a work permit is issued. The inspector should note all hazards in the area and be able to discuss emergency procedures with the permit applicant. The minimum information to be contained on this permit includes the following:

- Date
- Work to be performed
- Period for which the permit is valid
- Location of the job
- Type of fire suppression equipment available
- Inspection of area before work begins
- Authorization signature of individual in charge
- Notification of when a fire watch is assigned
- Final signature of the individual after the work has been completed

Fire Watch

Workers must maintain a fire watch for at least 30 minutes after any welding or thermal cutting operation is completed. This procedure is most critical when:

- Combustible contents are closer than 35 feet (11.7 m) from the point of operation.
- Adjacent areas having wall or floor openings are within a 35-foot (10.6 m) radius.
- Exposed combustible materials are in adjacent areas, especially concealed spaces, walls, and floors.
- Anytime a fire could spread due to conduction or radiation.

The inspector is directed to NFPA® 1, *Fire Code*, NFPA® 51B, *Standard for Fire Prevention During Welding, Cutting, and Other Hot Work,* and IFC requirements.

Fire watch personnel must be trained in the use of fire extinguishers and in procedures to warn occupants and summon the fire department in the event of a fire. Fire extinguishers that are appropriate for the level of hazard present must be available at all times when welding or thermal cutting operations are in progress.

Flammable Finishing Operations

Model fire codes establish requirements for flammable finishing, which involves the application of either flammable or combustible liquids or combustible powders onto wood, metal, or plastic surfaces **(Figure 9.60)**. The most common flammable finishing methods are:

- **Spray finishing**. In spray finishing, the primary hazard comes from pressurizing a flammable or combustible liquid and discharging it through a small orifice. This pressure creates an aerosol, which consists of numerous small droplets with little mass and a large surface area that is easily ignited.
- **Powder coating**. Powder coating involves the suspension of electrically charged combustible powders like polyester resin and applying them to a metal or plastic part with a negative electrical charge. The hazard in this process is flash fire or dust explosion.
- **Immersion coating**. Immersion coating involves the placement of a suspended article into an open dip tank of flammable or combustible liquids. Constant vapor production as the liquid evaporates increases the risk of fire.

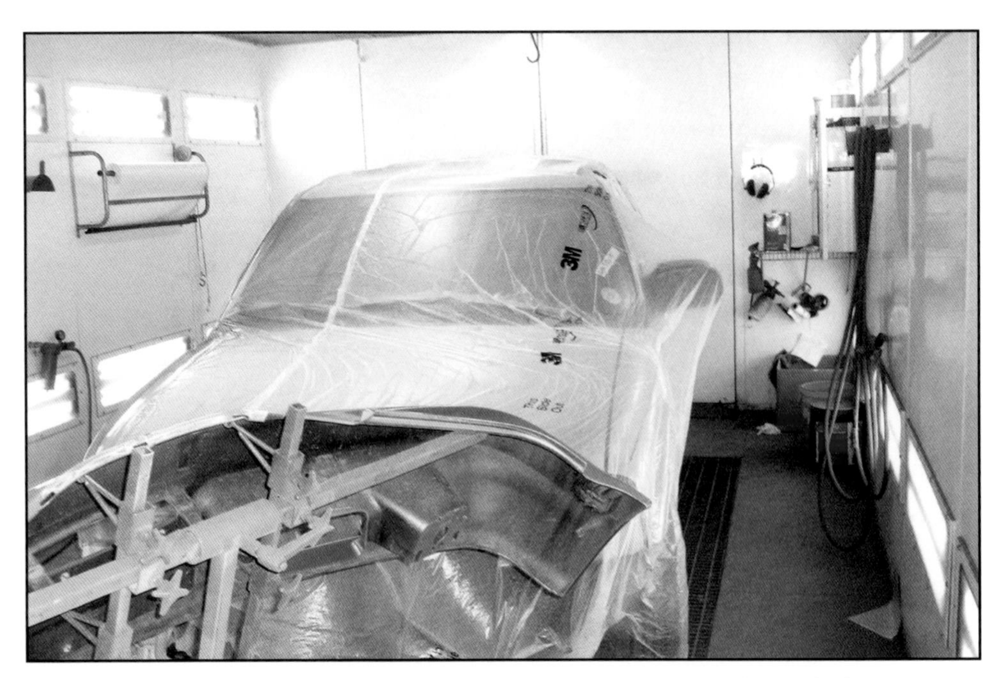

Figure 9.60 A finishing operation like this auto body shop presents hazards due to flammable liquids and combustible powders that have become atomized and are hanging in the air.

NOTE: More information on dust hazards is found later in this chapter.

Regardless of the type of the flammable finishing process, model fire codes specify one or more of the following controls to limit the likelihood of ignition and to provide protection to the occupants and the building if ignition occurs:

- The use of mechanical ventilation to maintain the atmosphere below specified flammable liquids.
- The installation of noncombustible equipment or noncombustible enclosures to limit the spread of fire beyond the flammable finishing area.
- The specification of hazardous (classified) electrical location boundaries in and around the spray finishing process to confirm the selected electrical equipment does not constitute an ignition source.
- An automatic fire extinguishing system in and around the flammable finishing process.
- Administrative controls limiting the amount of fuel in the flammable finishing area, and control of ignition sources including the prohibition of hot work.
- Interlocks to stop the flammable finishing process if mechanical ventilation is lost or the automatic fire extinguishing system activates.

The model fire code and applicable NFPA® standards generally exempt the application of water-based coatings, the use of coatings supplied from disposable aerosol containers, and the manual application of flammable liquids over small areas using brushes or rollers from regulation.

Spray Finishing

The most common flammable finishing operation is spray finishing. Spray finishing utilizes compressed air or hydraulic energy to atomize flammable or combustible liquids into small droplets. Hydraulic systems are termed airless spray finishing because it does not utilize compressed air. Airless systems operate at pressures ranging at 100-600 PSIG so the piping system must be engineered for these pressures. Compressed air systems operate between 5-20 PSIG.

Figure 9.61 A spray finishing booth must be enclosed and have exhaust ducts that discharge outside.

Model fire codes require that spray finishing be performed in either a spray booth or spray room **(Figure 9.61)**. Model fire codes classify a spray booth as an appliance. It is a noncombustible enclosure constructed of carbon, galvanized, or stainless steel panels with a plenum. The **plenum** is connected to a noncombustible exhaust duct that discharges outside the building.

Plenum — An enclosed space in a building or appliance designed to efficiently capture and discharge or circulate air.

The mechanical ventilation system must be designed to maintain the atmosphere inside the spray booth or spray room at an atmosphere less than 25 percent of the lower flammable limit for the most volatile flammable liquid that is applied. Generally, acetone is used because it exhibits one of the highest evaporation rates of all the industrial solvents used in flammable finishing. Generally speaking, a conventional spray booth designed for a passenger automobile is designed to exhaust at a rate of 8-11,000 cubic feet/minute.

A fire inspector will want to confirm this air exhaust rate. A mechanical inspector will commonly require an air balance test to confirm the exhaust fan is properly sized and functioning correctly. The air balance test will confirm that the mechanical ventilation system meets the manufacturer's specified flow rate.

An important concern to the inspector is verifying that the electrical equipment installed inside the spray booth is designed for hazardous (classified) locations. The interior space of spray rooms or booths is illuminated with electric lamps. Fire codes and NFPA® 33, *Standard for Spray Application Using Flammable or Combustible Material*, require that these lamps be listed for use in Class I, Division 1 atmosphere. This is a National Electrical Code® classification for electrical equipment installed in hazardous atmospheres where flammable vapors are present.

With few exceptions, all spray booths and spray rooms require some form of an automatic fire extinguishing system. Acceptable methods of protection include automatic sprinkler protection and dry chemical fire extinguishing systems.

Another key fire protection feature in any spray booth or room is the interlock between the spraying equipment and the mechanical ventilation. NFPA® 33, *Standard for Spray Application Using Flammable or Combustible Material* and model fire codes require the mechanical ventilation system be interlocked so that spraying cannot be performed unless the ventilation system is operating.

Powder Coating

Powder coating is a process in which metal parts are coated with a powdered polyester resin. The powder is ionized with positive electrons so it will adhere to the metal. After coating, the powder is backed to form a durable finish on the material surface. The powder is applied in a ventilated enclosure, which can be a spray booth or a spray room.

The primary hazard of powder coating is the ignition of the combustible powder inside the room or space where it is applied. Fire inspectors entering these installations need to recognize the following hazards and certify that the required engineering controls are in use:

- Verify that powder application equipment is listed for combustible powders. Because powdered polyester resin is a combustible resin, the equipment must be listed for hazardous locations under the jurisdiction's electrical code.
- A powder collection system is provided. The powder collector may be part of the booth, or an independent device that is ducted to a dust collector.
- An automatic fire extinguishing system is provided in the spray space.
- The powder coating booth is constructed using noncombustible materials.

Immersion Coating

In immersion coating (dipping and coating) operations, parts are immersed into liquids that may be combustible or have toxic qualities **(Figure 9.62)**. The potential for hazards comes from the combustible liquids and from the equipment used for these operations. This equipment, which increases the fire risk, includes:

- Flammable liquid storage tanks
- Conveyor pumps
- Heaters and heat exchangers
- Agitators
- Ventilation and exhaust equipment

To lower the potential of an incident, immersion coating operations should be located away from other process areas and never located near an egress area. The dip tanks and reservoirs should be contained in 1-hour fire-rated construction, and ignition sources identified and removed if possible. Splash boards, drains, and overflow protection devices are all production and safety components that should be included in evaluating dipping operations.

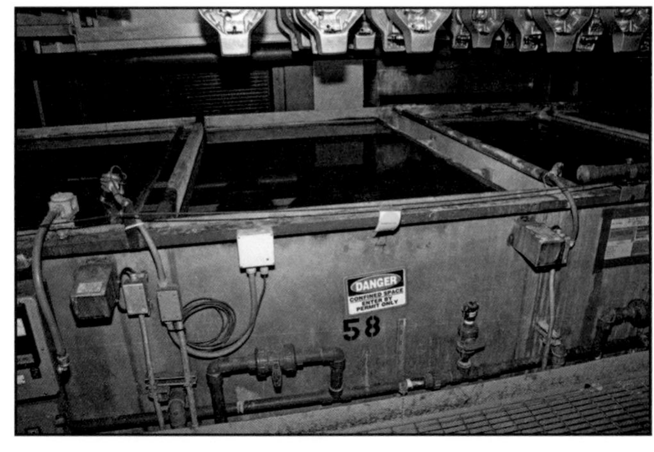

Figure 9.62 Dip tank operations use toxic and combustible liquids to coat parts with paint or other chemicals in industrial production.

In some cases, it may be necessary to use suspended nonflammable curtains that shield an operation. Where these processes are confined, explosion-venting construction can be used. Ventilation and exhaust of the areas are important to remove hazardous vapors that could lead to an explosion.

Housekeeping is important to limit accumulations of combustible materials. Maintenance is also needed to keep buildup of combustible residues to a minimum. Regular inspections of the process area, equipment, and storage of commodities associated with the process will limit the occurrence of incidents in a facility.

Quenching Operations

Quenching is the immersion of a metal part in a quench medium. The mechanical means immerses a part, moves it through the quench medium, and then removes it from the quench medium (**Figure 9.63, p. 360**). There are several mechanical components involved in a quenching process, such as:

- Elevators
- Conveyors
- Hoists
- Cranes

Mineral oils are commonly used as a quench medium. Regardless of the medium used, the properties of the medium must remain fluid and stable. Two types of quenching are heated and unheated.

Quenching operations must follow several requirements. These operations should be located in fire-resistive buildings away from the main production areas. A well-designed ventilation system must be employed to remove the collection of explosive atmospheres in the processing area. The area must be protected by an automatic sprinkler system.

Quenching operations should be designed where conveyor systems automatically stop in the event of a fire. All portions of the quench process should be interlocked so that they all stop operating.

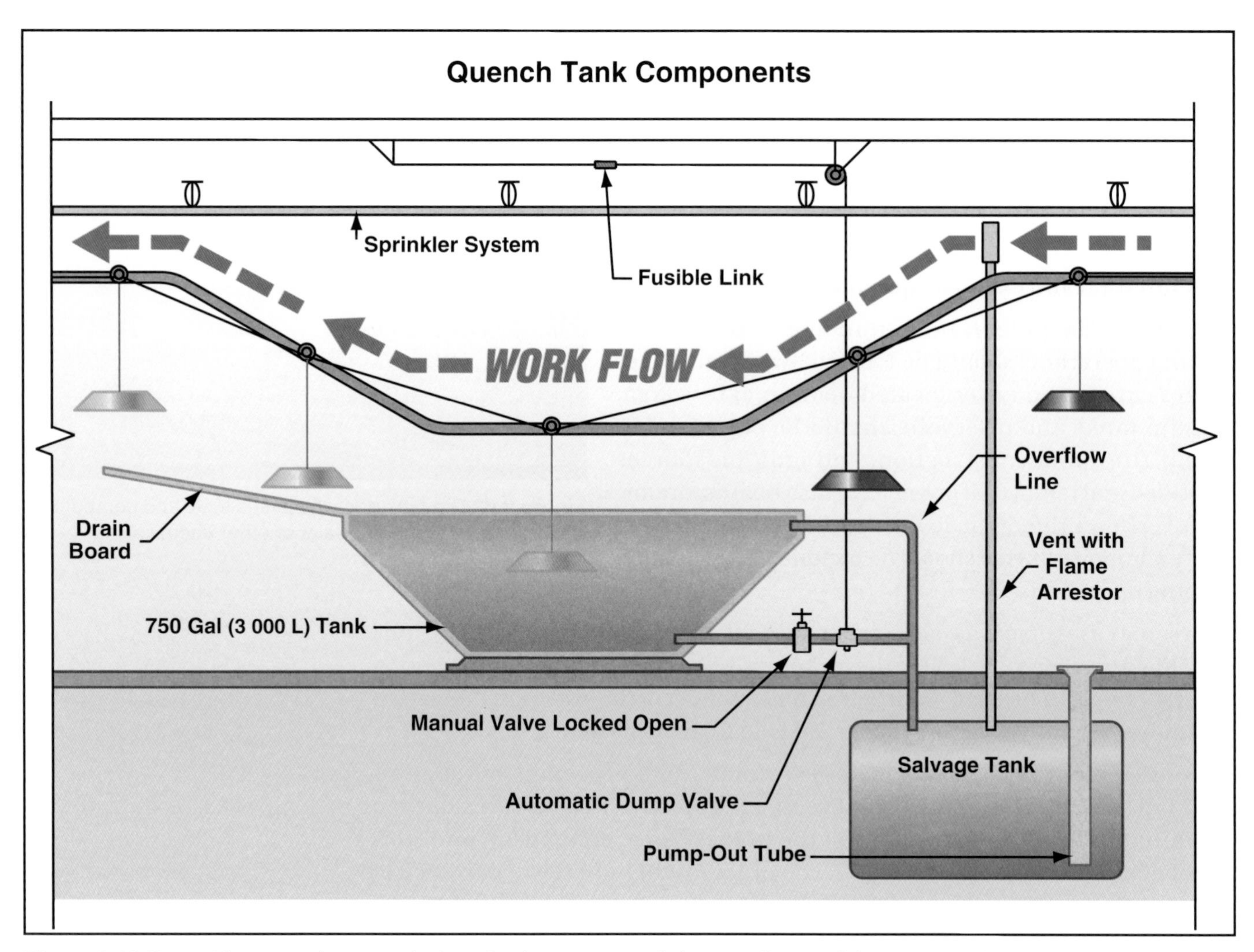

Figure 9.63 Quenching operations are designed to immerse metals in a medium and then remove them in an automated assembly line.

NOTE: All aspects of the quenching process should be electrically grounded in accordance with NFPA® 77, *Recommended Practice on Static Electricity.*

Hazard reduction around a quenching process includes elimination of:

- Open flames
- Spark-producing equipment or processes
- Equipment whose exposed surfaces exceed the autoignition temperature of the quenching medium.

These hazards should be completely enclosed if they cannot be removed. It is important that workers be trained in handling flammable liquids in order to lessen hazards.

Overflow is a significant issue regarding tanks used for quenching operations. Inspectors should be mindful of a number of tank-related considerations, such as:

- Tanks should be located at grade level so that flammable liquids will not overflow onto lower levels. Quenching tanks should never be placed in basements because of the potential hazard from fire and the increased risk to emergency responders.

- All tanks should be designed with enough freeboard (distance from the liquid surface to the top of the tank when the tank is full) to keep the tank from overflowing. The freeboard should never be less than 6 inches (150 mm).
- All tanks should have drains to protect from overflow, particularly those that contain 150 gallons (600 L) of quench oil or more. Any tank over 500 gallons (2 000 L) must have a bottom drain that opens automatically or manually in the event of a fire.
- Tanks should be built within dikes with drains capable of containing the contents of the tank and piping capable of moving it into specialized holding tanks.
- Installation of drain boards and automatic-closing covers can prevent overflow due to automatic sprinkler system discharges.
- Fire extinguishers of several types must be available for extinguishing quench mediums.

Dry Cleaning Operations

Dry cleaning is the process of removing dirt, grease, and stains from clothing or other textiles by using solvents. A facility where clothing is dropped off and returned would be classified as a business occupancy. A facility with on-site cleaning may be classified differently, such as a hazardous occupancy.

Often dry cleaning is not thought of as a hazardous process because it involves the basic task of cleaning clothes. However, the dry cleaning process involves the use of numerous chemical, fire, and environmental hazards. The primary fire hazard with dry cleaning is the presence of all the elements necessary for uncontrolled fires: fuels, ignition sources, and oxygen.

In earlier years, perchlorethylene was the primary solvent used in dry cleaning operations. Perchlorethylene reduced fire hazards but introduced environmental concerns. Newer solvents in common use today have fewer environmental concerns but greater flammability hazards.

Dry cleaning plants are classified by the type of solvents used in the process **(Table 9.1)**. For example, a Type II plant uses Type II solvents. Because Class I solvents have flash points below 100°F (40°C), Type I plants are prohibited. Small quantities of Class I solvents may be stored and used in the other classifications of plants.

Table 9.1
Classifications of Dry Cleaning Plants and Solvents

Plant Classification	Cleaning Solvent Classification
Type I	**Class I** — Liquids having a flash point below 100°F (38°C)
Type II	**Class II** — Liquids having a flash point at or above 100° F (38°C) and below 140°F (60°C)
Type IIIA	**Class IIIA** — Liquids having a flash point at or above 140°F (60°C) and below 200°F (93°C)
Type IIIB	**Class IIIB** — Liquids having a flash point at or above 200°F (93°C)
Type IV	**Class IV** — Liquids that are classified as nonflammable; dry cleaning not conducted by the public
Type V	**Class IV** — Liquids that are classified as nonflammable; dry cleaning conducted by the public

Based on information contained in the International Fire Code® (IFC®).

NOTE: NFPA® 32, *Standard for Drycleaning Plants,* also contains extensive information on occupancies that contain dry cleaning operations.

During the plans review process, an inspector should establish that the fire protection requirements of the locally adopted code are met. During inspections, an inspector must verify the following **(Figures 9.64 a and b)**:

- The quantity of Class I solvents is limited and properly stored.
- The dry cleaning plant type and solvent used match.
- Routine maintenance is performed to prevent accumulation of fluff, lint, or waste that could ignite or cause a fire to spread rapidly.
- All containers with flammable or combustible petroleum-based solvents are properly stored, transported, and used.
- A No Smoking policy is in place, enforced, and appropriate signs are posted.
- The correct size and number of portable fire extinguishers are available and personnel are trained in their use.
- Fire protection equipment, including detection and alarm equipment and automatic sprinklers, are operational.

Dust Hazards

Dust consists of suspended particulates in the air. Given the correct mixture of oxygen and an ignition source, combustible dust suspended in air and in sufficient quantity can create a fire, conflagration, or **explosion**. A dust explosion can be severe enough to destroy an entire facility. The U.S.

Explosion — physical or chemical process that results in the rapid release of high pressure gas into the environment.

Figures 9.64 a and b The inspector must verify that dry cleaning solvents are stored properly and that fire protection equipment is operable.

Chemical Safety and Hazard Investigation Board (CSB) identified 281 combustible dust incidents between 1980 and 2005 that led to the deaths of 119 workers, injured 718, and extensively damaged numerous industrial facilities.

Definition of Combustible Dust

Combustible dust is defined as a solid material composed of distinct particles or pieces, regardless of size, shape, or chemical composition, which presents a fire or deflagration hazard when suspended in air or some other oxidizing medium over a range of concentrations. Combustible dusts are often organic material or metal dusts that are finely ground into very small particles, fibers, chips, chunks, flakes, or a small mixture of these.

Source: OSHA Hazard Communication Guidance for Combustible Dusts.

Even materials that do not burn in larger pieces (such as aluminum or iron), given the proper conditions, can form combustible or explosive dust. Materials that can form combustible dusts are used in a wide range of industries and processes, such as agriculture, chemical manufacturing, pharmaceutical production, furniture, textiles, fossil fuel power generation, recycling operations, and metal working and processing, which includes 3D welding — a form of 3D printing. **(Figure 9.65, p. 364)**.

In order for a dust explosion to occur, the following five conditions must be present **(Figure 9.66, p. 365)**:

1. Combustible dust must be suspended in air.
2. The particle concentration must be within its explosive range.
3. An ignition source must be present.
4. The dust must be in a confined space.
5. Oxygen content must be capable of supporting combustion.

If one of the above five elements is missing, an explosion cannot occur.

Dust explosions occur in series **(Figure 9.67, p. 366)**. The first explosion usually is not as severe as subsequent explosions. The first explosion stirs dust that has settled on ledges, walls, equipment, or in areas of low travel. The second introduction of particles to the air generally results in larger and stronger explosions. Many processes can produce the potential for a dust explosion.

- **West Pharmaceutical Services, Kinston, NC, January, 2003: 6 dead, 38 injured, including two firefighters.**

The explosion disabled the building's sprinkler system and ignited fires throughout the facility, which produced rubber products for medical use. An investigation revealed that combustible dust from a plastic raw material – polyethylene dust – had accumulated on hidden surfaces above the production area. Because the facility produced supplies for medical use, crews continuously cleaned dust from visible areas. However, polyethylene dust was also drawn upward and accumulated above a ceiling installed over the rubber production area. The dust gradually built up on ceiling tiles, beams, conduits, and light fixtures to form an explosion hazard.

Combustible Dust

Does your company or firm process any of these products or materials in powdered form?

If your company or firm processes any of these products or materials, there is potential for a "Combustible Dust" explosion.

Agricultural Products
Egg white
Milk, powdered
Milk, nonfat, dry
Soy flour
Starch, corn
Starch, rice
Starch, wheat
Sugar
Sugar, milk
Sugar, beet
Tapioca
Whey
Wood flour

Agricultural Dusts
Alfalfa
Apple
Beet root
Carrageen
Carrot
Cocoa bean dust
Cocoa powder
Coconut shell dust
Coffee dust
Corn meal
Cornstarch
Cotton
Cottonseed
Garlic powder
Gluten
Grass dust
Green coffee
Hops (malted)
Lemon peel dust
Lemon pulp
Linseed
Locust bean gum
Malt
Oat flour
Oat grain dust
Olive pellets
Onion powder
Parsley (dehydrated)
Peach
Peanut meal and skins
Peat
Potato
Potato flour
Potato starch
Raw yucca seed dust
Rice dust
Rice flour
Rice starch
Rye flour
Semolina
Soybean dust
Spice dust
Spice powder
Sugar (10x)
Sunflower
Sunflower seed dust
Tea
Tobacco blend
Tomato
Walnut dust
Wheat flour
Wheat grain dust
Wheat starch
Xanthan gum

Carbonaceous Dusts
Charcoal, activated
Charcoal, wood
Coal, bituminous
Coke, petroleum
Lampblack
Lignite
Peat, 22%H_2O
Soot, pine
Cellulose
Cellulose pulp
Cork
Corn

Chemical Dusts
Adipic acid
Anthraquinone
Ascorbic acid
Calcium acetate
Calcium stearate
Carboxy-methylcellulose
Dextrin
Lactose
Lead stearate
Methyl-cellulose
Paraformaldehyde
Sodium ascorbate
Sodium stearate
Sulfur

Metal Dusts
Aluminum
Bronze
Iron carbonyl
Magnesium
Zinc

Plastic Dusts
(poly) Acrylamide
(poly) Acrylonitrile
(poly) Ethylene (low-pressure process)
Epoxy resin
Melamine resin
Melamine, molded (phenol-cellulose)
Melamine, molded (wood flour and mineral filled phenol-formaldehyde)
(poly) Methyl acrylate
(poly) Methyl acrylate, emulsion polymer
Phenolic resin
(poly) Propylene
Terpene-phenol resin
Urea-formaldehyde/ cellulose, molded
(poly) Vinyl acetate/ ethylene copolymer
(poly) Vinyl alcohol
(poly) Vinyl butyral
(poly) Vinyl chloride/ ethylene/vinyl acetylene suspension copolymer
(poly) Vinyl chloride/ vinyl acetylene emulsion copolymer

Dust Control Measures

The dust-containing systems (ducts and dust collectors) are designed in a manner (i.e., no leaking) that fugitive dusts are not allowed to accumulate in the work area.

The facility has a housekeeping program with regular cleaning frequencies established for floors and horizontal surfaces, such as ducts, pipes, hoods, ledges, and beams, to minimize dust accumulations within operating areas of the facility.

The working surfaces are designed in a manner to minimize dust accumulation and facilitate cleaning.

Ignition Control Measures

Electrically-powered cleaning devices such as vacuum cleaners, and electrical equipment are approved for the hazard classification for Class II locations.

The facility has an ignition control program, such as grounding and bonding and other methods, for dissipating any electrostatic charge that could be generated while transporting the dust through the ductwork.

The facility has a Hot Work permit program.

Areas where smoking is prohibited are posted with "No Smoking" signs.

Duct systems, dust collectors, and dust-producing machinery are bonded and grounded to minimize accumulation of static electrical charge.

The facility selects and uses industrial trucks that are approved for the combustible dust locations.

Prevention Measures

The facility has separator devices to remove foreign materials capable of igniting combustible dusts.

MSDSs for the chemicals which could become combustible dust under normal operations are available to employees.

Employees are trained on the explosion hazards of combustible dusts.

Protection Measures

The facility has an emergency action plan.

Dust collectors are not located inside of buildings. (Some exceptions)

Rooms, buildings, or other enclosures (dust collectors) have explosion relief venting distributed over the exterior wall of buildings and enclosures.

Explosion venting is directed to a safe location away from employees.

The facility has isolation devices to prevent deflagration propagation between pieces of equipment connected by ductwork.

The dust collector systems have spark detection and explosion/ deflagration suppression systems.

Emergency exit routes are maintained properly.

Occupational Safety and Health Administration
U.S. Department of Labor

www.osha.gov • (800) 321-OSHA • TTY (877) 889-5627

Figure 9.65 A wide range of industries and processes can produce combustible dusts. *Source: OSHA.*

- **Imperial Sugar Company, Wentworth, GA, 2008: 14 dead, 36 injured.**

On February 7, 2008, a series of sugar dust explosions at the Imperial Sugar manufacturing facility, which converted raw cane sugar into granulated sugar, resulted in 14 worker fatalities. Eight workers died at the scene and six others eventually succumbed to their injuries.

The resulting investigation determined that the first dust explosion initiated in the enclosed steel belt conveyor located below the sugar silos. Recently installed steel cover panels on the belt conveyor had allowed explosive concentrations of sugar dust to accumulate inside the enclosure. The first explosion ignited the sugar dust that had accumulated on the floors and elevated horizontal surfaces, lofted the dust, and propagated more dust explosions through the buildings. Secondary dust explosions occurred throughout the packing buildings, parts of the refinery, and the bulk sugar loading buildings.

- **Lakeland Mills, Prince George, BC, 2012: 2 dead, 22 injured.**

On April 23, 2012, a massive explosion in the Lakeland Mills sawmill killed two workers, injured 22, and destroyed the mill. The Prince George fire department had warned the Lakeland plant about dangerous dust accumulations and the need for a fire safety plan. Eleven days before the blast, an engineer had inspected the plant and warned of the risk of an explosion. The explosion at the Lakeland Mills sawmill occurred three months after a similar blast killed two others at Babine Forest Products near Burns Lake, BC.

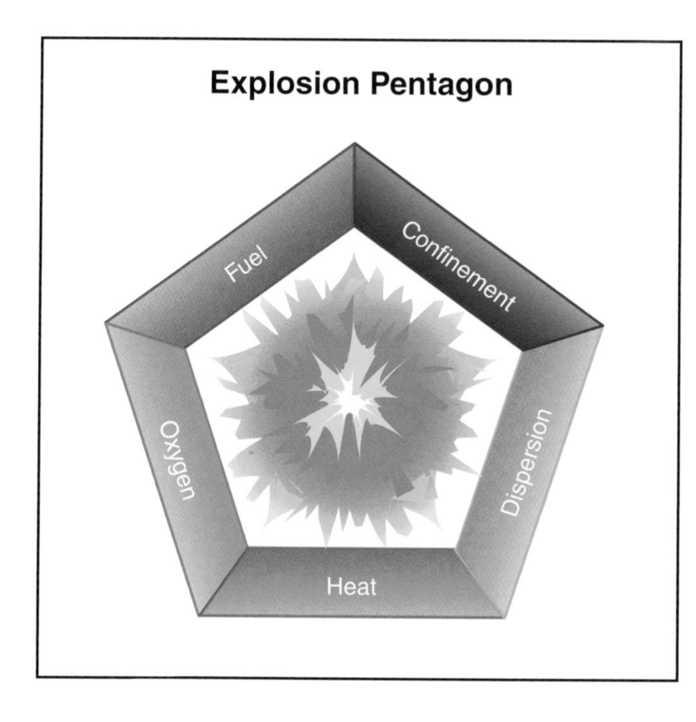

Figure 9.66 These five conditions are necessary for a dust explosion to occur.

Process Hazard Analysis

NFPA® 654, *Standard for the Prevention of Fire and Dust Explosions from the Manufacturing, Processing, and Handling of Combustible Particulate Solids,* is an all-encompassing standard on how to design a safe dust collection system. The standard requires that process hazard analysis be conducted for

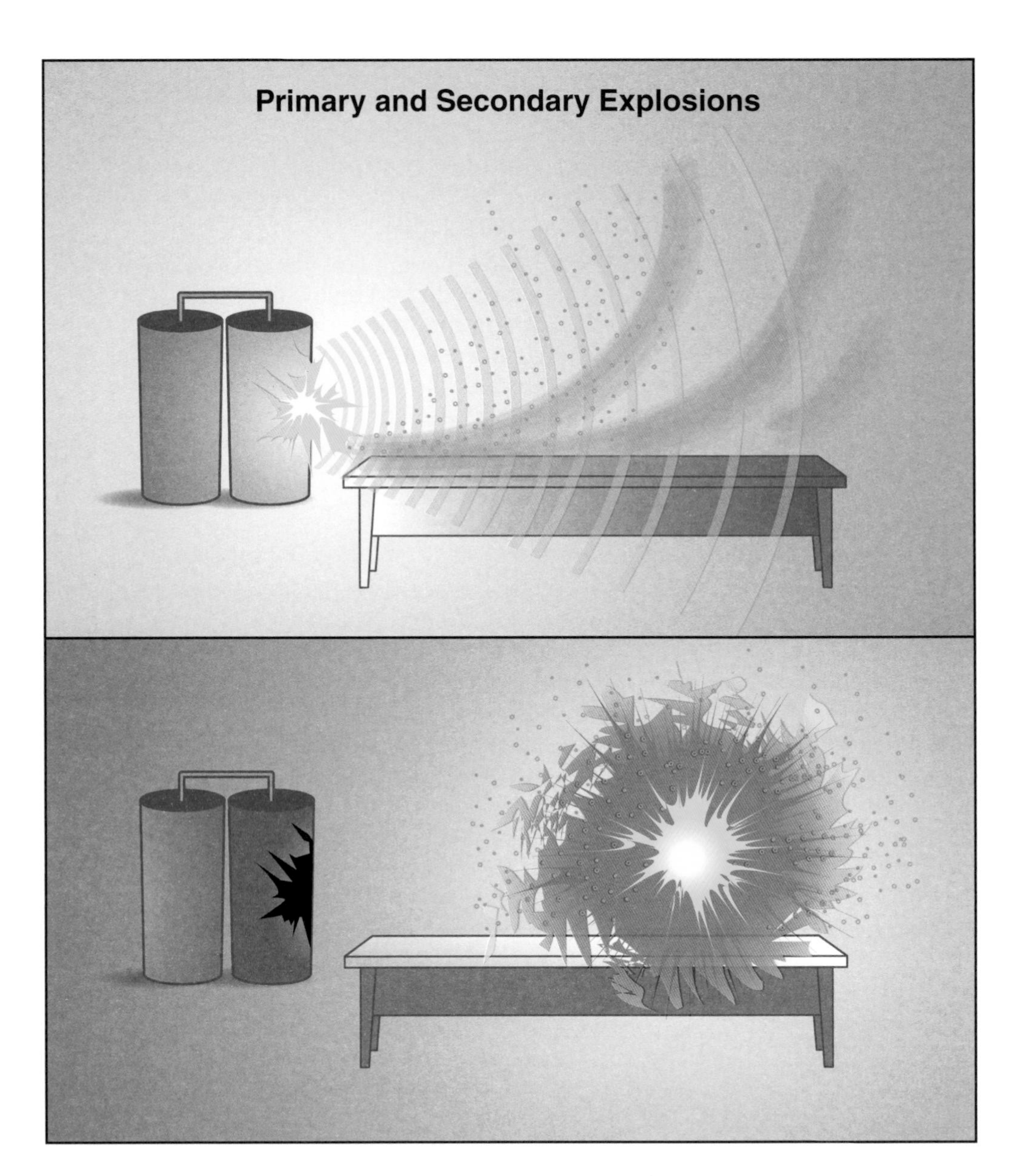

Figure 9.67 Primary dust explosions disperse combustible dusts in a facility. Secondary explosions occur as those dusts are ignited, resulting in a much larger explosion.

processes handling powders and bulk solids that present a fire or explosion hazard. This standard can assist the inspector as a reference when preparing to conduct inspections. As always, the inspector must also be familiar with the locally adopted codes.

In process hazard analysis (PHA), a qualified testing agency should test all combustible dusts that may produce a dust explosion to determine the following data:

- Particle size (surface area-to-mass ratio)
- Moisture content as received and dried
- Minimum dust concentration to ignite
- Available oxygen in atmosphere
- Minimum energy required for ignition
- Maximum rate of pressure rise at various concentrations
- Minimum dust layer ignition temperature
- Maximum explosion pressure at optimum concentration

Employers who use materials or processes capable of producing combustible dusts must evaluate those operations and tasks where dusts are present or may be generated and provide the required safety and handling information to their employees. They must determine generally applicable control measures, such as appropriate engineering controls, work practices, or personal protective equipment, and include that information on the SDS. An SDS is required for each hazardous chemical used, and it must be readily accessible to workers.

Classification and Grouping

NFPA® 499, *Recommended Practice for the Classification of Combustible Dusts and of Hazardous (Classified) Locations for Electrical Installations in Chemical Process Areas,* describes how dusts are grouped and classified. Combustible dusts are in one of the following groups:

- ***Group E*** — Metal dusts including aluminum, magnesium, their commercial alloys, and other dusts whose particle size, abrasiveness and conductivity present hazards in the use of electrical equipment.
- ***Group F*** — Volatile dusts containing more than 8 percent carbonaceous materials based upon ASTM D-3175, *Standard Test Method for Volatile Matter in the Analysis Sample of Coal and Coke.* These materials include coal, carbon black, charcoal, and coke dusts present in manufacturing and power production facilities.
- ***Group G*** — Dust atmospheres produced in flour, grain, wood, plastic, and chemical processing operations **(Figure 9.68)**.

Figure 9.68 Manufacturing operations that use wood are a common source of dust hazards. *Courtesy of Rich Mahaney.*

Additionally, combustible dusts are classified by the location where they are found. Combustible dusts comprise Class II Locations, described in NFPA 70®, *National Electrical Code®*, and include two divisions as follows:

- ***Class II, Division 1*** — Combustible dust is in the air under normal operating conditions and in sufficient quantities to produce an explosive or ignitable mixture; mechanical failure or abnormal operation of machinery might cause such an ignitable mixture to be produced; Group E combustible dusts are in quantities sufficient to be hazardous; dust accumulations are greater than 1/8 inch (3 mm) thick under normal conditions

- ***Class II, Division 2*** — Combustible dust is produced due to an abnormal operation and is present in quantities to produce explosive or ignitable mixtures; dust accumulations have occurred but are normally insufficient to interfere with normal operation of electrical equipment; accumulation of combustible dust on or in the vicinity of electrical equipment could lead to its ignition by abnormal operation or failure of the electrical equipment.

Dust Controls

Facilities that manufacture or process materials that can cause dust hazards must employ some type of dust control, whether passive or active. Passive means of dust control include providing enclosures for conveyor belts or reducing the speeds at which materials are moved through the facility on conveyor belts or through chutes.

Active dust controls, known as air material separators, are usually mechanical dust-collection systems capable of collecting 99.5 percent of the dust and storing it in bins outside the facility.

- NFPA® 68, *Standard on Explosion Protection by Deflagration Venting*
- NFPA® 69, *Standard on Explosion Prevention Systems*

Preventive measures taken for dust collection and reduction must be constantly monitored, and cleaning programs must be performed daily or as needed:

- Dust continually picked up by manual sweeping, vacuuming, or other means and placed in appropriate storage bins or containers **(Figure 9.69)**.
- Metal of all types collected and separated so that no metal chips are introduced into the material-handling or processing equipment.
- Electrical equipment in the plant needs to be compliant with NFPA 70®, National Electrical Code® and listed for this environment and use. Motors in particular may need to be nonsparking and intrinsically safe **(Figure 9.70)**.
- All equipment grounded to minimize static electricity.
- Smoking restricted to designated areas.
- Fire protection equipment provided throughout all buildings as required by locally adopted fire and building codes and the referenced NFPA® standards.

NOTE: Large facilities may have their own dust control programs that the inspector should review.

Fire Protection

Fire protection in all process industries should be in accordance with locally adopted codes. The building should have fire extinguishers in accordance with NFPA® 10, *Standard for Portable Fire Extinguishers*. Each facility must have a written emergency action plan that includes, but is not limited to, the following information:

- Means of notification for occupants in the event of fire and explosion
- Preplanned evacuation assembly area
- Person designated to notify emergency responders including the fire department

Figure 9.69 Dust collectors in a woodworking facility. *Courtesy of Rich Mahaney.*

- Facility layout drawings showing egress routes, hazardous chemical locations, and fire protection equipment
- Location of safety data sheets (SDSs) or material safety data sheets (MSDSs) for hazardous chemicals
- Emergency telephone number(s)
- Emergency response duties for occupants

The keys to preventing fires in materials-handling facilities are controlling dust and ignition sources **(Figure 9.71)**. In a woodworking facility, conveyors and pneumatic systems remove fine dust. The conveyors are magnetized to remove any trapped metal pieces that could spark and become an ignition source. Pneumatic systems, such as cyclones for dust collection, must be carefully designed so that the air being returned to the plant does not contain any dust that may lead to a combustible mixture with air. In the sections that follow, three of the most common facilities that fire inspectors will encounter are described: grain facilities, woodworking and processing facilities, machine shops, and manufacturing facilities.

Figure 9.70 Electrical equipment needs to be intrinsically safe to reduce fire and explosion hazards. *Courtesy of Rich Mahaney.*

Figure 9.71 Dusts must not be allowed to accumulate near power sources. *Courtesy of Rich Mahaney.*

Grain Facilities

It is important that dust-collecting systems be located outside the grain elevator **(Figures 9.72 a and b, p. 370)**. Buildings in grain-handling operations should be built in accordance with the building code used in the jurisdiction as well as NFPA® 61, *Standard for the Prevention of Fires and Dust Explosions in Agricultural and Food Processing Facilities.* Life safety issues for a grain-handling facility should comply with the appropriate locally adopted fire and building codes.

Figure 9.72 Dust collectors outside the facility. *Courtesy of Rich Mahaney.*

A method to prevent the escape of dust into surrounding areas must be provided at:

- Belt loaders
- Belt discharge or transfer points
- Trippers
- Turnheads
- Distributors
- Unfiltered vents

Woodworking and Processing Facilities

Although wood dust presents an explosion hazard similar to the other materials, it also presents a significant fire hazard. The most common place for fires to occur in woodworking operations is at the dust hogger, a machine that turns scrap wood into splinters or chips. Small pieces of metal, called tramp, wear off the machine during the milling operation and accumulate in the hog trap. The tramp can become hot enough to ignite wood on contact. This fire can then spread into other portions of the machinery, including the dust collection system. These fires are often hard to control because they burrow through piles of sawdust. The fires also may be located within an elaborate system of ducts, storage bins, and other associated equipment.

Machine Shops and Manufacturing Facilities

Metal dusts such as those composed of various ferrous and nonferrous metals, such as aluminum and magnesium particles, are commonly produced in machine shops and manufacturing facilities **(Figure 9.73)**. Fine metal dust has an enormous explosive potential. An inspector visiting one of these facilities must address several issues to confirm the safety of those handling this material.

Figure 9.73 Processes that produce metal shavings and metal dusts must be carefully controlled to minimize the presence of spark-producing hazards.

The inspector should review the company's policies on smoking and open flames. No smoking, open flames, electric- or gas-cutting or welding equipment, or spark-producing operations should be allowed where metal dusts exist. Employers need to enforce strict rules that employees may not carry lighters, matches, and smoking materials into processing or handling areas.

Employees working in a metal dust processing or handling area should wear flame resistant personal protective clothing. It is imperative that workers be trained in emergency procedures appropriate for the processes being conducted. Workers need to be familiar with the emergency PPE they must use in the event of a fire or explosion.

Inspection Procedures

An inspector should confirm that all of the following criteria are met when inspecting facilities:

- Warning signs must be posted for areas containing inert fire protection systems as well as explosion protection systems.
- Spark-producing portable power tools and propellant-actuated tools should not be used where combustible dust is present. When it is necessary to use spark-producing tools, all dust-producing machinery in the area must be turned off.
- All equipment, floors, and walls must be carefully cleaned, and all dust accumulations in the area removed.
- Machinery that causes vibrations that could dislodge dust in the area of work should also be turned off until the work is completed. An inspection must be conducted at the end of the job to verify that any spark-producing tools or debris that could enter equipment are removed from the premises.

- All work areas must be kept clean and uncluttered.
- Packaging, such as sacks, boxes, uninstalled machinery or parts, or other supplies, are stored away from areas where the only other combustible material is the agricultural commodity being stored.
- Miscellaneous storage must be placed so it does not impede facility housekeeping or fire suppression operations.

Torch-Applied Roofing Materials

Torch-applied roofing materials pose a serious fire hazard to roofing contractors and building owners. Sometimes the hazards are obvious, such as torching to a combustible deck or near flammable liquids. Other concerns are less obvious, such as torching around drains or penetrations where flames can be drawn into a building. Roofing contractors must instruct employees in the following:

- Never torch directly to combustible decks or materials.
- Never torch to areas that cannot be seen fully.
- Do not use torches near vents or air intakes.
- Never use a torch to heat a propane tank that begins to frost on the outside.
- Have appropriate fire extinguishers within easy reach at all times whenever working with torch-applied roofing materials. Fire-watch inspections must be conducted for at least two hours after the work has been completed and the last torch has been turned off.

Asphalt and Tar Kettles

Asphalt and tar kettles are typically trailer-mounted devices used to heat and dispense asphalt or tar for use on roads and roofs **(Figure 9.74)**. The asphalt or tar carried in a trailer-mounted tank or kettle is heated using an LPG burner or similar heating device. Because this equipment can overheat a flammable or combustible substance, kettles have the potential to cause serious fires. If the hot contents spill, they can ignite other fuels and spread fire rapidly as a mass of flaming liquid.

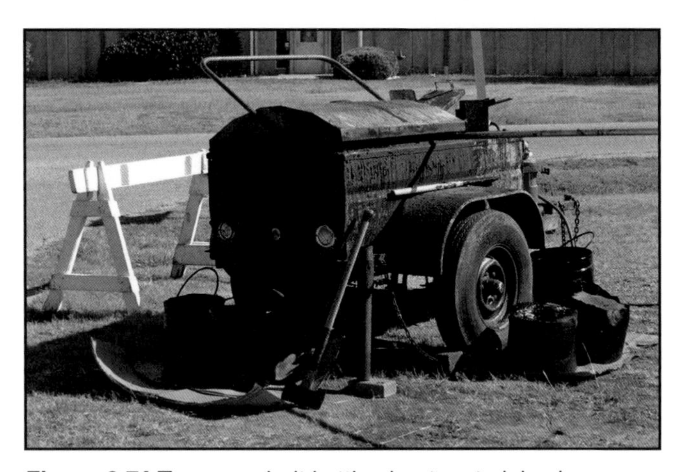

Figure 9.74 Tar or asphalt kettles heat materials above fire point and therefore present a serious fire hazard. Inspectors must closely observe how these kettles are used at work sites.

The adopted fire code will provide requirements for the safe use of asphalt and tar kettles. An inspector must be aware of fire protection and safety practices and procedures associated with these kettles so that the severity of these hazards can be reduced. The kettles should have limit controls, tight-fitting lids, and other safety devices to prevent fire or spills.

Kettles must never be operated inside a building or on its roof. The area in which the kettle is operating should be identified by traffic cones, barriers, and other suitable means as described in the adopted code.

At least one employee who is knowledgeable about the operation being conducted and about any potential hazards should constantly attend the kettle. The employee must be within a reasonable distance of the kettle (this distance varies from code to code) and have the kettle within sight. A portable fire extinguisher, rated at 40-B:C must be located within 25 feet (7.5 m) of the kettle. If the kettle is being used for roofing, an additional 3-A:40-B:C extinguisher must be available.

LPG cylinders used to supply heating elements on the trailer must be secured to prevent turnover. Regulators are required on all cylinders. LPG containers for roofing kettles must not be used inside any building. All kettles must have an approved, functional, and visible temperature gauge that indicates the temperature of the material being heated. They must also have a lid over the product tank that should be closed in the event of a fire. At no point should water be used to extinguish these types of fires.

Distilleries

The primary hazards in craft distilling are fire and explosion. Fire can occur when vapors from flammable organic compounds such as ethanol are released from leaks in tanks, casks, and equipment, such as transfer pumps, pipes, and flexible hoses **(Figure 9.75)**. Dust from processing grain and combustion from wood floors, casks, and racks can also cause fires or explosions. A vapor explosion can occur if enough vapors are released in an enclosed space with ignition sources present. Ignition sources include:

- Open flames
- Torch cutting and welding
- Sparks (static, electrical, and mechanical)
- Hot surfaces
- Heat from friction
- Radiant heat
- Combustible dust

Figure 9.75 Fire and explosions can occur in distilleries due to flammable vapors. *Courtesy of Rich Mahaney.*

The production of beer and spirits produces solutions of ethanol, which is a highly flammable liquid. Raw materials for fermentation and mashing processes involve the handling, storage, and milling of wheat and barley, which generate flammable dust. Grain roasting and drying require huge quantities of fuel, which is usually natural gas. Hence, all types of flammable materials (vapor, dust, and gas) necessary for an explosion are present in beer and spirits manufacturing facilities.

The inspector will need to be familiar with locally adopted codes. Hazardous materials requirements in fire codes may not apply to distilleries, but the hazardous materials requirements in the building codes may apply **(Figure 9.76, p. 374)**. An occupancy that exceeds the maximum allowable quantities (MAQs) for storage may become a hazardous occupancy that will need to meet fire protection, sprinkler, and separation requirements.

NOTE: See Chapter 10, Hazardous Materials, for more information on Maximum Allowable Quantities.

The best way to prevent a distillery fire is to control flammable vapors and ignition sources.

- Note venting processes so that flammable vapors inside the building do not accumulate.

Figure 9.76 The hazardous materials portions of building codes may apply in addressing hazards caused by distillery operations. *Courtesy of Rich Mahaney.*

- Make sure the electrical system in rooms where distilling or blending flammable liquids is done conforms to the National Electrical Code® Class I Division 1 Hazardous Location for Group D Flammable Liquids.
- Check for proper ground and bonding technique when pouring ethanol from the storage container to the still container and when decanting large amounts of finished product or byproduct.
- Note that heaters and natural gas appliances that use pilot lights at least 10 feet away from the pouring and distilling areas.
- Verify that fire sprinkler systems meet the fire jurisdiction's requirements for extinguishing an ethanol distillery fire.

Source: Distillery information adapted from Oregon OSHA.

Chapter Summary

Inspectors are expected to immediately recognize common fire hazards in their communities. Although there are numerous individual hazards that can cause a fire or contribute to a higher risk of a fire event, categorizing some hazards into general groups can help an inspector learn what to observe and note. Although a single community may not have all or even most of the categories included in this chapter, an inspector must have a familiarity with each.

The most effective approach for an inspector is to research the types of occupancies in the community, the types of processes that occur there, and the building and fire code requirements for each facility before performing inspections. In addition, inspectors must be familiar with the types of permits that are locally required before facilities can engage in hazardous processes. These permits determine the fire safety requirements that must be present at the site where the process occurs.

Review Questions

1. What are some examples of unsafe behaviors?
2. Name some of the unsafe behaviors inspectors may encounter when dealing with flammable and combustible liquids.
3. Name some of the actions that must be taken to control hazards when handling compressed/liquefied gas cylinders.
4. What are some of the signs that could indicate electrical hazards are present?
5. What kinds of information should an inspector be prepared to verify when onsite at a storage facility or retail outlet?
6. What are the two main hazards associated with recycling plants?

1. Identify fire and life safety hazards presented by heating, ventilation, and air conditioning systems.
2. List some of the conditions inspectors must look for when inspecting cooking equipment like ventilation hoods.
3. Name some of the special considerations inspectors need to be aware of when it comes to industrial furnaces and ovens.
4. What conditions should an inspector be alert to when inspecting facilities that use industrial trucks?
5. Name some of the provisions required for tents.
6. What are some of the operations that can create hazardous atmospheres?
7. Name the hot work fire safety issues inspectors should primarily be concerned with.
8. Name some of the controls model fire codes require for flammable finishing processes.
9. What are the three types of flammable finishing operations for which model fire codes specify requirements?
10. How can hazards around quenching processes be reduced?
11. What are some of the items an inspector must verify when inspecting dry cleaning operations?
12. What are the five conditions that must be present for a dust explosion to occur?
13. What are some of the conditions an inspector should be aware of in a distillery?

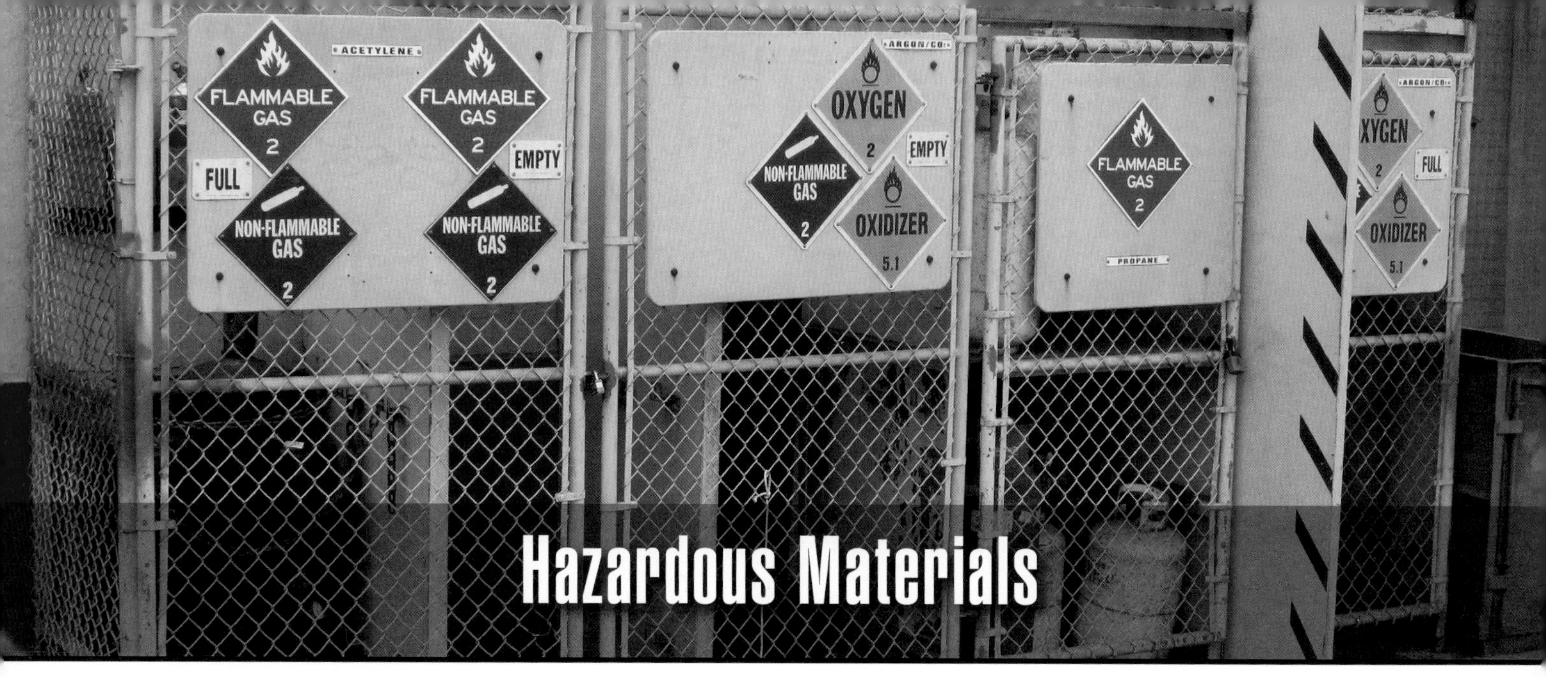

Photo courtesy of Rich Mahaney.

Chapter Contents

Case History 379

Application of Hazardous Materials Regulations 379

- Product Containment 381
- Pressure Relief 381
- Fire Protection 382
- Exemptions 382

Classification of Hazardous Materials ... 385

- Physical Hazard Materials 386
- Health Hazard Materials 396

Identification of Hazardous Materials 401

- Safety Data Sheets 402
- Transportation Placards, Labels, and Markings 402
- Other Markings 411
- Resource Guidebooks 418
- Canadian Dangerous Goods System 421
- Mexican Hazard Communication System 427
- Piping Identification 430
- Cylinder Markings 430

Permissible Amount of Hazardous Materials in a Building 430

- Maximum Allowable Quantity (MAQ) per Control Area 431
- Control Areas 434

Nonbulk and Bulk Packaging 435

- Containers 435
- Bulk Packaging 438

Testing, Maintenance, and Operations... 449

Process Control 449

Unauthorized Discharge 450

Hazardous Materials Piping, Valves, and Fittings 451

Hazardous/High-Hazard Occupancies 452

- Categories 452
- Explosives 453
- Requirements 454

Engineering Controls for Hazardous Materials 455

- Spill Control and Secondary Containment 456
- Mechanical Ventilation System 458
- Automatic Sprinkler Protection 458

Chapter Summary 459

Review Questions 460

Key Terms

Certificate of Occupancy........................385
Cryogen 388
Deflagration 386
Detonation 436
Intermediate Bulk Container (IBC) 437
Maximum Allowable Quantity
Per Control 431
Mixture 399
Pyrophoric..393
Reactivity 382
Threshold Limit Value (TLV)...................397

NFPA® Job Performance Requirements

This chapter provides information that addresses the following job performance requirements of NFPA® 1031, *Standard for Professional Qualifications for Fire Inspector and Plan Examiner* (2014).

Fire Inspector I	Fire Inspector II
4.3.4	5.3.3
	5.4.6

Hazardous Materials

Learning Objectives

After reading this chapter, students will be able to:

Inspector I

1. Explain the application of hazardous materials regulations. (4.3.8, 4.3.13)
2. Identify some of the applicable codes and standards that apply to hazardous materials. (4.3.8, 4.3.13)
3. Explain the classification system used for hazardous materials. (4.3.12, 4.3.13)
4. Describe the classification and properties of physical hazard materials. (4.3.13)
5. Explain the classification of health hazard materials. (4.3.13)
6. Describe the code requirements for the marking of hazardous materials for identification by emergency responders. (4.3.13)
7. Describe code considerations for determining the permissible amount of hazardous materials within a building. (4.3.13)
8. Explain the requirements for storage and use of nonbulk and bulk packaging. (4.3.13)
9. Describe the code requirements for testing, maintenance and operation of equipment, containers and tanks. (4.3.8, 4.3.13)Inspector II

Inspector II

1. Describe an inspector's role in process controls. (5.3.6)
2. Identify an inspector's responsibility after an unauthorized discharge. (5.3.6, 5.3.9)
3. Describe requirements for piping, valves, and fittings that convey hazardous materials (5.3.9)
4. Recognize the classification system and requirements for hazardous and high-hazard occupancies. (5.3.6, 5.3.9)
5. Describe engineering controls required for hazardous materials. (5.3.9)

Chapter 10
Hazardous Materials

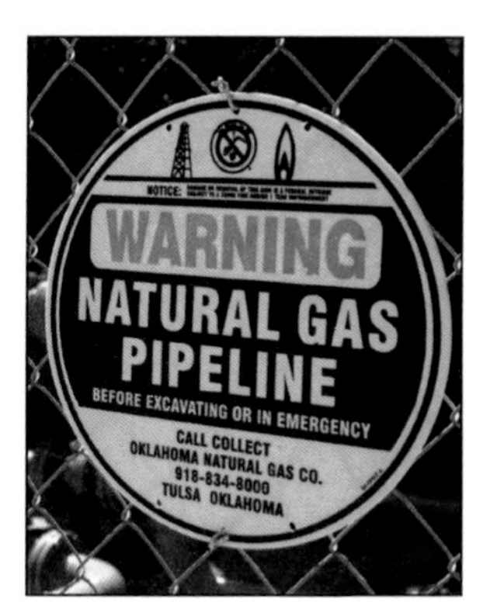

Case History

On April 9, 1998, 20 firefighters from a volunteer fire department responded to a fire in an 18,000-gallon propane tank located at a turkey farm in Iowa. Because the burning tank was venting, firefighters decided to allow the tank to burn itself out and protect exposed buildings. Some of the firefighters positioned themselves between the burning propane tank to water down nearby turkey sheds as the remaining firefighters performed other tasks.

About eight minutes after the firefighters arrived on scene, the tank suffered a BLEVE, exploding into four parts. One piece of the exploding tank struck two firefighters who were about 105 feet from the tank, killing them instantly. Six other firefighters and a deputy sheriff, who had arrived on scene just before the explosion, were also injured.

NIOSH investigators concluded that, to prevent similar incidents, fire departments should:

- Follow guidelines as outlined in published literature and guidebooks for controlling fire involving tanks containing propane.
- Adhere to emergency response procedures contained in 29 CFR 1910.120(q) - Emergency response to hazardous substance release procedures.
- Educate firefighters to the many dangers associated with a propane tank explosion, which is also known as a Boiling Liquid Expanding Vapor Explosion (BLEVE).

Model fire codes regulate hazardous materials to ensure their safe storage, handling, use and manufacturing. A release of a hazardous material may not cause a fire, but it may have a detonation, deflagration or release a toxic material that can incapacitate, injure, or kill people who are exposed to the compound. Because of these concerns, the codes and standards focus on several safety philosophies. This chapter explains application of hazardous materials regulations, classification of hazardous materials, identification, regulations, permissible amounts of hazardous materials allowed, types of packaging, unauthorized discharges, hazardous/high-hazard occupancies, and engineering controls for hazardous materials.

Application of Hazardous Materials Regulations

Hazardous materials present a challenging problem for fire inspectors because they commonly present multiple hazards. Physical hazard materials burn, accelerate burning, and either detonate or deflagrate. A release of a health hazard material can injure or incapacitate the public or emergency responders. An inspector examining the scope and purpose of the IFC or NFPA® 1 will find that over half of these codes are focused on hazardous materials or processes that use these materials. Examples include application of flammable finishes, fumigation, and motor vehicle fuel dispensing **(Figure 10.1)**. IFC Chapters 50 through 67 and NFPA® 400 focus on preventing and minimizing incidents involving the storage, use, handling, and dispensing of hazardous materials.

The regulations in model fire codes emphasize reporting the storage and use of hazardous materials above certain quantities. Model codes mandate the permit applicant to submit a properly prepared Hazardous Materials Management Plan or Hazardous Materials Inventory Statement to the fire code official under their authority to issue either a construction or operating permit. Businesses that fail to report the storage of hazardous materials are violating the jurisdiction's fire code. The burden of reporting rests with the business because they are bringing these unique hazards to the jurisdiction.

Hazardous waste materials can be chemical compounds that are mixtures of used industrial products. Changing chemicals from their useful functions to a waste can complicate their classification under the IFC or NFPA® 400, *Hazardous Materials Code*.

Hazardous materials responses generally require that responding personnel be specially trained and equipped to manage a chemical release. Such incidents may also require additional time to mitigate. The IFC and NFPA® 400 both require, as a condition of an operating permit, that a business designate a fire department liaison or emergency response coordinator. This individual is responsible for understanding the hazards of storing, using, and handling the materials. The liaison is responsible for explaining the site's emergency response procedures and has access to the facility's safety data sheets. This individual may also assist the fire department in planning an emergency response.

Figure 10.1 A fire involving several petroleum storage tanks is the type of incident the model codes and standards are intended to prevent. *Courtesy of Scott Stookey, International Code Council, Washington D.C.*

Product Containment

Confining a hazardous material in its container, tank, or vessel is of paramount importance. One of the bigger problems inspectors will find concerns the design and construction of storage tanks. Numerous incidents have occurred where the wrong construction materials or the wrong tank assembly methods have resulted in a hazardous materials release. In addition, businesses may try to reuse abandoned tanks to store hazardous materials. Once a design professional has evaluated used equipment, it can be used again safely. However, when the construction materials have not been compared to the hazards and chemistry of the stored materials, the results can include rapid and catastrophic container or tank failure. This concern extends to the piping system used to move the chemical through a facility or plant.

Inspectors need to be vigilant about small changes to plants and buildings where modifications occur to tanks, containers, or their piping systems. The inspector should ask how the stored material will affect the tank, piping, or the valves that contact the solid, liquid, or gas considered to be hazardous **(Figure 10.2)**.

Pressure Relief

Regardless of the method, pressure relief is an important element in any hazardous materials system that a fire inspector evaluates **(Figure 10.3)**. Hazardous materials stored in containers, cylinders, tanks, or pressure vessels are always assumed to be capable of being exposed to fire. An unwanted fire can pressurize a container and cause it to catastrophically fail and explode if it does not have a properly designed means of pressure relief. A pressure relief can be

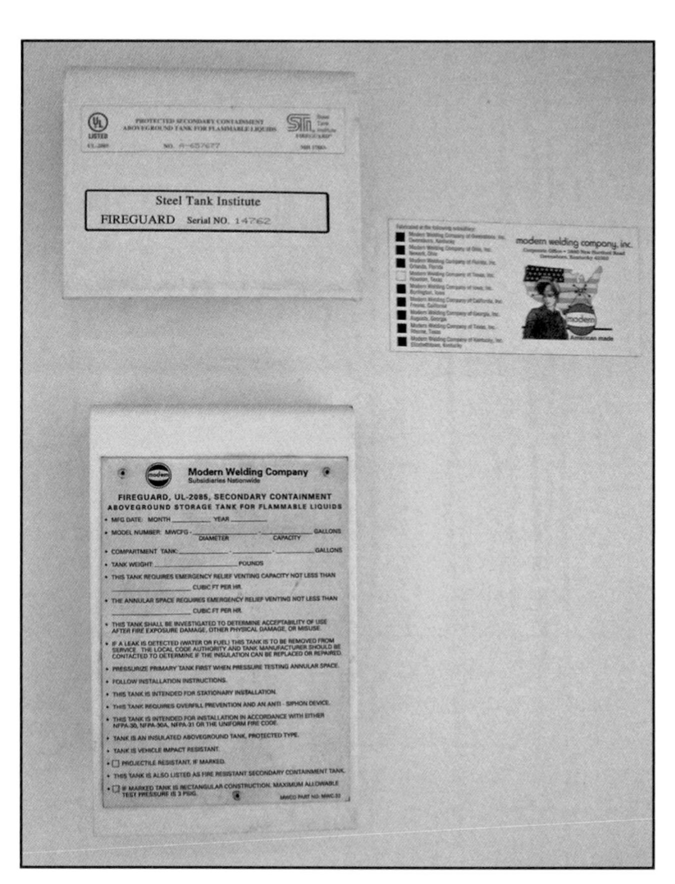

Figure 10.2 Inspectors must be vigilant when examining labels on tanks. *Courtesy of Scott Stookey, International Code Council, Washington D.C.*

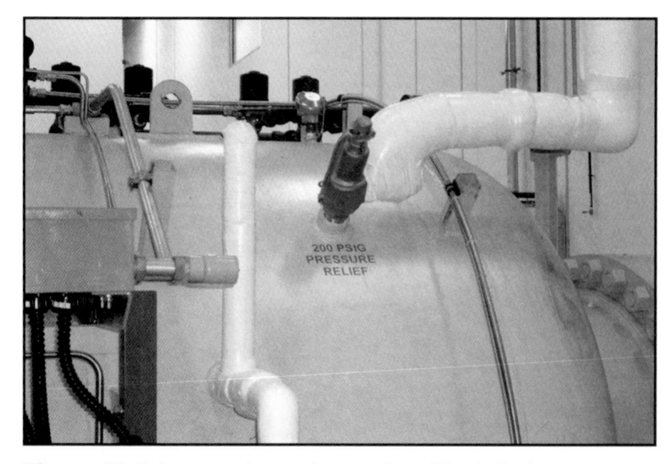

Figure 10.3 A properly engineered and installed means of pressure relief is required for many hazardous material systems. *Photograph courtesy of Scott Stookey, International Code Council, Washington, D.C.*

heat-activated or one that responds to an increase in pressure. It can also be a design where the container safely vents under controlled fire exposure.

Reactivity — Ability of two or more chemicals to react and release energy and the ease with which this reaction takes place.

Fire Protection

Many hazardous materials present flammable or **reactivity** hazards. NFPA® 400 and the IFC require certain levels of fire protection for many classes of hazardous materials. Protection can include locating the materials in fire-resistant enclosures such as gas cabinets or hazardous materials storage cabinets **(Figure 10.4)**. Based on the hazards of the stored material, fire-resistive construction may be required to separate rooms or areas where hazardous materials are stored or used from other areas of a building. In other cases, both model codes require that an automatic sprinkler system protect the building or area.

The goal of the codes is to protect the stored material from a fire and to limit the risk of its involvement. An inspector will need to rely on a particular NFPA® standard for a given class of hazardous materials to determine if the sprinkler discharge density and design area are correct for the stored materials (**Figure 10.5**).

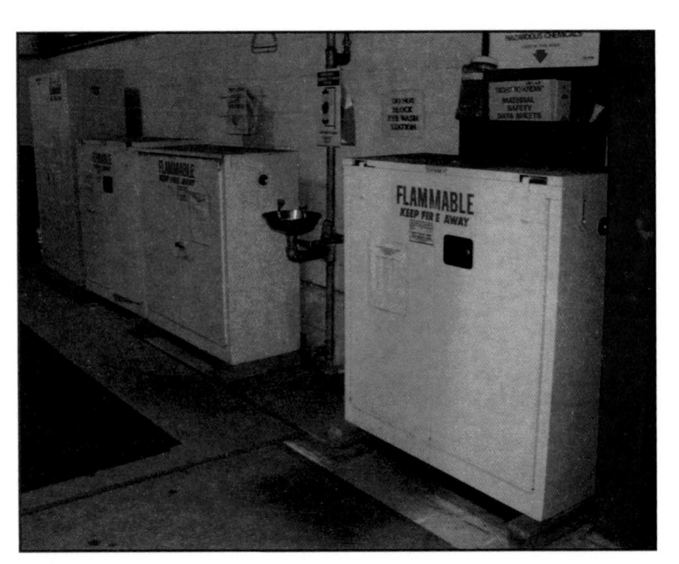

Figure 10.4 Loss of package integrity can result in the release of a hazardous material in storage. *Courtesy of Scott Stookey, International Code Council, Washington, D.C.*

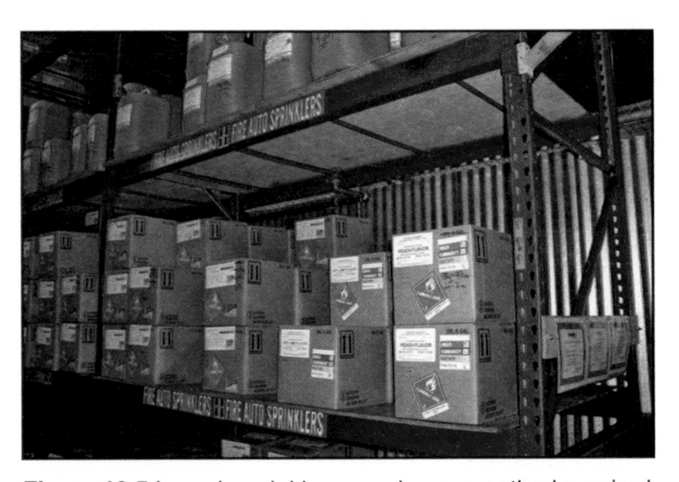

Figure 10.5 In-rack sprinklers may be one method required by the jurisdiction to provide fire protection for stored hazardous materials.

Exemptions

The hazardous materials provisions in the IFC and NFPA® 400 cover a variety of processes and uses. However, these codes do not cover all situations where hazardous materials are stored, used, or handled. In some cases, federal or provincial laws preempt their storage or use. In other cases, specific requirements in the model codes take precedence over the hazardous material requirements. It is important for inspectors to understand that certain activities or processes that involve hazardous materials may be exempt from regulation, including:

- Hazardous materials transportation
- Pesticides, fungicides, and rodenticides
- Certain building systems

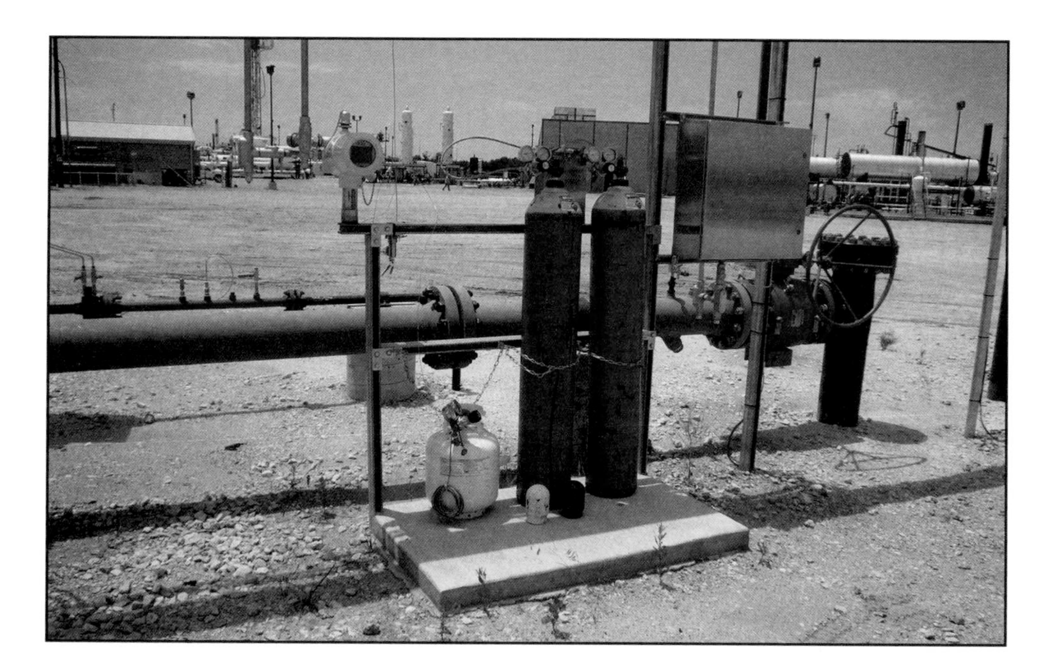

Figure 10.6 This natural gas pipeline is regulated by the U.S. Department of Transportation and is not regulated by the International Fire Code and NFPA® 400. *Courtesy of Scott Stookey, International Code Council, Washington, D.C.*

Hazardous materials transportation is outside the scope of the IFC and NFPA® 400. The Department of Transportation and its Pipeline and Hazardous Materials Safety Administration (PHMSA) regulate hazardous material transportation and its pipeline in the U.S. The IFC and NFPA® 400 regulate facilities that store, package, or use hazardous materials (**Figure 10.6**).

NOTE: See the **IFSTA Hazardous Materials for First Responder** manual for additional information.

Another exemption is the application of pesticides, fungicides, and rodenticides. The U.S. Environmental Protection Agency (EPA) under the Federal Insecticide, Fungicide, and Rodenticide Act (FIFRA), approves the use of these materials in the U.S. The exemption does not limit the inspector's authority to apply the fire code to the storage or manufacturing of these materials.

Certain building systems that store or use hazardous materials can be exempt from hazardous materials regulations. In the IFC, mechanical refrigeration, fuel oil storage, and stationary storage battery systems are generally exempt from the hazardous materials provisions **(Figure 10.7)**.

To accomplish the code's intent, fire inspectors must become familiar not only with the jurisdiction's adopted codes but also the standards applicable to hazardous materials storage, use, and handling. The IFC and NFPA® 400 adopt a number of standards by reference. These standards reflect the minimum requirements for the design and construction of containers, cylinders, tanks and piping systems. They also address proper marking

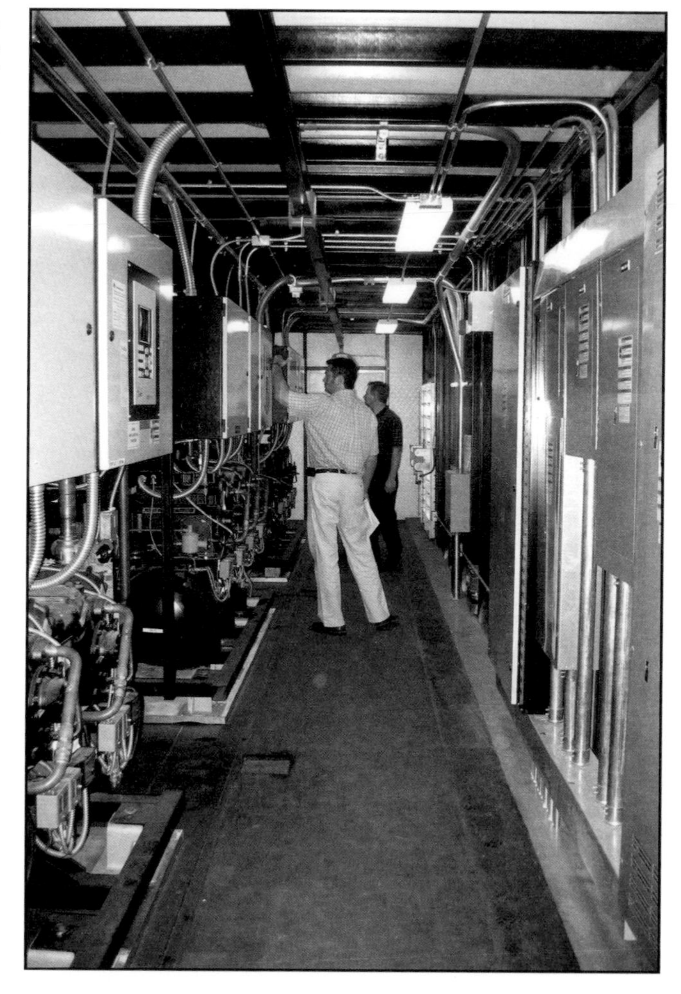

Figure 10.7 A refrigeration plant is exempt from the hazardous material requirements in a model fire code, but it is not exempt from other fire code requirements and the jurisdiction's mechanical code. *Courtesy of Scott Stookey, International Code Council, Washington, D.C.*

and identification of cylinders and hazardous materials piping. Other standards provide guidance to the fire inspector in ensuring the proper classification of gases. The following list summarizes the major standards that are commonly adopted and enforced for the storage, handling, and use of hazardous materials:

- **ICC**
  - — *International Fire Code* (IFC)
  - — *International Building Code* (IBC)
  - — *International Mechanical Code* (IMC)
- **NFPA**
  - — NFPA® 1, *Fire Code*
  - — NFPA® 30 and 30A, *Flammable & Combustible Liquid Code, Motor Vehicle Fuel Dispensing and Repair Garages Code*
  - — NFPA® 55, *Compressed Gases and Cryogenic Fluids Code*
  - — NFPA® 58, *Liquefied Petroleum Gas Code*
  - — NFPA® 91, *Standard for Exhaust Systems for Air Conveying of Vapors, Gases, Mists, and Noncombustible Particulate Solids*
  - — NFPA® 400, *Hazardous Materials Code*
- **API (American Petroleum Institute)**
  - — API 620 *Design and Construction of Large, Welded, Low-Pressure Storage Tanks*
  - — API 650 *Welded Steel Tanks for Oil Storage*
  - — API 653 *Tank Inspection, Repair, Alteration and Reconstruction*
- **ASME (American Society of Mechanical Engineers)**
  - — B31.3 *Process Piping*
  - — Boiler and Pressure Vessel Code, Section VIII, Division 1 and 2
- **ASTM** (American Society of Testing and Materials)
  - — Include flash point tests
- **CGA (Compressed Gas Association)**
  - — CGA C-7, *Guide to Preparation of Precautionary Labeling and Marking of Compressed Gas Containers*
  - — CGA S-1.1, S-1.2, S-1.3, *Pressure Relief Devices - Parts 1-3*
  - — CGA P-18, *Standard for Bulk Inert Gas Systems*
  - — CGA P-20, *Standard for Classification of Toxic Gas Mixtures*
  - — CGA P-23, *Standard for Categorizing Gas Mixtures Containing Flammable & Nonflammable Components*
- **Underwriters Laboratories (UL)**
  - — UL 58 *Steel Underground Storage Tanks for Flammable and Combustible Liquids*
  - — UL 142 *Steel Aboveground Storage Tanks for Flammable and Combustible Liquids*
  - — UL 1316, *Glass-Fiber-Reinforced Plastic USTs for Petroleum Products, Alcohols and Alcohol-Gasoline Mixtures*

— UL 2080 *Fire-Resistant Aboveground Storage Tanks*

— UL 2085 *Protected Aboveground Storage Tanks*

— UL 2245 *Vaulted Storage Tank Systems*

Classification of Hazardous Materials

Before a fire inspector can inspect hazardous materials, these materials must be assigned the proper classification using the criteria in the adopted fire code. Verifying proper chemical classification is similar to inspecting an existing building: generally, the inspector will review any property or permit information. One information source is the building's **Certificate of Occupancy** because it may indicate the occupancy classification, type of construction, building area, and whether it is protected by an automatic sprinkler system.

Certificate of Occupancy — Issued by a building official after all required electrical, gas, mechanical, plumbing, and fire protection systems have been inspected for compliance with the technical codes and other applicable laws and ordinances.

Proper hazard classification is the first step in regulating hazardous materials. Improper classification leads to the incorrect storage and use of materials. Safety Data Sheets (SDS) do not always provide enough information to properly classify hazardous materials under the International Fire Code (IFC) and NFPA® 400 classification systems (**Figure 10.8**). NFPA® 1 and the IFC give fire inspectors the authority to require a technical report and opinion prepared by an individual who is qualified to determine a chemical's hazard classification.

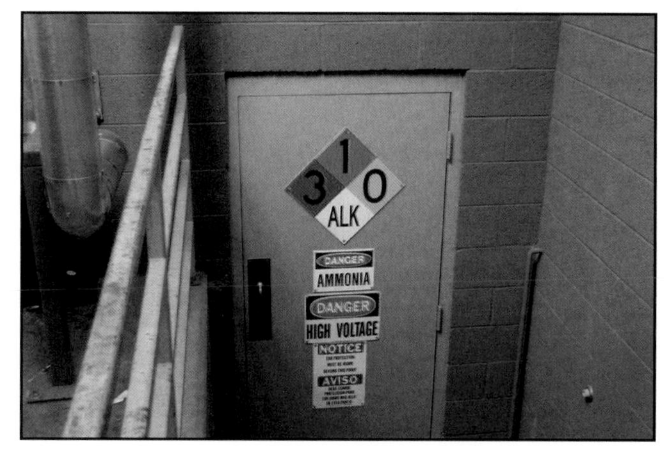

Figure 10.8 This machine room door clearly indicates the hazards inside. *Courtesy of Rich Mahaney.*

NOTE: More information about SDS is found in Chapter 9 and later in this chapter.

A single compound or hazardous material may have multiple hazards. For example, anhydrous chlorine, which is used to treat water and manufacture semiconductors, plastics, and many other compounds, is classified as an Oxidizing, Toxic, and Corrosive liquefied compressed gas by the IFC and NFPA® 55, *Compressed Gases and Cryogenic Fluids Code.*

When a material presents multiple hazards, fire codes require all of the hazards be addressed. For anhydrous chlorine, its storage and use would be required to comply with the IFC and NFPA® 55 requirements for a liquefied compressed gas that is classified as being an oxidizer, corrosive, and toxic.

Evaluating Hazards

In the context of fire codes, classification is an evaluation to determine the degree of hazard a material may present. Classification of hazardous materials can be challenging due to the numerous hazardous materials in the marketplace and in research. Inspectors should realize that the permit applicant is responsible for properly classifying the material.

NOTE: Inspectors are commonly asked to assist in classification because it may be faster and easier than trying to correct errors. They should exercise caution to avoid giving incomplete or inaccurate information.

Several resources available to assist inspectors in determining classifications are as follows:

- Hazardous Materials Expert Assistant software. The database contains about 8,000 hazardous materials and compounds and allows the user to search by chemical name, synonym, or its Chemical Abstract Service number.
- Appendix E in the IFC contains examples of hazardous materials classified using the code's criteria. Section E103 offers guidance on the steps to take to evaluate the hazards of a material.
- Wireless Information System For Emergency Responders (WISER), a free internet-based search engine developed by the National Library of Medicine. It is a compilation of a number of databases developed by several U.S. agencies.

Hazardous materials are broadly classified into two major categories: physical hazard materials and health hazard materials. The sections that follow describe both of these categories.

Detonation — (1) Supersonic thermal decomposition, which is accompanied by a shock wave in the decomposing material. (2) Explosion with an energy front that travels faster than the speed of sound. (3) High explosive that decomposes extremely rapidly, almost instantaneously.

Deflagration — (1) Chemical reaction producing vigorous heat and sparks or flame and moving through the material (as black or smokeless powder) at less than the speed of sound. A major difference among explosives is the speed of this reaction. (2) Intense burning, a characteristic of Class B explosives.

Physical Hazard Materials

Physical hazard materials are those materials that burn, accelerate burning, **detonation**, or **deflagration**. There are ten classes of physical hazard materials and within them a number of subcategories. Inspectors should understand that these subcategories will make a major difference in how the applicable code provisions are applied. The subcategories will also dictate the building's occupancy classification in the IBC and NFPA® 400.

Physical hazard materials include the following:

- Flammable and Combustible Liquids
- Compressed and Liquefied Compressed Gases
- Flammable Solids or Gases
- Organic Peroxides
- Oxidizers and Oxidizing Gases
- Pyrophorics
- Unstable (Reactive) Materials
- Water-Reactive Materials
- Cryogenic Fluids
- Explosives and Blasting Agents

Flammable and Combustible Liquids

Flammable and combustible liquids are the most common class of hazardous material that inspectors will encounter. A liquid is classified as either flammable or combustible when it contains carbon and hydrogen. Liquids are classified as either flammable or combustible based on their flash point and boiling point temperatures. The flash point is only one of a number of properties that must be evaluated and considered in assessing the overall flammability hazard of a liquid **(Table 10.1)**.

Table 10.1
Model Fire Code Classification of Flammable and Combustible Liquids

Flash Point Temperature (°F)	Boiling Point Temperature (°F)	Classification
Flammable Liquid		
Less than 73°F	Less than 100°F	Class I-A
Less than 73°F	Equal to or greater than 100°F	Class I-B
Equal to or greater than 73°F and less than 100°F	Not applicable	Class I-C
Combustible Liquid		
100 to 140°F	Not applicable	Class II
140 to 200°F		Class III-A
Greater than 200°F		Class III-B

Flammable and combustible liquids do not burn: the vapor they release can be ignited and burn. In comparison to ordinary combustibles, these liquids exhibit much higher heat release rates; as a result, fires involving these materials are far more difficult to control. When a flammable liquid container or other vessel leaks, the liquid will begin to evaporate, depending on the atmospheric temperature. Occupants should store flammable and combustible liquids such that their Lower Flammable Limit (LFL) is never exceeded.

NOTE: Refer to Chapter 3, Fire Behavior, for additional descriptions of flash points, upper flammable limits, and lower flammable limits.

Flammable and combustible liquids are widely available and have a variety of uses, including:

- Motor vehicle fuels
- Food preparation
- Lubricants
- Semiconductor fabrication
- Coatings
- Creating plastics

This wide availability is one reason why this class of hazardous materials is the most misused and has contributed to numerous injuries and deaths. Inspectors should be aware of the serious hazards associated with flammable and combustible liquids.

NOTE: Refer to Chapter 9 for additional information on flammable and combustible liquids.

Compressed and Liquefied Gases

Gases have numerous medical, industrial, and research applications. The model fire codes regulate their storage and use. Regulations address two major concerns:

- Physical and health hazards of the stored gas
- Design, construction, and protection of the compressed gas container or cylinder

Application of the model code provisions requires that a flammable compressed gas installation comply with the general provisions for compressed gas systems and the material-specific provisions for flammable gases. Gases are categorized into three groups depending on their physical state in containers under certain temperatures and pressures and their range of boiling points:

- Nonliquefied compressed gases
- Liquefied compressed gases
- Dissolved gases

Figure 10.9 Nested cylinders of flammable compressed gases. *Courtesy of Scott Stookey, International Code Council, Washington, D.C.*

Nonliquefied compressed gases. Compressed gases are those that do not liquefy at normal temperature and pressure (68°F at 14.7 PSI absolute [1 PSIG]) and under pressures as high as 10,000 PSIG (pounds per square inch gauge). Compressed gases can be liquid if they are cooled below their boiling point temperature, which converts them to cryogenic fluids. Examples of such gases include oxygen, helium, methane, and nitrogen.

Liquefied compressed gases. Liquefied compressed gases are those that become liquids at ordinary temperatures and pressures from 25 to 600 PSIG. Liquefied gases are elements or compounds that have boiling points relatively near atmospheric temperatures, ranging from approximately -130°F to 25-30°F (-90 to -1°C). Liquefied compressed gases would become solid at the low temperatures used for cryogenic fluids.

Rules for liquefied gases limit the maximum amount that can be put into a container to allow space for liquid expansion when ambient temperatures rise. Examples of liquefied compressed gases include anhydrous ammonia, propane, and carbon dioxide **(Figure 10.9)**.

Dissolved gases. A gas is dissolved when it is placed in a solution with another chemical to stabilize it. Acetylene is assigned to this category. Acetylene is supplied and stored dissolved in acetone, which renders it safe to transport and use with proper precautions.

Cryogenic Fluids

Cryogen — Gas that is converted into a liquid by being cooled below -150°F (-100°C).

The process of converting gases into liquid form through refrigeration is known as cryogenics. Cryogenic liquids, or **cryogens**, have a boiling point of -130°F (-90°C). Cryogenic liquids are also known as refrigerated liquids, especially while they are in transit. Common cryogenic liquids include:

- Nitrogen
- Oxygen

- Hydrogen
- Helium
- Argon
- Neon
- Krypton
- Xenon
- Liquefied natural gas (LNG)/methane
- Carbon monoxide

One advantage of using cryogenic liquids is the ability to modify a material's liquid-to-gas volume ratio. For example, 862 volumes of gaseous oxygen can be liquefied to a single volume by reducing the gas temperature below its boiling point. A cryogenic cylinder of liquid oxygen can hold 12 times more gas than a pressurized cylinder of oxygen **(Figure 10.10)**.

Besides the savings in storage space and weight they offer, cryogenic liquids are valuable simply because they are extremely cold. For example, liquid nitrogen is used to freeze liquids for emergency pipeline repairs, to harden gum-like materials such as plastics and cosmetics before they are ground, and for medical purposes.

The hazards associated with cryogenic liquids can be reduced to three categories:

- Inherent hazard of the particular gas, which may be intensified when it is in liquid form.
- High liquid-to-vapor ratio.
- Extremely low temperatures.

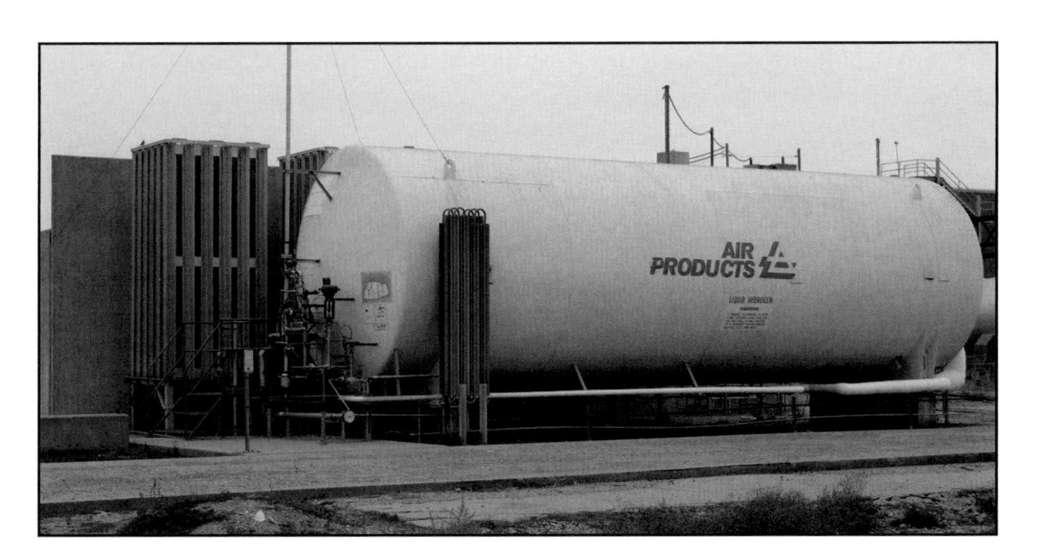

Figure 10.10 A liquid hydrogen tank. *Courtesy of Ron Moore, McKinney (TX) Fire Department.*

Cryogenic liquids and gases, such as hydrogen, methane, LNG (liquefied natural gas), and carbon monoxide, create a flammability hazard because they can burn or explode. Hydrogen is very easily ignited and particularly hazardous, forming flammable mixtures with air over a wide range of concentration (4 percent to 75 percent by volume).

If cryogenic liquids are trapped within a confining space such as a pipe or other container without adequate venting, the liquid will vaporize, expand, and cause a violent pressure explosion of the confining vessel. Liquid-to-vapor ratios can create extreme explosive conditions when the product is heated within the confining space. This situation is particularly true in piping.

A pressure-relief device should be installed on every length of pipe between two shutoff valves. All pipes carrying cryogenic liquids must be designed so that they slope up from the container to avoid the possibility of trapping fluids inside them. Pipes and associated fittings must be manufactured from:

- Stainless steel
- Aluminum
- Copper
- Monel® nickel-copper alloy

Because of their extremely low temperatures, all cryogenic liquids can inflict severe burns (similar to severe frostbite) upon contact with exposed skin. When small amounts are spilled on skin, cryogenic liquids tend to move across the flesh quickly. In large amounts, they cling to skin because of their low surface tension.

Inhaling cryogenic gases can severely damage the respiratory tract. Vapors can damage the eyes by causing the water in the eyes to freeze. In all situations where exposure to cryogenic liquids is possible, an inspector must wear appropriate personal protective equipment (PPE).

When cryogenic liquids are released into the atmosphere, they will refrigerate any moisture in the air and create a visible fog. The fog normally extends over the entire area that contains cryogenic vapors.

Flammable Solids or Gases

Flammable solids have an ignition temperature of less than 212°F (100°C). According to the IFC, flammable solids are any solids other than explosives that are capable of causing a fire through the following actions:

- Friction
- Absorption of moisture
- Spontaneous chemical reaction
- Retained heat

A variety of materials are loosely categorized as flammable solids:

- Metal powders
- Readily combustible solids that ignite by friction
- Self-reactive materials that undergo a strong exothermic decomposition
- Explosives that are wetted to suppress their explosive properties

Another type of flammable solid includes spontaneously combustible materials. Inspectors should be familiar with the following two primary types of spontaneously combustible materials:

- **Pyrophoric materials** — Liquids, solids, or gaseous materials that, even in small quantities and without external ignition sources, can ignite within 5 minutes after coming into contact with air.
- **Self-heating materials** — Materials that have the potential to self-heat when they come in contact with air.

NOTE: Pyrophorics are described in greater detail later in this chapter.

Another classification includes the dangerous-when-wet materials. These materials, when they come in contact with water, are likely to become spontaneously flammable or produce flammable or toxic gas. Magnesium phosphide is an example of a dangerous-when-wet material. Materials such as these must not be stored in locations or in such a manner where they can become combustible or dangerous.

Flammable solids that are metals require the placement of a Class D fire extinguisher close to the hazard. An inspector should refer to NFPA® 10, *Standard for Portable Fire Extinguishers,* for the requirements on extinguisher sizes and travel distances. Additionally, the inspector should consult the SDS to determine the appropriate fire extinguishing agent according to the manufacturer's specifications.

Flammable gases can be compressed, liquefied compressed, or dissolved. Examples include:

- Ethane
- Hydrogen
- Isobutene
- Propane

NFPA® 55 and the IFC classify a gas as flammable when it meets either of the following conditions:

- It is ignitable at atmospheric pressure when in a mixture of 13 percent or less by volume in air.
- It has a flammable range of at least 12 percent in air at atmospheric pressure, regardless of its lower limit.

Organic Peroxides

Organic peroxides only exist as solids or liquids. They release energy in the form of heat and are used to introduce energy into chemical reactions so more useful compounds can be created.

Organic peroxides present fire and reactivity hazards. The degree of the hazard depends on their classification. As they react, all organic peroxides exhibit one or all of the following hazards:

- Sensitive to heat
- Releases heat upon decomposition
- Forms free radicals upon decomposition
- Introduction of a contaminant can initiate an uncontrollable reaction
- Decomposition products can be flammable or toxic

Table 10.2
Hazard Classification of Organic Peroxides

Hazard Classification	Decomposition Rate	Reactivity Hazard	Burning Rate
Explosive	Rapid	Detonation	Not Applicable
I	Rapid	Deflagration	Not Applicable
II	Moderate	Severe	Very Rapid
III	Not Applicable	Moderate	Rapid
IV	Not Applicable	NA	Minimal

All organic peroxides have a limited shelf life, usually one year or less. The shelf life can be reduced if recommended temperature limits are not maintained. Once an organic peroxide reaches its shelf life, the material should be disposed of in accordance with the manufacturer's recommendations. An expired organic peroxide can become unstable.

Model fire codes assign hazard classes for organic peroxides. The lower the Roman numeral classification, the greater the decomposition, reactivity, and burning rate hazards the organic peroxide presents **(Table 10.2)**.

Oxidizers and Oxidizing Gases

Solid and liquid oxidizers and oxidizing gases do not burn: they accelerate the rate of burning. An oxidizer yields oxygen or other chemical compounds that promote or initiate combustion.

If an oxidizer is heated or contaminated, it can initiate a self-sustained reaction and can result in an explosion or violent decomposition. Oxidizers can also be:

- Corrosive
- Unstable (reactive)
- Water reactive
- Toxic (in some cases)

Fire code-compliant storage and use of oxidizers is designed to separate them from incompatible hazardous materials **(Figure 10.11)**. Oxidizers should never have contact with any petroleum-based materials because the reaction will be immediate and violent.

Oxidizers are unique in their naming convention. When fire inspectors encounter a chemical that ends in the letters *ate* or *ite*, or if the beginning of a chemical begins with the letters *per*, the hazardous material is probably an oxidizer. Examples of oxidizers following this naming convention include:

- Hydrogen *per*oxide
- Calcium hypochlor*ite*
- Potassium *per*chlor*ate*

- Ammonium nitr*ate*
- *Per*chloric acid

Solid and liquid oxidizers are divided into four hazard categories based on their ability to accelerate burning, their rate of decomposition, and if their decomposition can ignite combustible (Class A) materials. The higher the Arabic numeral rating, the greater the hazard the solid or liquid oxidizer presents. **Table 10.3** summarizes the oxidizer classifications found in model fire codes:

Oxidizing gases can support and accelerate combustion more than air does. They can exist as compressed or liquefied gases and oxygen can be converted into an oxidizing cryogenic fluid. Cryogenic oxygen is commonly used for respiratory therapy and can be found in home health care liquid oxygen containers.

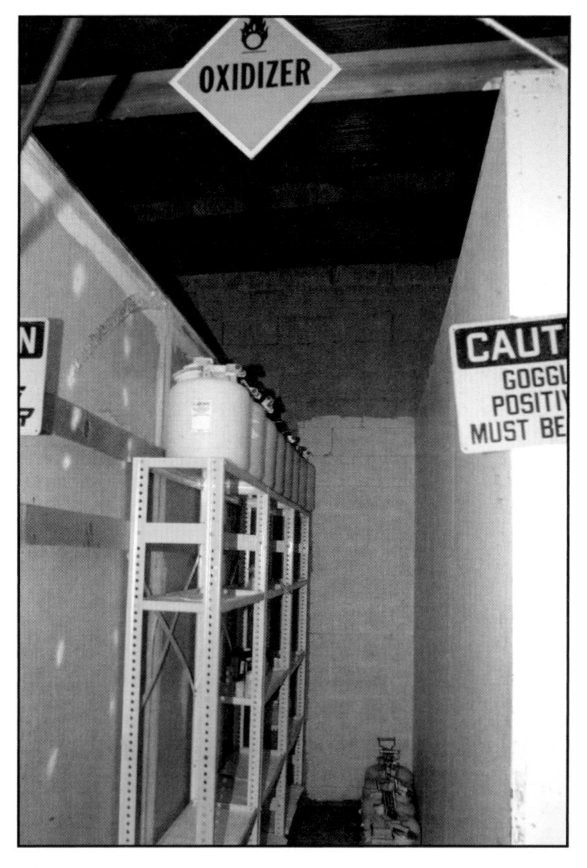

Figure 10.11 Noncombustible construction allows oxidizers to be stored inside this structure.

Pyrophorics

Pyrophorics are a category of solid, liquid, or gaseous hazardous materials that are a significant physical hazard. A material is classified as pyrophoric when it autoignites in air at temperatures of 130°F (55°C) or less. Pyrophorics do not require an ignition source – simply releasing the material in an atmosphere with air or oxygen causes autoignition. Because air contains moisture, pyrophorics ignite upon release in air. Some pyrophorics are also water reactive. Pyrophorics also commonly present other physical and health hazards, including classification by the IFC and NFPA® 400 as Toxic or Highly Toxic materials.

Pyrophoric — Material that ignites spontaneously when exposed to air. *Also known as* Air-Reactive Material.

While some pyrophoric gases and liquids exist, most pyrophoric materials are solid metals. Some isotopes of plutonium and uranium are pyrophoric. They are used in the manufacturing of semiconductors and in the synthesis of certain pharmaceuticals and specialty chemical compounds. Examples of pyrophoric materials include:

- Arsine
- Diborane
- Silane
- Sodium-Potassium (Na-K) alloys

Table 10.3
Hazard Classifications of Solid and Liquid Oxidizers

Hazard Classification	Burning Rate	Decomposition	Ignition of Combustibles
4	Explosive	Explosive	Spontaneous ignition occurs
3	Severely increased	Vigorous, self sustained	Yes
2	Moderate increase	None	May cause ignition
1	Slight increase	None	Does not ignite combustibles

Because of the inherent hazard of autoignition, the model codes require automatic sprinkler protection in buildings housing pyrophoric materials. NFPA® 1 and the IFC prohibit the storage or use of any pyrophoric material in an unsprinklered building.

In the case of alkali metals such as Na-K or pyrophoric radioactive isotopes, the application of water can grow and spread the fire. Because special agents are needed to control and extinguish fires involving pyrophoric materials, fire inspectors should seek technical assistance when evaluating fire extinguishing materials for them.

Requirements for pyrophoric gases are found in NFPA® 55 and NFPA® 318, *Standard for the Protection of Semiconductor Fabrication Facilities*. Model fire codes also reference Compressed Gas Association Standard G-13, *Storage and Handling of Silane and Silane Mixtures*.

Unstable (Reactive) Materials

Unstable (reactive) materials are solids, liquids, or gases. They can react adversely due to changes in temperature, pressure, or mechanical or thermal shock. Most oxidizers and organic peroxides are also classified as unstable (reactive) materials. The stabilizing chemical is called a diluent. If this stabilizing chemical is lost, the material begins to undergo a reaction that can cause a fire or uncontrollable reaction.

Unstable (reactive) materials can be found in the manufacturing of plastics, including expanded foams and urethane compounds. Blowing agents that create foam plastics are commonly classified as unstable (reactive). Examples of unstable (reactive) materials include:

- Hydrogen peroxide in concentrations greater than 52 percent by weight
- Ethylene oxide
- Unsaturated polyester resin

Unstable (reactive) materials are classified based on the type of uncontrolled reaction they are capable of producing and how the reaction can be initiated by changes in temperature or pressure. A thermal shock that occurs if the material is not cooled at a proper rate during a polymerization can produce a reaction. **Table 10.4** summarizes the hazard classifications for unstable (reactive) materials.

Water-Reactive Materials

A water-reactive material can react violently or explosively if it comes in contact with moisture or is improperly mixed with water. Model fire codes regulate water-reactive solids and liquids. There are no water-reactive gases. Many pyrophoric metals are also water reactive. Water-reactive materials are categorized into three different hazard classes which are summarized in **Table 10.5**. Examples of water-reactive materials include:

- Trimethylaluminum
- Calcium carbide
- Sulfuric acid
- Sodium hydroxide

Table 10.4
Model Fire Code Hazard Classifications for Unstable (Reactive) Materials

Hazard Classification	Type of Uncontrolled Reaction	Temperature and Pressure at the Time the Uncontrolled Reaction Occurs	Material Sensitive to Shock
4	Detonation or explosive decomposition or reaction	Normal	Yes; mechanical and thermal
3	Detonation or explosive decomposition or reaction	Heated under confinement or a strong initiating source is introduced	Yes; mechanical and thermal at a elevated temperature or pressure
2	Violent chemical change	Elevated	No
1	Normally stable but can become unstable	Elevated	No

Table 10.5
Model Fire Code Hazard Classifications for Water-Reactive Materials

Hazard Classification	Hazards of Reaction
3	Reacts explosively with water, without heat or confinement.
2	Reacts violently with water or has the ability to boil water. This class of material produces flammable, toxic or other hazardous gases, or evolves enough heat to autoignite or cause ignition of combustibles upon exposure to water or moisture.
1	Reacts with some release of energy, but not violently.

Explosives and Blasting Agents

Explosives and blasting agents are materials capable of producing a sudden, violent expansion of gases that may be accompanied by a shock or pressure wave. Explosives exist in solid or liquid states.

The explosives classification system in the IFC and NFPA® 495 is based on the UNDMGC and U.S. DOT regulations. All explosive materials that meet the UNDMGC criteria are designated as Hazard Class 1 materials. Explosives and blasting agents are further categorized based on the type of hazard they present.

NOTE: The Inspector II portion of this chapter contains a table of explosives categories.

Health Hazard Materials

Model codes recognize health hazard materials as those where a single brief exposure to the hazardous material can result in death, injury, or incapacitation. The toxicity limits in the model fire codes are lower than those in U.S. Department of Transportation (DOT) and Transport Canada (TC) regulations. For example, under the model fire codes, sulfur dioxide is neither a toxic nor highly toxic gas; under DOT and TC rules, it is a toxic inhalation hazard (TIH). This example illustrates why a DOT label or placard cannot always be used as a basis for classifying hazardous materials in fire codes.

Highly Toxic and Toxic Materials

Materials classified as toxic or highly toxic include substances capable of producing serious illness or death once they enter the bloodstream. For example, anesthetics and narcotics (ethers) reduce the muscular powers of vital bodily functions, resulting in loss of consciousness, suppressed breathing, and ultimately death. Irritants produce some type of local inflammation or irritate the skin, respiratory membranes, or eyes.

Toxic or highly toxic materials may enter the body through inhalation, ingestion, absorption (contact), or injection **(Figure 10.12)**. It is imperative that inspectors become aware of the possible toxic effects of a material and make sure that they are wearing appropriate PPE.

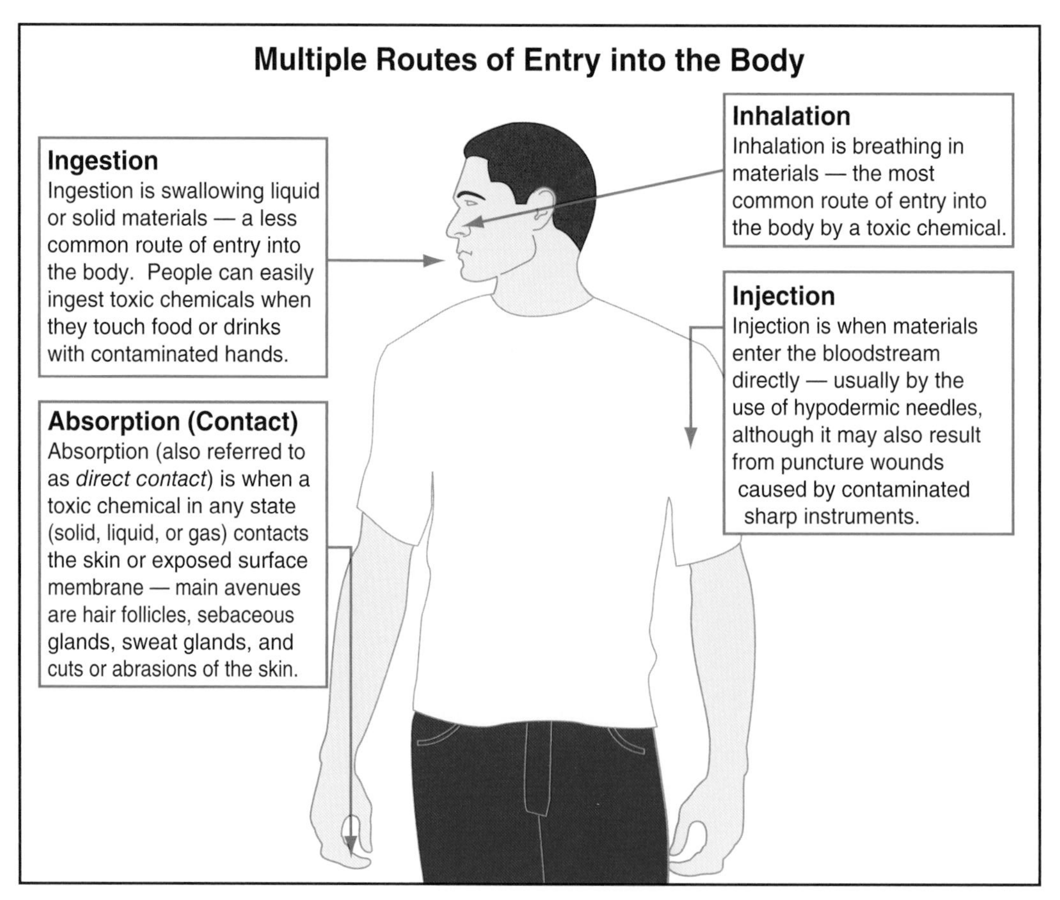

Figure 10.12 Highly toxic materials have four means of entering the body as shown in the illustration.

WARNING!

During inspections of facilities where toxic materials are being manufactured, stored, or shipped, an inspector must have appropriate personal protective equipment (PPE) and training in its use when there is any possibility of being exposed to these products.

Threshold Limit Value (TLV) — Concentration of a given material that may be tolerated for an 8-hour exposure during a regular workweek without ill effects.

Inspectors must also consider the **threshold limit value (TLV)** of potentially toxic materials. This value is the concentration of a given toxic material that the body generally can tolerate without ill effects. The American Conference of Governmental Industrial Hygienists (ACGIH) and the National Institute of Occupational Safety and Health (NIOSH) publish TLVs. These values are useful when an inspector needs to rate the maximum allowable concentration of a toxic material when evaluating proper ventilation.

Classification of health hazard materials as either Toxic or Highly Toxic is based on a material's acute toxicity LC_{50} or LD_{50} value. The values in this method of measurement represent the lowest exposure that will produce an adverse consequence to the body.

- *LD_{50}:* Refers to the ingested dose of a given substance that was lethal to 50 percent or more of the test population when they swallowed or ate the substance
- *LC_{50}:* Refers to the concentration in the air of a given substance that killed 50 percent or more of the test population when they inhaled or absorbed the vapors, fumes, or mists of the substance

Two sources for hazardous material toxicity data are:

- Registry of Toxic Effects of Chemical Substances (RTECS)
- *Sax's Dangerous Properties of Industrial Materials*

Examples of Highly Toxic materials include:

- Fluorine
- Hydrogen cyanide
- Methyl ethyl isocyanic acid
- White phosphorus

Examples of Toxic materials include:

- Chlorine
- Aniline
- Sulfuric acid
- Chromium trioxide

Corrosives

Corrosives exist in all three physical states. They are extensively used in water and wastewater treatment, metal finishing, and extraction and processing of minerals. They are also building block chemicals used in the manufacturing of a variety of products, such as:

- Food
- Pharmaceuticals
- Plastics
- Other chemicals like sulfur dioxide or ammonium nitrate

The classification of corrosive in the model fire codes is assigned when the material causes irreversible alteration or visible destruction to human skin. At higher concentrations, corrosives are a health hazard because contact on skin or eyes can cause burns. The classification criteria used in the fire codes is the same as the UNDMGC criteria. A material classified as corrosive by DOT or TC has the same classification in the fire codes (**Figure 10.13**).

WARNING!

Avoid contact with corrosives or corrosive spills. These materials rapidly dehydrate water from the skin while simultaneously reacting with proteins to inflict burns. Materials like sodium hydroxide attack and destroy proteins in the skin and are absorbed into the bloodstream, causing injury to the central nervous and circulatory systems.

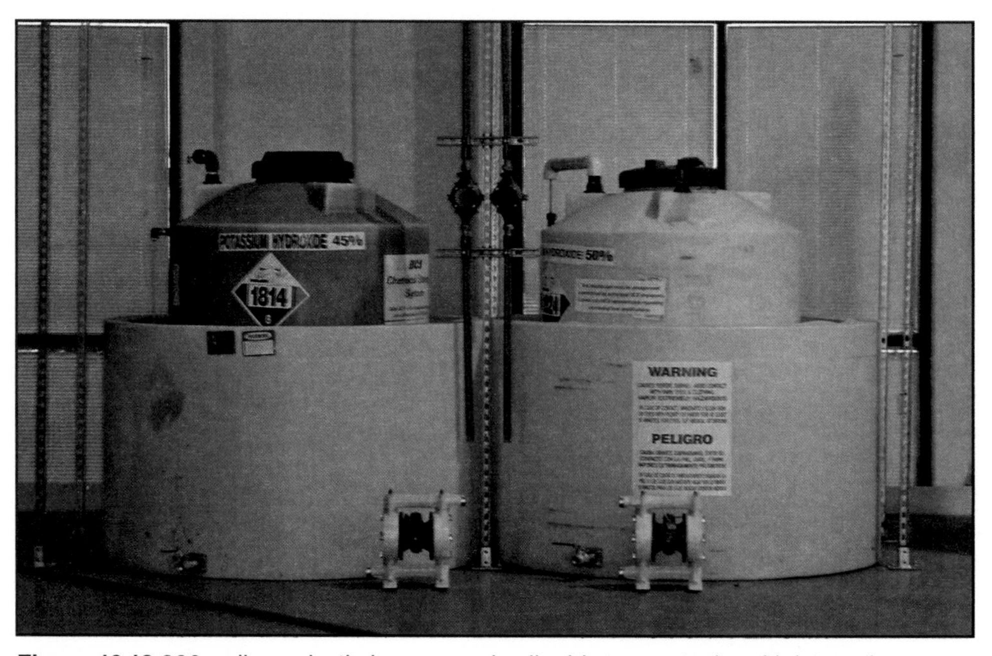

Figure 10.13 300-gallon polyethylene corrosive liquid storage tanks with integral secondary containment. *Courtesy of Scott Stookey, International Code Council, Washington, D.C.*

Mixtures

A **mixture** is a substance that contains two or more materials that are not chemically combined. Mixing chemicals can change their chemical and physical properties like their boiling or melting point. It can also change the hazard classification of a material. A classic example is aerosols. Aerosols are packaged by mixing a propellant with a base product, which may or may not be flammable.

Mixture — Substance containing two or more materials not chemically united.

Mixing two chemicals can create a new compound with its own unique hazards. For example, when metal is plated with chrome, the process starts by mixing the solid oxidizer chromium trioxide with water. This mixture creates a solution of chromic acid, which is not an oxidizer but a corrosive. An inspector evaluating such a process would need to understand the hazards of solid oxidizers and corrosive liquids.

Incompatible Materials

Model fire codes define incompatible materials as two materials that are mixed and cause a reaction that generates the following:

- Heat
- Fumes
- Gases
- Byproducts that are hazardous to life or property

Mixing hazardous materials that are not chemically compatible may produce irritating, noxious, or toxic gases, or generate uncontrolled pressures in containers or tanks that result in a fire or an explosion.

The model fire codes address incompatible hazardous materials in storage. Identifying incompatible storage requires knowing the hazard classification of the two materials, reviewing each product's SDS, and comparing the reactivity information to determine their chemical compatibility.

Examples of hazard classes always assumed to be incompatible include:

- Flammable or combustible liquids + oxidizers
- Oxidizers + organic peroxides
- Acids + bases

Table 10.6, p. 400, is a matrix that fire inspectors may use to identify incompatible hazardous materials storage. When storage arrangements are questionable, the matrix can help inspectors determine if they should require additional information to confirm compatibility of the stored hazardous materials.

Chemical Compatibility

A free computer application for determining if two hazardous materials are chemically compatible is available from the National Oceanic and Atmospheric Administration (NOAA). The application allows users to evaluate binary (two-chemical) mixtures. The database also evaluates a chemical's compatibility with air and water.

Table 10.6
Chemical Compatibility Matrix

	Corrosives	Explosives	Flammable Gas	F&C Liquid	Flammable Solid	Highly Toxic	Inert Gas	Organic Peroxide	Oxidizer	Oxidizing Gas	Pyrophoric	Unstable (Reactive)	Water Reactive
Corrosives	▿	⯃	▽	▽	▽	■	●	▽	▽	▽	▽	▽	▽
Explosives	⯃	▿	⯃	⯃	⯃	⯃	⯃	⯃	⯃	⯃	⯃	⯃	⯃
Flammable Gas	▽	⯃	●	▽	▽	■	●	▽	▽	▽	●	●	▽
F&C Liquid	▽	⯃	▽	●	▽	■	●	▽	▽	▽	▽	▽	▽
Flammable Solid	▽	⯃	▽	▽	●	■	●	▽	▽	▽	▿	▽	▽
Highly Toxic	■	⯃	■	■	■	●	●	■	■	■	■	■	■
Inert Gas	●	●	●	●	●	●	●	●	●	●	●	●	●
Organic Peroxide	▽	⯃	▽	▽	▽	■	●	▿	▽	▽	▽	▽	▽
Oxidizer	▿	⯃	▽	▽	▽	■	●	▿	▿	▿	▽	▽	▽
Oxidizing Gas	▿	⯃	▽	▽	▽	■	●	▿	▿	●	▽	▽	▽
Pyrophoric	▽	⯃	●	▽	●	■	●	▿	▿	▿	●	▽	▽
Unstable (Reactive)	▽	⯃	▽	▽	▽	■	●	▿	▿	▿	●	●	▽
Water Reactive	▽	⯃	▽	▽	▽	■	●	▿	▿	▿	▿	▿	●

⯃ Incompatible with all hazardous materials. Store the material in model code compliant explosive magazine.

▽ Chemically incompatible. Provide separation in accordance with the jurisdiction's adopted fire code.

▿ Compatibility between each stored hazardous material should be confirmed.

■ Store is a code compliant hazardous materials storage cabinet, gas cabinet, exhausted enclosure or area separated by 1 hour fire-resistance rated construction.

● Compatible hazardous materials.

Figure 10.14 Fuel oil tanks stored in the interior of a building require an enclosure like this one constructed in accordance with NFPA® 30 to separate the tank safely from the rest of the occupancy.

When incompatible storage is identified, the model fire codes offer several compliance options. Hazardous materials that are incompatible are considered separated if:

- They are separated by a distance of 20 feet (6 m) or more.
- One of the materials is stored in an approved storage cabinet, exhausted enclosure, or gas cabinet **(Figure 10.14)**.
- The materials are separated by a noncombustible line-of-sight barrier.

All of the codes exempt hazardous materials packages weighing less than 5 pounds (2.5 kg) or with a volume of 0.5 gallons (2 L) or less.

Identification of Hazardous Materials

The markings used to identify container contents and the information contained in hazardous materials resource guides enable emergency responders to act quickly and accurately. These marking can also assist the inspector in determining proper storage or handling procedures during inspections. For detailed information on hazardous materials, SDS sheets must also be present where hazardous materials are stored, used, or transported.

Inspectors must understand placards and material numbering systems that DOT, Transport Canada (TC), Mexican Secretariat for Communications and Transport (SCT), international governing bodies, and other agencies use. Inspectors must be familiar with the UN classification system and DOT placard and markings system and know how to use the *Emergency Response Guidebook (ERG)*, the *National Institute for Occupational Safety and Health (NIOSH) Pocket Guide to Chemical Hazards*, or the *Hazardous Materials Guide for First Responders* from the U.S. Fire Academy (USFA).

NOTE: For additional information on hazardous material identification and response, consult the IFSTA **Hazardous Materials for First Responders** and **Awareness Level Training for Hazardous Materials** manuals.

Safety Data Sheets

An SDS, formerly known as material safety data sheet (MSDS), is a detailed information bulletin prepared by the manufacturer or importer of a chemical to describe or give information about hazards **(Figure 10.15)**. Inspectors can acquire an SDS from the following:

- Material's manufacturer, supplier, and/or shipper
- Emergency response center such as CHEMTREC® (Chemical Transportation Emergency Center)
- Facility hazard communication plan
- Attached to shipping papers and/or containers

SDS sheets are being used worldwide and must include the sixteen sections shown in **Table 10.7.** Inspectors may still see some SDSs that were developed to American National Standards Institute (ANSI) standards, OSHA standards, or Canadian standards.

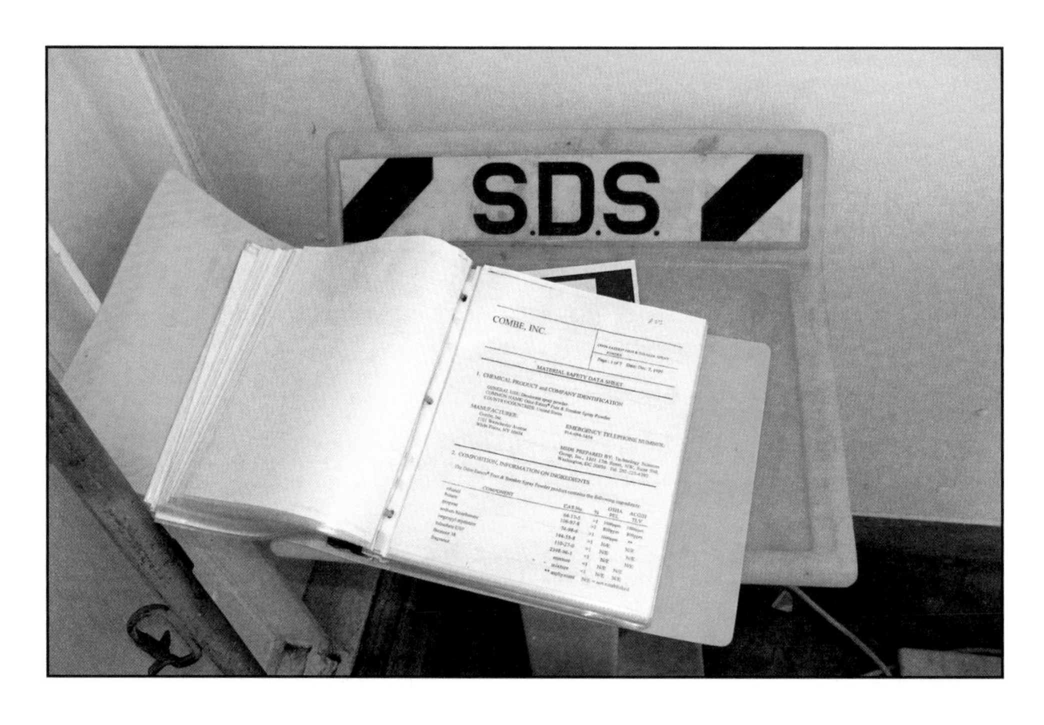

Figure 10.15 The Occupational Safety and Health Administration (OSHA) mandates that safety data sheets (SDS) be maintained on site for all products used or stored at a facility.

Transportation Placards, Labels, and Markings

As a result of the North American Free Trade Agreement (NAFTA) of 1994 and other agreements, trucks carrying products that include hazardous materials move freely among the U.S., Canada, and Mexico. To help regulate this movement of vehicles and materials, the NAFTA partners have all adopted the *UN Recommendations* regarding transportation placards, labels, and markings for identifying hazardous materials or dangerous goods. With few country-specific variations, the majority of the placards, labels, and markings used to identify these materials are very similar in each country. These recommendations enhance safety by improving the capabilities for accurate recognition of hazardous materials and eliminating duplicate or inconsistent marking and labeling systems.

Table 10.7
Information Disclosed on a U.S. Safety Data Sheet

Section	Catagory	Information
Section 1	Identification	Product identifier; manufacturer or distributor name, address, phone number; emergency phone number; recommended use; restrictions on use.
Section 2	Hazard(s) identification	All hazards regarding the chemical; required label elements.
Section 3	Composition/information	Information on chemical ingredients; on ingredients trade secret claims.
Section 4	First-aid measures	Important symptoms/effects, acute, delayed; required treatment.
Section 5	Fire-fighting measures	Suitable extinguishing techniques, equipment; chemical hazards from fire.
Section 6	Accidental release measures	Emergency procedures; protective equipment; proper methods of containment and cleanup.
Section 7	Handling and storage	Precautions for safe handling and storage, including incompatibilities.
Section 8	Exposure controls/personal protection	OSHA's Permissible Exposure Limits (PELs); ACGIH Threshold Limit Values (TLVs); and any other exposure limit used or recommended by the chemical manufacturer, importer, or employer preparing the SDS where available as well as appropriate engineering controls; personal protective equipment (PPE).
Section 9	Physical and chemical properties	Chemical's characteristics.
Section 10	Stability and reactivity	Chemical stability and possibility of hazardous reactions.
Section 11	Toxicological information	Routes of exposure; related symptoms, acute and chronic effects; numerical measures of toxicity.
Section 12	Ecological information*	
Section 13	Disposal considerations*	
Section 14	Transport information*	
Section 15	Regulatory information*	
Section 16	Other information	Includes the date of preparation or last revision.

*Note: Since other Agencies regulate this information, OSHA will not be enforcing Sections 12 through 15(29 CFR 1910.1200(g)(2)).

UN Hazard Classes

Under the UN system, there are nine classes for hazardous materials. These classes enable an inspector to readily categorize hazardous materials based on the risks they pose:

Class 1: Explosives

Class 2: Gases

Class 3: Flammable and Combustible Liquids

Class 4: Flammable Solids, Spontaneously Combustible Materials, and Dangerous-When-Wet Materials

Class 5: Oxidizers and Organic Peroxides

Class 6: Poison (Toxic) and Poison Inhalation Hazard

Class 7: Radioactive Materials

Class 8: Corrosive Materials

Class 9: Miscellaneous Dangerous Goods

Each of the nine hazard classes has a specific placard that identifies the class of material and assists an inspector in identifying the hazards associated with the product **(Figure 10.16)**. A material's hazard class is indicated either by its class (or division) number or name. The hazard class or division number must be displayed in the lower corner of placards corresponding to the primary hazard class of a material.

UN Commodity Identification Numbers

The UN has also developed a system of four-digit identification numbers used in conjunction with illustrated placards in North America. Each hazardous material is assigned a unique four-digit number. This number is often displayed on placards, labels, orange panels, and/or white diamonds for materials being transported in cargo tanks, portable tanks, tank cars, or other containers and packages **(Figure 10.17, p. 406)**. On orange panels, letters *UN* for United Nations (which means it is recognized for international transportation) may precede the number **(Figure 10.18, p. 406)**.

The yellow-bordered pages section in the *Emergency Response Guidebook (ERG)* provides a key to the four-digit identification numbers (see *Emergency Response Guidebook* section). Therefore, inspectors can use the 4-digit UN identification number and the *ERG* to determine appropriate response information. The four-digit UN identification number also appears on shipping papers, and it should match the numbers displayed on the exteriors of tanks or shipping containers. In North America, UN numbers are usually found on the following container packages:

- Rail tank cars
- Cargo tank cars
- Portable tank cars
- Bulk packages
- Vehicle containers containing large quantities (at least 8,820 lbs or 4 400 kg) of hazardous materials
- Certain nonbulk packages (for example, poisonous gases in specified amounts)

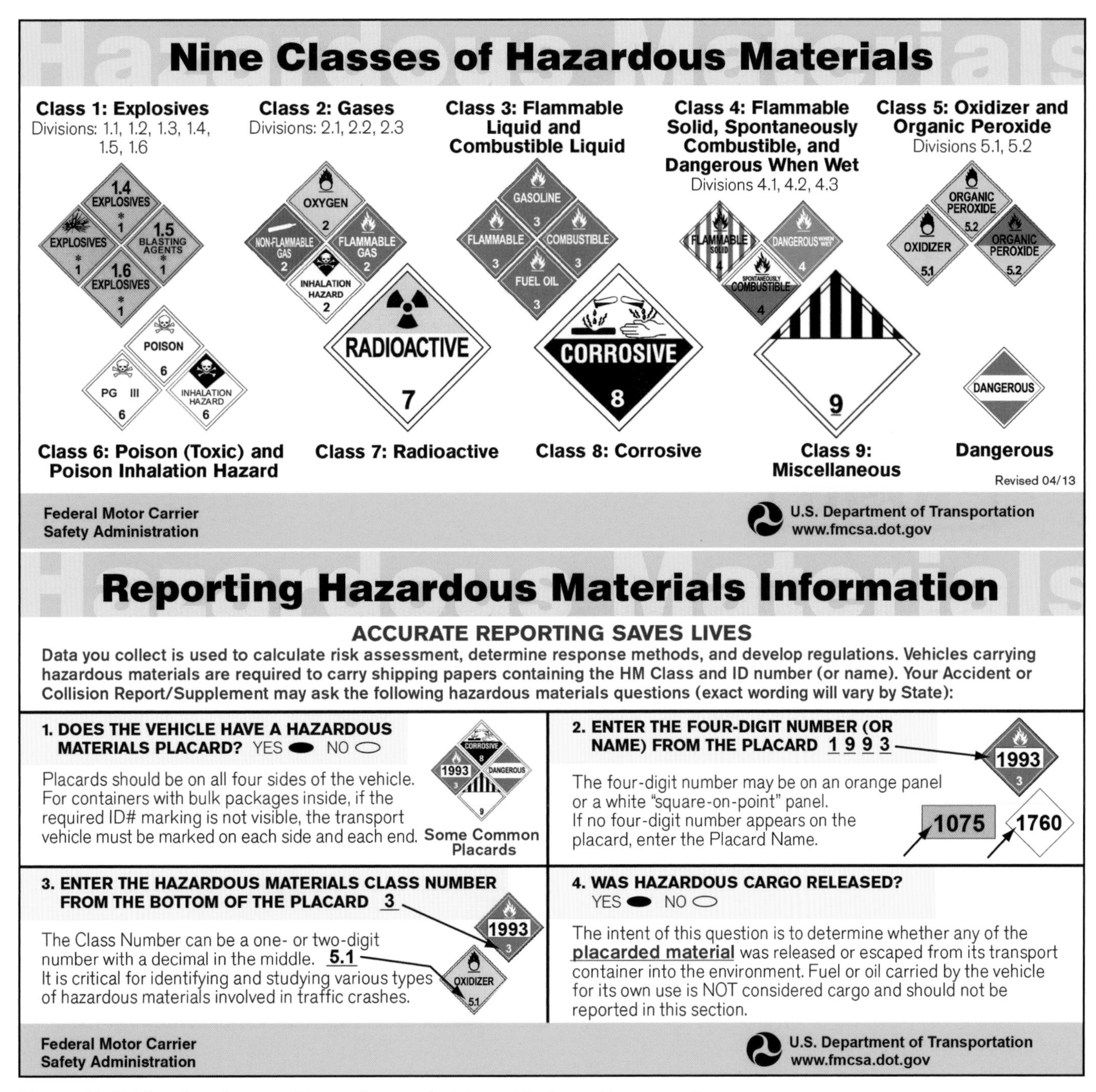

Figure 10.16 The nine classes of hazardous materials and their markings are shown.

DOT Placards, Labels, and Markings

Although the UN system serves as the basis for most DOT regulations, DOT regulations address some additional categories of substances, including other regulated materials (ORM-Ds), materials of trade (MOTs), and fumigated loads. The inspector must record information about these substances during an evaluation of a facility or industrial location.

Although a basic understanding of the UN system enables an inspector to quickly recognize most hazardous materials or dangerous goods, knowledge of the following standard systems used in the U.S. is important in order to accurately recognize these hazards:

Figure 10.17 These UN identification number placards provide a quick reference to identify hazardous materials.

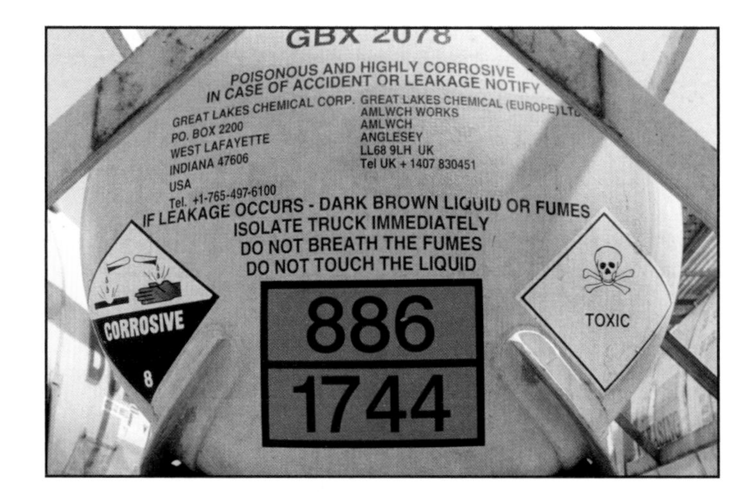

Figure 10.18 Sometimes the 4-digit ID number will be displayed in black lettering on an orange box. *Courtesy of Rich Mahaney.*

DOT placard. Diamond-shaped, color-coded sign that shippers provide to identify the materials in transportation containers **(Figure 10.19)**. Placards are found on a number of containers, including:

— Bulk packages

— Rail tank cars

— Cargo tank vehicles

— Portable tanks

— Unit loading devices containing hazardous materials over 640 cubic feet (18 m^3) in capacity

— Certain nonbulk containers

Some important facts to remember regarding placards are as follows:

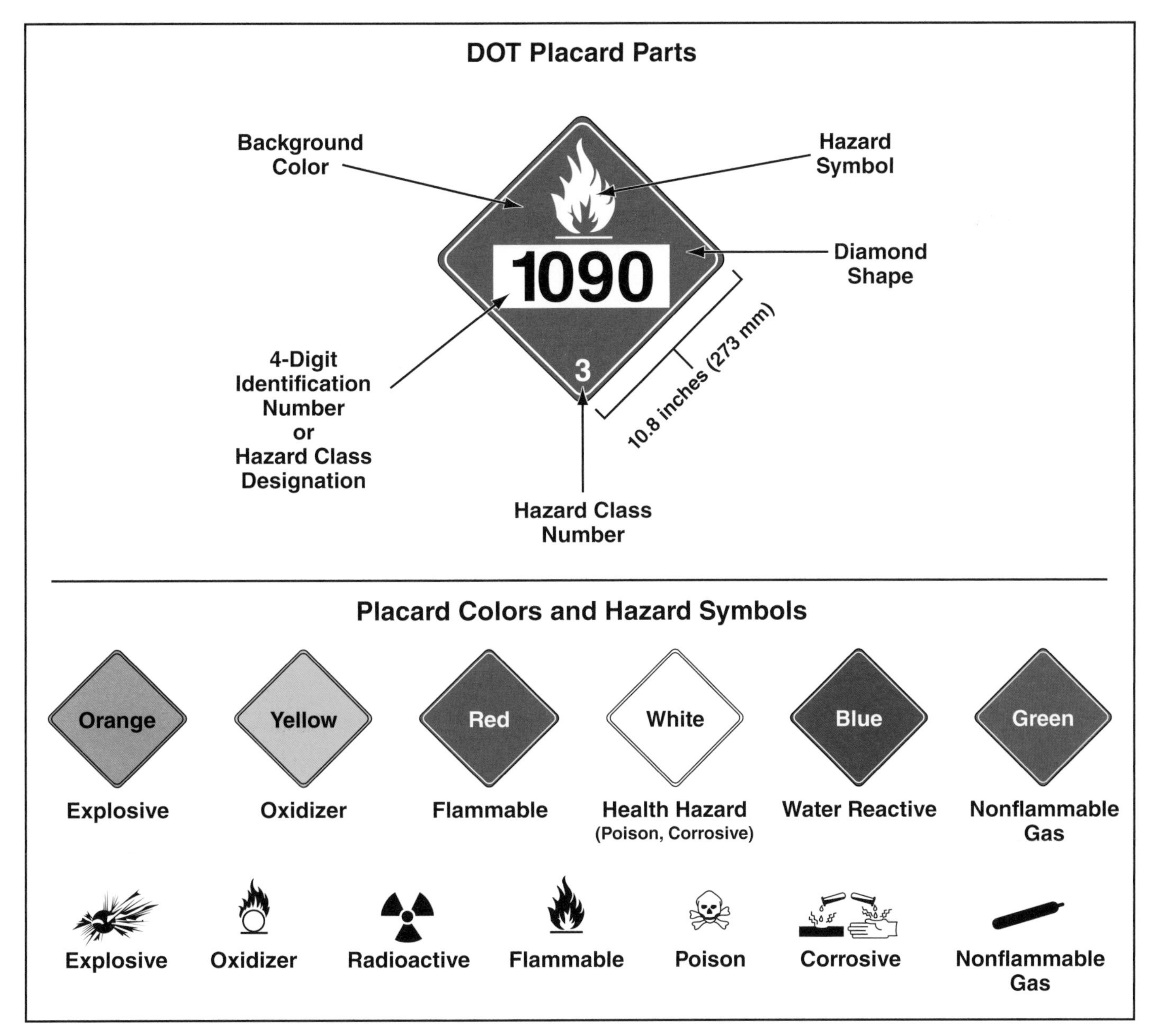

Figure 10.19 U.S. Department of Transportation placards present an easily understood, visual description of the hazards presented by the materials in the marked container.

- Shippers are required to provide placards. Drivers may not know what they are carrying or may have varying degrees of information about the hazardous materials in their vehicles.
- A placard is not required for shipments of infectious substances, other regulated materials (ORM-Ds), materials of trade (MOTs), limited quantities, small-quantity packages, radioactive materials, or combustible liquids in nonbulk packaging.
- Some private agriculture and military vehicles may not have placards, even though they are carrying significant quantities of hazardous materials. For example, farmers may carry fertilizer, pesticides, and fuel between fields or to and from their farms without any placard.

- The hazard class or division number corresponding to the primary or subsidiary hazard class of a material must be displayed in the lower corner of a placard **(Figures 10.20 a and b)**.
- Other than the Class 7 or *DANGEROUS* placard, text indicating a hazard (for example, the word *FLAMMABLE*) is not required. Text may be omitted from the oxygen placard only if the specific ID number is displayed.

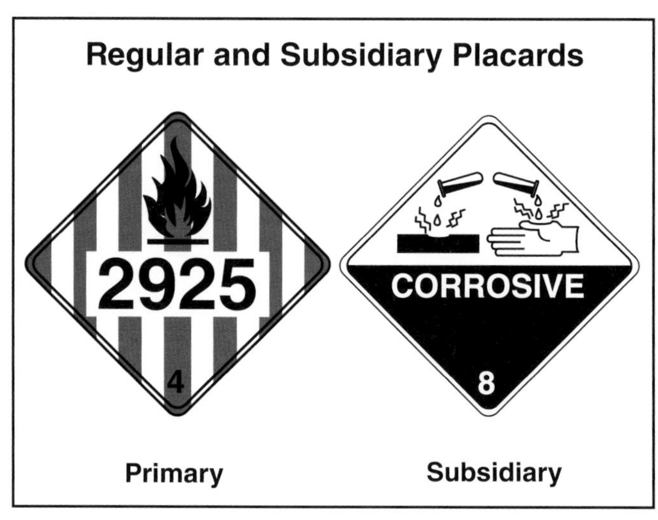

Figure 10.20 a Both primary and subsidiary placards must have the hazard class or division number displayed.

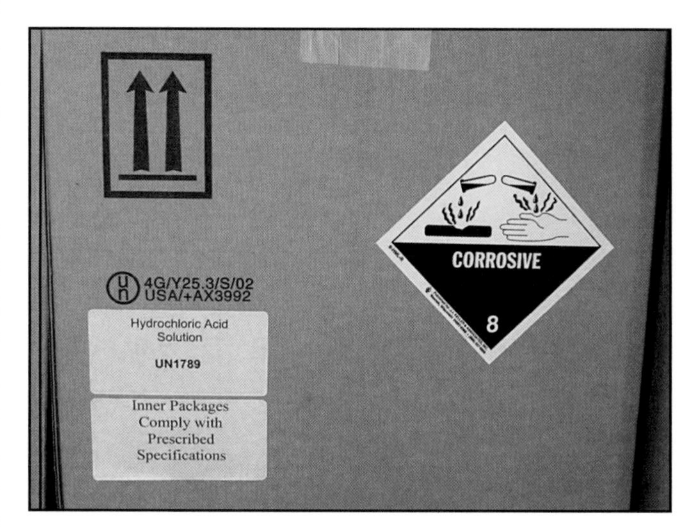

Figure 10.20 b DOT labels are not limited to bulk vehicle containers. They can appear on nonbulk drums, boxes, bags, and other small containers.

DOT label. Printed matter on a 3.9-inch (100 mm), square-on-point diamond that may or may not have written text identifying the hazardous material within the packaging; communicates the hazards posed by the material in case the package falls from the transport vehicle and spills its contents. Each label hazard class is assigned an appropriate pictogram as well as a division number. Packaging contains a primary label for materials that meet the definition of more than one hazard class. **Table 10.8** contains more information and examples of unique DOT labels.

DOT-required labels provide information similar to that conveyed by vehicle placards. Unlike placards that may only be used on the outside of large transport containers, labels are used on nonbulk packaging, such as:

- Drums
- Boxes
- Bags
- Other small containers that are normally located inside facilities

Table 10.8
Unique U.S. DOT Labels

Subsidiary Risk Labels	
Subsidiary risk labels may be used for the following classes: Explosives, Flammable Gases, Flammable Liquids, Flammable Solids, Corrosives, Oxidizers, Poisons, Spontaneously Combustible Materials, and Dangerous-When-Wet	
Class 1: Explosives	
	**Explosive Subsidiary Risk Label**
Class 3: Flammable Liquid	
	**Flammable Liquid Label** — Marks packages containing flammable liquids. *Examples:* gasoline, methyl ethyl ketone
Class 6: Poison (Toxic), Poison Inhalation Hazard, Infectious Substance	
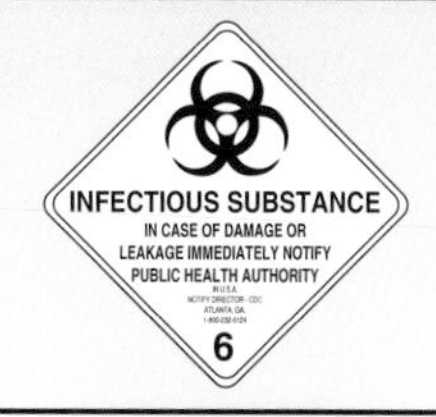	**Infectious Substances Label** — Marks packages with infectious substances (viable micro-organism, or its toxin, which causes or may cause disease in humans or animals). This label may be used to mark packages of Class 6.2 materials as defined in 49 *CFR* 172.432. *Examples:* anthrax, hepatitis B virus, *escherichia* coli (E. coli)
	**Biohazard Label** — Marks bulk packaging containing a regulated medical waste as defined in 49 *CFR* 173.134(a)(5). *Examples:* used needles/syringes, human blood or blood products, human tissue or anatomical waste, carcasses of animals intentionally infected with human pathogens for medical research

Continued

Used in this manner, an inspector can immediately identify hazardous materials or dangerous goods. A package may have several labels placed side by side, with the primary hazard label located on the left and the secondary or subsidiary class label on the right **(Figure 10.21, p. 411)**. DOT regulations require that subsidiary labels have their class numbers displayed.

NOTE: For additional information and regulations regarding the use of labels, refer to Title 49 (Transportation) of the *Code of Federal Regulations (CFR)*.

DOT marking. Descriptive name, identification number, weight, or specification that includes instructions, cautions, or UN marks (or a combination thereof) required on outer packaging containing hazardous materials or goods. The markings found on packaging are described in **Table 10.9, p. 412**.

Table 10.8 (continued)
Unique U.S. DOT Labels

Class 7: Radioactive Materials

Packages of radioactive materials must be labeled on two opposite sides, with a distinctive warning label. Each of the three label categories — RADIOACTIVE WHITE-I, RADIOACTIVE YELLOW-II, or RADIOACTIVE YELLOW-III — bears the unique trefoil symbol for radiation.

Class 7 Radioactive I, II, and III labels must always contain the following additional information:

- Isotope name
- Radioactive activity

Radioactive II and III labels will also provide the transport Index (TI) indicating the degree of control to be excercised by the carrier during transportation. The number in the transport index box indicates the maximum radiation level measured (in mrem/hr) at one meter from the surface of the package. Packages with the Radioactive I label have a Transport Index of 0.

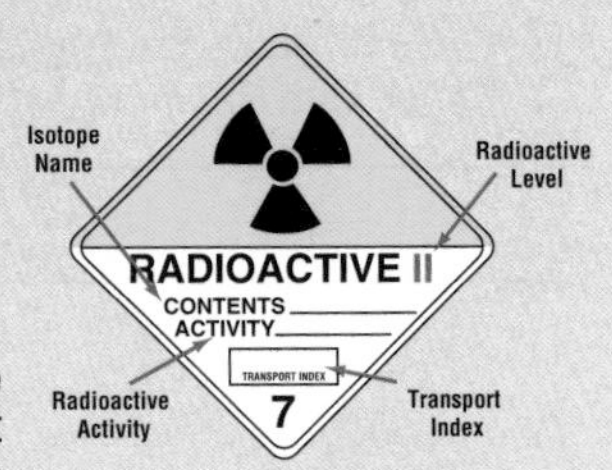

	**Radioactive I Label** — Label with an all-white background color that indicates that the external radiation level is low and no special stowage controls or handling are required.
	**Radioactive II Label** — Upper half of the label is yellow, which indicates that the package has an external radiation level or fissile (nuclear safety criticality) characteristic that requires consideration during stowage in transportation.
	**Radioactive III Label** — Yellow label with three red stripes indicates the transport vehicle must be placarded RADIOACTIVE.
	**Fissile Label** — Used on containers of fissile materials (materials capable of undergoing fission such as uranium-233, uranium-235, and plutonium-239). The Criticality Safety Index (CSI) must be listed on this label. The CSI is used to provide control over the accumulation of packages, overpacks, or freight containers containing fissile material.
	**Empty Label** — Used on containers that have been emptied of their radioactive materials, but still contain residual radioactivity.

Aircraft Labels

	**Danger - Cargo Aircraft Only** — Used to indicate materials that cannot be transported on passenger aircraft.

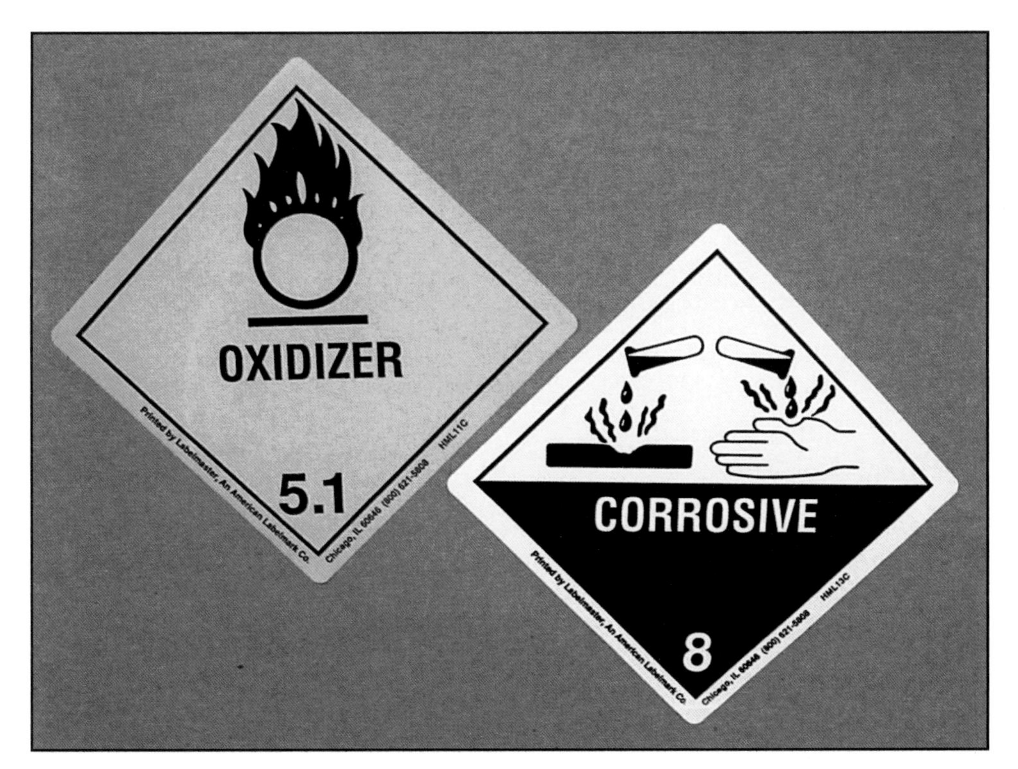

Figure 10.21 A container may be marked with several hazardous materials labels. The primary label is on the left and subsidiary labels are to the right. In this photo, for example, the primary label is the oxidizer label.

Inspectors must pay particular attention in the receiving areas of a warehouse, plant, or business to verify that hazardous materials are marked correctly and that they are stored in the manner required. Additionally, they must verify that the exterior markings on the structures are current and reflect the hazards present.

Illegal Shipments

Unfortunately, improperly marked, unmarked, and otherwise illegal shipments are common. These shipments often include incompatible products, illegal products, and waste products shipped and disposed of without permits. An inspector must be wary of unmarked or mismarked containers.

Other Markings

Not all markings apply to hazardous materials while in transit. OSHA's Hazard Communications Standard (HCS) requires employers to identify hazards in the workplace and train employees how to recognize those hazards. It also requires employers to certify that all hazardous material containers are labeled, tagged, or marked correctly, along with appropriate hazard warnings. The standard does not specify what identification system (or systems) to use, leaving that up to individual employers.

Table 10.9
Unique U.S. DOT Markings

Marking	Description
HOT	**Hot Marking** — Has the same dimensions as a placard and is used on elevated temperature materials. *Note:* Bulk containers of molten aluminum or molten sulfur must be marked MOLTEN ALUMINUM or MOLTEN SULFUR, respectively.
	Marine Pollutant Marking — Must be displayed on packages of substances designated as marine pollutants. *Examples:* cadmium compounds, copper cyanide, mercury based pesticides
INHALATION HAZARD	**Inhalation Hazard Marking** — Used to mark materials that are poisonous by inhalation. *Examples:* anhydrous ammonia, methyl bromide, hydrogen cyanide, hydrogen sulfide
DANGER DO NOT ENTER	**Fumigant Marking** — Warning affixed on or near each door of a transport vehicle, freight container, or railcar in which the lading has been fumigated or is undergoing fumigation with any material. The vehicle, container, or railcar is considered a package containing a hazardous material unless it has been sufficiently aerated so that it does not pose a risk to health and safety.
	Orientation Markings — Markings used to designate the orientation of the package. Sometimes these markings will be accompanied by words such as "this side up."
CONSUMER COMMODITY ORM-D	**ORM-D** — Used on packages of ORM-D materials. *Examples:* consumer commodities, small arms cartridges

Continued

Emergency responders and inspectors may encounter a variety of different (and sometimes unique) labeling and marking systems in their jurisdictions. Other identification marking requirements are as follows:

- ***Chemical manufacturers and importers*** — Required under OSHA HCS to provide appropriate labels on their product containers, including the following:
  - — Name of the product
  - — Manufacturer's contact information
  - — Precautionary hazard warnings
  - — Directions for use and handling
  - — Names of active ingredients
  - — First aid instructions
  - — Other pertinent information

Table 10.9 (concluded)
Unique U.S. DOT Markings

Marking	Description
OVERPACK	**Inner Packaging** — Used on authorized packages containing hazardous materials being transported in an overpack as defined in 49 *CFR* 171.8 and 49 *CFR* 173.25 (a) (4).
	Excepted Quantity — Excepted quantities of hazardous materials. The "*" must be replaced by the primary hazard class, or when assigned, the division of each of the hazardous materials contained in the package. The "**" must be replaced by the name of the shipper or consignee if not shown elsewhere on the package.
UN3373	**Category B Biological Substances** — Diagnostic and clinical specimens that do not cause permanent disability or life threatening or fatal disease to humans or animals when exposure occurs.
keep away from heat	**Keep Away From Heat Marking** — Used for aircraft transportation of packages containing self-reactive substances of Division 4.1 or organic peroxides of Division 5.2.
...kg max	**Marking of IBCs** — For IBCs not designed for stacking, the figure "0" and the symbol for IBCs not capable of being stacked must be displayed. For IBCs designed for stacking, the maximum permitted stacking load applicable (in kilograms) when the IBC is in use must be included with the symbol for IBCs capable of being stacked.

- ***Military services (U.S. and Canada)*** — Provide their own marking systems for hazardous materials and chemicals in addition to DOT and TC markings on transports.
- ***Pipeline companies*** — Must provide markers where pipelines cross under (or over) roads, railroads, and waterways. Many types of hazardous materials in liquid form, particularly petroleum varieties, are transported across both the U.S. and Canada in an extensive network of pipelines, most of which are buried underground.
- ***NFPA® 704 System*** — Provides number ratings for rapidly identifying the presence of hazardous materials and their potential severity based on health, flammability, instability, and related hazards.

NOTE: The NFPA® 704 identification system is described on page 416.

Figure 10.22 Manufacturer's labels must also display information about the container's hazardous contents. The terms *CAUTION*, *WARNING*, *DANGER*, and *poison* all have specific meanings on manufacturer's labels and should never be ignored.

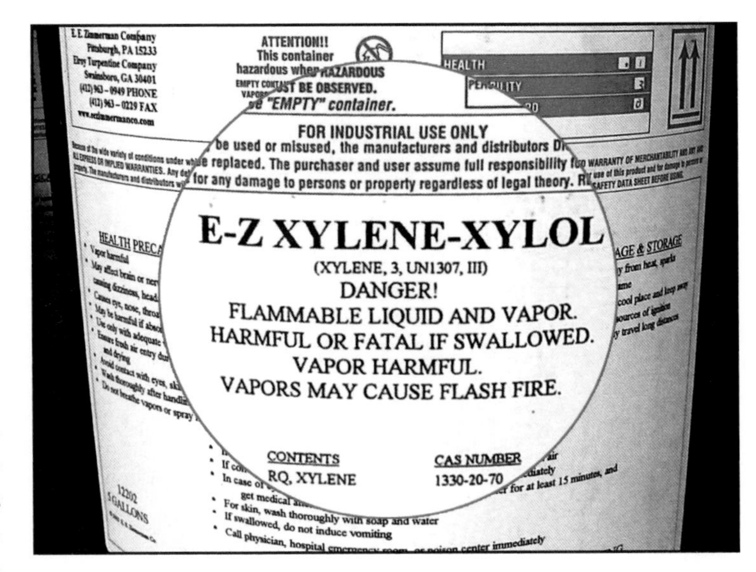

Manufacturers' Warnings

Under the U.S. Federal Hazardous Substances Act (FHSA), labels on products destined for consumer households must incorporate one of the following signal words to indicate the degree of hazard associated with the product **(Figure 10.22)**:

- ***CAUTION*** — May have minor health effects (such as eye or skin irritation)
- ***WARNING*** — Moderate hazards that have significant health effects or flammability
- ***DANGER*** — Highest degree of hazard (used on products that have potentially severe or deadly effects); also used on products that explode when exposed to heat
- ***POISON*** — Use in addition to DANGER on the labels of highly toxic materials

Military Markings

These markings are used on fixed facilities, but they may also be seen on military vehicles (although they are not required). The military placard system is not necessarily uniform; some buildings and areas that store hazardous materials may not be marked due to security reasons. **Table 10.10** provides the U.S. and Canadian military markings for explosive ordnance and fire hazards, chemical hazards, and PPE requirements.

CAUTION

The military ships some hazardous materials and chemicals by common carrier. In these instances, materials are not required to be marked with U.S. Department of Transportation (DOT) or Transport Canada (TC) markings.

Pipeline Markings

Markers must be in sufficient numbers along a pipeline to identify the pipe's location. However, pipeline markers do not always mark the exact location of the pipeline. Pipeline markers in the U.S. and Canada include the signal words

Table 10.10
U.S. and Canadian Military Symbols

Symbol	Fire (Ordnance) Divisions
1	**Division 1: Mass Explosion** Fire Division 1 indicates the greatest hazard. This division is equivalent to DOT/UN Class 1.1 Explosives Division **Also, this exact symbol may be used for:** **Division 5: Mass Explosion — very insensitive explosives (blasting agents)** This division is equivalent to DOT/UN Class 1.5 Explosives Division
2	**Division 2: Explosion with Fragment Hazard** This division is equivalent to DOT/UN Class 1.2 Explosives Division **Also, this exact symbol may be used for:** **Division 6: Nonmass Explosion — extremely insensitive ammunition** This division is equivalent to DOT/UN Class 1.6 Explosives Division
3	**Division 3: Mass Fire** This division is equivalent to DOT/UN Class 1.3 Explosives Division
4	**Division 4: Moderate Fire — no blast** This division is equivalent to DOT/UN Class 1.4 Explosives Division

Symbol	Chemical Hazards
"Red You're Dead"	**Wear Full Protective Clothing (Set One)** Indicates the presence of highly toxic chemical agents that may cause death or serious damage to body functions.
"Yellow You're Mellow"	**Wear Full Protective Clothing (Set Two)** Indicates the presence of harassing agents (riot control agents and smokes).
"White is Bright"	**Wear Full Protective Clothing (Set Three)** Indicates the presence of white phosphorus and other spontaneously combustible material.

Continued

Table 10.10 (concluded)

Symbol	Chemical Hazards
	Wear Breathing Apparatus Indicates the presence of incendiary and readily flammable chemical agents that present an intense heat hazard. This hazard and sign may be present with any of the other fire or chemical hazards/symbols.
	Apply No Water Indicates a dangerous reaction will occur if water is used in an attempt to extinguish the fire. This symbol may be posted together with any of the other hazard symbols.
Symbol	**Supplemental Chemical Hazards**
G	**G-Type Nerve Agents** — persistent and nonpersistent nerve agents *Examples: sarin (GB), tabun (GA), soman (GD)*
VX	**VX Nerve Agents** — persistent and nonpersistent V-nerve agents *Example: V-agents (VE, VG, VS)*
BZ	**Incapacitating Nerve Agent** *Examples: lacrymatory agent (BBC), vomiting agent (DM)*
H	**H-Type Mustard Agent/Blister Agent** *Example: persistent mustard/lewisite mixture (HL)*
L	**Lewisite Blister Agent** *Examples: nonpersistent choking agent (PFIB), nonpersistent blood agent (SA)*

CAUTION, WARNING, or DANGER (representing an increasing level of hazard) and contain information describing the transported commodity and the name and emergency telephone number of the carrier **(Figure 10.23)**.

NFPA® 704 System

The NFPA® 704 system provides a widely recognized method for indicating the presence of hazardous materials at commercial, manufacturing, institutional, and other fixed-storage facilities. This system is commonly required by local ordinances for all occupancies that contain hazardous materials. It is designed to alert emergency responders to health, flammability, instability, and related hazards (specifically oxidizers and water-reactive materials). These hazards may present short-term, acute exposures resulting from a fire, spill, or similar emergency. The NFPA® 704 system is *not* designed for transportation or general public use or for the following hazards:

- Nonemergency occupational exposures
- Explosives and blasting agents, including commercial explosive materials
- Chronic health hazards
- Etiologic agents and other similar hazards

Specifically, the NFPA® 704 system uses a rating system of numbers from *0* to *4*. The number *0* indicates a minimal hazard, whereas the number *4* indicates a severe hazard. The rating is assigned to three categories:

- Health
- Flammability
- Instability

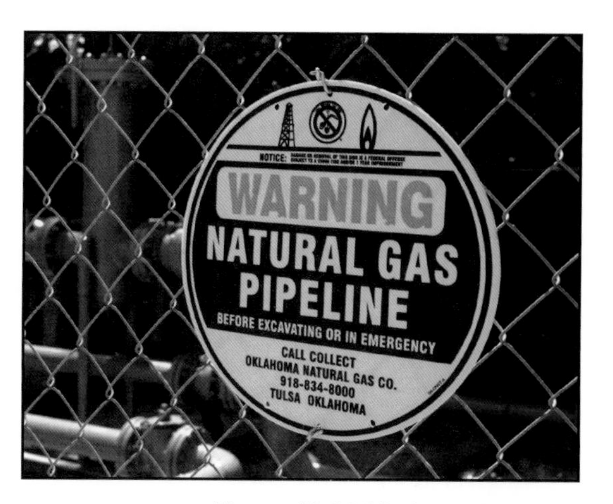

Figure 10.23 Markers along pipelines must include signal words such as *WARNING* and must be frequent enough along pipelines to identify its location.

Special hazards are located in the six o'clock position and have no specified background color; however, white is most commonly used **(Figure 10.24)**. Only two special-hazard symbols are presently authorized for use in this position by NFPA®: *W̶* and *OX* (respectively, indicating unusual reactivity with water or that the material is an oxidizer). However, inspectors may see other symbols in the white quadrant on old placards, including the trefoil radiation symbol. If more than one special hazard is present, multiple symbols may be seen.

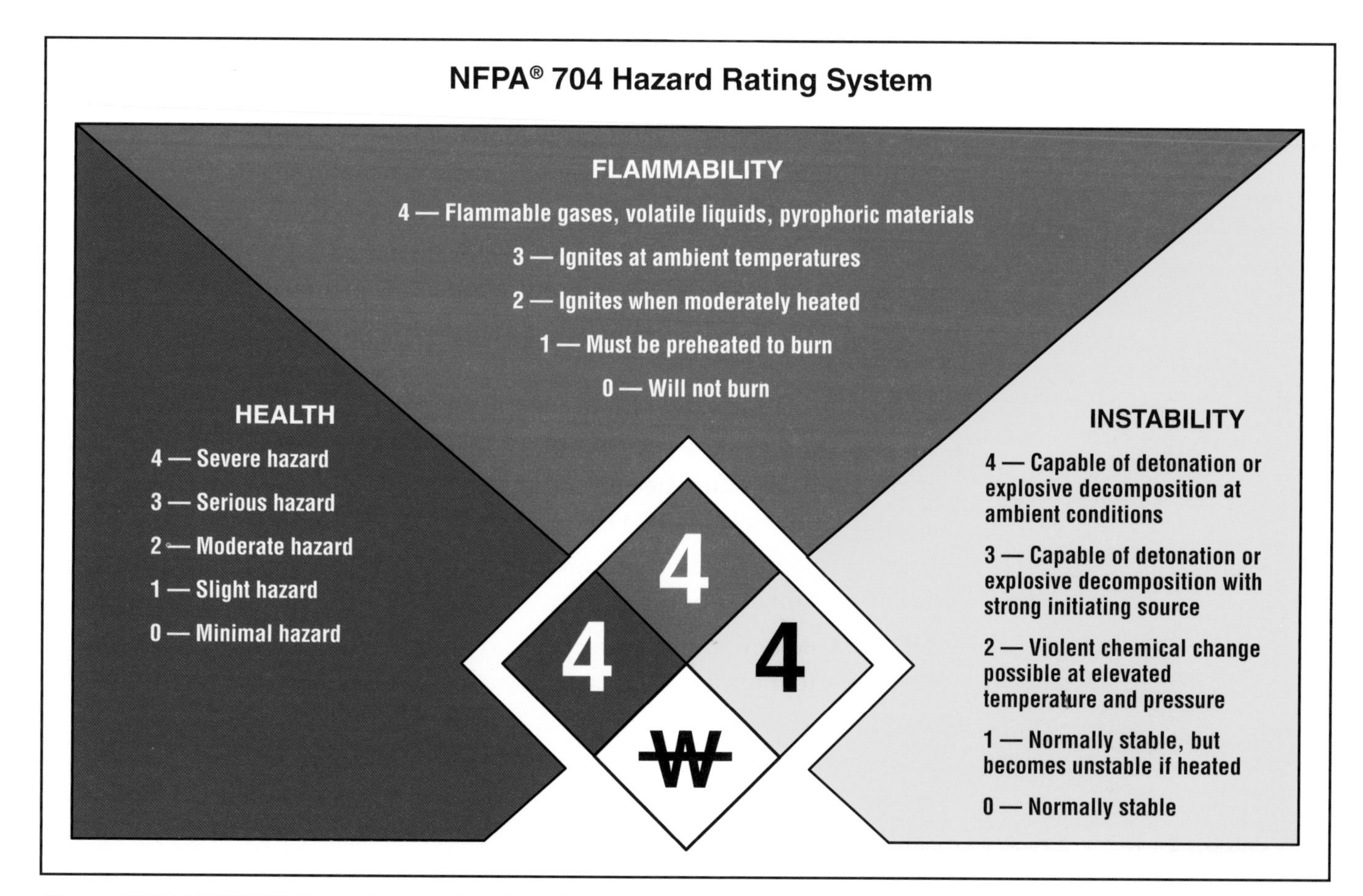

Figure 10.24 NFPA® 704 includes a coded placard system with quadrants numbered *0* through *4* in the upper three quadrants and a lettering system in the remaining quadrant. This illustration shows a key to the numerical ratings on these placards.

NFPA® 704 Limitations

When NFPA® 704 markings are placed on a structure rather than an individual container, the information they provide has limits. The markings do not identify which chemical(s) are present or indicate exactly where they are located. Inspectors need to examine container markings, employee information, company records, or preincident surveys to positively identify the materials.

Resource Guidebooks

Several references are widely used to identify hazardous materials:

- *Emergency Response Guidebook (ERG)*
- *NIOSH Pocket Guide to Chemical Hazards (NPG)*
- *Hazardous Materials Guide for First Responders* from the U.S. Fire Administration

In addition, the *Hazardous Materials Information Resource System* (HMIRS) is available for inspectors who deal with U.S. government facilities.

Emergency Response Guidebook

The current *ERG* was developed jointly by TC, DOT, and the Secretaría de Comunicaciones y Transportes (Secretariat for Communications and Transport or SCT) of Mexico. The *ERG* is primarily a guide to aid emergency responders in quickly identifying the hazards of materials involved in an emergency incident. This knowledge will also help protect them and the general public during the initial response phase. The *ERG* is also a useful resource for inspectors in identifying hazardous materials/dangerous goods during inspections.

NOTE: The *ERG* does not address all circumstances that may be associated with hazardous materials/dangerous goods. It is primarily designed for use at hazardous materials/dangerous goods incidents occurring on highways or railroads.

Inspectors can use the *ERG* in several ways:

- Identify the four-digit UN identification number on a placard or on shipping papers and then find the appropriate guide in the yellow-bordered pages of the guidebook **(Table 10.11)**.
- Find the name of the material involved (if known) in the blue-bordered pages section of the guidebook.

NOTE: Inspectors must exercise care when using this method because many chemical names differ only by a few letters, so exact spelling is important.

- Identify the transportation placard of the material and then use the three-digit guide code associated with the placard in the Table of Placards and Initial Response Guide to Use On-Scene located in the front of the *ERG*.

Table 10.11
Emergency Response Guidebook Contents

ERG Pages	Identification	Description	Purpose	Special Notations or Markings
D Number Index (Yellow Pages)	Index list of hazardous materials in numerical order of ID number	Four-digit UN/NA ID number of the material followed by its assigned Initial Action Guide number (orange pages) and then the material's name	Enables emergency responders to quickly identify the Initial Action Guide (orange pages) to consult using the ID number of the material involved	When a material in the yellow index is highlighted, it means that the material releases gases that are toxic inhalation hazard (TIH) materials. These materials require the application of additional emergency response distances.
Material Name Index (Blue Pages)	Index list of hazardous materials in alphabetical order of material name	Name of the material followed by its assigned Initial Action Guide number (orange pages) and then the four-digit ID number	Enables emergency responders to quickly identify the Initial Action Guide (orange pages) to consult using the name of the material involved	When a material in the blue index is highlighted, it means that the material releases gases that are TIH materials. These materials require the application of additional emergency response distances.
Initial Action Guides (Orange Pages)	List of guides that address a group of materials that possess similar chemical and toxicological characteristics	Guide title identifies the general hazards of the hazardous material addressed. Guides are presented in a two-page format: • Left-hand page provides safety-related information. • Right-hand page provides emergency response guidance and activities for fire situations, spill or leak incidents, and first aid.	Provides emergency responders with important safety recommendations and general hazard information	Each guide is divided into three main sections: 1. ***Potential Hazards*** — Describes potential hazards that the material may display in terms of fire/explosion and health effects upon exposure. The highest potential is listed first, which allows emergency responders to make decisions regarding the protection of the emergency response team as well as the surrounding population. 2. ***Public Safety*** — Outlines suggested public safety measures based on the situation at hand. — Provides general information regarding immediate isolation of the incident site and recommended type of protective clothing and respiratory protection — Lists suggested evacuation distances for small and large spills and fire situations (fragmentation hazard) — Directs the reader to consult the tables on the green pages listing TIH materials and water-reactive materials when the material is highlighted in the yellow or blue pages

Continued

Table 10.11 (Concluded)

ERG Pages	Identification	Description	Purpose	Special Notations or Markings
Initial Action Guides (Orange Pages)				*3. **Emergency Response*** — Addresses emergency response areas, including precautions for incidents involving fire, spills, or leaks along with first aid. — Lists several recommendations under each area to further assist the responder in the decision-making process — Gives information on general guidance for first aid before seeking medical care
Table of Initial Isolation and Protective Action Distances (Green Pages)	Table listing TIH materials by ID number — includes certain chemical warfare agents and water-reactive materials that produce toxic gases upon contact with water	Provides distances divided by small and large spills as well as night and day	Provides two different types of recommended safe distances: • Initial isolation distances • Protective action distances	Highlights TIH materials for easy identification in both numeric (yellow pages) and alphabetic (blue pages) indexes

UN/NA = United Nations/North America

ID = Identification

TIH = Toxic inhalation hazard

Using Placards and Container Profiles in the *ERG*

Another method of using the *ERG* involves using the placards and container profiles provided in the white pages in the front of the book. Inspectors can identify placards (without the four-digit ID numbers) or container shapes and then reference the guide number to the orange-bordered page provided in the circle nearest the placard or container shape.

NIOSH Pocket Guide to Chemical Hazards

The *NIOSH Pocket Guide to Chemical Hazards (NPG)* is a reference source of general industrial hygiene information. The *NPG* provides key information and data in abbreviated or tabular form for over 600 chemicals or groups of chemicals that may be encountered in the work environment. The *NPG* is designed for occupational safety personnel, employers, and employees, but it can also provide a quick reference for inspectors. The *NPG* includes the following pieces of information for individual chemicals:

- Chemical names, synonyms, trade names, conversion factors, Chemical Abstract Service (CAS) number, Registry of Toxic Effects of Chemical Substances® (RTECS®), and DOT numbers

- NIOSH recommended exposure limits (RELs)
- OSHA permissible exposure limits (PELs)
- NIOSH immediately dangerous to life or health (IDLH) values
- Physical description of the agent with chemical and physical properties
- Measurement methods
- Personal protection and sanitation recommendations
- Respirator recommendations
- Information on health hazards, including route, symptoms, first aid, and target organ information

Hazardous Materials Guide for First Responders

The *Hazardous Materials Guide for First Responders* is the result of a study performed by USFA to determine resources that are available to emergency responders. The study indicated that available resources did not meet the specific needs of the Awareness or Operational Levels. As a result, this guide contains material not found in the *ERG* or *NPG*. The guide contains seven sections:

Section 1: Indexes: Alphabetical Material Name Index and a UN/NA Number Index

Section 2: Specific Material Guides: specific recommendations for 430 commonly encountered materials

Section 3: Materials Summary Response Table: summary information for 1,422 less commonly encountered materials

Section 4: DOT Placards, Chart 10

Section 5: Silhouettes of Rail Cars, Tank Trucks, and Chemical Tanks

Section 6: General Approach to a Hazardous Materials Incident

Section 7: Glossary of Terms and Abbreviations

Hazardous Materials Information Resource System

The Hazardous Materials Information Resource System (HMIRS) is a U.S. Department of Defense (DoD) automated system the Defense Logistics Agency developed. HMIRS is the central repository for SDSs for the U.S. Government military services and civil agencies. It also contains other valuable information including hazard communications (HAZCOM) warning labels and transportation information. HMIRS includes data for hazardous materials that the federal government purchased through the DoD and civilian agencies.

Canadian Dangerous Goods System

In Canada, the federal or provincial government regulates a number of products designated as dangerous goods. Municipal inspectors are not normally empowered to enforce these regulations. In some cases, a regulation is referred to and it becomes a part of provincial legislation. Local bylaws, however, may address the storage, use, or transport of these same products for local purposes.

Inspectors must be aware of and familiar with other regulations to avoid conflicts between federal, provincial, and local laws. Inspectors should also be prepared to notify the proper agency to address a hazard or dangerous commodity when they encounter one. **Table 10.12** contains information on the placards, labels, and markings that TC uses to indicate various classifications of dangerous goods.

NOTE: Inspectors can obtain federal and provincial legislation from government web sites. Federal legislation can be found at the Department of Justice Canada web site. The provincial Queen's Printer web site usually contains provincial legislation.

The *Controlled Product Regulations* under the *Hazardous Products Act* establishes the Workplace Hazardous Materials Information System (WHMIS). This system is designed to reduce the risk from hazardous products in the workplace. Health Canada and provincial agencies administer WHMIS. The provincial agency responsible for occupational health and safety (OHS) normally regulates WHMIS within each province. As a result, inspectors may not have authority to apply these regulations on their own. WHMIS is applicable to federal government employees through the *Canadian Labour Code (CLC),* administrated by Human Resources and Skills Development, Canada.

Provincial OHS regulations typically place the responsibility on employers to safeguard the health and safety of their workers who may be exposed to WHMIS-controlled products. Employers must see that the following requirements are met:

- Controlled products used, stored, handled, or disposed of in the workplace are properly labeled.
- SDS are readily available to workers.
- Workers receive education and training in the safe storage, handling, usage, and disposal of controlled products in the workplace.

The WHMIS regulations require suppliers, distributors, and importers to provide labels and SDSs for a WHMIS-controlled product that they sell for use in Canadian workplaces. Controlled products are any products that:

- Burn readily
- Explode
- Produce toxic reactions, allergies, infectious diseases, or dangerous reactions

WHMIS generally requires all containers holding a controlled product to display a WHMIS label with specific information on it, although there are exceptions **(Figure 10.25, p. 427)**. A critical component of WHMIS is that an SDS must be available on site for each controlled product to allow employees to identify hazards in the storage, use, and disposal of the product.

The supplier of a controlled product must label all controlled-product containers with a Supplier Label. When the controlled product is moved from the supplier container to another unmarked container for use at a worksite, the new container must be labeled with a Workplace Label. **Table 10.13, p. 428** shows the WHMIS labels and hazard classes.

NOTE: The Workplace Label does not require the distinctive crosshatch border that is required on the Supplier Label.

Table 10.12
Canadian Transportation Placards, Labels, and Markings

Class 1: Explosives	
1.1 1 Placard and Label	**Class 1.1** — Mass explosion hazard
1.2 1 Placard and Label	**Class 1.2** — Projection hazard but not a mass explosion hazard
1.3 1 Placard and Label	**Class 1.3** — Fire hazard and either a minor blast hazard or a minor projection hazard or both but not a mass explosion hazard
1.4 * 1 Placard and Label	**Class 1.4** — No significant hazard beyond the package in the event of ignition or initiation during transport * = Compatibility group letter
1.5 * 1 Placard and Label	**Class 1.5** — Very insensitive substances with a mass explosion hazard
1.6 * 1 Placard and Label	**Class 1.6** — Extremely insensitive articles with no mass explosion hazard
Class 2: Gases	
2 Placard and Label	**Class 2.1 — Flammable Gases**
2 Placard and Label	**Class 2.2 — Nonflammable and nontoxic Gases**

Continued

Table 10.12 (Continued)	
Class 2: Gases (continued)	
2 Placard and Label	Class 2.3 — Toxic Gases
1005 2 Placard and Label	Anhydrous Ammonia
2 Placard and Label	Oxidizing Gases
Class 3: Flammable Liquids	
3 Placard and Label	Class 3 — Flammable Liquids
Class 4: Flammable Solids, Substances Liable to Spontaneous Combustion, and Substances that on Contact with Water Emit Flammable Gases (Water-Reative Substances)	
FLAMMABLE SOLID 4 Placard and Label	Class 4.1 — Flammable Solids
4 Placard and Label	Class 4.2 — Substances Liable to Spontaneous Combustion
4 Placard and Label	Class 4.3 — Water-Reactive Substances
Class 5: Oxidizing Substances and Organic Peroxides	
5.1 Placard and Label	Class 5.1 — Oxidizing Substances

Continued

Table 10.12 (Continued)	
Class 5: Oxidizing Substances and Organic Peroxides (continued)	
Placard and Label	**Class 5.2 — Organic Peroxides**
Class 6: Toxic and Infectious Substances	
Placard and Label	**Class 6.1 — Toxic Substances**
Label Only	**Class 6.2 — Infectious Substances** Text: INFECTIOUS In case of damage or leakage, Immediately notify local authorities AND INFECTIEUX En cas de Dommage ou de fuite communiquer Immédiatement avec les autorités locales ET CANUTEC 613-996-6666
Placard Only	**Class 6.2 — Infectious Substances**
Class 7: Radioactive Materials	
Label and Optional Placard	**Class 7 — Radioactive Materials** **Category I** — White RADIOACTIVE CONTENTS............................CONTENU ACTIVITY................................ ACTIVITÉ
Label and Optional Placard	**Class 7 — Radioactive Materials** **Category II** — Yellow RADIOACTIVE CONTENTS............................CONTENU ACTIVITY................................ACTIVITÉ INDICE DE TRANSPORT INDEX
Label and Optional Placard	**Class 7 — Radioactive Materials** **Category III** — Yellow RADIOACTIVE CONTENTS............................CONTENU ACTIVITY................................ACTIVITÉ INDICE DE TRANSPORT INDEX

Continued

Table 10.12 (Concluded)	
Class 7: Radioactive Materials (continued)	
RADIOACTIVE 7 **Placard**	**Class 7 — Radioactive Materials** The word RADIOACTIVE is optional.
FISSILE CRITICALITY SAFETY INDEX 7 **Placard**	**Class 7 — Fissile** CRITICALITY SAFETY INDEX
Class 8: Corrosives	
8 **Placard and Label**	**Class 8 — Corrosives**
Class 9: Miscellaneous Products, Substances, or Organisms	
9 **Placard and Label**	**Class 9 — Miscellaneous Products, Substances, or Organisms**
Other Placards, Labels, and Markings	
DANGER	**Danger Placard**
UN3373	**Biological Substances Sign**
	Elevated Temperature Sign

Continued

Table 10.12 (Concluded)

Other Placards, Labels, and Markings (continued)	
DANGER This unit is under fumigation with / Cette unité est sous fumigation au (Name of fumigant) / (Nom du fumigant) Applied on / Depuis le Date / Date Time / Heure DO NOT ENTER / DÉFENSE D'ENTRER	**Fumigation Sign** Text is in both English and French
	Marine Pollutant Sign

A Supplier Label must show all text in English and French, have a WHMIS- hatched border, and give the following information:

- Product identifier (common name of product)
- Supplier identifier (name of company that sold it)
- Statement that an MSDS/SDS is available
- Hazard symbols (pictures of the classifications)
- Risk phrases (words that describe the main hazards of the product)
- Precautionary measures (how to work with the product safely)
- First aid measures (what to do in an emergency)

A Workplace Label, however, only shows the following information:

- Product identifier (common name of product)
- Information on safe handling

Product K1 / Produit K1

Danger **Fatal if swallowed.** **Causes skin irritation.**	**Danger** **Mortel en cas d'ingestion.** **Provoque une irritation cutanee.**
Precautions: Wear protective gloves. Wash hands thoroughly after handling. Do not eat, drink or smoke when using this product. Store locked up. Dispose of contents/containers in accordance with local regulations. IF ON SKIN: Wash with plenty of water. If skin irritation occurs: Get medical advice or attention. Take off contaminated clothing and wash it before reuse. IF SWALLOWED: Immediately call POISON CENTRE or doctor. Rinse mouth.	**Conseils:** Porter des gants de protection. Se laver les mains soigneusement après manipulation. Ne pas manger, boire ou fumer en manipulant ce produit. Garder sous clef. Éliminer le contenu/récipient conformément aux règlements locaux en vigueur. EN CAS DE CONTACT AVEC LA PEAU: Laver abondamment à l'eau. En cas d'irritation cutanée: Demander un avis médical/consulter un médecin. Enlever les vêtements contaminés et les laver avant réutilisation. EN CAS D'INGESTION:Appeler immédiatement a un CENTRE ANTIPOISON ou un médecin. Rincer la bouche.

Figure 10.25 Workplace Hazardous Materials Information (WHMIS) labels must be completed on all controlled products in Canadian workplaces.

Mexican Hazard Communication System

The Official Mexican Standards (*Normas Oficiales Mexicanas* or *NOMs*) augment the Mexican Regulation for the Land Transport of Hazardous Materials and Wastes. The Mexican SCT is responsible for publishing and maintaining the *NOMs*. Mexico's equivalent to the Hazard Communication Standard (HCS) is NOM-018-STPS-2000. It, too, requires employers to see that hazardous chemical substances in the workplace are appropriately and adequately labeled.

The system is essentially an adoption of the NFPA® 704 marking system and *Hazardous Material Identification Guide (HMIG)*. The *HMIG* is a labeling system developed and marketed by Lab Safety Supply Inc. An inspector must remember, however, that employers in Mexico may use any alternative system that complies with the objectives of the standard.

Table 10.13
GHS/WHMIS Symbols and Hazard Classes

Symbol	Description
	Exploding bomb (for explosion or reactivity hazards).
	Flame (for fire hazards).
	Flame over circle (for oxidizing hazards).
	Gas cylinder (for gases under pressure).
	Corrosion (for corrosive damage to metals, as well as skin, eyes).
	Skull and crossbones (can cause death or toxicity with short exposure to small amounts).
	Health hazard (may cause or suspected of causing serious health effects).
	Exclamation mark (may cause less serious health effects or damage the ozone layer*).
	Environment* (may cause damage to the aquatic environment).
	Biohazardous infectious materials (for organisms or toxins that can cause diseases in people or animals).

*The GHS system also defines an Environmental hazards group. This group(and its classes) was not adopted in WHMIS 2015. However, you may see the environmental classes listed on labels and Safety Data Sheets (SDSs). Including information about environmental hazards is allowed by WHMIS 2015.

NOM-026-STPS-1998, "Signs and Colors for Safety and Health," authorizes the use of some International Organization for Standardization (ISO) safety symbols from ISO-3864, "Safety Colors and Safety Signs," on signs to communicate hazard information **(Table 10.14)**. General caution symbols in Mexico are triangular rather than round like those in Canada or rectangular as in the U.S.

Table 10.14
Sample ISO-3864 Type Symbols*

Corrosive	Explosive	Flammable	Toxic/ Poisonous
Biological Hazard	**Radiation**	**Oxidizer**	**Irritant**

* ISO = International Organization for Standardization. This table is not comprehensive.

Mexican transportation placards, labels, and markings are based on the *UN Recommendations* and use the same hazard classes and divisions. In fact, Mexican and Canadian placard and labeling systems are virtually identical. However, because international regulations allow the insertion of text (other than class or division number) in the space below the symbol as long as the text relates to the hazard, placards and labels in Mexico will have text in Spanish **(Figure 10.26)**.

Figure 10.26 Mexican labels are nearly the same as Canadian labels. The text is in Spanish; however, English-speaking inspectors should rely on the symbols on the labels to identify the hazards indicated.

Additionally, information provided on markings will also be written in Spanish. English-speaking inspectors along the U.S./Mexican border and along routes frequented by trucks from Mexico should understand the following common Spanish hazard warning terms:

- Peligro (danger)
- Gases Flamables (flammable gas)
- Radioactivo (radioactive)
- Liquido Criogenico (cryogenic liquid)

Some differences between the Mexican transportation regulations and the U.S. HMR include the following:

- Official Mexican regulations do not authorize the use of *DANGEROUS* placards.
- Package markings are consistent, except that the proper shipping name is provided in Spanish in addition to English.

- The *HOT* mark used for elevated temperature in the U.S. is not authorized in Mexico.
- The Mexican standard regarding the classification of flammable liquids does not incorporate provisions for combustible liquids. Combustible liquid requirements only apply in the U.S.

Piping Identification

In addition to identifying packages and areas where hazardous materials are stored, model fire codes also require identification of the piping that conveys hazardous materials. Such identification enables plant personnel to trace piping in the event of a leak or a spill and stop the event by closing the valves. Pipe labels must identify the contents and direction of flow. Labels are required at a maximum spacing of 25 feet (7.5 m), at each change of piping direction, and on both sides of floor or ceiling penetrations **(Figure 10.27)**.

Cylinder Markings

Many gas consumers believe that the color of a cylinder indicates its hazards, but a cylinder's color does not indicate which gas it may store. The only way to identify a cylinder's contents is by labeling or marking the cylinder. Cylinders require markings in accordance with Compressed Gas Association (CGA) Standard C-7, *Guide to the Preparation of Precautionary Labeling and Marking of Compressed Gas Containers*. Labels are located on the shoulder or wall of the cylinder. CGA Standard C-7 requires that the marking state the name of the hazardous material, its DOT and UN hazardous material identification label and hazard division identification number, which is used in the *ERG*. If the compressed gas is an inhalation hazard, warning statements about this hazard are required, and its reportable quantity (RQ) if such a value is assigned by the U.S. Department of Transportation (DOT) **(Figure 10.28)**. If the amount of hazardous material released due to an accident exceeds the RQ, its release must be reported to the United States Coast Guard National Response Center.

Permissible Amount of Hazardous Materials in a Building

Once a hazardous material hazard classification(s) is assigned, its storage situation must be defined. Model fire codes consider three situations for hazardous materials:

- Storage
- Use — Closed system
- Use — Open system

A material in a package that is static and not moving is defined as storage. Hazardous materials storage can present its own hazards. For example, containers of compressed gases and cryogenic fluids contain materials that are stored at high pressures and create stored kinetic energy. Loss of container integrity will result in the release of a hazardous material **(Figure 10.29)**.

When a hazardous material is moved from its original container, it is considered *use*. A use-closed system does not allow for the release of the hazardous particulates or vapor. A piping system transporting a chemical

Figure 10.27 Pipe identification must indicate its contents and the direction of flow. *Courtesy of Scott Stookey, International Code Council, Washington, D.C.*

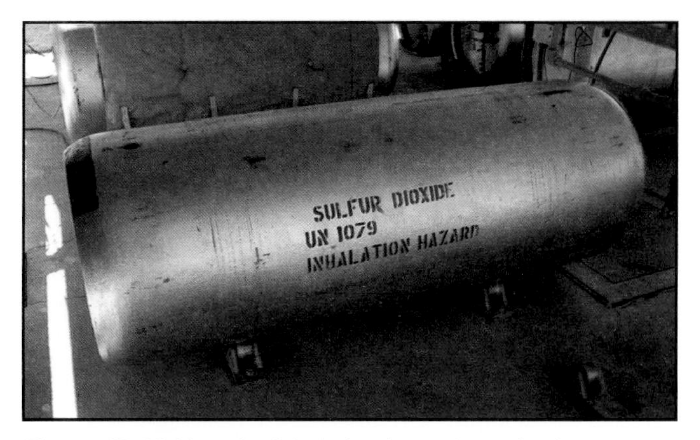

Figure 10.28 Note the inhalation hazard required on this container of sulfur dioxide. *Courtesy of Ron Moore, McKinney (TX) Fire Department.*

from a storage tank to machinery by pumping the liquid is an example of a use-closed system. Under the model fire code provisions, the connection of compressed gas cylinders to any distribution system is classified as a use-closed system **(Figure 10.30, p. 432)**.

Use-open systems include liquids dispensed from a container or intermediate bulk containers. In a use-open system, vapors or dusts are liberated to the atmosphere. A use-open system has a much higher potential for such unintended consequences as fire or chemical exposure to persons. When liquids are dispensed, the model fire codes require several controls to prevent their uncontrolled release, including automatic shutoff valves and a means of spill control **(Figure 10.31, p. 432)**.

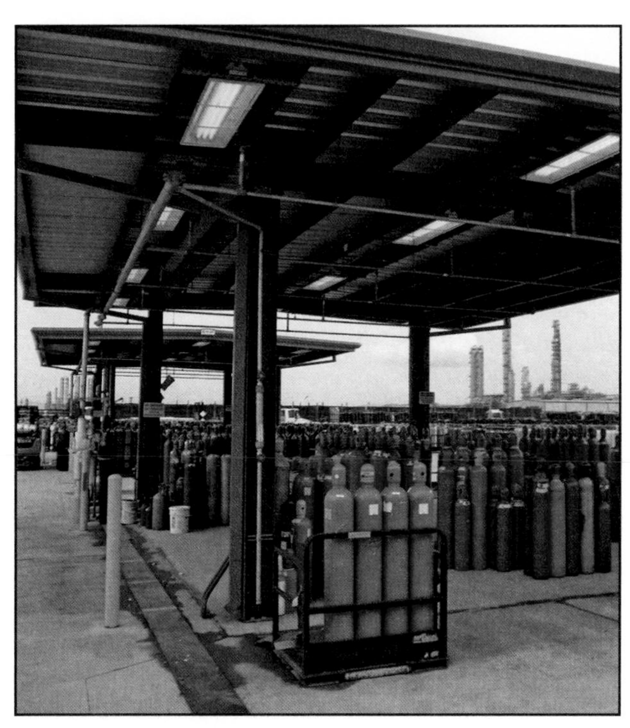

Figure 10.29 Loss of package integrity can result in the release of a hazardous material in storage. *Courtesy of Scott Stookey, International Code Council, Washington, D.C.*

Maximum Allowable Quantity (MAQ) per Control Area

Model fire codes set limits for the amount of hazardous materials allowed inside a building. The **maximum allowable quantity per control area** is the largest amount of a particular class of hazardous materials permitted inside a building without changing the occupancy to a High Hazard use.

MAQs in NFPA® 400 and the IFC are determined based on:

- The building's occupancy classification
- The material's physical state
- The material's situation
- The classification of the hazardous material

The code official must have the material hazard classification, physical state, and the amount of material that is in storage and use to determine the MAQ. When the MAQ is exceeded, the tables in the model fire codes will stipulate the Hazardous (High Hazard) occupancy classification that is required to accommodate the amount of hazardous materials in storage and use.

Maximum Allowable Quantity Per Control Area — Maximum amount of a hazardous material to be stored or used within a control area inside a building or an outdoor control area; maximum allowable quantity per control area is based on the material state (solid, liquid, or gas) and the material storage or use conditions (Source: *International Fire Code®*, 2015 edition).

Figure 10.30 Use-closed systems do not liberate vapors or dusts to the atmosphere. *Courtesy of Scott Stookey, International Code Council, Washington, D.C.*

Figure 10.31 Dispensing is classified as a use-open system and presents a higher risk of a fire or chemical exposure to persons. *Courtesy of Scott Stookey, International Code Council, Washington, D.C.*

Users of MAQ tables in the model fire codes should recognize the importance of understanding the various footnotes in these tables. These footnotes establish certain requirements or conditions that allow an increase of hazardous materials inside buildings without requiring a change of occupancy. In other cases, they set forth limitations or prohibitions for certain classes of hazardous materials.

Most buildings are not protected by automatic sprinkler systems. The hazardous material provisions in the model fire code recognize this and allow an MAQ increase when materials are isolated by noncombustible enclosures. These enclosures offer increased resistance to fire by providing a layer of noncombustible material between the hazardous material and an external fire. Listed flammable liquid or hazardous material cabinets provide at least 10 minutes of fire resistance based on the ASTM E-119 fire test **(Figure 10.32)**.

When hazardous materials are isolated by noncombustible enclosures or the building is protected by an approved automatic sprinkler system, the model fire codes typically double the MAQ without a change of occupancy.

Figure 10.32 An approved flammable liquid or hazardous materials storage cabinet can be used to increase the MAQ per control area for most classes of hazardous materials.

This increase is permitted because model fire codes only consider the ramifications of a single event. For example, if a fire occurs inside an unsprinklered building where the hazardous material is stored in an exhausted enclosure, it is expected that the enclosure will reduce the potential for a release due to fire exposure.

In the case of a fire in a sprinklered building, it is expected that the sprinkler system will control the fire and prevent its spread to the stored hazardous materials. The philosophy of the model fire codes is that adding the benefits of individual protection features can safely increase the MAQ without changing the classification of the occupancy **(Figure 10.33)**.

Model fire codes recognize that certain classes of hazardous materials present such a high risk of fire or explosion that they are not allowed in unprotected buildings. Classes of hazardous materials that warrant this level of protection include:

- Class 4 Oxidizers
- Detonable Unstable (Reactive) materials
- Explosives (excluding Division 1.4G consumer fireworks)
- Pyrophoric materials

When these or other classes of hazardous materials require automatic sprinkler protection, inspectors should review their adopted fire code to verify that the required discharge density and design area is selected. Most NFPA® standards governing hazardous materials have specific sprinkler design area requirements that exceed the minimum requirements in NFPA® 13 **(Figure 10.34, p. 434)**.

Figure 10.33 Adding sprinkler protection can increase the MAQ of an area. *Courtesy of Scott Stookey, International Code Council, Washington, D.C.*

Control Areas

Figure 10.34 For a material like silane, a pyrophoric gas, the material cannot be stored or used inside a building that is not protected by an automatic sprinkler system. *Courtesy of Scott Stookey, International Code Council, Washington, D.C.*

If an occupancy does not have hazardous material storage cabinets, gas cabinets, or similar storage and is not protected by an automatic sprinkler system, the model fire codes offer another option to increase the MAQ: control areas. A control area is a physical boundary inside a building where hazardous materials can be stored and used as long as the MAQ is not exceeded. For the purposes of applying hazardous material regulations, all buildings are considered a single control area. Control areas can be:

- An area bounded by the floor
- Exterior walls
- Roof of a building
- Areas of a building separated by fire-resistant rated construction

The number of control areas inside a building and their MAQ is influenced by the building's occupancy. For a given class of hazardous material, different occupancies will have different MAQs. For example, the IFC has MAQ values for Group M (Mercantile) occupancies that are different when compared to Factory, Business, or Institutional occupancies.

Model codes have established MAQs based on the loss history of certain hazardous materials in certain occupancies. MAQs are also based on concerns for occupant and firefighter safety. For example, the IFC has specific limits for amount of Toxic gas that can be stored or used outside laboratories in Group B occupancies. The reason for this limit is that occupants of a Group B business office normally would not expect a material like anhydrous chlorine in such a space.

Construction requirements for control areas are in the jurisdiction's adopted building code. Requirements for control areas are deferred to the building code because they can require the construction of wall or horizontal assemblies meeting minimum fire assembly ratings.

Consider a building in which the owner wants to store 150 pounds of Class II organic peroxides, which has a MAQ of 50 pounds per control area. The owner does not want to utilize hazardous material cabinets or automatic sprinkler protection. Separating the building into three control areas, each separated by 1-hour or greater fire-resistance rated construction, and limiting the amount of storage to 50 pounds of Class II organic peroxides is an acceptable design. This arrangement is acceptable because it is highly unlikely that a fire in a single control area would spread beyond the control area of origin because the fire-resistive construction compartmentalizes the building.

Nonbulk and Bulk Packaging

Packaging for hazardous materials is categorized into two broad categories by DOT and TC: Nonbulk and Bulk. These categorizations are used in NFPA® 30 and NFPA® 400 to define requirements for storage and use of hazardous materials. The IFC uses a different approach for defining hazardous material packages as:

- Containers
- Cylinders
- Portable containers
- Stationary tanks

Fire inspectors must know the differences in these terms and how they relate to hazardous materials storage and use.

Nonbulk packaging is categorized as any package that is designed to safely contain the following:

- 119 gallons (475 L) or less of liquids
- 882 pounds (440 kg) or less of solids
- Compressed gas cylinders with a water capacity of 1,000 pounds (500 kg) or less.

Water capacity (w.c.) is the accepted measurement of the available volume of compressed gas cylinders. The reason for expressing available volume in this manner is that gases have different densities based on their molecular weight. Gas cylinders are generally filled by weight. The cylinder's empty (termed *tare*) weight is combined with the density of the packaged gas to prevent overfilling of cylinders **(Figure 10.35)**.

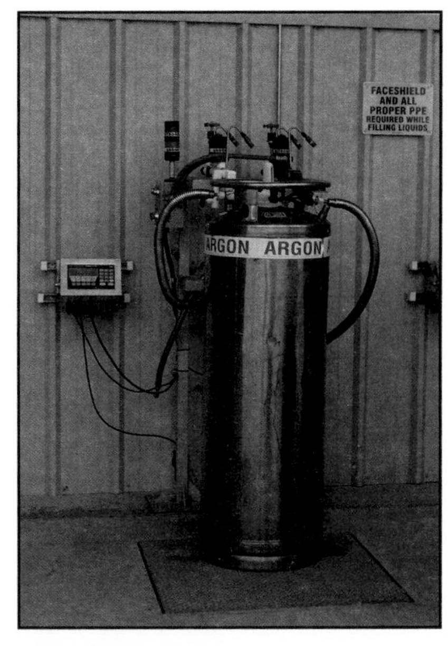

Figure 10.35 Compressed and liquefied compressed gas cylinders and cryogenic fluid containers are filled by weight based on their water capacity. *Courtesy of Scott Stookey, International Code Council, Washington, D.C.*

Nonbulk packages containing liquefied compressed gases, cryogenic fluids, or liquids with a high vapor pressure must contain a means of venting. The same hold true for flammable liquids stored in portable tanks. The vented products cannot create a flammable mixture in air or a poisonous or asphyxiating atmosphere **(Figure 10.36)**.

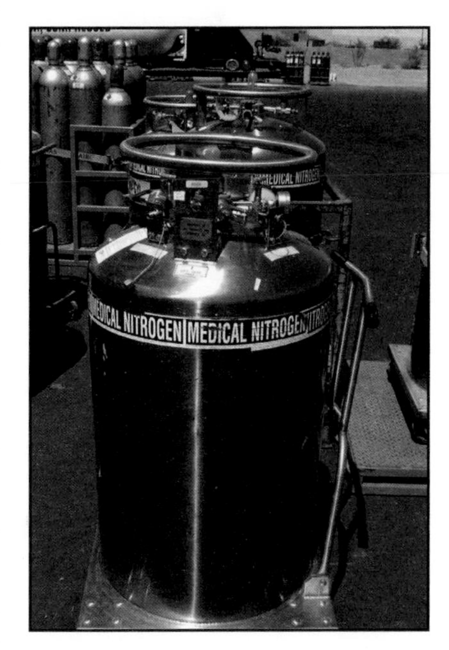

Figure 10.36 Cryogenic fluid containers are designed to constantly vent so the container's internal pressure remains within acceptable limits. *Courtesy of Scott Stookey, International Code Council, Washington, D.C.*

Containers

Code requirements for containers address packaging of flammable and combustible liquids. NFPA® 30 has specific limits on the volume and type of containers based on the liquid's classification and its material of construction. Additional requirements for containers and their construction are also established based on the building's occupancy.

All model fire codes exempt liquid fuels stored in motor vehicle fuel tanks from regulation. Another exemption is the storage of distilled spirits in containers if their volume does not exceed certain limits.

The construction requirements for containers, along with the classification of the stored hazardous materials, have a significant effect on the design and performance of automatic sprinkler systems. For example, consider an automatic sprinkler system designed to protect the storage of Class IC flammable liquids. The automatic sprinkler design is based on the flammable liquids being packaged in carbon steel containers. If the metal packaging is changed to plastic packaging without appropriate modifications to the automatic sprinkler system, the most likely outcome will be total loss of the building **(Figure 10.37, p. 436)**.

Figure 10.37 Containers are selected based on the stored hazardous material and its physical state. *Courtesy of Scott Stookey, International Code Council, Washington, D.C.*

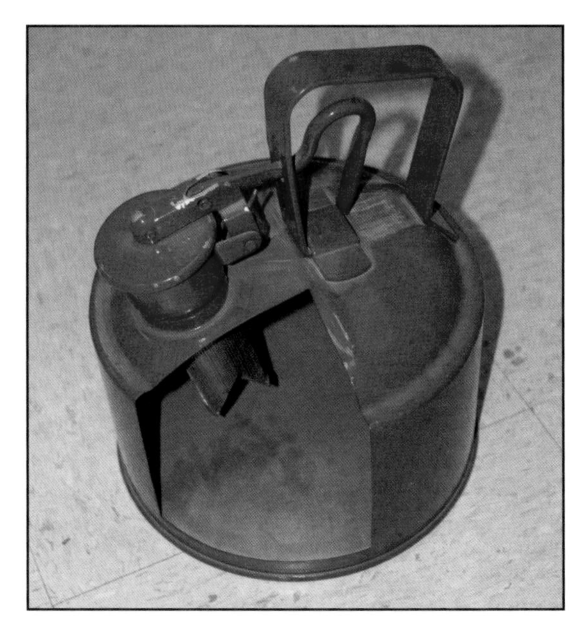

Figure 10.38 Self-closing safety cans feature a manual lever that automatically returns to the closed position when the level is released.

Safety Cans

An inspector most often encounters flammable or combustible liquids stored in containers that are a maximum of 5 gallons (20 L) or less in size. Although there are other acceptable methods of storing these small amounts, the safest containers are approved safety cans. Safety cans are constructed to reduce the chance of leakage or container failure. They are also designed to virtually eliminate vapor release from the container under normal conditions. Safety cans use self-closing lids with vapor seals and contain a flame arrester in the dispenser opening **(Figure 10.38)**. The self-closing lid also acts as a pressure-relief device when the can is heated.

Safety Can Approvals

Safety cans should be approved by Underwriters Laboratories, Inc. (UL), Underwriters' Laboratories of Canada (ULC), or FM Global. Canadian regulations require packaging and portable tanks to meet the specific design and construction criteria of the following documents:

- Transportation of Dangerous Goods Regulations
- Canadian Standards Association (CSA) B376-M, *Portable Containers for Gasoline and Other Petroleum Fuels*
- CSA B306-M, *Portable Fuel Tanks for Marine Use*
- Underwriters' Laboratories of Canada/Other Recognized Document (ULC/ORD) C30, *Safety Containers*
- Section 6 of CSA B620, *Highway Tanks and Portable Tanks for the Transportation of Dangerous Goods*

Intermediate Bulk Containers

Intermediate Bulk Containers (IBCs), also known as totes, are a common form of DOT authorized nonbulk packaging for liquids or solids. IBCs are designed for volumes between 119 to 793 gallons (475 L to 3 150 L). IBCs are constructed of:

- Carbon or stainless steel
- Aluminum
- High-density polyethylene
- Multiple-layer fiberboard box with a plastic liner

Intermediate Bulk Container (IBC) — Rigid (RIBC) or flexible (FIBC) portable packaging, other than a cylinder or portable tank, that is designed for mechanical handling with a maximum capacity of not more than three 3 cubic meters (3 000 L, 793 gal, or 106 ft^3) and a minimum capacity of not less than 0.45 cubic meters (450 L, 119 gal, or 15.9 ft^3) or a maximum net mass of not less than 400 kilograms (882 lbs).

With the exception of multiple-layer fiberboard IBCs, which are only allowed for single trips, all IBCs are subject to requalification every 30 months. NFPA® 30 prohibits the use of rigid plastic IBCs for the storage and dispensing of Class I flammable liquids and further limits the use of fiberboard IBCs to Class III-B liquids.

IBCs are designed for gravity dispensing. Model fire code requirements can vary on gravity dispensing but generally require a means of spill control. This method is prohibited for dispensing Class I-A flammable liquids and Highly Toxic liquids. Gravity dispensing of the remaining classes of flammable and combustible liquids is allowed when an automatic-closing valve is provided.

IBCs require a means of pressure relief. The design of the pressure relief is based on the classification of the stored hazardous material. NFPA® 30 has specific requirements for the design of pressure relief devices installed on IBCs for flammable and combustible liquids.

Inspectors need to be aware of IBCs inside buildings. Depending on the liquid's hazard classification, the storage of certain liquids will automatically change the building to a High-Hazard or Hazardous occupancy. For example, a structure may have an MAQ of 120 gallons for Class IC flammable liquids. Placing one 550-gallon IBC filled with Class IC liquids inside that building would exceed the by greater than two times the allowed MAQ in any building not classified as High Hazard/Hazardous occupancy **(Figure 10.39)**.

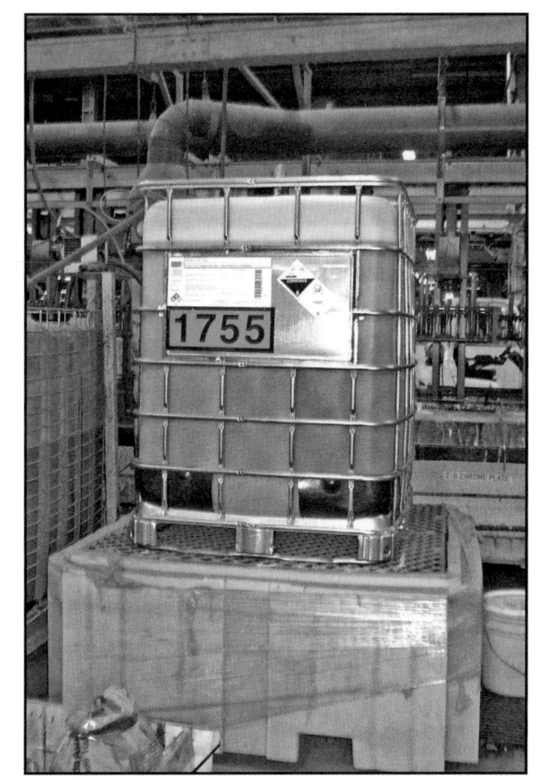

Figure 10.39 380-gallon high-density polyethylene IBC. *Courtesy of Scott Stookey, International Code Council, Washington, D.C.*

Cylinders

A compressed gas container is a pressure containment vessel. The container must be properly engineered to safely store and dispense the gas based on its physical or health hazards. Compressed gas cylinders are designed and constructed in accordance with U.S. DOT requirements and approved standards. Unless they are disposable, compressed gas containers are subject to periodic testing and examination. Stationary pressure vessels must be constructed in accordance with the *Boiler and Pressure Vessel Code* (BPVC) requirements for unfired pressure vessels.

All compressed gas containers, cylinders, or tanks have common and standard features:

- The size of a cylinder or container is limited. If it is a tank, like a stationary pressure vessel, its size is generally not limited.

- The volume of gas that can be stored is directly proportional to the density of the gas molecules. The lighter the gas molecules, the greater the volume of gas that can be stored when compared to heavier gas molecules stored in the same volume.
- Cylinders, containers, and tanks always have a circular cross-sectional area. They are not constructed using other geometric shapes because of the imposed forces in the corners of rectangular-shaped containers. They are designed to operate at pressures above 25 PSIG.
- Proper storage and use of compressed gas containers is dependent on the material contained within, construction, and intended use. Cylinders are designed for either a vertical or horizontal orientation—certain specially designed cylinders can be stored and used in either orientation.
- With the exception of a limited number of physical or health hazard materials, all compressed gas containers are equipped with a pressure relief device. The pressure relief device protects the cylinder from an explosion resulting from excessive heating of its contents.
- Valve fittings are manufactured with unique, material-specific connections to prevent the connection of the gas cylinder to the wrong pipe or distribution system **(Figure 10.40)**.
- Cylinders require markings that identify the standard to which they are constructed, that they are qualified for service, their contents, and the basic hazards of the stored gas.

Bulk Packaging

Bulk packaging refers to a packaging, other than that on a ship or barge, in which materials are loaded with no intermediate form of containment. This packaging type includes a transport vehicle or freight container, such as a cargo tank, railcar, or portable tank **(Figures 10.41 a and b)**. Bulk packaging also includes stationary tanks.

NOTE: Because hazardous materials transportation is outside the scope of the IFC and NFPA® 400, bulk packaging in this manual will focus on stationary tanks.

Stationary tanks used to store hazardous materials are permanently installed, either underground or aboveground. There are similarities in all forms of storage tanks, but the differences in designs and application require fire inspectors to understand the various types of tanks available. Depending on the design, stationary tanks are either listed by a nationally recognized testing laboratory or they must be approved by the fire official. The most common stationary tanks are those used for storing flammable and combustible liquids.

Stationary tanks are either shop fabricated or field erected. A shop-fabricated tank is assembled from metallic or nonmetallic materials, tested, factory certified, and shipped to the installer. All underground storage tanks and most aboveground storage tanks are shop-fabricated because they are less expensive than the field-erected types. Model fire codes require shop-fabricated tanks for storing flammable and combustible liquids to be listed. If they are constructed as process vessels, they may be listed or approved, depending on their design pressure and function **(Figures 10.42 a and b, p. 439)**.

Figure 10.40 The cylinder valve connection is designed to ensure that the cylinder is not connected to an incompatible piping system. The valve shown has a spring-loaded safety relief valve. *Courtesy of Scott Stookey, International Code Council, Washington, D.C.*

Figures 10.41 a and b Portable containers include intermediate bulk containers, which exist in a wide variety of styles and sizes. Containers may be constructed of (a) hard plastic or (b) metal and used to transport solids or liquids. *Courtesy of Rich Mahaney.*

Figures 10.42 a and b a Horizontal cylindrical atmospheric AST. b Rectangular atmospheric AST. *Both photographs courtesy of Scott Stookey, International Code Council, Washington, D.C.*

Field-erected tanks are commonly found in refineries and petroleum terminals but are also found at:

- Marine ports
- Chemical packaging plants

- Manufacturing plants
- Power plants (where heavy fuel oils are stored to provide energy to boilers or steam turbines)

They are engineered specifically for the product they will store and for the site conditions. They are not limited in diameter or height, and it is very common for tanks to hold ten to hundreds of millions of gallons (liters) of product. Field-erected tanks are always installed aboveground and are usually approved by the fire official **(Figure 10.43)**.

Stationary tanks are also categorized by their design pressure. The applicable engineering standard used as the basis for design and approval governs its design pressure. NFPA® standards for flammable and combustible liquids and liquefied petroleum gases prohibit installing an underground storage tank aboveground and aboveground storage tanks being installed underground. **Table 10.15** summarizes the classification of storage tanks by pressure ratings.

Shop-Fabricated Aboveground Storage Tanks

Fire inspectors commonly encounter atmospheric aboveground storage tanks (ASTs). ASTs can range in volume from 60 to 60,000 gallons (240 L to 240 000 L). A variety of installations use ASTs, including:

- Flammable and combustible liquids storage areas
- Lubrication storage tanks in vehicle repair garages
- Fuel tanks for diesel-drive fire pumps and generators
- Manufacturing plants

ASTs for flammable and combustible liquids are required by NFPA® 30 and the IFC to be listed. The standard considered universal by tank manufacturers and AHJs is UL 142. It also addresses specialized designs like ASTs used as foundations for generators (termed generator sub-base ASTs) or ASTs with integral secondary containment **(Figure 10.44)**.

UL 142 requires that any shop-fabricated AST storing flammable or combustible liquids have a permanent nameplate. This nameplate is important to fire officials because it indicates the application that the tank is designed for and any special installation requirements. The nameplate also indicates the standard the tank was constructed to. It also stipulates the required flow rate for tanks equipped with emergency vents **(Figure 10.45, p. 442)**.

UL 142 requires that all atmospheric ASTs be constructed of either carbon or stainless steel. Aluminum is not allowed because it has a much lower melt point when compared to steel.

ASTs are constructed with a normal and an emergency vent. These vent openings are provided to protect the tank from excessive vacuum or positive pressure. The normal vent protects the tank from collapse or overpressurization when liquid is introduced or withdrawn. A normal vent is also required for all underground storage tanks **(Figure 10.46, p. 442)**. NFPA® 30 and the IFC require that the normal vent on tanks storing Class I flammable liquids have a pressure/vacuum (PV) vent. A PV vent is designed so that:

- Air is not introduced into the storage tank unless product is being removed
- Flammable vapors remain inside the tank unless it is being filled.

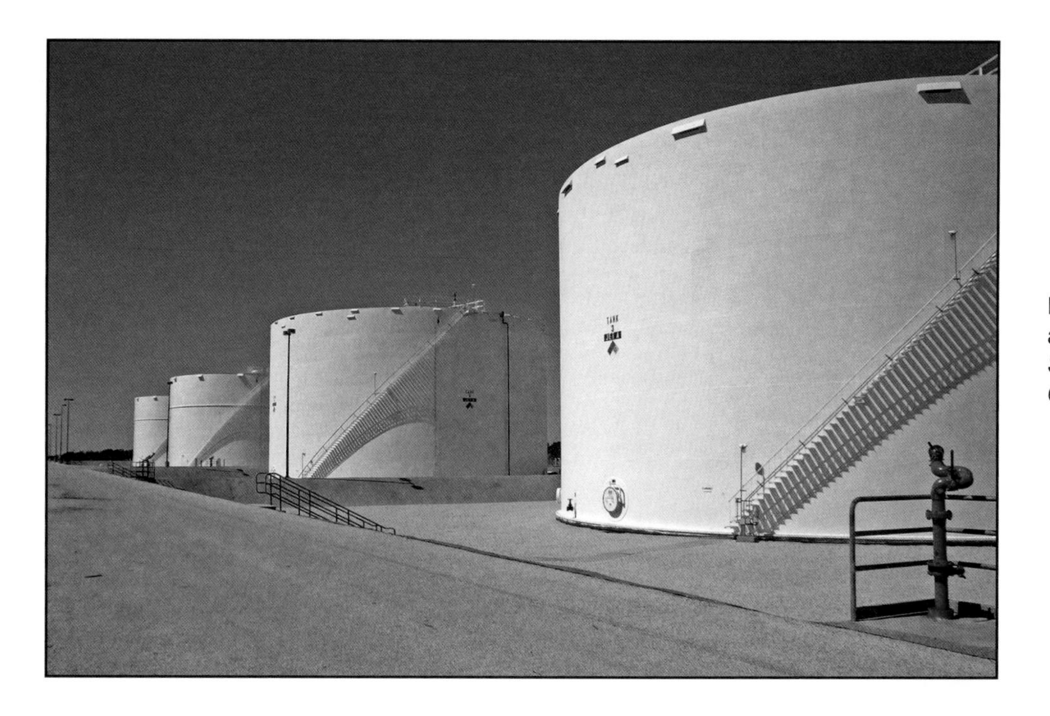

Figure 10.43 Field-erected atmospheric AST. *Courtesy of Scott Stookey, International Code Council, Washington, D.C.*

Table 10.15
Storage Tank Types by Pressure Rating

Type of Tank	Permitted Installation Location	Design Pressure
Atmospheric	Aboveground or Underground	Atmospheric pressure to 1.0 PSIG during normal operation. Atmospheric pressure to 2.5 PSIG during emergency venting or relief. Limited to -0.5 PSIG vacuum pressure.
Low Pressure	Aboveground	1.0 to 15.0 PSIG under normal operating and emergency relief conditions. May be designed to operate at vacuum pressure.
Pressure Vessel	Aboveground or Underground	> 15 PSIG under normal operating and emergency relief conditions. May be designed to operate at vacuum pressure.

Figure 10.44 Generator sub-base atmospheric AST. *Courtesy of Scott Stookey, International Code Council, Washington, D.C.*

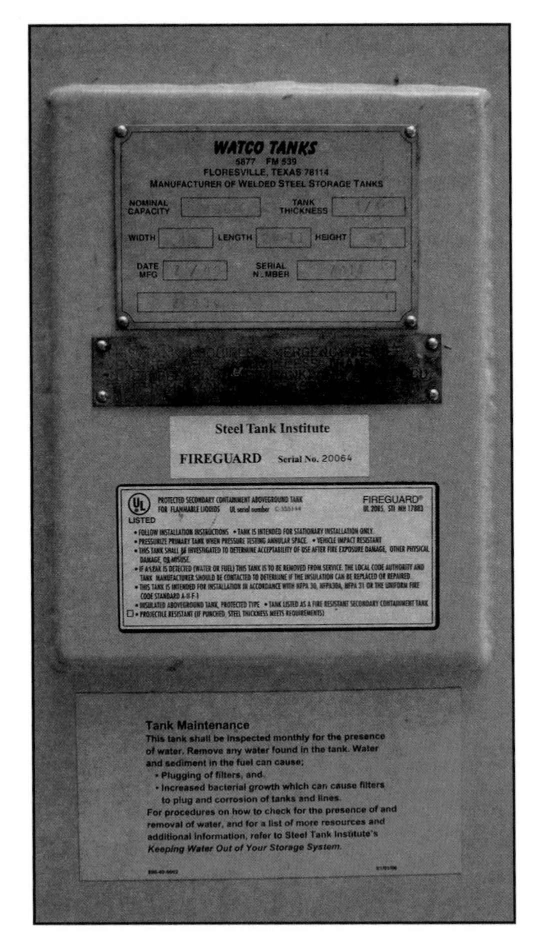

Figure 10.45 Shop-fabricated nameplate. The nameplate indicates the tank's standard of construction and special information, such as the emergency vent flow rate. *Courtesy of Scott Stookey, International Code Council, Washington, D.C.*

Figure 10.46 A normal vent ensures the tank's internal pressure is not drastically changed when product is added or removed. *Courtesy of Scott Stookey, International Code Council, Washington, D.C.*

The PV vent must terminate at least 12 feet (3.5 m) above grade, and 5 feet (1.5 m) from lot lines that can be built upon **(Figure 10.47)**.

The most important safety feature on any AST is the emergency vent, which is designed to prevent the tank from exploding if it becomes involved in fire. The emergency vent serves as a pressure relief device to safely exhaust the vapor generated when fire heats a tank's contents. An emergency vent is not required for underground storage tanks because they have no fire risk.

Emergency vents are designed to open when subjected to a few ounces of pressure. The designs vary based on whether the AST is shop-fabricated or field-erected. An inspector's primary concerns for emergency vents on ASTs are:

- Is the tank equipped with an emergency vent?
- Is the vent operational?
- Is it properly sized?
- Has the vent been modified or obstructed so it is no longer functional? **(Figures 10.48 a and b)**

Sizing of emergency vents is performed by the tank designer based on the NFPA® 30 requirements. Emergency vents are sized based on the following:

- Surface area of the tank
- Tank orientation — vertical or horizontal AST
- Tank geometry — cylindrical or rectangular

Field-Erected Aboveground Storage Tanks

Field-erected ASTs are fabricated from plate carbon steel and stainless steel. The steel sheets are erected into a structurally safe tank onsite. Model codes do not limit the diameter or height of field-erected ASTs, so

Figure 10.47 Pressure/vacuum vent. *Courtesy of Scott Stookey, International Code Council, Washington, D.C.*

Figures 10.48 a and b a Pallet-type emergency vent. b Shear-pin type emergency vent. *Both photos courtesy of Scott Stookey, International Code Council, Washington, D.C.*

they commonly are designed to store millions of gallons of flammable and combustible liquids.

Tanks designed to operate at atmospheric pressure are designed to API 650, *Welded Tanks for Oil Storage*. Tanks that operate at a pressure between 2-15 PSIG or that store liquids at temperatures above 200°F (93°C) are classified as low pressure field-erected ASTs and constructed to API 620, *Design and*

Construction of Large, Welded, Low-Pressure Storage Tanks. Inspectors will need to review the manufacturer's nameplate affixed to the tank to determine the standard of construction that was used as the basis for its design and construction **(Figure 10.49)**. API Standard 653, *Tank Inspection, Repair, Alteration, and Reconstruction,* requires that major modifications or repairs to field-erected ASTs be noted on a permanent repair nameplate affixed to the storage tank **(Figure 10.50)**.

Field-erected ASTs are constructed to American Petroleum Institute (API) standards. Field- erected ASTs are constructed as cone roof, open floating roof, or internal floating roof tanks **(Figure 10.51)**. The type of roof selected is based on the petroleum product the tank will store.

Storage in cone roof tanks is generally limited to combustible liquids because they produce fewer vapor emissions. A typical fixed-roof tank consists of a cylindrical steel shell with a cone- or dome-shaped roof that is permanently

Figure 10.49 Field-erected ASTs are constructed on the owner's site. Courtesy of *Scott Stookey, International Code Council, Washington, D.C.*

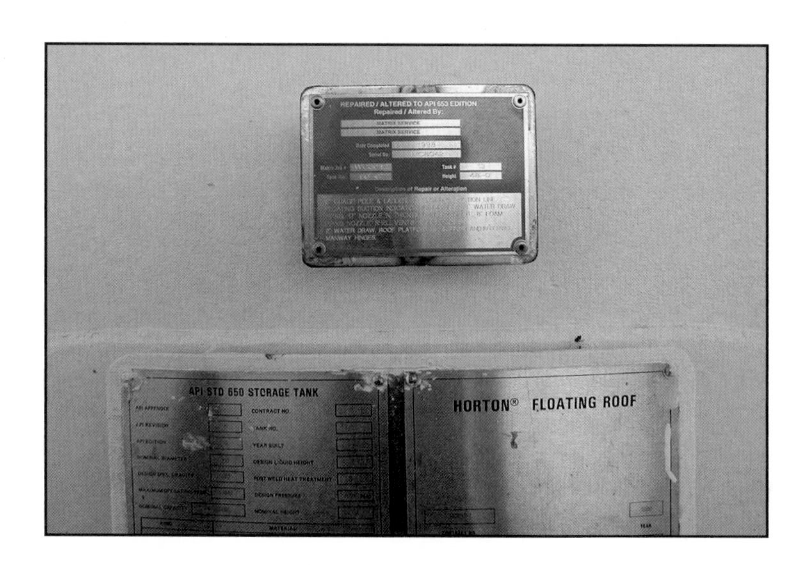

Figure 10.50 Field-erected AST nameplates. The bottom nameplate displays the basic information about the tank when it was constructed. The top nameplate documents repairs that were performed on the AST. Courtesy of *Scott Stookey, International Code Council, Washington, D.C.*

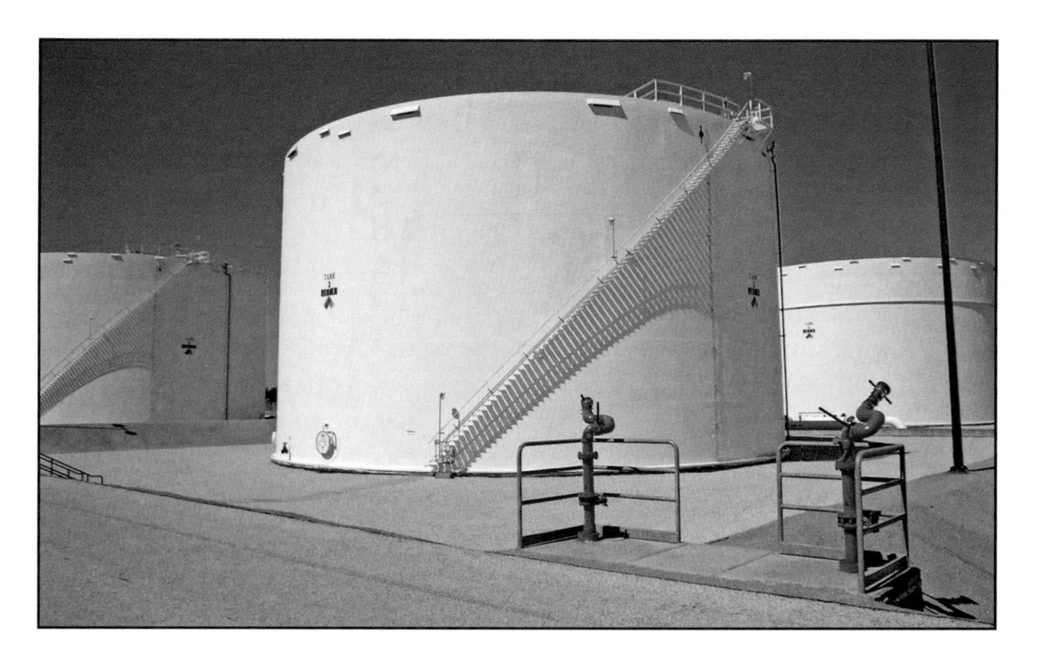

Figure 10.51 Cone roof storage tanks. Courtesy of *Scott Stookey, International Code Council, Washington, D.C.*

affixed to the tank shell. New tanks utilize-welded construction and are liquid- and vapor-tight. Older tanks may be of riveted or bolted construction and may not be vapor-tight.

NOTE: **Appendix G** contains information about floating roof tanks.

Fire protection system requirements for field-erected ASTs will vary based on the tank's volume, contents, and design. The need for fire protection systems is also influenced by the tank's location and the capabilities of fire department. In rural areas, a storage tank may not require fire protection because it does not present an exposure risk. Conversely, a field-erected AST in an industrial area may require fire protection because the fire department lacks the necessary resources to control and suppress a fire. Inspectors should review their locally adopted codes to determine when fire protection for field-erected ASTs is required **(Figure 10.52)**.

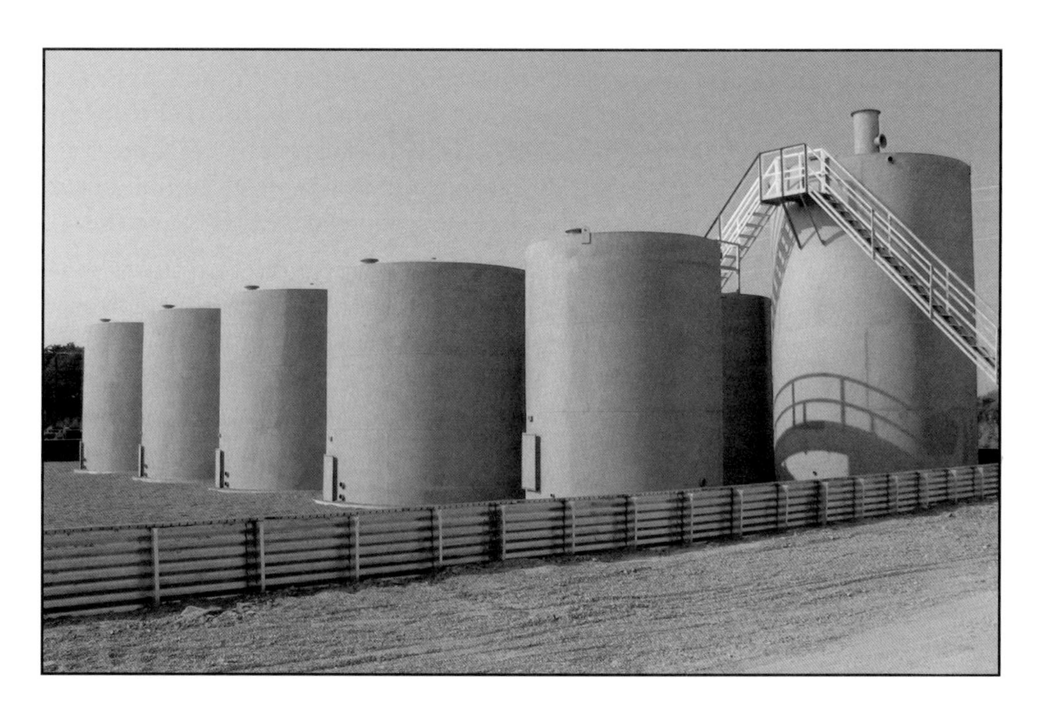

Figure 10.52 Tank fire protection can include foam makers and chambers installed at the tank roof or the installation of ground-mounted monitor nozzles, which can be used to cool foam makers prior to the application of foam. *Courtesy of Scott Stookey, International Code Council, Washington, D.C.*

The tank owner or a representative is required to perform a monthly visual inspection of the exterior of each field-erected AST to identify any potential sources of possible leaks using "tell-tale" fittings built into the storage tank. An authorized inspector certified to API 653 must perform an external examination every five years. At an interval no greater than 20 years, the tank must be removed from service and internally inspected.

Underground Storage Tanks

Underground storage tanks (USTs) are designed to operate at atmospheric pressure. Of all the available storage tank designs, USTs offer the highest degree of fire safety because the tank is installed in the ground and backfilled with sand, gravel, or other clean fill. With the exception of lightning strikes, USTs are safe from most fire exposures.

Model fire code-compliant USTs are listed. They are constructed of carbon steel or glass-fiber reinforced plastic. Carbon steel USTs are constructed to UL 58, *Standard for Steel Underground Tanks for Flammable and Combustible Liquids*, and glass-fiber reinforced USTs are constructed to UL 1316, *Glass-Fiber-Reinforced Plastic Underground Storage Tanks for Petroleum Products, Alcohols, and Alcohol-Gasoline Mixtures.*

It is a violation of model fire codes to attempt to install an aboveground tank underground. Unlike ASTs, USTs are designed to resist the dead load of water-saturated soil on the front (heads) and side (shell). Listed USTs are designed for:

- Storage of gasoline
- Diesel
- Bio-fuels
- Alcohol-blended fuels **(Figure 10.53)**

UST openings are located on top of the tank and are prohibited below the tank's liquid level. USTs require openings for a normal vent, filling the tank with liquid, and openings to accommodate pumps to supply dispensers. Model fire codes require electronic monitoring of the tank's liquid level to prevent the tank from being overfilled and product reconciliation.

Product reconciliation is one method used for verifying the liquid-tightness of the UST. It is a measurement of the product that comes into the UST from the tanker and compares it to the amount of petroleum fuel sent to consumers via the dispensers. If the measurements deviate beyond a set limit, an audible alarm activates at the control panel. Reconciliation is required daily **(Figure 10.54)**.

Model fire codes do not require USTs to have secondary containment. However, many state environmental laws do require it, along with electronic monitoring of the secondary containment space. Inspectors will need to review their applicable state environmental laws to determine any special UST design or installation requirements. The model codes require overfill protection and spill control for all USTs.

Pressure Vessels

Pressure vessels are engineered stationary containers designed to store liquids, cryogenic fluids, or gases at pressures of 15 PSIG or greater. With certain exceptions, any unfired vessel over 6 inches (150 mm) in diameter and more than 2 feet (600 mm) in length that operates at 15 pounds PSIG or greater and is not part of a piping system is classified as a pressure vessel by the *Boiler and Pressure Vessel Code* (BPVC). Pressure vessels can be constructed for either a horizontal or vertical orientation. The design, construction, and examination requirements for pressure vessels that are used for hazardous materials are contained in Section VIII, Division 1 of BPVC **(Figure 10.55)**.

The BPVC requires marking pressure vessels with a nameplate that demonstrates it was constructed in accordance with its requirements. The official code symbol for all pressure vessels is a shamrock with the letter "U" inside it. In addition to the BPVC code symbol, pressure vessel nameplates also require the following markings:

- Manufacturer's name preceded with the words "certified by"
- Maximum allowable working pressure (MAWP), expressed in PSIG at the design temperature, expressed in degrees Fahrenheit (°F)

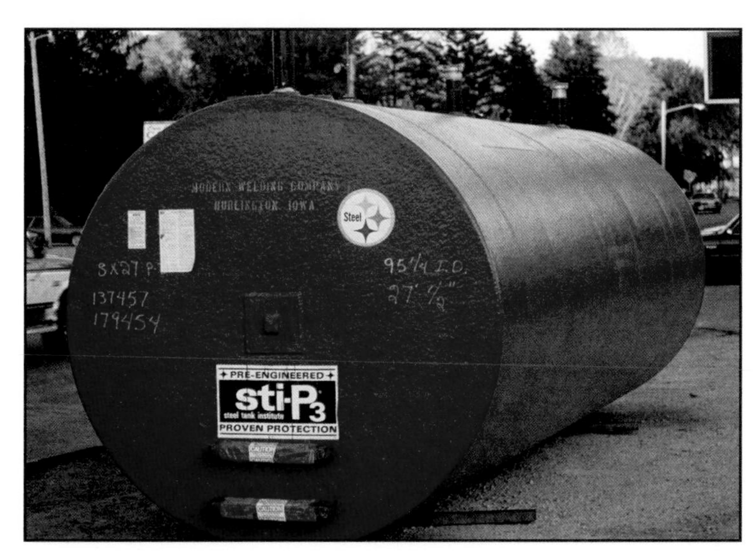

Figure 10.53 Carbon steel UST. Courtesy of *Scott Stookey, International Code Council, Washington, D.C.*

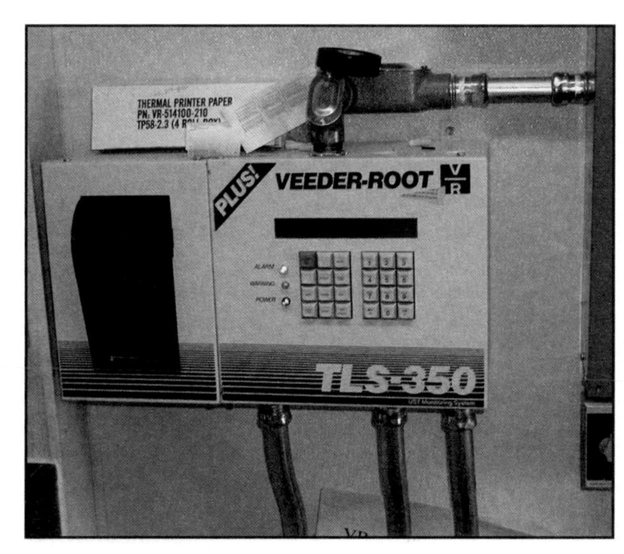

Figure 10.54 Electronic tank reconciliation system. This particular system also monitors the UST leak detection system. Courtesy of *Scott Stookey, International Code Council, Washington, D.C.*

Figure 10.55 Horizontal pressure vessels for compressed argon, nitrogen and oxygen. Courtesy of *Scott Stookey, International Code Council, Washington, D.C.*

- Minimum design metal temperature (MDMT) expressed in °F at a given pressure (PSIG)
- Manufacturer's serial number for the pressure vessel
- Year of construction
- Required abbreviations specified by the BPVC

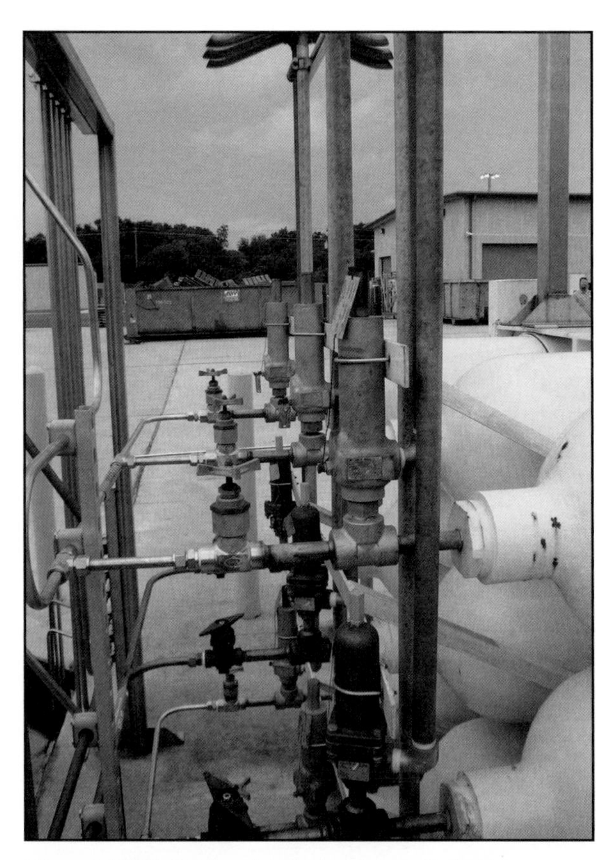

Figure 10.56 Spring-loaded safety relief valve. Courtesy of *Scott Stookey, International Code Council, Washington, D.C.*

The BPVC requires that a pressure vessel be equipped with a pressure relief device to relieve the rise of internal pressure due to an abnormal condition or emergency, including fire exposure. Pressure relief devices are either safety relief valves or non-reclosing relief devices such as burst discs. Safety relief valves open at a predetermined pressure below the design pressure of the vessel and close automatically when the pressure falls below the opening pressure. Burst discs (also termed as rupture discs) cannot close upon operation and are replaced after activation. The operating pressure of the relief device, regardless of type, generally does not exceed the MAWP of the pressure vessel **(Figure 10.56)**.

Case History

On December 3, 2004, the Houston (TX) Fire Department responded to a fire and explosion at a Polyethylene Wax Processing Facility operated by Marcus Oil and Chemical. The incident involved the catastrophic failure of a 50,000-pound, 12-foot-diameter by 50-foot-long pressure vessel (25 000 kg, 3.5 m by 15 m). The vessel was propelled 150 feet (45 m) into an adjacent building and metal debris caused structural damage to residential dwellings up to ¼ mile (0.4 km) away. The blast overpressure broke glazing within a 2 mile (3 km) radius of the site.

The U.S. Chemical Safety Board (CSB) investigation determined that operators had pressurized the tank with nitrogen gas containing 18 percent oxygen instead of the intended concentration of not more than eight percent oxygen. Furthermore, the pressure vessel was not constructed in compliance with the BPVC. It was improperly modified by the cutting of a hole and welding in of heating coils by plant personnel who were not trained or certified in pressure vessel construction or modification. The vessel was not tested following the modification to verify that it could withstand the system operating pressure (theoretically calculated at 80 pounds PSIG), and was not equipped with a safety relief device.

Testing, Maintenance, and Operations

Fire inspectors evaluating hazardous materials will encounter processes and equipment used to facilitate their safe storage and handling. When inspecting hazardous materials, it is important that the mechanical integrity and the tightness of the system are maintained to prevent an unauthorized discharge. Because certain activities like dispensing and use of liquids are fairly common, inspectors must have a good understanding of the hazards involved and how the model fire codes regulate use. This section will address testing, maintenance, and operations involving hazardous materials.

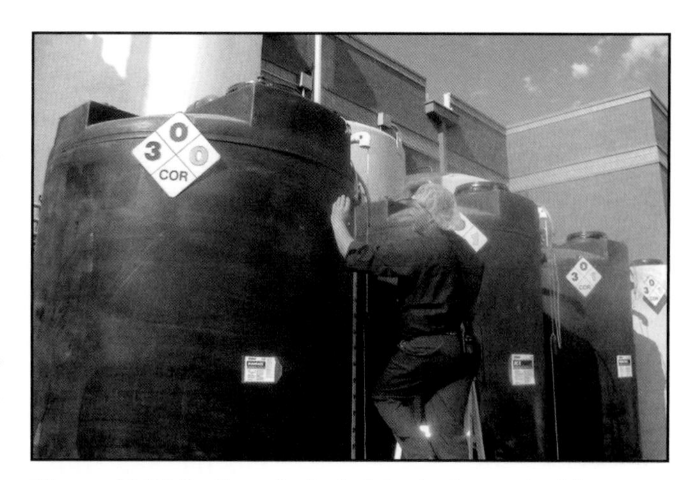

Figure 10.57 Testing of a leak detector is required for aboveground storage tanks with integral containment that cannot be visually monitored. Courtesy of *Scott Stookey, International Code Council, Washington, D.C.*

IFC and NFPA® 400 requirements for the integrity of tanks and piping include verifying that tank vents are operational and that tanks and piping are free of any leaks. Leaking tanks, pipes, and fittings must be removed from service. Damaged or nonfunctional tank or piping components should be repaired or replaced with approved components installed in accordance with the manufacturer's instructions. Tanks with integral secondary containment that cannot be visually monitored require an electronic means of leak detection, which requires periodic testing **(Figure 10.57)**.

Gravity dispensing of liquids is a common use of hazardous materials. Gravity dispensing occurs when liquids are flowed from containers or IBCs using the weight of the stored liquid to flow from the container when a valve is opened. If the valve is left open, or if it fails because of improper material selection or a lack of maintenance, the entire contents can be spilled.

For gravity dispensing to be safely accomplished, the arrangement of the dispensing system must anticipate small liquid losses. Before dispensing can occur, a self-closing automatic valve that requires an operator is required. The receiving container must comply with jurisdiction's fire code. A means of spill control is also required.

NOTE: Refer to Chapter 9 for additional information about safely dispensing liquids.

Process Control

Manufacturing processes regularly use two or more hazardous materials at specified intervals in order to complete a controlled reaction. If the materials are not added properly, the process is not adequately cooled, or the materials are not agitated, the material may rapidly decompose, resulting in a fire or explosion.

The model fire codes require that such systems be properly designed by a competent design professional. The IFC and NFPA® 400 require that hazardous materials follow a proper path during the manufacturing processes. This can vary from using complex controls that measure the weight or flow of materials to using specialized threaded or mechanical connections so incompatible materials are not accidentally mixed. These codes also require that the process must not exceed the temperature or pressure limits established by the designer. Because processes like these can become complicated, inspectors may require a Technical Report and Opinion to clarify the process or its potential hazards. An authorized person who can demonstrate understanding of the process, its hazards, and necessary controls regarding the correct rate, pressure, and temperature of processes should prepare the report **(Figure 10.58)**.

Unauthorized Discharge

The IFC and NFPA® 400 define an unauthorized discharge as a release or emission of a hazardous material that does not conform to the requirements of the jurisdiction's fire code or applicable public health and safety regulations **(Figure 10.59)**. A leaking valve on an 8-inch (200 mm) diameter liquid withdrawal pipe installed on a 200,000 barrel field-erected AST is an unauthorized release of a material that is not consistent with NFPA® 30 or the IFC. Under the IFC and NFPA® 400, this can require mitigation (clean up) of the spill by the property owner.

When an unauthorized discharge occurs, an inspector's responsibility is to work to prevent additional incidents. To this end, a stop-use order may be issued stating that the equipment cannot be returned to service until the required corrections are made. In some cases, the hazardous material will need to be removed from the system, which presents its own hazards. The plan of action should consider the required steps so that repairs or improvements can be safely accomplished without creating another incident.

Figure 10.58 Control of the process by trained and competent operators is an important consideration by an inspector evaluating a process using hazardous materials. Courtesy of *Scott Stookey, International Code Council, Washington, D.C.*

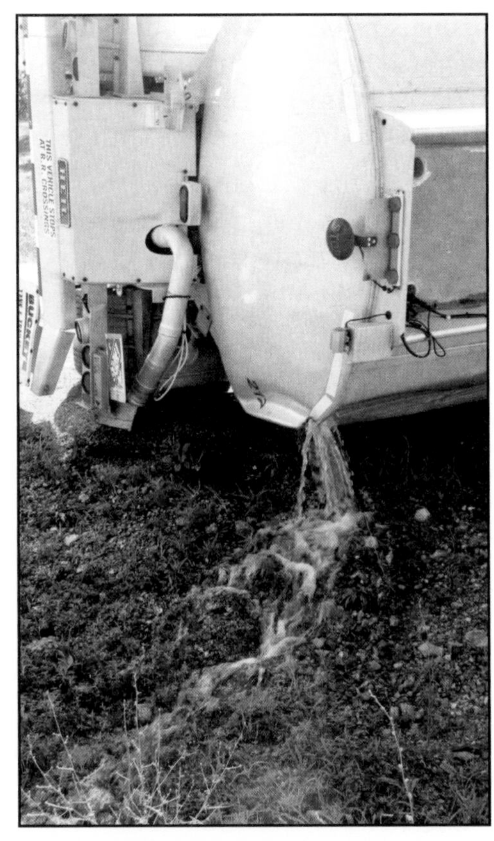

Figure 10.59 Incidents involving liquids may be more difficult to mitigate than those involving solids. Liquids may also emit vapors, which behave like gases.

Hazardous Materials Piping, Valves, and Fittings

Pipes, valves, and fittings are components of piping systems that convey hazardous materials from cylinders and tanks to a point of use. The IFC and NFPA® 400 have extensive requirements for piping systems design, erection, and testing. Piping systems conveying hazardous materials are required to be constructed of materials with adequate strength and durability that are chemically compatible with the liquid or solid being conveyed **(Figure 10.60)**. Model fire codes require hazardous materials piping systems be constructed to ASME B31.3. The inspector needs to verify that the installation matches the approved plans.

In piping systems conveying flammable, pyrophoric, toxic, and corrosive liquids and gases, model fire codes prescribe the installation of an excess flow control valve or other means of flow control. Excess flow control valves are designed to stop the flow of a liquid or gas at a predetermined setting or if the pipe is severed. Excess flow control valves are selected based on the pipe diameter and the material being conveyed. A key inspection issue for excess flow control valves is pipe diameter changes. In a piping system requiring excess flow control valves, such a valve needs to be installed at each location where the pipe diameter changes **(Figure 10.61)**.

Figure 10.61 The brass object is an excess flow control valve. It is a safety valve designed to close if the transfer hose is severed on this chlorine tank car. Courtesy of *Scott Stookey, International Code Council, Washington, D.C.*

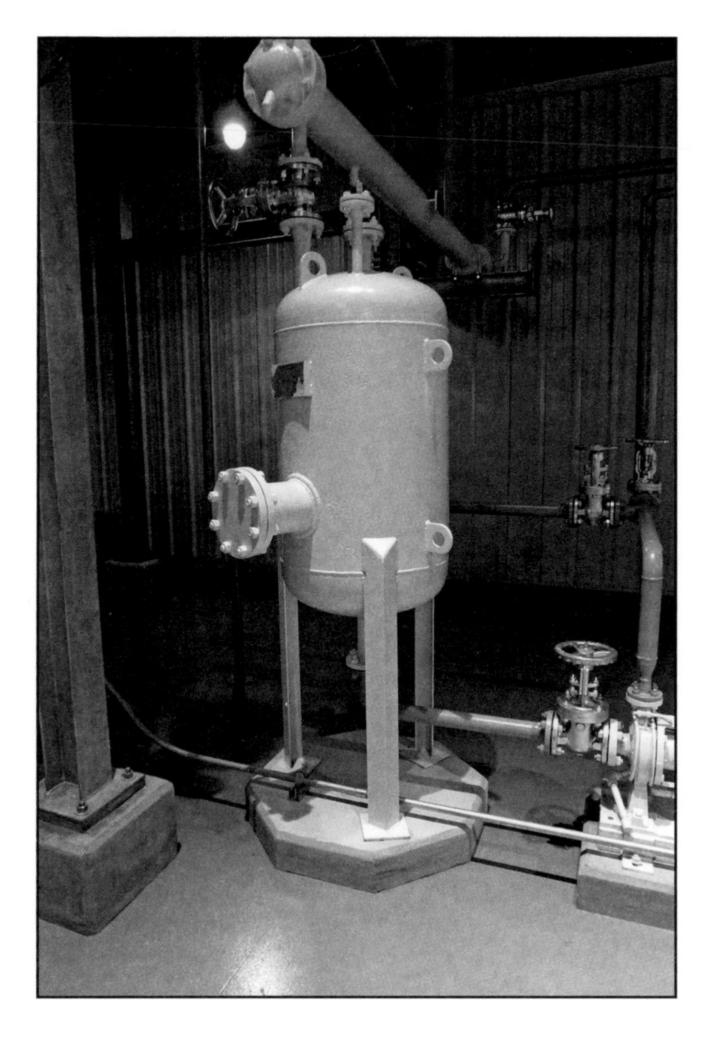

Figure 10.60 A properly designed piping system is erected using the appropriate materials that are chemically compatible for the system's design temperature and pressure. Courtesy of *Scott Stookey, International Code Council, Washington, D.C.*

Hazardous/High-Hazard Occupancies

The amount of hazardous materials in a building may exceed the MAQ even with such engineering controls as automatic sprinkler protection or exhausted enclosures. In these cases, the IBC and NFPA® 400 require the building to comply with its applicable Hazardous (Group H) or High-Hazard occupancy classification. In comparison to other occupancies in the IBC or NFPA® 101, *Life Safety Code*®, hazardous occupancies represent some of the most challenging building uses. This is because their design and construction are based on the classification of hazardous materials that are stored or used and will change as these change.

Hazardous occupancies are assigned to one or more of four different categories, based on the physical or health hazards of the stored materials. Generally speaking, the lower the occupancy number, the greater the hazard the building presents to the occupants and emergency responders.

Categories

Group H-1/High Hazard Level 1 occupancy classifications are assigned to occupancies that present a detonation hazard. This occupancy classification is assigned to buildings that store explosives, blasting agents, or other hazardous materials. These buildings represent the highest level hazard of all physical hazardous materials **(Figure 10.62)**.

Group H-2/High Hazard Level 2 occupancies store materials that present a deflagration hazard. A deflagration is a rapid, energetic release of energy. Group H-2 occupancies also present a risk from accelerated burning of the material. They include buildings or spaces storing the following:

- Flammable gases
- Pressurized flammable or combustible liquids
- Pyrophorics
- Higher-reactivity oxidizers and organic peroxides **(Figure 10.63)**

Group H-3/High Hazard Level 3 occupancies store or use hazardous materials that readily support combustion or present other physical hazards. Level 3 occupancies are designed to store:

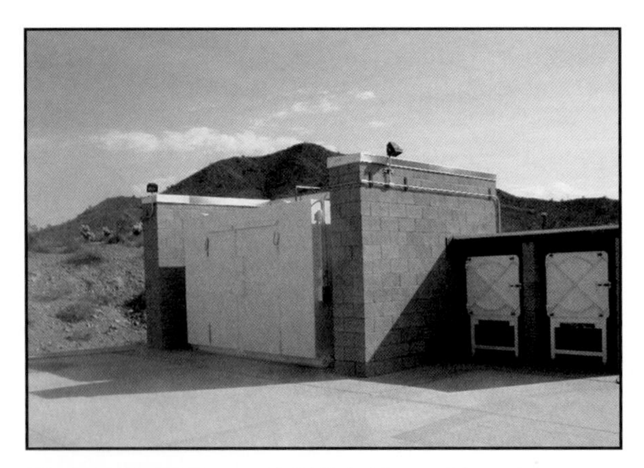

Figure 10.62 Group H-1/High Hazard Level 1 occupancy. Courtesy of *Scott Stookey, International Code Council, Washington, D.C.*

Figure 10.63 Group H-2/High Hazard Level 2 occupancy. Courtesy of *Scott Stookey, International Code Council, Washington, D.C.*

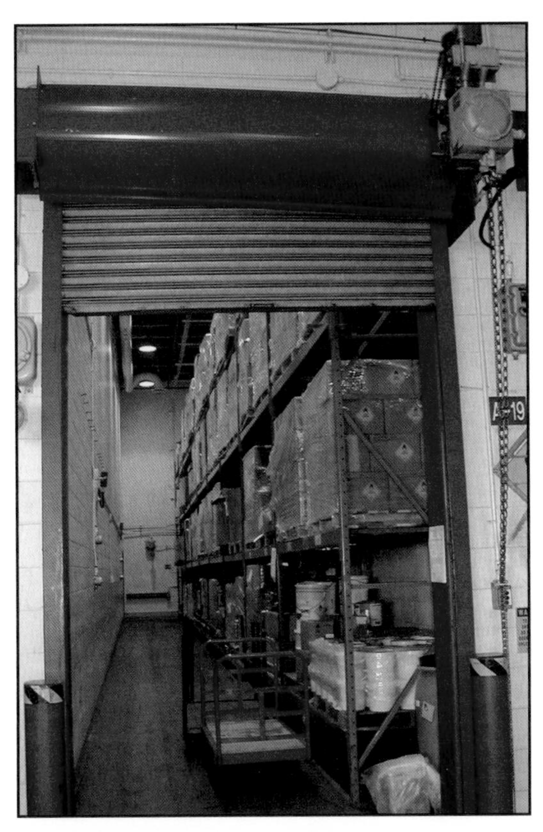

Figure 10.64 Group H-3/High Hazard Level 3 occupancy. Courtesy of *Scott Stookey, International Code Council, Washington, D.C.*

Figure 10.65 Group H-4/High Hazard Level 4 occupancy. Courtesy of *Scott Stookey, International Code Council, Washington, D.C.*

- Unpressurized flammable and combustible liquids
- Lower reactivity oxidizers and organic peroxides
- Flammable solids
- Oxidizing gases **(Figure 10.64)**

Group H-4/High Hazard Level 4 occupancies are designed for the storage of health hazard materials. This occupancy classification is for corrosives, and toxic and highly toxic materials exceeding the MAQ **(Figure 10.65)**.

NOTE: The IBC and NFPA® 400 require a dual hazardous occupancy classification in buildings where the MAQ for physical and health hazard materials are exceeded.

Explosives

An explosion can result in either a detonation or deflagration. A detonation is a heat-producing reaction characterized by a shock wave in the material. The reaction zone progresses through the material at a rate greater than the speed of sound. The principal heating mechanism is one of shock compression. A deflagration is also a heat-producing reaction and occurs when a flammable gas, vapor or combustible dust is rapidly oxidized. The reaction zone progresses through the unburned material at a rate less than the speed of sound.

Some industries designate explosive materials as either primary or secondary explosives. Primary explosives are very sensitive materials, meaning they can be easily detonated and produce a fast pressure wave. These are commonly used in detonators, which are devices used for initiating a detonation and include lead azide or lead styphnate. Secondary explosives will detonate when initiated by a shock wave created by a detonator **(Figure 10.66)**. They normally will not detonate when heated or ignited. Examples include RDX (Cyclotrimethylenetriamine) and PETN (Pentaerthyritol Trinatrate). **Table 10.16** shows the levels of explosives classifications.

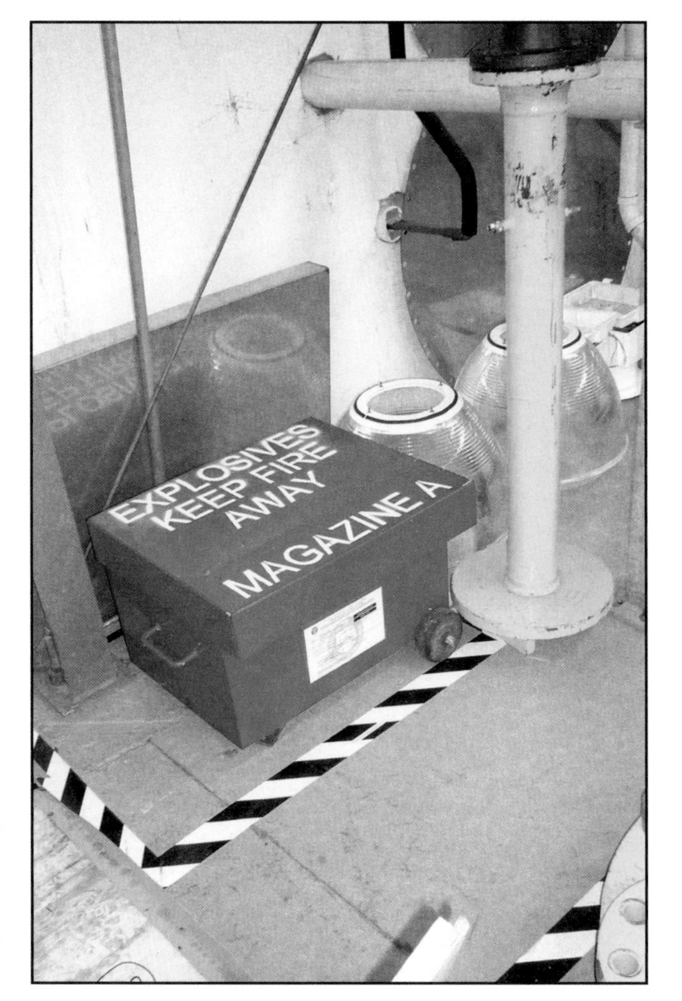

Figure 10.66 Type III explosives daybox. Courtesy of *Scott Stookey, International Code Council, Washington, D.C.*

Requirements

Hazardous occupancies have requirements common to the four occupancy classes. These requirements are consistent between the model building codes and include:

- Group H-1/High Hazard Level 1 occupancies are not permitted in mixed occupancy buildings. They require a separate, stand-alone building without a basement because of the detonation hazard.
- These occupancies require an automatic sprinkler system complying with NFPA® 13. These occupancies must also comply with the sprinkler requirements in the particular NFPA® hazardous material standard.
- Continuous mechanical ventilation to remove vapors or gases released because of an accidental spill or unauthorized discharge to exhaust to a post-chemical treatment system or outside the building.
- Fire-resistive separation when a Hazardous/High Hazard occupancy is located in a mixed occupancy building.
- Under the IBC, additional seismic safety features and special inspections are required for Group H occupancies based on the physical or health hazard classification of the stored materials.
- Means of egress provisions for Group H/High-Hazard Level occupancies are more restrictive when compared to the other occupancies. Hazardous occupancies always require panic hardware regardless of occupant load. The common path of egress travel distances is reduced, and exit access travel distances cannot be increased by installing an automatic sprinkler system.

Table 10.16
UNDMGC/US DOT Explosives Hazard Classification

DIVISION	HAZARDS	EXAMPLES
1.1	Explosives with a mass explosion hazard. A mass detonating hazard is one which affects almost the entire charge instantaneously. It includes substances that can be caused to detonate by means of a blasting cap when confined or will transition from deflagration to detonation when confined or unconfined.	Dynamite, trinitrotoluene (TNT), RDX, PETN, Composition 4 (C-4)
1.2	Explosives that have a projectile hazards but not a mass explosion hazard	Nondetonating encased explosives, military specification ammunition
1.3	Explosives that have a fire hazard and either a minor blast or projectile hazard, or both, but is not capable of a mass explosion. The primary hazard is radiant energy or violent burning, or both. Can deflagrate when confined.	Smokeless powder, display fireworks, perforating charges used in crude oil/natural gas production, explosive bolts, nitrocellulose wetted with not less than 25% methanol, by weight.
1.4	Explosives that pose a minor explosion hazard. The explosive effects are largely confined to the package and no projection of fragments of appreciable size or range is expected. An internal fire must not cause virtual instantaneous explosion of the almost the entire contents of the package.	Consumer fireworks, explosive actuators, signal or aerial flares, power device cartridges
1.5	Blasting agents that are very insensitive explosives. The division is comprised of substances that have a mass explosion hazard but are so insensitive that there is very little chance of detonation under normal conditions or transport. Blasting agents are a mass detonation hazard	Ammonium nitrate combined with fuel oil
1.6	Extremely insensitive articles which do not have a mass explosion hazard. Articles classified as Division 1.6 exhibit a negligible chance of accidental initiation or propagation.	Military ordnance

Engineering Controls for Hazardous Materials

Model fire codes prescribe specific administrative and engineering controls in Group H/High-Hazard occupancies. Secondary containment is required in most hazardous occupancies for many classes of liquid and solid physical and health hazard materials. Mechanical ventilation systems that capture and exhaust vapors generated in the event of a spill or a gas release are required in all hazardous occupancies. Automatic sprinkler protection in hazardous occupancies requires special design considerations. This section reviews the more common engineering controls employed in hazardous occupancies.

Spill Control and Secondary Containment

Hazardous occupancies storing liquid or solids require spill control or secondary containment. They are required to contain an accidental spill and keep it within the occupancy. Secondary containment is prescribed when the outdoor MAQ is exceeded, and in all cases of exterior flammable and combustible liquid storage. Spill control or secondary containment is not required for compressed or liquefied compressed gases.

Using the floor surface of the hazardous occupancy will satisfy the requirements for spill control. Use of the building floor is a permissible method if the container volume is less than 55 gallons (220 L) of liquids or 550 pounds (275 kg) of solids. The IFC and NFPA® 400 require that the floor be liquid tight by using concrete or asphalt construction. Containment pallets can also satisfy the requirement for spill control. Containment pallets are designed to capture the contents of a single 55-gallon (220 L) or smaller container **(Figure 10.67)**.

Model fire codes require secondary containment for the following:

- **Liquids** — When the container volume is greater than 55 gallons (220 L) or the aggregate volume is greater than 1,000 gallons (4 000 L).
- **Solids** — When the individual container weight is greater than 550 pounds (275 kg) or the aggregate weight of hazardous materials is more than 10,000 pounds (5 000 kg).

Model fire codes offer several compliance options for secondary containment. In all cases, the design must provide a liquid-tight method of containing the spill from the largest container, portable tank, or stationary AST. Designs may utilize raised curbs or foundations constructed with a depression with sufficient depth to contain the contents **(Figure 10.68)**.

Another code compliance option is a drainage system that captures a spill and drains it to an outdoor containment basin or a storage tank. A key consideration in the design of a drainage system is it must not allow the mixing of incompatible hazardous materials **(Figure 10.69)**.

Figure 10.67 Containment pallet. Courtesy of *Scott Stookey, International Code Council, Washington, D.C.*

Figure 10.68 Curbed secondary containment structure in a Group H occupancy. Courtesy of *Scott Stookey, International Code Council, Washington, D.C. (Photograph courtesy of Scott Stookey, International Code Council, Austin TX)*

Figure 10.69 Drainage scupper in a Group H occupancy. The scupper terminates into a dedicated drain that terminates into an exterior containment structure. Courtesy of *Scott Stookey, International Code Council, Washington, D.C.*

Model fire codes require monitoring of a secondary containment or drainage system. In most cases, this can be accomplished by visual surveys. An electronic means of monitoring may be necessary for drainage systems where all parts of the system cannot be visually monitored.

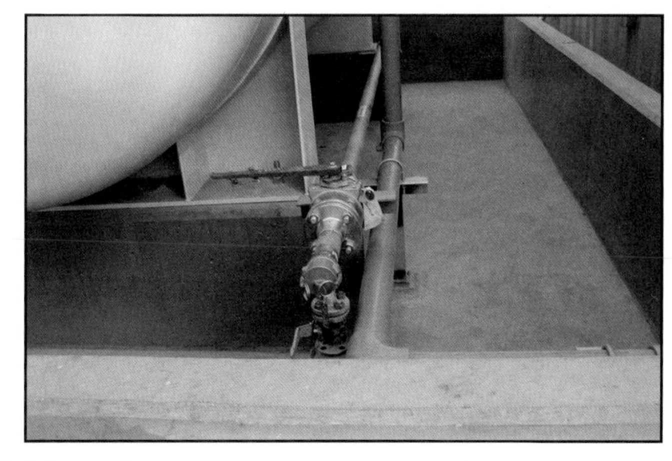

Figure 10.70 Outdoor secondary containment structure for hazardous materials storage. Courtesy of *Scott Stookey, International Code Council, Washington, D.C.*

An indoor secondary containment or drainage system is designed to collect the volume of the largest container or tank plus the discharge from an operating automatic sprinkler system. Model fire codes require that the system be capable of containing a minimum of 20 minutes of sprinkler water flow over the design area of the automatic sprinkler system or the area of hazardous occupancy, whichever is smaller.

With the exception of flammable and combustible liquids, model fire codes require secondary containment of hazardous materials outdoor storage when the outdoor MAQ is exceeded **(Figure 10.70)**. The secondary containment design for hazardous materials must contain the volume of the largest container or storage tank and contain the volume of rainfall produced in a 24-hour/25-year storm.

Storm Data

The National Oceanic and Atmospheric Administration (NOAA) isopluvial maps that show 24-hour/25-year storm data. *Pluvial* is a meteorological term meaning the amount of precipitation falling in one day or other specified period that is likely to be equaled or exceeded only once in a century. Isopluvial maps for each U.S. state are available at the NOAA Hydrometeorological Design Studies Center.

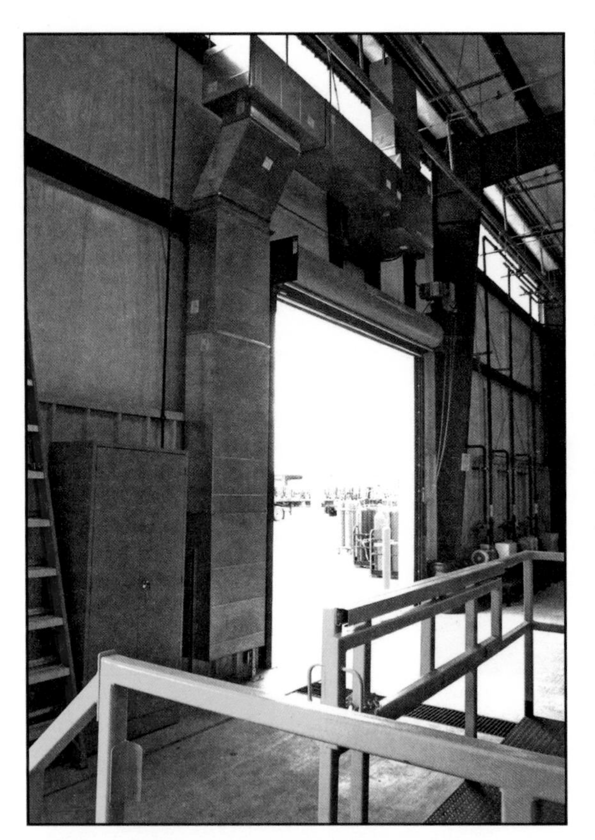

Figure 10.71 A mechanical ventilation system in a Group H/High-Hazard use is designed to capture vapors, gases or fumes from a leak or accidental release. Courtesy of *Scott Stookey, International Code Council, Washington, D.C.*

Mechanical Ventilation System

Group H/High Hazard occupancies must have mechanical ventilation systems. These systems provide adequate air movement to exhaust fumes, vapors, or gases created by a container, piping, or valve or component leak. Mechanical ventilation is not required for flammable solids.

Hazardous mechanical exhaust systems are designed and constructed in accordance with the IMC or NFPA® 91. The design of the system must provide enough air at a sufficient velocity to capture gases or vapors. A design issue is the vapor density of the stored materials. Materials with a vapor density greater than one require ventilation openings within 12 inches (300 mm) of the floor. Lighter-than-air vapors or gases require ventilation openings within 12 inches (300 mm) of the highest point of the room. Mechanical ventilation systems in hazardous occupancies are designed to provide a minimum air flow rate of one cubic foot per minute per square foot (cubic meter per minute per square meter) over the area of occupancy or space **(Figure 10.71)**.

The selection of the duct and fan air foils (blades) is another consideration. Nonmetallic ducts may be required for corrosive materials. Materials like glass-fiber reinforced plastic or chlorinated PVC are acceptable but must be constructed of materials with limited flame spread and smoke production characteristics. These duct materials may be required to be listed for specific applications. In most cases, the duct and air foils are of noncombustible construction.

Unless they are designed to remove contaminants, hazardous mechanical exhaust systems cannot recirculate air into the building or space. Model fire codes require a post-release treatment system for ventilation systems designed to capture highly toxic or toxic vapors or gases. These systems are designed to reduce the toxicity of the released material.

To assist inspectors in determining the adequacy of emergency vents in stationary tanks, UL 142 requires the nameplate on all shop-fabricated ASTs to indicate the required emergency vent flow rate. The flow rate is expressed in standard cubic feet/hour (SCFH).

To verify that the emergency vent is properly sized, the inspector should compare the required flow rate posted on the nameplate to the flow rate marked on the emergency vent **(Figure 10.72)**. NFPA® 30 requires all emergency vents to be permanently marked with their design flow rate. If the flow rate on the emergency vent equals or exceeds the prescribed flow rate on the tank nameplate, the installation complies with NFPA® 30. If the emergency vent flow rate is less than the value on the tank's nameplate, the installation is not in compliance with any model fire code.

Automatic Sprinkler Protection

Group H/High-Hazard occupancies require automatic sprinkler protection. Depending on the occupancy classification and locally adopted codes, sprinkler protection may be required throughout the building or only within the

Figure 10.72 An inspector may be required to verify the size of vent openings on a tank. Courtesy of *Scott Stookey, International Code Council, Washington, D.C.*

Figure 10.73 Horizontal barrier with in-rack sprinklers in a Group H flammable liquids warehouse. Courtesy of *Scott Stookey, International Code Council, Washington, D.C.*

hazardous occupancy. The model fire codes generally do not dictate the type of sprinkler system, only that it must comply with NFPA® 13 requirements. NFPA® 13 requires specialized automatic sprinkler designs be installed for certain classes of physical hazard materials, such as:

- Flammable and combustible liquids
- Compressed gases that are classified as pyrophoric, flammable or oxidizers
- Solid and liquid oxidizers and organic peroxides

Of all the physical hazards, flammable and combustible liquids present the most variables in the design of an automatic sprinkler system.

One method of satisfying the NFPA® 30 sprinkler design requirements is installing in-rack sprinklers with a horizontal barrier. A horizontal barrier is used in rack storage to accelerate the activation of the in-rack sprinklers. The construction requirements for horizontal barriers and the sprinkler design requirements must be strictly adhered to so the fire is controlled before it begins to compromise the container **(Figure 10.73)**.

Chapter Summary

The ability to distinguish hazardous materials or dangerous goods by their properties and identify the materials or goods by the markings on the containers is basic to an inspector's duties. This chapter provided information that will help an inspector apply the requirements contained in the model building and fire codes for storing, handling, transporting, dispensing, and using the materials or goods. Information was also given for federal markings used to identify the materials or goods while in transit. This information can be used to generate an inventory of the type and size of the hazards present in the community. An inspector must be proficient in reading the placards, markings, and labels of all types of hazardous materials or dangerous goods and using reference guidebooks.

During the course of an inspection, an inspector may encounter materials that may individually or in combination pose a risk or hazard. It is essential for an inspector to recognize these potential hazards and have them removed, modify the way hazardous materials are stored, or provide for measures that reduce risks to acceptable levels. The inspector must also be familiar with the maximum allowable quantities of hazardous materials that are permitted based on occupancy type and the materials themselves. Hazardous and High-hazard occupancies require special attention because of the materials involved and potential for a catastrophic event.

This chapter provided information on the types of storage containers and packaging used to store hazardous materials and dangerous goods in North America. Also included were inspector concerns related to unsafe behaviors that may result from improper storing, handling, dispensing, transporting, using, and disposing of hazardous materials.

Review Questions

1. Name some of the ways in which hazardous materials regulations are applied.
2. List some of the codes and standards that apply to hazardous materials.
3. What are some of the resources available to assist inspectors in determining hazardous material classifications?
4. What makes flammable and combustible liquids the most "misused"?
5. Describe the three categories of gases.
6. What hazards are associated with cryogenic liquids?
7. What actions can make flammable solids capable of causing a fire?
8. Explain the hazard classifications of organic peroxides.
9. What are the hazard categories of solid and liquid oxidizers based upon?
10. Name some examples of pyrophoric materials.
11. What do unstable materials react to?
12. List the states in which explosives and blasting agents exist.
13. How are highly toxic and toxic materials classified?
14. How are corrosives classified in fire codes?
15. How does mixing two or more materials impact hazard classification?
16. What is required to identify the appropriate storage classification of incompatible materials?
17. List some of the resources inspectors must be familiar with and know how to use in order to identify hazardous materials.
18. Compare and contrast DOT placards, labels, and markings.
19. What are some of the other types of markings inspectors might encounter?
20. List some of the references used to identify hazardous materials.

21. Why is it important for inspectors to be aware of different aspects of the Canadian Dangerous Goods System?
22. What are some of the differences between markings in the Mexican Hazard Communication System and those found in the U.S. and Canada?
23. Describe the requirements for marking piping and cylinders.
24. What must a code official have to determine the MAQ?
25. Describe examples of control areas.
26. List examples of containers used for nonbulk and bulk packaging.
27. List types of stationary tanks that may be used for nonbulk and bulk packaging.
28. What are some of the things inspectors need to verify when inspecting hazardous materials storage?

1. What can an inspector require to clarify processes?
2. What can an inspector do in response to a unauthorized discharge?
3. List some of the requirements for piping, valves, and fittings used to convey hazardous materials.
4. Describe the classification categories for hazardous and high-hazard occupancies.
5. What is the purpose of spill control and secondary containment?
6. List some of the requirements for mechanical ventilation systems.
7. List some of the requirements for automatic sprinkler systems in hazardous occupancies.

Water Supply Distribution Systems

Chapter Contents

Case History **465**

Water Supply System Components **466**

System Design.. 466

Water Supply Sources... 468

Means of Moving Water 470

Distribution Systems 472

Water Supply Testing **477**

Fire Hydrant Inspections 480

Pitot Tube and Gauge 481

Fire Flow Test Computations 482

Required Residual Pressure 484

Fire Flow Test Procedures 485

Available Fire Flow Test Results Computations 490

Chapter Summary **493**

Calculations **493**

Review Questions **498**

Key Terms

Butterfly Valve 474
Cavitation 485
Combination System 470
Direct Pumping System 470
Distribution System 466
Flow Hydrant 485
Gate Valve 474
Gravity System 470
Normal Operating Pressure 486
Pitot Tube 480
Residual Pressure 479
Static Pressure 479
Steamer Connection 477
Test Hydrant 485
Water Main 472

NFPA® Job Performance Requirements

This chapter provides information that addresses the following job performance requirements of NFPA® 1031, *Standard for Professional Qualifications for Fire Inspector and Plan Examiner* (2014).

Fire Inspector I

4.3.16

Water Supply Distribution Systems

Learning Objectives

After reading this chapter, students will be able to:

Inspector I

1. Identify components of public water supply systems. (4.3.16)
2. Identify characteristics of private water supply systems. (4.3.16)
3. Explain water supply testing. (4.3.16)
4. Explain fire hydrant inspections. (4.3.16)
5. Describe how to use a pitot tube and gauge to take flow readings. (4.3.16)
6. Explain how to use flow test computations. (4.3.16)

Chapter 11
Water Supply Distribution Systems

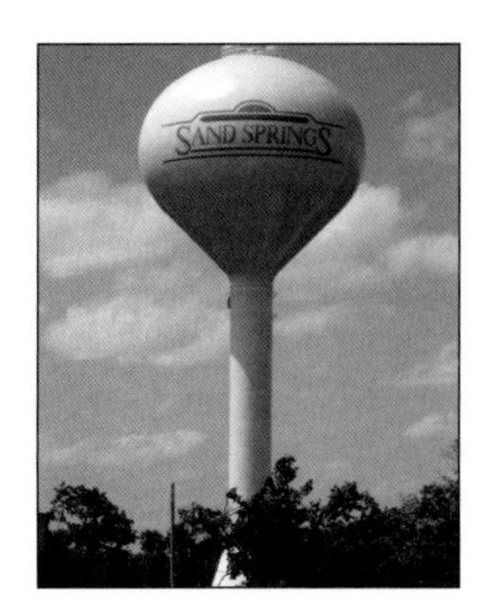

Case History

An August, 2011 fire in a residential neighborhood in Seattle, Washington, resulted in total loss of a structure and minor injuries to several firefighters. Arriving crews encountered significant problems with establishing a reliable water supply. The first and closest hydrant failed to provide enough water pressure, and the second hydrant, farther away, didn't work at all. Finally, a third hydrant, 11 blocks away, worked. Nine pumpers were required to keep water pumping to the house. Because it took so long to flow water to the burning structure, the house was a total loss.

Although the city of Seattle conducts regular hydrant inspections, most of the supply problems traced back to the area's unincorporated days, with smaller mains installed by a county water district. Property owners had been asked twice to pay for upgrades, and twice said no. After the fire, upgrades were ordered to bring most of the city up to modern-sized mains.

I

An adequate water supply and distribution system is necessary for agricultural, industrial, and domestic use. It is also one of the most important tools firefighters use to control and extinguish fire **(Figure 11.1)**. Water-based fire-suppression systems, including automatic sprinklers, standpipes, and other types of systems, are useless without a water supply distribution system. Inspectors must be familiar with the types of water supply distribution systems in their communities in order to verify that the systems can adequately handle emergency situations.

NOTE: Sprinklers and standpipes are described in Chapter 12.

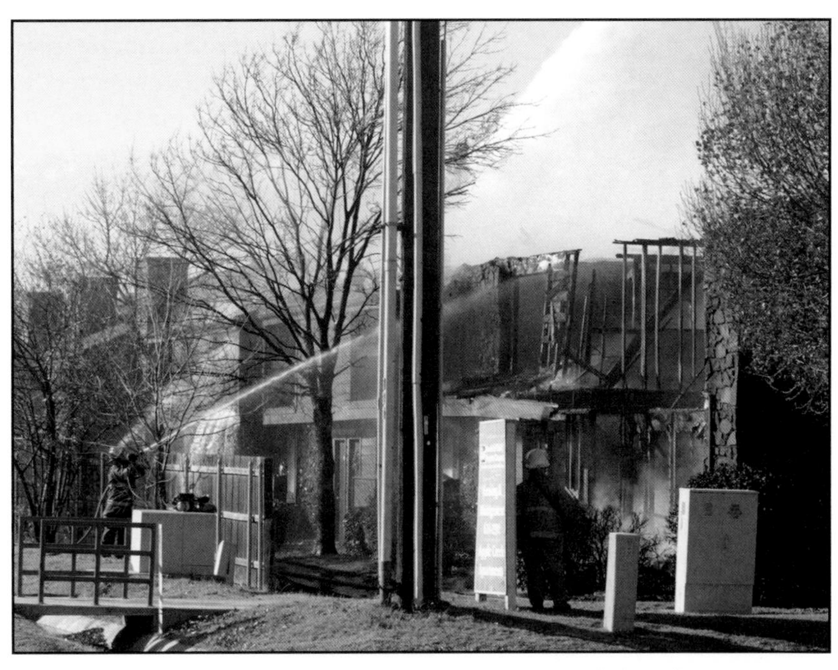

Figure 11.1 Fire fighting operations such as this one rely heavily on an adequate water supply and distribution system. Without this system or without its functioning optimally, fires like this might not be extinguished.

Although the responsibility for the water supply distribution system may rest with various water suppliers, an inspector should understand the basic principles, components, and design of water supply distribution systems. Inspectors must also be aware of the fire inspection division's authority to test and inspect public and private water supply distribution systems. Inspectors can usually obtain information on the local water supply distribution system or network from the water department.

This chapter provides information for an understanding of water supply distribution system components, including water sources, processing, movement, and delivery, plus hydrants and valves. The chapter also includes information on flow testing, hydrant inspection and maintenance, and water supply capabilities analyses.

Water Supply System Components

There are two basic types of water supply systems in North America: public and private systems. Public water supply distribution systems are usually a function of local government, such as a department of municipal government or a state-authorized water district. The local water department is usually a separate city utility whose main function is to provide sanitary (potable) water that is safe for human use. An elected board generally governs a state-authorized water district.

A private water supply system may provide water under contract to a municipality, region, or single property. Private water supply distribution systems may be specific to an industrial facility or complex (such as a refinery), or supply water to a residential subdivision. The facility owner/occupant of a private water supply system is responsible for the inspection, testing, and maintenance of the system.

An inspector may encounter private water supply systems when performing site inspections or plans review for new construction. The inspector may also be required to determine that private water supply distribution systems have been properly tested, inspected, and maintained. Inspectors must be familiar with the basic principles of any private water supply distribution systems that are within their jurisdictions.

System Design

The design of public and private water supply systems varies from region to region. However, all systems have the following basic components **(Figure 11.2)**:

- **Water supply source(s)** — Lakes, reservoirs, ponds, rivers, streams, aquifers, wells, or springs
- **Processing or treatment facilities** — Purification or desalination plants
- **Means of moving the water** — Water pumps
- **Water distribution and storage systems** — Storage tanks, control valves, piping systems, and hydrants

NOTE: A reservoir can serve as a water source and as part of the **distribution system**.

Distribution System — That part of an overall water supply system that receives the water from a water source and delivers it by means of pumps or gravity to the area to be served.

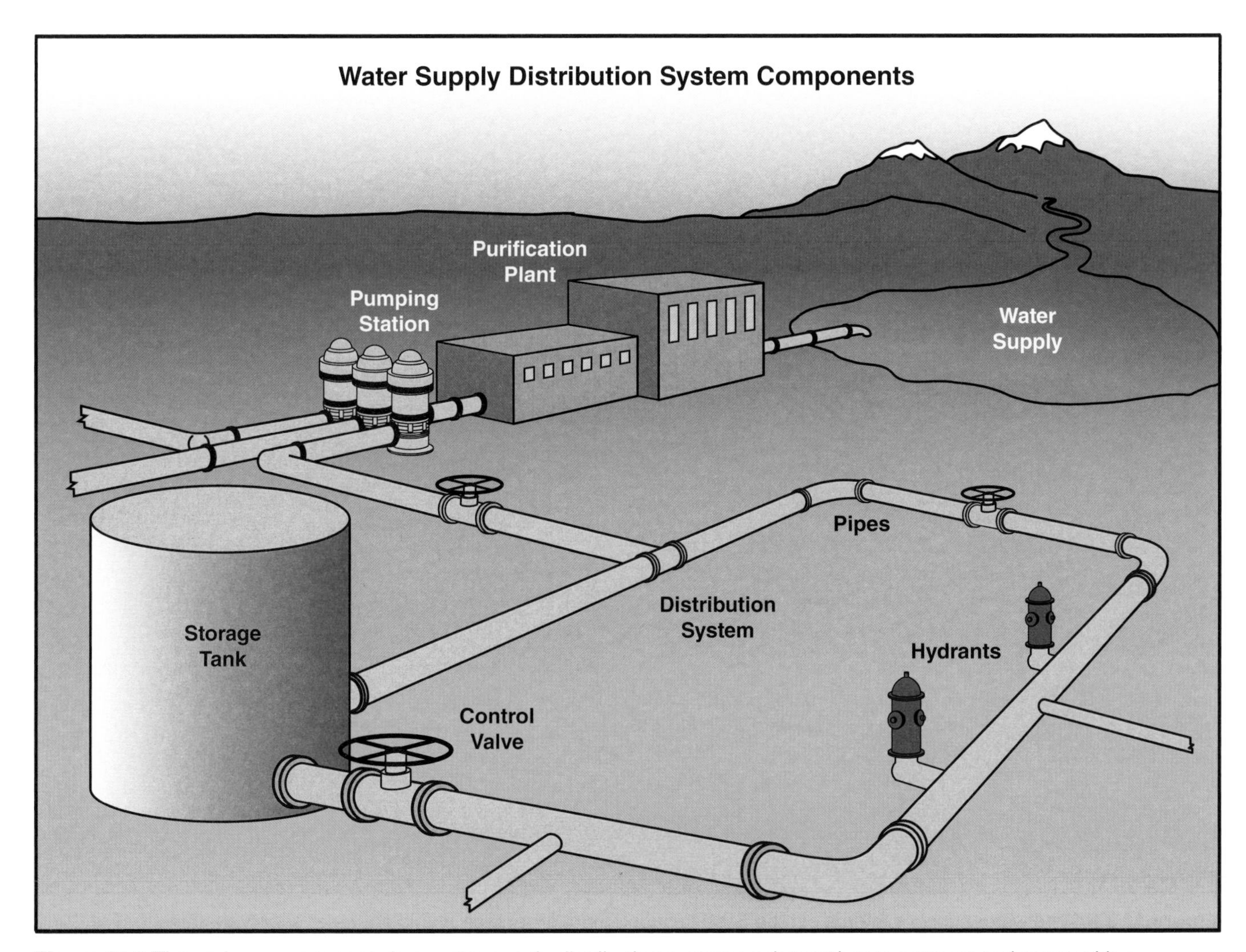

Figure 11.2 The various components in a water supply distribution system work together to connect end users with a municipality's water supply.

Fire Protection and Water Distribution Systems

The inspector should be aware that not all water distribution systems have been designed and installed with fire protection as a primary concern or a primary intent of the system. Providing water for fire protection may be a secondary goal; as a result, the system may not be adequate for this use. In addition, many systems have not been updated to keep pace with the needs of modern fire protection systems.

Inspectors must be familiar with the design and reliability of private water supply distribution systems in their jurisdiction **(Figure 11.3, p. 468)**. Well-maintained distribution systems may provide a reliable source of water for fire protection purposes. Small capacity or poorly maintained private water supply distribution systems should not be relied upon to provide all the water necessary for adequate fire protection. Reviewing documentation is the best method for evaluating the capabilities of the system.

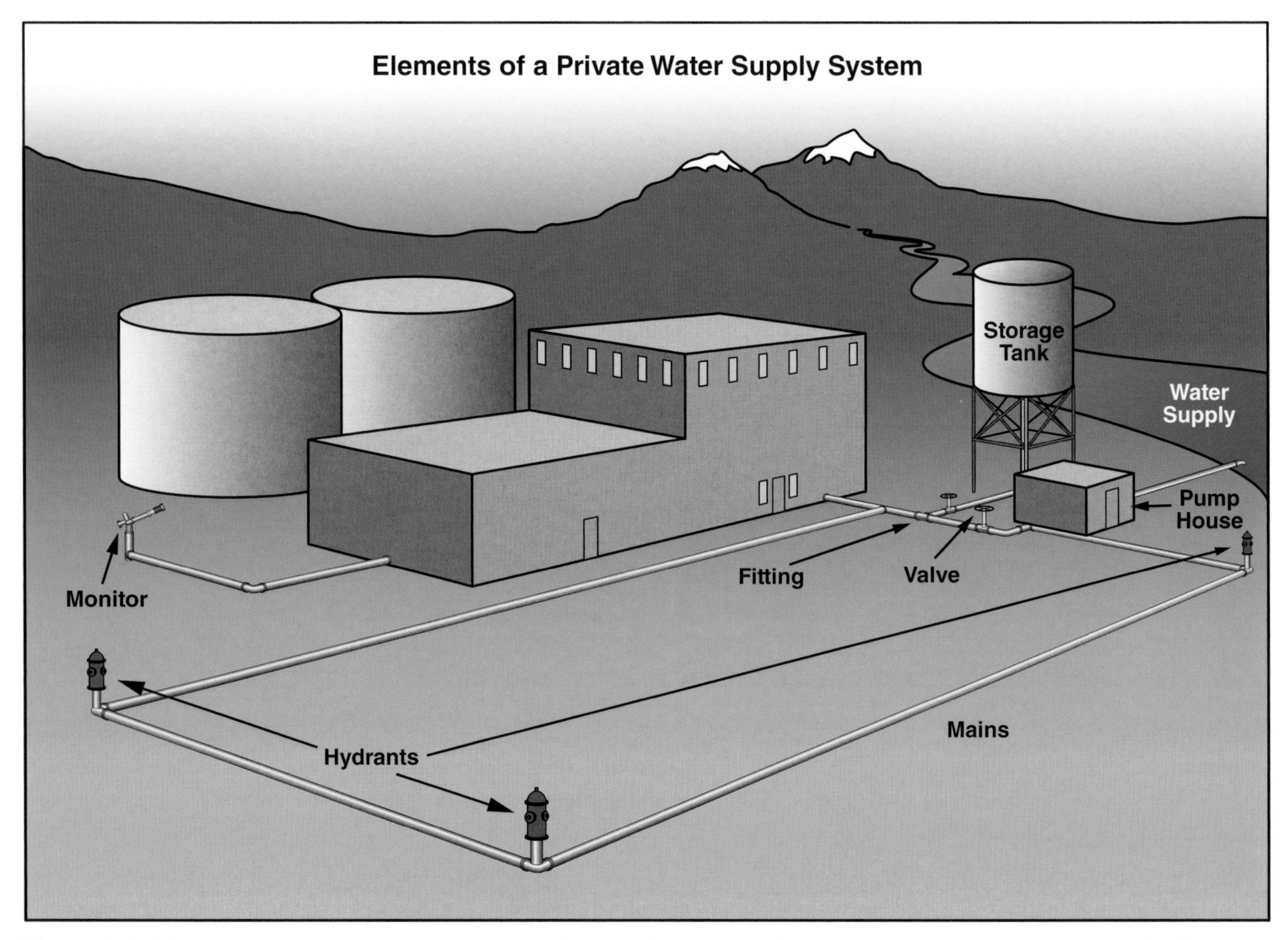

Figure 11.3 Private water supply systems can be used to supplement or replace municipal water supply systems for fire protection.

Historically, many significant fire losses can be traced, at least in part, to the failure of a private water supply distribution system that municipal fire departments have used at an incident. Problems, such as the discontinuation of electrical service to a facility whose fire protection system is supplied by electrically driven fire pumps, have resulted in disastrous losses.

Water Supply Sources

Water originates at the source and is moved to the treatment or processing facility before reaching the distribution system for use. The point of use may be an individual residence, commercial property, private water supply system, or public fire hydrant. Water sources may include:

- Reservoirs (impounded water)
- Suction tanks
- Pressure tanks
- Gravity tanks

These sources may be independent of a public water distribution system or may supplement it.

Before it can be used, most water must be processed to remove impurities that can be harmful to humans, animals, and plants. Most water purveyors operate water treatment plants, and most that depend on ocean water operate desalination plants and treatment plants. Once water has been used in most municipalities, it enters sanitary sewer systems that take it to another type of treatment plant that removes human waste and impurities before discharging it back into rivers, lakes, and oceans.

Private water supply distribution systems commonly receive their water from a municipal water supply distribution system separated from the private system by a water meter and check valve or an on-site water supply, such as a well or lake. In some cases, the private system may have its own water supply source independent of the municipal water distribution system, such as a well, impounded water supply, or river.

Reservoirs

A reservoir may be an open lake or pond or an enclosed structure similar to a tank. Water may originate from the lake, pond, or well and be pumped into the reservoir **(Figure 11.4)**.

Suction Tanks

Suction tanks are located at ground level and provide a water supply source for fire pumps. Tanks are sized based on the demand and system needs.

Pressure Tanks

Pressure tanks can be used for limited private fire protection services. They contain water under air pressure that is released when the system requires it. Pressure tanks should be provided with low/high water-level and air-pressure supervision gauges or devices **(Figure 11.5)**.

Figure 11.4 Pumping stations facilitate the movement of water from a natural source, such as a lake into a reservoir.

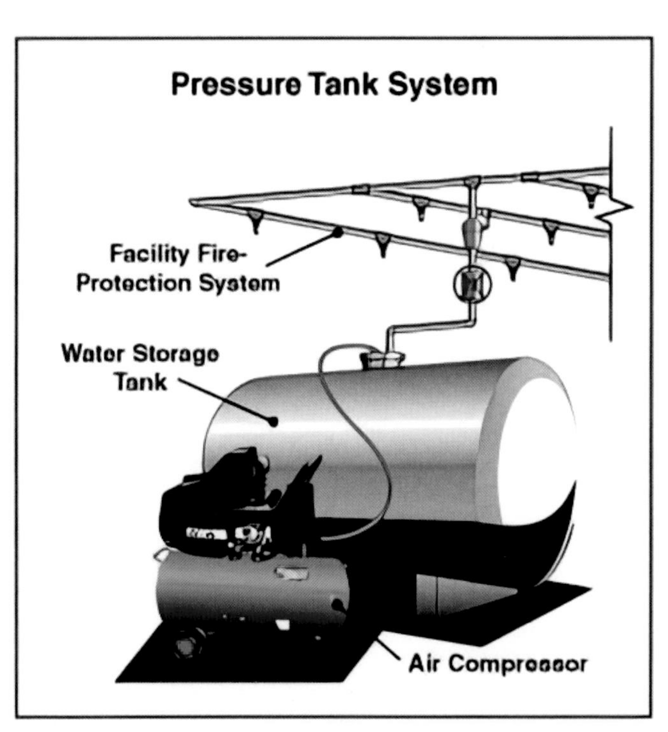

Figure 11.5 Air pressure moves water from the storage tank through the facility fire-protection system.

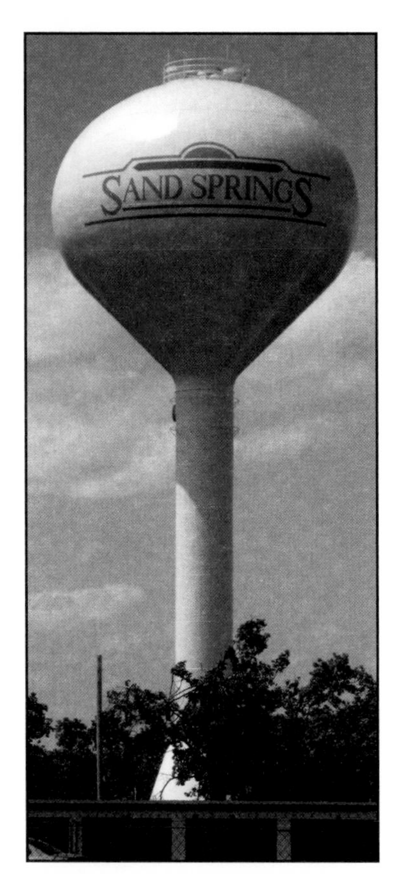

Figure 11.6 Elevated water storage tanks use gravity to create water pressure throughout the system. *Courtesy of Sand Springs (OK) Fire Department.*

Gravity Tanks

Gravity tanks are used to stabilize or balance the pressures on distribution systems at times of peak demand. The tanks are elevated above the point of demand, either on a tower, top of a structure, or high terrain **(Figure 11.6)**.

Once the demand on a system becomes so great that the pumps cannot keep up, the system pressure will begin to drop. When the pressure drops to a point where it cannot keep the tank full, the tank will begin to add water to the system. Elevated storage tanks become an important fire protection feature during peak demand periods. It is only necessary to keep the tanks full and the valves open so that water will flow when needed. Gravity tanks can require considerable maintenance and often require protection against freezing.

Elevated tanks have several problems in providing fire protection water supplies. The most important of those problems is the limitation of pressure available from the tank. A tank has to be 100 feet (30 m) high to generate a pressure of only 43 psi (301 kPa). This pressure is not adequate to meet many modern fire protection system designs.

For example, early suppression fast response (ESFR) sprinklers used for warehouse protection require at least 50 psi (350 kPa) at the most hydraulically demanding sprinkler. System demands for other high-challenge occupancies also require tanks to be at impractical heights to provide adequate pressure. Therefore, it is more useful to use ground-level storage and a fire pump for the water supply redundancy required at many highly protected facilities.

Means of Moving Water

Water must be moved from its initial source to treatment facilities and from there to the point where it is distributed and used. Three methods for moving water through the system are gravity, direct pumping, or a combination system using both direct pumping and gravity. A **gravity system** delivers water from the source to the distribution system without pumping equipment **(Figure 11.7)**. The difference in the height of the water source and the point of use creates elevation pressure, also known as elevation head pressure. The elevation pressure forces the water throughout the water distribution system.

When elevation pressure cannot provide sufficient pressure to meet the needs of the community, one or more pumps installed in the distribution system create the pressure within the distribution system. This system for moving water is called a **direct pumping system** (**Figure 11.8**).

Most communities use **combination systems** that consist of both gravity tanks and the direct pumping process to provide adequate pressure. Water is pumped into the distribution system and elevated storage tanks that provide gravitational pressure **(Figure 11.9, p. 472)**. When the consumption demand is greater than the rate at which the water is pumped, water flows from the storage tanks into the distribution system. When demand is less, water is pumped into the storage tanks.

Gravity System — Water supply system that relies entirely on the force of gravity to create pressure and cause water to flow through the system. The water supply, which is often an elevated tank, is at a higher level than the system.

Direct Pumping System — Water supply system supplied directly by a system of pumps rather than elevated storage tanks.

Combination System — Water supply system that is a combination of both gravity and direct pumping systems. It is the most common type of municipal water supply system.

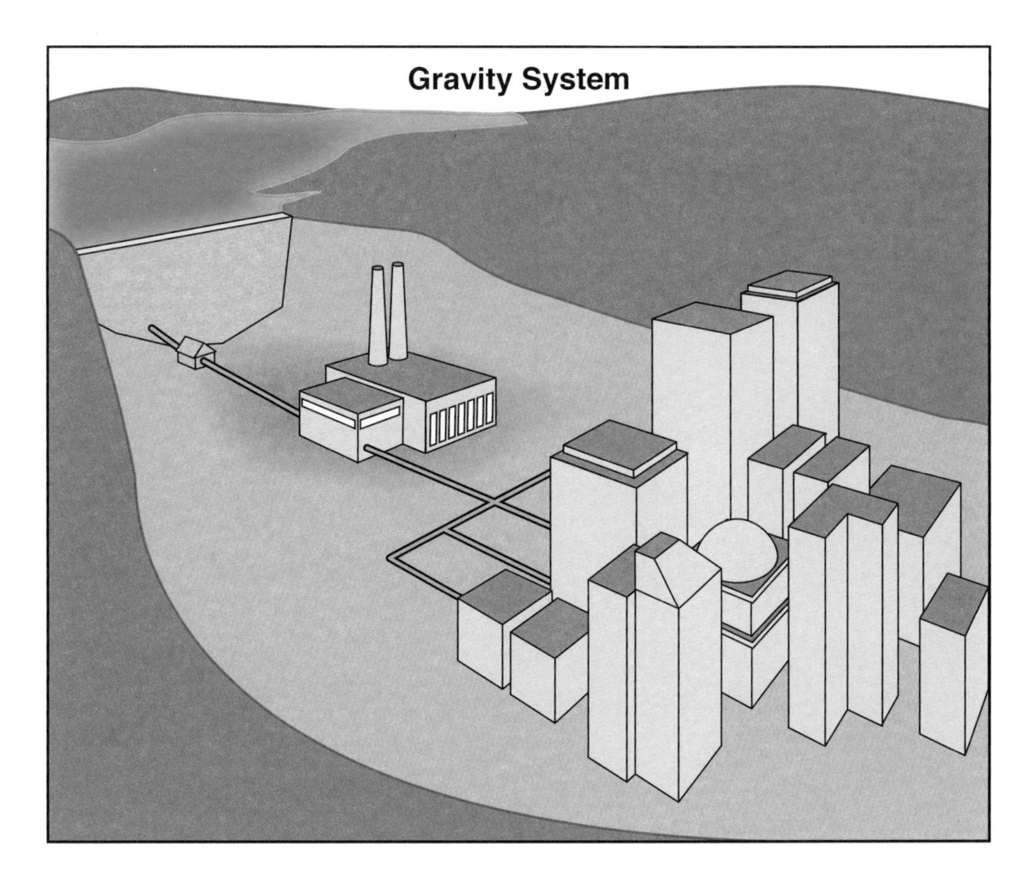

Figure 11.7 A gravity system uses the height of the water source to create pressure instead of relying on pumps.

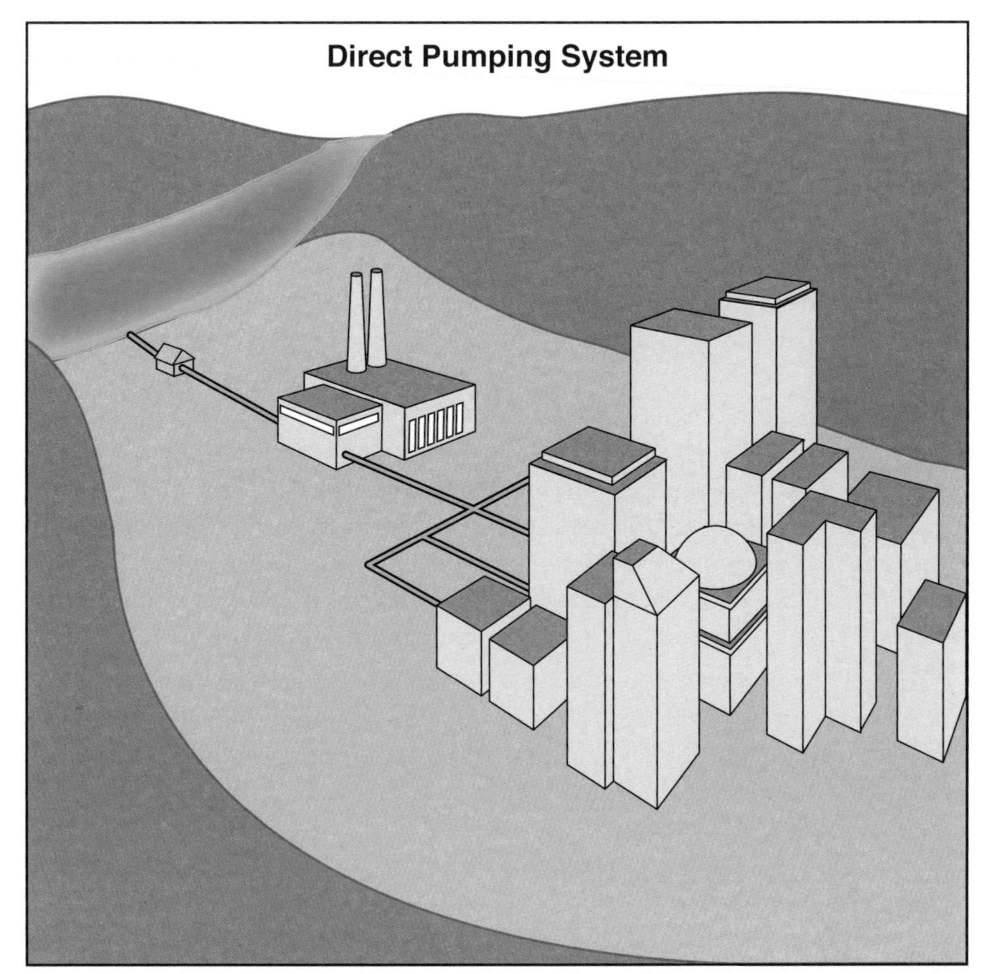

Figure 11.8 When the surface water source lacks the required elevation to generate pressure, a series of pumps moves the water.

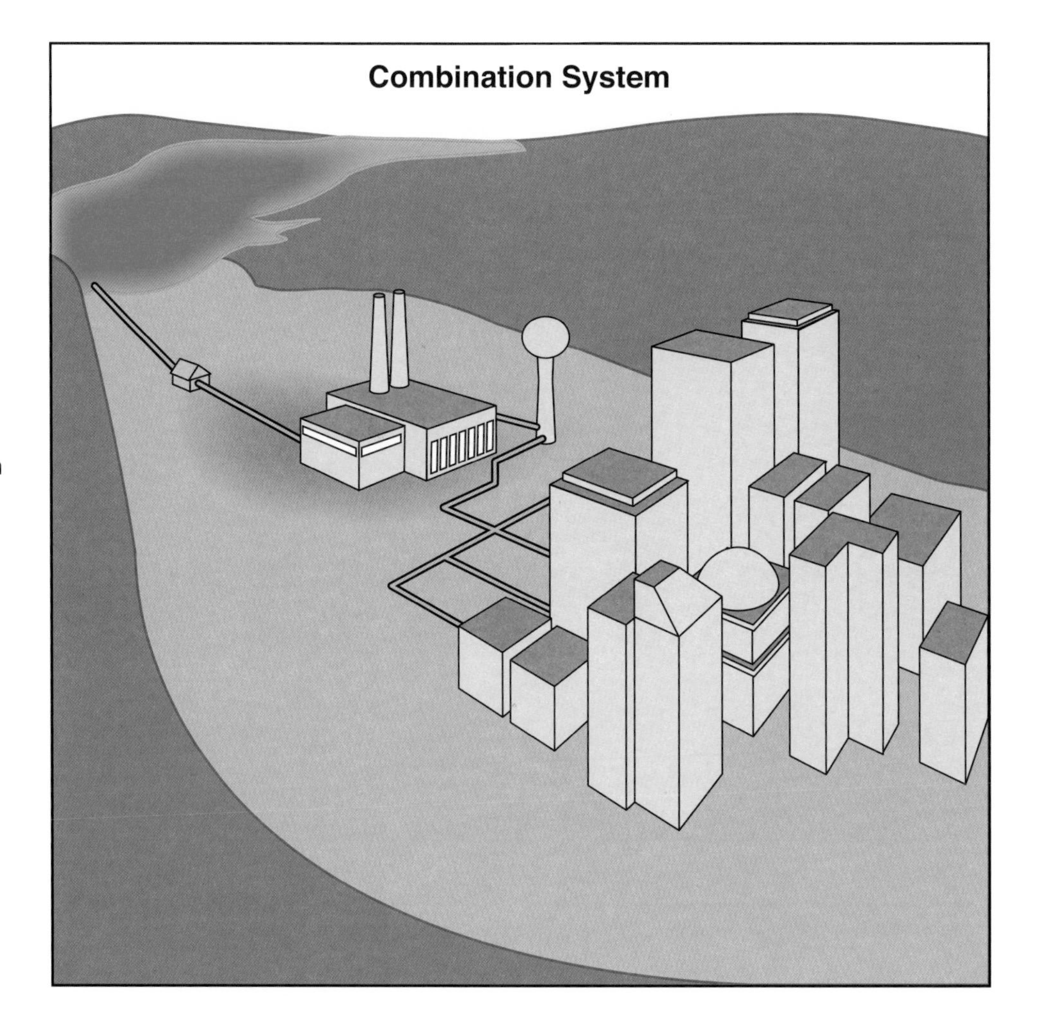

Figure 11.9 A combination system uses elevated storage tanks to accommodate variable water usage rates.

Distribution Systems

The water distribution system consists of a network of pipes, storage tanks, control valves, and hydrants throughout the community or service area that carry the water under pressure to the points of use. The sections that follow describe each of these system components.

Piping

Water Main — A principal pipe in a system of pipes for conveying water, especially one installed underground.

The distribution system receives water from the pumping station/treatment facility and delivers it throughout the service area. Fire hydrants, gate valves, elevated storage, and reservoirs are secondary parts of the distribution system. The term *grid* or *gridiron* describes the interlocking network of **water mains** that compose a water distribution system. A water distribution system consists of three types of water mains loosely referred to as follows **(Figure 11.10)**:

- **Primary feeders** — Large pipes, also known as arterial mains, with relatively widespread spacing. These mains convey large quantities of water to various points in the distribution system and supply smaller secondary feeder mains. Fire hydrants are rarely attached directly to these mains.
- **Secondary feeders** — Intermediate pipes that interconnect with the primary feeder lines to create a grid. Control valves can isolate each secondary feeder.

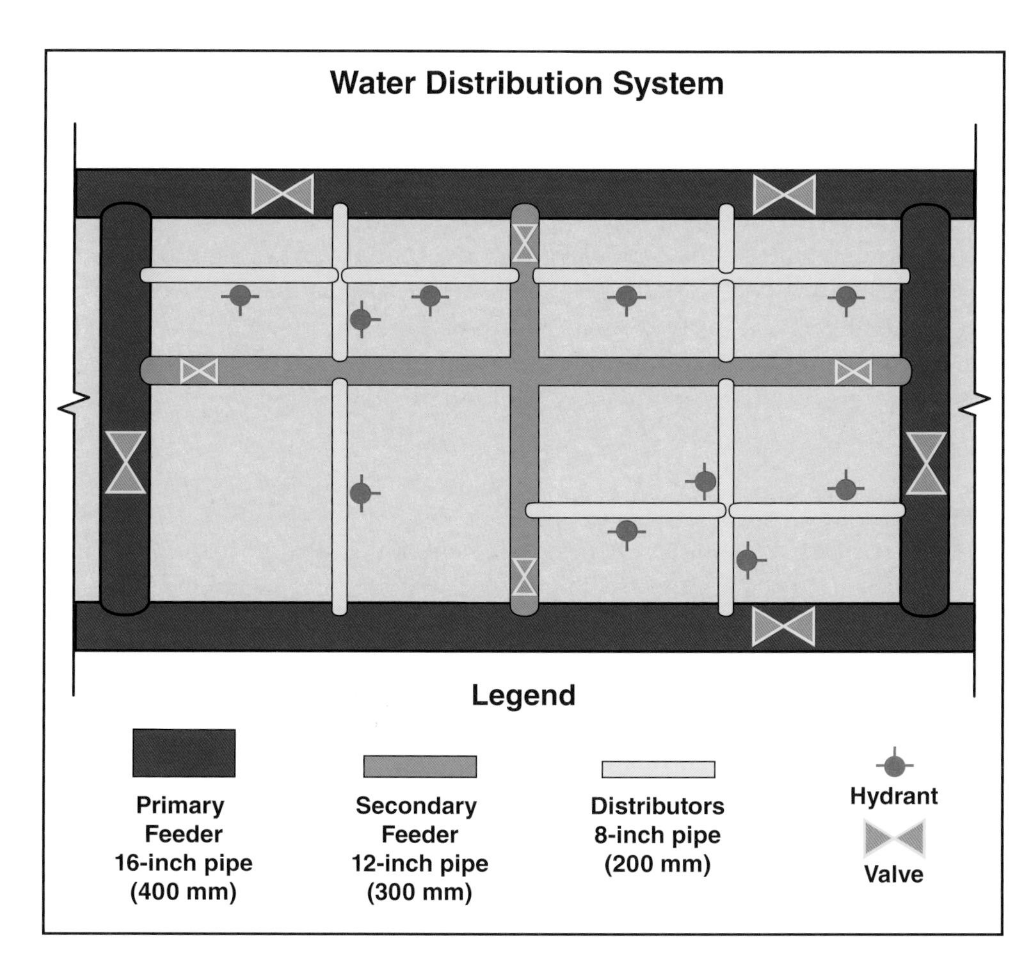

Figure 11.10 This illustration shows how the various feeders and distributors in a water distribution system work together.

- **Distributors** — Small water mains that serve individual fire hydrants and commercial and residential consumers. Distributors may form an intermediate grid between secondary feeders or may be dead-end lines with the hydrant or supplied property at the end of the line.

Water distribution systems are generally designed using hydraulic calculation software that is intended to provide constant pressure range and reliable delivery throughout the system. The grid or loop system is designed to provide constant pressure or flow when pipes or the grid need repair. Another advantage of the grid system is that high demand in one area does not reduce water flow in other areas. Dead-end lines can result in reduced water flows.

Connections to the piping system provide access to the water supply system. These connections may be through water flow control valves and flowmeters at the point where customers gain water from the system or through fire hydrants that are used for fire protection.

Private fire protection system piping, valves, and fire hydrants are located on private property and maintained by the property owner **(Figure 11.11)**. Generally, these components are similar to the ones found on public water supply distribution systems. Trained facility employees or individuals certified to perform construction and repairs are responsible for exercising the system control valves annually.

Figure 11.11 Water towers store water at elevated pressure and are used when pumps supplying a water distribution system can no longer maintain the demand.

Some distribution systems maintain separate piping for fire protection and domestic or industrial services. This separation differs from most municipal water supply distribution systems in which fire hydrants are connected to the same mains that supply water for domestic or industrial use.

Separate distribution systems are cost-prohibitive for most municipal applications but are economically practical in many private applications. There are a number of advantages to having separate piping arrangements in a private water supply distribution system. The property owner has control over the water supply source, and neither system (fire protection or domestic/industrial services) is affected by service interruptions to the other system.

Storage Tanks

Water distribution systems may have storage tanks located throughout the system. These storage tanks are usually constructed of steel or concrete.

Control Valves

Control valves are installed throughout the water distribution system. They are generally located on the small line at the point where it attaches to the large line. Control valves are used to isolate the flow of water to the following:

- Individual hydrants or properties
- Sections of distribution lines, secondary feeders, and primary feeders
- The entire system

Control valves are spaced so that a minimum length of the water distribution system will be out of service if it is necessary to close a valve. Workers close valves to initiate repairs to the system or to isolate a water main break. Valves should be exercised at least once a year to verify that they are in good working condition.

Control valves for water distribution systems are broadly divided into indicating and nonindicating types **(Figures 11.12 a and b)**. An indicating valve shows whether the valve is open, closed, or partially closed. Valves in private fire protection systems are usually of the indicating type. The post indicator valve (PIV) and the outside stem and yoke (OS&Y) valve are two common indicator valves. Chapter 12, Water-Based Fire Suppression Systems, describes these valves in more detail. The nonindicating control valve has no means for visibly determining if the valve is open or closed.

Gate Valve — Control valve with a solid plate operated by a handle and screw mechanism; rotating the handle moves the plate into or out of the waterway.

Butterfly Valve — Control valve that uses a flat circular plate in a pipe that rotates 90 degrees across the cross section of the pipe to control the flow of water.

Valves in water systems are usually the nonindicating type and located underground. In treatment plants and pump stations, control valves are located aboveground. Covers located in or near the street indicate access to control valves. If a buried valve is properly installed, the valve is operated through a valve box with a special valve key **(Figure 11.13)**. Water purveyors commonly keep the valve key, although it could also be carried on fire apparatus.

Control valves may be gate valves or butterfly valves. A **gate valve** is usually the non-rising stem type. As the operator turns the valve nut, the gate either rises or lowers to control the water flow. A **butterfly valve** usually has a rubber or rubber-composition seat that is bonded to the valve body. The valve disk rotates 90 degrees to open or close the valve.

Figures 11.12 a and b Valves provided in a water distribution system can be indicating (a) and nonindicating (b) varieties.

Backflow Preventers

In addition to the waterflow control valve, the water supply line may also have a water flowmeter and a backflow preventer. The water flowmeter determines the quantity of water that the facility uses for billing purposes. The backflow preventer prohibits potentially contaminated water from flowing into the public water system.

Two sources of water may serve some private facilities for fire protection: one from the municipal system and another from a private source. In many cases, the private water source for fire protection provides nonpotable (not for drinking) water. If this is the case, officials need to take adequate measures to prevent contamination caused by backflow of nonpotable water into the municipal water supply distribution system. Some jurisdictions do not allow the interconnection of potable and nonpotable water supply systems, which means that the protected property is required to maintain two separate distribution systems.

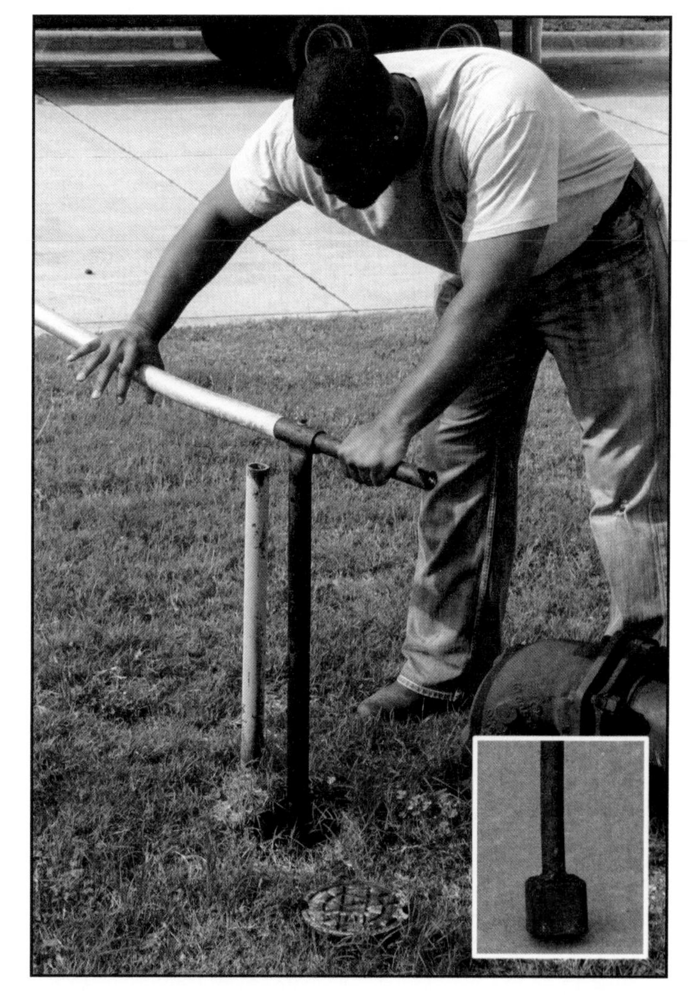

Figure 11.13 Nonindicating valves in public water systems are usually located underground and are operated through a valve box activated with special valve key like the one shown.

Fire Hydrants

The locally adopted code determines the type and location of fire hydrants in the system. The two main types of modern fire hydrants are:

- **Dry-barrel** — Designed for use in climates that have freezing temperatures; the control valve is on the distribution line located below the frost line **(Figure 11.14)**. The stem nut used to open and close the control valve is located on top of the hydrant. Water flows into the hydrant only when a firefighter operates the stem nut. Any water remaining in a closed dry-barrel hydrant drains through a small drain valve that opens at the bottom of the hydrant when the main valve approaches a closed position. A hydrant wrench used to turn the stem opens or closes the valve.
- **Wet-barrel** —Designed for use in mild climates where temperatures remain above freezing; water remains in the hydrant at all times. Compression valves are usually at each outlet, but there may be another control valve in the top of the hydrant to control the water flow to all outlets **(Figure 11.15)**. This hydrant type features the valve at the hose outlet.

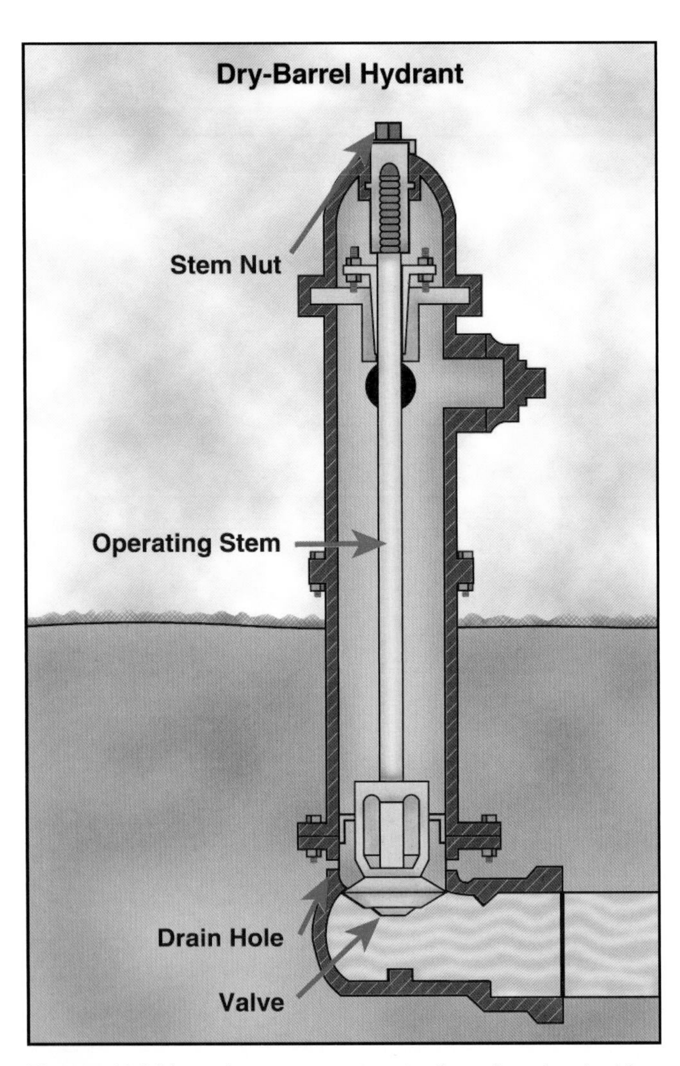

Figure 11.14 In order to prevent water from freezing inside the hydrant, dry-barrel hydrants do not have water inside them unless water is flowing.

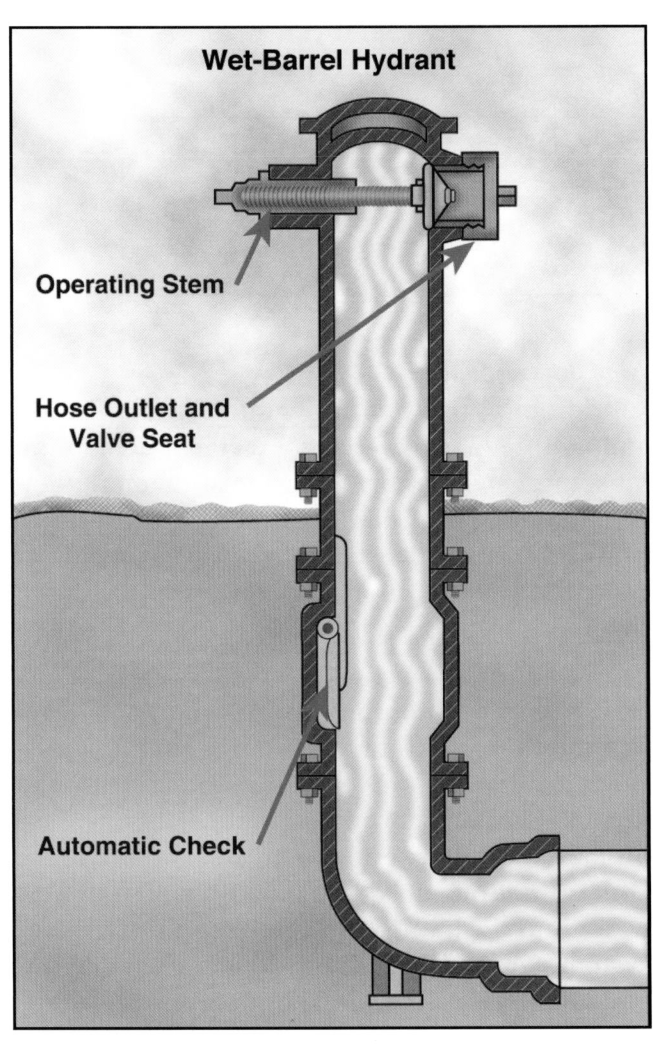

Figure 11.15 Wet-barrel hydrants have water under pressure inside them at all times.

Regardless of the location, design, or type, hydrant discharge outlets are considered *standard* if they contain the following two components:

1. At least one large (4- or 4½-inch [100 mm or 115 mm]) outlet, often referred to as the **steamer connection** or pumper outlet connection.
2. Two hose outlet nozzles for 2½-inch (65 mm) couplings **(Figure 11.16)**.

Steamer Connection — Large-diameter outlet, usually 4½ inches (115 mm), at a hydrant or at the base of an elevated water storage container.

Hydrant inspections should occur at least once a year. In some jurisdictions, fire companies perform these inspections. Records of hydrant inspections and maintenance performed should be maintained **(Figure 11.17, p. 478)**. Inspectors need accurate reports of observations and tests to evaluate the operational readiness of a water supply distribution system.

Fire hydrants may be painted to designate ownership of the hydrant and to indicate their flow capacity. Hydrants may be painted violet to indicate that they contain nonpotable water. **Table 11.1, p. 479,** illustrates the colors of the bonnets (tops) and discharge caps on municipal hydrants based on NFPA® 291, *Recommended Practice for Fire Flow Testing and Marking of Hydrants.*

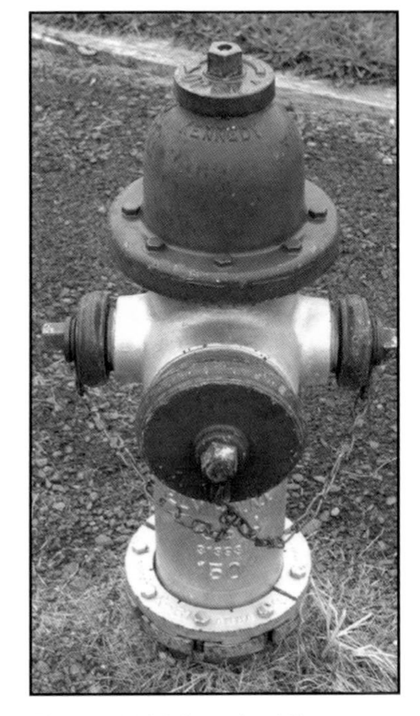

Figure 11.16 Standard fire hydrant discharge outlets must include at least one large outlet and two hose outlets.

Although the water purveyor may be responsible for maintenance on public hydrants, fire departments depend on hydrants working during a fire incident. As a result, periodic inspections, maintenance, and flow testing must be performed. Inspectors should note signs of damage or disrepair and report them. A qualified individual must inspect private hydrants and provide documentation for review, according to NFPA® 25 and ISO guidelines.

Water Supply Testing

Inspectors assigned to the fire inspection division, building department, or water department may be assigned the task of analyzing the water supply for the municipality or service area. They may perform this analysis periodically, when the water distribution system is altered or expanded, or when applications are made for new construction. Inspectors may also be required to witness the testing of private water supply distribution systems or review fire protection system documents based on these tests.

Fire Flow vs. Water Flow

The terms fire flow and water flow are often used interchangeably. This is incorrect:

- Water flow is a general term used to express how much water is flowing for a given time duration.
- Fire flow is the required volume of water (expressed in gallons per minute [GPM] at a residual pressure of 20 PSIG for a minimum flow duration. The flow duration can range from 2-6 hours. Fire flow is a legal requirement in all model fire codes.

The basis for the 20 PSIG value stems from American Water Works Association requirements for municipal water supply systems. The 20 PSIG value ensures that when water flow is great (such as during a defensive fire) and thousands of GPM are flowing, the pressure does not drop below a point where it is possible to collapse a water main.

HYDRANT RECORD

LOCATION ______ HYDRANT NO. ______

POSITION ______ MAKE ______

INSTALLED ______ TYPE ______ TURNS TO OPEN ______ R. ______ L. ______

SIZE OF LEAD ______ SIZE OF MAIN ______

VALVE IN LEAD ______ FT. ______ TURNS TO OPEN ______ R. ______ L. ______

BENCH MARK ______ ELEV. ______

PRESSURE TESTS

DATE	STATIC PRESSURE	FLOW PRESSURE	GPM	DATE	STATIC PRESSURE	FLOW PRESSURE	GPM

REMARKS

______

______

______

RECORD OF MAINTENANCE

WORK PERFORMED ______ DATE ______

______

Flowed

Lubricated

Cap Gasket Replaced

Bonnet Gasket Replaced

Valve Leather Replaced

Drain Valve Replaced

Cap Replaced

Lead Valve Operated

Painted

Raised

Moved

Figure 11.17 A hydrant record is used to record maintenance and inspections of fire hydrants.

Table 11.1
Classifications and Markings of Municipal Fire Hydrants

Classification	Fire Flow	Barrel Color	Top and Nozzle Cap Colors	Pressure
Class AA	1,500 gpm (5 680 L/min) or greater	Chrome Yellow	Light Blue	20 psi (140 kPa)
Class A	1,000–1,499 gpm (3 785–5 675 L/min	Chrome Yellow	Green	20 psi (140 kPa)
Class B	500–999 gpm (1 900–3 780 L/min)	Chrome Yellow	Orange	20 psi (140 kPa)
Class C	500 gpm (1 900 L/min) or less	Chrome Yellow	Red	20 psi (140 kPa)

Based on information given in *NFPA® 291, Recommended Practice for Fire Flow Testing and Marking of Hydrants, 2007*.

Because an adequate water supply is essential to the operation of water-based fire suppression systems, inspectors must be able to analyze the water supply. This analysis may involve flow testing hydrants on the water distribution system or reviewing data provided by the municipal water department or water supply provider. Fire flow tests include the measurement of **static pressure**, **residual pressure**, and the formulas and calculations used to determine available water from these tests **(Figure 11.18)**.

Residual Pressure — Pressure remaining at a given point in a water supply system while water is flowing.

Static Pressure — Pressure at a given point in a water system when no water is flowing.

Fire flow tests determine the rate of fire flow available for fire suppression at various locations within the distribution system. An inspector should remember, however, that hydrant pressure varies between periods of peak demand and minimal demand. Calculations or graphical analyses from fire flow tests provide the following information:

- Amount of water flow from individual hydrants
- Water flow pressures
- Quantity of water available at any pressure
- Pressure available across a wide range of flows

Knowing the capacity of a water distribution system is essential when fire companies prepare preincident plans. Test results may

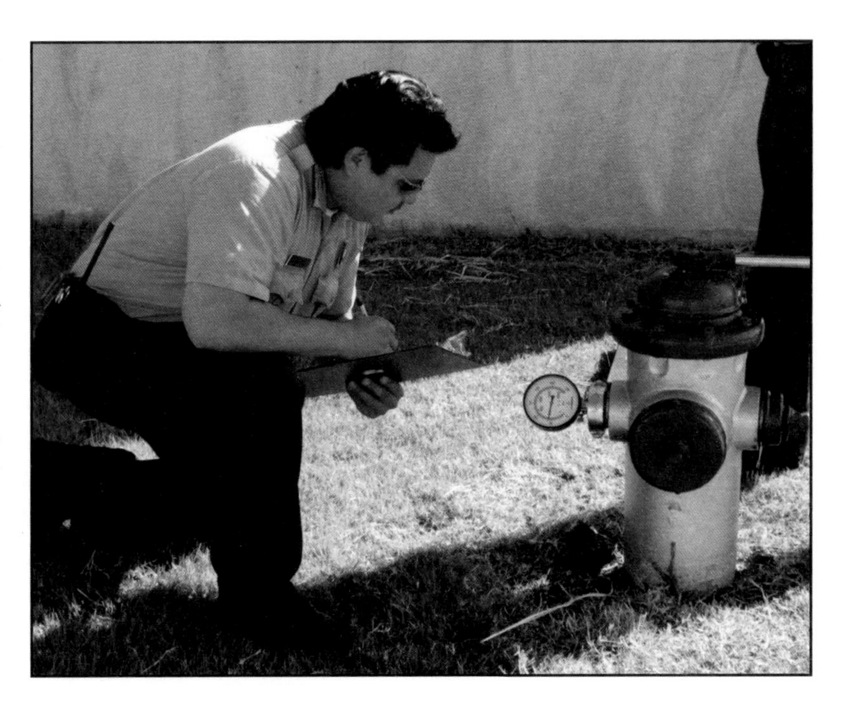

Figure 11.18 An inspector measures static and residual pressures to determine the amount of water available for fire protection.

also indicate weak points in a water distribution system. Water department personnel can use this information to plan improvements in an existing system and design extensions to newly developed areas.

Tests that are repeated at the same locations semiannually may reveal a loss in the carrying capacity of water mains, closed or partially closed valves, and a need for improving certain arterial mains. Flow tests should be conducted after any extensive water main improvements, following the construction of water line extensions, and on a regular predetermined basis on existing portions of the distribution system.

Pitot Tube — Instrument that is inserted into a flowing fluid (such as a stream of water) to measure the velocity pressure of the stream; commonly used to measure flow. A pitot tube functions by converting the velocity energy to pressure energy that can then be measured by a pressure gauge. The gauge reads in units of pounds per square inch (psi) or kilopascals (kPa).

Fire Hydrant Inspections

To perform fire flow tests and analyze the results, an inspector must have the ability to:

- Operate a **pitot tube** and gauge (an instrument used to determine the fire flow from a hydrant).
- Calculate how much water is flowing from a hydrant.
- Determine the residual pressure required for the water distribution system when water is flowing from the system.
- Perform the test, adhere to test precautions, determine possible obstructions in the system, and analyze the results based on computations.

Fire hydrant inspections should be performed on an annual basis to monitor the physical condition of all hydrants in the response area. Inspections may also be performed as part of the fire flow tests.

Hydrants should be inspected before fire flow tests begin. Inspectors or emergency response companies may perform these inspections as part of their normal duties. The materials needed to complete the inspection of fire hydrants include a/an:

- Notebook
- Gauging device for checking discharge outlet threads (a female coupling for the various discharge outlet sizes may be used)
- Approved lubricant
- Small, flat brush
- Gate valve key
- Pressure gauge and a tapped hydrant cap **(Figure 11.19)**
- Appropriate pitot tube and gauge
- Hydrant wrench
- Water stream diffuser

During hydrant inspections, inspectors should observe the following conditions:

- Check for any obstructions near the hydrants, such as sign posts, utility poles, shrubbery, or fences **(Figure 11.20 a)**.
- Check the hydrant outlet(s) to verify that they face the proper direction and that there is clearance between the outlet and surrounding ground **(Figure 11.20 b)**. The clearance between the bottom of the butt (discharge outlet) and the grade should be at least 15 inches (380 mm).

Figure 11.19 A pressure gauge for testing hydrant flow is one piece of equipment that is necessary to accurately test water pressure.

- Check for mechanical damage to the hydrant, such as dented outlets or rounded (stripped) stem nuts.
- Check the condition of the paint for rust or corrosion. Verify that the discharge outlet caps are not painted shut.
- Check water flow by having the hydrant fully opened.
- Check to see that a dry-barrel hydrant drains once the stem valve is closed.

Pitot Tube and Gauge

An inspector must be familiar with the use of a pitot tube and gauge for determining hydrant flow rates **(Figure 11.21)**. Using a pitot tube and gauge to take a flow reading is not difficult, but it must be done properly to obtain accurate readings. The procedure for using a pitot tube and gauge is as follows:

Step 1: Open the petcock (small faucet or valve used to control the flow of liquids) on the pitot tube and verify that the air chamber is drained. Then close the petcock.

Step 2: Edge the blade into the stream with the small opening or point centered in the stream and held away from the hydrant butt or nozzle approximately one-half the diameter of the opening. For example, with a 2½-inch (65 mm) hydrant butt, this distance is 1¼ inches (32 mm). The pitot tube blade should be parallel to the outlet opening with the air chamber kept above the horizontal plane passing through the center of the stream. This position increases the efficiency of the air chamber and helps avoid needle fluctuations.

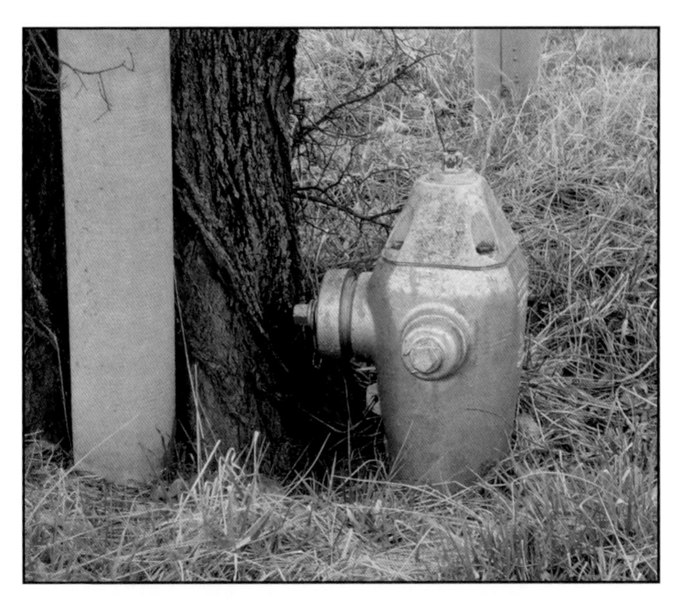

Figure 11.20 a Obstructed hydrants are difficult for responding firefighters to access and use. Inspectors must note and enforce correction of obstructed hydrants.

Figure 11.20 b A hydrant like this that lacks sufficient clearance between the outlet and the ground needs to be reported for correction. *Courtesy of Dennis Marx.*

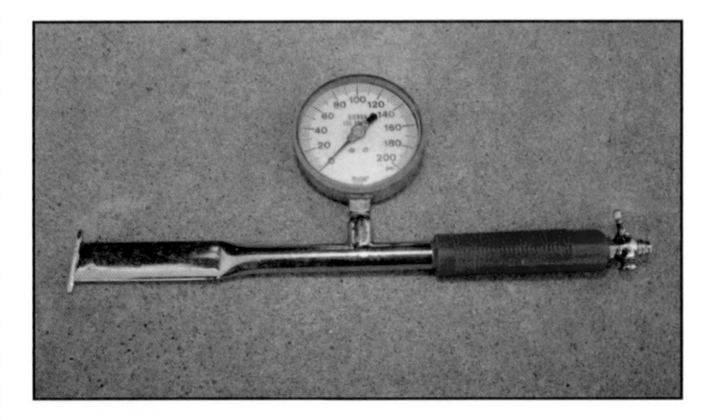

Figure 11.21 Using a pitot tube and gauge for determining hydrant flow rates is a skill that all inspectors must learn and be able to perform.

Step 3: Check the velocity pressure reading from the gauge. If the needle is fluctuating, read and record the value located in the center between the high and low extremes.

Step 4: After the test is completed, open the petcock and be certain that all water drains from the assembly before storing it.

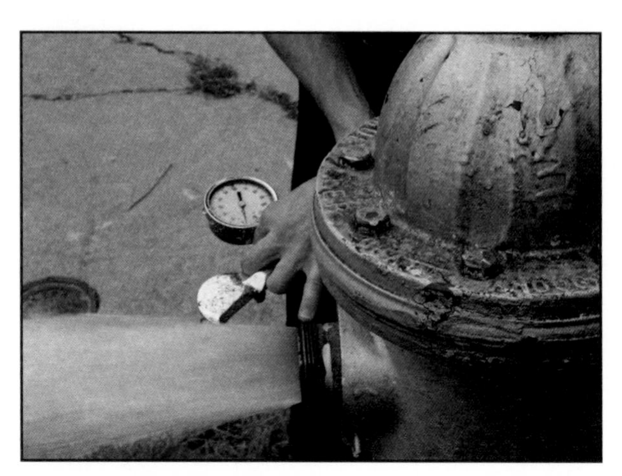

Figure 11.22 This inspector is demonstrating the proper method for using a pitot tube and gauge to measure flow at a hydrant.

Figure 11.22 shows a good method of holding a pitot tube and gauge in relation to a hydrant outlet or nozzle. Note that the operator grasps the pitot tube just behind the blade with the first two fingers and thumb of the left hand, while the right hand holds the air chamber. The little finger of the left hand rests upon the hydrant outlet or nozzle tip to steady the instrument. Unless some effort is made to steady the pitot tube, the movement of the water will make it difficult to get an accurate reading.

Another method of holding the pitot tube is illustrated in **Figure 11.23 a**. The left-hand fingers are split around the gauge outlet, and the left side of the fist is placed on the edge of the hydrant orifice or outlet. The blade can then be sliced into the stream in a counterclockwise direction **(Figure 11.23 b)**. The right hand once again steadies the air chamber.

Fire Flow Test Computations

The easiest way to determine how much water is flowing from a hydrant outlet(s) is to refer to prepared tables for nozzle/outlet discharge. Jurisdictions may choose to develop their own tables based on the flow pressures that are common to their area. These tables are computed by using the following flow formulas (both Customary System and International System of Units):

$$\text{Gallons per minute (gpm)} = (29.83) \times C_d \times d^2 \times \sqrt{P} \quad (1)$$

In SI

$$\text{Liters per minute (L/min)} = (0.0667766) \times C_d \times d^2 \times \sqrt{P} \quad (2)$$

Where: *Cd = Coefficient of discharge*

d = Actual diameter of the hydrant or nozzle orifice in inches (mm)

P = Pressure in psi (kPa) as read at the orifice

The constant *29.83 (0.0667766)* is derived from the physical laws relating water velocity, pressure, and conversion factors that conveniently leave the answer in gallons per minute (L/min). This formula was derived by assigning a coefficient of *1.0* for an ideal frictionless discharge orifice. An actual hydrant orifice or nozzle will have a lower coefficient of discharge, reflecting friction factors that slow the flow velocity. The coefficient will vary with the type of hydrant outlet or nozzle used.

When using a hydrant orifice, the operator will have to feel the inside contour of the hydrant to determine which one of the three types of hydrant outlets is being used **(Figure 11.24)**. Refer to the manufacturer's recommendations for determining the coefficient of discharge for a specific nozzle.

The flow formulas also depend on the actual internal diameter of the outlet or nozzle opening being used. A ruler with a scale that measures to at least 1/16th of an inch (1.5 mm) should be used to measure the diameter of the outlet or nozzle opening.

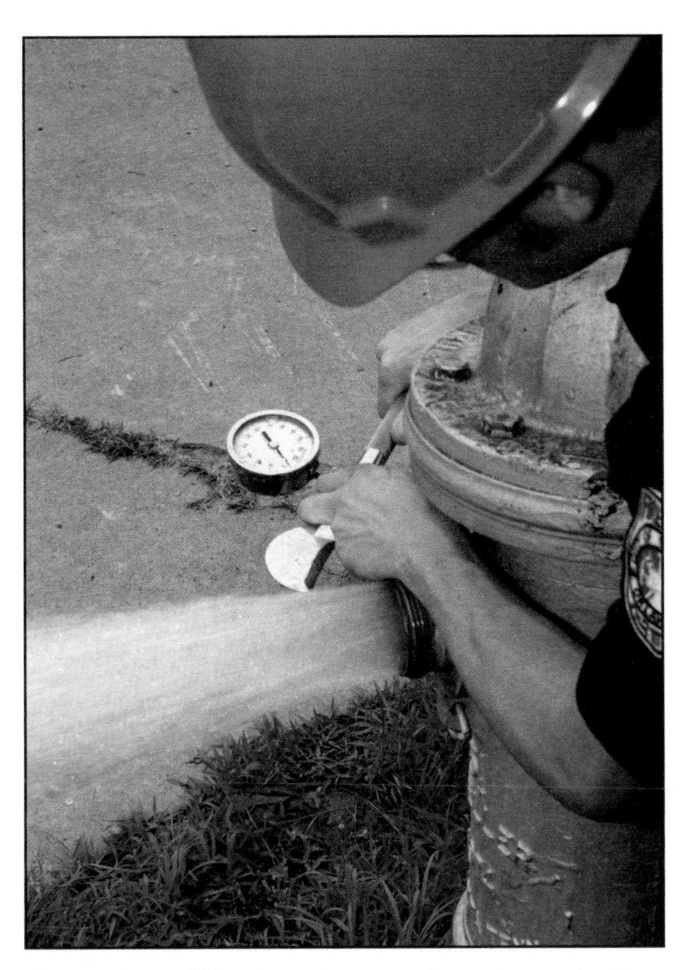

Figure 11.23 a This photo shows an alternative method for holding a pitot tube and gauge when testing hydrants.

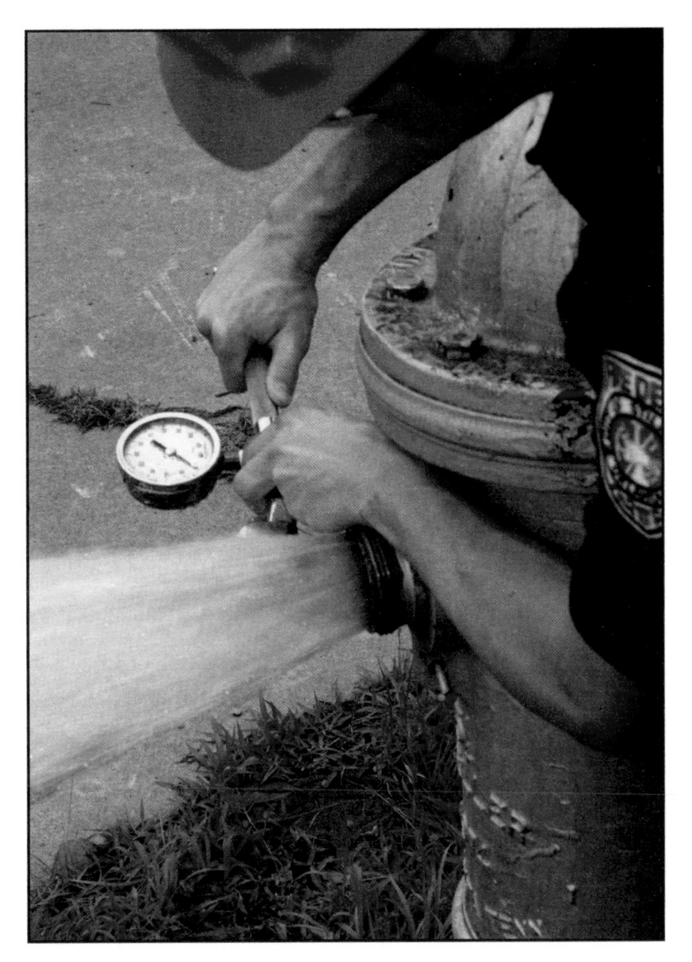

Figure 11.23 b With the pitot tube and gauge stabilized against the hydrant with his left hand, this inspector uses a counterclockwise motion to slice the blade into the water stream with his right hand.

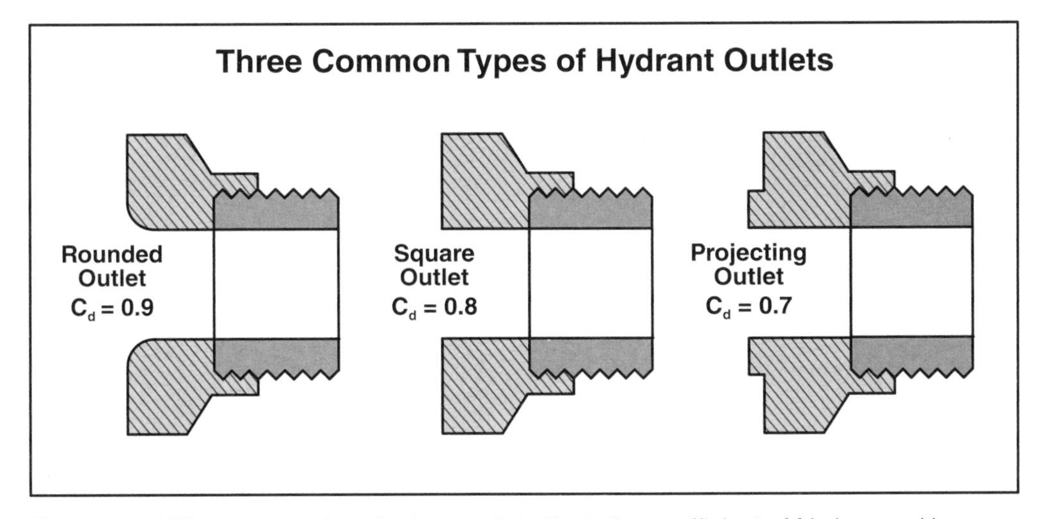

Figure 11.24 The contour of the hydrant outlet affects the coefficient of friction used in hydraulic calculations.

Flow Calculations

Assuming a 2½-inch (65 mm) hydrant outlet that has an actual diameter of 2⁷⁄₁₆ inches (2.44 inches [62 mm]) with a *C* factor of 0.80 and a flow pressure of 10 psi (69 kPa) read from the pitot gauge is used, the waterflow equation would read as follows:

$$gpm = 29.83 \times C_d \times d2 \times \sqrt{P}$$

$$gpm = 29.83 \times 0.80 \times (2.44)^2 \times \sqrt{10}$$

$$gpm = 449.28 \text{ or } \approx 450$$

in SI

$$L/min = 0.0667766 \times C_d \times d^2 \times \sqrt{P}$$

$$L/min = 0.0667766 \times 0.80 \times (62)^2 \times \sqrt{69}$$

$$L/min = 1\ 705.78 \text{ or } \approx 1\ 700$$

Table 11.2
Correction Factors for Large Diameter Outlets

Velocity Pressure	Factor
2 psi (13.8 kPa)	0.97
3 psi (20.7 kPa)	0.92
4 psi (27.6 kPa)	0.89
5 psi (34.5 kPa)	0.86
6 psi (41.4 kPa)	0.84
7 psi (48.3 kPa) or over	0.83

Generally, 2½-inch (65 mm) outlets should be used to conduct hydrant flow tests because the stream from a large hydrant outlet (4 to 4½ inches [100 mm to 115 mm]) contains voids; the entire stream of water is not solid. For this reason, the listed formula alone will not give accurate results for flows using large outlets.

If it is necessary to use the large outlets, use a correction factor to give more accurate results. Multiply the flow (as determined by Formulas 1 or 2 given previously) by one of the factors shown in **Table 11.2**, corresponding to the velocity pressure measured by the pitot tube and gauge. See **Example 1: Water Flow Calculation**, at the end of this chapter.

These formulas allow the computation of total flow from the flowing hydrants when performing an area fire-flow test. The formulas also indicate the flow from the hydrant only at the time of the test. High-risk areas or locations with large, daily fluctuations in demand may necessitate multiple tests to determine the minimum flow available.

Required Residual Pressure

Fire protection engineers have established 20 psi (140 kPa) as the minimum required residual pressure when computing the available water for area flow-test results. This residual pressure can be defined as enough pressure to overcome friction inside any one of the following:

- Short 6-inch (150 mm) branch pipe
- Hydrant
- Apparatus intake hose

Residual pressure must also allow a safety factor to compensate for gauge error. The Environmental Protection Agency (EPA) and many state health departments require the 20 psi (140 kPa) minimum to prevent external ground (surface) water or other contaminants and pollutants from being drawn into the distribution system at pipe connection points.

Pressure differentials can collapse a water main or create **cavitation** (implosion of air pockets drawn into fire pumps connected to the system). More frequently, a fire apparatus operating at a low system pressure may draw the entire capacity of the system at a location. If a discharge valve is turned off too quickly, a water hammer (sudden surge in pressure) will be generated and may be transferred to the water main, resulting in damaged or broken mains, connections, or other system components.

Cavitation — Condition in which vacuum pockets form due to localized regions of low pressure at the vanes in the impeller of a centrifugal pump and cause vibration, loss of efficiency, and possible damage to the impeller.

Fire Flow Test Procedures

During a fire flow test, the static pressure and the residual pressure should be taken from a fire hydrant (commonly called the **test hydrant**) as close as possible to the location requiring the test results. The **flow hydrants** are those hydrants where pitot gauge readings are taken to find their individual flows. These readings are then added to find the total flow during the test.

Test Hydrant — Fire hydrant used during a fire flow test to read the static and residual pressures.

Flow Hydrant — Fire hydrant from which the water is discharged during a hydrant fire flow test.

In general, the test hydrant should be located between the flow hydrant and the water supply source when flow testing a single hydrant. The flow hydrant should be downstream from the test hydrant **(Figures 11.25 a and b)**. The actual direction of flow in the system is difficult to determine. Water department personnel should be consulted for assistance. When flow testing multiple hydrants, the test hydrant should be centrally located relative to the flow hydrants.

NOTE: Water is actually never discharged from a test hydrant; rather, a pressure gauge cap is placed on the discharge and the hydrant is fully opened.

The procedure for conducting an available water test is as follows:

Step 1: Locate personnel at the test hydrant and all flow hydrants to be used.

Step 2: Remove a hydrant cap from the test hydrant and attach the pressure gauge cap with the petcock in the open position. After checking the

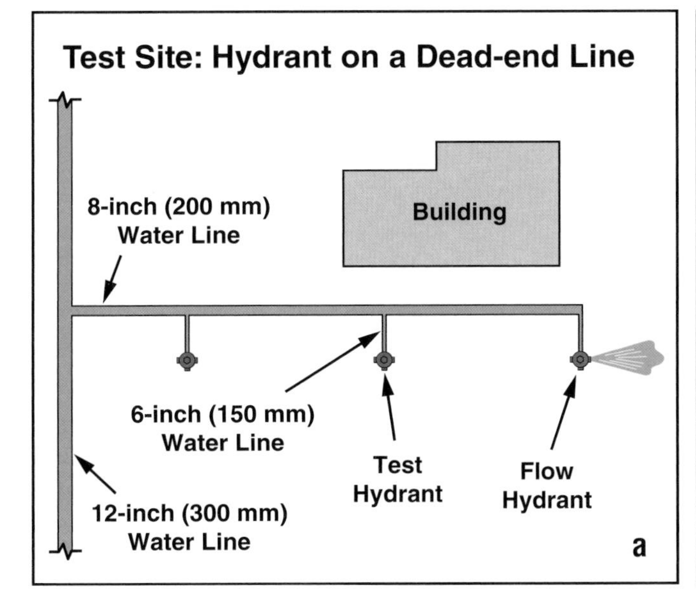

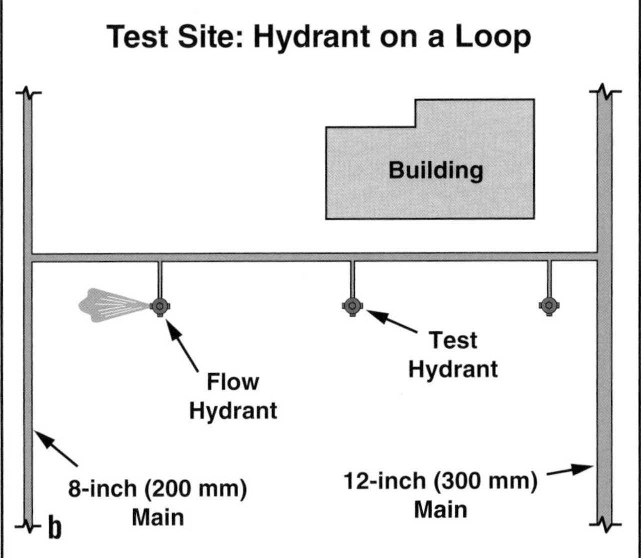

Figures 11.25 a and b The test method that is selected depends on the location of the hydrant in the water supply system. (a) When the hydrant is on a dead-end system, the flow hydrant is located downstream from the test hydrant. (b) On a looped system, the flow hydrant is located in the direction of the flow. In this example, it is downstream from the larger water main and toward the smaller main.

other caps for tightness, slowly open the hydrant by turning the operating nut several turns. Once the air has escaped and a steady stream of water is flowing, close the petcock and fully open the hydrant **(Figure 11.26)**.

Step 3: Read and record the static pressure as seen on the pressure gauge.

Step 4: Removes the cap(s) from the outlet(s) on the flow hydrants. Check and record the hydrant coefficient and the actual inside diameter of the orifice when using a hydrant outlet. If a nozzle is placed on the outlet, check and record its coefficient and the diameter of the nozzle orifice.

Step 5: Open flow hydrants as necessary and read and record the pitot gauge reading of the velocity pressures **(Figure 11.27)**. The individual at the test hydrant simultaneously reads and records the residual pressure. **NOTE:** The residual pressure should not drop below 20 psi (140 kPa) during the test; if it does, the number of flow hydrants must be reduced.

Step 6: Slowly close the flow hydrant to prevent water hammer in the water mains. After checking for proper drainage, replace and secure all hydrant caps. Report any hydrant defects.

Normal Operating Pressure — Amount of pressure that is expected to be available from a hydrant, prior to pumping.

Step 7: Check the pressure gauge on the test hydrant for a return to **normal operating pressure**, then close the hydrant. Open the petcock valve to prevent a vacuum on the pressure gauge. Remove the pressure gauge cap. After checking for proper drainage, replace and secure the hydrant cap. Report any hydrant defects.

When testing the available water supply, determining the number of hydrants to open depends on an estimate of the flow available in the area. For example, a strong probable flow may require that several hydrants be opened to ensure accurate test results. Enough hydrants should be opened to drop the static pressure by at least 10 percent. If more accurate results are required, the pressure drop should be increased to 25 percent.

For example, if the static pressure is 80 psi (560 kPa), the residual pressure should be at least 72 psi (504 kPa). For more accurate results, the residual pressure may be dropped 25 percent, which would be to 60 psi (420 kPa). Graphical analysis or mathematical calculations can then determine the flow available at 20 psi (140 kPa).

Water mains may contain such low pressures that no flow pressure can be read on the pitot gauge. If this situation occurs, the flow orifice must be reduced by using a straight stream nozzle with tips smaller than 2½ inches (65 mm). The nozzle is placed on the hydrant outlet to increase the flow velocity to a point where the velocity pressure is measurable with the pitot gauge **(Figure 11.28)**. Using these straight stream nozzles will require an adjustment in the water flow calculation that must include the smaller nozzle diameter and corrected coefficient of friction for the nozzle tip.

Inspectors must take precautions when performing fire flow tests to ensure their own safety and that of the public. At the same time, inspectors must be familiar with the various types and causes of obstructions in the water distribution system.

Figure 11.26 When a pressure gauge cap is attached and the petcock is in the open position, water will flow freely through the cap and its flow will be designated on the accompanying gauge.

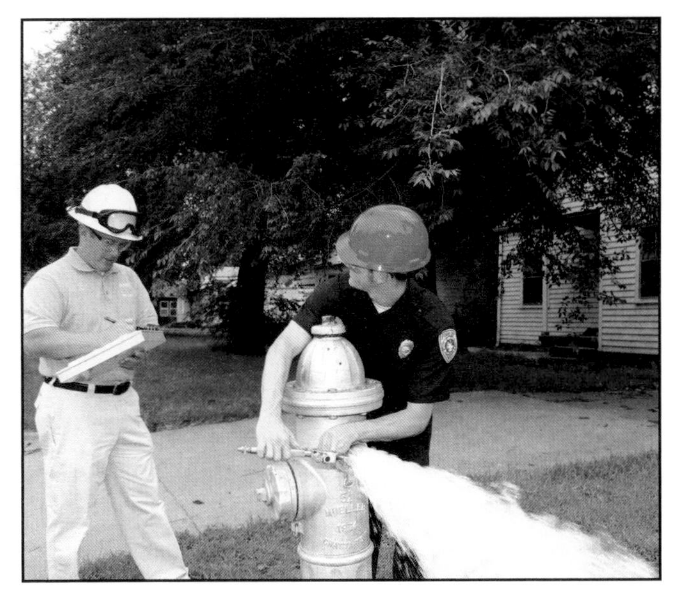

Figure 11.27 After flow has been established at a hydrant, an inspector should record the hydrant pressure as read from the gauge.

Precautions

Inspectors and firefighters must take certain precautions before, during, and after flow tests to avoid injuries to individuals participating in the test or to passersby. Minimizing damage to public and private property from the water discharge is crucial. Controlling pedestrian and vehicular traffic during all phases of the testing may require assistance from local law enforcement personnel. It may also be advisable to conduct flow tests in busy areas at nonpeak hours, such as early in the morning.

Before conducting a flow test, an inspector should notify a water department official. Opening hydrants may interfere with normal operating conditions in the water supply system. Additionally, if water service personnel are performing maintenance work in the immediate vicinity, the results of the flow test would not be typical for normal conditions.

The appropriate officials should also notify businesses in the area about the tests in order to reduce the number of false reports of possible water main breaks. Employing a notification process will also promote a positive working relationship between the fire department, water department, and the public.

During the hydrant testing procedures, everyone should take the following safety measures:

- Wear the following protective equipment:
  - Helmet
  - Gloves
  - Eye protection
  - Reflective safety vests

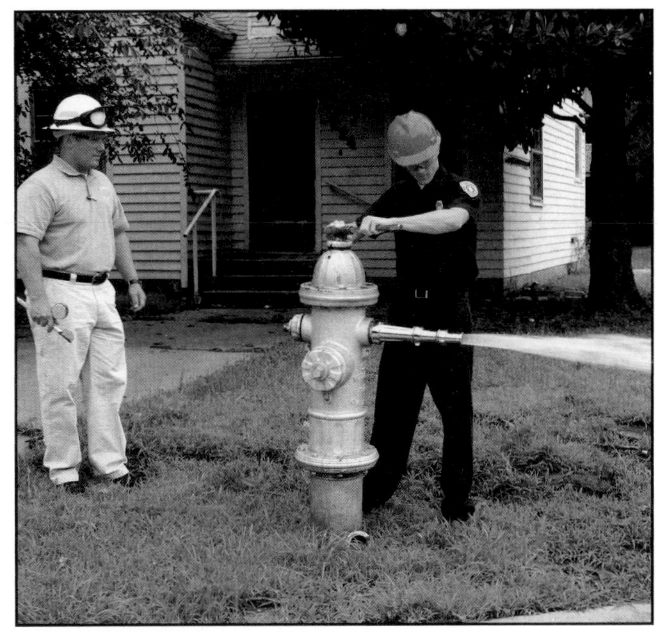

Figure 11.28 Adding a nozzle decreases flow and increases pressure to a measurable level.

Figure 11.29 Make sure that traffic safety cones or signs are placed on streets and roadways to divert traffic away from flow testing areas.

- Erect traffic safety signs on busy streets **(Figure 11.29)**.
- Tighten caps on hydrant outlets that are not being used.
- Do not stand in front of closed caps.
- Do not lean over the top of a hydrant when operating it.
- Do not flow water during freezing weather.

It is also important to take steps to protect the surrounding property. Property damage control measures include the following:

- Open and close hydrants slowly to avoid water hammer.
- Do not flow hydrants where drainage is inadequate.
- Check downstream to see where the water will flow.

Because flowing water across a busy street could cause an accident, take proper measures beforehand to slow or stop traffic. A good rule to follow is "*When in doubt, do not flow!*" If there are difficulties in conducting a flow test, consider solutions so that the test can be completed without disruptions or property destruction.

Protect Property

Always protect property from possible damage during a flow test. List any damage that does occur and prepare a report for the department safety officer.

WARNING!

Do not leave deep standing water resulting from a hydrant test unattended in retention basins in locations where children are playing. Wait until the water has been absorbed or drained before leaving the area. Do not allow children to play in runoff water.

Obstructions

After several years, fire flow tests may show a progressive reduction in the water flow capacity of the distribution system. An investigation of the situation usually reveals that the water mains were obstructed in some manner or that system components (including pumps) were worn or leaking. Reduced water flow may be the result of one or more of the following obstructions or conditions:

- **Encrustations** — Growths, crusts, tubercular lump corrosion, or rust on the inside walls of water mains. These can form around the inside wall of a water main, eventually narrowing the actual diameter of the pipe to a point where its capacity is greatly impaired. The following conditions are causes **(Figures 11.30 a and b)**:
  - — Chemicals in the water
  - — Biological or organism growth
  - — Biodegradation of water agents and pipe materials
  - — Progressive growth of rust deposits of iron pipes
  - — Accumulation of various salts due to oxidation
  - — Biological reactions produced by organisms present in most water supplies
- **Sedimentation deposits** — Sedimentary decay (mud, clay, or leaves), foreign matter other than sediment (stones, tools, wood, or lead), dead organisms, and decayed vegetation found chiefly in the bottom of a water main, especially in large, low-velocity portions of a system, and in dead-end idle flow parts of a system.
- **Malfunctioning valves** — Closed or partially closed
- **Malfunctioning pipes** — Leaking or broken pipe section or mains

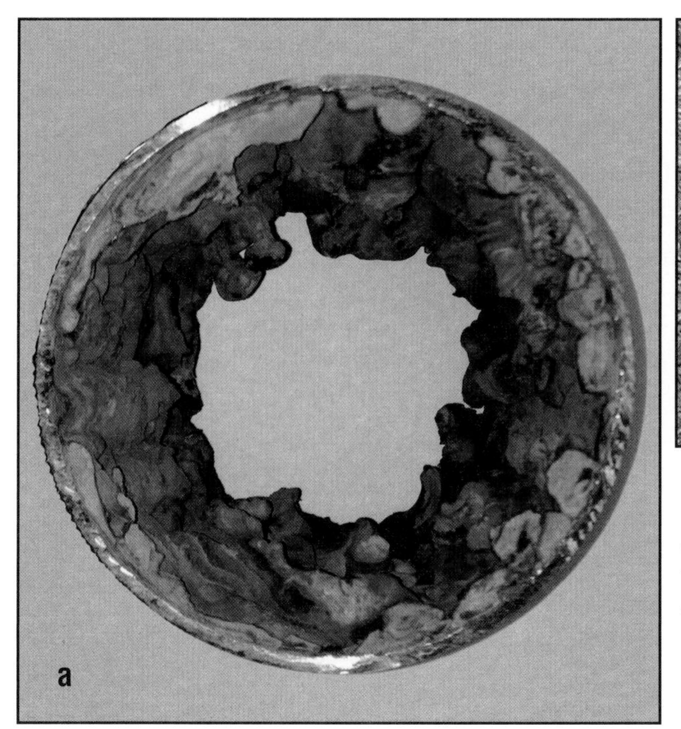

Figures 11.30 a and b Encrustations like the tubular corrosion in these pipes fittings can greatly reduce the flow through a pipe. *a courtesy of Brett Lacey; b courtesy of Steve Toth.*

- **Malfunctioning pumps** — Worn or damaged impellers or pressure adjustments set too low for the demand of the system
- **Foreign matter other than deposits** — Chunks of lead (from ball and spigot joints), boards, crowbars, tool handles, stones, and other materials
- **Increased friction loss** — Combination of encrustation, sedimentation, partially closed or closed valves, and foreign matter

Inspectors should be aware of the consequences of stuck or partially closed valves. If a valve is only partially closed, it would not be noticed during normal water usage. However, high-friction loss would reduce the water volume available for fire suppression systems or fire suppression operations. Routine testing can locate partially closed valves before they pose a problem.

Available Fire Flow Test Results Computations

Graphical analysis and mathematical computation are two ways to compute fire flow test results. Both methods are discussed in the sections that follow.

Graphical Analysis

The fire flow chart in **Figure 11.31** is a logarithmic scale developed to simplify the process of determining available water in an area. The chart is accurate to a reasonable degree if an inspector uses a fine-point pencil or pen when plotting results. The inspector can multiply or divide the figures on the vertical and/or horizontal scales by a constant if necessary to fit any situation.

The procedure for conducting a graphical analysis is as follows:

Step 1: Determine which gpm (L/min) scale to use.

Step 2: Locate and plot the static pressure on the vertical scale at 0 gpm (0 L/min).

Step 3: Locate the total waterflow measured during the test on the chart.

Step 4: Locate the residual pressure noted during the test on the chart.

Step 5: Plot the residual pressure above the total water flow measured.

Step 6: Draw a straight line from the static pressure point through the residual pressure point on the water flow scale.

Step 7: Read the gpm (L/min) available at 20 psi (140 kPa) and record the figure. This reading represents the total available water.

The following information boxes give examples of graphical analyses for water-flow tests using one and two outlets. Example calculations are in both Customary System units and International System of Units (SI). **See Example 2 (US): Flow Test for One Outlet** and **Example 3 (SI): Flow Test for One Outlet**, at the end of this chapter.

Figure 11.32, p. 492, shows the test results plotted for graphical analysis of the water supply. The static pressure of 50 psi (345 kPa) is plotted at 0 gpm (0 L/min). The residual pressure of 25 psi (173 kPa) is above the total measured flow of 892 gpm (3 532 L/min), Scale A.

NOTE: Inspectors must understand that pitot gauge pressures are never plotted on the graph; only the flow that corresponds to the pitot gauge pressures is used.

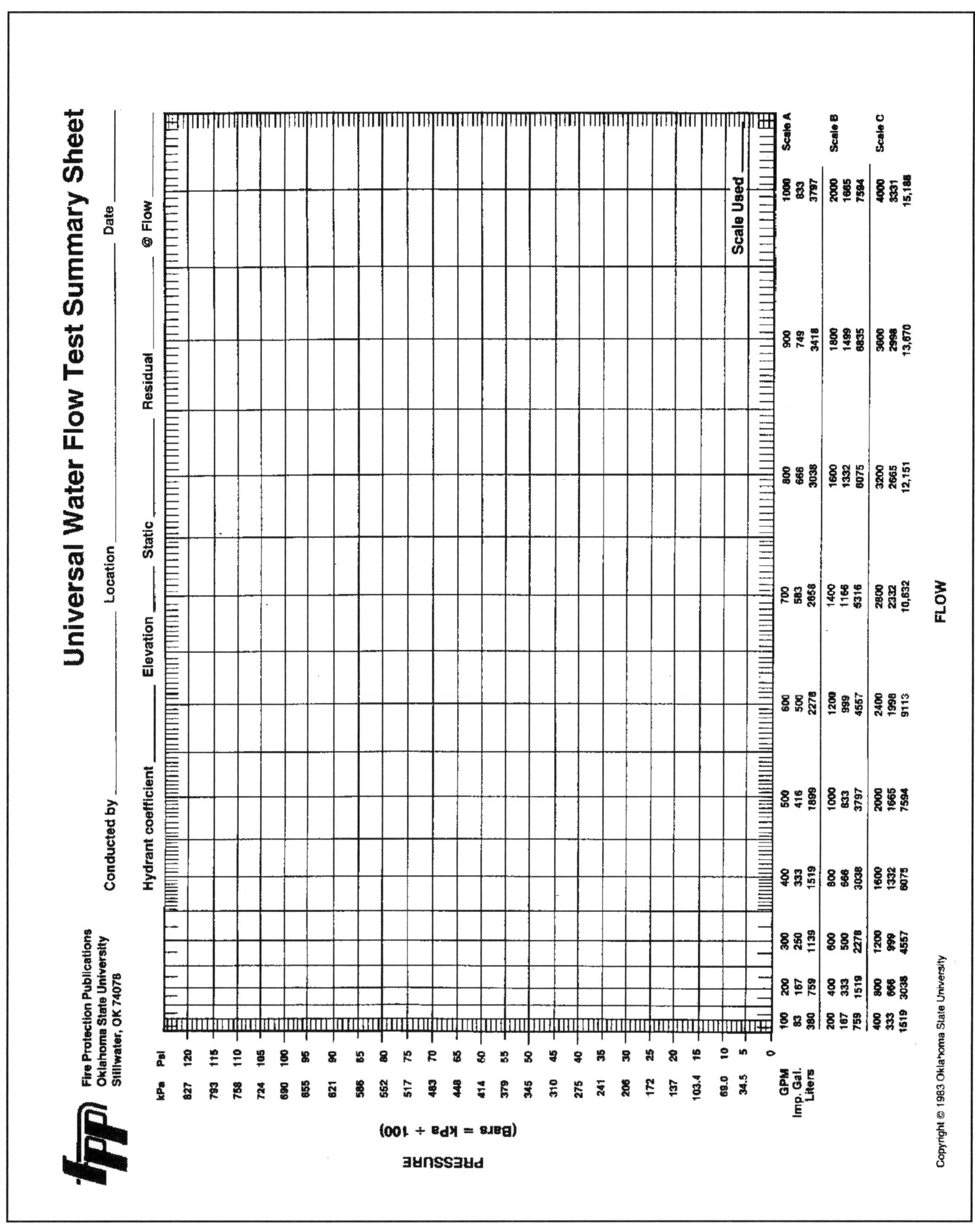

Figure 11.31 This Universal Water Flow Test Summary Sheet shows a logarithmic scale that has been developed to simplify the process of determining available water in an area.

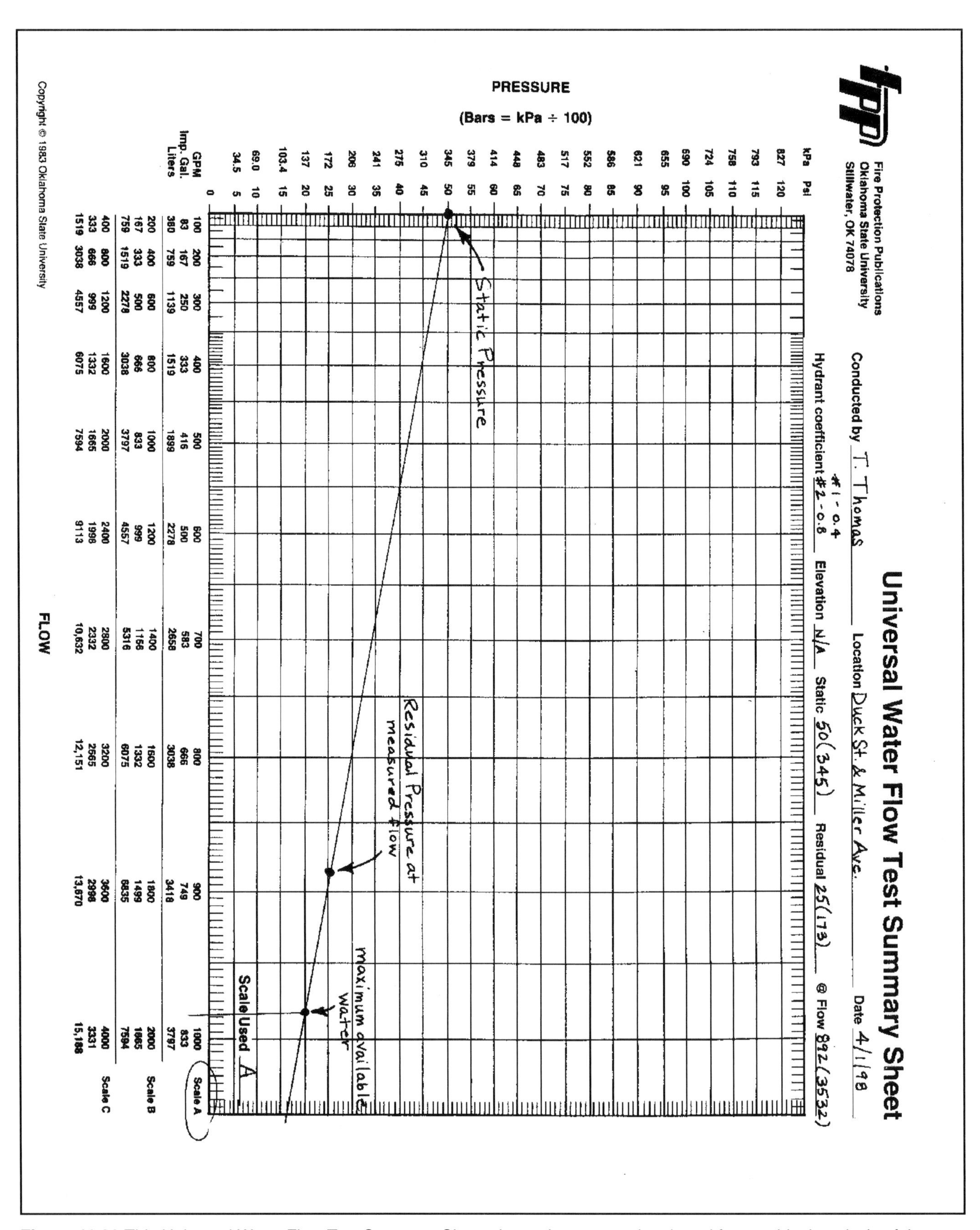

Figure 11.32 This Universal Water Flow Test Summary Sheet shows the test results plotted for graphical analysis of the water supply.

A line drawn through the static and residual pressure points represents the water supply at the test location. Inspectors should note that approximately 978 gpm (4 000 L/min) would be available at 20 psi (140 kPa). This figure represents the minimum desired intake pressure. See **Example 4 (US): Flow Test for Two Outlets**, and **Example 5 (SI): Flow Test for Two Outlets**.

These examples show that the waterflow scale must be changed, so that a line can be drawn down to the 20 psi (140 kPa) level **(Figure 11.33, p. 494)**. The available water rate at 20 psi (140 kPa) in this case would be approximately 1,970 gpm (7 450 L/min).

Mathematical Method

The mathematical method of determining fire flow uses a variation of the Hazen-Williams formula for determining available water. **Example 6, Mathematical Method for Determining Fire Flow**, shows how to calculate fire flow in this manner.

Although inspection personnel must understand these formulas and how the calculations are made, personnel more commonly use software to do them. Inspection personnel simply enter the information from the flow tests into the computer, and the available water supply is automatically determined. Several commercial water flow programs are available.

Chapter Summary

Adequate water supply systems are critical for agriculture, industrial, domestic and fire protection uses. Because water-based fire suppression in the form of water application or sprinkler systems are often the first line of defense against a fire, inspectors must know that the water supply systems in their districts are adequate. Water systems may be public or private, and the responsibility for seeing that the systems are inspected and maintained varies. Nonetheless, inspectors need to be aware of water supply systems and methods for testing the adequacy of water supplies.

Calculations

Example 1: Water Flow Calculation

From Table 11.2, a flow of 6 psi (41.4 kPa) through a 4-inch (100 mm) outlet is indicated as 1,050 gpm (3 974.68 L/min). However, tests have shown that only 84 percent of this quantity is actually flowing due to voids in the water stream. Accordingly, the actual flow is

$$1{,}050 \times 0.84 = 882 \text{ gpm}$$

In SI

$$3\,974.68 \times 0.84 = 3\,338.7 \text{ L/min}$$

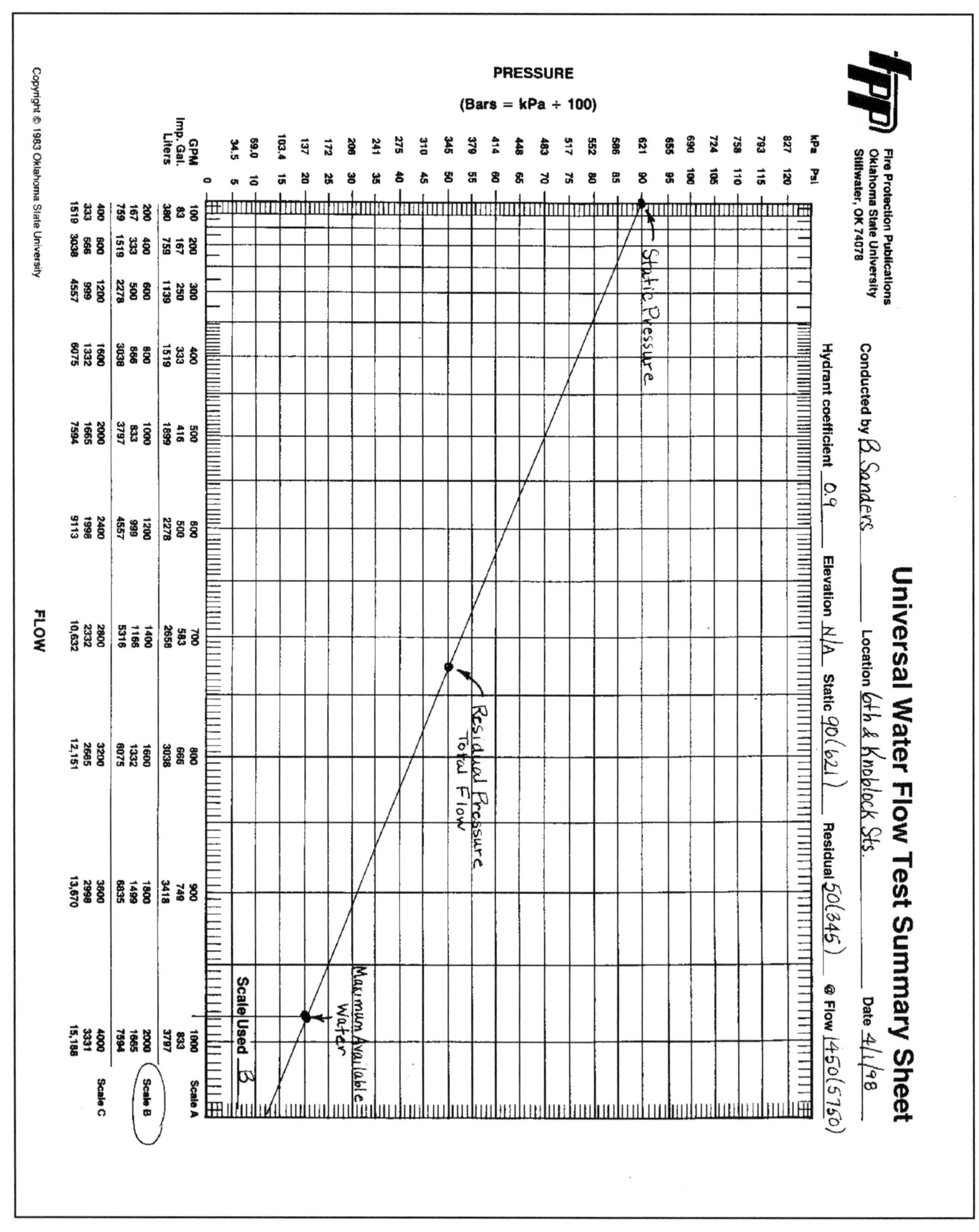

Figure 11.33 The waterflow scale changed so that a line can be drawn down to the 20 psi (140 kPa) level.

Example 2 (U.S.): Flow Test for One Outlet

Given:

Test Hydrant = 50 psi static and 25 psi residual

Flow Hydrant No. 1 = Using one 2½-inch outlet with C = 0.9, pitot gauge reading = 7 psi, and actual discharge diameter = 2.56 inches

$$(29.83)\ (0.9)\ (2.56)^2\ (\sqrt{7}) = 466 \text{ gpm}$$

Flow Hydrant No. 2 = Using one 2½-inch outlet with C = 0.8, pitot gauge reading = 9 psi, and actual discharge diameter = 2.44 inches

$$(29.83)\ (0.8)\ (2.44)^2\ (\sqrt{9}) = 426 \text{ gpm}$$

$$\text{Total Water Flow} = 466 + 426 = 892 \text{ gpm}$$

Example 3 (SI): Flow Test for One Outlet

Given:

Test Hydrant = 345 kPa static and 173 kPa residual

Flow Hydrant No. 1 = Using one 65 mm outlet with C = 0.9, pitot gauge reading = 48 kPa, and actual discharge diameter = 66.5 mm

$$(0.0667766)\ (0.9)\ (66.5)^2\ (\sqrt{48}) = 1\ 841 \text{ L/min}$$

Flow Hydrant No. 2 = Using one 65 mm outlet with C = 0.8, pitot gauge reading = 62 kPa, and actual discharge diameter = 63.5 mm

$$(0.0667766)\ (0.8)\ (63.5)^2\ (\sqrt{62}) = 1\ 691 \text{ L/min}$$

$$\text{Total Flow} = 1\ 841 + 1691 = 3\ 532 \text{ L/min}$$

Example 4 (U.S.): Flow Test for Two Outlets

Given:

Test Hydrant = 90 psi static and 50 psi residual

Flow Hydrant = Using two 2½-inch outlets with each C = 0.9, pitot gauge reading for each = 17 psi, and actual diameter = 2.56 inches

$$(29.83)\ (0.9)\ (2.56)^2\ (\sqrt{17}) = 725 \text{ gpm} \times \text{two outlets} = 1{,}450 \text{ gpm}$$

Example 5 (SI): Flow Test for Two Outlets

Given:

Test Hydrant = 621 kPa static and 345 kPa residual

Flow Hydrant = Using two 65 mm outlets both with C = 0.9, pitot gauge reading for each = 117 kPa, and the actual diameter = 66.5 mm

$$(0.0667766)\ (0.9)\ (66.5)^2\ (\sqrt{117}) = 2\ 875 \text{ L/min} \times \text{two outlets} = 5\ 750 \text{ L/min}$$

Example 6: Mathematical Method for Determining Fire Flow

$$Q_r = (Q_f \times h_r^{0.54}) \div h_f^{0.54} \qquad (3)$$

Where: Q_r = *Flow available at desired residual pressure*

Q_f = *Flow during test*

h_r = *Pressure drop to residual pressure (normal operating pressure minus required residual pressure)*

h_f = *Pressure drop during test (normal operating pressure minus residual pressure during flow test)*

The values for *hr* or *hf* to the 0.54 power are listed in **Table 11.3**. Most scientific calculators allow the user to determine these values without a set of complicated steps and procedures.

Using the values from Example 2, in addition to a normal operating pressure of 55 psi:

$$Q_f = 892\ gpm$$

$$h_r = 55\ psi - 20\ psi = 35\ psi$$

$$h_f = 55\ psi - 25\ psi = 30\ psi$$

Under *h* at 35:

$$h^{0.54} = 6.82$$

Under *h* at 30:

$$h^{0.54} = 6.28$$

Therefore:

$$Q_r = (892 \times 6.82) \div 6.28$$

$$Q_r = 969\ gpm$$

From values in Example 3:

$$Q_f = 3\,532\ L/min$$

$$h_r = 380\ kPa - 138\ kPa = 242$$

$$h_f = 380\ kPa - 173\ kPa = 207$$

Under *h* at 242:

$$h^{0.54} = 19.38$$

Under *h* at 207:

$$h^{0.54} = 17.81$$

Therefore:

$$Q_r = (3\,532 \times 19.38) \div 17.81$$

$$Q_r = 3\,843\ L/min$$

NOTE: When doing these calculations in SI, the figures obtained are often higher than those provided in **Table 11.3**. It will be necessary to use a calculator to determine $h^{0.54}$.

Table 11.3
Values for Computing Fire Flow Tests

h	$h^{0.54}$	h	$h^{0.54}$	h	$h^{0.54}$	h	$h^{0.54}$	h	$h^{0.54}$	h	$h^{0.54}$	h	$h^{0.54}$
1	1.00	26	5.81	51	8.36	76	10.37	101	12.09	126	13.62	151	15.02
2	1.45	27	5.93	52	8.44	77	10.44	102	12.15	127	13.68	152	15.07
3	1.81	28	6.05	53	8.53	78	10.51	103	12.22	128	13.74	153	15.13
4	2.11	29	6.16	54	8.62	79	10.59	104	12.28	129	13.80	154	15.18
5	2.39	30	6.28	55	8.71	80	10.66	105	12.34	130	13.85	155	15.23
6	2.63	31	6.39	56	8.79	81	10.73	106	12.41	131	13.91	156	15.29
7	2.86	32	6.50	57	8.88	82	10.80	107	12.47	132	13.97	157	15.34
8	3.07	33	6.61	58	8.96	83	10.87	108	12.53	133	14.02	158	15.39
9	3.28	34	6.71	59	9.04	84	10.94	109	12.60	134	14.08	159	15.44
10	3.47	35	6.82	60	9.12	85	11.01	110	12.66	135	14.14	160	15.50
11	3.65	36	6.93	61	9.21	86	11.08	111	12.72	136	14.19	161	15.55
12	3.83	37	7.03	62	9.29	87	11.15	112	12.78	137	14.25	162	15.60
13	4.00	38	7.13	63	9.37	88	11.22	113	12.84	138	14.31	163	15.65
14	4.16	39	7.23	64	9.45	89	11.29	114	12.90	139	14.36	164	15.70
15	4.32	40	7.33	65	9.53	90	11.36	115	12.96	140	14.42	165	15.76
16	4.47	41	7.43	66	9.61	91	11.43	116	13.03	141	14.47	166	15.81
17	4.62	42	7.53	67	9.69	92	11.49	117	13.09	142	14.53	167	15.86
18	4.76	43	7.62	68	9.76	93	11.56	118	13.15	143	14.58	168	15.91
19	4.90	44	7.72	69	9.84	94	11.63	119	13.21	144	14.64	169	15.96
20	5.04	45	7.81	70	9.92	95	11.69	120	13.27	145	14.69	170	16.01
21	5.18	46	7.91	71	9.99	96	11.76	121	13.33	146	14.75	171	16.06
22	5.31	47	8.00	72	10.07	97	11.83	122	13.39	147	14.80	172	16.11
23	5.44	48	8.09	73	10.14	98	11.89	123	13.44	148	14.86	173	16.16
24	5.56	49	8.18	74	10.22	99	11.96	124	13.50	149	14.91	174	16.21
25	5.69	50	8.27	75	10.29	100	12.02	125	13.56	150	14.97	175	16.26

Review Questions

1. What is the difference between public and private water supplies?
2. What are the common components of water supply systems?
3. Name the types of water sources.
4. What methods are used to move water?
5. List types of water distribution systems.
6. What can an inspector learn from testing water supplies?
7. List the tools needed to perform a fire hydrant inspection.
8. List the steps in the procedure for using a pitot tube and gauge.
9. List the physical laws from which the constant used in the flow test formula is derived.
10. Explain residual pressure.
11. Describe the procedure to conduct a flow test.
12. What precautions should an inspector take when conducting water flow tests?
13. What are some of the obstructions an inspector may encounter when conducting water flow tests?
14. List the steps involved in a graphical analysis.

Water-Based Fire Suppression Systems

Photo courtesy of Rich Mahaney.

Chapter Contents

Case History 503
I **Automatic Sprinkler Systems 504**
Basic Types and Design 505
Components 507
Residential Systems 518
Automatic Fire Suppression Systems 519
Water-Spray Fixed System 520
Water-Mist Systems... 521
Foam-Water Systems... 527
Standpipe and Hose Systems 525
Components 526
Classifications 526
Types 527
Water Supplies and Residual Pressure 529
Standpipe Hose Valves... 530
Pressure-Regulating Devices 530
Fire Department Connections.................................. 531
Stationary Fire Pumps 532
Types 532
Pump Drivers 535
Controllers 536
II **Evaluate System Components and Equipment 538**
Plans Review 539
Preacceptance Inspections 539
Acceptance Testing.. 540
Routine Inspections and Testing 540
Chapter Summary 557
Review Questions 558

Key Terms

Centrifugal Pump....................................532
Fusible Link...515
Head Pressure 533
High-Rise Building 532
Large Diameter Hose (LDH)...................532
OS&Y Valve 509
Post Indicator Valve (PIV) 509
Pressure-Maintenance Pump.................535
Pressure-Reducing Valve 531
Riser...511
Sprinkler..512
Thermocouple...526

NFPA® Job Performance Requirements

This chapter provides information that addresses the following job performance requirements of NFPA® 1031, *Standard for Professional Qualifications for Fire Inspector and Plan Examiner* (2014).

Fire Inspector I

4.3.5

Fire Inspector II

5.3.4

5.4.8

5.3.11

5.4.3

5.4.4

Water-Based Fire Suppression Systems

Learning Objectives

After reading this chapter, students will be able to:

Inspector I

1. Describe the components and basic operation of automatic sprinkler systems. (4.3.5)
2. Explain the operation of fixed fire suppression systems. (4.3.5)
3. Recognize types of standpipe and hose systems. (4.3.5)
4. Explain the components and operation of stationary fire pumps. (4.3.5)

Inspector II

1. Explain how to evaluate fire protection systems and equipment. (5.3.4, 5.3.8, 5.4.3, 5.4.4)
2. Explain inspection and testing of fire suppression systems and components. (5.3.4, 5.3.11, 5.4.4)

Chapter 12
Water-Based Fire Suppression Systems

Case History

Chicago, 2012: A woman died in a Chicago high-rise fire after the building elevator took her to the fire floor. It was believed that the victim died instantly when the doors opened. The fire also injured nine other people, including two firefighters.

The legal responsibility is complicated by Chicago's Home Rule status, which means that the city only enforces its own municipal standards. The building owners stated that they were following Chicago fire codes, which are less stringent than state codes. The Illinois fire marshal said that state code supersedes city code in matters of fire safety. The fire marshal cited the building with a number of violations, including lack of a fire alarm system, lack of automatic elevator recall, and lack of an automatic sprinkler system.

The city of Chicago has given older high-rises a three-year extension to retrofit buildings with modern fire safety equipment, which includes the most costly addition, a sprinkler system. Buildings exempt from sprinkler retrofitting requirements must submit to a fire safety inspection, install voice communication systems, and install one-hour fire rated doors in stairwells.

Inspectors must be familiar with all water-based fire suppression systems that they may encounter in their jurisdictions, including:

- Automatic sprinkler systems
- Water-spray fixed systems
- Water-mist systems
- Foam-water systems
- Standpipe and hose systems
- Fire pumps

This chapter describes each of these types of water suppression systems. Inspectors are responsible for understanding how the systems work, making visual observations regarding their installation and compliance with design documents, and inspecting the system so that the owner maintains the system in a ready condition in accordance with the locally adopted code.

NOTE: This chapter describes water-based suppression systems that control or contain hazardous conditions. Chapter 13 describes special-agent extinguishing systems, Chapter 14 describes detection systems.

Automatic Sprinkler Systems

Automatic sprinkler systems, in their basic forms, are used as the first line of defense against fires and have been in use for over one hundred years. The design, installation, testing, and inspection requirements for sprinkler systems composed the first standard written by NFPA® in 1896. That standard — NFPA® 13, *Standard for the Installation of Sprinkler Systems* — has been in continuous publication since that date.

Today, the automatic sprinkler system has proven to be an unsurpassed fire protection device. Fire loss data reveals that in buildings equipped with automatic sprinklers, about ninety-four percent of all fires were controlled or extinguished by the sprinkler system. For the remaining fires that were not controlled in sprinkler-equipped buildings, failure was due to human actions, including the following:

- Improper maintenance
- Inadequate or no water supply
- Incorrect design
- Obstructions
- Intentionally set fire

According to the National Fire Sprinkler Association (NFSA), there has never been a multiple loss of life (three or more deaths) due to fire or smoke in a sprinkler-equipped building. Deaths that have occurred were in cases where the victims were close to the fire (asleep in the room of origin) or as the result of an explosion.

The model building codes have required the installation of sprinkler systems in hotels, hospitals, and places of assembly for many years. According to NFPA® reports, the fire death rate per thousand in sprinkler-equipped hotels and motels is 1.6 as compared to 9.1 in hotels and motels without sprinkler systems.

For a fire suppression system to be effective, it must be reliable and automatic. It should also utilize a readily available and inexpensive extinguishing agent. The system should be able to discharge the agent directly on the fire or in a manner that floods the fire area before it can expand beyond the incipient phase. Automatic sprinkler systems meet this criterion for an effective fire suppression system.

An automatic sprinkler system consists of a series of discharge devices (called sprinklers) that are arranged to distribute enough water to either extinguish a fire or control its spread until firefighters arrive. A network of pipes supplies water to the sprinklers. Thermally sensitive devices such as fusible links keep most sprinklers (except deluge system sprinklers) closed. When heat from a fire affects these devices, sprinklers closest to the fire activate automatically. Deluge or preaction type sprinkler systems operate when an electronic detector or manual control device is activated. Most fires in sprinkler-equipped structures are controlled by the operation of one or two sprinklers.

NOTE: Sprinklers are described in more detail later in this chapter.

Basic Types and Design

Traditionally, the fire service has recognized four basic types of automatic sprinkler systems as defined in NFPA® 13:

- ***Wet-pipe sprinkler system*** — Continually charged with water under pressure that discharges immediately when heat from a fire activates one or more sprinklers (**Figure 12.1**).
- ***Dry-pipe sprinkler system*** — Continually charged with compressed air. When a sprinkler activates, the air is released, allowing the dry-pipe waterflow control valve to operate and charge the system with water. These systems are typically used in areas where freezing temperatures are likely to occur (**Figure 12.2**).

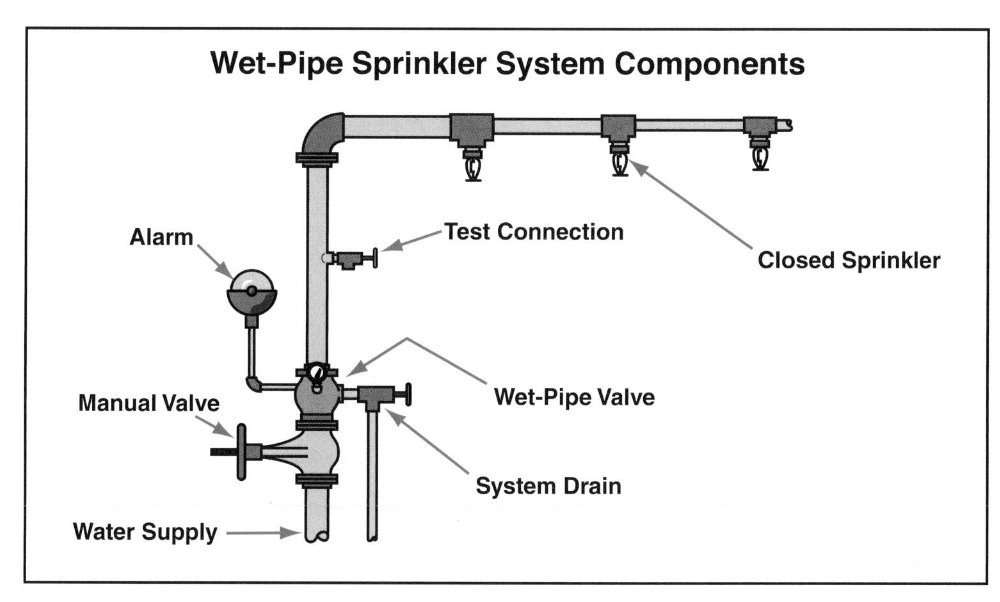

Figure 12.1 A wet-pipe sprinkler system includes several essential components.

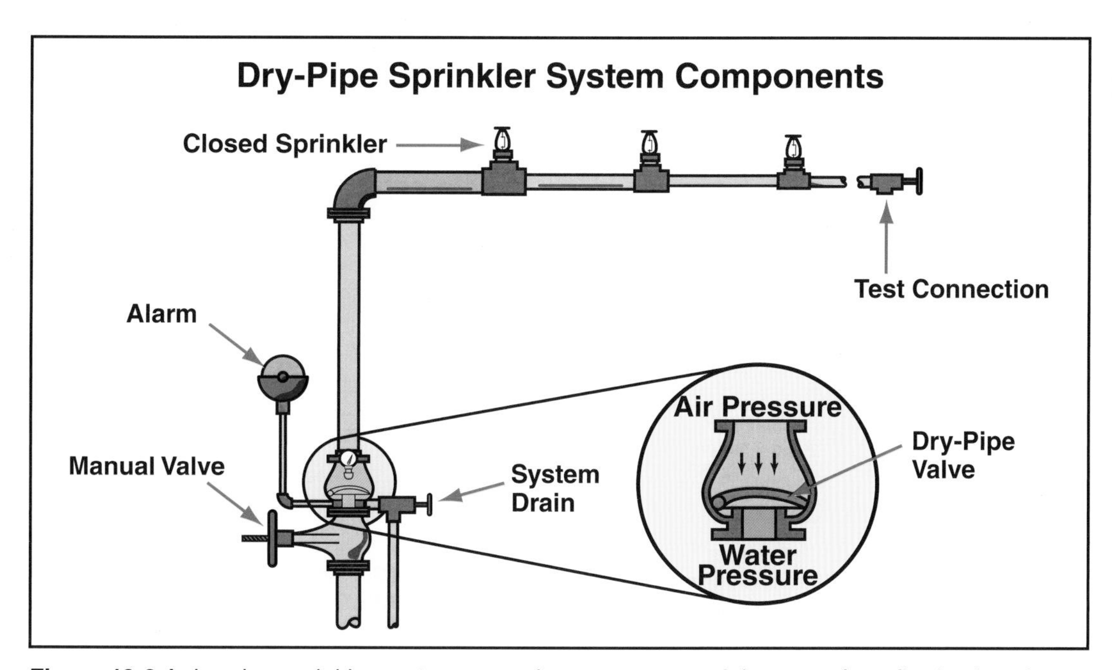

Figure 12.2 A dry-pipe sprinkler system uses air pressure to seal the water from flowing into the pipes before activation.

- ***Deluge sprinkler system*** — Consists of open sprinklers attached to unpressurized dry pipes. (There is no thermally sensitive device in a deluge system). The system activates when a detection device in the protected area senses a fire and opens the waterflow control valve to the system. All sprinklers discharge water simultaneously **(Figure 12.3)**.
- ***Preaction sprinkler system*** — Similar to a dry system, a preaction system is continually charged with air that may or may not be under pressure. The system only operates when a sprinkler opens due to a thermally sensitive device activating and one or more detection devices in the same area triggers the waterflow control valve to open **(Figure 12.4)**.

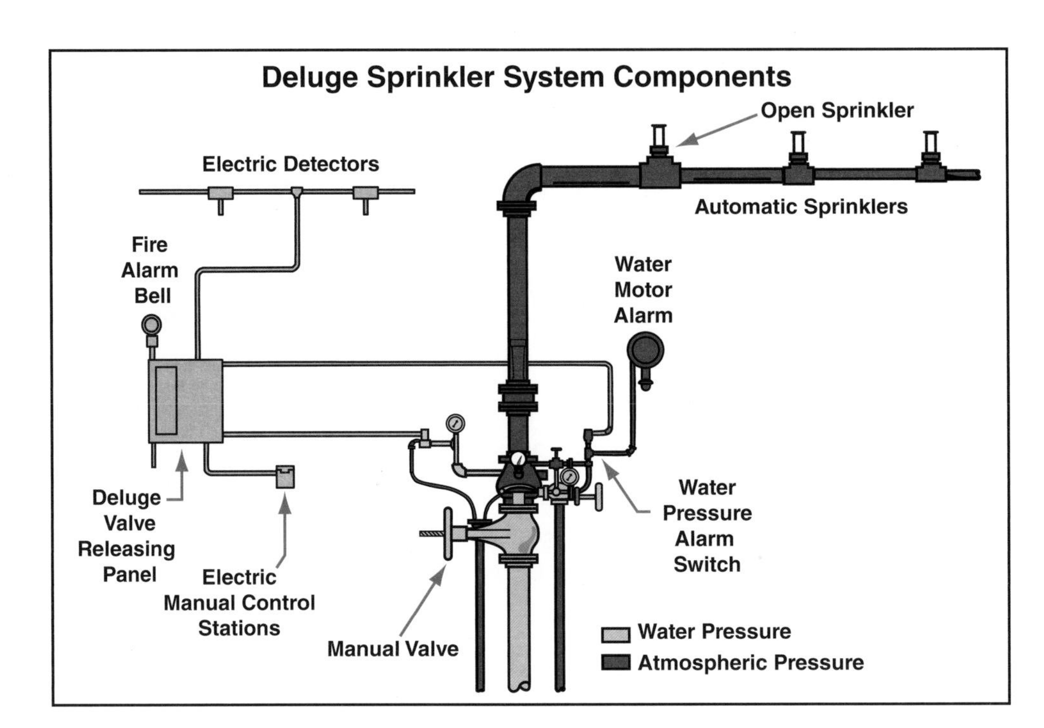

Figure 12.3 A deluge system is provided with sprinklers having no thermally sensitive device. When the system activates, all sprinklers discharge simultaneously.

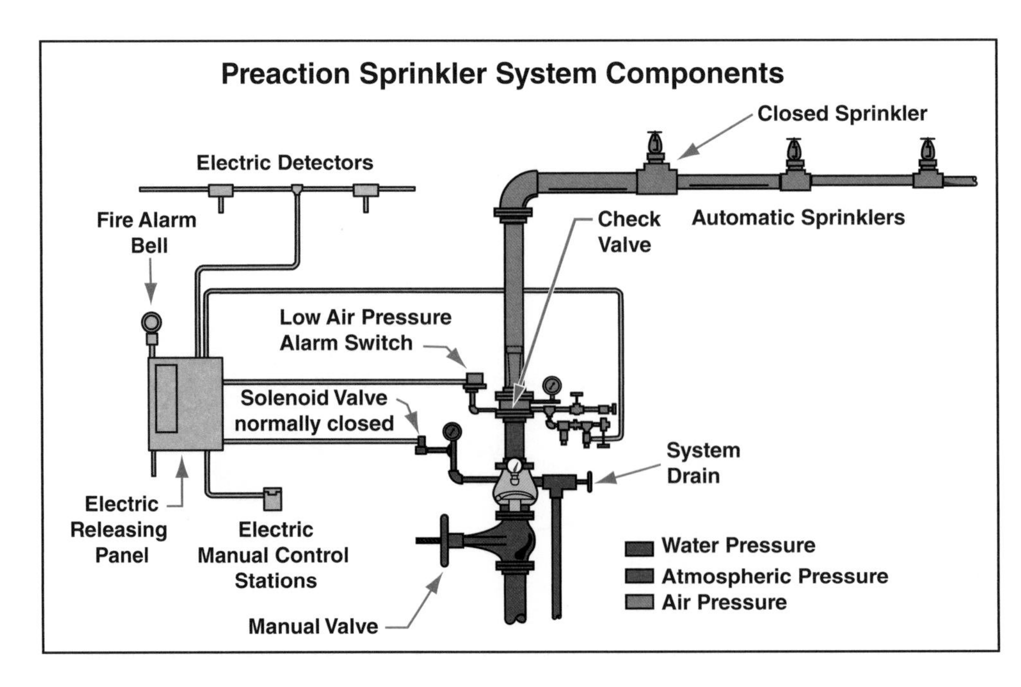

Figure 12.4 Preaction systems require sprinklers to be activated by smoke- or heat-detection systems before water will flow to the sprinklers.

Advances in sprinkler system design and technology, as well as ever-increasing types of hazards, have led to the addition of six more types of systems. These systems, which are variations of the four basic types, including the following:

- ***Antifreeze sprinkler system*** — Wet-pipe system that is continually charged with an antifreeze solution. When the system is activated, the antifreeze solution discharges, activating the waterflow valve and allowing water to flow to the open sprinkler(s). This system requires additional maintenance: the antifreeze solution must be tested annually and adjusted if not in compliance.
- ***Circulating closed-loop sprinkler system*** — Wet-pipe system that uses the sprinkler system to circulate water for non-fire-protection building services such as heating or cooling. It is a closed system in which water is not removed from the system unless the sprinklers are activated.
- ***Combined dry pipe and preaction sprinkler system*** — System that is continually charged with air under pressure combined with a detection system that controls the operation of the waterflow control valve. The detection system activates the waterflow control valve, the release of the pressurized air in the system, and the facility fire alarm. When the system is charged with water, activation of the individual sprinklers will discharge water. This system is used in special circumstances where an additional level of protection against false activation is required, such as in data centers or museums. It is also sometimes known as a double preaction system or double interlock system.
- ***Multicycle sprinkler system*** — System designed to be operated repeatedly in response to a detection device. The system shuts on and off based on the demand indicated by the detection device.

Several sprinkler design methods are looped systems and gridded systems. Gridded systems consist of parallel cross mains connected by multiple branch lines **(Figure 12.5)**. Any activated sprinklers will receive water from both mains. This type of system has the advantage of water flowing to the sprinklers from multiple directions.

Looped sprinkler systems consist of interconnected cross mains that provide multiple routes for water to reach any point in the system **(Figure 12.6)**. The branch lines are not interconnected. This system is a common design because of the advantage of water flow from multiple directions.

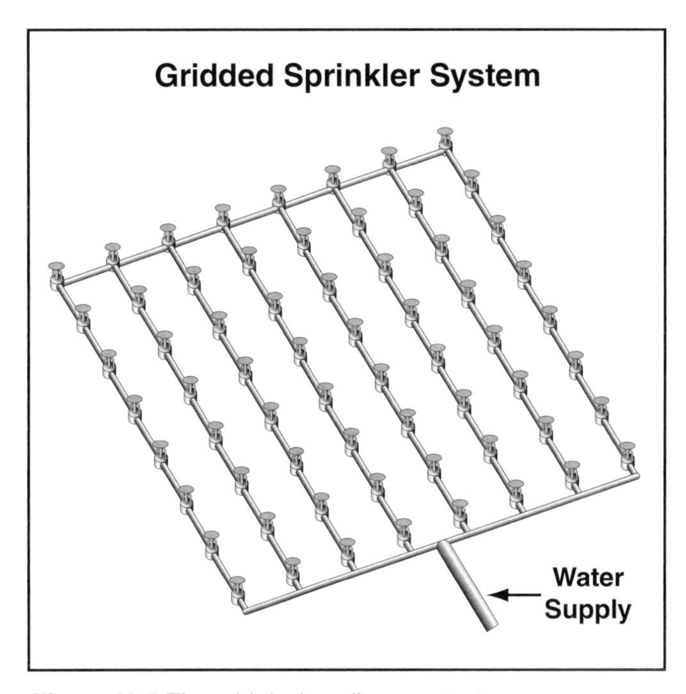

Figure 12.5 The grid design allows water to access sprinklers from multiple directions.

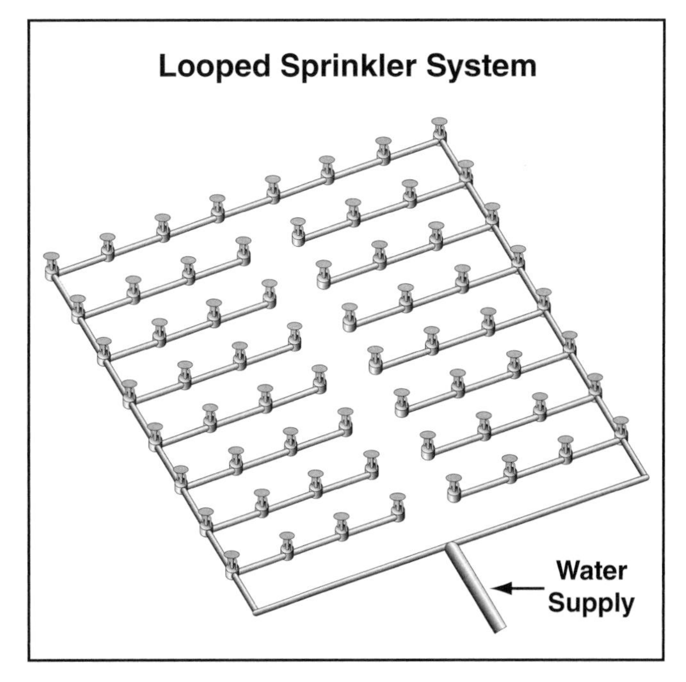

Figure 12.6 Looped systems also allow water flow from multiple directions; however, branch lines are not interconnected like they are in a grid system.

Components

Automatic sprinkler system components are designed to deliver water to a fire or hazard as quickly as possible. While components may vary based on the building's construction or the hazards being protected, all sprinkler systems have features that are common, including:

- Water supplies
- Waterflow control valves
- Operating valves
- Water distribution pipes
- Sprinklers
- Detection and activation devices

Water Supplies

Every automatic sprinkler system must have a water supply of adequate volume, pressure, and reliability. The hazard being protected, the occupancy classification, and fuel load conditions determine the minimum waterflow required for the system. In addition, a water supply must be able to deliver the required volume of water to the highest or most remote sprinkler in a structure while maintaining a minimum required residual (remaining) pressure in the system.

Sprinkler systems must have a primary water supply and may be required to have a secondary water supply. The primary water supply may come from a public or a private source. Often, a public water supply source is the only type of connection available. If no public supply is available, a private water supply will be necessary. Private water supplies may originate from impounded water sources, such as on-site ponds, reservoirs, wells, or storage tanks.

NOTE: Refer to Chapter 11, Water Supplies, for additional information on water supply sources.

In some instances, a second independent water supply is not only desirable but also required. Storage tanks may be used as secondary supply sources, although they may also be primary sources (such as with residential sprinkler systems). Fire pumps that take suction from large static water sources, lakes, reservoirs, or wells are also used as secondary sources of water supply. When properly powered and supervised, these pumps may be used as a primary water supply source. The inspector may be able to make a visual evaluation for degradation of water supply, such as noticing that a lake is down due to a drought year.

Fire department connections (FDCs) are also included as part of automatic sprinkler systems to boost the pressure of the primary water supply. A fire department pumper connected to the public/private water supply can pump water into the sprinkler system (riser) through the FDC.

NOTE: Additional information on FDCs is included later in this chapter.

Waterflow Control Valves

Every automatic sprinkler system is equipped with a main waterflow control valve or valves on either side of a check valve (backflow preventer) that prevent sprinkler water from flowing back into the water supply. Waterflow control valves are used to turn off or isolate the water supply to the system when it is necessary to perform maintenance, change sprinklers, or interrupt operation. These valves are located between the water supply and the sprinkler system. Waterflow control valves may also be located throughout the system to isolate specific zones. Municipal supply fire loops may not have indicating valves.

Waterflow control valves must be indicating-type valves that visually indicate whether they are open or closed **(Figures 12.7 a - c)**. Several common types of indicator control valves are used in automatic sprinkler systems:

- **Outside stem and yoke (OS&Y)** — Gate valve that has a yoke on the outside with a threaded stem or screw; the threaded portion of the stem is out of the yoke when the valve is open and inside the yoke when the valve is closed.
- **Post indicator valve (PIV)** — Underground gate valve that has a hollow metal post attached to the valve housing. The valve stem inside the post has a target on which the words OPEN or SHUT appear.
- **Wall post indicator valve (WPIV)** — Similar to a PIV except that it extends horizontally through the wall with the target and valve operating nut on the outside of the building.
- **Post indicator valve assembly (PIVA)** — Similar to the PIV except that it uses a butterfly valve, while the PIV uses a gate valve.

OS&Y Valve — Outside stem and yoke valve; a type of control valve for a sprinkler system in which the position of the center screw indicates whether the valve is open or closed. *Also known as* Outside Screw and Yoke Valve.

Post Indicator Valve (PIV) — Type of valve used to control underground water mains that provides a visual means for indicating "open" or "shut" positions; found on the supply main of installed fire protection systems. The operating stem of the valve extends above ground through a "post," and a visual means is provided at the top of the post for indicating "open" or "shut."

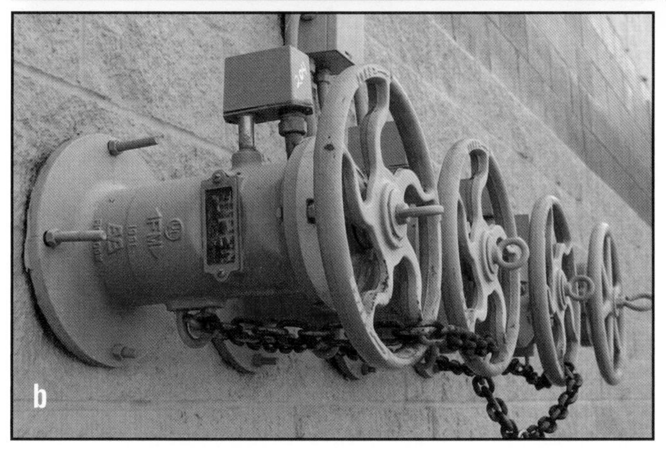

Figures 12.7 a-c (a) Outside sprinkler riser assembly with two zones; (b) Wall post indicator valves (WPIVs); (c) Post indicating valves (PIVs). *Photos a and b courtesy of Rich Mahaney.*

Operating Valves

In addition to the main waterflow control valves, automatic sprinkler systems employ such operating valves as alarm test valves, check valves, automatic drain valves, globe valves, and ball-drip valves **(Figures 12.8 a and b)**:

- **Alarm-test valves** — Located on the riser and used to flow water for testing the waterflow alarm; performs a function similar to the inspector's test valve that may be located at a remote part of the system piping.
- **Check valves** — Used to limit the flow of water to one direction; placed in the water supply line to prevent recirculation or backflow of water from the sprinkler system into the municipal water supply system.
- **Drain valves** — Used to drain water from piping when pressure is relieved in the pipe; may be main drain or auxiliary valves.
- **Trim valves** — May be globe, ball, or gate valves; used primarily as drains and test valves. With many packaged valves in use, ball valves are common. The inspector's test valve is usually a globe valve and may sometimes be found at a remote location within a sprinkler system.
- **Drip check or drip ball valves** — Used for drains and to keep water from freezing in pipes and damaging them.

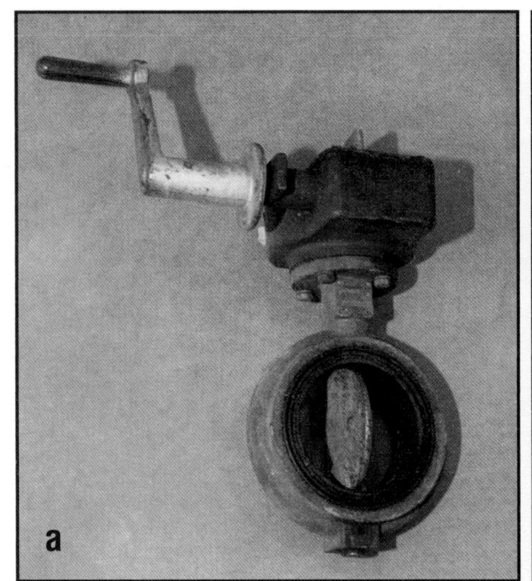

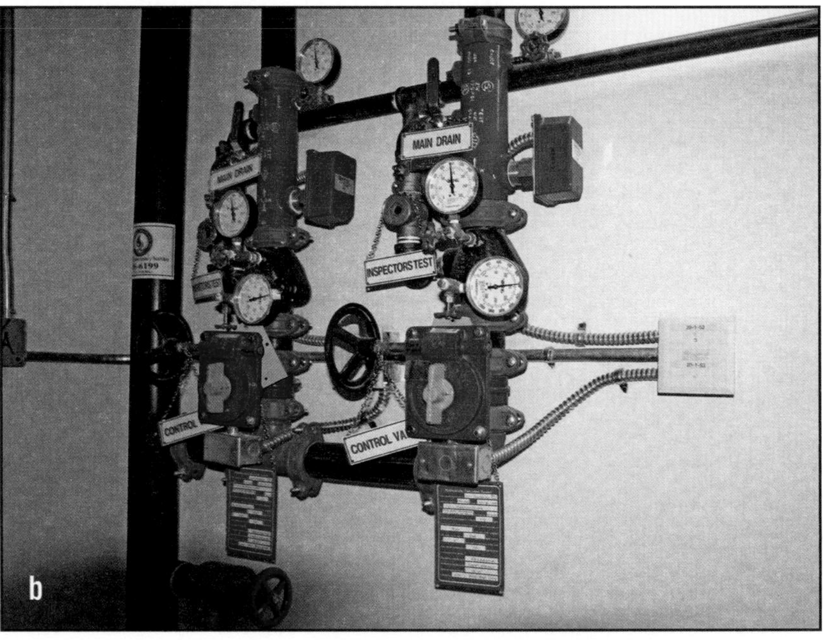

Figures 12.8 a and b (a) The butterfly valve rotates inside the waterway. (b) Sprinkler zone test and drain assemblies.

Water Distribution Pipes

An automatic sprinkler system consists of an arrangement of pipes of different diameters. The system starts with an underground water supply main that may originate from a public or private water supply. The underground supply main contains a check valve to prevent sprinkler water from flowing back (backflowing) and possibly contaminating the potable (drinkable) water supply.

System risers are vertical sections of pipe that connect the underground supply to the rest of the piping in the system. The riser has the system waterflow control valve and associated hardware that is used for testing, alarm activation, isolation, and maintenance.

Riser —Vertical water pipe used to carry water for fire protection systems above ground, such as a standpipe riser or sprinkler riser.

Risers supply the cross main that directly serves a number of branch lines **(Figure 12.9)**. Sprinklers are installed on the branch lines with nipple risers (short vertical sections of pipe). Hangers, rings, and clamps support the entire system, which may be pitched (sloped) to facilitate drainage. The requirements for supports spacing can be found in the applicable design specification, and the inspector should compare the design to the installation.

The arrangement of pipes is created by using one of two methods: pipe schedule tables or hydraulic calculations. Inspectors who are reviewing sprinkler plans should be aware of the method used to design the system.

Pipe schedule tables have been in use for over 100 years. The design is based on tables in NFPA® 13 that designate the maximum number of sprinklers that a given size of pipe can supply, sprinkler spacing, and occupancy classification. The pipe schedule table method is limited to light and ordinary hazards. Some sprinkler contractors may still use this method, but it is not as accurate as the hydraulic calculation method.

Hydraulic calculations were initially performed manually to design the systems, but now computer design programs make the process much easier, more accurate, and less expensive. Hydraulically designed systems are based on the type of occupancy to be protected, the type of hazard, the required

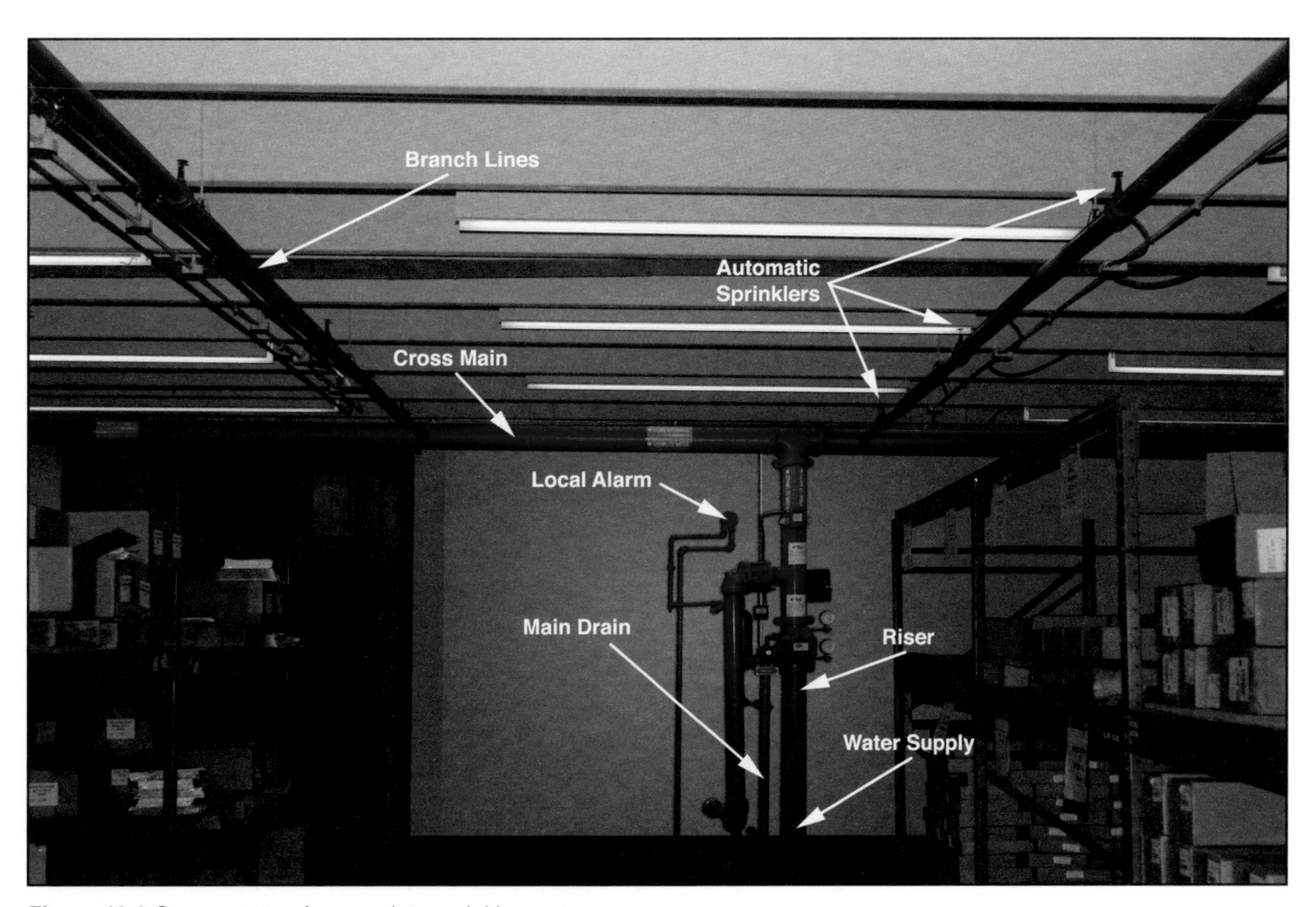

Figure 12.9 Components of a complete sprinkler system.

density (quantity of water to be discharged), and the minimum pressure for each operating sprinkler. The layout and diameter of the distribution pipes are determined from these design requirements.

In addition to being more accurate, the hydraulic calculation method can provide a cost-effective design that is not limited by the restrictions placed on the pipe schedule design. The system is designed based on the greatest demand at the greatest distance from the base of the waterflow control valve or system riser. Calculations may be made to protect all types of hazards, including extra hazards, with any type of piping and any type of sprinkler discharge device.

Sprinkler — Waterflow device in a sprinkler system. The sprinkler consists of a threaded nipple that connects to the water pipe, a discharge orifice, a heat-actuated plug that drops out when a certain temperature is reached, and a deflector that creates a stream pattern that is suitable for fire control. Formerly referred to as a sprinkler head.

NOTE: See Chapter 11 for an illustration of hydraulic calculation tables.

Sprinklers

The **sprinkler** is the actual discharge device that applies water (or foam) to a fire or other hazard. Sprinklers are basically small fog-type nozzles that emit a stream of water and break it into small droplets, spreading it over a large area and allowing for efficient steam conversion and cooling of the fire. A deflector, which is a small circular piece of metal mounted on the end of the sprinkler, creates the discharge pattern **(Figure 12.10)**.

Figure 12.10 As the seated cap is released, water begins to flow from the sprinkler.

Sprinklers are defined by their orientation or by NFPA® definitions. Orientation refers to the way the sprinkler is installed in the ceiling or wall. With some design variations, each type of sprinkler may be installed as follows **(Figures 12.11 a-e)**:

- **Concealed** — Recessed and covered with a removable decorative cover plate that releases when exposed to a specific level of heat.
- **Flush** — Mounted in a ceiling with the body of the sprinkler, including the threaded shank, above the plane of the ceiling.
- **Pendant** — Installed downward from the branch line so that water flowing from the sprinkler strikes the deflector and is distributed over the protected area.
- **Recessed** — Installed in a recessed housing within the ceiling of a compartment or space; all or part of the sprinkler other than the threaded shank is mounted in the housing.
- **Sidewall** — Mounted horizontally from a wall face so that the deflector causes the water to be distributed in an arc over the protected area while a portion of the spray is directed at the supporting wall face.
- **Upright** — Installed upward from a branch line so that water is discharged up against the deflector.

Figure 12.11 Sprinkler orientations include the following: (a) upright sprinkler; (b) pendant sprinkler; (c) sidewall sprinkler; (d) hidden/concealed sprinkler; (e) flush sprinkler.

The inspector needs to remember that the correct sprinkler needs to be installed in the correct location based on orientation and hazard type. It is not uncommon to find a pendant sprinkler installed where an upright one should be or to have a sprinkler with a high heat rating installed improperly. In systems with multiple types of sprinklers, the inspector should verify that the spare sprinkler box contains an adequate number of sprinklers of each type, along with the appropriate wrench or wrenches **(Figure 12.12, p. 514)**.

NFPA® has begun to define sprinklers by their design and performance characteristics rather than their orientation. NFPA® 13 lists various types of sprinklers. An inspector will need to be familiar with all types and their uses when performing both sprinkler plans review and facility inspections **(Figures 12.13 a - c, p. 514)**. Some common types of sprinklers are as follows:

- **Control mode specific application (CMSA) sprinkler** — A type of spray sprinkler capable of producing characteristic large water droplets and that is listed for its capability of providing fire control of specific high-challenge fire hazards.

Figure 12.12 An inspector should verify that sprinkler cabinets are stocked with at least one replacement sprinkler for each type used in the system.

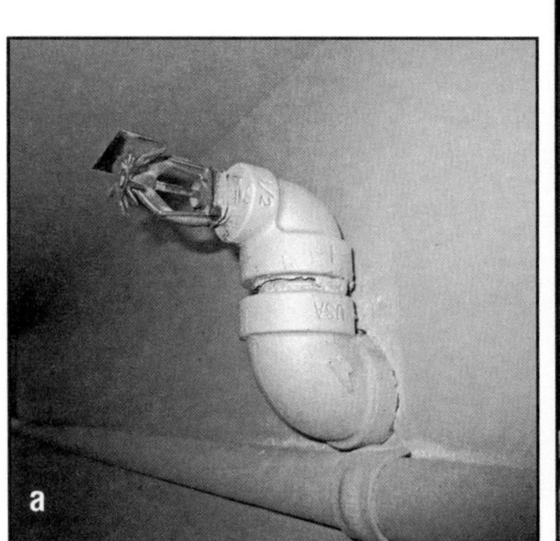

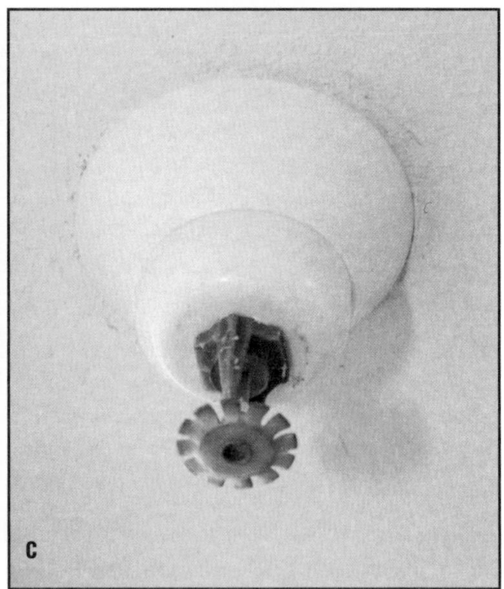

Figures 12.13 a - c Inspectors may encounter a variety of sprinkler types while performing their duties. Some examples include (a) a sidewall sprinkler (b) an upright sprinkler with a fusible link and (c) a pendant sprinkler. *Photo a courtesy of Rich Mahaney.*

- **Early suppression fast-response (ESFR)** — Designed to suppress fires in high-challenge hazards such as warehouses; developed as a result of the research into residential-type sprinklers that must react rapidly to protect life in the room of origin.
- **Extended coverage (EC)** — Designed to cover an area greater than 15 feet (5 m) x 15 feet (5 m).
- **Old-style/conventional** — Designed to direct 40 to 60 percent of its discharge in a downward direction. May be installed in an upright or pendent position, and the deflector may be altered according to installation. Inspectors may find these sprinklers in old facilities.

- **Open** — Designed with an open orifice and no thermal or heat-responsive element or plug. Most commonly found in deluge systems.
- **Quick-response early suppression (QRES)** — Designed to provide increased life safety in hotels, motels, and similar residential occupancies.
- **Quick-response extended coverage (QREC)** — Designed to provide increased life safety in hotels, motels, and similar residential occupancies; has a larger area of coverage than QRES sprinklers.
- **Residential (RES)** — Designed for fast response in residential occupancies where life safety in the room of origin is the primary concern; used in systems that are designed to meet the the requirements of
  - — NFPA® 13D, *Standard for the Installation of Sprinkler Systems in One- and Two-Family Dwellings and Manufactured Homes*
  - — NFPA® 13R, *Standard for the Installation of Sprinkler Systems in Low-Rise Residential Occupancies*.
- **Special** — Designed and listed to protect special hazards.
- **Standard spray (SS)** — Designed for use in general types of occupancies; replaced the old-style or conventional sprinklers; may be designed for use as upright (SSU) or pendant (SSP); take longer to activate than the RES, QREC, QRES, or ESFR sprinklers.
- **Nozzle** — Designed for use in situations that require a special water-discharge pattern, directional spray, or any other unusual discharge characteristics; similar to the type used in water-spray fixed systems.

Although a standard sprinkler may have an operating temperature of 165° F (74° C), the thermal lag of the **fusible link** may delay the operation of the sprinkler until the surrounding air temperature is considerably higher. By redesigning the fusible link, the sprinkler can respond more quickly and activate before conditions in the room become untenable.

Fusible Link — (1) Connecting link device that fuses or melts when exposed to heat; used in sprinklers, fire doors, dampers, and ventilators. (2) Two-piece link held together with a metal that melts or fuses at a specific temperature.

Within each sprinkler type, there may be multiple variations depending on the manufacturer and the hazard requirements. All sprinklers must meet the design criteria set forth by NFPA® and tested for the specific application for which they are intended.

An inspector may encounter other variations of basic sprinkler types during plans review or field inspections due to special conditions:

- **Corrosion-resistant** — Constructed of stainless steel or is factory coated with a corrosion-resistant material, such as wax, asphalt, lead, enamel, polyester, or Teflon®; usually installed in areas containing acids or caustic materials or processes.
- **Dry** — Attached to a dry pipe nipple that is separated from a wet branch line by a pressure seal; may be used in either dry or wet systems where the protected area may freeze, such as walk-in freezers in grocery stores.
- **Institutional** — For use in prisons, jails, or facilities that house the mentally ill; designed to be damage resistant with no removable parts.
- **Intermediate level or rack storage** — Used in high-rack storage spaces; shields are added to prevent the sprinkler from being affected by the operation of sprinklers at a higher level.

- **Ornamental or decorative** — Specially painted or plated by the manufacturer to match the décor of the surrounding area.
- **Pilot line detector** — A standard spray or thermostatic fixed temperature release device used as a detector to pneumatically or hydraulically release the main valve, controlling the flow of water into a fire protection system

Detection and Activation Devices

Sprinkler systems are activated in several ways:

- Thermal detectors in the sprinkler
- Smoke, heat, or rate-of-rise detectors in the protected area
- Manual controls

NOTE: See Chapter 14, Fire Detection and Alarm Systems, for further information on alarm systems.

Thermocouple — Device for measuring temperature in which two electrical conductors of dissimilar metals such as copper and iron are joined at the point where heat is applied.

Electronic detection and manual controls may also act as notification or alarm systems to warn building occupants. Detection and activation devices include the following alarms, detectors, and activators:

- **Sprinkler activation** — Occurs when **thermocouples** (devices for measuring temperatures) reach a preset temperature, releasing the sprinkler plug. Commonly used release mechanisms are fusible links, glass bulbs, and chemical pellets. Sprinklers are rated by their activation temperatures, which may be stamped on the sprinkler **(Figures 12.14 a-c)**.
- **Specific application sprinkler** — Activates based on the maximum temperature expected at the level of the sprinkler under normal conditions. An important consideration is the anticipated rate-of-heat release that a fire would produce in a particular area. These temperature ratings are given in **Table 12.1**.

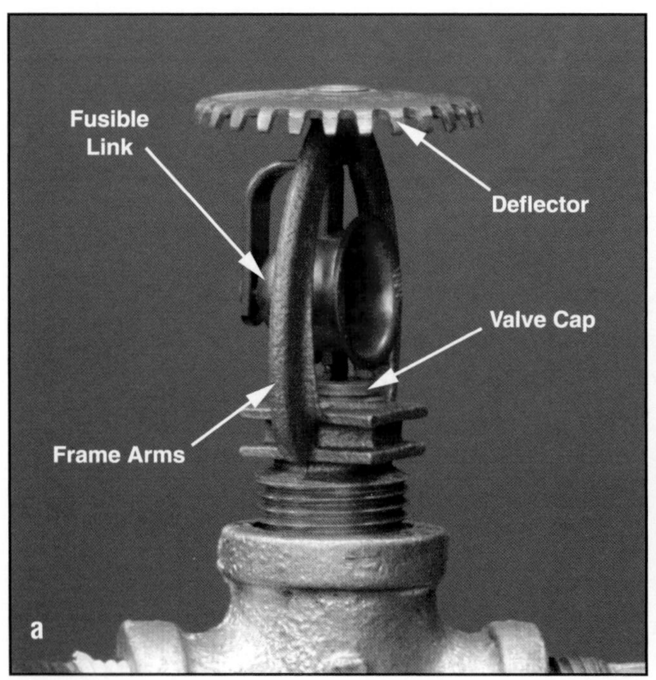

Figures 12.14 a - c (a) Fusible links, (b) glass bulbs, and (c) pellet heat sensors are all mechanisms designed to break or fail as a reaction to heat. When they fail, water is released from the sprinklers.

- **Electronic heat detector** — Activates preaction and deluge systems by smoke, heat, or rate-of-rise detectors in the protected area; may also act as notification or alarm system to warn occupants. When the detector senses an abnormal situation, the detector sends an electronic impulse to the fire control panel and the sprinkler waterflow control valve **(Figure 12.15)**. The valve opens, permitting water to enter the system. *Differences:*
  - *Preaction system:* The sprinklers only discharge water when the sprinkler thermocouples activate, either simultaneously or after the detector has activated.
  - *Deluge system:* The sprinklers will discharge water immediately.
- **Waterflow alarm** — Activates an alarm on the sprinkler waterflow control valve and area valves (indicating the specific point in the system where water is discharging). The waterflow alarm sounds a warning when a sprinkler activates and water begins to flow in the facility, on the exterior of the facil-

Table 12.1
Sprinkler Color Coding, Temperature Classification, and Temperature Rating

Color Coding	Temperature Classification	Temperature Rating	
		°F	°C
Uncoded or Black	Ordinary	135-170	57-77
White	Intermediate	175-225	79-107
Blue	High	250-300	121-149
Red	Extra High	325-375	163-191
Green	Very Extra High	400-475	204-246
Orange	Ultra High	500-575	260-302

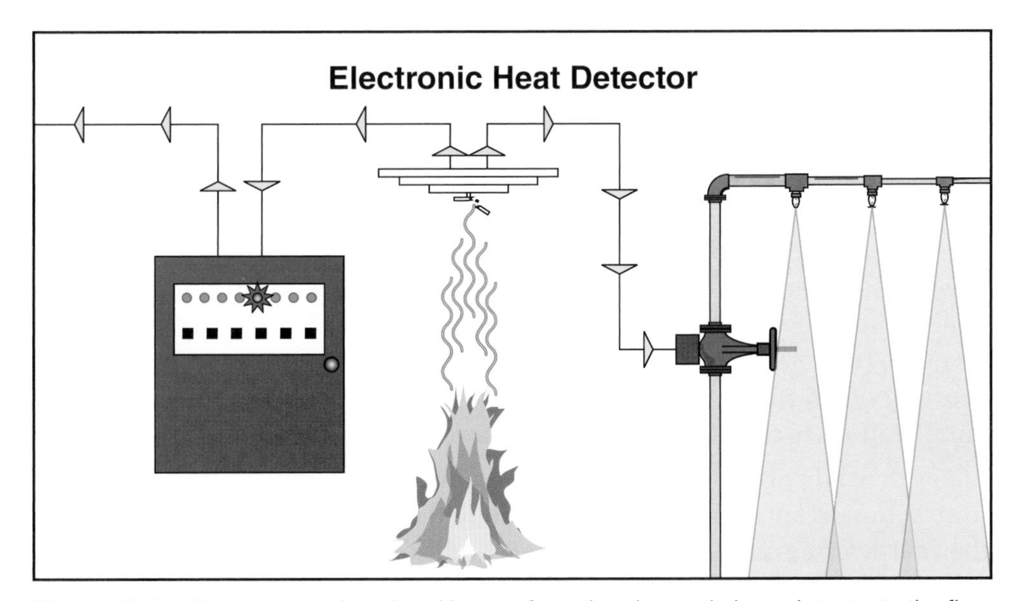

Figure 12.15 When activated, a signal is sent from the electronic heat detector to the fire-control panel and the fire protection system control valve.

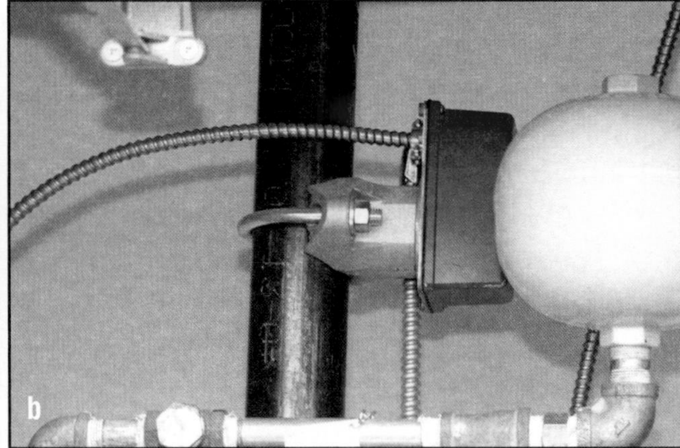

Figures 12.16 a and b (a) Water motor gongs and (b) electric sensors on pipes are types of waterflow alarms. These alarms sound to indicate that a sprinkler system has activated within a facility.

ity, and sometimes at a monitored alarm-control panel **(Figures 12.16 a and b)**. Waterflow alarms may also activate as part of the alarm system. These alarms can be mechanical in nature, activating by an impeller located in the pipe that directly spins a mechanical alarm. Alternatively, a water pressure or waterflow device can activate an audible and/or visual fire alarm system that can also transmit off site.

- **Manually activated system** — Depends on human intervention to operate the sprinkler system. In this case, a worker may activate the system before a fire occurs, such as in the case of a hazardous gas or liquid release. Some locations have manual systems for fire brigades to activate. These are also located in electrical rooms or data centers.

Residential Systems

Sprinkler systems installed according to NFPA® 13 standards focus on property conservation with life safety as a primary concern. Sprinkler systems installed according to NFPA® 13R, 13D, and IRC P2904 residential system standards address life safety by working to prevent flashover in the room of origin and enabling occupants to evacuate. This compromise was reached with the idea of reducing system cost, increasing the number of system installations, and the safety of the building occupants. As such, some of the control hardware, fire department connections, required system flows, and connections to the public or private water supply have been scaled back or eliminated based on the hazard.

Figure 12.17 Residential sprinklers are installed high on the walls to prevent fire from traveling above the sprinkler spray.

In recent years, some new NFPA® 13D and IRC P2904 compliant systems have been developed that are fully integrated into the domestic plumbing system. In these systems, the water flow to the plumbing fixtures also serves the sprinkler; therefore, there is no way to shut down the sprinkler system without shutting down the domestic system. It is also important to remember that modifications to the plumbing system may compromise the sprinkler system.

Residential sprinklers also have distribution patterns that are different from standard sprinklers. They are designed to discharge water to break the thermal layering and prevent flashover conditions, allowing for the occupants to evacuate **(Figure 12.17)**. Sprinkler distribution patterns vary according to manufacturer, so an inspector should consult the relevant standard for requirements.

Water Supply and Flow Rate Requirements

The water supply requirements for residential sprinklers are less than those for standard sprinkler systems. NFPA® 13D requires only 18 gpm (70 L/min) for any single sprinkler. When there are two or more sprinklers, each requires 13 gpm (50 L/min) as a minimum water supply. Economics is not the only reason for the reduction in water supply. Small domestic water supplies service many residential buildings. For example, some single-family dwellings may be supplied by wells. Although a ten-minute water supply may not completely control all fires, it will delay a flashover and allow time for the occupants to reach safety.

The water supply for residential sprinklers may be taken from several sources, which may include a connection to the public water system, an onsite pressure tank, or a storage tank with an automatic pump.

To be of value, any sprinkler system must be in service. As with a standard sprinkler system, inadvertent or deliberate closing of valves renders the system useless. When a residential sprinkler system is supplied from a public water system, using a single valve to control both the sprinklers and the domestic service virtually eliminates the possibility of the supply valve being closed. With this arrangement, the sprinklers cannot be turned off without turning off the household domestic supply (sinks and toilets).

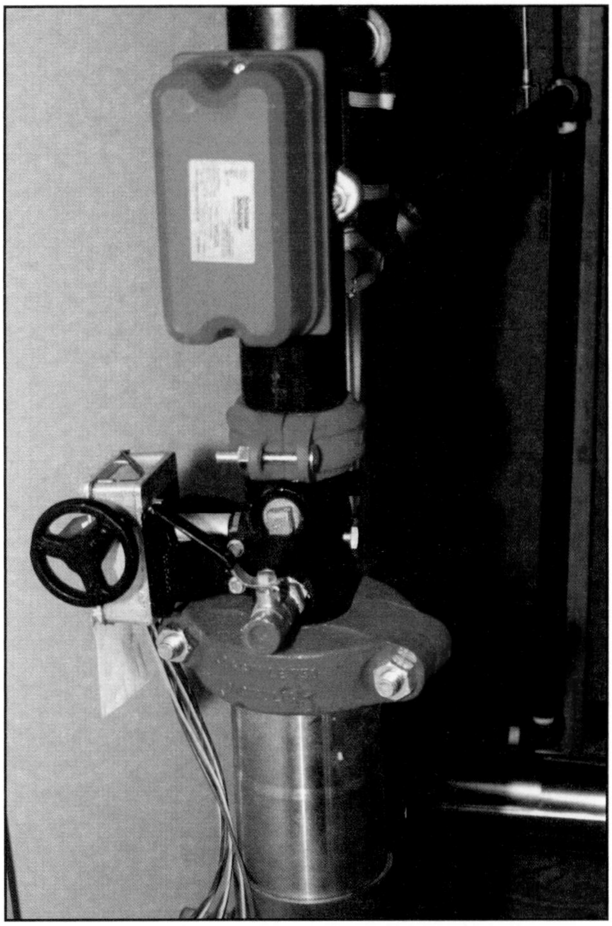

Figure 12.18 Control valves meeting NFPA® 13R are necessary parts of a water-based fire suppression system.

Even if the sprinkler system is viewed as unneeded, it is unlikely that a homeowner would be willing to go very long without the household water supply. Where plumbing or water department requirements do not permit this type of uninterrupted connection, NFPA® 13R permits the sprinkler valve to be supervised or simply locked in the open position **(Figure 12.18)**.

Spacing

Unless otherwise approved, spacing for sprinklers in residential systems is a maximum of 144 square feet (13 m^2) per sprinkler **(Figure 12.19, p. 520)**. The maximum spacing between sprinklers is 12 feet (3.5 m) with the maximum allowable distance of a sprinkler from a wall being 6 feet (2 m). Sprinkler manufacturers have produced a variety of sprinkler designs, and the spacing of sprinklers may be based upon the particular sprinklers that have been tested and listed. Some residential sprinklers, therefore, can be spaced to protect an area as large as 20 × 20 feet (6 m by 6 m). However, with this sprinkler spacing, the minimum discharge with one sprinkler operating is increased to 32.5 gpm (123 L/min) and 22.5 gpm (85 L/min) per sprinkler with two sprinklers operating.

Automatic Fire Suppression Systems

Other types of automatic fire suppression systems are similar to but are not considered to be automatic sprinkler systems. These include water-spray, water-mist, and foam-water systems.

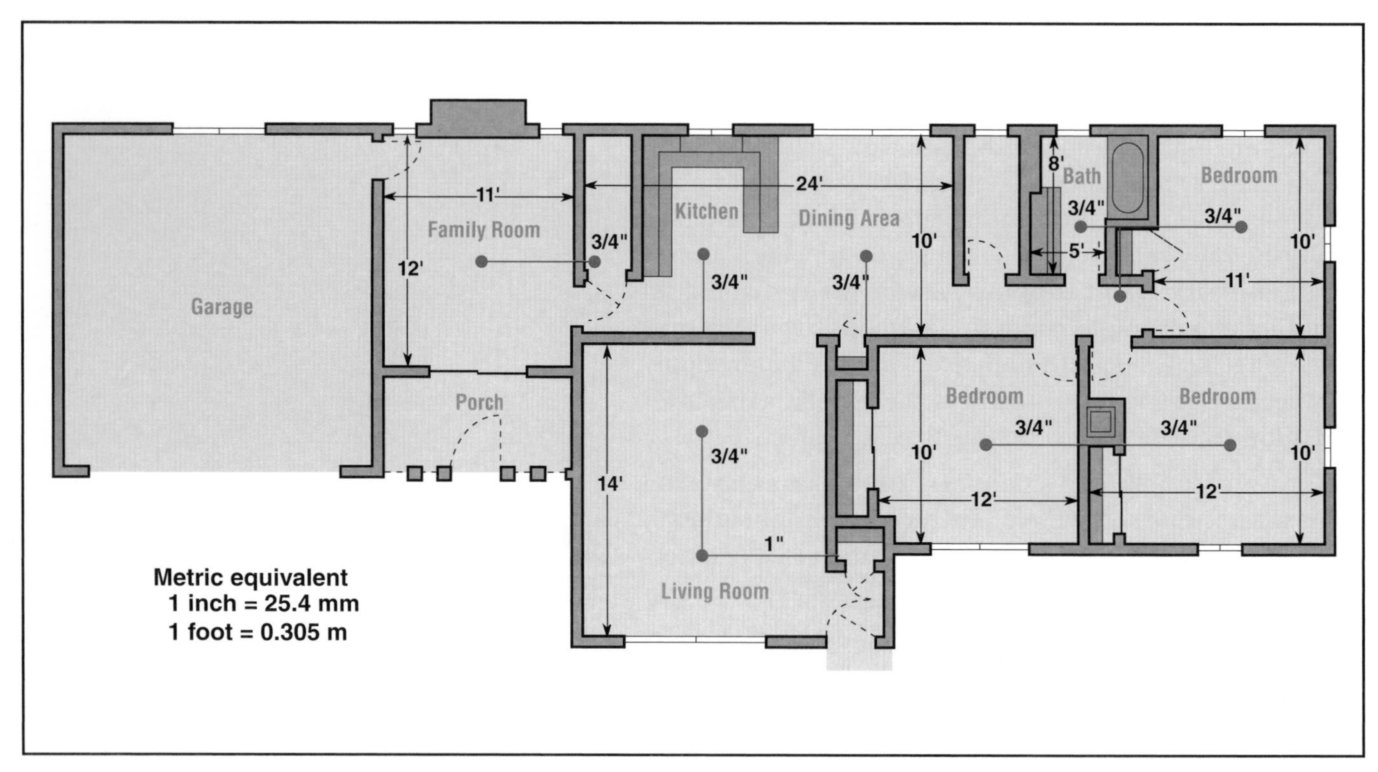

Figure 12.19 A common residential sprinkler system design.

Water-Spray Fixed System

The water-spray fixed system discharges water over the area or surface to be protected through an arrangement of pipes and nozzles. An automatic heat detection system or a manual activation system may be used to activate the water-spray system.

Components of a water-spray fixed system are similar to those of an automatic sprinkler system and include the following:

- Reliable water supply
- Piping
- Automatic or manual detection and activation devices
- Waterflow control valve
- Water-spray nozzles

A water-spray fixed system provides protection to specific hazards or hazardous processes by applying water droplets of a predetermined pattern, particle size, velocity, and density through specially designed nozzles. The system may be independent from or supplementary to other types of fire protection systems. General hazard categories include:

- Flammable gaseous and liquid materials
- Electrical equipment
  - — Transformers **(Figure 12.20)**
  - — Oil switches
  - — Motors
  - — Cable trays
  - — Cable runs

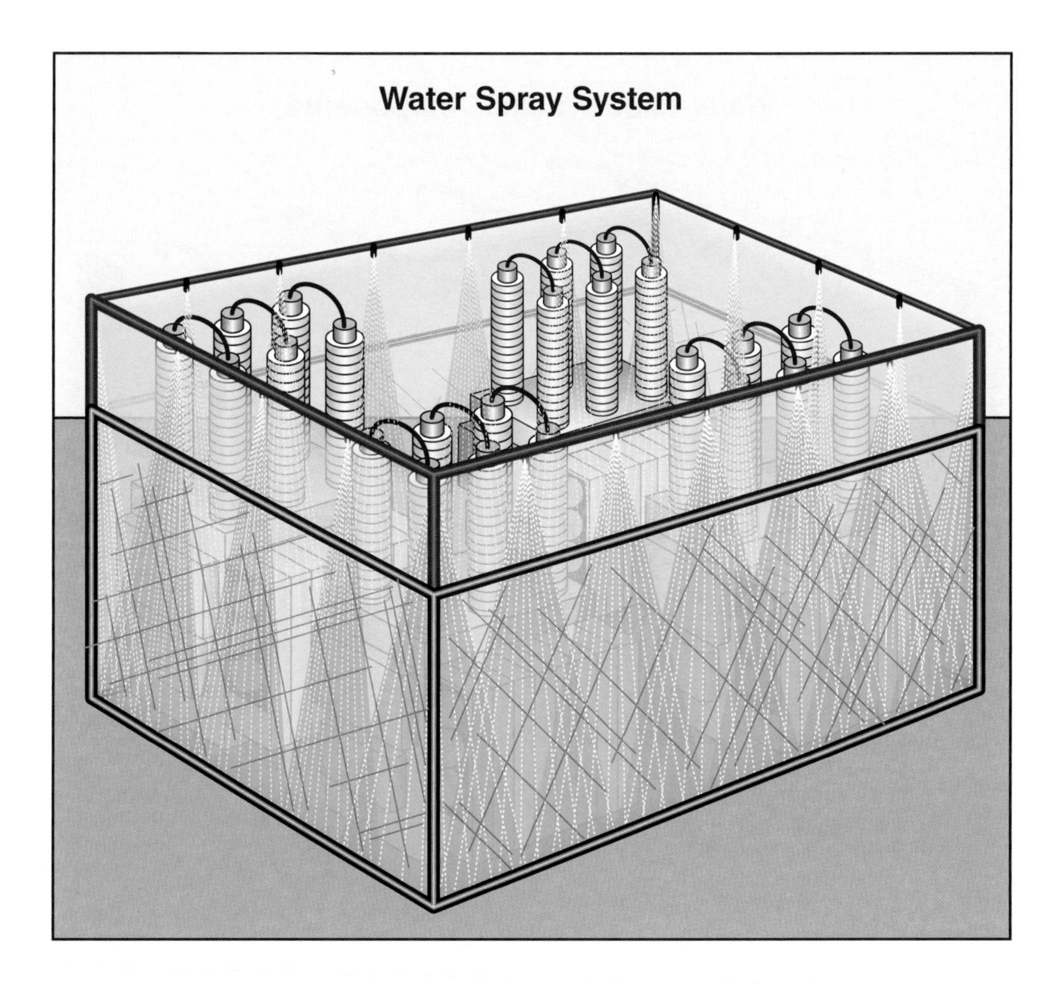

Figure 12.20 A fixed spray water system is used to protect a group of transformers.

- Ordinary Class A combustibles
- Certain hazardous solids
  - — Propellants
  - — Pyrotechnics
- Exposure protection
  - — Separation between hazards
  - — Means of egress

Water-spray fixed system design and installation are regulated by NFPA® 15, *Standard for Water-spray Fixed Systems for Fire Protection*. NFPA® 25, *Standard for the Inspection, Testing, and Maintenance of Water-Based Fire Protection Systems,* outlines testing, inspection, and maintenance requirements.

Water-Mist Systems

A water-mist system is similar to a water-spray fixed system, except that the water-mist system discharges a fine mist of water. The mist absorbs larger quantities of heat than water-spray or automatic sprinkler systems. In theory, the water-mist raises the humidity of the room enough to halt the combustion process. The very fine spray also controls or extinguishes fire by displacing oxygen and blocking radiant heat production **(Figure 12.21, p. 522)**. Because more fire can be controlled with less water damage, the water-mist system is considered a replacement for fixed fire suppression systems that used halogenated hydrocarbon agents.

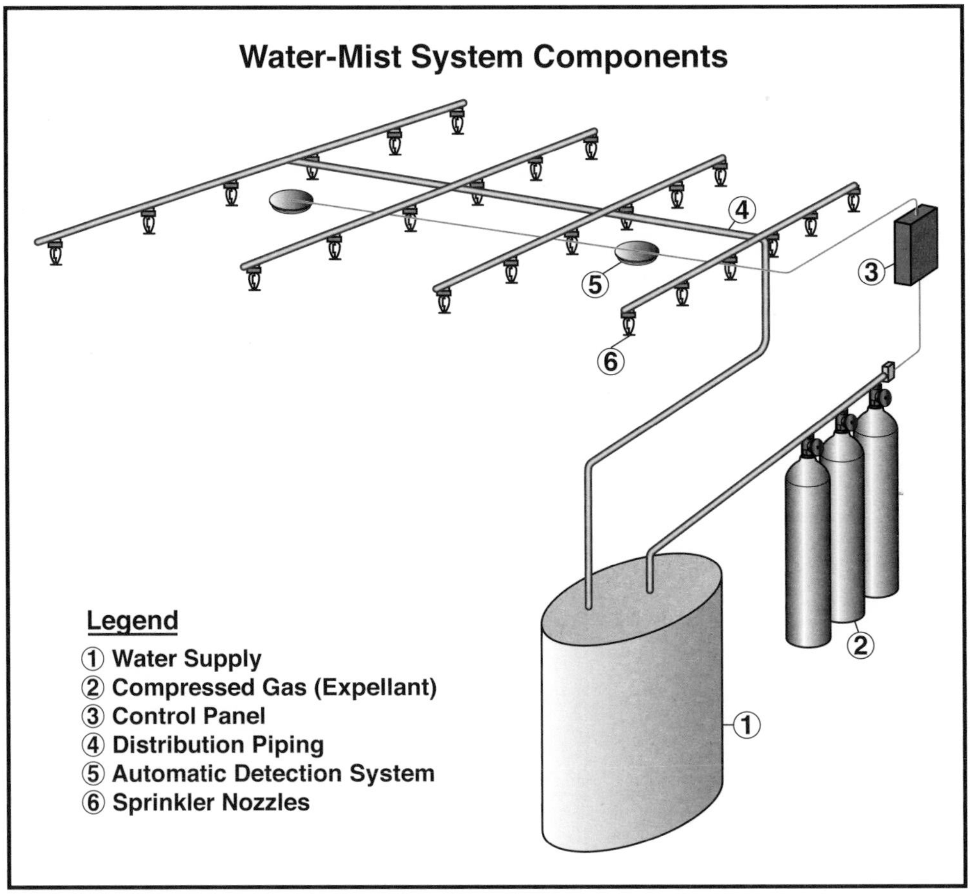

Water-Mist System Components

Legend

1. Water Supply
2. Compressed Gas (Expellant)
3. Control Panel
4. Distribution Piping
5. Automatic Detection System
6. Sprinkler Nozzles

Figure 12.21 Water-mist system components include expellant cylinders, water source, piping, detectors, control panel, and sprinklers.

Water-mist systems are currently used to protect the following types of hazards:

- Gas jet fires
- Computer equipment and rooms
- Flammable and combustible liquids
- Aircraft passenger cabins
- Marine vessels
- Ordinary Class A combustibles

These systems are designed to protect lives and property by extinguishing Class A and Class B fires, controlling fire temperatures in compartments, and preventing flashover with subsequent extension to other compartments in a structure. Continuing research also indicates that water-mist systems may be suitable for use in residential occupancies and flammable and combustible storage facilities.

Water-mist systems are designed to be operated at considerably higher pressures than standard sprinkler systems. The basic pressure ranges in which these systems operate are as follows:

- Low-pressure system — 175 psi (1 225 kPa) or less
- Intermediate-pressure system — 175 to 500 psi (1 225 kPa to 3 500 kPa)
- High-pressure systems — 500 psi (3 500 kPa) or greater

Compressed-air, nitrogen, or high-pressure water pumps create these higher pressures. Compressed air may be supplied to the system from storage cylinders or air pumps. Air pressure may be applied to the water through the water tube itself (single-fluid system) or through a second air tube to each spray nozzle (twin-fluid system).

In general, a water-mist system is composed of small-diameter, pressure-rated copper, or stainless-steel tubing. Small-diameter spray nozzles are spaced evenly on the tubing. Depending on the design of the system, the spray nozzles may be of the open or closed sprinkler variety. A product-of-combustion detection system activates the system, and may have a set amount of time to discharge. These systems should also have a means of manual operation if individuals discover the fire before the activation of the system.

The most common type of water-mist system works similarly to a traditional deluge sprinkler system. All of the spray nozzles in a particular room or zone are open, and when the detection devices activate, the water discharges from the spray nozzles.

NFPA® 25 lists requirements for inspecting, maintaining, and service testing water-mist systems. This information was previously contained in NFPA® 750, *Standard on Water-mist Fire Protection Systems*. Systems that utilize air cylinders require that the cylinders be hydrostatically tested on a regular basis. Empty cylinders must be tested before recharging if it has been more than five years since their last tests. Cylinders that have not been discharged should be emptied and tested every twelve years.

Inspectors should verify that the required replacement components such as extra spray nozzles are present in sufficient quantities. Inspectors should also try to verify that the owner/occupant or fire protection equipment company is servicing the system on a regular basis. Typical servicing functions include the following:

- Lubricating control valve stems
- Adjusting packing glands on valves and pumps
- Bleeding moisture and condensation from air compressors and air lines
- Cleaning strainers
- Replacing corroded or painted nozzles
- Replacing damaged or missing pipe hangers
- Replacing damaged valve seats or gaskets

Foam-Water Systems

Foam-water automatic sprinkler systems are designed to discharge a foam-water solution onto a fire. They are commonly employed to protect Class B fire hazards but may also be used for Class A hazards. When properly designed, they may also be used to suppress vapors created by a flammable liquid spill.

Foam-water systems are designed to a deliver a foam-water solution at the required design concentration over the discharge density and design area of the system. These systems can be wet-pipe, dry-pipe, preaction, or deluge systems. NFPA® 16, *Standard for the Installation of Foam-Water Sprinkler and Foam-Water-Spray Systems*, details the design and construction requirements of foam-water automatic sprinkler systems. The design professional establishes

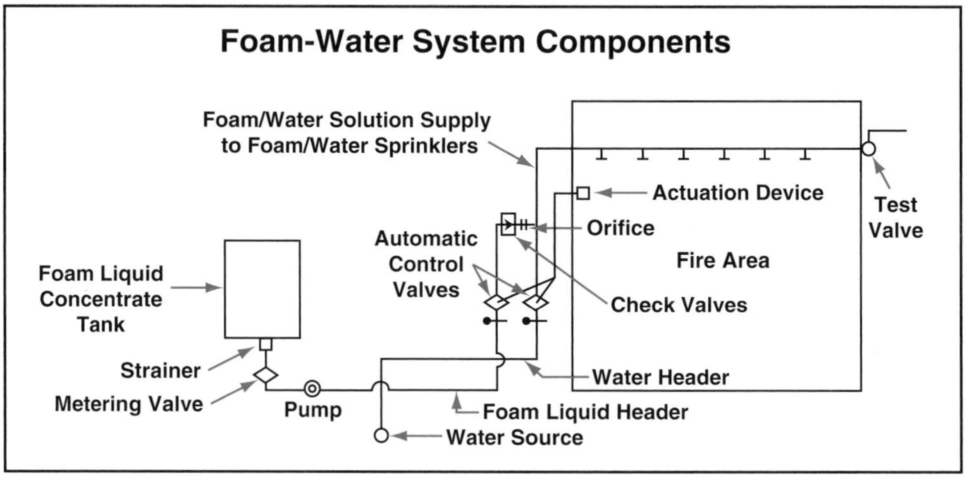

Figure 12.22 Foam-water system components are similar to a water-based sprinkler system with the addition of the foam liquid storage tank.

the required duration for foam application, but it must be for at least fifteen minutes to control and suppress the fire. The inspector should verify that the minimum required volume of foam concentrate and that the correct type of foam is being provided and maintained.

All foam-water systems require a proportioner. The proportioner may be an in-line valve or pump and is designed to inject and mix the foam with the water to create the foam-water solution at the correct concentration. NFPA® 25 requires an annual test of the proportioner to verify that it is properly mixing the foam-water solution and creating the specified concentration.

NOTE: See Chapter 13 for more information about foam proportioners.

In certain designs, the ceiling sprinklers are used as the detection feature when the foam-water sprinklers are pre-primed with foam solution. In other cases, a fire alarm and detection system is required to activate the foam-water system, especially if it is designed as a preaction or deluge system **(Figure 12.22)**.

Standpipe and Hose Systems

Standpipe and hose systems are designed to provide a means for rapidly deploying fire hose and operating fire streams on all levels of multistory structures and at remote points in large-area structures **(Figure 12.23)**. Depending on the type of standpipe system installed, it is intended to be used by firefighters, trained occupants, or both. A permanent water supply that is augmented through an FDC may supply the system, or it may be a stand-alone system that requires the fire department to charge the system with water through the FDC **(Figure 12.24)**. The system may also be part of or separate from an automatic sprinkler, water-spray, water-mist, or foam-water system.

Some building codes require operational standpipes during construction **(Figure 12.25)**. As construction progresses, the standpipe is extended to subsequent floors. The system may or may not have an attached water supply, but it must have appropriate FDCs, hose discharge connections, and such other necessary appliances as fire pumps to provide the required volume and pressure at the highest point of discharge.

It is very important for the inspector to verify the connection threads on all hose connections and the fire department connections **(Figure 12.26)**. Improper hose threads can cost valuable time at an emergency incident and may ultimately render the standpipe unusable until it is corrected.

As part of the maintenance requirements of NFPA® 25, these systems need to be flow and hydrostatic tested every five years. At a minimum, the Inspector I should request documentation of the last test to verify system readiness.

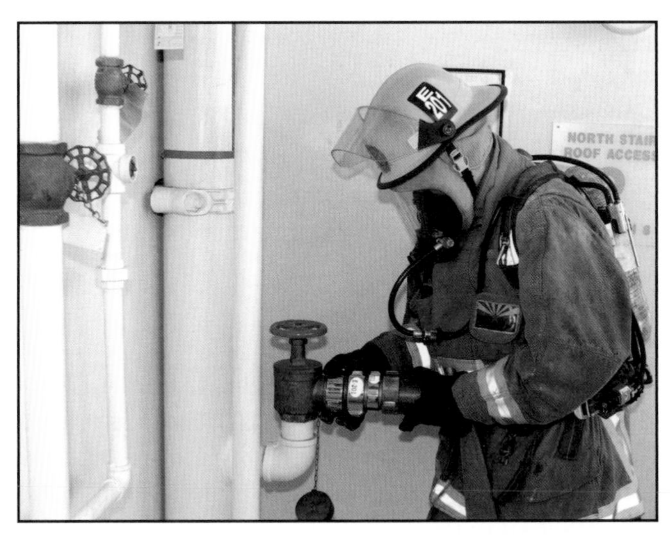

Figure 12.23 Standpipes allow firefighters to extend fire hoses and nozzles to any portion of the structure to extinguish a fire.

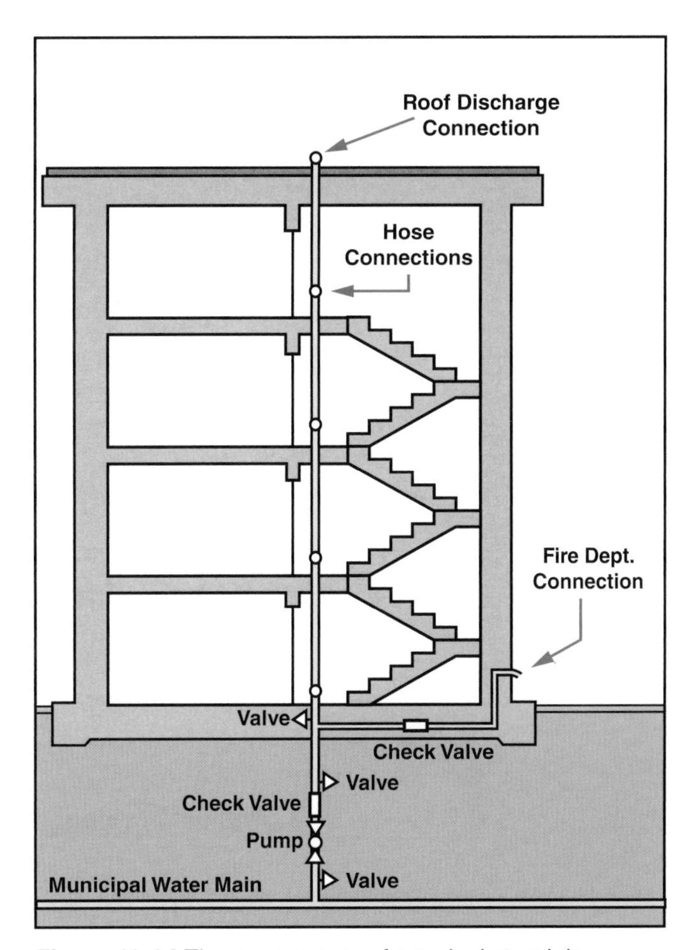

Figure 12.24 The components of a typical standpipe system.

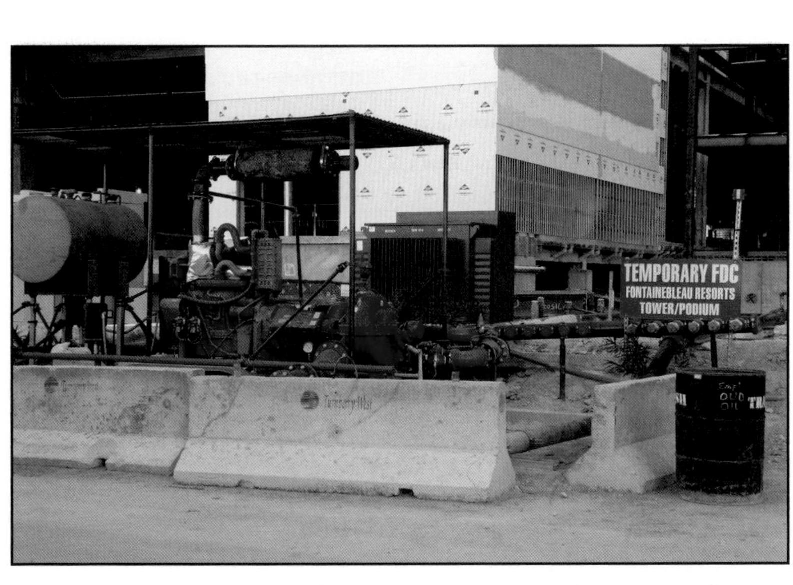

Figure 12.25 A temporary FDC may be required during construction.

Figure 12.26 It is important to verify that hose valve threads are in good condition.

Components

Inspectors should be familiar with standpipe components so they know what they are required to verify during an inspection. Standpipe system components are similar to sprinkler system components. They consist of the following:

- Hose stations (defines classification of system)
- Water supplies
- Waterflow control valves (similar to those used in automatic sprinkler systems)
- Risers (piping systems used to transfer water from the supply to the discharge)
- Pressure-regulating devices
- Fire department connections (FDCs)

Classifications

NFPA® 14, *Standard for the Installation of Standpipe and Hose Systems*, addresses the design and installation of standpipes. The standard establishes three classes of standpipe systems that are based on the intended use of the hose station or discharge outlet.

Class I: Firefighters

Class I standpipe systems are primarily for use by firefighters trained in handling large handlines (2½-inch [65 mm] hose) **(Figure 12.27)**. Class I systems must be capable of supplying effective fire streams during the more advanced stages of fire within a structure. Class I systems have 2½-inch (65 mm) hose connections or hose stations attached to the standpipe riser.

Class II: Trained Building Occupants

The Class II system is primarily designed for use by building occupants who are trained in its use or by fire department personnel. These systems are equipped with a 1½-inch (38 mm) hose, nozzle, and hose rack **(Figure 12.28)**.

There is some disagreement over the value of Class II standpipe and hose systems. The presence of the small hose may give a false sense of security to occupants and create the impression that they should attempt to fight a fire, even though the safer course would be to escape. In any case, fire department personnel cannot depend on Class II standpipe hose for fire control operations. The installation of Class II standpipes is declining in favor of Class I or Class III installations.

Class III: Combination

Class III standpipes combine the features of Class I and Class II systems. Class III systems have both 2½-inch (65 mm) hose connections for fire department personnel and 1½-inch (38 mm) hose and connections for use by the trained building occupants **(Figure 12.29)**. The design of the system must allow both Class I and Class II services to be used simultaneously. This is typically done by equipping the 2½-inch (65 mm) connection with a 1½-inch (38 mm) adapter. This arrangement is more popular than equipping the standpipe with specific 2½-inch (65 mm) connections with their own valves.

Figure 12.27 This Class I standpipe connection is designed to be used by firefighters, so features a larger hose connection.

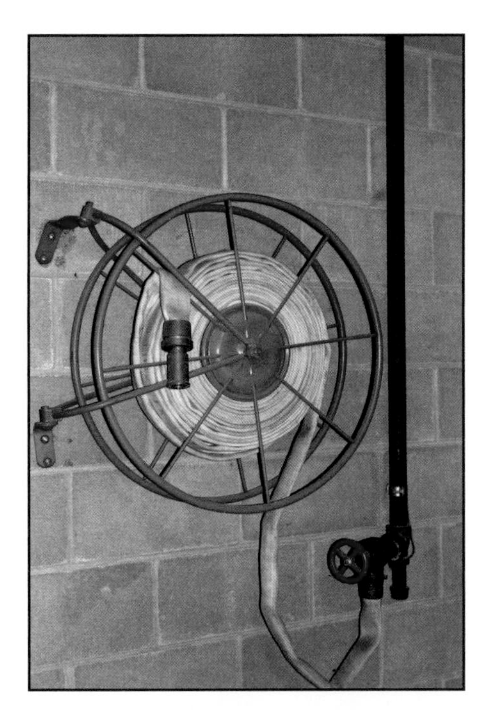

Figure 12.28 Class II hose stations are designed to be used by occupants in addition to firefighters.

Figure 12.29 Class III hose stations may or may not have an attached hose. Many jurisdictions remove the hose from these stations, leaving only the connections for firefighters responding to the scene.

Types

In addition to the three classes of standpipe and hose systems, NFPA® 14 lists the following five types of standpipe systems:

- **Automatic-wet** — Contains water in the system at all times and is attached to a water supply capable of supplying the system demand at all times. When the hose valve is opened, water is immediately available **(Figure 12.30, p. 528)**.
- **Automatic-dry** — Contains air or nitrogen under pressure and is permanently attached to a water supply. Water enters the system when a hose valve is opened **(Figure 12.31, p. 528)**.

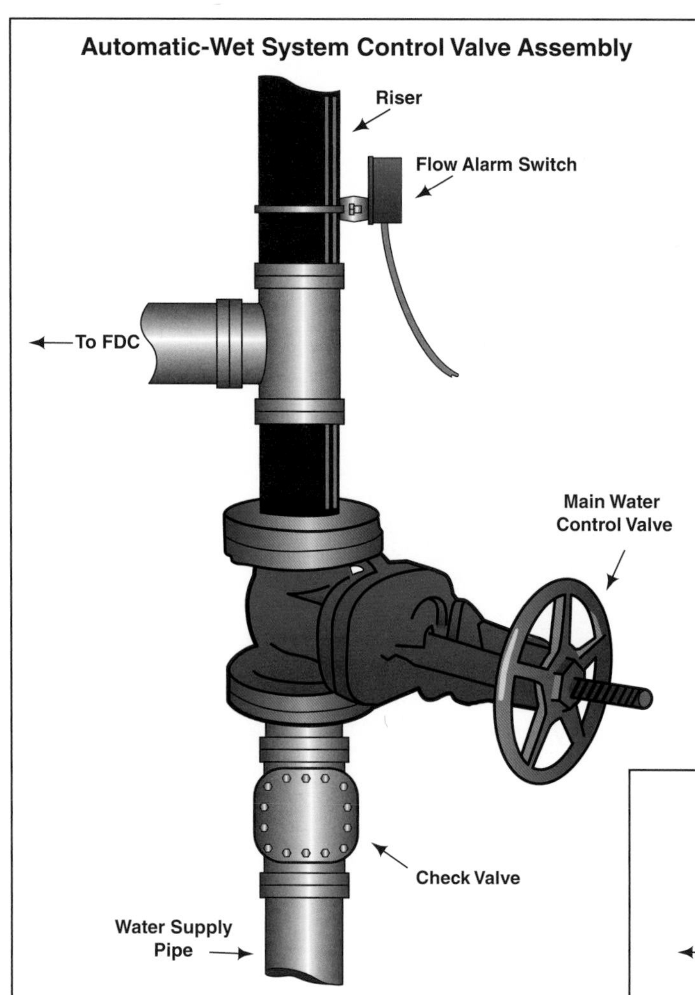

Figure 12.30 The automatic wet system contains water in the system at all times.

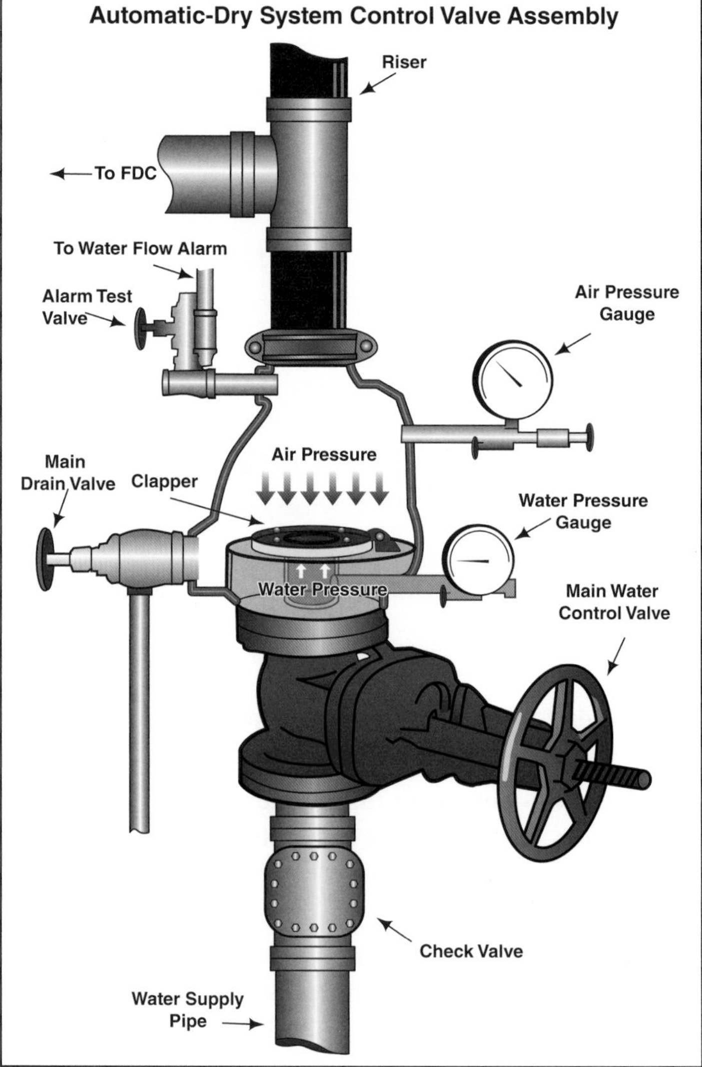

Figure 12.31 The automatic dry system allows water to enter the system when a hose valve is opened.

- **Semiautomatic-dry** — Standpipe with empty pipe that is connected to a permanent water supply. It uses a device, such as a deluge valve, to admit water into the system piping upon activation of a remote control device located at a hose connection.
- **Manual-dry** — Contains no water in the system and is not connected to the domestic water supply. Relies on the fire department to provide water through the FDC to meet system demand.

NOTE: Canadian building and fire codes do not accept dry standpipes as meeting code requirements.

- **Manual-wet** — Maintains water from a domestic fill connection in the piping to detect leaks in the system. However, no permanently connected water source is attached to the system, and the fire department must provide water through the FDC.

An automatic wet standpipe system with an automatic water supply is the most common type of standpipe. With this type of system, water is constantly available at the hose station. However, wet standpipe systems cannot be used in cold environments, so a dry system may be necessary. Automatic dry standpipe systems have the disadvantages of greater cost and maintenance requirements than other systems.

An inspector assessing standpipe systems must check for the following:

- Are standpipes accessible?
- Do valves work?
- Are the caps in place?
- Are doors closed and latched?

Water Supplies and Residual Pressure

The water supply systems for standpipe systems have specific requirements according to the adopted code. The amount of water required for a standpipe system depends on the occupancy and height of the building and the total number of standpipes in the building. Water supply requirements may be reduced by installed automatic sprinkler systems.

Some occupancies require secondary water supplies. The water is usually supplied from automatic fire pumps that augment the pressure provided from a municipal water main. The inspector should be aware of multiple water supplies and verify their operational readiness.

The current NFPA® 14 minimum requirement for residual pressure is 100 psi (700 kPa) at the fire hose outlet. However, this is a minimum and may not be adequate to supply a modern automatic nozzle on the end of an attack line connected to the topmost hose outlet. Because of this situation, some other building and fire codes require higher minimum residual pressures. An inspector should consult the code used in the local jurisdiction for minimum requirements in standpipe-protected occupancies.

The Inspector I should ask for inspection and maintenance records to determine operational readiness. The inspector should confirm that the water supply has not been changed or that all valves are accessible and have not been shut off or damaged.

Standpipe Hose Valves

Current practice is to locate fire hose valves so that any part of a floor is within 130 feet (40 m) of the standpipe hose connection. This distance allows any fire to be reached with 100 feet of hose (30 m), plus a 30-foot (10 m) fire stream **(Figure 12.32)**. Standpipes and their connections are most commonly located within rated stair enclosures so that firefighters have a protected point from which to begin an attack.

The actual hose connections can be located no more than 6 feet (2 m) from floor level. These connections must be plainly visible and not obstructed. Any caps over the connections must be easy to remove.

Buildings equipped with Class I or III systems may be required to have a 2½-inch (65 mm) outlet on the roof. This outlet is required when any of the following three situations are present:

- Combustible roof
- Combustible structure or equipment on the roof
- Exposures that present a fire hazard

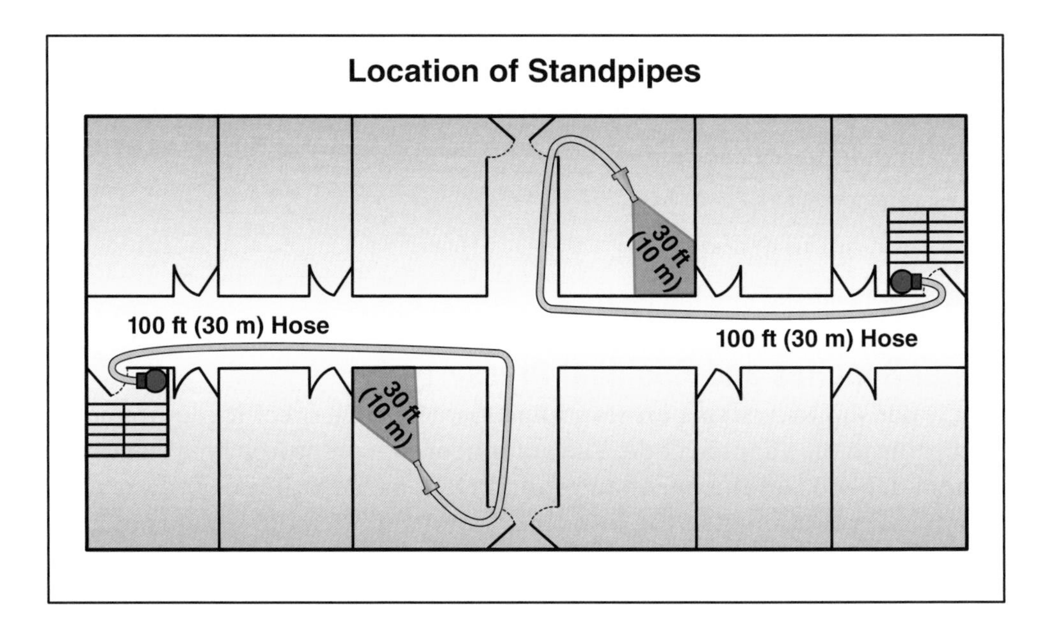

Figure 12.32 To gain complete coverage, standpipes must be properly located to be most effective for fighting fires.

Pressure-Regulating Devices

Where the discharge pressure at a hose outlet exceeds 175 psi (1 225 kPa), NFPA® 14 requires a pressure-regulating (restricting) device to limit the pressure to 100 psi (700 kPa), unless otherwise approved by the fire department. The use of a pressure-regulating device prevents pressures that make fire hose difficult or dangerous to handle. This device also enhances system reliability because it extends individual zones to greater heights. In some instances, it may improve system economy because its use may eliminate some pumps.

Pressure-regulating devices make the system design more complex. The three basic categories of pressure-regulating devices are as follows:

- **Pressure-restricting devices** — Consist of a simple restricting orifice inserted into the waterway. The amount of residual pressure drop through the orifice plate depends on the orifice diameter and available flow and pressure within the system. Each standpipe discharge connection is fitted with a restricting orifice with different sizes being required for each floor and application. They are limited to systems with 1½-inch (38 mm) hose discharges and 175 psi (1 225 kPa) maximum pressure. This device is not a preferred type because it does not control or reduce the water pressure in the system.

- **Pressure-control devices** — Preferred for managing excessive pressure and considered to be the most reliable method of pressure control. They use a pitot tube and gauge to read the pressure and automatically reduce the flow through the discharge. Some of the devices are field adjustable, and others are preset at the factory.
- **Pressure-reducing devices** — Preferred for managing excessive pressure; uses a spring mechanism that compensates for variations in pressure **(Figure 12.33)**. These mechanisms balance the available pressure within the system with the pressure required for hoseline use. Inspectors need to know that the second pipe is for testing of the PRVs.

NOTE: Refer to NFPA® 25 for specific testing requirements for these devices.

Figure 12.33 Pressure-reducing devices are used to manage excess pressure. These devices balance the available pressure within a system against the pressure needed for hoselines.

The inspector should verify that the required testing has been completed and that required **pressure-reducing valves** are installed and tested as specified.

Pressure-Reducing Valve — Valve installed at standpipe connection that is designed to reduce the amount of water pressure at that discharge to a specific pressure, usually 100 psi (700 kPa).

Standpipe systems that are equipped with pressure-regulating devices are designed so that they can be routinely tested. Systems must have dedicated drainage pipes with connections on each floor and a means for determining water flow.

A pressure-regulating device must be specified and/or adjusted to meet the pressure and flow requirements of the individual installation. For factory-set devices, the pressure-regulating device must be installed on the proper hose outlet. Installers must follow the manufacturer's instructions on making adjustments for field-adjustable devices. A pressure-regulating device that is not properly installed or adjusted for the required inlet pressure, outlet pressure, and flow may result in seriously reduced available flow and impaired suppression capabilities.

Fire Department Connections

Each standpipe system requires one or more FDCs through which a fire department engine can supply water into the system. Large buildings having two or more zones require an FDC for each zone **(Figure 12.34)**.

Standard requirements specify that there shall be no shutoff valve between the FDC and the standpipe riser. In multiple-riser systems, however, listed indicating valves are provided at the base of the individual risers. These are included so that the entire system, including other standpipes, is not taken out of service during maintenance.

Figure 12.34 A standpipe clearly marked to show the zones it services.

Large Diameter Hose (LDH) — Relay-supply hose of 3½ to 6 inches (90 mm to 150 mm) in diameter; used to move large volumes of water quickly with a minimum number of pumpers and personnel.

The hose connections to the FDC must adhere to the standards of the locally adopted code and have standard cap plugs or approved breakaway covers. Some jurisdictions require **large-diameter hose (LDH)** connections to supply standpipes. The hose coupling threads should conform to those used by the local fire department. See NFPA® 1963, *Standard for Fire Hose Connections,* and IFSTA's **Fire Hose Practices** manual for more information.

The FDC may also be protected with a locking intake cap that must be removed with a special key **(Figure 12.35)**. The local authority may regulate the use of these locking caps. A raised-letter sign on a plate or fitting reading *STANDPIPE* designates the FDC. If the FDC does not service the entire building, the sign must indicate which floors it does service.

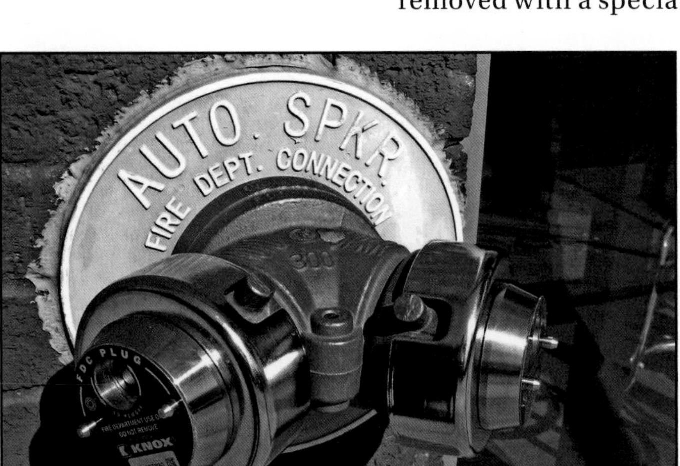

Figure 12.35 Make sure that locking FDCs are in good order. Standpipes are targets for vandalism and refuse. *Courtesy of Scott Strassburg.*

Stationary Fire Pumps

Inspectors must be familiar with the stationary (fixed) fire pumps that are found in many commercial, institutional, and industrial facilities. The main function of a fire pump is to increase the pressure of the water that flows through it. Usually, a fire pump is needed to supply a sprinkler or standpipe system because the available water supply source, such as an elevated tank or ground storage tank, does not have adequate pressure to meet the fire suppression system demand. **High-rise buildings** or areas with low static pressure on the public or private water supply may require the installation of fire pumps.

High-Rise Building — Building that requires fire fighting on levels above the reach of the department's equipment. The *International Building Code (IBC)* defines a high-rise building as one greater than 75 feet (23 m) in height, but other fire and building codes may define the term differently. *Also known as* High-Rise.

All stationary fire pumps and their installations must meet the requirements set forth in NFPA® 20, *Standard for the Installation of Stationary Pumps for Fire Protection.* However, numerous other NFPA® standards contain references to fire pumps, including standards on sprinklers, standpipes, water-spray, water-mist, and foam-water systems. The next section describes centrifugal pumps and pressure-maintenance pumps.

Types

According to NFPA® 20, a variety of fire pump types can be installed as fixed or stationary pumps for fire suppression systems **(Figure 12.36)**. Five types of **centrifugal pumps** are in use as stationary fire pumps:

Centrifugal Pump — Pump with one or more impellers that rotate and utilize centrifugal force to move the water. Most modern fire pumps are of this type.

- Horizontal split-case
- Vertical split-case
- Vertical inline
- Vertical turbine
- End suction

Horizontal Split-Case

The horizontal split-case pump is the most common type of fire pump found in stationary fire suppression systems. Sometimes referred to as a horizontal shaft pump, the drive shaft is on a horizontal plane with the pump on one end of the shaft and the driver (motor) on the other **(Figure 12.37)**. This pump is

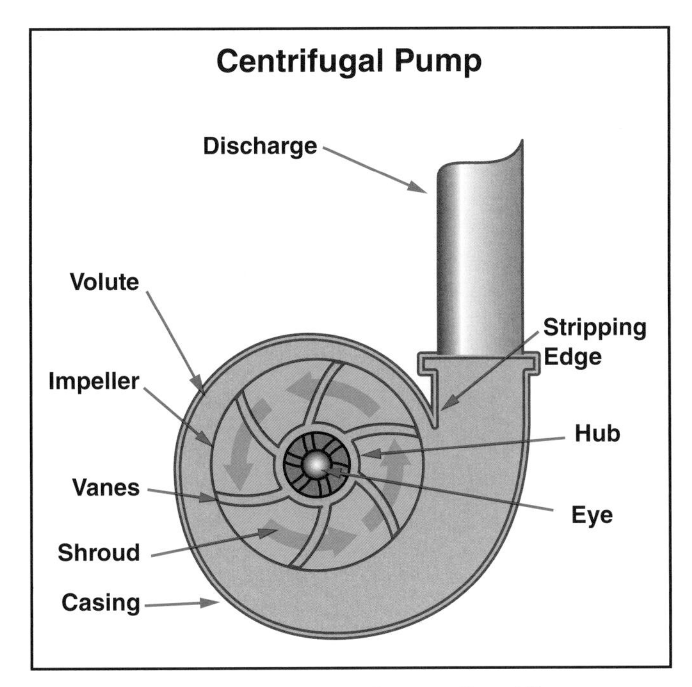

Figure 12.36 The major parts of a centrifugal fire pump.

Figure 12.37 A horizontal split-case pump boosts incoming pressure. The shaft is positioned horizontally with a pump at one end and a motor at the other.

used to boost the pressure from an incoming, pressurized water source such as a municipal water main or a facility water supply. Because it is not a self-priming pump (cannot draft water from a static supply source into the pump on its own), a horizontal split-case pump cannot be used to supply water from a static supply source such as a pond that has no positive **head pressure**.

Head Pressure — Pressure exerted by a stationary column of water, directly proportional to the height of the column.

Fire Pump Pressure Ratings

Inspectors should be aware that fire pump pressure ratings or capabilities are often expressed in terms of feet of head pressure rather than psi or kPa. A column of water 2.304 feet (0.7 m) high creates 1 psi (exactly 6.89 kPa) of pressure. Therefore, a pump rated at 231 feet (70.4 m) of head pressure will be equal to one rated at 100 psi (689 kPa).

NOTE: For ease of instruction, IFSTA generally uses a 1 psi = 7 kPa ratio when talking about pressure.

The most common gallon rating for the horizontal split-case pumps in use today are those in the 500 to 1,500 gpm (2 000 L/min to 6 000 L/min) range. However, pumps are available for flows as low as 150 gpm (600 L/min) and as high as 4,500 gpm (18 000 L/min). Unlike pumps on fire apparatus, stationary fire pumps are not rated at a particular pressure.

Vertical Split-Case

The vertical split-case pump is very similar to the horizontal split-case pump, except that the impeller shaft runs vertically **(Figure 12.38)**. This pump is almost always driven by an electric motor that sits on top of the pump. The main advantage of this pump is its compactness.

Figure 12.38 A vertical split-case pump has a vertical impeller shaft but serves the same function as a horizontal split-case pump.

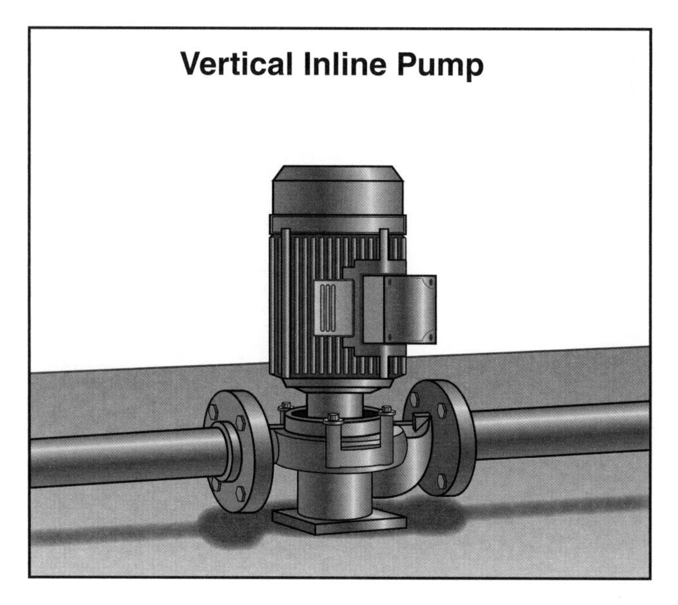

Figure 12.39 In a vertical inline pump, the driver or motor is located above the inline impeller.

Vertical Inline

The vertical inline pump is a single-stage pump designed to fit into the intake/discharge line with the driver located above the inline impeller **(Figure 12.39)**. The advantages of a vertical inline pump are the ease of installation as a replacement pump, the compact space required for the pump, and the ease of maintenance of the pump and driver. The pump has a capacity up to 1,500 gpm (6 000 L/min) and operating pressures up to 165 psi (1 150 kPa).

Vertical Turbine

The vertical turbine pump is very useful for lifting water from a source below the pump. Vertical turbine pumps are commonly used as well as pumps in nonfire-protection applications.

The vertical turbine pump impellers are actually located within the water supply source **(Figure 12.40)**. Water is drawn into the impeller and then discharged up through the impeller casing. Most of these pumps are multistage pumps. As the water exits one impeller, it enters the next and continues until it is discharged into the fire suppression system piping.

The volume capabilities of vertical turbine pumps are consistent with horizontal and vertical split-case pumps. However, vertical turbine pumps are available with discharge pressure ratings of up to 500 psi (3 500 kPa).

End Suction

The end suction pump is a variation of the horizontal split-case pump design **(Figure 12.41)**. End suction pumps have center line suction and discharge. They have pressure ratings from 40 to 150 psi (280 kPa to 1 050 kPa), along with flow ranges of 50 to 750 gpm (200 L/min to 3 000 L/min).

The advantages of the end suction pump are ease of installation, simplified piping arrangement, and reduced pipe strain. The pumps are self-venting, which eliminates the need for an automatic air-release valve that is normally installed to control overheating of the pump.

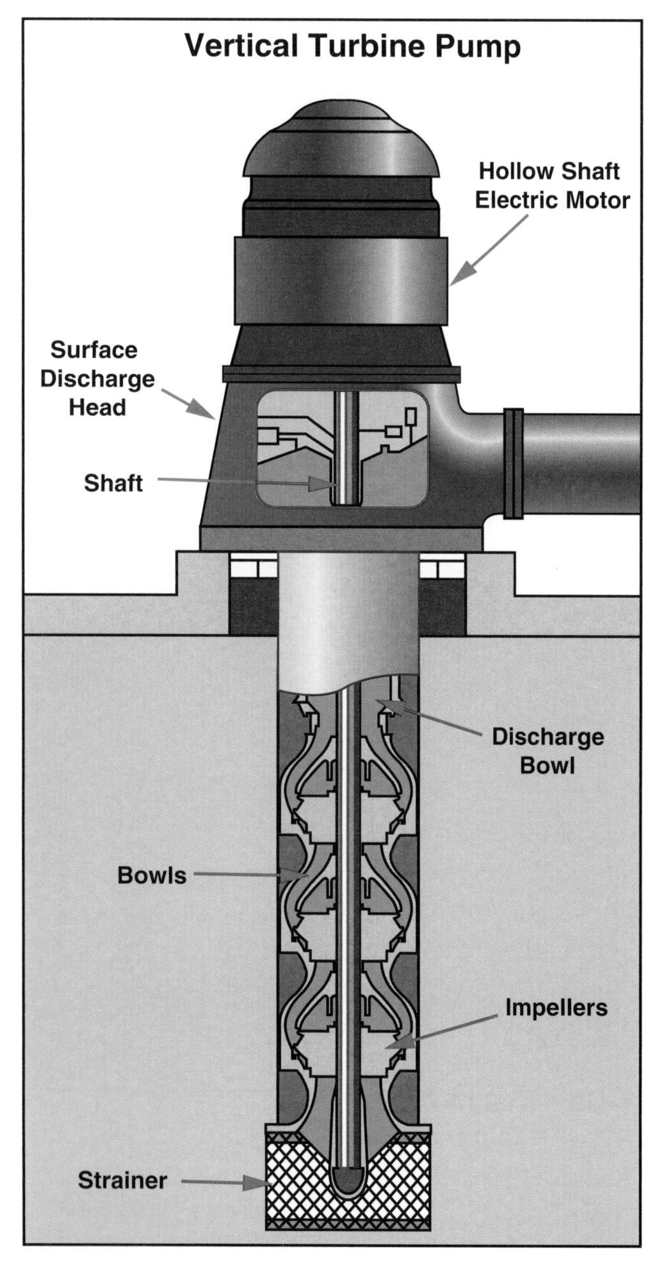

Figure 12.40 In a vertical turbine pump, the impellers are located in the water source.

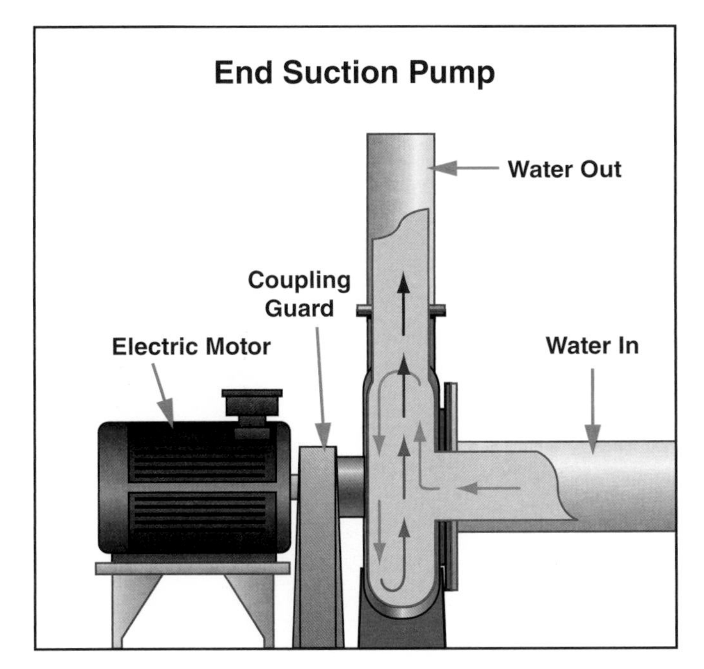

Figure 12.41 End suction pumps are single-stage pumps that have center line suction and discharge.

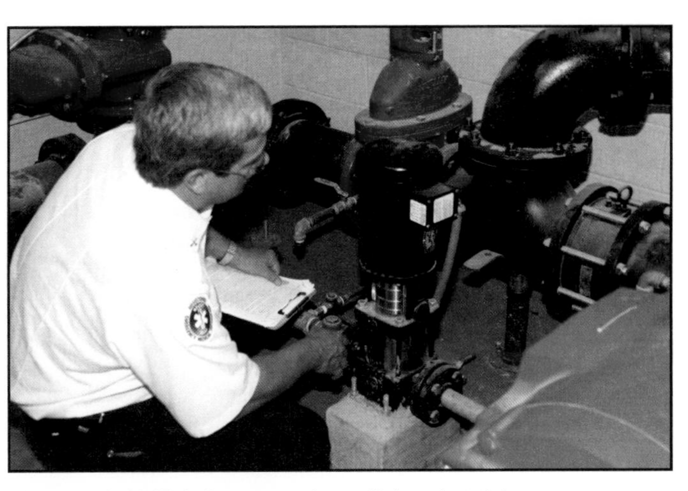

Figure 12.42 This inspector is verifying that this pressure maintenance pump is in working order and installed correctly.

Pressure-Maintenance Pump — Pump used to maintain pressure on a fire protection system in order to prevent false starts at the fire pump.

Pressure-Maintenance

Automatic sprinkler systems may require a small-capacity auxiliary pump. These small pumps are known as **pressure-maintenance pumps**, jockey pumps, or make-up pumps and are located in parallel with the primary fire pump. They are designed to maintain system pressure and prevent the larger main pump from starting repeatedly for a short period of time **(Figure 12.42)**. The design of these small-capacity, high-pressure pumps may be any one of the pump types mentioned previously or pump types listed in NFPA® 20.

Pump Drivers

The source of power that operates the fire pump is called the driver. Fire pumps are commonly powered by one of three types of drivers: electric motor, diesel engine, or steam turbine. Other types of engine drivers, such as gasoline, natural gas, and liquefied petroleum, have been used in the past but are not currently recognized in NFPA® codes.

Electric Motor Driver

An electric motor is the most common method for driving a fire pump. It is simple, reliable, and easily maintained. Electric motors used on fire pumps are not designed specifically for that purpose; however, all electric motors must meet the requirements of the National Electrical Manufacturers Association (NEMA).

The motor must have adequate horsepower (hp) to drive the fire pump. The pump capacity (gpm [L/min]), the net pressure (discharge pressure minus the incoming pressure), and the pump efficiency determine the required hp. For example, for a 1,000-gpm (4 000 L/min) pump rated at 100 psi (700 kPa), a motor of about 80 hp would be needed. Electric motors powerful enough to power fire pumps use a large amount of electricity and may require a larger electrical service to the building **(Figure 12.43, p. 536)**.

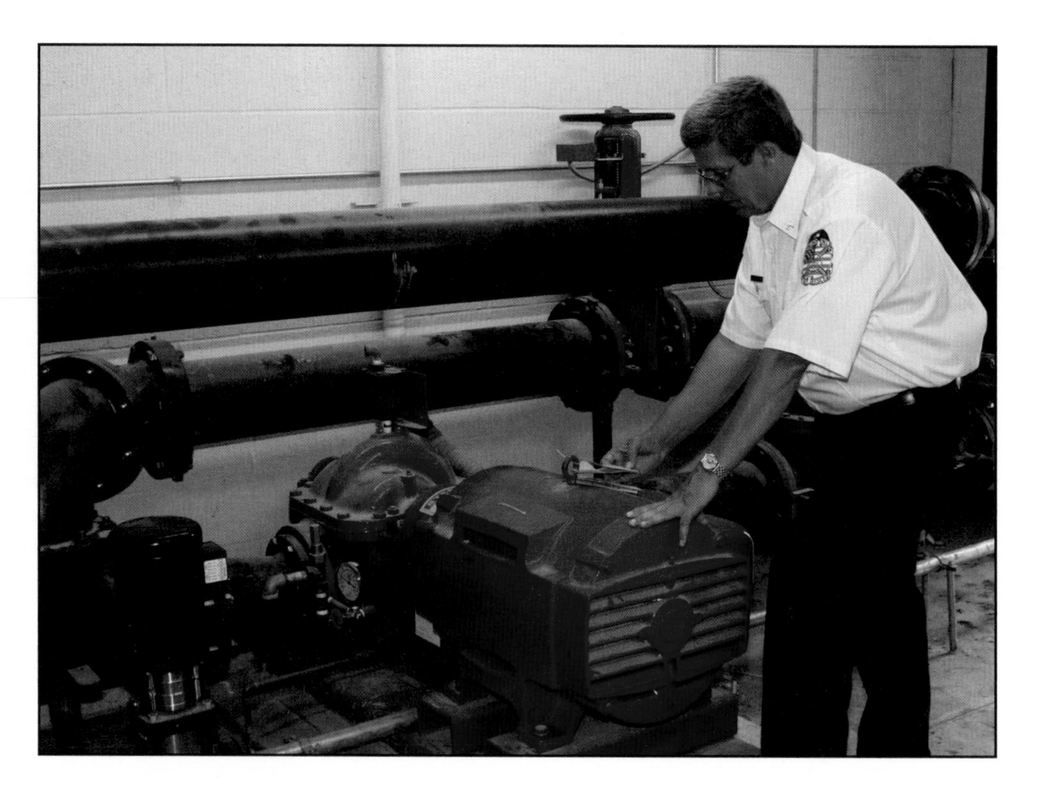

Figure 12.43 A large electric motor like this one being inspected draws a large amount of electricity to generate the 80 hp needed to activate the pump.

Diesel Engine Driver

A diesel engine is used where a driver independent of the local electrical power supply is needed. However, this engine is more complex and requires batteries to start; it also requires an on-site fuel supply. A diesel engine is generally more expensive and requires more maintenance than an electric motor. A diesel engine is required to be tested weekly by running it for at least thirty minutes.

Unlike an electric motor, the diesel engine used for fire pumps is tested and listed by testing laboratories **(Figure 12.44)**. Testing agencies require the diesel engine to be equipped with overspeed shutdown devices, tachometers, oil-pressure gauges, and temperature gauges.

The engine must have a closed-circuit-type cooling system and a fuel supply to provide at least 1 gallon (4 L) per hp. For example, a 185-hp motor requires a 185-gallon (740 L) fuel tank. Diesel engines that are contained in a room or other enclosure must also have an adequate flow of air through the room to promote proper combustion and removal of exhaust fumes.

Steam Turbine

The third type of fire pump driver is the steam turbine. Steam turbines are not as common as the electric or diesel drivers. Steam turbines provide steam pressure to drive both horizontal and vertical split-case pumps directly. When an uninterruptible supply of steam is available in sufficient quantities and at sufficient pressure, steam-driven pumps are feasible options. Otherwise, it is more economical to use electric- or diesel-driven equipment.

Controllers

A stationary fire pump starts automatically whenever the fire suppression system it supplies operates, and it is frequently designed to stop automatically. The fire pump controller accomplishes this action **(Figure 12.45)**. NFPA® 20

Figure 12.44 Testing agencies have specific requirements, such as overspeed shutdown devices for diesel engine powered pumps like this one.

Figure 12.45 The automatic pump starts and stops in a fire suppression system are controlled by a fire pump controller like the one shown.

describes the requirements for electric and diesel pump controllers. Additional requirements for testing and maintenance are a consideration when pumps are connected to an alternate power supply such as an engine-driven generator.

Electric Motor

Most fire pumps are designed to start when a drop in pressure occurs in the fire suppression system **(Figure 12.46)**. A pressure-sensing switch within the electric motor controller detects the drop in pressure resulting from the flow of water. The pressure switch then energizes a circuit that closes the contacts for the motor circuit and starts the fire pump motor. When the water stops flowing in the system, the pressure switch detects the resulting increase in pressure and interrupts the motor circuit, thus turning off the pump.

A pressure switch is adjustable and must be properly adjusted for the individual fire suppression system. If a pressure switch is set improperly, the pump may not start when it is needed or stop when the system is turned off. The pressure at which the pressure switch is set to start the fire pump must be lower than the pressure in the system. The pressure at which the pressure switch stops the fire pump must be less than the churn (no flow) pressure of the fire pump.

The fire pump controller also contains a provision for starting and stopping the pump manually. The controller contains other operating features, including a circuit breaker, a power-available indicating lamp, and a running period timer. The function of the

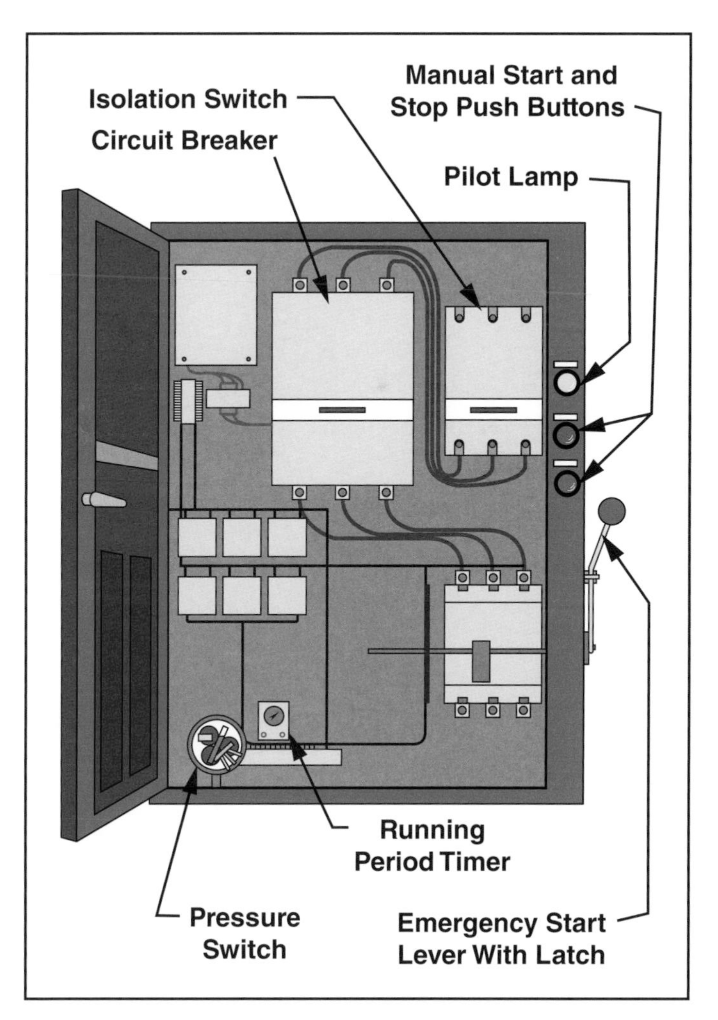

Figure 12.46 All electric fire pump controllers have the same basic features.

running timer is to keep the fire pump motor running for a minimum period of time once the motor has started. This action eliminates rapid opening and closing of the main motor contacts, which could result from system pressure fluctuations.

Diesel Motor

A diesel fire pump controller is more complicated than an electric motor controller. An electric motor controller basically opens and closes an electric circuit for the motor **(Figure 12.47)**. A diesel engine controller closes the circuit for the starting motor on the diesel engine. In addition, it monitors and contains alarms for the following conditions:

- Low engine oil pressure
- High engine coolant temperature
- Failure to start
- Engine overspeed shutdown
- Battery failure

Figure 12.47 The diesel fire pump controller is designed to indicate a number of problems that require attention.

One of the largest fire losses in many years occurred in 1995 at the Madden Textile Plant in Georgia. Although no lives were lost in the fire, the resulting property loss was valued at $200,000,000 — enough to place the company into bankruptcy. A contributing factor to the high loss was the failure of two diesel-driven fire pumps to operate properly. The cause of the failure was lack of adequate inspection, testing, and maintenance.

Evaluate System Components and Equipment

NFPA® 25 contains detailed lists of inspection, testing, and maintenance requirements for all types of water-based fire suppression systems. The standard also includes charts indicating the frequency for performing these tasks. The owner or occupant is responsible for seeing that trained and experienced personnel perform these tasks. The owner/occupant may contract with third-party testing and maintenance organizations to perform these tasks. The AHJ, regardless of governmental level or agency, is responsible for verifying that the required testing and maintenance are complete and that the required records are in order.

An inspector will have the following four opportunities to review, inspect, and witness tests to water-based fire suppression systems:

1. During the plans review and approval process that is required for all fire suppression systems.
2. During the construction phase when the equipment is being installed; an inspector will make periodic preacceptance inspections during installation.
3. During the acceptance test that is required of all systems.
4. During the life of the system, an inspector makes periodic visual inspections, witnesses annual tests, or reviews the records of such inspections and tests.

CAUTION

NEVER personally operate, adjust, manipulate, alter, or handle any fire suppression system devices or equipment during situations other than emergencies or planned training sessions. Always allow facility personnel to operate fire suppression system devices during the inspection and testing of these systems. Inspectors should only witness testing, inspection, and maintenance or accept appropriate documentation that required inspection, testing, and maintenance has been conducted by the owner or owner's representative.

Plans Review

The first contact an Inspector II may have with a new fire suppression system will be during the plans review phase **(Figure 12.48)**. In general, the inspector must determine that the proposed fire suppression system:

- Is appropriate for the type of occupancy, hazard, and construction type
- Is correctly designed
- Meets NFPA® and adopted fire and building code requirements
- Contains all necessary documentation to permit an accurate assessment of the design

NOTE: Chapter 15, Plans Review and Field Verification, describes the plans review process in more detail.

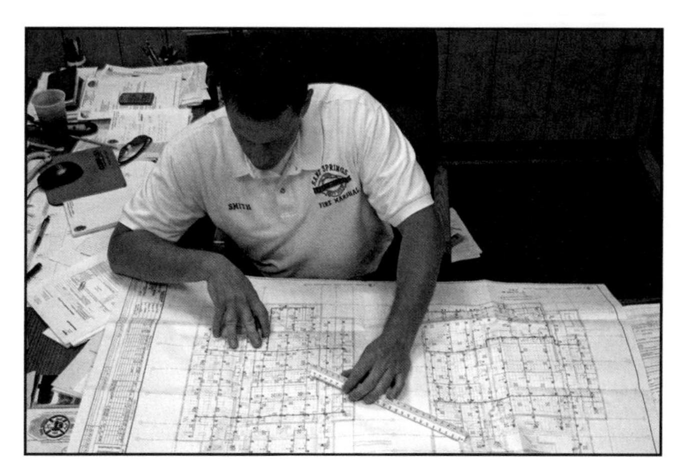

Figure 12.48 An Inspector II may need to evaluate fire protection systems during plans review.

Preacceptance Inspections

Once the inspector reviews and corrects plans for the system and the AHJ issues a permit for the installation of the system, the inspector will be ready to perform preacceptance inspections during the installation. Although the contractor should maintain a stamped (certified) copy of the plans on site, inspectors should bring copies of the plans and other documentation to the inspection site.

During installation of all fire protection systems, the inspector should compare the components of the system to those shown on the plans or listed in the construction documents. The contractor must submit any changes, alterations, or substitutions for review and approval as meeting or exceeding the original design.

An inspector should particularly notice pipe hangers. The hangers must match the requirements shown on the plans and in the documents. In particular, correct hanger spacing is essential to prevent damage to the system during operation.

Acceptance Testing

Acceptance tests are performed on all water-based fire extinguishing systems when the installation is complete. The installation contractor or the owner/occupant's trained maintenance personnel performs acceptance tests. Depending on the local building or fire code requirements, representatives of the building department or fire department may be required to witness the tests. **Table 12.2** gives acceptance test information for each type of system.

CAUTION

Due to waterflow quantities and pressures, electric voltage required by fire pumps, and high noise levels, inspectors must wear appropriate protective equipment, including gloves, eye and hearing protection, and hard hat or helmet.

Routine Inspections and Testing

Water-based fire suppression systems must be inspected and tested regularly to verify that they are operational. It is the responsibility of the owner/operator to have the inspections and tests performed and to maintain accurate records. It is the inspector's responsibility to review the records and sometimes witness the inspections and tests. This is done so the inspector can verify that the system is installed and maintained according to the original design and the applicable standards.

The inspector should be aware that modifications in a structure can have an impact on the sprinkler/water-based suppression system's effectiveness. Inspectors should pay close attention to the following situations:

- Changes in occupancy
- Commodity changes
- Water supply degradation
- Sprinkler changes
- Structural modifications, such as adding or removing walls

Table 12.2
Acceptance Testing Procedures

Fire-Suppression System	NFPA Standard References	Test Procedures
Automatic Sprinkler System **Inspectors:** Witness acceptance tests. **Installation Firm Representative:** Conducts acceptance tests.	• NFPA® 13, *Standard for the Installation of Sprinkler Systems.* • NFPA® 20, *Standard for the Installation of Stationary Pumps for Fire Protection.* • NFPA® 24, *Standard for the Installation of Private Fire Service Mains and Their Appurtenances.*	***1. Flush underground connections.*** • Flush underground mains and lead-in connections before mains are connected to the sprinkler piping. • Continue flushing until all debris and sediment is removed from the line and water is clear. ***2. Hydrostatic test system piping (wet-pipe system)*** • Ensure that the system will be able to handle the pressure if a pumper connects to the fire department connection (FDC). • Hydrostatically test all piping (including underground piping) at not less than 200 psi (1 400 kPa) for 2 hours. • Test the system at 50 psi (350 kPa) above the normal static pressure if the normal static pressure exceeds 150 psi (1 050 kPa). • Check for visible leakage while the system is pressurized; there should be none. • Check for drops in pressure. • Conduct underground supply pipe testing before completion of the interior sprinkler assembly. ***3. Hydrostatic test system piping (dry-pipe system)*** • Perform in the same manner as described for hydrostatic testing of wet-pipe sprinkler systems. • Test for 24 hours with not less than 40 psi (280 kPa) air pressure in freezing weather. • Locate and correct air leaks if there is a loss of more than 1½ psi (10 kPa). • Conduct hydrostatic test using water when the weather warms.
Water Spray Fixed System **Inspectors:** Witness acceptance tests. **Installation Firm Representative:** Conducts acceptance tests.	• NFPA® 13, *Standard for the Installation of Sprinkler Systems.* • NFPA® 15, *Standard for Water Spray Fixed Systems for Fire Protection.*	***1. Test similar to automatic sprinkler system requirements.*** • Flush all underground mains and lead-in piping before connecting the main to the system. • Hydrostatically test system piping in accordance with NFPA® 13 requirements. ***2. Make observations.*** • Water spray system actuation valve must operate within 40 seconds of a heat-detection device sensing a fire. • Ultrahigh-speed water spray systems must operate in less than 100 milliseconds of heat-detector activation. • Operation of the discharge nozzles must be visually observed. • Nozzles should *not* be obstructed or plugged to prevent the creation of the proper discharge pattern. They should be properly positioned. • Discharge patterns should *not* be obstructed or prevented from covering the protected area. ***3. Take pressure readings.*** • Take a pressure reading at the most remote nozzle on the system to determine that the piping is *not* obstructed. • Take a second pressure reading at the water-flow control valve. • Compare both readings to the manufacturer's design criteria.

Continued

Table 12.2 (Continued)

Fire-Suppression System	NFPA Standard References	Test Procedures
Water Mist System **Inspection Personnel (who are familiar with what is required):** Witness acceptance tests. **Installing Contractor or Building Owner Representative:** Performs acceptance tests. NOTE: When feasible, full-scale operational tests of the system should be conducted.	NFPA® 750, *Standard on Water Mist Fire Protection Systems.*	***1. Thoroughly flush all underground water supply piping and lead-in connections to the riser before they are connected to the water mist system.*** ***2. Visually and operationally inspect for correct installation and operation of all mechanical and electrical components of the system (includes any product-of-combustion detection system that may be used to activate the water mist system).*** ***3. Hydrostatically test all piping and tubing systems as described in the following requirements:*** • Low-pressure systems should be able to maintain a pressure of 200 psi (1 400 kPa) for 2 hours. • Intermediate- and high-pressure systems should be able to maintain 150 percent of their normal working pressure for 10 minutes and then for 110 minutes at normal working pressure. • Dry-pipe and preaction water mist systems should be subjected to an air-leakage test. • Water mist systems should be able to maintain 40 psi (280 kPa) of air pressure with no more than 1½ psi (10 kPa) of leakage for a 24-hour period.
Foam-Water System **Inspectors:** Witness acceptance tests. **Installation Firm Representative:** Conducts acceptance tests.	• NFPA® 16, *Standard for the Installation of Foam-Water Sprinkler and Foam-Water Spray Systems.* • NFPA® 13, *Standard for the Installation of Sprinkler Systems.*	***1. Test similar to automatic sprinkler systems requirements for the sprinkler portion of the system, including the following items:*** • Water-flow control valves • Activation • Piping • Sprinklers ***2. Test to determine the following:*** • System flow pressures • Actual discharge capacity • Consumption rate of foam-producing materials ***3. Determine operation of foam proportioning equipment by flow tests.*** NOTE: ***If it is not possible to test the foam discharge, a water discharge test may be permitted.*** • Discharge foam from a single system. • Discharge foam simultaneously from the maximum number of systems expected to operate. • Continue the discharge until a stabilized discharge of foam is obtained. • Conduct test at a minimum flow equal to the flow of the most remote four sprinklers. NOTE: The percentage of foam concentrate must be greater than the manufacturer's listed percentage rate.

Continued

Table 12.2 (Continued)

Fire-Suppression System	NFPA Standard References	Test Procedures
Standpipe and Hose System **Inspectors:** Witness acceptance tests. **Installation Firm Representative:** Conducts acceptance tests.	NFPA® 14, *Standard for the Installation of Standpipes and Hose Systems.*	***Perform the following tests and inspections when installation is complete:*** • Flush and flow-test the system to remove any construction debris and ensure that there are no obstructions. NOTE: This testing also ensures that the system is capable of flowing the required amount of water at the required minimum pressure. • Hydrostatically test at a pressure of at least 200 psi (1 400 kPa) for 2 hours to ensure the tightness and integrity of fittings. NOTE: If the normal operating pressure is greater than 150 psi (1 050 kPa), the system should be tested at 50 psi (350 kPa) greater than its normal pressure. • Perform a flow test on systems equipped with an automatic fire pump at the highest outlet to ensure that the fire pump will start when the hose valve is opened. • Test the fire pump to ensure that it will deliver its rated flow and pressure. • Inspect all devices used to ensure that they are listed by a nationally recognized testing laboratory. • Check hose stations and discharge connections to ensure that they are in cabinets within 6 feet (1.8 m) from the floor and positioned so that the hose can be attached to the valve without kinking. • Inspect each hose cabinet or closet for a conspicuous sign that reads *FIRE HOSE and/or FIRE HOSE FOR USE BY OCCUPANTS OF BUILDING.* • Check fire department connections (FDCs) for the proper fire department thread and a sign that reads *STANDPIPE* with a list of the floors served by that connection. • Check a dry standpipe for a sign that reads *DRY STANDPIPE FOR FIRE DEPARTMENT USE ONLY.*
Stationary Fire Pump **Inspectors from the Fire, Building, Water, Electrical, or Mechanical Departments:** Witness acceptance tests. **Manufacturer Representatives of the fire pump and its components (pump, engine, controller, and transfer switch):** Witness acceptance tests.	NFPA® 20, *Standard for the Installation of Stationary Pumps for Fire Protection.*	***1. Pump must meet the following three standard performance criteria during the acceptance test:*** • Must develop not more than 140 percent of the rated net pressure at shutoff or churn. • Must develop at least the rated net pressure while delivering the rated flow. • Must develop at least 65 percent of the rated net pressure while delivering 150 percent of the rated flow. ***2. List of basic equipment needed to conduct an acceptance test:*** • Pitot tube and gauge • Method for measuring pump speed • Voltmeter • Ammeter

Continued

Table 12.2 (Continued)

Fire-Suppression System	NFPA Standard References	Test Procedures
Stationary Fire Pump **Owner/Occupant or Representative:** Witness acceptance tests. **Installation Contractor:** Performs acceptance tests. **Service Factor —** Described as *power to spare;* that is, a number that is multiplied by the horsepower (hp) rating of a pump to equal the actual hp rating of the motor; the higher number generally means that the pump has more power (less likely to overload and overheat) as compared to the same pump with a lower number.	NFPA® 20, *Standard for the Installation of Stationary Pumps for Fire Protection.*	***3. Procedures:*** ***Manually controlled pumps*** Manually start and stop at least 10 times with the pump running at least 5 minutes each time. ***Automatically controlled pump*** • Conduct at least 10 automatic operations plus 10 manual operations with the pump running at least 5 minutes in each cycle. • Test an automatic controller if it starts the pump in response to a fire-protection system operation such as a fire-detection system. ***Electric-driven pump*** • Conduct one test with all hoselines open to see whether the pump will reach rated speed under full load without pulling excess current and releasing or breaking the circuit. **NOTE:** All of these multiple-operation tests determine whether the starting mechanism is operating properly. • Keep the pump in operation during all testing procedures for no less than 1 hour. • Notice the temperature of the pump bearings and the pump itself. None of the components should become hot to the touch. • Use the voltage and current measured to evaluate other acceptance criteria. **NOTE:** An electric motor should have a nameplate displaying the service factor (power to spare), full-load current rating, and rated voltage. • Conduct essentially the same test procedure for the vertical-shaft electric-driven pump. ***Diesel-driven pumps*** Test the same as for electric-driven pumps. Voltage and current readings are not necessary. ***During testing*** • Do not exceed the full-load current rating except as allowed by the service factor. • Do not exceed the ratio of the measured current in amperes to the full-load current rating at any time during the test. • Ensure that the measured voltage is never more than 5 percent below or more than 10 percent above the rated voltage. ***Data Collection*** Use the data collected when the test has been completed to construct performance curves that are compared with the manufacturer's certified curves. Use the net pressure in constructing the pressure versus flow curve. **NOTE:** If the performance curve falls very close to the characteristic curve, the velocity pressures should be considered. An increase or decrease in pipe size will cause pressure changes because of the change in water velocity. But for most practical applications, these pressure changes can be ignored.

Continued

Table 12.2 (Concluded)		
Fire-Suppression System	**NFPA Standard References**	**Test Procedures**
Stationary Fire Pump		***Results*** • Compare the actual performance of the pump to the certified shop test curves provided by the manufacturer. • Do not accept the installation if the pump does not meet or exceed the characteristic curves or malfunctions in any way. • Require the installing contractor, in conjunction with the equipment manufacturer, to make the installation comply with the standard.
Private Water Supply System **1. Private Fire Service Mains** **Inspectors:** Witness acceptance tests. **Installation Firm Representative:** Conducts acceptance tests. **2. Water Supply Tanks** **Owner and the Installation Contractor:** Performs inspections.	• NFPA® 24, *Standard for the Installation of Private Fire Service Mains and Their Appurtenances.* • NFPA® 22, *Standard for Water Tanks for Private Fire Protection.*	***1. Procedures:*** • Hydrostatically pressurize both underground and overhead piping for 2 hours to 200 psi (1 400 kPa) or 50 psi (350 kPa) above the maximum static pressure, whichever is larger. **NOTE:** Any leakage constitutes failure for the overhead piping. The underground pipe is permitted to leak a little (just a few quarts [liters] per hour) depending on the length of pipe and the number and type of valves and gaskets. • Flush the underground pipe before connection to the fire-protection system piping to remove any accumulated debris in the piping. **NOTE:** If foreign materials are not flushed from the piping before connection to the fire-protection system, these materials will damage or obstruct the system control valves, piping, and discharge devices, which can have a negative affect on the effectiveness of the system **NOTE:** The required flow rate for flushing pipes depends on the diameter of the underground pipe. However, the minimum flow rate is a velocity of 10 feet per second (3 m/sec). ***2. Water Supply Tanks*** Provide results of the inspection to the authority having jurisdiction.

The inspector must consult historical records detailing what the system is designed to protect. This knowledge will help the inspector determine whether to require additional protection for hazards or occupants, or whether the sprinkler system itself is adequate for the current situation.

Before performing any inspection, witnessing a test, or reviewing records, the inspector should take the following steps:

- Review the records of previous inspections and identify the make, model, and type of equipment, including the area protected by the system.
- Determine whether the occupancy classification has changed.
- Review building permits for the site to determine whether any approved alterations have been made to the structure or facility.
- Wear appropriate clothing for dirty locations such as attics and basements. Protective clothing may be necessary for certain manufacturing areas.
- Obtain permission from the owner/occupant before performing any inspection.

Inspectors should consult NFPA® 25, which contains specific inspection and testing requirements, before inspecting or witnessing the testing of any water-based fire suppression system. The inspector should verify that someone has informed the alarm monitoring company prior to starting or witnessing any test. This will avoid an unnecessary emergency response.

The sections that follow contain general items an inspector should look for during scheduled building inspections. For convenience, these sections follow a pattern that an inspector would take when inspecting an occupancy. Because most water-based fire suppression systems are based on the design of automatic sprinkler systems, this section contains the majority of the applicable information. The remaining sections focus on information specific to that particular type of system.

Automatic Sprinkler Systems

Any scheduled building inspection should also include automatic sprinkler systems. Generally, an inspector begins an inspection of a water-based fire suppression system in the system riser room **(Figure 12.49)**.

Every sprinkler system is equipped with various water-control valves and operating valves. Valves are used to shut off the water, drain the system, prevent recirculation of water, and serve other needs of the system. An inspector should first verify that all valves controlling water supplies to the sprinkler system and within the system (sectional valves) are open at all times. An inspector who finds a valve closed should report the condition to the responsible party and to the fire department. Examine control valves as follows:

- Verify that the indicating control valve is opened fully and secured or otherwise supervised in an approved manner (tamper switches, chained and padlocked in the open positions) **(Figure 12.50)**.
- Check the indicating control valve to confirm that it is in operating condition and is accessible and undamaged. The indicating control valve is used to shut off the water supply to the system when it is necessary to replace sprinklers or to perform other maintenance. These valves are located between the sources of the water supply and the sprinkler system.

NOTE: If there is a permanent ladder to access elevated valves, check to see that it is in good condition.

- Check the PIVs to verify that the operating wrench is in place. Verify that the PIV's target (open/shut sign) is properly adjusted and that the cover glass is in place and clean **(Figure 12.51)**. Verify that the PIV bolts are tight and the barrel casing is intact.
- Verify that the FDC is not blocked and is free of obstruction, the swivels rotate freely, and caps are in place **(Figure 12.52)**. If locking caps are in place, verify that appropriate keys are provided. Use a fire department male coupling to verify that the FDC female threads are compatible.
- Verify that automatic sprinkler protection is provided throughout the occupied areas of the building or area based on the locally adopted code.
- Verify that the inlet and outlet gauges on the sprinkler riser are indicating that water or water and air pressure are available in the automatic sprinkler system.

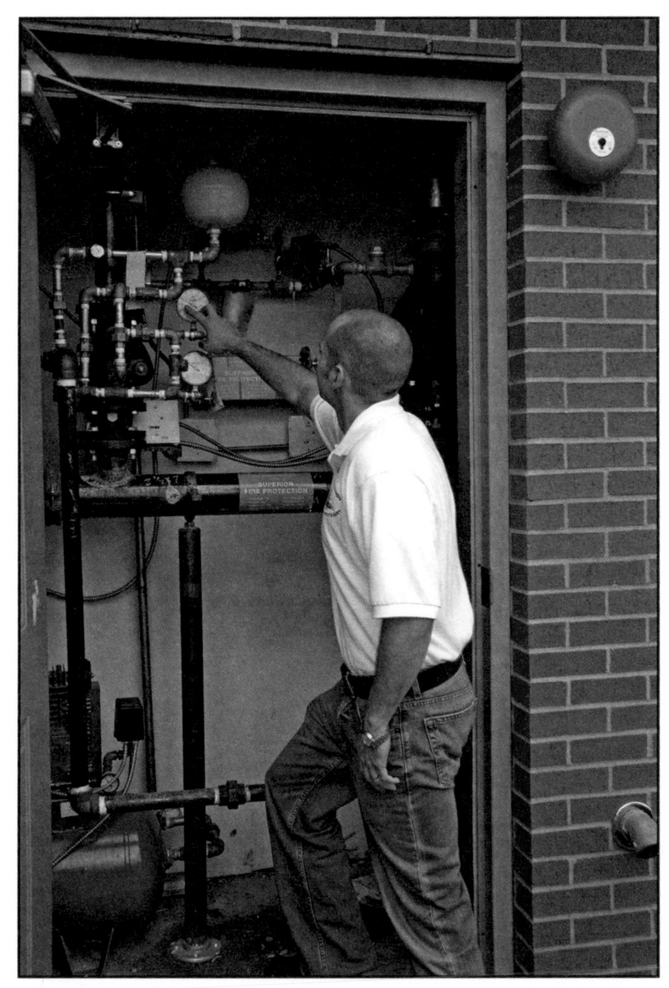

Figure 12.49 Inspectors should check fire suppression system pressure to verify that the system is at the correct pressure. If it is not, the inspector should direct that action be taken, including draining the system when the pressure is too high.

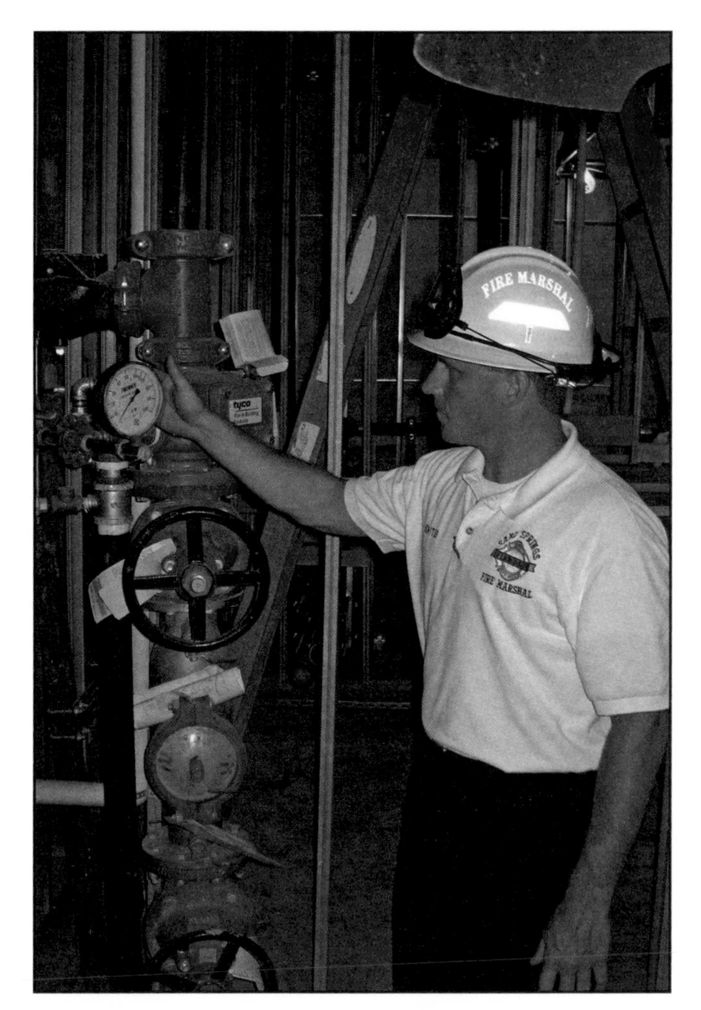

Figure 12.50 Verify that the indicating control valve is opened fully.

Figure 12.51 An inspector verifies that the post indicator valve (PIV) is readable and properly adjusted.

Figure 12.52 Make sure that the FDC is free from obstructions and that the caps are in place. *Courtesy of Rich Mahaney.*

- In buildings or areas protected by a wet-pipe sprinkler system, verify that piping is protected from freezing by approved methods in accordance with locally adopted codes.
- In buildings or areas protected by a dry-pipe sprinkler system, note that the riser room is heated, the air compressor is functional, and the system piping is pressurized.
- Verify that sprinklers are free from damage, corrosion, and paint, and are not subject to recall **(Figures 12.53 a and b)**.
- Review that the inspection, testing and maintenance report has been completed in accordance with NFPA® 25. Reports that are incomplete may require re-test of the system or inspection emphasis in certain areas.

Each sprinkler system riser has a main drain. The primary purpose of the main drain is to drain water from the system for maintenance purposes **(Figure 12.54)**. NFPA® 25 mandates quarterly main drain tests. The inspector should verify the inspection of the system and witness the alarm test in conjunction with the main drain test. The system pressure gauge frequently indicates a higher pressure than the supply pressure gauge due to pressure surges being trapped by a check valve.

Risers that are 4 inches (100 mm) or larger in diameter are equipped with a 2-inch (50 mm) main drain. The main drain test is also useful for detecting impairments, such as closed valves, obstructions, or gradual deterioration in the water supply. To perform a main drain test, the inspector should use the following steps:

1. Observe and record the pressure on the system side gauge at the system riser **(Figure 12.55)**.

Figures 12.53 a and b Sprinklers need to be inspected to verify that they are in good condition and free from corrosion.

NOTE: On a system using an alarm check valve, take the pressure readings from the lower gauge because erroneously high static pressures can exist above the valve.

2. Have a building representative fully open the main drain.

NOTE: The main drain will usually discharge outside the building; check the area to make sure it is clear.

3. Observe and record the pressure drop.
4. Have the building representative close the 2-inch (50 mm) main drain slowly.
5. Observe and record the final static pressure. If it is not the same as the initial static pressure, it is likely that pressure was trapped in the system.

The inspector should compare the new readings to previously recorded readings. If there are significant differences, there may be a partially closed supply valve or an obstruction in the supply line. Another cause can be faulty backflow preventers.

During an inspection of a dry-pipe sprinkler system, inspectors should verify the following conditions:

- All indicating control valves are open and properly supervised in the open position.

Figure 12.54 The main drain on a riser.

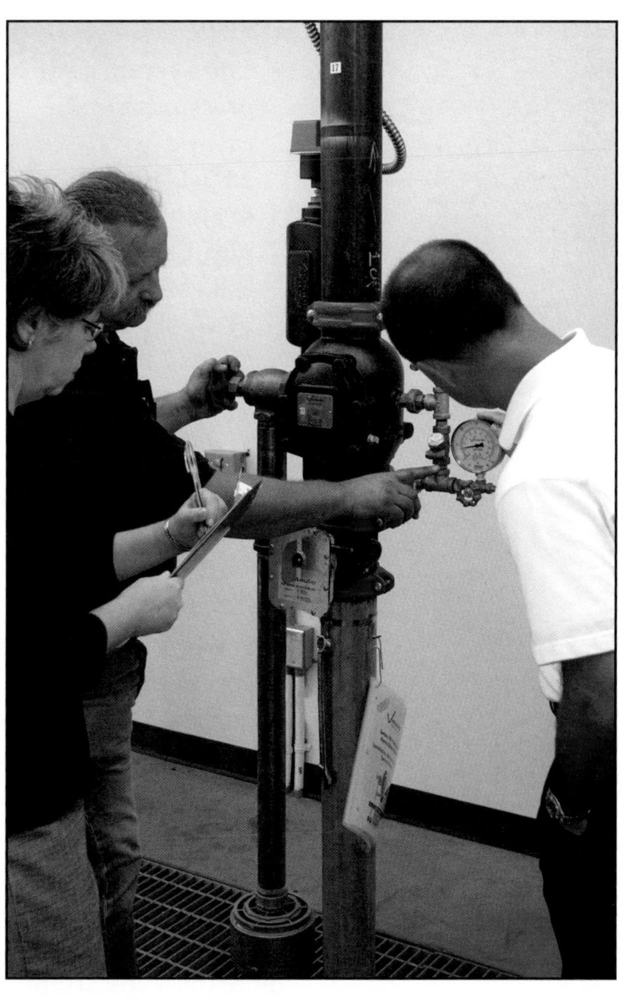

Figure 12.55 Record the pressure on the gauge at the system riser.

- Air-pressure readings correspond to previously recorded readings.
- The ball drip valve moves freely and allows trapped water to seep out of the FDC.
- The velocity drip valve located beneath the intermediate chamber is free to move and allow trapped water to seep out. Inspectors can check this valve by instructing the building representative to lift a push rod that extends through the drip valve opening. Where an automatic drip valve is installed, the velocity drip valve can be checked by moving the push rod located in the valve opening.
- Any drum drips are drained to eliminate the moisture trapped in the low areas of the system.
- The priming water is at the correct level. If necessary, personnel can drain water by opening the priming water test level valve until air begins to escape **(Figure 12.56)**.

NOTE: If the system is equipped with a quick-opening device, opening the priming water test line could trip the system.

Figure 12.56 The priming water must be at the correct level.

- The system's air pressure is maintained at 15 to 20 psi (105 to 140 kPa) above the trip point and no air leaks are indicated by a rapid or steady air loss. If inspectors note excessive air pressure, they should have the system drained.
- The system air compressor is approved for sprinkler system use, well-maintained, operable, and of sufficient size.

The main drain test for deluge and preaction sprinkler systems serves the same purpose as it does on wet- and dry-pipe systems. The test is conducted in the same manner as done for the wet- and dry-pipe systems.

During the inspection of an automatic sprinkler system, the inspector observes the alarm trip test performed by a representative of the sprinkler company of the owner/occupant **(Figure 12.57)**. The inspector *never* performs this test due to liability issues.

When planning an alarm trip test on a dry-pipe system, the inspector should be aware that this procedure can take from two to four hours to complete. The time needed depends on the amount and size of the piping in the system and the capacity of the air compressor to pressurize the system. Also, old valves may prove more difficult to reset because of leaking seats or worn parts, and more than one attempt may be necessary.

With electronically supervised equipment, the inspector may want to have personnel notify the alarm-monitoring organization before they conduct any tests. Another method is to notify the monitoring company after the first portion of the test to verify that the company handles its responsibility. After the tests are complete, the alarm-monitoring organization should confirm that the alarm equipment functions properly.

The inspector should inspect the sprinkler cabinet located near the waterflow control valve to verify that the required number of replacement sprinklers is available. The NFPA® sprinkler standards list the appropriate supply of extra sprinklers that the protected premises must keep on hand. Standards specify a minimum of one sprinkler for each type and temperature rating, as well as a sprinkler wrench.

After the inspector finishes examining the sprinkler system riser room, the next steps are to continue through the remainder of the building, noting the condition and location of sprinklers, hangers, and piping **(Figure 12.58)**. The inspector should pay particular attention to any temporary or permanent changes that could obstruct or alter sprinkler discharge patterns.

Also note whether all sprinklers are clean, undamaged, unobstructed, and free of corrosion or paint; if not, they must be replaced. Guards that protect against mechanical damage may also need to be added or repaired **(Figure 12.59)**. Carefully examine sprinklers in buildings subject to high temperatures, particularly in areas that

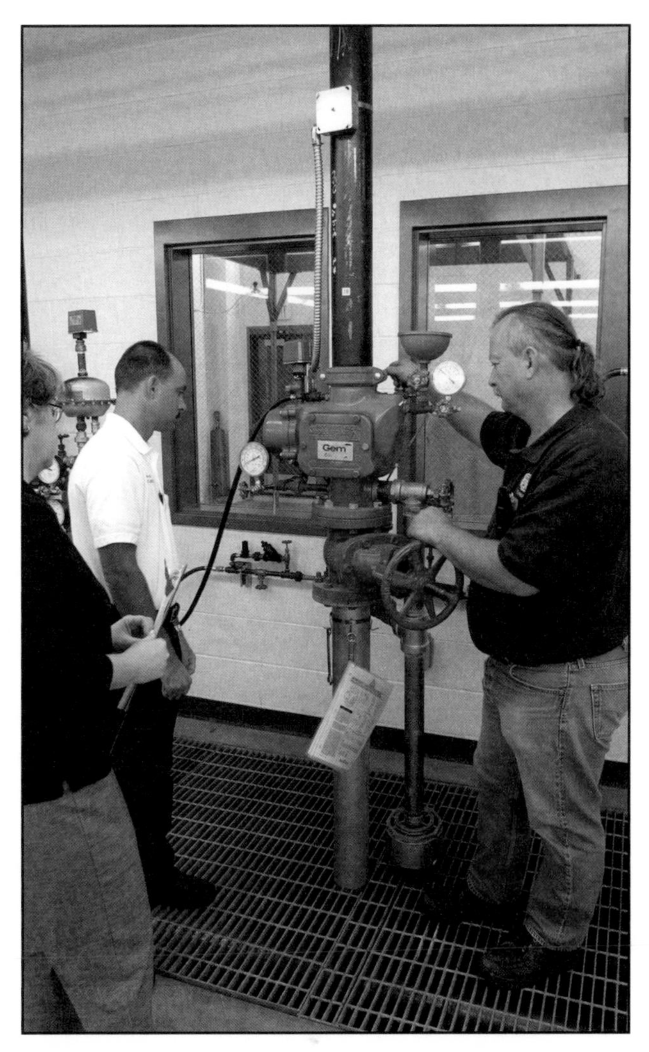

Figure 12.57 The main drain valve should be closed slowly to prevent accidental tripping of the valve .

Figure 12.58 Check the condition of sprinkler hangers and piping.

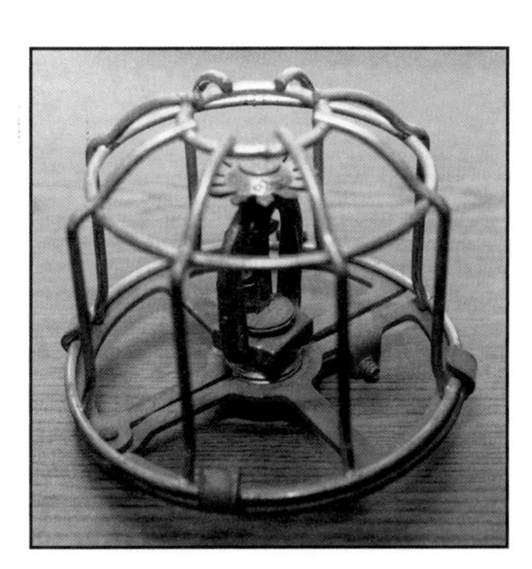

Figure 12.59 A protective cage can help avoid damage to a sprinkler.

have changes in occupancy, fire hazards, or mechanical equipment (such as heating and lighting). Any changes may require the installation of different types of rated sprinklers.

Any sprinkler showing evidence of weakness or damage needs to be replaced with a sprinkler of the same type and temperature. A creeping or sliding apart of the fusible link (cold flow) or leakage around the sprinkler orifice indicates a weak sprinkler. *Cold flow* is the distortion of a material caused by the repeated heating of a fusible link to near its operating temperature **(Figure 12.60)**. Using frangible bulb sprinklers or a higher temperature rated sprinkler will eliminate cold-flow problems.

Partitions, stock, lights, or other objects should not obstruct the distribution of water discharge from sprinklers, and the discharge area should be free of hanging displays. A minimum clearance of 18 to 36 inches (450 mm to 900 mm), measured from the deflector, should be maintained under sprinklers, depending on type **(Figure 12.61)**.

Inspect all sprinkler piping and hangers to determine that they are in good condition. Check for corrosion and physical damage to verify that there are no leaks in the pipes or fittings. Report loose sprinkler hangers. Note whether hangers are at proper intervals and match the original installation plans. Do not permit occupants to use sprinkler piping as a support for ladders, stock, ceiling grids, or other materials **(Figure 12.62)**.

Residential Sprinkler Systems

In sprinkler systems designed for one- and two-family dwellings, the water supply source is generally the same as the domestic water supply. This system works because the water supply requirements are substantially less for residential systems. An NFPA® 13D system requires a minimum ten-minute supply of

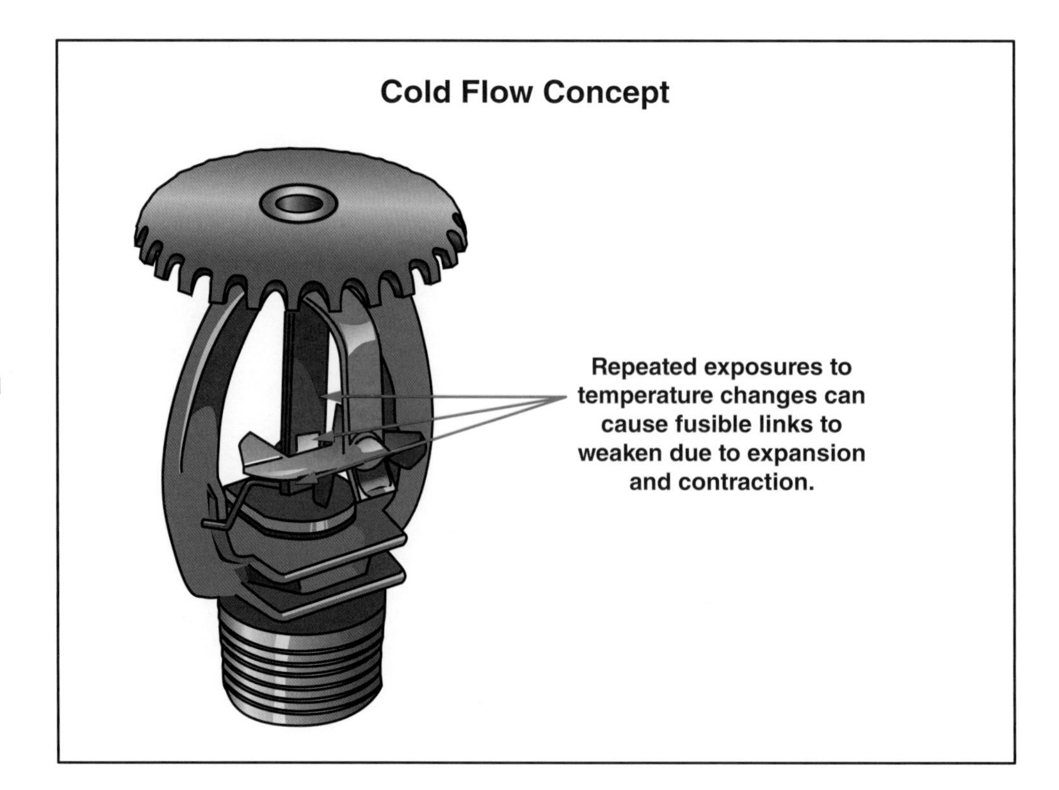

Figure 12.60 Repeated exposure to temperature changes can cause fusible links to weaken due to expansion and contraction.

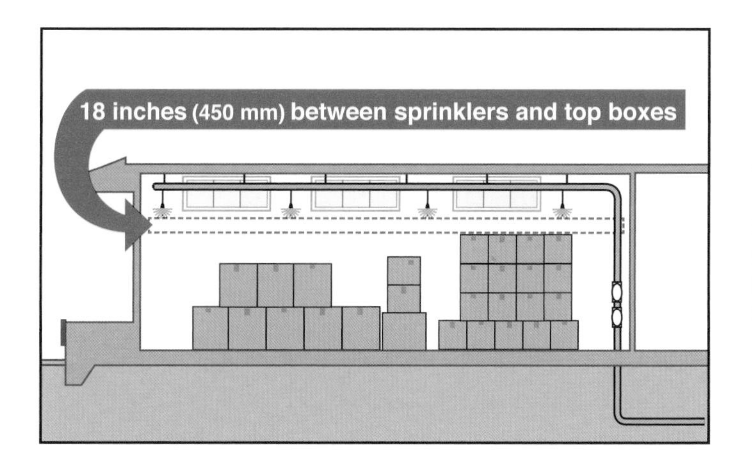

Figure 12.61 There should be at least 18 inches (450 mm) of clearance between sprinklers and stored materials. Some sprinklers require more clearance.

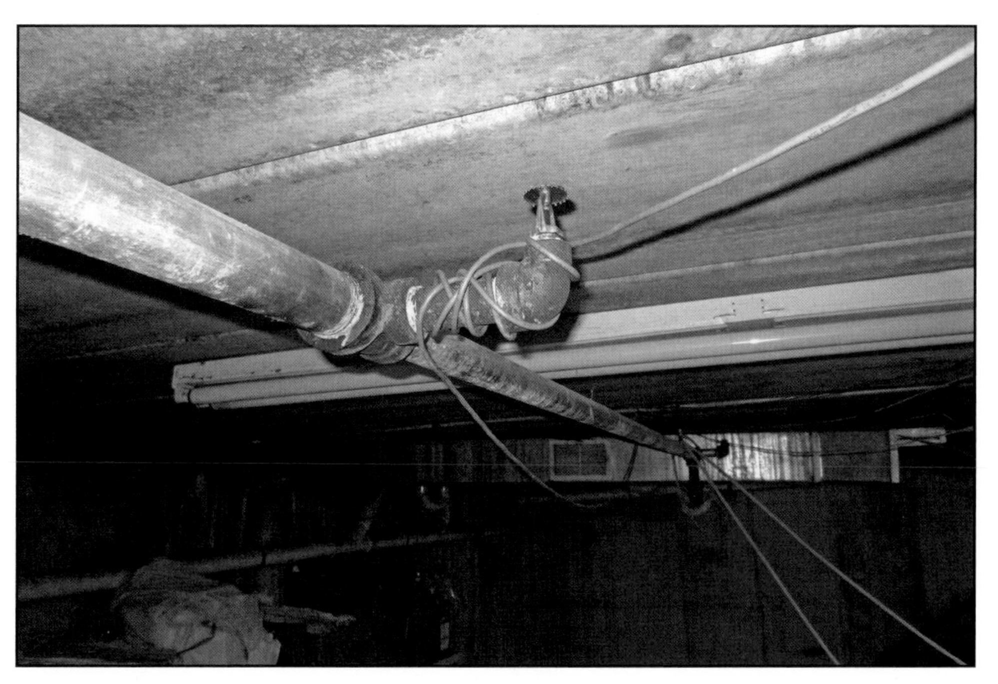

Figure 12.62 Using sprinkler piping to support wiring or other materials is unacceptable and must be corrected immediately. *Courtesy of Rich Mahaney.*

stored water. If structures are less than 2,000 square feet (180 m^2) in area and no more than one story in height, the minimum quantity of water needed is based on the flow rate of two sprinklers that are operating for seven minutes. Structures equipped with an NFPA® 13R system require that four sprinklers operate for thirty minutes. A major difference between residential sprinklers and standard sprinklers is their sensitivity or speed of operation. Residential sprinklers operate more quickly than standard sprinklers.

Any type of piping, such as steel, copper, and plastic, that is listed by an approved testing agency may be used for residential sprinkler systems, although the minimum acceptable size is ½-inch (12.5 mm). The inspector should be aware of the design criteria for piping. For example, piping may be listed for a sprinkler application but have limitations, such as exposed plastic piping.

Sprinkler coverage in residential systems is not as extensive as in standard commercial automatic sprinkler systems. The placement and spacing of residential sprinklers inside of dwelling and sleeping units is based on the sprinkler's listed area of coverage. A single residential sprinkler is permitted to protect its listed area of coverage. In some design standards, sprinklers can be omitted from the following areas in one- and two-family dwellings:

- Bathrooms not over a certain size
- Small closets not over a certain size
- Garages
- Porches
- Carports
- Uninhabited attics
- Entrance hallways

Local codes may be amended to remove some or all of these exemptions. Therefore, the inspector must be familiar with local amendments to the adopted fire and building codes.

As the area coverage increases, the required flow also increases. The area of coverage is also dependent on the ceiling being flat or sloped. In all cases, the area of coverage for any residential sprinklers cannot exceed 400 square feet (35 m^2).

Water-Spray Fixed Systems

The periodic inspections and tests of water-spray fixed systems are similar to those performed on automatic sprinkler systems. Periodic inspections of water-spray systems include all systems components, such as:

- Valves
- Detection devices
- Piping
- Nozzles
- Strainers
- Water supply

During periodic inspections, inspectors should review owner/occupant inspection reports to note whether the frequency of inspections, tests, and maintenance meet the minimum requirements found in NFPA® 25. When making an inspection, check the piping and fittings for the following conditions:

- Mechanical damage
- Leaks
- Corrosion
- Misalignment of parts
- Missing components
- Damaged or missing pipe hangers

General items the inspector should look for on water-spray fixed systems include the following:

- Inspect nozzles to verify that they are operable; free from corrosion, paint, or accumulations of dirt; and properly aligned for the protected area.

- Verify that the sprinkler cabinet contains replacement sprinklers and wrenches.
- Inspect the strainers on the mainline and nozzles for obstructions. Clean strainers and remove any debris to permit the proper flow of water.
- Verify that the water supply is dependable and that all waterflow control valves are in the locked and open position.
- Inspect the fire pumps, tanks, and connections to the public water supply in accordance with the requirements for those types of systems.

An inspector should witness operational tests to determine that the system's response time is less than forty seconds. Observe that the discharge pattern is correct for the surface being covered.

Foam-Water Systems

Parts of the foam-water system that are similar to those of other water-based fire protection systems should be tested and inspected in accordance with those system requirements. The system component that is special to the foam-water system is the foam proportioner.

During an inspection of a foam-water system, an inspector should verify that the control valves are in the appropriate position for the type of proportioner in use. The valves may be in the open or closed position. The inspection requirements for the various proportioners are as follows:

- **Standard pressure proportioner** — Verify that the ball drip valves are open and operable and that no corrosion is present on the foam concentrate storage tanks.
- **Bladder tank proportioner** — Check the operation of the waterflow control valves, look for corrosion on the exterior of the storage tank, and look for the presence of foam in the water around the bladder.
- **Line proportioner** — Inspect the strainers to verify that they are not obstructed, look for corrosion on the exterior of the storage tank, and verify that the pressure vacuum vent is operational.
- **Standard and inline balanced-pressure proportioners** — On these two types of proportioners, the inspector should verify the following:
  - — Strainers are not obstructed
  - — Pressure vacuum vent is operational
  - — Gauges are operational
  - — Sensing line valves are open
  - — Power is available to the foam pump
- **Orifice plate proportioner** — Same inspection requirements as the previous balanced-pressure proportioners, except that there are no sensing line valves in the system.

Figure 12.63 Inspectors should make certain that all hose stations are in working order and will perform as anticipated in case of an emergency.

Standpipe and Hose Systems

As with all fire protection systems, standpipe and hose systems require testing and inspection at regular intervals **(Figure 12.63)**. Building management personnel should make a visual inspection at least once a month. Because interior fire suppression operations depend on standpipe and hose systems, the fire department must also inspect them at regular intervals. A fire protection system contractor or the owner/occupant's employees should perform testing of standpipes and hose systems. Employees must be trained and certified to perform standpipe tests.

Standpipes and Sprinklers

Many standpipe risers supply water for automatic sprinkler systems as well as supplying water for hose streams. In a building equipped with automatic sprinklers and standpipes, the required flow and pressure are based on the most demanding sprinkler system and standpipe system according to hydraulic calculations.

Based on the requirements of the IFC, standpipes are required to be installed in all buildings, except R3 occupancies where the highest story is located more than thirty feet above the lowest level of fire department vehicle access or where the lowest level is more than thirty feet below the highest level of fire department vehicle access.

Inspectors should inspect standpipe and hose systems for the following conditions:

- All water supply valves are sealed (locked) in the open position.
- A fire pump in operating condition with power available.
- Individual hose valves are free of paint, corrosion, and other impediments.
- Individual hose valves are operable.
- Hose valve threads are not damaged.
- Hose valve wheels are present and not damaged.
- Hose cabinets are accessible.
- Hose (when present) is in good condition, is dry, and is properly positioned on the rack or reel.
- Discharge outlets in dry systems are closed.
- Dry standpipe is drained of moisture.
- The FDC is not blocked and is free of obstruction, the swivels rotate freely, and caps are in place. If locking caps are in place, verify that appropriate keys are provided. Use a fire department male coupling to make sure that the FDC female threads are compatible **(Figure 12.64)**.

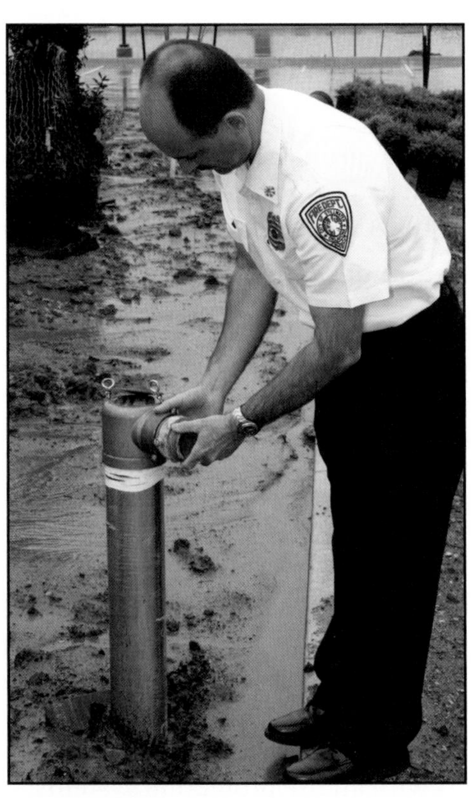

Figure 12.64 Connecting a male coupling to an FDC is a quick way to verify that the threads are not damaged.

- Water supply tanks are filled to the proper level.
- If the system is equipped with pressure-regulating devices, those devices are tested as the manufacturer requires.
- Dry-pipe systems are hydrostatically tested every five years.

Stationary Fire Pumps

The recommended inspection frequency for fire pump operation is weekly. Pumps should be operated from automatic starts, if so equipped, and brought to full speed while pumping a substantial stream. Building maintenance personnel or contracted fire protection specialists most commonly perform pump tests. Depending on local code requirements, inspectors may be required to witness the tests on a periodic basis.

Evaluate the electrical power for electrically driven fire pumps. Visually check the fire pump control panel to verify that the circuit breaker or disconnect is closed and that the power indicating light is on. If the pump takes water under pressure (such as from a city water main), check the incoming pressure gauge to verify that water is available.

The inspector must remember that simply checking the incoming pressure will not disclose an obstruction such as a partially closed valve in the water supply. A flow test is necessary to check for obstructions.

Chapter Summary

Water-based fire suppression systems have proven their value in controlling, containing, and preventing fires in many types of occupancies. Statistics show both the life safety and property protection value of these systems. When such systems fail, the cause is generally human error. To prevent failures, it is up to fire inspectors to determine that water-based fire suppression systems are designed, installed, tested, and inspected properly.

While the property owner/occupant is responsible for performing the various tests and inspections that are listed in the NFPA® standards, it is the fire inspector who must verify the performance of the tests and inspections. This verification occurs through report reviews, witnessing tests and inspections, and performing periodic inspections in the presence of owner/occupants.

An inspector must be familiar with the types of water-based fire suppression equipment, types of tests required for the equipment, and intervals at which the tests and inspections must occur. This chapter provided an overview of those requirements. The inspector should stay up to date on any changes in the NFPA® standards or locally adopted building and fire codes as well as technological changes in the various types of equipment.

Review Questions

1. Name basic types of automatic sprinkler systems.
2. List the major components included in automatic sprinkler systems.
3. What are some of the characteristics of residential sprinkler systems?
4. How does a water-spray fixed system operate?
5. Explain how a water-mist system works.
6. Describe how a foam-water system works.
7. Describe standpipe and hose systems and the standard components.
8. Differentiate between the standpipe classifications.
9. What are the types of standpipe and hose systems?
10. What are some of the inspection requirements for water supplies, standpipe hose valves, regulating devices, and fire department connections?
11. List the types of fire pumps that can be installed as stationary pumps for fire suppression systems.
12. Differentiate between the types of fire pump drivers.
13. Describe different types of controllers found on stationary fire pump systems.

1. What does an inspector need to do before performing an inspection, witnessing a test, or reviewing records?
2. Describe the steps involved in inspecting an automatic sprinkler system.
3. What are some of the conditions inspectors should check or verify when inspecting residential sprinkler systems?
4. Name some of the water-spray fixed system components inspectors need to check during an inspection.
5. What foam-water system components should be checked during an inspection?
6. What conditions should inspectors look for in standpipe and hose systems?
7. What are some of the inspection requirements for stationary fire pumps?

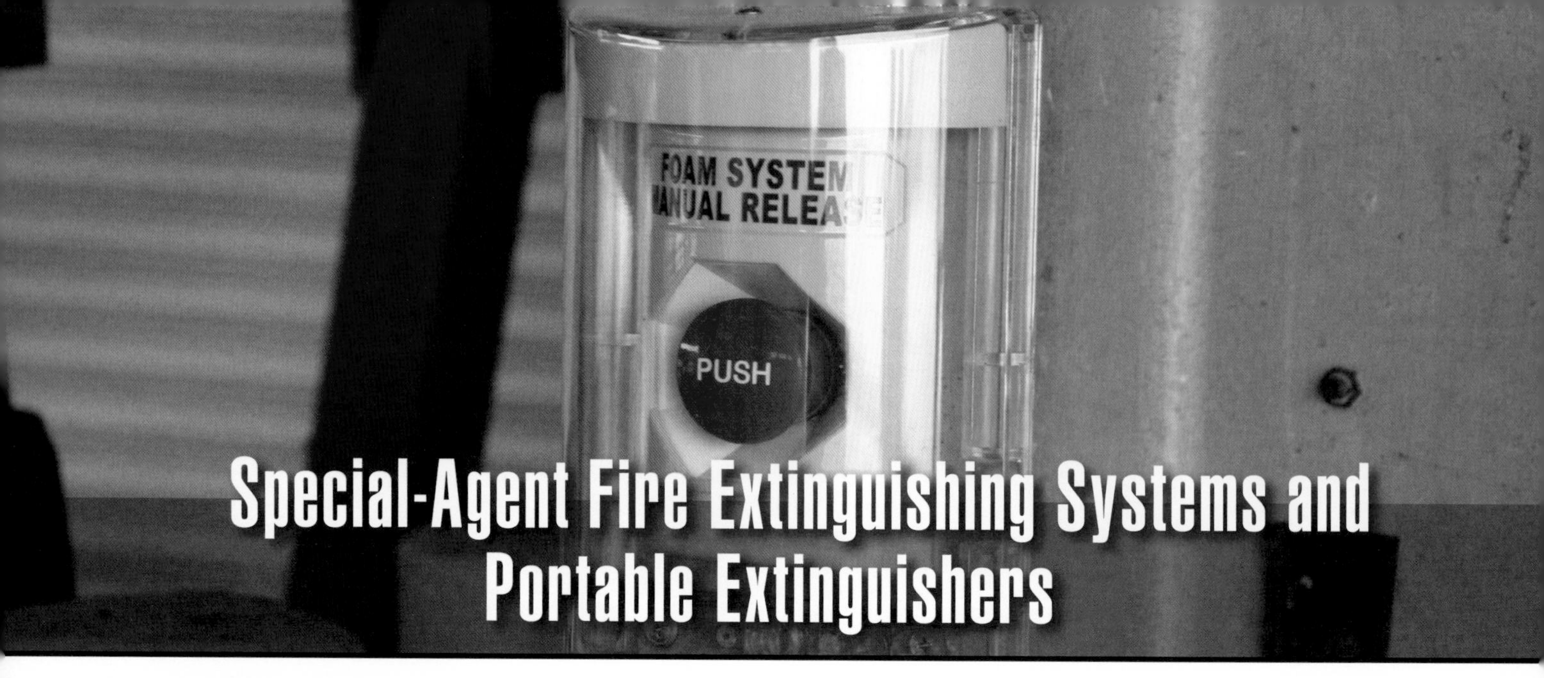

Special-Agent Fire Extinguishing Systems and Portable Extinguishers

Chapter Contents

Case History 563

I Special-Agent Fire Extinguishing Systems 564

- Fire Hazard Classification 565
- Dry-Chemical Systems 565
- Dry Powder Systems 569
- Wet-Chemical Systems 571
- Clean-Agent Systems 572
- Carbon Dioxide Systems 574
- Foam Systems 578

Portable Fire Extinguishers 584

- Classification Systems 585
- Rating Systems 585
- Agents 586
- Types 590
- Installation and Placement 592

II Inspection and Maintenance of Special-Agent Fire Extinguishing Systems 594

- Dry-Chemical Systems 595
- Wet-Chemical Systems 595
- Clean-Agent Systems 596
- Carbon Dioxide Systems 596
- Foam Systems 596

Fire Extinguishers: Selection, Location, Training, and Inspection 597

- Nature of the Hazard 597
- Extinguisher Size/Treavel Distance 599
- Training 600
- Inspection and Maintenance 600

Chapter Summary 603

Review Questions 603

chapter 13

Key Terms

Asphyxiant 575
Expellant Gas..568
Hydrostatic Test.......................................595
Proportioning..579
Total Flooding System 573

NFPA® Job Performance Requirements

This chapter provides information that addresses the following job performance requirements of NFPA® 1031, *Standard for Professional Qualifications for Fire Inspector and Plan Examiner* (2014).

Fire Inspector I	Fire Inspector II
4.3.5	5.4.3
4.3.7	5.4.4

Special-Agent Fire Extinguishing Systems and Portable Extinguishers

Learning Objectives

After reading this chapter, students will be able to:

Inspector I

1. Describe the components and operation of fixed fire suppression systems. (4.3.5)
2. Explain how to determine the operational readiness of portable fire extinguishers. (4.3.7)

Inspector II

1. Identify the appropriate evaluation and testing methods for special-agent fire extinguishing systems. (5.4.3, 5.4.4)
2. Describe proper selection, distribution, inspection, and maintenance of portable fire extinguishers. (5.4.3, 5.4.4)

Chapter 13
Special-Agent Fire Extinguishing Systems and Portable Extinguishers

Case History

The City of Sequim (WA) increased its fire safety inspections with a goal of bringing city restaurants in compliance with current fire and life safety codes. As a result, several restaurants in the city were ordered to repair or replace their duct and hood systems:

- An inspection of the hood and duct system in the Sunshine Café revealed holes in the sides of the hood covered in grease. The grease had been building up in a void space, meaning any fire on the stove could have had disastrous consequences. The owners shut down for several months, upgraded their fire protection system, and reopened.
- Las Palomas Mexican Restaurant was also found to have holes in its hood system. A previous fire had drained the older, noncompliant fire suppressant and the owners opted to close and upgrade the system.
- The Moon Palace was also ordered to upgrade its fire suppression system.

When fires occur, the most effective way to protect occupants is to extinguish the fire in its incipient stage. To provide this level of protection, codes require fire extinguishing systems in some types of occupancies and portable fire extinguishers in most types of public occupancies.

Chapter 12, Water-Based Fire Suppression Systems, described automatic sprinkler systems that apply water to the fire during its incipient stage. In some occupancies, however, water-based fire suppression systems are ineffective. These systems may not have an adequate water supply or the water that is used will cause additional, unnecessary damage to the building or contents. There may be various design limitations to water-based fire suppression systems, or the system may not effectively suppress a fire.

Occupancies where water-based fire suppression systems are ineffective require special-agent fire extinguishing systems. This chapter describes fixed and portable special-agent fire extinguishing systems and their classifications, operation, components, agents, and inspection and testing requirements. This chapter also describes portable fire extinguishers, their classifications, types, components, installations, inspection, and maintenance requirements.

Special-Agent Fire Extinguishing Systems

Special-agent fire extinguishing systems are fixed or mobile fire suppression systems that are used in locations where standard automatic sprinkler systems are not the most effective way to address a particular fire risk **(Figures 13.1 a and b)**. These locations include areas that may contain the following:

- Flammable and combustible liquids
- Water-reactive metals or chemicals
- Combustible metals that are flammable solids
- Food preparation equipment
- File storage or archives
- Electronic Information Technology (IT) equipment
- Electrical transformers and switches
- High-value content such as museums

While water-based fire extinguishing systems can have an almost unlimited supply of water, special-agent fire extinguishing systems must operate with a specific, limited, quantity of agent. Water-based fire extinguishing systems only need to control or contain a fire until the fire department arrives. For a special-agent extinguishing system to be considered successful, it must completely extinguish the fire.

Figures 13.1 a and b Special agent extinguishers are located in places such as machine shops and service stations. *b courtesy of Rich Mahaney.*

Because special-agent fire extinguishing systems are designed to protect a specific type of hazard or location, a variety of systems exist. The following sections contain descriptions about the following alternative fire extinguishing systems and their classifications:

- Dry chemical
- Dry powder
- Wet chemical
- Clean Agent
- Carbon dioxide (CO_2)
- Foam

Fire Hazard Classification

Special-agent fire extinguishing systems are classified by the type of fire they will extinguish. Labels may be affixed to the extinguishing agent storage tanks to indicate the class of fire for which the system is approved.

NOTE: Portable fire extinguishers are also classified and labeled according to the type of fire they will extinguish.

As a review from Chapter 3, the fire classifications are as follows:

- **Class A** — Involves ordinary, solid, combustible materials such as wood, cloth, paper, rubber, and many plastics.
- **Class B** — Involves flammable and combustible liquids and gases such as gasoline, oil, lacquer, paint, mineral spirits, and alcohol.
- **Class C** — Involves energized electrical equipment where the electrical nonconductivity of the extinguishing agent is of major importance; materials involved are either Class A (wiring insulation) or Class B (lubricants), and they can be extinguished once the equipment is de-energized.
- **Class D** — Involves combustible metals such as aluminum, magnesium, potassium, sodium, titanium, and zirconium (particularly in their powdered forms); may require special extinguishing agents or techniques
- **Class K** — Involves oils and greases normally found in commercial cooking kitchens and food preparation facilities using deep fryers; through a process known as saponification, extinguishing agents turn fats and oils into a soapy foam that extinguishes a fire.

NOTE: Class K is the most recently added classification of fire designated by NFPA®. It reflects the development and use of cooking oils that have extremely high combustion temperatures that require special extinguishing agents. In Europe this classification is designated as Class E.

Dry-Chemical Systems

A dry-chemical extinguishing system is used wherever rapid fire extinguishment is required and where reignition of the burning material is unlikely. This system is most commonly used to protect the following areas:

- Flammable liquid storage rooms
- Dip tanks
- Paint spray booths (**Figure 13.2, p. 566**)
- Exhaust duct systems

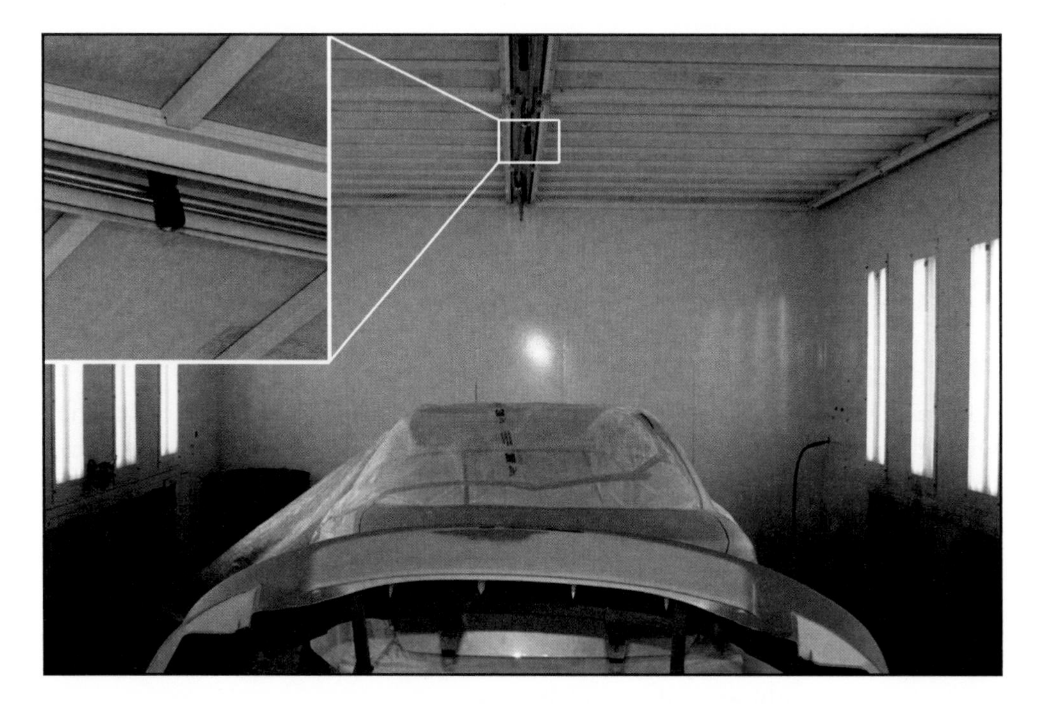

Figure 13.2 Dry-chemical systems like this one are found in paint spray booths to protect against the ignition of flammable vapors. Note the individual nozzle in the inset photo.

The Inspector I will be required to perform the following on dry-chemical fire extinguishing systems:

- Review installation plans
- Review manufacturers' specification sheets (cut sheets)
- Inspect new installations
- Witness acceptance tests of those new installations
- Perform periodic inspections of existing systems

The following sections will cover the following aspects of dry-chemical extinguishing systems:

- Application methods
- Agents
- Components

Application Methods

All dry-chemical systems must meet the requirements set forth in NFPA®17, *Standard for Dry Chemical Extinguishing Systems*. The two application methods for dry-chemical extinguishing systems are fixed system and handheld hoseline.

Fixed System. The fixed system application method consists of the agent storage tanks, expellant storage tanks, a heat-detection and activation system, piping, and nozzles. There are two main types of fixed systems:

—Local application (the most common type): Discharges agent onto a specific surface such as the cooking area in a restaurant kitchen **(Figure 13.3)**.

—Total flooding: Introduces a concentration of agent into a closed area, such as a spray paint booth.

Figure 13.3 Restaurants are a common location for a fixed special-agent system.

Handheld hoseline. The handheld hoseline application method requires trained personnel to apply the dry chemical from hose stations that are connected directly to the agent and expellant storage containers. Hose stations contain hoseline and nozzle assemblies that can reach all portions of the protected area. Some types of hazards such as fuel loading docks, aircraft hangars, outdoor parking areas for aircraft, and flammable liquids storage rooms, may require the installation of handheld hoselines.

Agents

Dry-chemical agents are used in situations where water would be ineffective or reactive with the burning materials. These agents extinguish the fire by separating the oxygen from the fuel, thereby inhibiting the chemical chain reaction necessary for combustion. The fire extinguishing agents used in fixed dry-chemical systems are generally the same agents that are used in portable fire extinguishers.

There are some disadvantages to the use of dry-chemical agents. A dry-chemical system is not recommended for an area that contains sensitive electronic equipment because some dry-chemical agents may become corrosive when exposed to moisture. These agents discharge a cloud of chemicals that leave a residue, creating a cleanup problem. The chemical residue has insulating characteristics that hinder the operation of the equipment unless extensive cleanup is performed.

Some of the dry-chemical extinguishing agents in use are:

- **Sodium bicarbonate** — Also known as ordinary dry chemical and is effective on Class B and Class C fires. When evaluated against an equal weight of carbon dioxide (CO_2), sodium bicarbonate is twice as effective for Class B fires. Sodium bicarbonate has a very rapid control capability against flaming combustion and has some effect on surface fires in Class A materials. In this connection, it has been used successfully on textile machinery where the concentration of fine textile fibers can produce a surface fire. Sodium bicarbonate used in fire extinguishing systems is chemically treated to be water-repellent and free-flowing.

- **Potassium bicarbonate** — Also known as Purple-K to differentiate it from other dry chemicals; has properties and applications similar to sodium bicarbonate. It is most effective on Class B and Class C fires. On a pound-for-pound basis, potassium bicarbonate can extinguish a fire twice the size that the same amount of sodium bicarbonate can. Potassium bicarbonate is also treated to be water-repellent and free-flowing.
- **Monoammonium phosphate** — Also known as multipurpose dry-chemical (pale yellow in color), and is effective on Class A, Class B, and Class C fires. It acts similar to other dry chemicals on flammable liquid fires. Using a combination of extinguishing methods, it quickly extinguishes flaming combustion. On ordinary combustible materials, the monoammonium phosphate melts, forming a solid coating that extinguishes the fire by smothering. Because monoammonium phosphate can have a corrosive effect on unprotected metals, corrosion can form around extinguisher system nozzles, piping, and agent containers.

Expellant Gas — Inert gas that is compressed and used to force extinguishing agents from a portable fire extinguisher; nitrogen is a commonly used expellant gas.

Components

Dry-chemical extinguishing systems consist of the following components:

- **Storage container for agent and/or expellant gas** — May contain both the agent and the pressurized expellant gas (either nitrogen or carbon dioxide), or they may be stored in separate containers **(Figure 13.4)**. A pressure gauge attached to the storage container indicates the stored pressure of the container **(Figure 13.5)**. Storage containers range from 30 to 100 pounds (15 kg to 50 kg). Although rare, some storage containers may hold as much as 2,000 pounds (1 000 kg) of agent. Containers must be located as close to the discharge point as possible. They must also be in an area that maintains a temperature range from -40°F to 120°F (-40°C to 50°C).
- **Piping to carry the agent and gas** — Engineered to account for the unique flow characteristics of the agent. The pipe configuration including the proper size, number of bends and fittings, and pressure drop (friction loss) must be calculated into the design **(Figure 13.6)**.
- **Nozzles** — Attached to a system of fixed piping to deliver the agent to the hazard. No standard nozzle designs exist; each system manufacturer has its own designs and nozzles must be listed in accordance with NFPA® 17.
- **Actuating mechanism** — Releases agent into the piping system in response to the activation of a fire detection system; in many cases it occurs after the melting of fusible links but may be activated by smoke, optical, or other forms of fire detection **(Figure 13.7)**. The fire detection device triggers a mechanical or electrical release that in turn triggers the flow of agent and expellant gas. Systems that have automatic actuation should also be equipped with audible warning signals to facilitate prompt evacuation of the area. Exiting the area quickly lessens problems caused by reduced visibility or breathing difficulties due to exposure to the discharged agent **(Figure 13.8, p. 570)**. The majority of fixed systems must also be capable of manual release and equipped with automatic fuel or power shutoffs.

Figure 13.4 Where multiple agents may be needed to extinguish different classes of fire at the same site, extinguishers may be housed in the same area but stored separately.

Figure 13.5 Inspect pressure gauges on storage containers of dry-chemical systems.

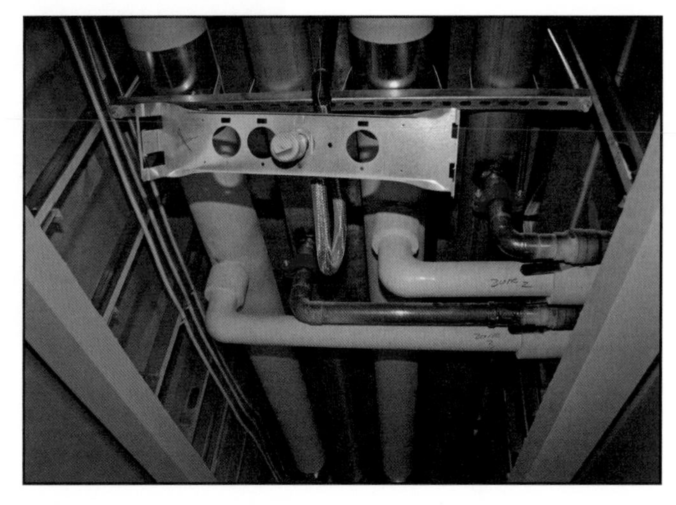

Figure 13.6 The inspector may need to check the building plans to verify that system piping is correct.

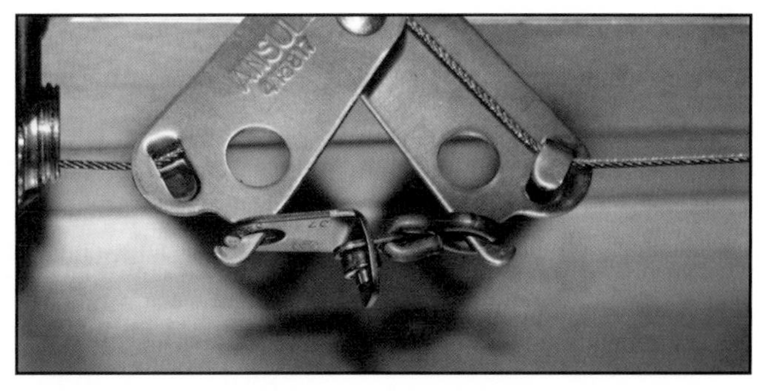

Figure 13.7 Like many automatic sprinkler systems, dry-chemical systems also use fusible links like this one.

Detailed maintenance of these systems is the responsibility of fire protection system companies. However, the owner/occupant representatives or fire inspectors should be aware of changes in hazards and be able to evaluate these systems for the following conditions:

- Mechanical damage or corrosion
- Proper orientation (aim) of nozzles **(Figure 13.9, p. 570)**
- Proper pressures on stored-pressure containers

Dry Powder Systems

Extinguishing agents for combustible metals, also known as dry powders, are designed to extinguish Class D fires involving aluminum, magnesium, sodium, and potassium. Because no single agent is

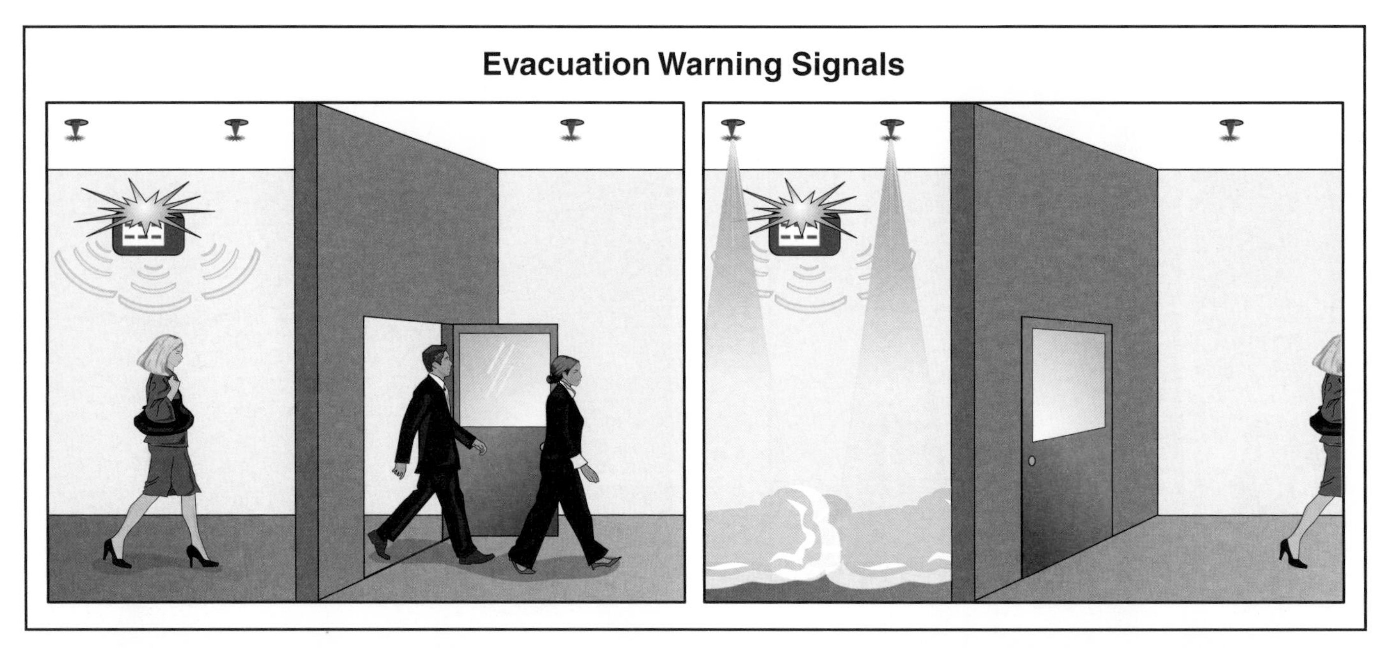

Figure 13.8 An audible warning system is used to alert occupants so that prompt evacuation from the affected area is possible.

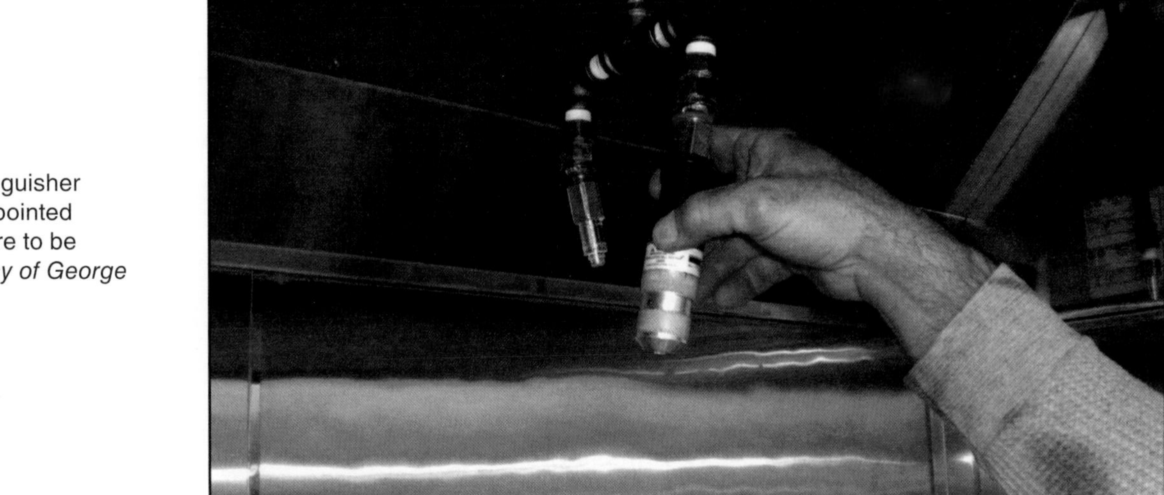

Figure 13.9 Extinguisher nozzles must be pointed correctly if they are to be effective. *Courtesy of George Apple.*

effective on all combustible metals, the extinguishing agent must be carefully chosen for the hazard metal being protected. Metal fires produce extreme heat and require a long period for complete extinguishment. The more common Class D agents that inspectors may find in their jurisdiction are as follows:

- **NA-X®** — Sodium carbonate-based agent with additives to enhance its flow; designed specifically for use on sodium, potassium, and sodium-potassium alloy fires (NOT suitable for use on magnesium fires). The extinguishing agent forms an encasing crust or cake on the burning material, which causes an oxygen deficiency, interrupts the combustion process, and thereby extinguishes the fire. Application can be from fixed systems, by hand from pails, or from portable extinguishers. NA-X® is listed by Underwriters Laboratories Inc. (UL) for use on burning materials at fuel temperatures up to 1,400°F (760°C).

- **MET-L-X®** — Sodium chloride (salt) based agent intended for use on magnesium, sodium, and potassium fires; contains additives to enhance flowing and prevent caking in the extinguisher **(Figure 13.10)**. It extinguishes metal fires by forming a crust on the burning metal. The agent is applied from the fire extinguisher to first control the fire and then it is applied more slowly to bury the fuel in a layer of the powder. The agent is stable when stored in sealed containers. It is nonabrasive and has no known toxic effects.
- **LITH-X®** — Graphite-based agent that extinguishes fires by conducting heat away from the fuel after a layer of the powder has been applied to the fuel; can be used on several combustible metals. It was developed to control fires involving lithium but can also be used to extinguish magnesium, zirconium, and sodium fires. Unlike other dry powders, it does not form a crust on the burning metal.

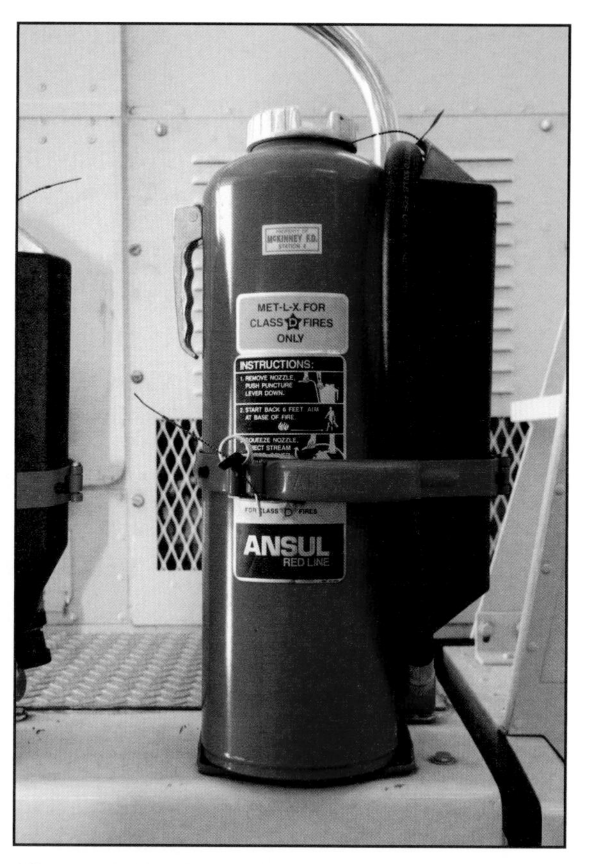

Figure 13.10 MET-L-X® extinguishers like this one release sodium carbonate-based agents that interrupt the combustion process by creating a crust on top of the burning fuel.

Wet-Chemical Systems

A wet-chemical system is most effective on fires in commercial cooking equipment that produce grease-laden vapors, such as those using solid fuels or equipment cooking with heated oils. The nature of the wet chemical is such that it reacts with animal or vegetable oils and forms a soapy foam **(Figure 13.11)**. A wet-chemical agent extinguishes oil fires by fuel removal, cooling, smothering, and flame inhibition. The primary difference between dry-chemical and wet-chemical systems is the type of agent used.

Extinguishing System Standards for Cooking Equipment

Cooking equipment must be protected by automatic fire extinguishing systems that meet the requirements of Underwriters Laboratories Inc. (UL) Standard 300, *Fire Testing of Fire Extinguishing Systems for Protection of Restaurant Cooking Areas* and/or Underwriters' Laboratories of Canada ULC/ORD-C1254.6, *Testing of Restaurant Cooking Area Fire Extinguishing System Units*.

NFPA® 17A, *Standard for Wet Chemical Extinguishing Systems*, became effective for all equipment manufactured after November, 1994. Inspectors need to be aware of the cooking medium used in equipment because it may not be appropriate for the system design. These systems should meet the requirements of UL 300.

Figure 13.11 Wet-chemical agent in a fixed system.

In recent years, tests and fire analyses have indicated that dry-chemical systems do not meet the criteria for extinguishing fires involving deep-fat fryers. Although an inspector will still encounter these types of systems protecting deep-fat fryers and other cooking equipment, the newer Class K agents (actually UL 300 system) are being used to protect this type of hazard.

Wet-chemical fire extinguishing agents are typically composed of water and either potassium carbonate, potassium citrate, or potassium acetate. The agent is delivered to the hazard area in the form of a spray.

Clean-Agent Systems

Clean agents are in a general category of fire extinguishing agents that effectively leave no residue. Clean-agent fire extinguishing systems are effective on Class A, Class B, and Class C fires and will not conduct electricity. Clean agents are approved by the U.S. Environmental Protection Agency (EPA) as nonharmful to the atmosphere (ozone safe). NFPA® 2001, *Standard on Clean Agent Fire Extinguishing systems,* describes the requirements for total flooding and local application clean-agent extinguishing systems.

NOTE: Total flooding systems are described later in this chapter.

Stored as liquefied compressed gas, clean agents convert to a gas when exposed to air. The gas extinguishes the fire by displacing the oxygen and disrupting the fire tetrahedron. These systems are not appropriate for occupied areas. When activated, they release agents that are potentially unhealthy or that will cause asphyxiation.

Some typical applications for clean-agent fire extinguishing systems include the following:

- Computer rooms
- Telecommunications facilities **(Figure 13.12)**
- Clean (manufacturing) rooms
- Document archives and art storage rooms
- Laboratories
- Contents that are delicate or sensitive to water or other corrosives

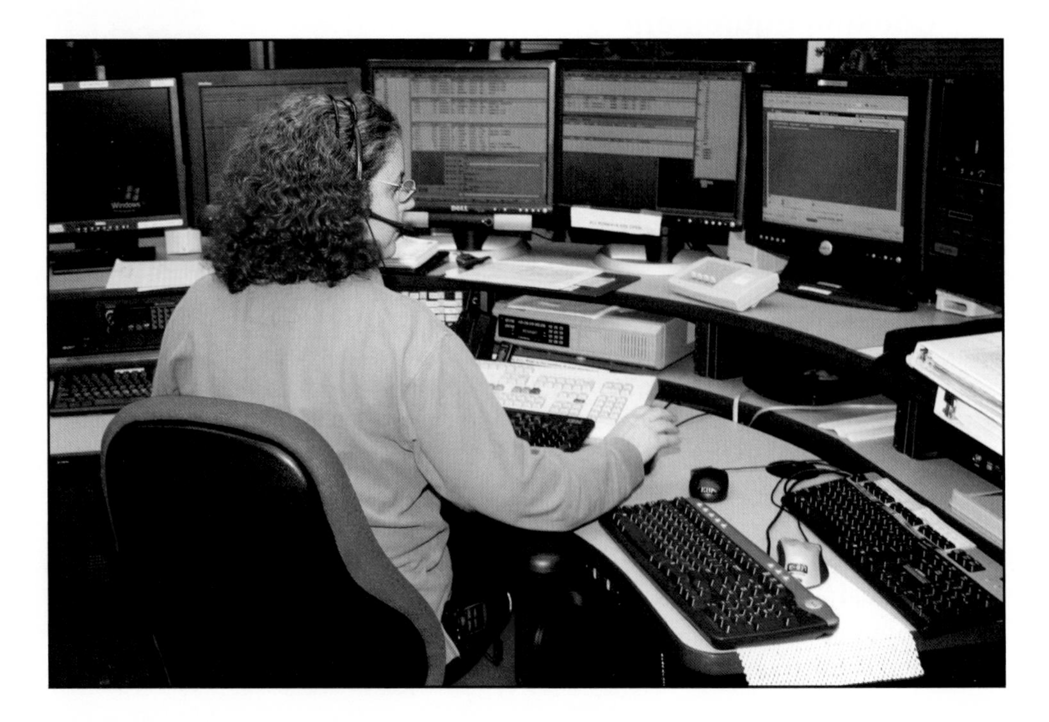

Figure 13.12 Rooms like this telecommunications center include large amounts of electrical equipment and are perfect candidates for clean-agent fire-extinguishing systems.

Agents

One of the first groups of clean agents developed included halogenated fire extinguishing agents. This group contains atoms from one of the halogen series of chemical elements: fluorine, chlorine, bromine, and iodine. The halogenated agents are principally effective on Class B and Class C fires. The term *halon* has been commonly used to describe this group of agents.

While halon production has been discontinued, many of these systems can remain in service until they need to be replaced or upgraded due to system discharge or the need to recharge the system. Therefore, the inspector should be familiar with their operation, regulations, and hazards.

Use of Halon Agents

Because halon has proven to be harmful to humans and the earth's ozone layer, international restrictions on its use were adopted with the Montreal Protocol in 1987. The agreement directed a phase-out of existing halon agents and forbade their manufacture after January 1, 1994.

Locations where halon-agent use is deemed to be essential may be granted an exemption from the phase-out plan. Halon fire extinguishing systems installed before the Montreal Protocol may remain in use until such time as they are discharged on a fire or the gas leaks and must be replaced. The criteria for this exemption are as follows:

- Halon-agent use is necessary for human health and safety or critical for the functioning of society.
- No technically or economically feasible alternatives are available.
- All feasible actions must be taken to minimize emissions during use.

Two types of halons are still in use: Halon 1211 (bromochlorodifluoromethane) and Halon 1301 (bromotrifluoromethane). Halon 1211 is most commonly found in portable fire extinguishers. Halon 1301 is used in some portable fire extinguishers but is more commonly found in fixed **total flooding systems**.

Total Flooding System — Fire suppression system designed to protect hazards within enclosed structures; foam is released into a compartment or area and fills it completely, extinguishing the fire.

NOTE: Halon materials may be recycled. Recycling reduces the material's volume by approximately ten percent each time it is recycled. The sale and storage of halon materials from the existing stockpiles is still permitted, so an inspector may find a facility with large amounts stored.

Several categories of clean agents are now commercially available: halocarbon agents (include hydrochlorofluorocarbon [HCFC] and hydrofluorocarbon [HFC]) and other inert (nonreactive) gas agents such as CO_2. Common halon replacement agents include the following:

- **Halotron®** — Halocarbon agent that becomes a rapidly evaporating liquid when discharged; leaves no residue and meets Environmental Protection Agency (EPA) minimum standards for discharge into the atmosphere. The agent does not conduct electricity back to the extinguisher operator, making it suitable for Class C fires.
- **FM-200™** — Halocarbon agent that leaves no residue and is not harmful to humans or the environment; does require significantly more agent to achieve extinguishment than Halon 1301.

- **Inergen®** — Blend of three naturally occurring gases: nitrogen, argon, and carbon dioxide; is environmentally safe and does not contain a chemical composition like many other proposed halon alternatives. It is stored in cylinders near the facility under protection.
- **ECARO-25™** — Hydrofluorocarbon-based agent that is nonconductive, noncorrosive, residue free, has zero ozone-depletion potential (ODP), and is environmentally preferred to halon. According to the manufacturer, it can be used in an existing halon system with only the replacement of the storage container and discharge nozzles. Third-party testing indicates that it is safe for human exposure in very short durations.
- **FE-36™** — Developed by DuPont™ to replace both Halon 1301 in local application systems and Halon 1211 in portable fire extinguishers. It consists of hydrofluorocarbon HFC-236fa. Suggested applications include computer and telecommunications rooms, museums and archives, hospitals, auto racetracks, and on aircraft.

Components

Clean-agent fire extinguishing systems may be fixed systems that are designed for local application or total flooding agent distribution. These system components include actuation devices, agent storage containers, piping, and discharge nozzles.

Carbon Dioxide Systems

Carbon dioxide (CO_2) is a clean agent that has proven effective at extinguishing fires involving many types of combustible materials. It is less effective at extinguishing fires involving nitrates that contain available oxygen, and not suitable for fires involving metals. The limitations of CO_2 are related to health effects associated with its use as well as restrictions imposed by the combustible material itself.

CO_2 fire extinguishing systems resemble clean-agent fire extinguishing systems. All systems must adhere to the requirements as described in NFPA® 12, *Standard on Carbon Dioxide Extinguishing Systems.* CO_2 fire extinguishing systems have been used to extinguish fires involving the following materials or equipment:

- Flammable and combustible liquids
- Electrical equipment and energized equipment, including power plants and large standby generators
- Flammable gases
- Other combustibles including cellulose materials

As delivered, CO_2 is extremely cold, approaching -110°F (-80°C), and can freeze exposed skin. The agent, however, has a limited cooling effect on a fire. CO_2's primary mechanism of extinguishment is accomplished through oxygen removal or smothering. The limited cooling effects of CO_2 occur when it is applied directly to the burning material.

Although CO_2 fire extinguishing systems were almost phased out by the growth of halon and subsequent clean-agent systems in the 1970s and 1980s, there has been a resurgence of new CO_2 systems to replace halon-based systems.

A CO_2 fire extinguishing system can protect a wide variety of hazards through total flooding or local fire protection applications. Some types of hazardous facilities that CO_2 systems can protect include the following:

- Automobile manufacturing facilities
- Industrial plants **(Figure 13.13)**
- Refineries/chemical plants
- Paint and coating operations
- Food/agricultural processing plants
- Pharmaceutical manufacturing facilities
- Power plant and large standby generators
- Large printing presses

NOTE: Handheld hoseline and standpipe systems are also used although they are not common and not addressed in this manual.

Figure 13.13 Notice the red CO_2 discharge horns over this hazard.

Personnel Safety

The most serious problem involving CO_2 systems, especially total flooding systems, is personnel safety. Carbon dioxide is an **asphyxiant**; eliminating oxygen from a fire also eliminates breathable oxygen from the atmosphere. Total flooding systems are designed to deliver at least a 34 percent concentration of CO_2 into an enclosed area. Concentrations this high are lethal to humans and have resulted in people being killed during the operation of these systems.

Asphyxiant — Any substance that prevents oxygen from combining in sufficient quantities with the blood or from being used by body tissues.

For safety reasons, total flooding systems must be provided with predischarge alarms as well as discharge alarms. However, alarms alone are not enough to guarantee personnel safety. All affected personnel must be educated about the dangers of CO_2 and trained in proper emergency procedures related to system discharge **(Figure 13.14, p. 576)**. Failure of alarms to operate or inadequate training could also result in fatalities.

Figure 13.14 Warning signs used within and adjacent to a space protected by a carbon dioxide (CO_2) extinguishing system. It is imperative that employees know the dangers of carbon dioxide and what to do if the system discharges.

Room integrity is critical if total flooding applications are to be effective. Inspectors must be able to evaluate the integrity of a room equipped with a total flooding system. Room integrity includes such system components as automatic door closers, door gaskets, and shut-down air handling equipment, including closing dampers. NFPA® 2001 requires annual room integrity tests. Inspectors should make a visual inspection of the room on follow-up inspections, looking for obvious issues or changes in the room integrity and order the room tested if necessary.

Local application systems are not as dangerous to personnel as total flooding systems. In local application systems, CO_2 is delivered directly onto the fire as opposed to filling an enclosure with the gas. Danger to personnel is also reduced if the local application system is located outdoors or in a large building.

Components

The components of CO_2 systems are similar to those of clean-agent fire extinguishing systems. These components include actuation devices, agent storage containers, piping, and discharge nozzles. The three means of actuation for CO_2 systems are as follows:

- **Automatic operation** — Initiated by a listed form of smoke or fire detection device. These actuation methods trigger control valves on the CO_2 supply that allow the agent to enter the system and discharge. See Chapter 14, Fire Detection and Alarm Systems, for additional information on smoke and fire detectors.
- **Normal manual operation** — Initiated by a person manually operating a control device and putting the system through its complete cycle of operation, including predischarge alarms **(Figure 13.15)**.
- **Emergency manual operation** — Used only when the other two actuation modes fail; causes the system to discharge immediately and without any warning to individuals in the area.

Figure 13.15 Much like fire alarm systems, carbon dioxide systems may have manual pull stations that workers can use to activate the system in case of emergency.

CO_2 systems exist as either high-pressure or low-pressure systems **(Figures 13.16 a and b)**. In a high-pressure system, the CO_2 is stored in standard DOT-approved cylinders at a pressure of about 850 psi (5 950 kPa). A low-pressure system is designed to protect much larger hazards. The liquefied CO_2 in these systems is stored in large, refrigerated tanks at 300 psi (2 100 kPa) at a temperature of 0°F (-18°C).

In either a high-pressure or a low-pressure system, the containers are connected to the discharge nozzles through a system of fixed piping. Nozzles for total flooding systems may be the high- or low-velocity types. However, high-velocity nozzles promote better agent disbursement throughout the

Figures 13.16 a and b *a* High pressure system. *b* Low-pressure carbon dioxide stored in a large, refrigerated tanks. *photo b courtesy of Chemetron Fire Systems.*

entire area. Local application nozzles are typically the low-velocity type, which reduces the possibility of splashing the burning product when the agent contacts it.

Foam Systems

A foam fire extinguishing system is used when water alone may not be an effective fire extinguishing agent. These locations include but are not limited to the following:

- Flammable liquids processing or storage facilities
- Aircraft hangars
- Rolled paper- or fabric-storage facilities **(Figure 13.17)**

Foam extinguishes a fire by one or more methods including the following:

- Smothering — Prevents air and flammable vapors from combining
- Separating — Intervenes between the fuel and the fire
- Cooling — Lowers the temperature of the fuel and adjacent surfaces
- Suppressing — Prevents the release of flammable vapors

When applied at the correct rate of discharge, foam suppresses fire by forming a chemical layer that is resistant to air on the liquid surface and absorbs heat. The water in the foam solution absorbs additional heat, which suppresses further production of vapors and protects against re-ignition.

The type of system and foam used depend on the hazard being protected. The sections that follow include a description of the following:

- Types of foam systems
- Foam generation
- Foam proportioning rates
- Foam expansion rates
- Foam concentrates
- Proportioners
- Inspection and testing

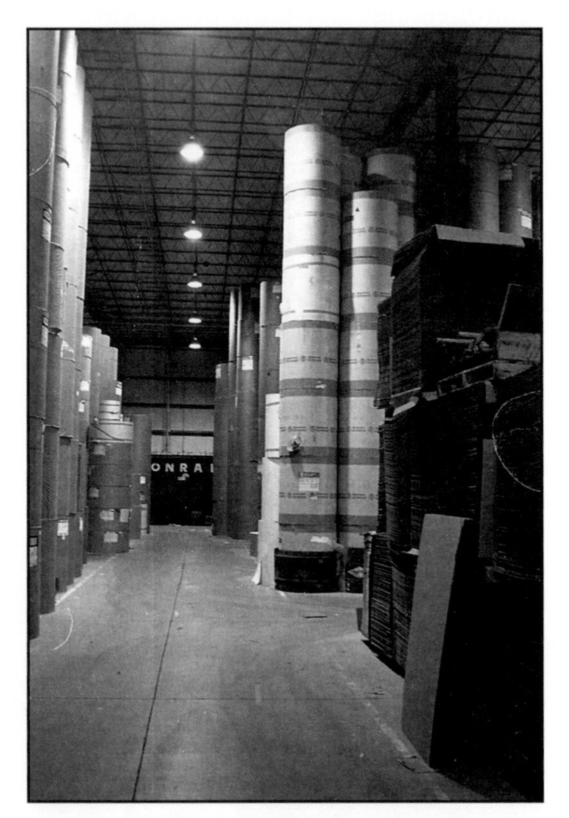

Figure 13.17 Rolled paper storage facilities are frequently protected by foam systems.

Types of Foam Systems

A foam fire extinguishing system must have an adequate water supply, foam concentrate supply, piping system, proportioning equipment, and foam discharge devices. According to NFPA® 11, *Standard for Low-, Medium-, and High-Expansion Foam*, there are four types of foam fire extinguishing systems:

- **Fixed** — Complete installation that is piped from a central foam station; automatically discharges foam through fixed delivery outlets to the pro-

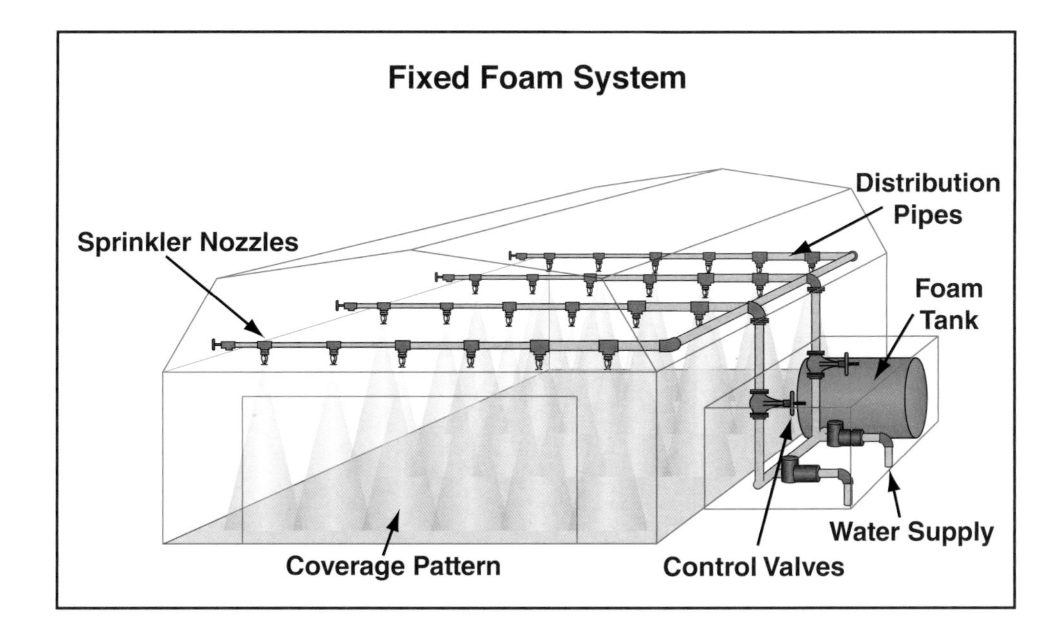

Figure 13.18 In a fixed foam system, foam is discharged directly from fixed outlets.

tected hazard **(Figure 13.18)**. If a pump is required to increase pressure on the system, it is usually permanently installed.

- **Semifixed** — Foam discharge piping is in place but not attached to a permanent source of foam; requires a separate mobile foam-solution source, usually a fire brigade or fire department pumper/truck. It is primarily used on flammable liquid storage tanks. This type of system is found in settings that involve several similar hazards such as petroleum refineries.
- **Mobile** — A wheel-mounted, foam fire extinguishing system; may contain a premixed foam solution, or have water supply connection capability; can be self-propelled or towed by a vehicle.
- **Portable** — Hand-transported foam-producing equipment, such as hose, nozzle, proportioner, and other materials.

Foam Generation

Most fire extinguishing foam concentrates in use today are the mechanical type, which means they must be **proportioned** with water and mixed with air (aerated) before they can be used. Before describing types of foam concentrates and the foam-making process, it is important to understand the following terms:

Proportioning — Mixing of water with an appropriate amount of foam concentrate

- **Foam concentrate** — Raw foam liquid before the introduction of water and air; usually shipped in 5-gallon (20 L) buckets or 55-gallon (220 L) drums. Foam concentrate for fixed systems is stored in large fixed tanks that can hold 500 gallons (2 000 L) or more.
- **Foam proportioner** — Device that introduces the correct amount of foam concentrate into the water stream to make the foam solution.
- **Foam solution** — Homogeneous mixture of foam concentrate and water before the introduction of air.
- **Foam** (also known as finished foam) — Completed product once air is introduced into the foam solution.

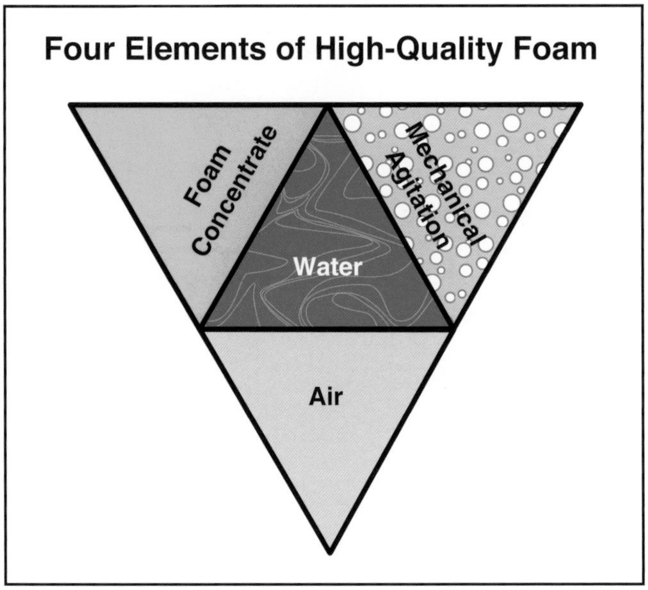

Four Elements of High-Quality Foam

Foam Concentrate

Mechanical Agitation

Water

Air

Figure 13.19 All four elements in the illustration are required in the correct proportions to create high-quality foam.

Four elements are necessary to produce high-quality fire fighting foam: foam concentrate, water, air, and mechanical agitation **(Figure 13.19)**. All of these elements must be present and blended in the correct proportions. Removing any element will result in either no foam or poor-quality foam.

There are two stages in the formation of foam:

- **Proportioning** — Water is mixed with foam liquid concentrate to form a foam solution. The foam solution then passes through the system piping, traveling to the foam distribution nozzle or sprinkler.
- **Aeration** — When the foam solution reaches the distribution nozzle or sprinkler, it is aerated, changing the foam solution into finished foam.

Proportioning equipment and foam nozzles or sprinklers are engineered to work together. Using a foam proportioner that is not hydraulically engineered and matched with the foam distributor nozzle or sprinkler (even if the two are made by the same manufacturer) can result in unsatisfactory quality foam or no finished foam at all. An inspector may need to become familiar with a variety of types of foam production equipment to inspect or analyze a foam fire extinguishing system.

Foam Proportioning Rates

Finished foam is 94 to 99½ percent water. Class B low-expansion foams in use today are designed to be used at 1-percent, 3-percent, or 6-percent concentrations. In general, foams are available at the following rates:

- Foam designed for hydrocarbon (petroleum-based organic compound) fires are used at 1-percent to 6-percent concentrations.
- Polar solvent fuels (lacquers, ketones, or alcohols) require 3-percent or 6-percent concentrates, depending on the particular brand being used.
- Medium- and high-expansion foams are typically used at 1-percent, 1½-percent, 2-percent, or 3-percent concentrations.

Foam Expansion Rates

Depending on its purpose, foam concentrate is designed for low, medium, and high expansion rates **(Figure 13.20)**:

- **Low** — Has a small air/solution ratio, generally in the area of 7:1 to 20:1; used primarily to extinguish fires involving liquid fuels; also used for vapor suppression on unignited liquid fuel spills. Low-expansion foam is most effective when the temperature of the liquid fuel does not exceed 212°F (100°C). If the fuel temperature exceeds this figure, much higher foam expansion rates will be required in order to cool the fuel below this temperature.
- **Medium** — Typically has expansion ratios between 20:1 and 200:1; used when rapid vapor suppression is needed. It can be used indoors or outdoors on either solid or liquid fuels. It will produce more foam with less water than the low-expansion type foams.

Classifications and Expansion Rates of Foam

Classification	Rate
Low Expansion	Less than 20:1
Medium Expansion	Between 20:1 and 200:1
High Expansion	Greater than 200:1

Figure 13.20 Foam expansion may be categorized into three classes that are used under specific circumstances.

- **High** — Generally has expansion ratios of 200:1 to about 1,000:1; useful as a space-filling agent in aircraft hangars and such hard-to-reach spaces as basements, mine shafts, and other subterranean areas, and bulk-baled commodities such as paper, rags, and cardboard **(Figure 13.21)**. Steam dilution caused by the vaporization of the foam in heated areas displaces gas and smoke, thus cooling the environment and extinguishing confined-space fires.

Figure 13.21 High-expansion foam dumps like the one pictured are useful wetting agents where fire load is high and potential fires are expected to grow rapidly. *Courtesy of the United States Air Force.*

Before inspecting a foam fire extinguishing system, it is important to check listings from UL and NFPA® 11, *Standard for Low-, Medium-, and High-Expansion Foam*, to determine recommended uses. It is also important to check the manufacturer's information sheets to determine proper storage of foam, its shelf life, and how to dispose of it without damaging the environment.

Foam Concentrate Types

Because foam concentrates must match the fuel to which they are applied, it is important to identify the type of fuel the fire extinguishing system is protecting. Foams designed for hydrocarbon fires will not extinguish polar solvent fires, regardless of the concentration at which they are used. However, foams that are designed for polar solvent fires may be used on hydrocarbon fires.

Fire extinguishing foam concentrate is manufactured with either a synthetic or protein base. Synthetic-based foam is made from a mixture of detergents. Protein-based foams are derived from either plant or animal matter. Some foam concentrates are a combination of the two. The following are types of foam concentrates an inspector may encounter in fixed fire suppression systems:

- Fluoroprotein foam
- Film forming fluoroprotein (FFFP) foam
- Aqueous film forming foam (AFFF)
- Alcohol-resistant aqueous film forming foam (AR-AFFF)

Proportioners

An inspector should be familiar with the basic concepts of foam proportioners to be able to evaluate all system components. The foam proportioners in common use for fixed systems include the following:

- **Balanced pressure proportioner** — Has a foam concentrate line connected to each fire pump discharge outlet or to the system riser. A foam concentrate pump (separate from the main pump) supplies the concentrate line from a fixed storage tank. This pump provides pressure equal to the pressure at which the fire pump is supplying water to the riser **(Figure 13.22)**. Because the foam concentrate and water are being supplied at the same pressure and the sizes of the discharges are proportional, the foam is proportioned correctly. This proportioner is one of the most reliable methods of foam proportioning. Advantages are:
  - — Ability to monitor the demand for foam concentrate and adjust the amount of concentrate being supplied
  - — Ability to discharge foam concentrate from some outlets and plain water from others at the same time; a single fire pump can supply both foam concentrate and water discharges.

  The discharge orifice of the foam concentrate line is adjustable at the point where it connects to the system riser. For example:
  - — If 3 percent foam is used, the foam concentrate discharge orifice is set to 3 percent of the total size of the water discharge outlet.
  - — If 6 percent foam is used, the foam concentrate discharge orifice is set to 6 percent of the total size of the riser.

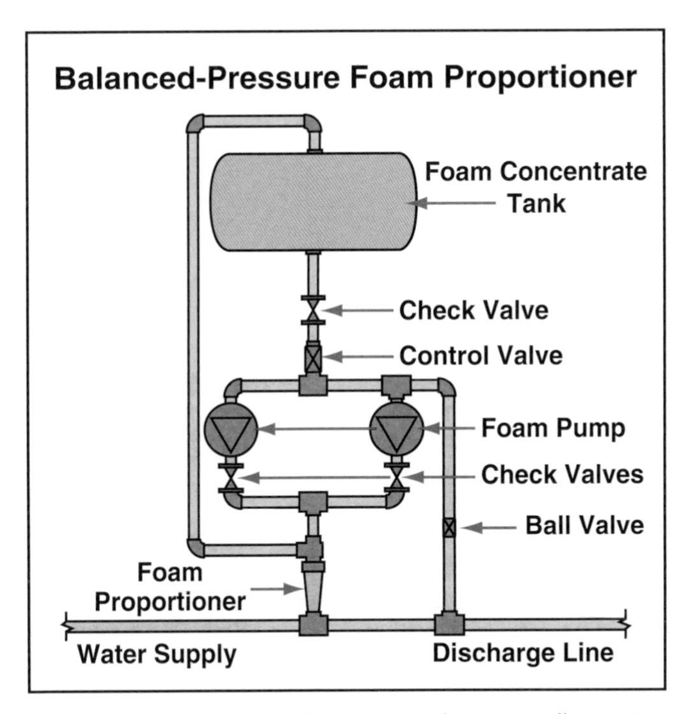

Figure 13.22 A balanced foam proportioner supplies water and foam at the same pressure to ensure the creation of high-quality foam.

- **Around-the-pump proportioner** — Has a small return line (bypass) from the discharge side of a fire pump back to the intake side of the pump; an inline eductor is positioned on the bypass line **(Figure 13.23, p. 584)**. This proportioner is rated for a specific flow and should be used at this rate, although it does have some flexibility. For example, a proportioner designed to flow 500 gpm (2 000 L/min) at a 6-percent concentration will flow 1,000 gpm (4 000 L/min) at a 3-percent rate. This automatic proportioner is especially useful when there is low water pressure or when a motor is not available for a separate foam concentrate pump. It is the most common type of built-in proportioner installed in mobile fire apparatus and some fixed system applications. Disadvantages are:
  - — Pump cannot take advantage of incoming pressure. If the inlet water supply is any greater than 10 psi (70 kPa), the foam concentrate will not be able to enter the pump intake.
  - — Pump must be dedicated solely to foam operation; this proportioner does not allow plain water and foam concentrate to be discharged from the pump at the same time.

Figure 13.23 Around-the-pump proportioners have a bypass line "around-the-pump" that is responsible for adding foam concentrate to the discharge stream.

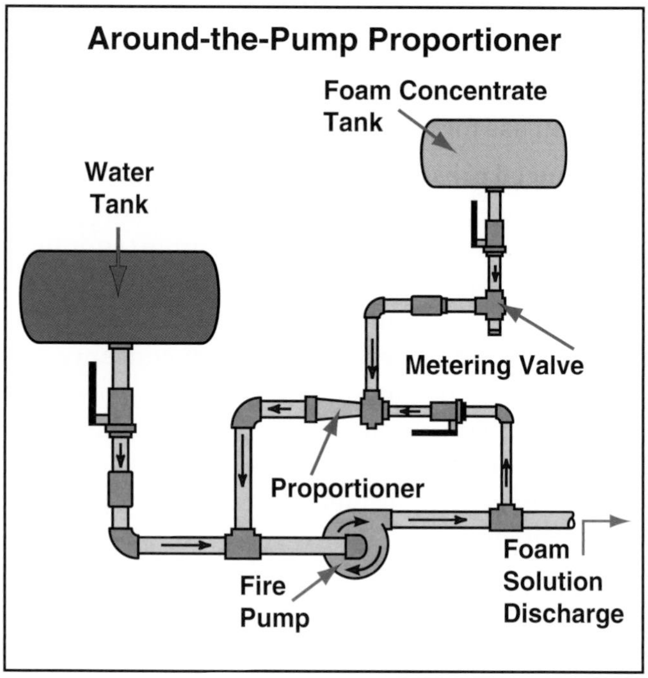

- **Pressure proportioning tank system** — Consists of one or two foam concentrate tanks that connect to both the water supply and foam solution lines of the overall system; designed so that a small amount of water from the supply source is pumped into the concentrate tank(s). This water volumetrically displaces the concentrate, forcing it into the foam solution line where it is mixed with discharge water. The system allows for automatic proportioning over a wide range of flows and pressures and does not depend on an external power source. However, the system is limited by the size of the concentrate tank.
- **Coupled water motor-pump proportioner** — Consists of two positive-displacement rotary-gear pumps mounted on a common shaft; the larger pump is for water and the smaller one is for the foam concentrate. As water flows through the larger pump, it causes the smaller pump to turn and draft foam concentrate from the foam tank. The correct foam/water solution results because the pumps are sized in proportion to each other. This type of proportioner is used in fixed-system applications. It is limited to two sizes, both designed for 6-percent proportioning rates.

Portable Fire Extinguishers

In many situations, a portable fire extinguisher provides the first line of defense against an incipient fire. The value of a fire extinguisher lies in the speed with which it can be used. The person using the extinguisher must understand how it operates and be physically capable of operating it. To be effective, a fire extinguisher must be:

- Readily visible and accessible
- Suitable for the hazard being protected
- Properly maintained **(Figure 13.24)**

- Of sufficient size to control an incipient fire
- Used in appropriate wind and weather conditions to provide effective application on exterior fires

NFPA® 10, *Standard for Portable Fire Extinguishers,* contains requirements for extinguisher installation and placement. Because locally adopted building and fire codes may be more restrictive, the inspector should also refer to locally adopted codes.

The sections that follow contain the following information:

- The way portable fire extinguishers are classified, rated, inspected, and maintained
- Types of extinguishing agents
- Proper selection, installation, and location for extinguishers
- Description on training occupants to use portable fire extinguishers

Figure 13.24 A fire extinguisher that is low in air pressure needs servicing immediately. *Courtesy of Rich Mahaney.*

Portable Extinguishers and Fixed Systems

People often mistakenly assume that additional portable fire extinguishers can substitute for required fixed fire extinguishing systems. One or several portable fire extinguishers should never be considered a substitute or replacement for an automatic fire extinguishing system.

Classification Systems

No single portable fire extinguisher is suitable for use on every type of burning fuel. Fire extinguishers are classified by the type of fire they will extinguish based on the five classifications of fires (A, B, C, D, and K). Labels are affixed to the extinguishers to indicate the class of fire for which the extinguisher is approved.

Rating Systems

Portable fire extinguishers are rated according to their intended use and extinguishing capability for the five classes of fire. The type and amount of extinguishing agent and the extinguisher's design determine the amount of fire that can be extinguished for a particular class of fire. This information is displayed on the front faceplate of the extinguisher using an alphanumeric classification system.

Multiple letters or numerical-letter ratings are used for portable fire extinguishers that are effective on more than one class of fire. Class A and Class B extinguishers also receive a numerical rating that precedes the letter. This rating designates the size of fire the extinguisher can be expected to extinguish when used by an untrained operator **(Table 13.1, p. 586)**.

Table 13.1
Portable Fire Extinguisher Ratings

Class	Ratings	Explanations
A	1-A through 40-A	1-A (1¼ gallons [5 L] of water) 2-A (2½ gallons [10 L] of water)
B	1-B through 640-B	Based on the approximate square foot (square meter) area of a flammable liquid fire a non-expert can extinguish
C	No extinguishing capability tests	Tests are to determine non-conductivity
D	No numerical ratings	Tested for reactions, toxicity, and metal burnout time
K	No numerical rating	Tested to ensure effectiveness against 2.25 square feet (0.2 m^2) of light cooking oil in a deep fat fryer

Occupants who need to use a fire extinguisher are usually under stress and in a hurry, so a labeling system that includes both an icon and the alphanumeric rating is desirable. Fire extinguisher rating labels are designed to enable a user to identify the type of fire extinguisher without having to understand the A-B-C-D-K rating system. For a complete discussion regarding the rating of portable fire extinguishers, refer to the **IFSTA Fire Protection, Detection, and Suppression Systems** manual.

To simplify the process of matching extinguishers with types of fires, several methods for identifying extinguishers by using symbols or icons have been developed. NFPA® 10 recognizes two methods of extinguisher recognition: the pictorial system and the letter-symbol system **(Table 13.2)**.

Agents

Portable fire extinguishers use many different types of extinguishing agents. Each extinguishing agent may be able to control one or more classes of fire, but one agent cannot extinguish all classes of fire. The agents are the same as those used in the dry-chemical fixed fire extinguishing and handheld hoseline systems mentioned previously. The following list highlights the more common fire extinguishing agents and extinguishers.

- **Water** — Liquid used to extinguish fires in Class A materials primarily by cooling the burning fuel. The most common size of water extinguisher is the 2½-gallon (10 L) model; maximum size that is considered portable is the 5-gallon (20 L) unit. Characteristics are:
  - — Water-type fire extinguishers are relatively easy to maintain but subject to freezing and must be kept in a heated area unless an approved antifreeze agent is added to the water.

Table 13.2
Pictorial and Letter Systems

Class Name	Letter Symbol	Image Symbol	Description
Class A or Ordinary Combustibles	A — Ordinary Combustibles		Includes fuels such as wood, paper, plastic, rubber, and cloth.
Class B or Flammable and Combustible Liquids and Gases	B — Flammable Liquids		Includes all hydrocarbon and alcohol based liquids and gases that will support combustion.
Class C or Electrical	C — Electrical Equipment		This includes all fires involving energized electrical equipment.
Class D or Combustible Metals	D — Combustible Metals		Examples of combustible metals are magnesium, potassium, titanium, and zirconium.
Class K or Kitchen	K — Cooking Oils		Includes unsaturated cooking oils in well-insulated cooking appliances located in commercial kitchens.

Reproduced with permission from Wayne State University, Detroit, MI.

— Fire extinguishing capability of an extinguisher is limited by the amount of water carried in a portable unit.

Water limitations are as follows:

— Ineffective by itself on most Class B flammable liquids fires

— Conducts electricity, making it dangerous to use on Class C electrical fires

— Reactive with some chemicals and Class D metals

— Can cause heated cooking oils to splatter, spreading a fire to unaffected areas

- **Carbon dioxide (CO_2)** — Colorless, noncombustible gas that is heavier than air; extinguishes primarily through a smothering action by establishing a gaseous blanket between the fuel and the surrounding air **(Figure 13.25, p. 588)**. These extinguishers are suitable for Class B and Class C fires. Characteristics of CO_2 extinguishers:

— Store CO_2 in a liquid state, which allows more agent to be stored in a given volume, at a pressure of about 840 psi (5 900 kPa)

— CO_2 has a white cloudy appearance when discharged from the extinguisher because of small dry-ice crystals formed by the condensation of surrounding water vapor that is carried along in the gas stream.

Limitations of CO_2 extinguishers are:

— Limited value when used on Class A fires that are located deep within the burning material. These fires can rekindle after the CO_2 dissipates into the atmosphere and normal oxygen returns.

— Difficult to project CO_2 very far from the extinguisher discharge horn because of its gaseous nature

— Although CO_2 is very cold when discharged, its temperature has a minimal effect in cooling or extinguishing a fire because it lacks the cooling effects of water-based fire extinguishing agents.

— Characteristically discharges with a loud noise that may startle an untrained operator

— There may be a discharge of static electricity that can shock the operator when operated in areas of low humidity.

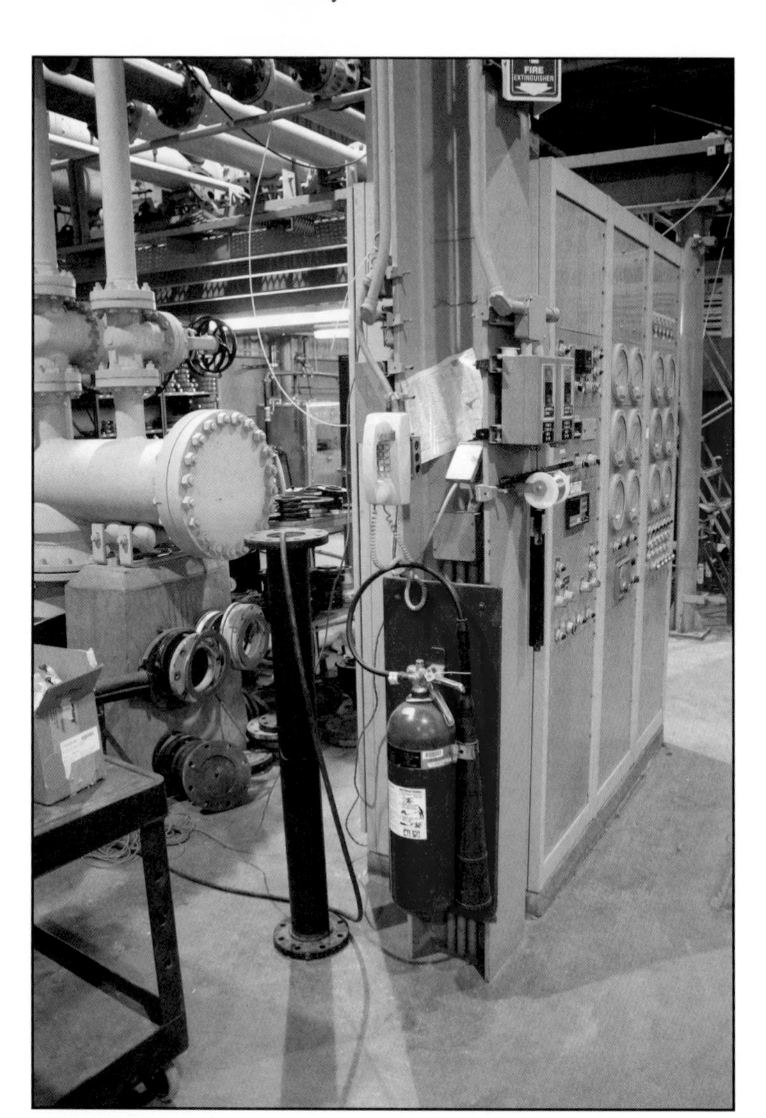

Figure 13.25 This CO_2 extinguisher is used to handle fire in Class B and C fuels.

> **WARNING!**
> Carbon dioxide (CO_2) is an asphyxiant. Do not use a CO_2 extinguisher in a confined space without supplied-air respiratory protection.

- **Foam** — Concentrate added to water to increase its effectiveness; may contain aqueous film forming foam (AFFF) or film forming fluoroprotein (FFFP) foam solutions **(Figure 13.26)**. They are suitable for both Class A and Class B fires. Most commonly, the foam concentrate is premixed with water in the extinguisher and discharged through a special aerating nozzle. Effectiveness features are:
  - Effective on Class A fuels because the foam agent is mixed with water and can cool the fuel.
  - Penetrates deeply into tightly packed fuels like fabrics or bales of materials because the foam breaks the surface tension of water.
  - Very effective on flammable liquid fires because of the double effect of a foam blanket and a surface film to exclude air from the fuel.
- **Dry chemical** — Very small solid particles (not gases or liquids) that can be projected more effectively from an extinguisher nozzle than can gaseous agents. Dry chemicals do not dissipate into the atmosphere as rapidly as gases; therefore, they are especially suitable for controlling fires outdoors. Frequently used agents include:
  - Sodium bicarbonate
  - Potassium bicarbonate
  - Monoammonium phosphate (multipurpose)
  - Urea potassium bicarbonate
  - Potassium chloride

Figure 13.26 AFFF is effective at extinguishing liquid fuel fires because the foam blanket both cools the fuel and creates an oxygen barrier to inhibit fire production.

- **Wet chemical** — Solution composed of water and either potassium carbonate, potassium citrate, or potassium acetate. The agent is delivered to the hazard area in the form of a spray. Extinguishers are intended for use with Class K fixed systems or commercial kitchens that have deep fat fryers using vegetable or animal fats. An inspector should determine the type of agent in use in the fixed system and verify that the portable extinguishers mounted in the kitchen are compatible **(Figure 13.27)**.
- **Clean agent** — Agent that leaves no residue when discharged. Two halons are still in use in portable fire extinguishers. Halon 1211 is most commonly found in portable fire extinguishers. Halon 1301 is used in some portable fire extinguishers, but is more commonly found in fixed-system applications. Portable extinguishers using halon replacements contain inert gases such as argon or DuPont FE-36™. Extinguishers are available in a variety of capacities and may cover 5 to 25 square feet (0.45 m² to 2.25 m²).

CAUTION

It is very important to use the correct extinguishing agent on a fire. Using the wrong agent can be dangerous and can result in a fire not being extinguished, a violent reaction, or both.

Types

Portable fire extinguishers are also classified by the methods used to expel the extinguishing agent. These classification types include the following:

- **Stored-pressure extinguisher** — Contains an expellant gas and extinguishing agent in a single chamber **(Figure 13.28)**. The pressure of the gas forces the agent out through a siphon tube into the chamber, valve, and nozzle assembly. Though simple to use, this extinguisher usually requires special charging equipment for pressurization. Licensed distributors usually perform needed services. These extinguishers are normally found in office buildings, department stores, or even private residences where a high-use factor is not expected. The air-pressurized water (APW) extinguisher is one of the most common types of stored-pressure extinguishers.
- **Cartridge-operated extinguisher** — Stores the expellant gas in a separate cartridge attached to the side of the agent cylinder or tank **(Figure 13.29)**. During activation, the expellant gas (carbon dioxide or nitrogen) is released into the agent cylinder. The pressure of the gas forces the agent into the application hose. Discharge is controlled by a handheld nozzle/lever. No pressure gauge is provided. These extinguishers are found in industrial operations such as paint spraying or solvent manufacturing facilities where they may be used frequently.
  - Inspecting: Verify that the expellant gas cartridge has not been activated; weigh to be sure that it has adequate gas.
  - Recharging: Replace the gas cartridge and fill the agent cylinder (does not require special equipment; may be performed in-house).

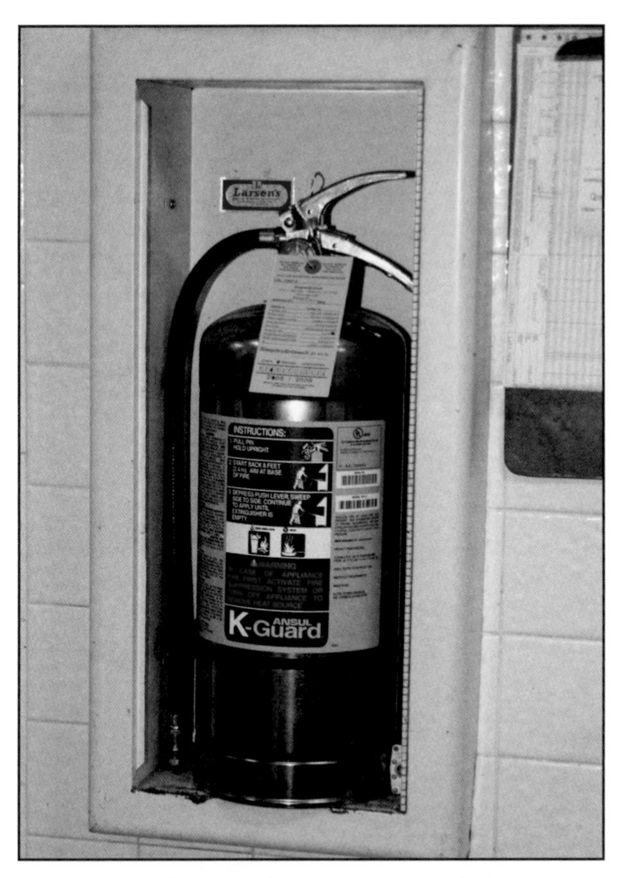
Figure 13.27 Extinguishers mounted in kitchens with deep fat fryers should be Class K and clearly marked as such so that they are used appropriately.

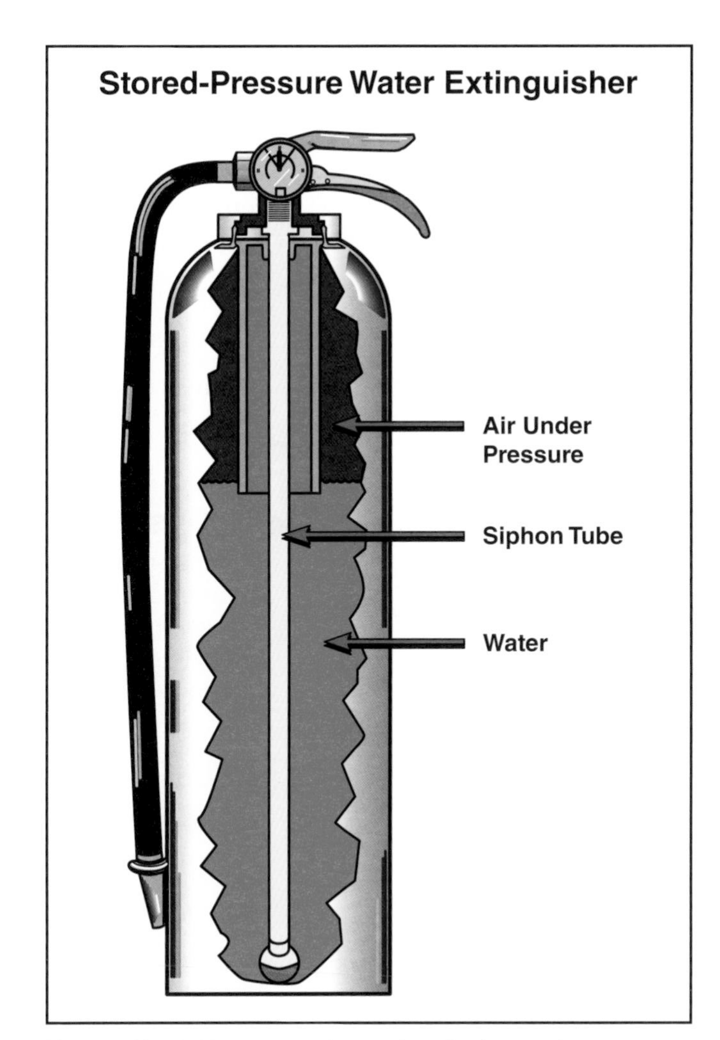

Figure 13.28 Air pressure is used to discharge the water from the extinguisher.

Figure 13.29 Cartridge-operated extinguishers house the expellant gas in a separate cartridge attached to the side of the agent tank.

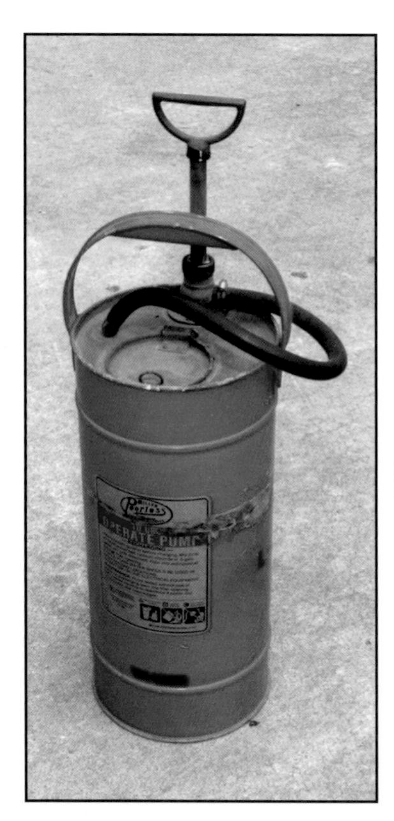

Figure 13.30 This pump-type water extinguisher depends on the manual operation of the pump to expel water from the tank.

- **Pump-operated extinguisher** — Discharges its agent by the manual operation of a pump **(Figure 13.30)**. Because only water can be used in this type of extinguisher, its primary advantage is that it can be refilled from any available water source. Pump-operated extinguishers are not generally found outside fire departments or fire brigades. Maintenance is extremely simple, consisting mainly of checking the following:
  - — Extinguisher is full.
  - — Pump is operational.
  - — Hose and nozzle have not suffered any mechanical damage.

Obsolete Extinguishers

Fire inspectors may occasionally encounter fire extinguishers that are out of production and no longer suitable for use. Operating these extinguishers, even when used as directed, could result in injury or death to the user. The types of obsolete fire extinguishers inspectors are more likely to encounter are:

- Inverting-type
- Soldered or riveted shell soda-acid
- Chemical foam
- Cartridge-operated water
- Loaded stream

In 1982, the Occupational Safety and Health Administration (OSHA) ordered that all these extinguishers be removed from service. An inspector who finds an obsolete extinguisher should require that it be removed from service immediately and replaced with an extinguisher that meets the requirements specified in NFPA®10.

Installation and Placement

Proper extinguisher placement is an essential but often overlooked aspect of fire protection. Properly placed fire extinguishers should exhibit the following:

- Visible and marked with legible signage
- Not blocked by storage or equipment
- Near points of egress or ingress
- Near normal paths of travel
- Placed in proper physical environment for extinguisher **(Figure 13.31)**

Properly securing extinguishers to the structure will avoid injury to building occupants and damage to the extinguisher. Some examples of improper mounting would be an extinguisher placed where it protrudes into a path of travel or one that is sitting on top of a workbench with no mount at all. To minimize potential problems, extinguishers are frequently placed in cabinets or wall recesses to protect both the extinguisher and people who might walk into them **(Figure 13.32)**. If an extinguisher cabinet is placed in a fire-rated wall, the cabinet must have the same fire-resistance rating as the wall assembly.

Figure 13.31 Extinguishers placed outside can become rusted due to exposure. *Courtesy of Rich Mahaney.*

Figure 13.32 Easy-to-open wall-mounted cabinets are designed to protect passersby and prevent damage to extinguishers.

An extinguisher must also be placed so that it is easy to access. An extinguisher that is placed too high presents a lifting and safe handling hazard. Standard mounting heights specified for extinguishers are as follows:

- Extinguishers with a gross weight not exceeding 40 pounds (18 kg) should be installed so that the top of the extinguisher is not more than 5 feet (1.5 m) above the floor **(Figure 13.33)**.
- Extinguishers with a gross weight greater than 40 pounds (18 kg), except wheeled types, should be installed so that the top of the extinguisher is not more than 3½ feet (1 m) above the floor.
- Clearance between the bottom of the extinguisher and the floor should never be less than 4 inches (100 mm)

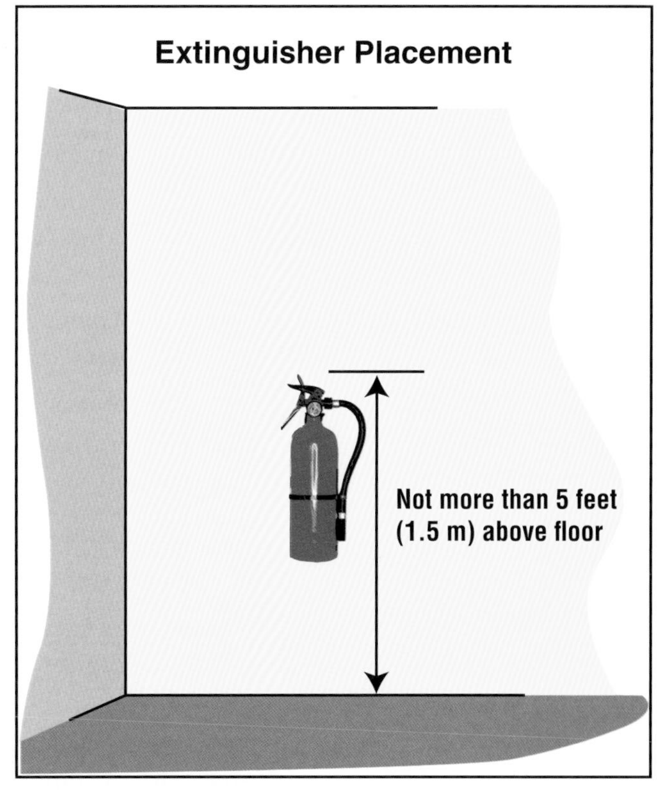

Figure 13.33 Extinguishers that weigh less than 40 pounds (18 kg) should be installed with the top of the extinguisher not more than 5 feet (1.5 m) above the floor.

Physical environment is very important to extinguisher reliability. The greatest concern is the temperature of the environment:

- Water-based extinguishers must be located where freezing is not possible because testing laboratories evaluate them at temperatures between 40°F and 120°F (4°C and 49°C).

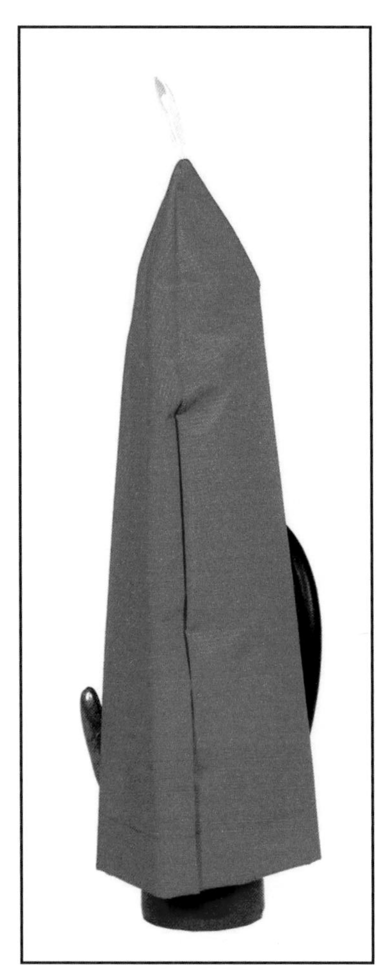

Figure 13.34 An extinguisher may need a protective cover to avoid damage and corrosion. *Courtesy of Tyco Products.*

- Other types of extinguishers can be installed where the temperature is as low as -40° F (-40°C).
- Specialized extinguishers are available for temperatures as low as -65°F (-54°C). Extinguishers using plain water can be provided with antifreeze recommended by the manufacturer.

Other environmental factors that may adversely affect an extinguisher's effectiveness include snow, rain, and corrosive fumes. A corrosive atmosphere may be present not only in an industrial environment but also in marine applications where extinguishers are exposed to saltwater spray. In outdoor installations, an extinguisher can be protected with a plastic bag or placed in a cabinet **(Figure 13.34)**. For marine applications, extinguishers are available that have been listed for use in a saltwater environment.

Inspection and Maintenance of Special-Agent Fire Extinguishing Systems

An Inspector II reviews proposed installation of different kinds of fire protection systems, including portable fire extinguishers. The inspector will be required to recognize proper selection, location, distribution, inspection, and maintenance of these systems.

Dry-Chemical Systems

In addition to being familiar with the agents and operation of dry-chemical systems, the Inspector II is required to perform the following additional duties:

- Evaluate the application methods, agents, components
- Evaluate inspection and testing requirements

A qualified individual should verify to the inspector that the dry-chemical system is maintained and serviced as required by NFPA® 17. This standard recommends that qualified personnel perform the following procedures monthly:

- Check all parts of the system to verify that they are in their proper locations and connected.
- Inspect manual actuators for obstructions.
- Inspect the tamper indicators and seals to note that they are intact **(Figure 13.35)**.
- Check the maintenance tag or certificate for proper placement.
- Examine the entire system for any obvious damage.
- Check the pressure gauges to verify that they read within their operable ranges **(Figure 13.36)**.

The inspector should be able to review records of inspections and maintenance. Any problems that are noted should be corrected immediately. In most cases, this situation requires notification of the organization responsible for the maintenance of the system.

Figure 13.35 Dry-chemical activation stations like this one should have an intact tamper seal when inspected.

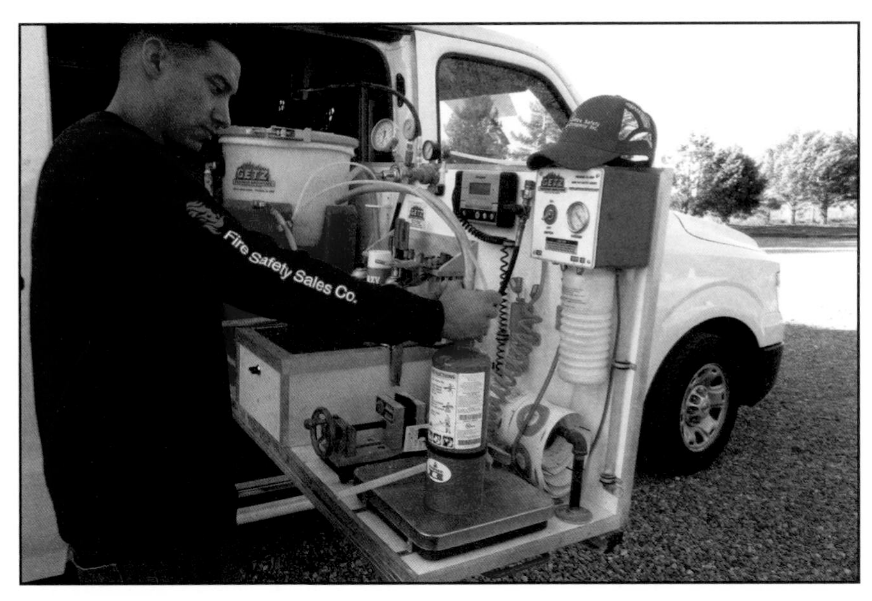

Figure 13.36 A dry-chemical extinguisher being weighed as part of servicing. *Courtesy of Fire Safety Sales.*

Hydrostatic Test — Test method that uses water under pressure to check the integrity of pressure containment components such as pipes or cylinders.

Dry-chemical agent storage containers (cylinders) that are less than 150 pounds (75 kg) must be **hydrostatically tested** every 12 years, so the inspector should verify that this testing is up-to-date. Larger storage cylinders have no hydrostatic test requirements. The inspector should also check any auxiliary functions such as audible/visible alarms, power shutdowns for process equipment, and functionality of ventilation fans.

Wet-Chemical Systems

The inspection and maintenance procedures for wet-chemical extinguishing systems are basically the same as those used for dry-chemical extinguishing systems. NFPA® 17A, *Standard for Wet Chemical Extinguishing Systems,* contains detailed requirements for listed UL 300 wet-chemical systems.

During an inspection, the inspector must verify that facility or fire protection systems company personnel have properly performed system maintenance and have conducted all manufacturer-required system checks **(Figure 13.37)**. The inspection should review the following:

- System parts are in their correct location.
- Manual actuators are unobstructed.
- Tamper indicators and seals are intact.
- Maintenance tags are in place and up-to-date.
- Obvious damage is noted.
- Gauges are within operational limits.
- Equipment modifications or repairs are noted.

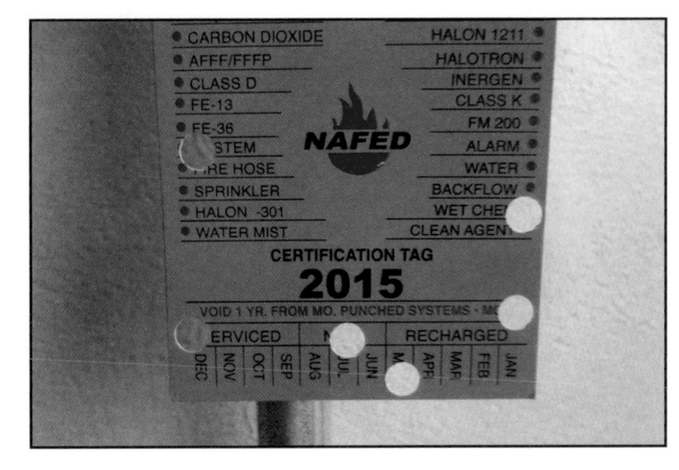

Figure 13.37 Verify that system checks are up to date. *Courtesy of Rich Mahaney.*

Clean-Agent Systems

Inspection and testing procedures for clean-agent fire extinguishing systems can be found in NFPA® 2001, *Standard on Clean Agent Fire Extinguishing Systems* and in NFPA® 12A, *Standard on Halon 1301 Fire Extinguishing Systems*. An annual inspection by qualified personnel is required for all system components. The quantity and pressure of the clean agent must be checked semiannually. Records on all tests and inspections must be maintained and available for review by the inspector.

To properly inspect clean-agent systems, inspectors must know the types of agents used in the systems, system components, and inspection and testing requirements. In addition to the information provided in the sections that follow, inspectors should consult the system manufacturer's instructions.

Clean-agent storage containers must meet U.S. Department of Transportation (DOT) and Transport Canada (TC) requirements and be hydrostatically tested every 5 years. Hoses used for local application must be pressure tested annually.

In addition to the system inspections and tests, the protected enclosure must be inspected annually. The inspector must determine that the integrity of the space has not been compromised by penetrations that would permit the agent to escape during a discharge.

Carbon Dioxide Systems

Because of the relative complexity of CO_2 systems, only fire suppression system contractors who are licensed representatives of the system manufacturer should perform maintenance and testing. Nonetheless, inspectors need to be familiar with the requirements for servicing the system, which include inspecting for the following:

- Physical damage to components
- Excessive corrosion
- Change in hazard
- Enclosure integrity
- Up-to-date records to document that required tests, inspections, and maintenance have occurred

Agent cylinders should be checked semiannually and changed if necessary. Hydrostatic testing requirements for storage containers are the same as those for clean-agent fire extinguishing system containers.

Foam Systems

Foam fire extinguishing systems are highly complex systems that require specially trained personnel to inspect, service, and test them. When inspecting these systems, the requirements outlined in NFPA® 25, *Standard for the Inspection, Testing, and Maintenance of Water-Based Fire Protection Systems,* as well as those procedures described by the manufacturer of the system must be followed. Because some state governments may modify standards, inspectors need to check those regulations as well. Inspectors should do the following:

- Check the concentrate tank for signs of sludge, damage, or deterioration. Verify that inspections have been performed and documented.

- Semiannually: Check all valves and alarms attached to the system.
- Annually: Check foam concentrates, foam equipment, and foam proportioning systems. Verify that qualitative tests are performed on the concentrate to certify that no contamination is present.

NOTE: For more information on foam and foam systems, consult IFSTA's **Fire Protection, Detection, and Suppression Systems** and **Principles of Foam Fire Fighting** manuals.

Fire Extinguishers: Selection, Location, Training, and Inspection

Fire extinguishers must be properly located or distributed throughout a building so that they are readily available during an emergency. The main factors that influence proper selection and distribution of fire extinguishers are:

- Nature of the hazard to be protected
- Extinguisher size
- Travel distances

Additional factors in the selection and distribution of fire extinguishers are:

- Potential severity (size, intensity, and rate of fire spread) of any resulting fire
- Personnel available to operate the extinguisher, including their physical abilities, emotional characteristics, and any training they may have in extinguisher use
- Environmental conditions that may affect the use of the extinguisher, such as temperature, winds, and presence of toxic gases or fumes
- Any anticipated adverse chemical reactions between the extinguishing agent and the burning material
- Any health and occupational safety concerns, such as exposure of the extinguisher operator to heat, products of combustion, and extinguishing agent
- Inspection and service required to maintain the extinguishers

Nature of the Hazard

The specific nature of the hazard dictates the type, size, number, and location of fire extinguishers. Some hazards may require an additional extinguisher beyond the distribution required by the locally adopted code. The general method for locating or distributing extinguishers described in NFPA® 10 and used nationally consists of classifying occupancies as light (low) hazard, ordinary (moderate) hazard, or extra (high) hazard. Fire extinguisher distribution is specified on the basis of these classifications.

- **Light-hazard occupancy** — One in which the amount of ordinary combustible material or flammable liquids present is such that an incipient fire of small size may be expected. Such occupancies include classrooms (but not necessarily all parts of a school), churches, and assembly halls **(Figure 13.38, p. 598)**. Some parts of an occupancy that are classified as light hazard may actually be ordinary hazard. Examples of this situation are the shop or storage areas in a school.

- **Ordinary-hazard occupancy** — One in which the amount of ordinary combustibles and flammable liquids present would likely result in an incipient fire of moderate size. In such occupancies, fire growth would not be so rapid as to be beyond the control of fire extinguishers if the fire were discovered quickly. Such occupancies include mercantile storage and display, light manufacturing facilities, parking garages, and warehouses with storage below 12 square feet (1.08 m^2) not classified as extra hazard **(Figure 13.39)**.
- **Extra-hazard occupancy** — One in which the amount of ordinary combustible materials and flammable liquids present are high and a rapidly spreading fire may develop. Examples include automotive repair shops, painting facilities, manufacturing operations that use flammable liquids, restaurants with deep fat fryers, and locations with high-piled storage of combustibles **(Figure 13.40)**.

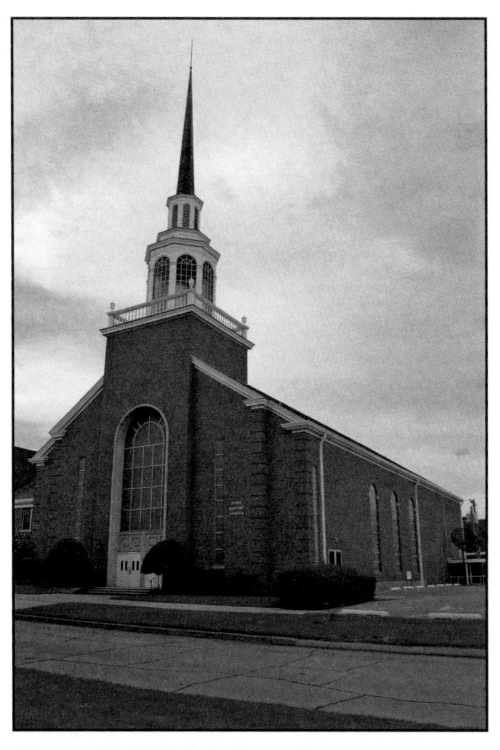

Figure 13.38 A light-hazard occupancy does not contain large amounts of combustibles.

Figure 13.39 An ordinary-hazard occupancy, like this strip mall, contains enough combustibles and flammable liquids that an incipient fire would grow more quickly. *Courtesy of Ron Moore, McKinney (TX) Fire Department.*

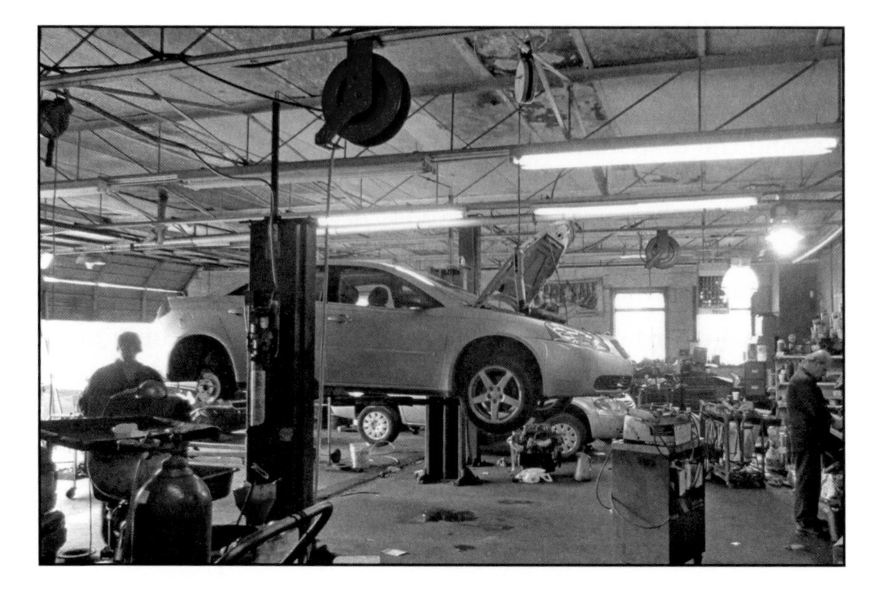

Figure 13.40 Fires can be expected to grow rapidly in an extra-hazard occupancy. *Courtesy of Rich Mahaney.*

Extinguisher Size/Travel Distance

The quantity of burning fuel that a fire extinguisher can extinguish determines its size. The ability to control the class of fire is converted into the maximum area of floor space each extinguisher can protect. See the following descriptions and examples:

- **Class A fuels** — For each occupancy classification, NFPA® 10 recommends the minimum size extinguisher needed and the maximum floor area to be protected by the extinguisher for Class A fuels. **Table 13.3** defines the maximum area per unit of agent that Class A extinguishers of a given rating can protect. In all occupancies, the maximum travel distance to an extinguisher for Class A hazards is 75 feet (25 m).
- **Class B fuels** — Travel distance to the hazard is the most important factor in determining the distribution of Class B extinguishers. Flammable liquid fires develop very rapidly and occur in a variety of situations that are fundamentally different from a fire control standpoint. Two situations deserve special consideration:
  - Spill fire where the flammable liquid does not have depth. According to NFPA® 10, anything less than ¼ inch [6 mm] is considered to be without depth.
  - Flammable liquids with depth (¼-inch [6 mm] at least), such as dip tanks

Table 13.3
Portable Fire Extinguisher Requirements for Class A Fire Hazards

Occupancy Hazard	Minimum Single Extinguisher Rating	Maximum Floor Area Per Unit of A	Maximum Floor Area Per Extinguisher	Maximum Travel Distance To Extinguisher	Notes
Light	2-A	3,000 ft² (278.70 m²)	11,250 ft² (1 045.15 m²)	75 ft (22.86 m)	Two 1-A rated water-type extinguishers may be substituted for one 2-A rated extinguisher
Ordinary	2-A	1,500 ft² (139.35 m²)	11,250 ft² (1 045.15 m²)	75 ft (22.86 m)	None
Extra	4-A	1,000 ft² (92.90 m²)	11,250 ft² (1 045.15 m²)	75 ft (22.86 m)	Two 2.5 gallon (9.46 L) water-type extinguishers may be substituted for one 4-A extinguisher

Based on information in Chapter 9, International Fire Code, 2006 edition

Individual hazards involving flammable liquids with depth are often protected by fixed extinguishing systems. These systems lessen the requirements for portable fire extinguishers in the area but do not eliminate the need for them. A spill fire could occur beyond the effective reach of a fixed system and require the use of portable extinguishers.

- **Class C fuels** — No special spacing rules are required for Class C hazards because fires involving energized electrical equipment usually involve Class A or Class B fuels.

- **Class D fuels** — Placement and distribution cannot be generalized. Determining extinguisher placement for Class D fuels involves making an analysis of the specific metal, the amount of metal present, the configuration of the metal (solid or particulate), and the characteristics of the extinguishing agent. NFPA® 10 recommends only that the travel distance for Class D extinguishers not exceed 75 feet (25 m).
- **Class K fuels** — Fire is always present in the working environment of commercial kitchens. Employees in such areas are responsible for maintaining appropriate cooking temperatures to ensure safety. Because employees are in various levels of training and because fire hazards may be in the restaurant seating area (light hazard), NFPA® 10 has assigned a more restrictive distance requirement. In areas where Class K fires are likely, the maximum travel distance from the hazard to the extinguisher is reduced to 30 feet (10 m).

Training

The effectiveness of portable fire extinguishers is limited by the ability of the building occupants to use them. While the inspector is not responsible for this type of training, the inspector can inquire about how well and how often the occupants are trained in extinguisher use. Training is a service that the fire department can provide if no other source is available. The inspector can recommend that training be given to the occupants and can provide information on who to contact.

Inspection and Maintenance

In most occupancies, fire extinguishers are used so infrequently they tend to be ignored until a fire occurs. Without regular inspections, however, they can be:

- Stolen, misplaced, or obstructed **(Figure 13.41)**
- Damaged as a result of being struck by a vehicle such as a forklift truck **(Figure 13.42)**
- Low in pressure for a variety of mechanical reasons
- Used on a fire and replaced on its mount without being recharged
- Clogged with solid agent that obstructs the discharge hose or nozzle

Periodic fire extinguisher inspections are very important to fire and life safety in a facility. An industrial complex may have hundreds of extinguishers; simply checking to verify that they have not been stolen or vandalized is an important part of plant protection. Portable fire extinguisher inspections are usually performed by building or facility personnel but may be performed by fire suppression or inspection personnel **(Figure 13.43)**. It is important for inspectors to consult local or state licensing/certification requirements for servicing.

NFPA® 10 recommends monthly extinguisher inspections. The owner/occupant is responsible for keeping accurate records of these inspections. The extinguisher inspection tag should feature the inspector's name or initials; bar code readers may also be used to record the inspection. The inspection tag also provides the owner/occupant with chronological data to verify compliance with codes and insurance requirements **(Figure 13.44)**.

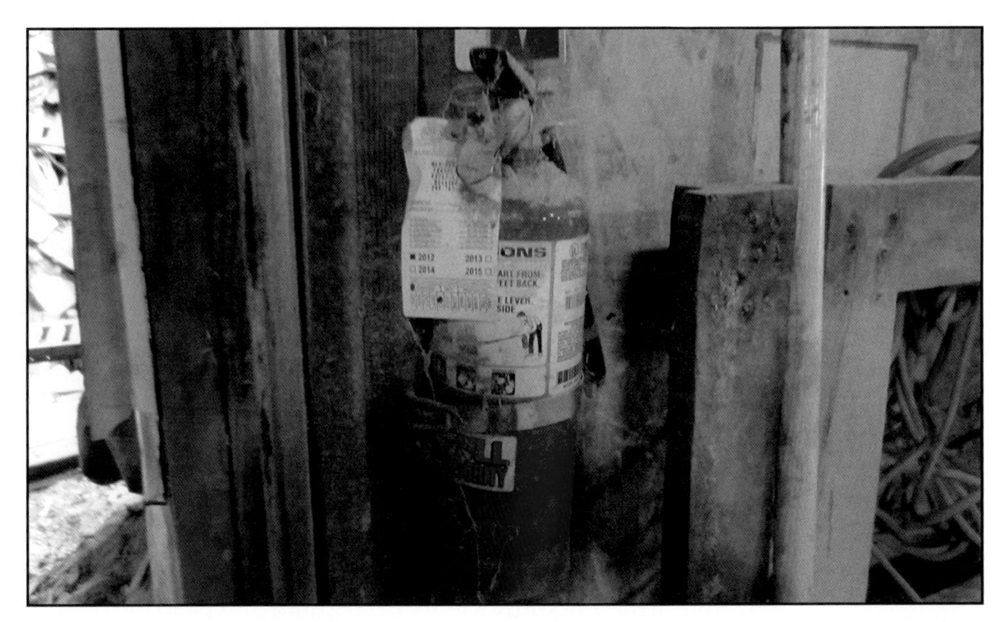

Figure 13.41 This extinguisher is nearly obscured by dust. *Courtesy of Rich Mahaney.*

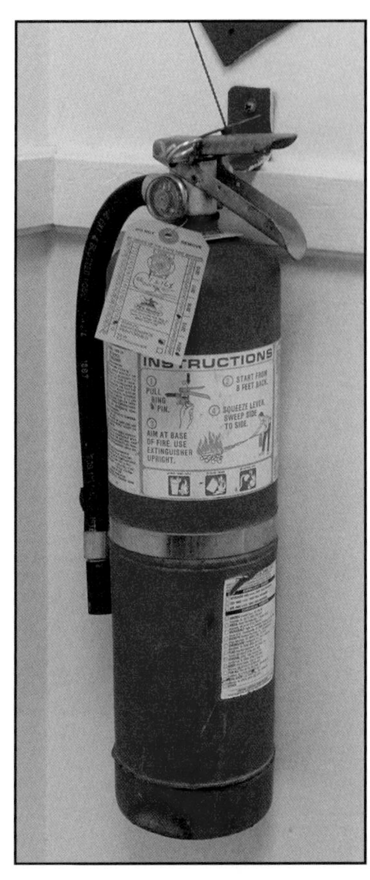

Figure 13.42 Damaged extinguishers need to be taken out of service. *Courtesy of Rich Mahaney.*

Figure 13.43 As with any other part of a fire suppression system, fire extinguishers must be periodically inspected.

In addition to the inspection tag, an extinguisher may have a verification of service collar tag. The service collar tag is made of polyethylene or aluminum and is tightened down against the opening of the extinguisher body. The purpose of this tag is to verify that the extinguisher was actually opened and discharged during maintenance. The service collar tag will have maintenance data impressed into the plastic for the inspector's records. This tag must not be damaged.

While the owner/occupant is responsible for seeing that the monthly inspection performed, the inspector is responsible for verifying that the inspection has occurred. The fire inspector should also look for the same items that the owner/occupant's personnel did:

- Note that extinguisher is in its proper location.
- Verify that access to the extinguisher is not obstructed by boxes, clothing, or storage items.
- Check that the extinguisher is suitable for the hazard protected.

- Check the inspection tag to determine if maintenance is due.
- Examine the nozzle or horn for obstructions.
- Verify that lock pins or tamper seals are intact **(Figure 13.45)**.
- Check for signs of physical damage.
- Check that the extinguisher is fully charged with expellant and agent.
- Check that the pressure gauge indicates proper operating pressure **(Figure 13.46)**.
- Check collar tag for current information and/or damage.
- Check that required signage is in place **(Figure 13.47)**.
- Check to see if the operating instructions on the extinguisher nameplate are legible.

Fire extinguisher maintenance should be performed whenever an inspection reveals the need for maintenance or the unit is due for periodic maintenance required by the manufacturer. Personnel who are trained and certified by the extinguisher manufacturer should perform the actual maintenance. For more information on maintenance intervals and fire extinguisher inspection, see NFPA® 10.

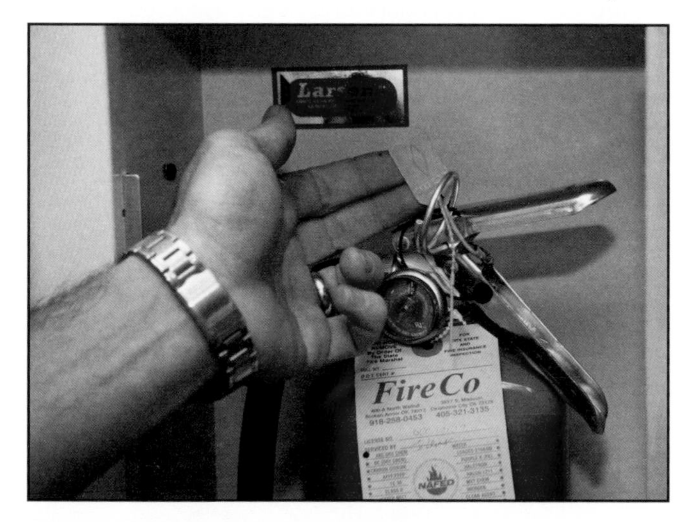

Figure 13.45 Check lock pins or tamper seals during inspections to verify that they are in place and intact.

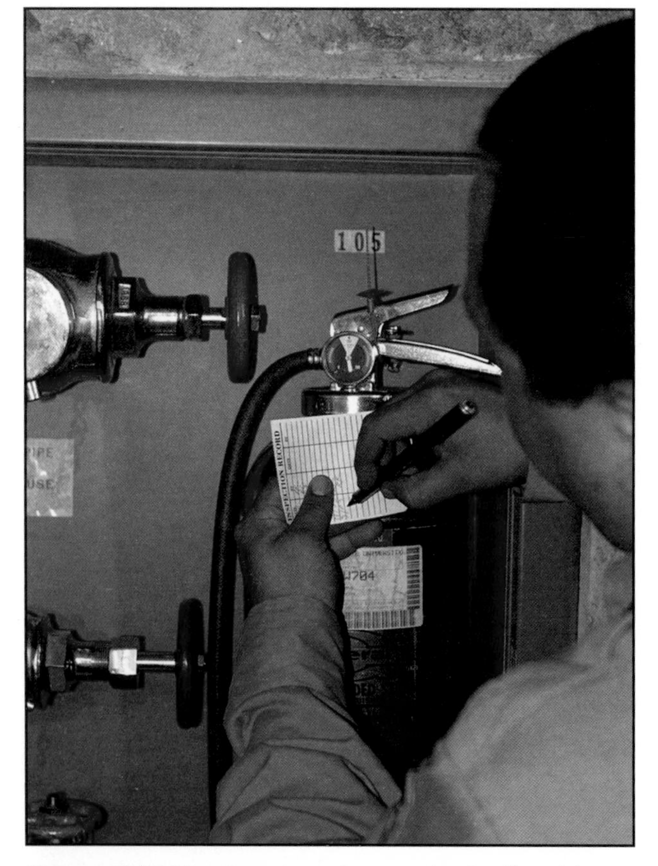

Figure 13.44 Keeping accurate records of extinguisher inspections is very important.

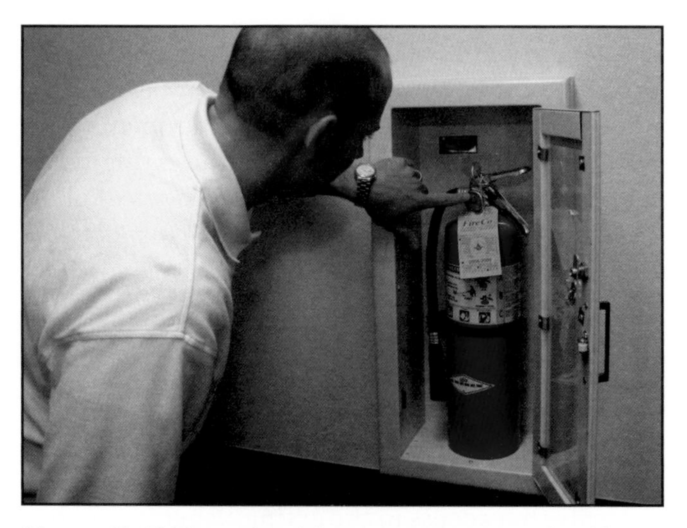

Figure 13.46 The extinguisher pressure gauge should indicate the proper pressure for the type of extinguisher being inspected.

Chapter Summary

Special-agent extinguishing systems and portable fire extinguishers are major components of fire and life safety systems. This chapter provided the fire inspector with an overview of the types of special-agent systems and portable fire extinguishers that might be found during a building or facility fire inspection. The inspector should become familiar with the specific types of special-agent fire extinguishing systems that are found locally as well as the locally adopted codes and standards that govern the distribution and installation of portable fire extinguishers.

Figure 13.47 Equally important as inspecting the extinguishers themselves is verifying that signage indicating the location of the extinguisher is in place and clearly visible.

Review Questions

1. Define the different fire classifications.
2. Name types of dry-chemical application methods.
3. List the components in a dry-chemical extinguishing system.
4. Identify the common agents used in dry powder systems.
5. Differentiate between dry-chemical systems and wet-chemical systems.
6. What components are included in a clean-agent fire extinguishing system?
7. Why are CO_2 systems dangerous to personnel?
8. List the means of actuation for CO_2 systems.
9. Explain how foam proportioners operate.
10. Describe the pictorial symbols used for the different classes of fire.
11. What should properly placed fire extinguishers exhibit?
12. How does an inspector determine if an extinguisher is operational and ready for use?

1. Describe the inspection requirements for dry-chemical systems.
2. What is included in an inspection of wet-chemical systems?
3. List some of the items an inspector should verify when inspecting clean-agent systems.
4. What should inspectors look for when inspecting CO_2 systems?
5. What items do inspectors need to verify when inspecting foam fire suppression systems?
6. What are the main factors that influence the proper distribution of portable fire extinguishers?
7. What should an inspector look for when inspecting portable fire extinguishers?

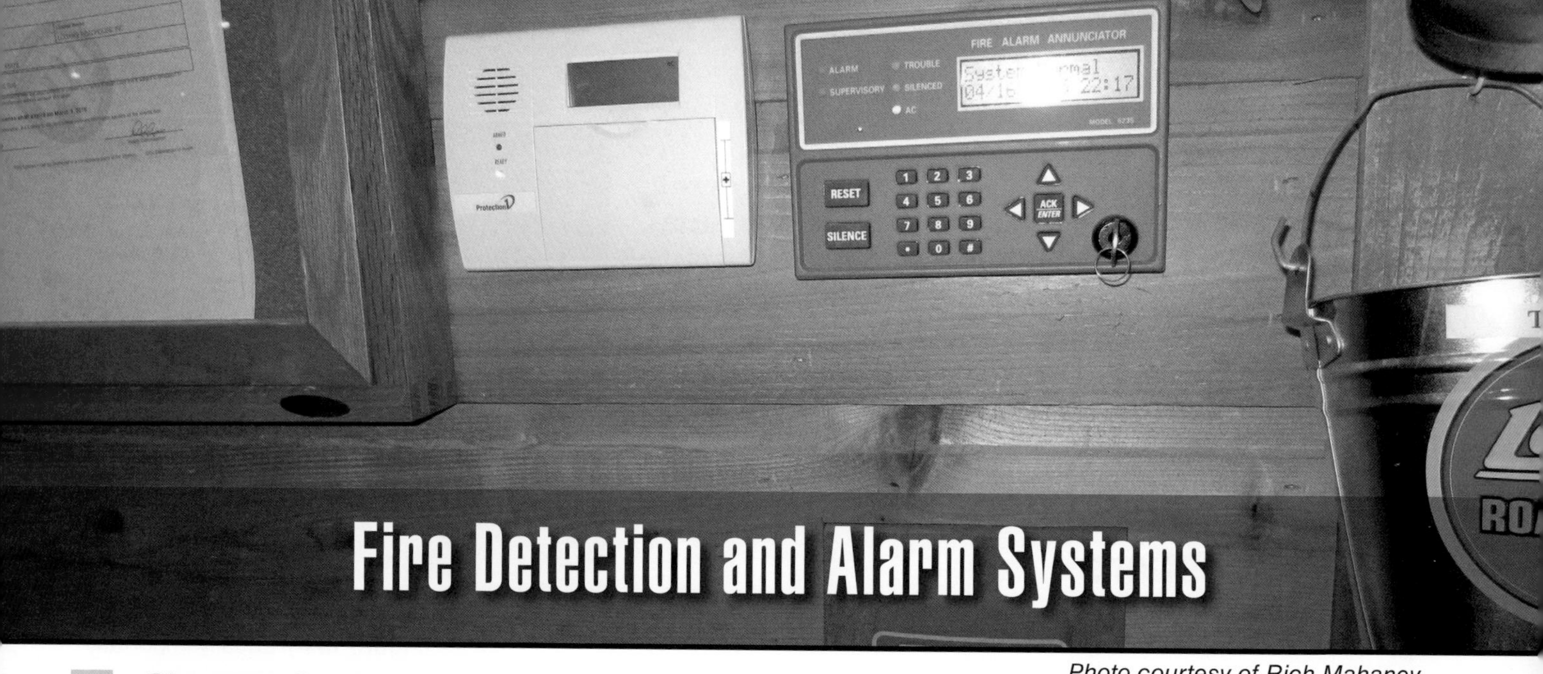

Fire Detection and Alarm Systems

Photo courtesy of Rich Mahaney.

Chapter Contents

Case History 607

Fire Alarm System Components .... 608 (I)

Fire Alarm Control Units 608

Primary Power Supply 609

Secondary Power Supply 609

Initiating Devices 610

Notification Appliances 610

Additional Alarm System Functions 612

Alarm Signaling Systems 613

Protected Premises Systems (Local) 614

Supervising Station Alarm Systems 617

Public Emergency Alarm Reporting Systems 621

Emergency Communications Systems 621

Automatic Alarm-Initiating Devices 623

Fixed-Temperature Heat Detectors 624

Rate-of-Rise Heat Detectors 627

Smoke Detectors 629

Flame Detectors 633

Combination Detectors 634

Sprinkler Waterflow Alarm-Initating Devices 634

Manually Actuated Alarm-Initiating Devices 635

Inspection and Testing 637

Inspection Considerations for Fire Alarm Control Units 638

Inspection Considerations for Alarm-Initiating Devices 639

Inspection Considerations for Alarm Signaling Systems 641

Inspecting, Testing, and Evaluating Fire Detection and Alarm Systems .... 642 (II)

Acceptance Testing 642

Occupant Notification Devices 643

Chapter Summary 647

Review Questions 647

chapter 14

Key Terms

Acceptance Test 642
Alarm Signal 613
Central State System 618
Fire Alarm Control Unit (FACU) 608
Ionization Smoke Detector 632
Mass Notification System (MNS) 622
Photoelectric Smoke Detector 630
Proprietary Alarm System 629
Protected Premises System 614
Public Emergency Alarm Reporting System 621
Remote Receiving Station 619
Service Test 637
Smoke Damper 612
Smoke Detector 629
Supervisory Signal 613
Thermistor 628
Trouble Signal 613

NFPA® Job Performance Requirements

This chapter provides information that addresses the following job performance requirements of NFPA® 1031, *Standard for Professional Qualifications for Fire Inspector and Plan Examiner* (2014).

Fire Inspector I	Fire Inspector II
4.3.6	5.3.4
	5.4.3
	5.44

Fire Detection and Alarm Systems

Learning Objectives

After reading this chapter, students will be able to:

Inspector I

1. Identify fire alarm system components. (4.3.6)
2. Explain types of alarm-signaling systems. (4.3.6)
3. Explain types of automatic alarm-initiating devices. (4.3.6)
4. Describe manual alarm-initiating devices. (4.3.6)
5. Describe service testing and inspection methods for fire detection and alarm systems. (4.3.6)

Inspector II

1. Explain methods to evaluate fire detection and suppression system equipment for life safety, property conservation and hazards. (5.3.4, 5.4.3)
2. Describe inspection and testing methods for fire detection and suppression system equipment. 5.4.4

Chapter 14
Fire Detection and Alarm Systems

Case History

Athens, Georgia, 2014: An investigation by the student newspaper at the University of Georgia revealed numerous expired fire extinguishers on campus. A total of 57 fire extinguishers in the labs, classrooms, and hallways of 19 buildings were checked; 46 of the extinguishers were expired and had not been checked in more than a year. Annual checks for many of the more than 8,000 extinguishers on campus had slowed considerably because of difficulties in contracting with an outside company. After the story came to light, a company was quickly contracted to start the process of checking, servicing, and replacing extinguishers. The Safety Division was also working to implement an electronic database to keep track of the extinguishers on campus.

History has proven that early detection of a fire and the signaling of an appropriate alarm remain significant factors in preventing large losses due to fire. Properly installed and maintained fire detection and alarm systems can help to increase the survivability of occupants and emergency responders while decreasing property losses (**Figure 14.1**).

Together with automatic fire suppression systems, fire detection and alarm systems are part of the active fire protection systems found in many occupancies. To this end, adopted building and/or fire codes may require the installation of fire detection and alarm systems. These systems usually require installation and maintenance by trained individuals.

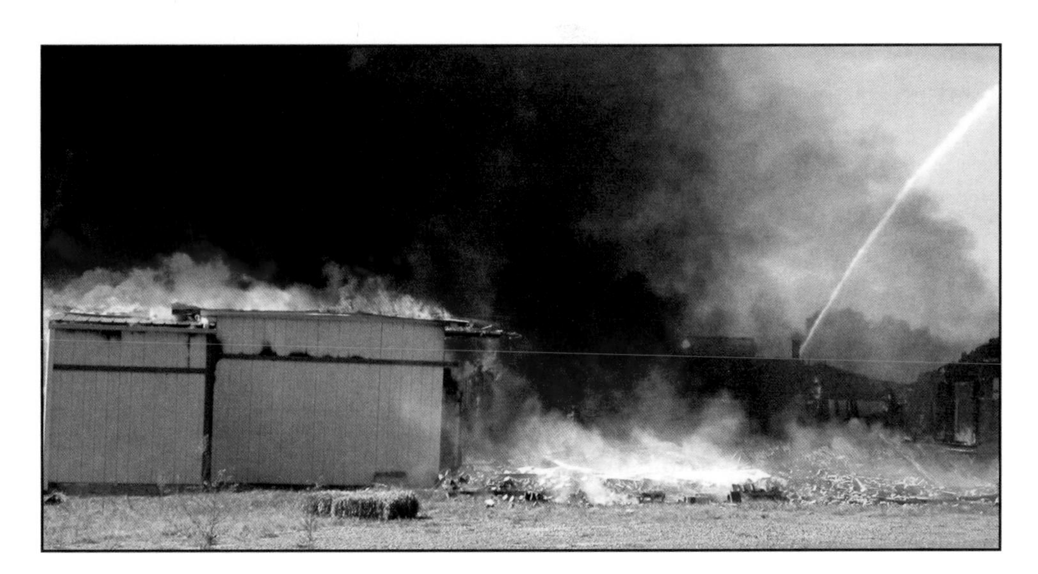

Figure 14.1 Large-loss fires can often be prevented when there is little or no delay between detection and alarm transmission.

This chapter provides information on the fundamental components of fire detection and alarm systems. Addressed in more detail are fire alarm control units, detection and alarm system components, types of signals, alarm-initiating devices, and notification appliances. This chapter also highlights the procedures that fire inspectors or other personnel should follow while inspecting and testing these systems. Also described is the importance of preparing and maintaining accurate records regarding the installation, testing, modification, and maintenance of fire detection and alarm systems.

Fire Alarm System Components

Modern detection and signaling systems vary in complexity from those that are simple to those that incorporate advanced detection and signaling equipment. Such systems are typically designed and installed by qualified individuals as determined by the AHJ. The design, installation, and approval of a fire detection and alarm system may also require acceptance testing by regulatory agencies before new buildings are occupied or the system is placed in service.

The design and installation of the fire detection and alarm system should conform to applicable provisions of NFPA® 70, National Electrical Code®, and NFPA® 72, National Fire Alarm and Signaling Code, and locally adopted codes and ordinances. Other standards also apply to the installation of these systems and are addressed later in this chapter within the discussions of the various types of systems. Each of the following sections highlights a basic component of a fire detection and alarm system.

Fire Alarm Control Unit (FACU) — The main fire alarm system component that monitors equipment and circuits, receives input signals from initiating devices, activates notification appliances, and transmits signals off-site. Formerly called the *fire alarm control panel (FACP)*.

Fire Alarm Control Units

The **fire alarm control unit (FACU)**, formerly called the fire alarm control panel (FACP), contains the electronics that supervise and monitor the integrity of the wiring and components of the fire alarm system. The FACU basically serves as the brain for the alarm system **(Figure 14.2)**. It receives signals from alarm-initiating devices, processes the signals, and produces output signals that activate audible and visual appliances. The FACU also transmits signals to an off-site monitoring station when provided. Power and fire alarm circuits are connected directly into this panel. In addition, the remote auxiliary fire control units and notification appliance panels are considered to be part of the fire alarm system and are connected and controlled.

Controls for the system are located in the FACU **(Figure 14.3, p. 610)**. The FACU can also perform other functions, such as:

- Providing two-way firefighter communication
- Providing remote annunciator integration
- Controlling elevators, HVAC, fire doors, dampers, locks, or other fire protection features

The FACU can also provide public address messages and mass notifications alerts through prerecorded evacuation messages or independent voice communications.

NOTE: Some fire alarm control units are designed for both security and fire protection. In these types of systems, fire protection is engineered into the system to assume the highest priority.

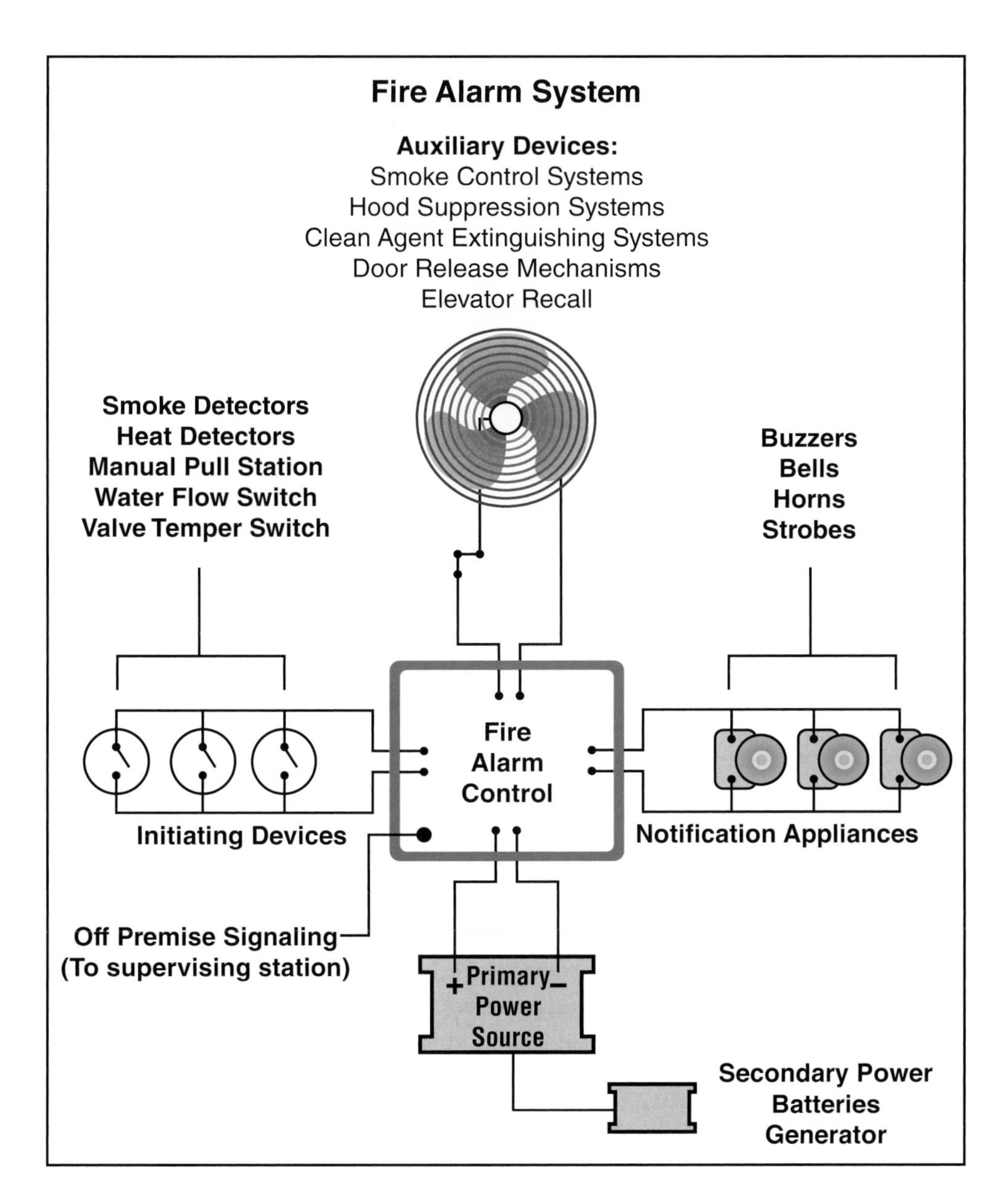

Figure 14.2 This schematic shows the different components of a fire alarm control unit (FACU), the central hub of an alarm system.

Primary Power Supply

The primary electrical power supply usually comes from the building's main power connection to the local utility provider. In rare instances where electrical service is unavailable or unreliable, an engine-driven generator can provide the primary power supply. If such a generator is used, either a trained operator must be on duty 24 hours a day or the system must contain multiple engine-driven generators. One of these generators must always be set for automatic starting. The FACU must supervise the primary power supply and signal an alarm if the power supply is interrupted **(Figure 14.4, p. 610)**.

Secondary Power Supply

All fire alarm systems must have a secondary power supply. This requirement is designed so that the system will be operational even if the main power supply fails. The secondary power supply must be capable of providing normal, (nonalarm) standby conditions capacity and power to fully operate an alarm condition. The time period requirements for secondary power operation capa-

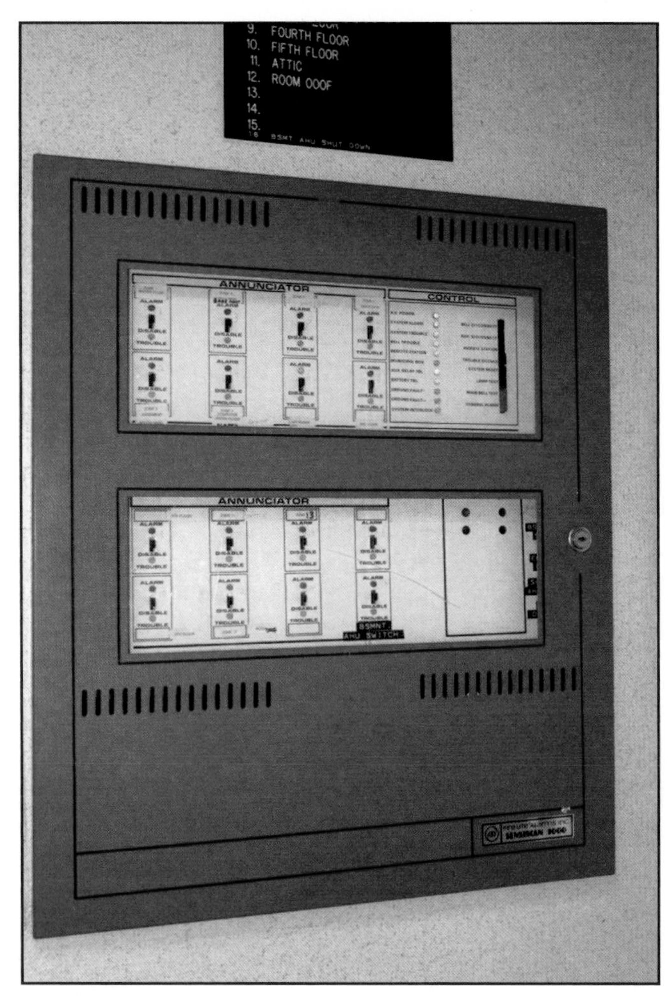

Figure 14.3 An FACU can monitor alarms, control elevators, and public address messages.

Figure 14.4 On the power supply circuit board, one switch should be permanently labeled as the FACU (sticker in the photo) to ensure that anyone doing maintenance on the circuit board does not deactivate that switch and create a safety hazard.

bilities vary and can be found in NFPA® 72. Secondary power sources can consist of batteries with chargers, engine-driven generators with a storage battery, or multiple engine-driven generators, of which one must be set for automatic starting **(Figure 14.5)**.

Initiating Devices

A fire detection system consists of manual and automatic alarm-initiating devices that are activated by the presence of fire, smoke, flame, or heat **(Figure 14.6)**. The devices then send a signal to the FACU using one of two methods: a hard-wire system or a generated signal conveyed by radio wave over a special frequency to a radio receiver in the panel. Both automatic and manual alarm-initiating devices are addressed in more detail in the next sections and include but are not limited to the following devices:

- Manual pull stations
- Smoke detectors
- Flame detectors
- Heat detectors
- Combination detectors
- Waterflow devices

Notification Appliances

Audible notification signaling appliances are the most common types of alarm-signaling systems used for signaling a fire alarm in a structure. Once

Figure 14.5 A backup battery, like the one under the circuit board in the photo, should be available to all components of a fire detection system if primary power is unavailable. *Courtesy of Ron Moore, McKinney (TX) Fire Department.*

Figure 14.6 A ceiling-mounted fire alarm speaker and strobe light combination unit.

an alarm-initiating device is activated, it sends a signal to the FACU, which then processes the signal and initiates actions. The primary action initiated is usually local notification, which can take the form of:

- Bells
- Buzzers
- Horns
- Speakers
- Strobe lights
- Other warning appliances

Depending on the system's design, the local alarm may either activate a single notification appliance, notification appliances within a specific zone, designated floor(s), or the entire facility. Notification appliances fall under the following categories **(Figures 14.7 a-c, p. 612)**:

- **Audible** — Approved sounding devices, such as horns, bells, or speakers, that indicate a fire or emergency condition.
- **Visual** — Approved lighting devices, such as strobes or flashing lights, that indicate a fire or emergency condition.
- **Textual** — Visual text or symbols indicating a fire or emergency condition.
- **Tactile** — Indication of a fire or emergency condition through sense of touch or vibration.

NOTE: Audible appliances are described in more detail in **Appendix H**.

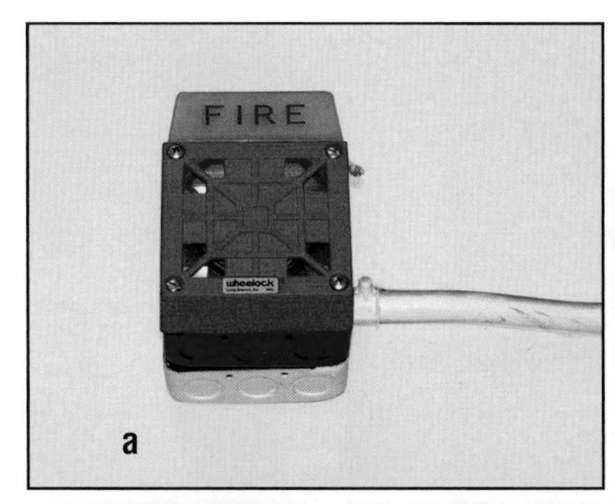

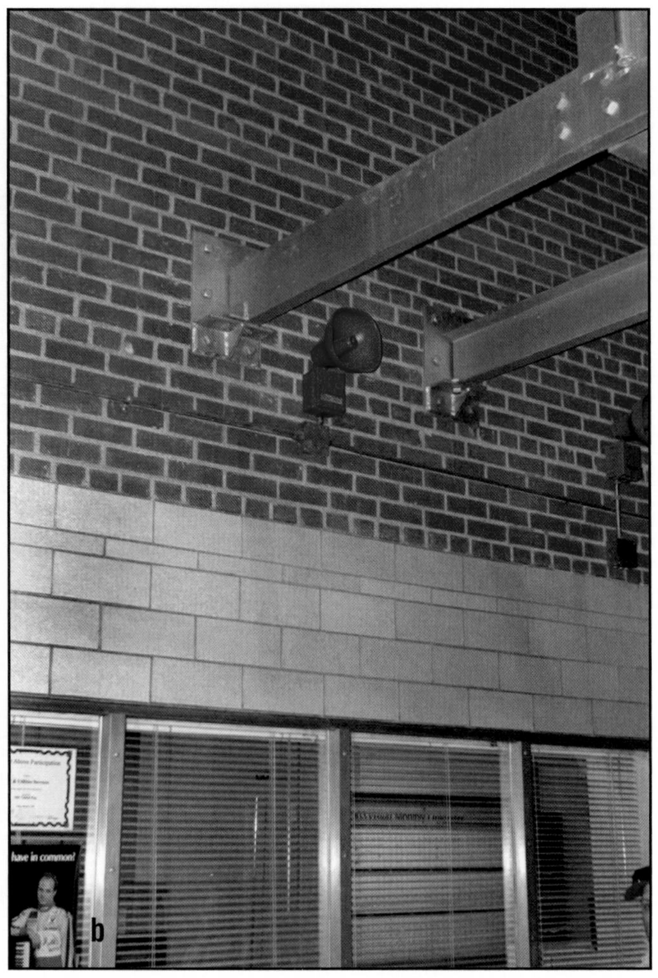

Figures 14.7 a-c Notification devices include bells, horns, strobe lights, and speakers.

Additional Alarm System Functions

Building codes have special requirements for some types of occupancies in case of fire. In these cases, the fire detection and alarm system can be designed to initiate the following actions:

Smoke Damper — Device that restricts the flow of smoke through an air-handling system; usually activated by the building's fire alarm signaling

- Turn off the heating, ventilating, and air-conditioning (HVAC) system
- Close **smoke dampers** and/or fire doors **(Figure 14.8)**
- Pressurize stairwells and/or operate smoke control systems for evacuation purposes
- Unlock doors along the path of egress
- Provide elevator recall to the designated floor and prevent normal operations **(Figure 14.9)**.
- Operate heat and smoke vents
- Activate special fire suppression systems, such as preaction and deluge sprinkler systems or a variety of special-agent fire extinguishing systems

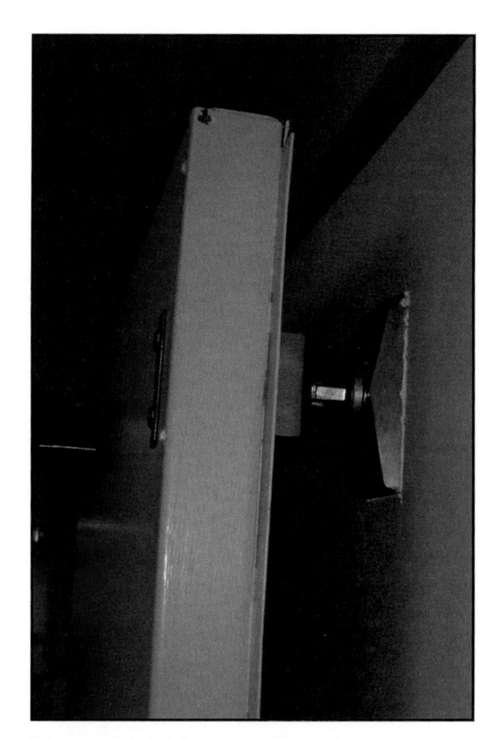

Figure 14.8 Magnetic door closures like this one are designed to remain open during normal building activity and release when alarms are activated to close the door.

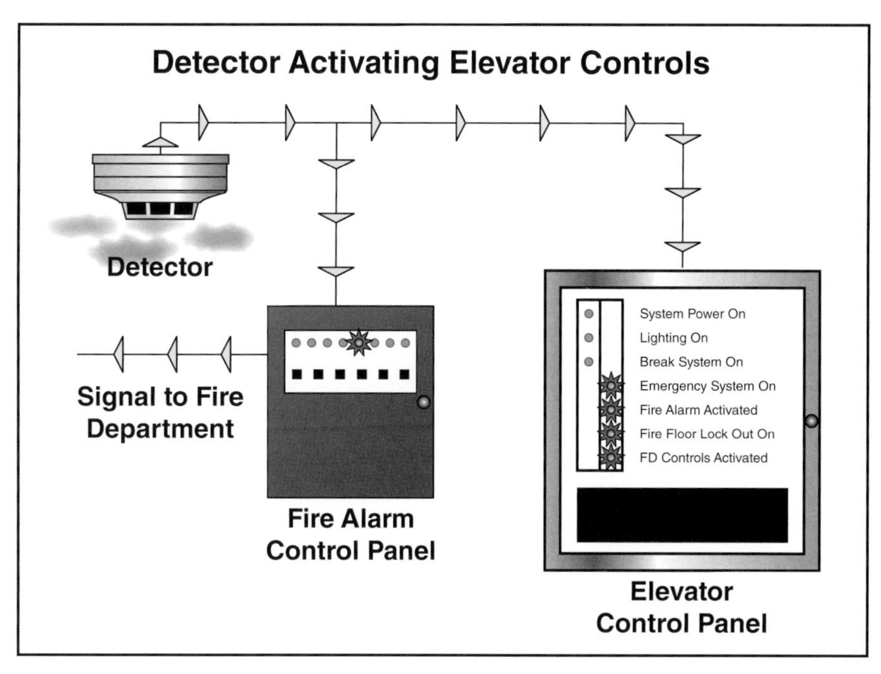

Figure 14.9 In this scenario, the activation of the fire detection system has caused an override of the elevator controls so that the elevator can be used by firefighters.

Alarm Signaling Systems

Fire detection and alarm systems are designed to receive certain types of signals from devices and perform an action based upon the type of signal received. Some signals may indicate a fire condition, while others may indicate that a device on the system needs to be serviced. The FACU should be programmed to respond to different signal types in an appropriate manner.

Fire detection and alarm systems are equipped with three types of specialty signals, depending on the type and nature of the alarm they are reporting:

- An **alarm signal** is a warning of a fire emergency or dangerous condition that demands immediate attention. Locally adopted codes may require fire alarm signals from systems monitored by a supervising station to notify the responding fire department. Activation of smoke detectors, manual pull stations, waterflow switches, and other fire extinguishing systems are all initiating devices that send fire alarm signals.
- A **supervisory signal** indicates an off-normal condition of the complete fire protection system. Supervisory signals also include a returned-to-normal signal, meaning that the condition has been resolved. These signals are used to monitor the integrity of the fire protection features of the system.
- A **trouble signal** indicates a problem with a monitored circuit or component of the fire alarm system or the system's power supply. Each signal must be audibly and visually displayed at the FACU in a distinct manner that differentiates one type of signal from another. Trouble conditions include loss of primary power or failure or removal of an initiating device, such as a smoke detector.

Alarm Signal — Signal given by a fire detection and alarm system when there is a fire condition detected.

Supervisory Signal — Signal given by a fire detection and alarm system when a monitored condition in the system is off-normal.

Trouble Signal — Signal given by a fire detection and alarm system when a power failure or other system malfunction occurs.

NOTE: A trouble signal indicates a problem with the fire detection and alarm system. A supervisory alarm indicates a problem with an accessory of the fire alarm system.

Tamper Switches

Tamper switches are devices used to supervise indicating valves installed in a fire protection water supply system. These switches monitor water shutoff valves that supply the sprinkler system. If a water shutoff valve within the water-based fire protection system — such as an automatic fire sprinkler system, standpipe, static tank, or fire pump — is closed, a supervisory signal is annunciated at the FACU and transmitted to a supervising station.

A simple fire alarm system may only sound a local evacuation alarm. A more complex system may sound a local alarm, activate building services, and notify fire and security agencies to respond. The type of system required depends upon the type of occupancy of the building and is affected by the following factors:

- Level of life safety hazard
- Structural features of the building
- Hazard level presented by the contents of the building
- Availability of fire suppression resources, such as water supply, hydrants, and automatic sprinkler systems
- State and local code requirements

Inspectors should be able to recognize each type of system and understand how each system operates. This recognition is important when performing inspections or conducting preincident planning. Several types of systems include the following:

- Protected premises (local)
- Supervising station alarm systems
- Public emergency alarm reporting system
- Emergency communications systems

Both emergency communications systems and parallel telephone systems may be found in buildings with certain occupancies or building types. Mass notification systems are a special type of emergency communications systems that may be found as a part of a building's alarm system to provide specific and detailed instructions to a building's occupants. NFPA® 72 contains the requirements for all fire alarm and protective signaling systems and should be consulted for further information. The following section describes the major systems.

Protected Premises System — Alarm system that alerts and notifies only occupants on the premises of the existence of a fire so that they can safely exit the building and call the fire department. If a response by a public safety agency (police or fire department) is required, an occupant hearing the alarm must notify the agency.

Protected Premises Systems (Local)

A **protected premises system** is designed to provide notification to building occupants only on the immediate premises **(Figure 14.10)**. Where these systems are allowed, there are no provisions for automatic off-site reporting.

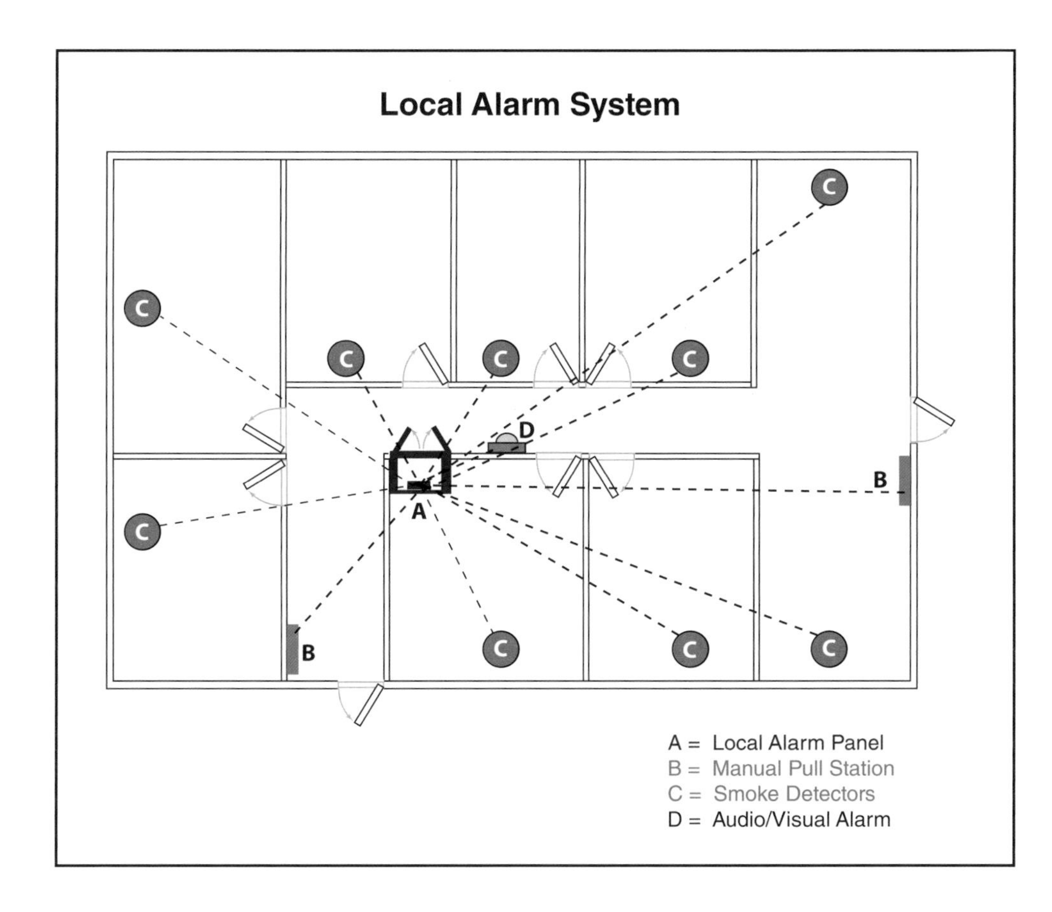

Figure 14.10 A local alarm system alerts occupants of one building to an incident occurring on that property.

The protected premises system can be activated by manual means, such as a pull station, or by automatic devices, such as smoke detectors. A protected premises system may also be capable of annunciating a supervisory or trouble condition to ensure that service interruptions do not go unnoticed.

Presignal Alarms

Presignal alarms are unique systems that may be employed in locations such as hospitals where occupants need assistance to evacuate in a safe and orderly manner. This presignal is usually a separate signal that is recognizable only by personnel who are familiar with the system. The presignal may be a recorded message over an intercom, a soft alarm signal, or a pager notification. Depending on the policies of the occupancy and local code requirements, emergency personnel may elect to handle the incident without sounding a general alarm. Emergency responders may sound the general alarm after investigating the problem, or the general alarm will sound automatically after a certain amount of time has passed and the fire alarm control unit has not been reset.

Conventional Alarm Systems

A conventional system is the simplest type of protected premises alarm system. When an alarm-initiating device, such as a smoke detector, sends a signal to the FACU, all of the alarm-signaling devices operate simultaneously **(Figure 14.11, p. 616)**. The signaling devices usually operate continuously until the FACU is reset. The FACU is incapable of identifying which initiating device

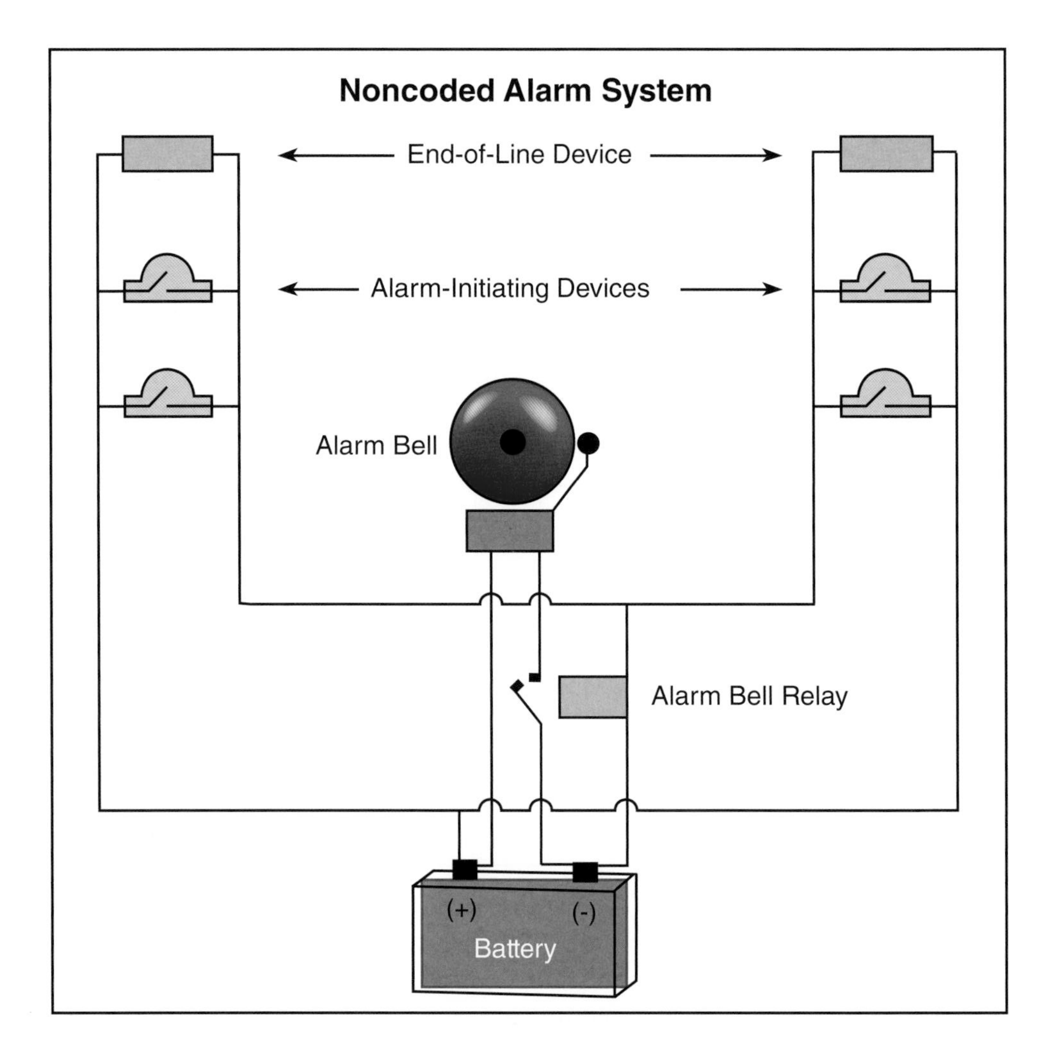

Figure 14.11 In conventional alarm systems, all the individual alarm appliances operate at the same time.

triggered the alarm; therefore, building and fire department personnel must walk around the entire facility and visually check to see which device was activated. These systems are only practical in small occupancies with a limited number of rooms and initiating devices.

An FACU serves the premises as a local control unit. This system is found in occupancies that use the alarm signals for other purposes. In the past, schools sometimes used the same bells for class change as for fire alarms. The FACU enables the fire alarm to have a sound that is distinct from class bells, eliminating confusion as to which type of alarm is sounding. Modern codes do not allow systems such as these; however, older systems that do are still encountered.

Zoned Conventional Alarm Systems

Fire-alarm system annunciation enables emergency responders to identify the general location, or zone, of alarm device activation. In this type of system, an annunciator panel, FACU, or a printout visibly indicates the building, floor, fire zone, or other area that coincides with the location of an operating alarm-initiating device **(Figure 14.12)**.

Alarm-initiating devices in common areas are arranged in circuits or zones. Each zone has its own indicator light or display on the FACU. When an initiating device in a particular zone is triggered, the notification devices are activated, and the corresponding indicator is illuminated on the FACU. This signal gives responders a better idea of where the problem is located.

NOTE: A zone is a defined area within the protected premises.

An annunciator panel may be located remotely from the FACU, often in a location designated by the fire department. Such an installation may be found at the driveway approach to a large residential retirement complex, for example. This type of annunciator panel usually has a map of the complex coordinated to the zone indicator lamps. Arriving firefighters use the information provided on the annunciator panel to locate the building involved. Another type of annunciator panel may be found in the lobby area of a building. It will have a graphic display of the involved area.

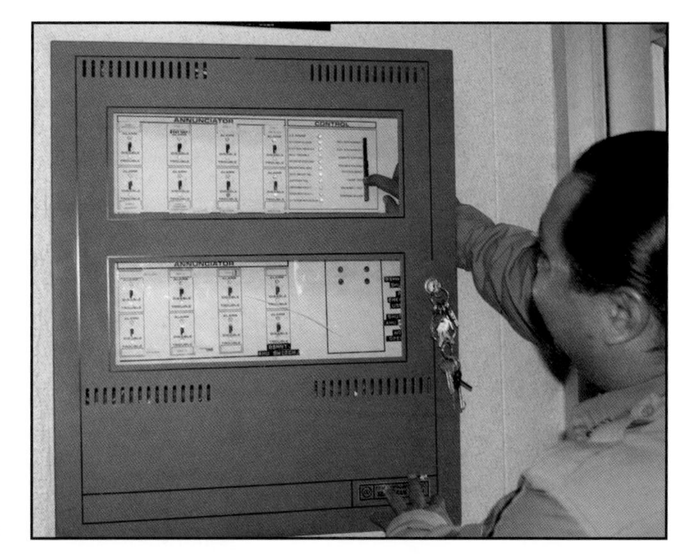

Figure 14.12 Zoned systems are designed to help emergency responders quickly identify the location of a fire.

Older Coded Fire Alarm Systems

A coded system may also be equipped with its own indicator lamp on the FACU, a signal-coding device that is placed into the circuit. This device causes the signaling devices to sound in a unique pattern for each zone. This pattern enables employees or fire department personnel to determine the problem zone by listening to the pattern of the alarms. This is not permitted in new installations, but may exist in older installations.

Usually, the audible pattern consists of a series of short rings, a brief pause, and a second series of short rings, followed by a long pause. The cycle then repeats. Most systems are designed to sound the alarm for the first zone that comes in and then disregard all subsequent zones that initiate alarm signals.

Figure 14.13 Addressable alarms display the location of each initiating device.

Addressable Alarm Systems

Addressable alarm systems display the location of each initiating device on the FACU and an annunciator panel if provided (**Figure 14.13**). This connection enables emergency responders to pinpoint the specific device that has been activated. Addressable systems reduce the amount of time that it takes to respond to emergency situations. These systems also allow repair personnel to quickly locate and correct malfunctions in the system.

Supervising Station Alarm Systems

Fire alarm systems are required by model fire codes to be monitored at a constantly attended location. For buildings that are not constantly attended by qualified personnel, initiating device signals are required to be transmitted to a supervising station. A supervising station is a facility that receives signals from a protected premises fire alarm system and where the signal is processed by personnel.

NFPA® 72 designates supervising stations as:

- **Central** — A central supervising station is an independent business that is also listed by a nationally recognized testing laboratory. A central station is recognized as the most reliable type of supervising station.
- **Proprietary** — A proprietary supervising station is a supervising station under the same ownership as the buildings protected by the fire alarm systems. At a proprietary supervising station, personnel are constantly in attendance to supervise and investigate fire alarm system signals.
- **Remote** — A remote supervising station is not listed and operates as a business. Personnel are in attendance at all times to supervise and investigate signals.

Central Station System — Alarm system that transmits a signal to a constantly attended location (central station) operated by an alarm company. Alarm signals from the protected property are received in the central station and are then retransmitted by trained personnel to the fire department alarm communications center.

Central Station System

A **central station system** is a listed supervising station that monitors the status of protected premises alarm systems and provides inspection, testing, and maintenance through contracted services **(Figure 14.14)**.

Typically, a central station is a company that sells its services to many customers. When an alarm is activated at a particular client's location, central station employees receive that information and contact the fire department and representatives of the occupancy. The alarm systems at the protected property and the central station are most commonly connected by dedicated telephone lines. All central station systems should meet the requirements set forth in NFPA® 72. When meeting the listing requirements, central stations must be listed by UL.

The primary difference between a central station system and a proprietary system is that the receiving point for alarms in a central station system is located outside the protected premises and is monitored by a contracted service **(Figure 14.15)**. The external receiving point is called the central station.

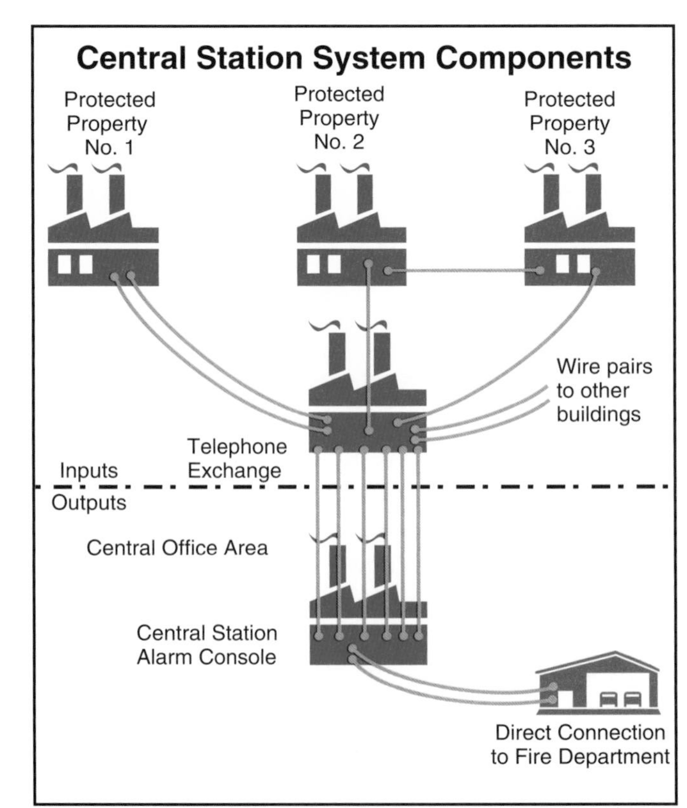

Figure 14.14 Multiple properties can be monitored in a central station system.

Figure 14.15 Central station receiving sites are housed in an off-site location that receives alarms and routes them to the responding fire department.

Proprietary System

A **proprietary system** is used to protect large commercial and industrial buildings, high-rise structures, and groups of commonly owned facilities, such as a college campus or industrial complex in single or multiple locations. Each building or area has its own system that is wired into a common receiving point that the facility owner owns and operates. The receiving point must be in a separate structure or a part of a structure that is remote from any hazardous operations **(Figure 14.16, p. 620)**.

Proprietary Alarm System — Fire alarm system owned and operated by the property owner.

The receiving station of a proprietary system is continuously staffed by trained personnel who can take necessary actions upon alarm activation. The operator should be able to automatically summon a fire department response through the system controls or by using the telephone. Many proprietary systems and receiving points are used to monitor security functions in addition to fire and life safety functions. Modern proprietary systems can be complex and have a wide range of capabilities, including:

- Coded-alarm and trouble signal indications
- Building utility controls
- Elevator controls
- Fire and smoke damper controls

Remote Receiving System

A listed supervising station that monitors the status of protected premise alarm systems through contracted services is called a **remote receiving station**. Remote receiving stations do not provide inspection, testing, or maintenance services.

Remote Receiving System — System in which alarm signals from the protected premises are transmitted over a leased telephone line to a remote receiving station with a 24-hour staff; usually the municipal fire department's alarm communications center sending an alarm signal to the FACU.

Parallel Telephone and Multiplexing Systems

A parallel telephone system consists of a dedicated circuit between each individual alarm box or protected property and the fire department telecommunications center. NFPA® 72 requires that these telephone systems are not used for any other purpose than to relay alarms. These systems are generally not found today due to the existence of private monitoring firms.

Multiplexing systems allow the transmission of multiple signals over a single line. This type of system allows the alarm initiating devices to be identified individually, as in an addressable system, or in a group through the interaction of the fire alarm control unit with each independent device. Remote devices, such as relays, can be controlled over the same line to which initiating and indicating devices are connected. This connection greatly reduces the amount of circuit wiring needed for large applications.

The control panels for multiplexing systems can range from the simple and relatively inexpensive to the sophisticated and costly. Some multiplex systems have the added advantage of being able to test the performance of the devices, reducing manpower requirements for preventive maintenance.

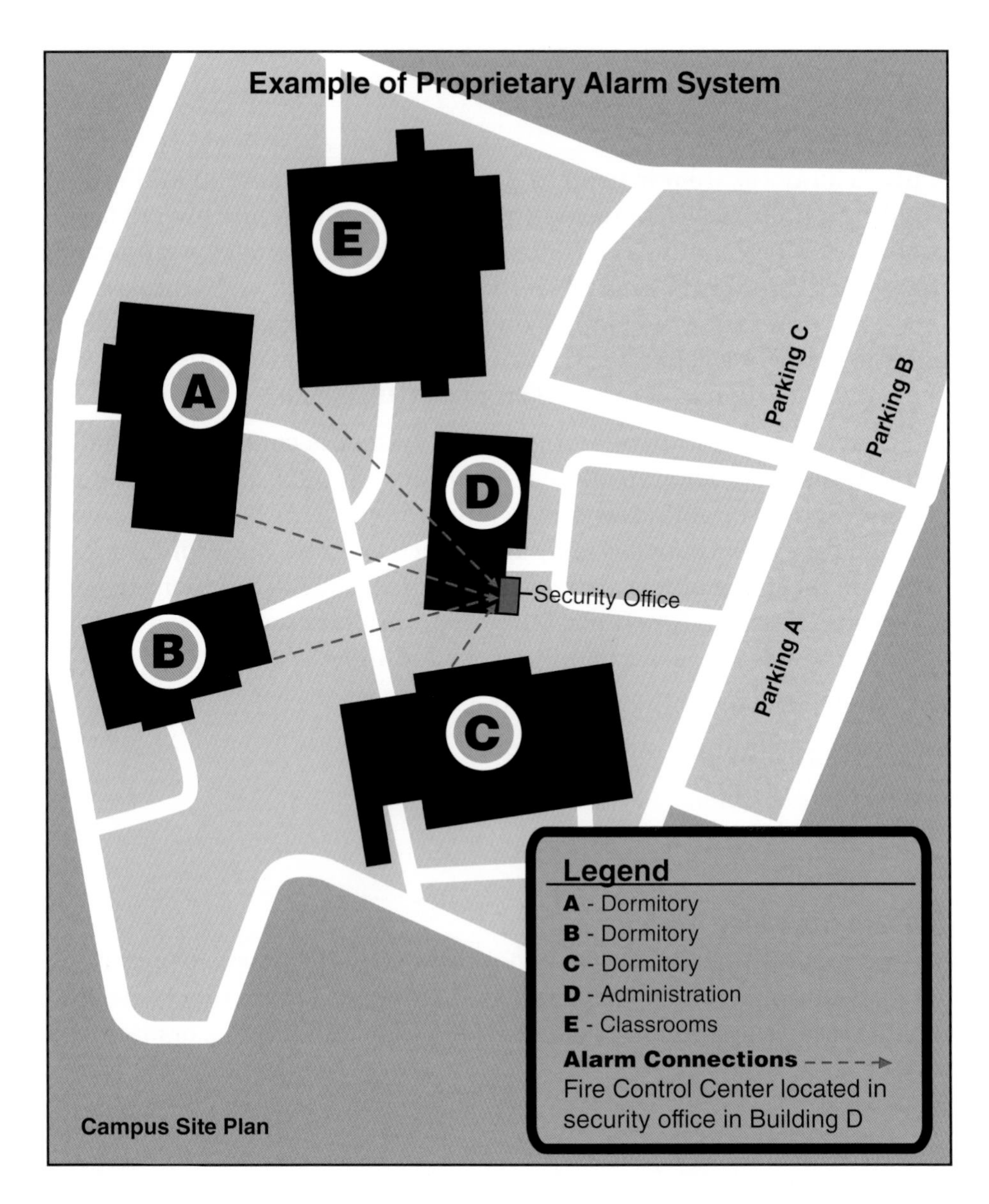

Figure 14.16 A proprietary alarm system is used to monitor several commonly owned facilities, such as a number of university buildings, with one receiving point.

Depending on local requirements, the fire department may approve other organizations to monitor the remote system. In some small communities, the local emergency services telecommunications center monitors the system **(Figure 14.17)**. This arrangement is particularly common in communities that have volunteer fire departments whose stations are not continuously staffed. In these cases, emergency services telecommunications personnel must be aware of the importance of these alarm signals and trained in the actions that must be taken upon alarm receipt.

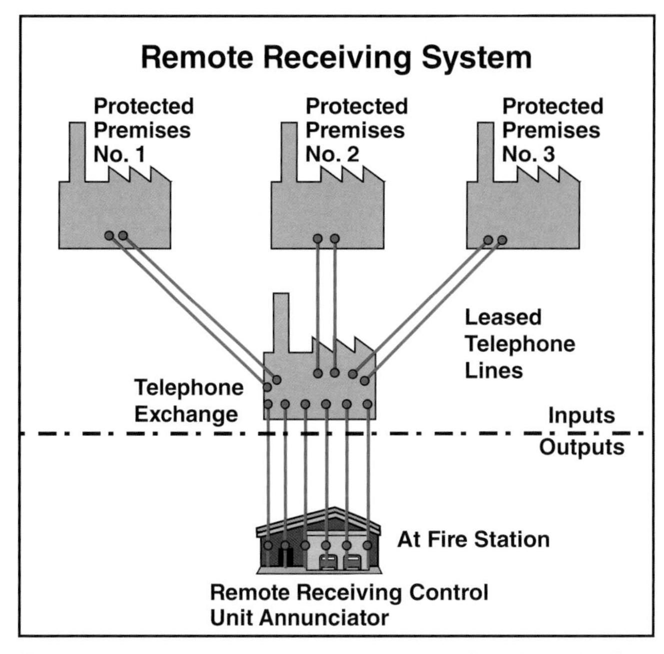

Figure 14.17 A remote receiver connects directly to the fire department dispatch center.

Public Emergency Alarm Reporting Systems

In some communities, fire alarm signals from a protected premises are transmitted directly to the fire department. Instead of being connected to the fire department telecommunications center through a municipal fire alarm box system, the **public emergency alarm reporting system** is connected by another means, usually a leased telephone line. Where permitted, a radio signal over a dedicated fire department radio frequency may also be used.

Public Emergency Alarm Reporting System — System that connects the protected property with the fire department alarm communications center by a municipal master fire alarm box or over a dedicated telephone line.

NOTE: For more information, see the IFSTA **Fire Protection, Detection, and Suppression Systems** manual.

A local energy system has its own power source and does not depend on the supply source that powers the entire municipal fire alarm system. In these systems, initiating devices can be activated even when the power supply to the municipal system is interrupted. However, interruption may result in the alarm only being sounded locally and not being transmitted to the fire department telecommunications center. The ability to transmit alarms during power interruptions depends on the design of the municipal system.

A shunt system is electrically connected to an integral part of the municipal fire alarm system and depends on the municipal system's source of electric power. When a power failure occurs in this type of system, an alarm indication is sent to the fire department communications center. NFPA® 72 allows only manual pull stations and waterflow detection devices to be used on shunt systems. Fire detection devices are not permitted on a shunt system.

Emergency Communications Systems

An emergency communications system is a supplementary system that may be provided in facilities in conjunction with detection and alarm signaling systems. The purpose of emergency communications systems is to provide a reliable communication system for occupants and firefighters. This system may either be a stand-alone system or it may be integrated directly into the overall fire detection and alarm-signaling system. System types include voice notification, two-way communication, and mass notification.

Voice Notification Systems

A one-way voice notification system warns building occupants that action is needed and tells them what action to take. This type is most commonly used in high-rise buildings, places of assembly, and educational occupancies. Occupants can be directed to move to areas of refuge in the building, leave the building, or stay where they are if they are in an unaffected area.

Two-Way Communication Systems

This system is most helpful to fire suppression personnel who are operating in a building, particularly in high-rise structures that interfere with portable radio transmissions. These two-way emergency communication systems use either intercom controls or special telephones **(Figure 14.18)**. Emergency phones are connected in the stairwells and other locations as required by the AHJ, NFPA® 72, or the adopted building code. These phones enable firefighters to communicate with the Incident Commander at the Fire Command Center. Most building codes require these systems in high-rise structures.

Other Emergency Communication Systems

The International Fire Code requires new and existing buildings to be provided with approved radio coverage for emergency responders. Radio coverage inside these buildings must be equivalent to the existing public safety communication capabilities outside the building. Fire department radio transmissions may be augmented by portable or fixed radio repeaters, bi-directional amplifiers, or a leaky coax. Radio repeaters operate by boosting or relaying fire department radio signals in buildings that may shield or disrupt normal high-frequency radio transmissions due to the weakness of these higher frequencies. Leaky coax systems are similar. Rather than boosting or relaying radio signals, these systems simply increase the transmitting capability in these building types by creating a more effective (virtual) antenna that can improve radio communications.

Mass Notification System (MNS) — System that notifies occupants of a dangerous situation and provides information and instructions.

Mass Notification Systems (MNS)

The purpose of a **mass notification system (MNS)** is to provide emergency communications to a large number of people on a wide-scale basis **(Figure 14.19)**. This communication can be directed to the occupants of a building or even an entire community. The events of September 11, 2001, as well as school shootings and other incidents, have provided evidence of the need for this type of system.

While the military was the first to implement this technology, today public and private facilities use it. Mass notification systems may be incorporated into an emergency communications system. Those individuals designing this type of system must take into consideration the building being protected as well as the needs of the occupants. When installed, mass notification systems may have a higher priority and override the fire alarm based on risk analysis. Specifications for mass notification systems are included in NFPA 72® and should be consulted for more information.

Figure 14.19 One-way voice communication systems are commonly used to provide occupants in a structure with instructions during an emergency.

Figure 14.18 An intercom or other two-way system may be used when it is necessary for occupants to respond to officials giving them instructions during an emergency.

Automatic Alarm-Initiating Devices

Automatic alarm-initiating devices, commonly called *detectors*, continuously monitor the atmosphere of a building, compartment, or area. When certain changes in the atmosphere are detected, such as a rapid rise in heat, the presence of smoke, or a flame signature, a signal is sent to the FACU. These signals originate a change-of-state condition like sprinkler water flow, the presence of smoke in a room or area, operation of a manual fire alarm box, or the loss of power to a fire pump.

Avoiding Nuisance Alarms

An inspector should remember that products of combustion may be present when there is no emergency condition. For example, flame detectors may activate if a welder strikes an arc in a monitored area, or smoke detectors may activate due to excessive moisture or atmospheric particles such as dust. These possibilities force fire protection system designers to take into account the normal activities that take place in any given protected area. They then must design a detection system that minimizes the chances of an accidental activation **(Figure 14.20)**.

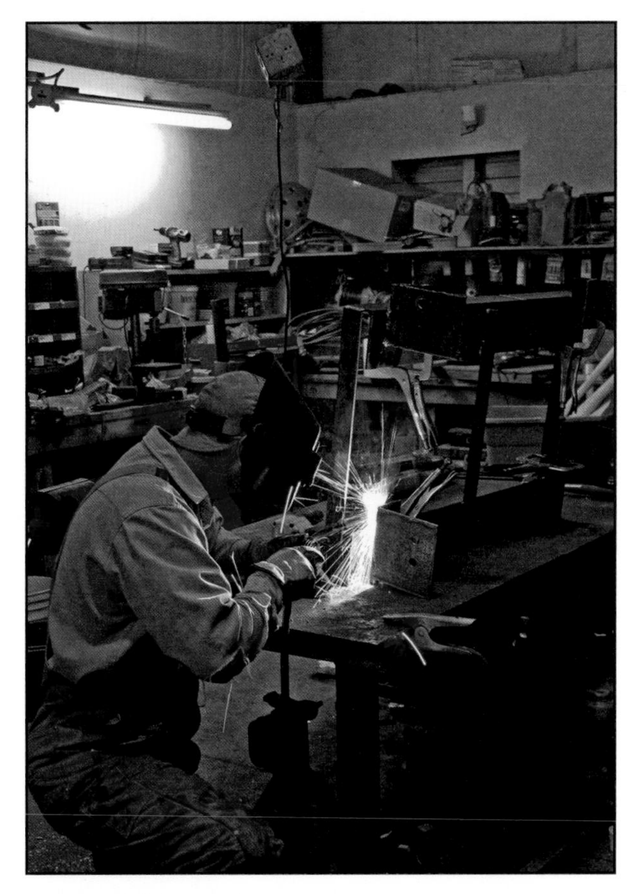

Figure 14.20 Detection systems must take into account expected activities in order to minimize accidental activations.

The basic types of automatic initiating devices are those that detect heat, smoke, flame, and water flow. The two categories of heat detectors are fixed temperature and rate of rise. Devices that are a combination of these basic types are also available. The sections that follow describe the use of these devices:

- Fixed-temperature heat detectors
- Rate-of-rise heat detectors
- Smoke detectors
- Flame detectors
- Combination detectors
- Sprinkler waterflow alarm-initating devices

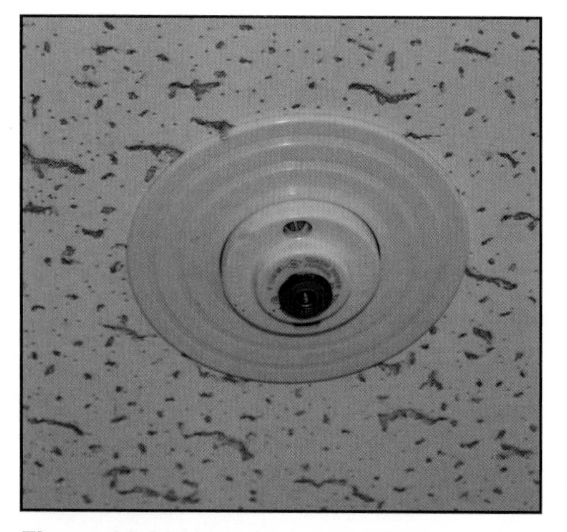

Figure 14.21 Heat detectors set to predetermined temperature ratings should be installed in ceilings in areas that are expected to accumulate heat.

Fixed-Temperature Heat Detectors

Fire detection systems using heat detection devices are among the oldest still in service. They are relatively inexpensive compared to other types of systems, and they are the least prone to nuisance alarms. They are, however, typically slower to activate under fire conditions than other types of detectors.

To be effective, heat detectors must be properly placed where heat is expected to accumulate (**Figure 14.21**). Heat detectors must also be selected at a temperature rating that will give at least a small margin of safety above the normal ceiling temperatures expected in a particular area. According to NFPA® 72, heat-sensing fire detectors must be color-coded and marked with their listed operating temperatures. See **Table 14.1** for a list of colors and temperatures for these detectors.

Because heat is a product of combustion, devices that use one of the following three primary principles of physics can detect the presence of heat:

1. Heat causes expansion of various materials.
2. Heat causes melting of certain materials.
3. Heated materials have thermoelectric properties that are detectable.

Table 14.1
Heat-Sensing Fire Detector Color Coding, Temperature Classification, and Temperature Rating

Color Coding	Temperature Classification	Temperature Rating °F	Temperature Rating °C
Uncolored	Low	100–134	39–57
Uncolored	Ordinary	135–174	58–79
White	Intermedite	175–249	80–121
Blue	High	250–324	122–162
Red	Extra High	325–399	163–204
Green	Very Extra High	400–499	205–259
Orange	Ultra High	500–575	260–302

All heat-detection devices operate on one or more of these principles. The various styles of fixed-temperature devices or detectors used in fire detection systems are addressed in the sections that follow.

Fusible Links/Frangible Bulbs

Although fusible links and frangible bulbs are usually associated with automatic suppression systems, they are also used in fire alarm systems. The operating principles of links and bulbs are the same in either type of system; only their application differs.

Fusible links are used to hold a spring device in the detector in the open position. When the melting point of the fusible link is reached, it melts and drops away. This action causes the spring to release and touch an electrical contact that completes a circuit and sends an alarm signal **(Figure 14.22)**. In order to restore the detector, the fusible link must be replaced.

A frangible bulb is also inserted into a detection device to hold two electrical contacts apart, much like that described for the fusible link. As the temperature increases, the liquid in the bulb expands **(Figure 14.23)**. The expanded liquid compresses the air bubble in the glass, and the bulb fractures and falls away. The contacts close to complete the circuit and send the alarm. In order to restore the detector, either the frangible bulb or the entire detector must be replaced.

Bimetallic Heat Detector

A bimetallic heat detector uses two types of metal with different heat-expansion ratios that are bonded. When subjected to heat, one metal expands faster than the other and causes the combined strip to arch. The amount the strip arches depends on the characteristics of the metals, amount of heat they are exposed to, and degree of arch present when in normal positions. All of these factors are calculated into the design of the detector.

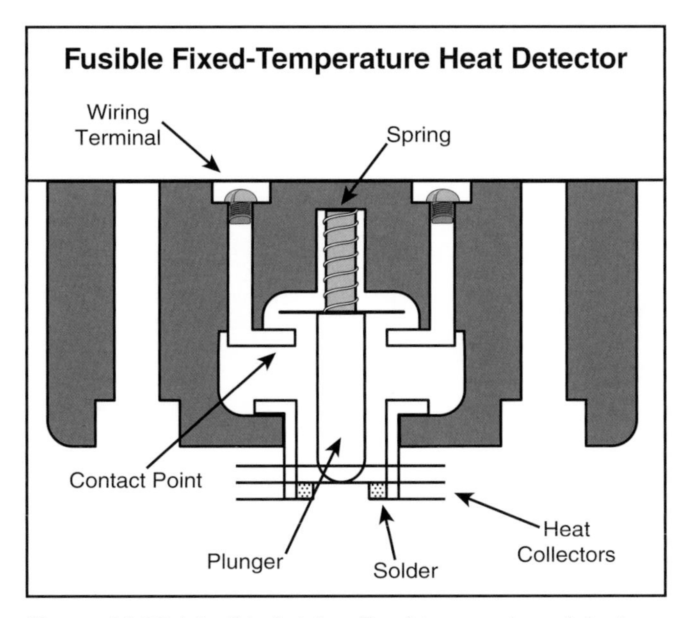

Figure 14.22 A fusible link is a fixed-temperature detector that uses solder with a known melting point to separate a spring from the contact points.

Figure 14.23 Frangible bulb detectors are designed to activate when the glass bulb breaks in response to heat.

A bimetallic strip may be positioned with one or both ends secured in the device. When positioned with both ends secured, a slight bow is placed in the strip. When heated, the expansion causes the bow to snap in the opposite direction. Depending on the design of the device, this action either opens or closes a set of electrical contacts that in turn sends a signal to the FACU **(Figure 14.24)**. Most bimetallic detectors are the automatic-resetting type. They need to be checked, however, to verify that they have not been damaged.

Continuous-Line Heat Detector

Most of the detectors described in this chapter are the spot style; that is, they detect conditions only at the spot where they are located. However, one style of heat detection device, the continuous-line device, can be used to detect conditions over a wide area.

Two models of continuous-line heat detectors are available: One model consists of a conductive metal inner core cable that is sheathed in stainless steel tubing **(Figure 14.25)**. The inner core and sheath are separated by an electrically insulating semiconductor material, which keeps the core and sheath from touching but allows a small amount of current to flow between them. The insulation is designed to lose some of its electrical-resistance capabilities at a predetermined temperature anywhere along the line. When the heat at any given point reaches the resistance-reduction point of the insulation, the amount of current transferred between the two components increases. This increase results in an alarm signal being sent to the FACU. This heat-detection device restores itself when the level of heat is reduced.

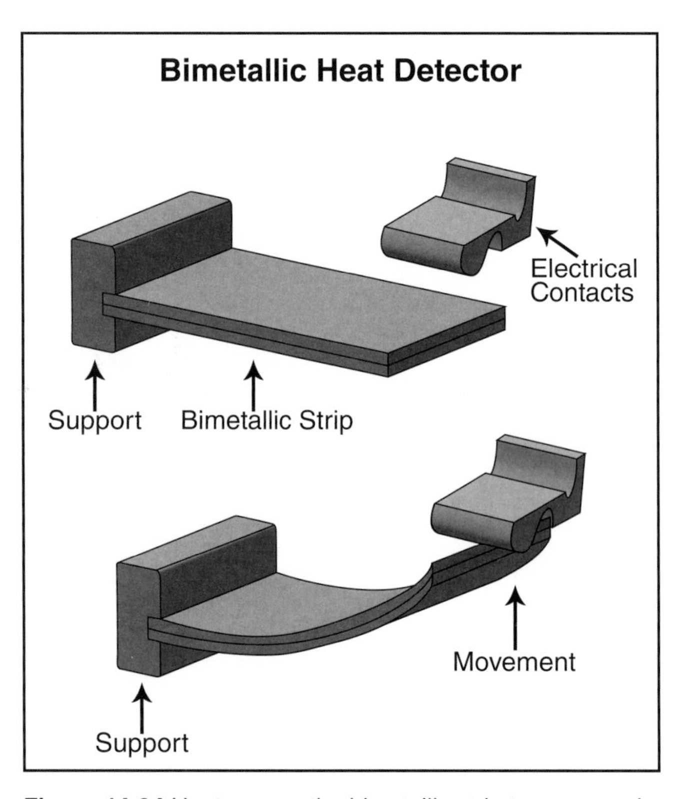

Figure 14.24 Heat causes the bimetallic strip to move and activate the alarm.

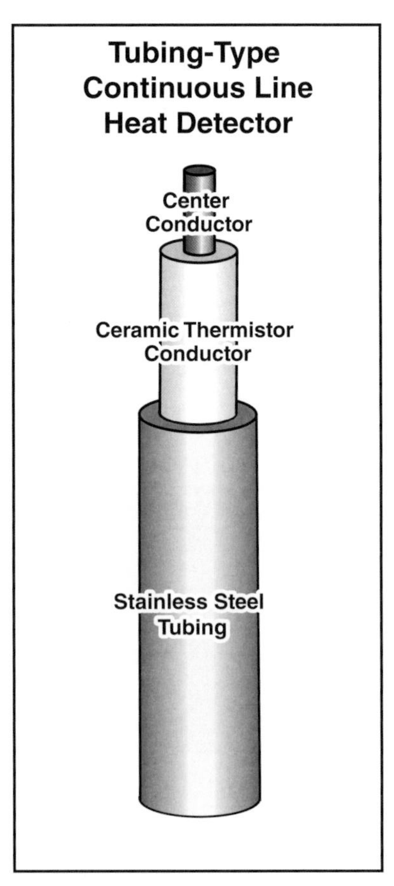

Figure 14.25 Continuous line detectors detect extreme temperatures at any point along the line of the cable.

A second model of continuous-line heat-detection device uses two wires that are each insulated and bundled within an outer covering. When the melting temperature of each wire's insulation is reached, the insulation melts and allows the two wires to touch, which completes the circuit and sends an alarm signal to the FACU (**Figure 14.26**). To restore this continuous-line heat detector, the fused portion of the wires must be removed and replaced with new wire.

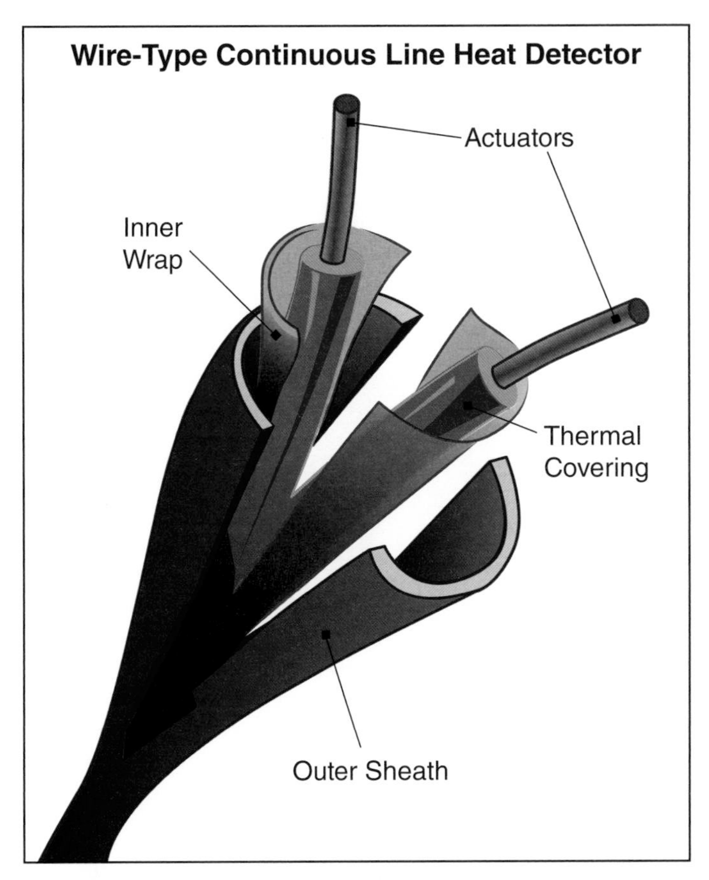

Figure 14.26 Heat causes the insulation on the wires to melt, allowing the wires to touch and complete the circuit.

Rate-of-Rise Heat Detectors

A rate-of-rise heat detector operates on the principle that fires rapidly increase the temperature in a given area. These detectors respond at substantially lower temperatures than fixed-temperature detectors. Typically, rate-of-rise heat detectors are designed to send a signal when the rise in temperature exceeds 12° to 15°F (7°C to 8°C) degrees per minute because temperature changes of this magnitude are not expected under normal, nonfire circumstances.

Most rate-of-rise heat detectors are reliable and not subject to nuisance activations. However, they can occasionally be activated under nonfire conditions. For example, if a rate-of rise detector is placed near a garage door in an air-conditioned building, an influx of summer air when the door opens will rapidly increase the temperature around the heat detector, causing it to activate. Avoiding such improper placement of a heat detector prevents nuisance activations.

All rate-of-rise heat detectors automatically reset. The following different styles of rate-of-rise heat detectors are currently in use:

- Pneumatic rate-of-rise line heat detector
- Pneumatic rate-of-rise spot heat detector
- Rate-compensation heat detector
- Electronic spot-type heat detector

Pneumatic Rate-of-Rise Line Heat Detector

A pneumatic rate-of-rise line heat detector can monitor large areas of a building. Line heat detectors consist of a system of metal pneumatic tubing arranged over a wide area of coverage (**Figure 14.27, p. 628**).

The space inside the tubing acts as a pressurized air chamber that allows the contained air to expand as it heats. These heat detectors contain a flexible diaphragm that responds to the increase in pressure from the tubing. When an area being served by the tubing experiences a temperature increase, the air pressure increases and the heat-detection device operates.

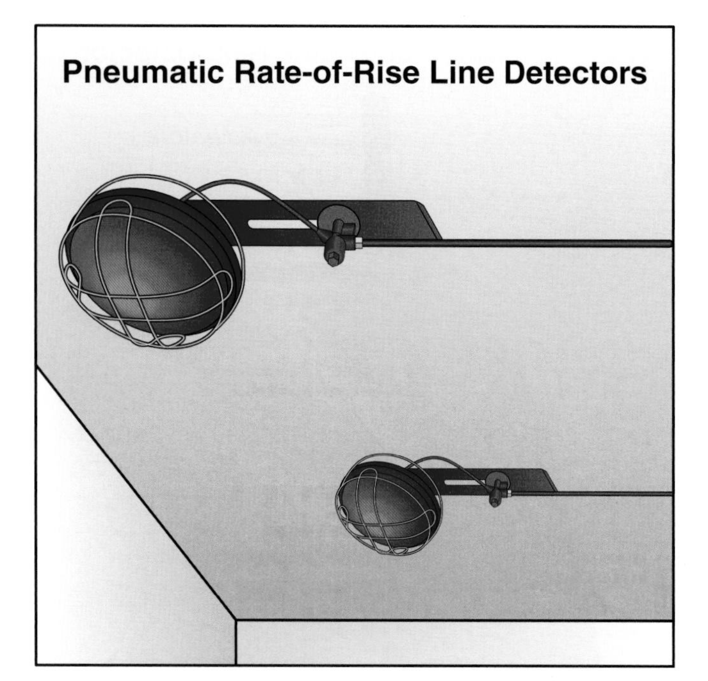

Figure 14.27 Line heat detectors used in pneumatic rate-of-rise systems depend on the change in temperature and increase in pressure of the air in the tubing to activate the alarm system.

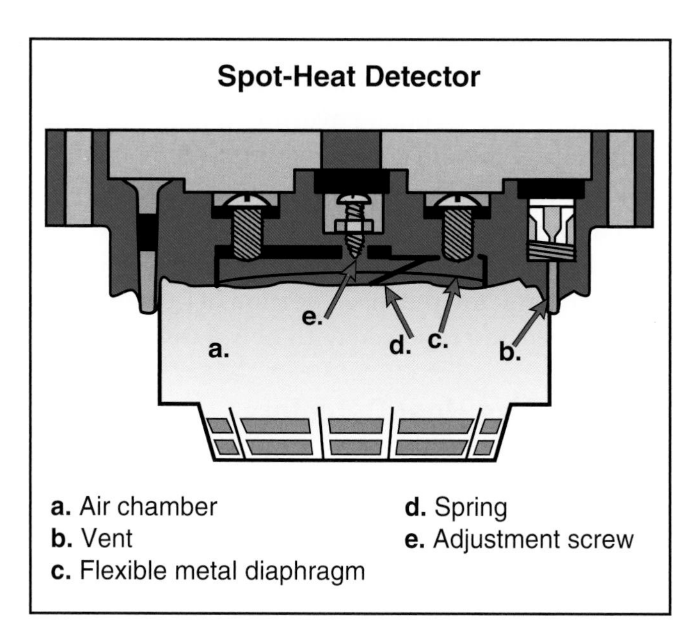

Figure 14.28 A spot detector monitors a specific location and sends a signal to the alarm panel.

Pneumatic Rate-of-Rise Spot Heat Detector

A rate-of-rise spot heat detector operates on the same principle as the pneumatic rate-of-rise line heat detector. The major difference between the two is that the spot heat detector is self-contained in one unit that monitors a specific location **(Figure 14.28)**. Alarm wiring extends from the detector back to the FACU.

Rate-Compensation Heat Detector

This heat detector is designed for use in areas that are subject to regular temperature changes, but at rates that are slower than those of fire conditions. Rate-compensation heat detectors contain an outer bimetallic sleeve with a moderate expansion rate. This outer sleeve contains two bowed struts that have a slower expansion rate than the sleeve. The bowed struts have electrical contacts. In the normal position, these contacts do not touch. When the detector is heated rapidly, the outer sleeve expands lengthwise. This expansion reduces the tension on the inner strips and allows the contacts to meet, thus sending an alarm signal to the FACU.

If the rate of temperature rise is fairly slow, such as 5°F to 6°F (2°C to 3°C) per minute, the sleeve expands at a slow rate that maintains tension on the inner strips. This tension prevents unnecessary system activations.

Thermistor — Semiconductor made of substances whose resistance varies rapidly and predictably with temperature.

Electronic Spot-Type Heat Detector

An electronic spot-type heat detector consists of one or more **thermistors** that produce a marked change in electrical resistance when exposed to heat. The rate at which thermistors are heated determines the amount of current that is generated **(Figure 14.29)**. Greater changes in temperature result in larger amounts of current flowing and activation of the alarm system.

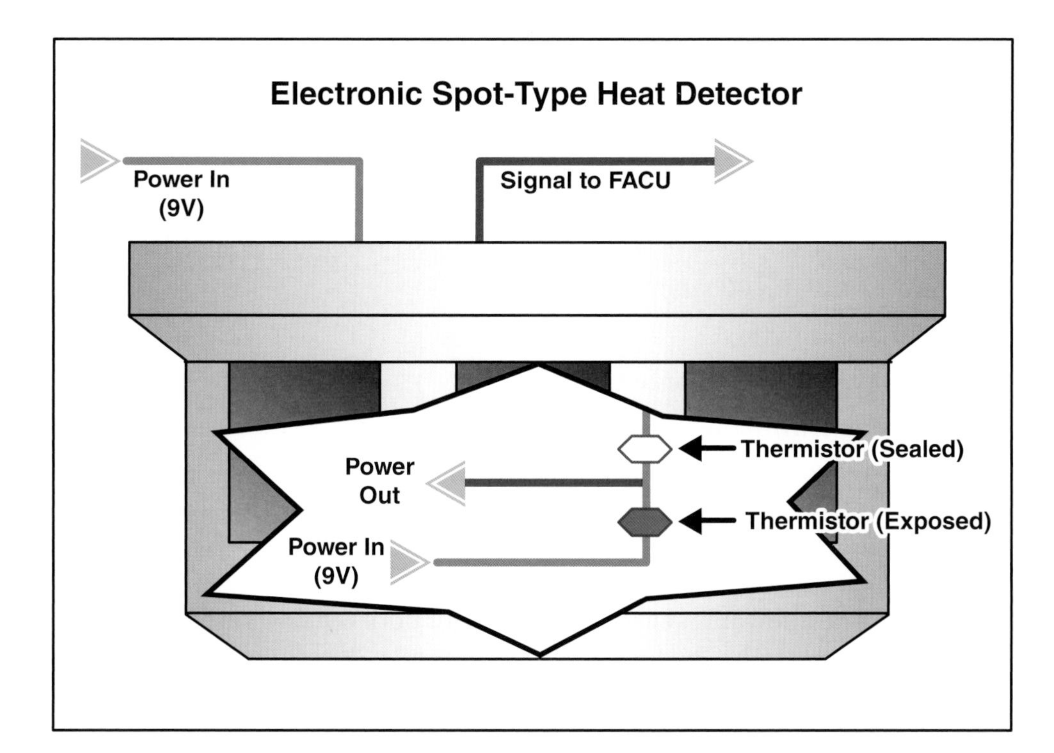

Figure 14.29 The rate at which the temperature of the internal thermistors increases determines the amount of current that is generated to activate an alarm signal.

These heat detectors can be calibrated to function as rate-of-rise (approximately 15°F [8°C] per minute) detectors and function at a fixed temperature. Heat detectors of this type are designed to bleed or dissipate small amounts of current, which reduces the chance of a small temperature change activating an alarm.

Smoke Detector — Alarm-initiating device designed to actuate when visible or invisible products of combustion (other than fire gases) are present in the room or space where the unit is installed.

Smoke Detectors

Smoke detectors serve the purpose of early detection, notification, and reaction. Some detectors are also used to activate mechanical or electrical systems, such as dampers, doors, and electronic shutdown. Smoke detectors have evolved into two principal types of devices:

- A detector that provides early detection and reports back to an alarm panel to initiate evacuation alarms
- A detector that provides some type of signal to initiate one of the actions discussed above.

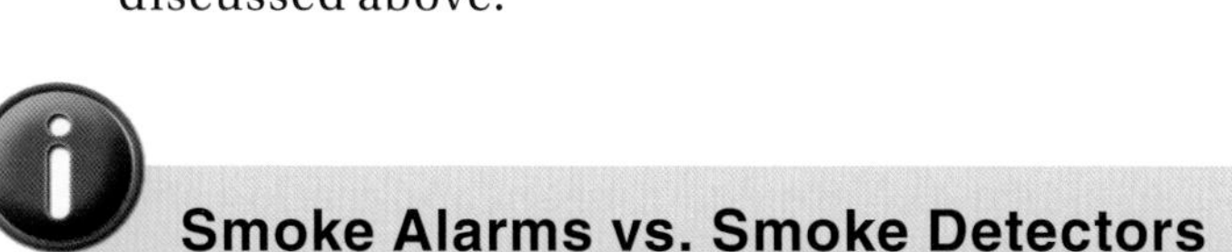

Smoke Alarms vs. Smoke Detectors

The terms *smoke alarm* and *smoke detector* are often used interchangeably. While this is common practice, it is technically incorrect. Smoke alarms are the devices typically installed in residential occupancies. These devices combine a smoke detector with a local notification appliance. When activated, smoke alarms emit an audible alarm to notify occupants of the presence of smoke.

Smoke detectors differ from smoke alarms in that they do not include a local notification appliance. When activated, smoke detectors send a signal to an FACU or a similar device. The FACU then initiates the alarm to notify occupants.

Smoke detection is the preferred automatic alarm device in such occupancy types as residences and health and institutional care facilities because smoke detectors sense the presence of a fire much more quickly than heat-detection devices. Because of the dangers of toxic fire gases, an early warning can mean the difference between a safe escape and no escape at all.

Many factors affect the performance of smoke detectors, including:

- Type and amount of combustibles
- Rate of fire growth
- Proximity of the detector to the fire
- Ventilation within the area involved

Smoke detectors and smoke alarms are tested, certified, and listed based on their performance by third-party testing services. Regardless of their principle of operation, all smoke detectors are required to respond to the same fire tests. Two basic methods of smoke detection are in use: photoelectric and ionization. The allowable sensitivity ranges for both types of smoke detectors are established by UL. The following sections also describe duct and video-based detectors.

Photoelectric Smoke Detector — Type of smoke detector that uses a small light source, either an incandescent bulb or a light-emitting diode (LED), to detect smoke by shining light through the detector's chamber: smoke particles reflect the light into a light-sensitive device called a photocell.

Photoelectric Smoke Detectors

Photoelectric smoke detection works on all types of fires and usually responds more quickly to smoldering fires than ionization smoke detection. Photoelectric smoke detection is best suited for areas containing overstuffed furniture and other areas where smoldering fires can occur.

A photoelectric device consists of a photoelectric cell coupled with a specific light source. The photoelectric cell functions in one of two ways to detect smoke: projected-beam application (obscuration) or refractory application (scattered).

The projected-beam application style of photoelectric detector uses a beam of light focused across the area being monitored onto a photoelectric-receiving device such as a photodiode. The cell constantly converts the beam into current, which keeps a switch open. When smoke interferes with or obscures the light beam, the amount of current produced is lessened. The detector's circuitry senses the change in current, and initiates an alarm when a current change threshold is crossed **(Figure 14.30)**.

Projected-beam application smoke detectors are particularly useful in buildings where a large area of coverage is desired, such as in churches, atriums, or warehouses. Rather than wait for smoke particles to collect at the top of an open area and sound an alarm, the projected-beam application smoke detector is strategically positioned to sound an alarm more quickly.

Projected-beam application smoke detectors need to be mounted on a stable stationary surface. Any movement due to temperature variations, structural movement, and vibrations can cause the light beams to misalign.

A refractory application photoelectric smoke detector uses a beam of light from a light-emitting diode (LED) that passes through a small chamber at a point distant from the light source. Normally, the light does not strike the photocell or photodiode. When smoke particles enter the light beam, light strikes the particles and reflects in random directions onto the photosensitive device, causing the detector to generate an alarm signal **(Figure 14.31)**.

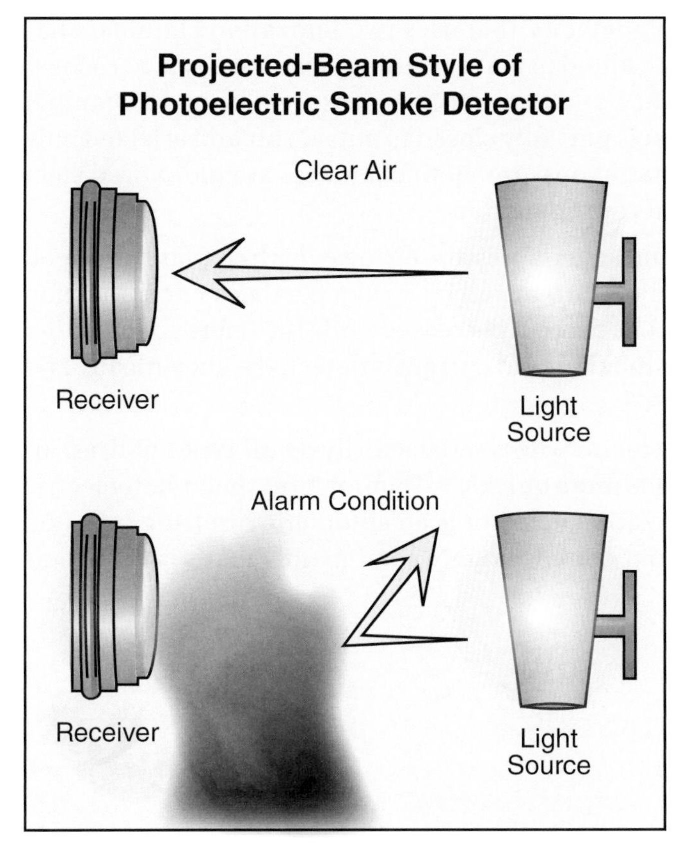

Figure 14.30 A projected-beam photoelectric detector activates when light is blocked by smoke from reaching a sensor.

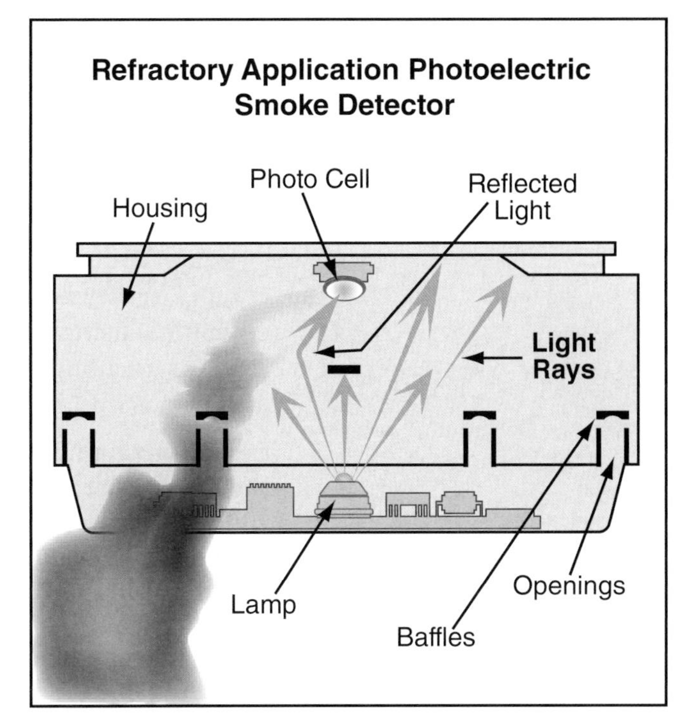

Figure 14.31 A refractory photoelectric smoke sends an alarm when light reaches a sensor after reflecting off smoke.

Ionization Smoke Detector — Type of smoke detector that uses a small amount of radioactive material to make the air within a sensing chamber conduct electricity.

Ionization Smoke Detectors

An **ionization smoke detector** contains a sensing chamber consisting of two electrically charged plates (one positive and one negative) and a radioactive source for ionizing the air between the plates. A small amount of Americium 241 that is adjacent to the opening of the chamber ionizes the air particles as they enter. The ionized particles free electrons from the negative electron plate and the electrons travel to the positive plate. Thus, a small ionization current measurable by electronic circuitry flows between the two plates.

Products of combustion, which are much larger than the ionized air molecules, enter the chamber and collide with the ionized air molecules. As the two interact, they combine and the total number of ionized particles is reduced. This action results in a decrease in the chamber current between the plates. When a predetermined threshold current is crossed, an alarm is initiated (**Figure 14.32**).

Changes in humidity and atmospheric pressure in the room can cause an ionization detector to malfunction and initiate a nuisance alarm. To compensate for the possible effects of humidity and pressure changes,

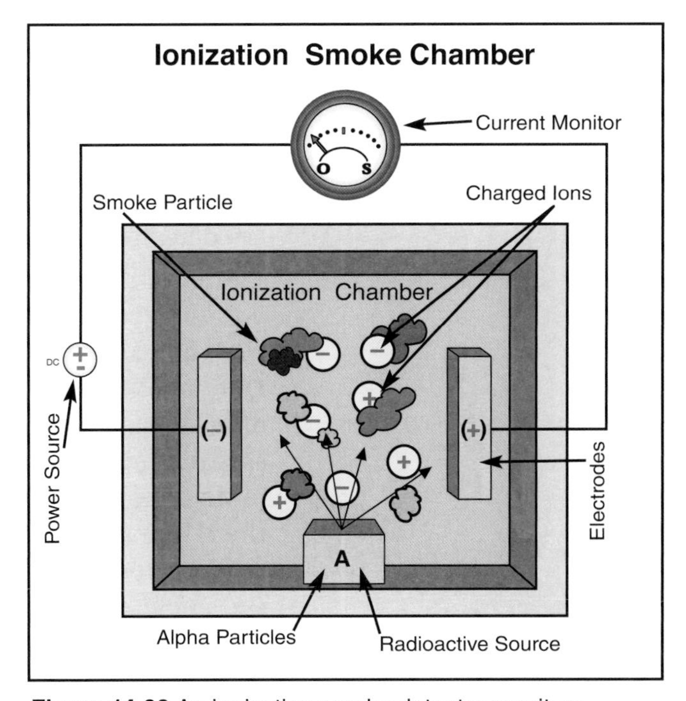

Figure 14.32 An ionization smoke detector monitors charged particles inside the detector.

a dual-chamber ionization detector that uses two ionization chambers has been developed and may be found in many jurisdictions. One chamber senses particulate matter, humidity, and atmospheric pressure. The other chamber is a reference chamber that is partially closed to outside air and affected only by humidity and atmospheric pressure. Both chambers are monitored electronically and their outputs are compared.

When the humidity or atmospheric pressure changes, both chambers respond equally to the change, but remain balanced. When particles of combustion enter the sensing chamber, its current decreases while the reference chamber remains unchanged. The imbalance in current is detected electronically and an alarm is initiated.

An ionization smoke detector works satisfactorily on all types of fires, although it generally responds more quickly to flaming fires than photoelectric smoke detectors. The ionization detector is an automatic resetting type and is best suited for rooms that contain highly combustible materials, such as the following:

- Cooking fat/grease
- Flammable liquids
- Newspapers
- Paint
- Cleaning solutions

An air-sampling smoke detector is a type of ionization detector that is designed to continuously monitor a small amount of air from the protected area for the presence of smoke particles. There are two basic types of air-sampling smoke detectors. The most common one is the cloud-chamber type **(Figure 14.33)**. This detector uses a small air pump to draw sample air into a high-humidity chamber within the detector. The detector then imparts the high humidity to the sample and lowers the pressure in the test chamber. Moisture condenses on any smoke particles in the test chamber, which creates a cloud inside the chamber. The detector triggers an alarm signal when the density of this cloud exceeds a predetermined level.

The second type of air-sampling smoke detector is composed of a system of pipes spread over the ceiling of the protected area **(Figure 14.34)**. A fan in the detector/controller unit draws air from the building through the pipes. The air is then sampled using a photoelectric sensor.

Duct Smoke Detectors

Duct smoke detectors are installed in the return or supply ducts or plenums of HVAC systems to prevent smoke and products of combustion from being spread throughout the building. Duct smoke detectors are specifically listed for installation within higher air velocities. Upon the detection of smoke, the HVAC system will either shut down or transition into a smoke-control mode. The detection of smoke in duct areas is sometimes difficult because the smoke can be diluted by the return air from other spaces or outside air. Duct smoke detectors are no substitute for other types of smoke detectors in open areas.

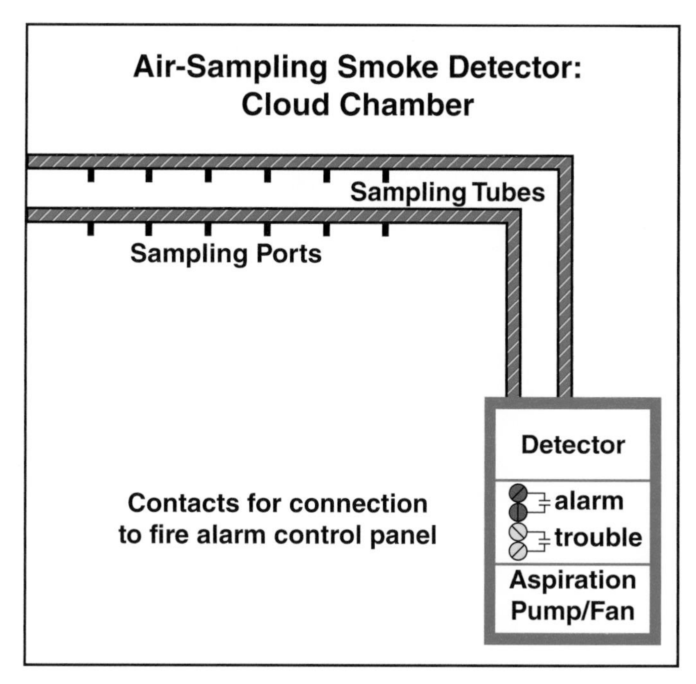

Figure 14.33 An aspiration pump draws samples of air into the air-sampling smoke detector, which activates when it detects smoke or other particles of combustion.

Figure 14.34 A similar type of air-sampling smoke detector uses pipes to draw air into the system. Activation depends on a photoelectric cell.

Video-Based Detectors

Video-based smoke and flame detection operates on the principle of detecting changes in a digital video image from a camera or a series of cameras. Images are transmitted from a closed-circuit television to a computer that looks for changes in the images. These cameras will work only in a lighted space. They also provide an image to an operator who is monitoring the system. These systems offer advantages in large, open facilities where there may be a delay in smoke movement and detection.

Flame Detectors

A flame detector is sometimes called a light detector. There are three basic types:

- Those that detect light in the ultraviolet wave spectrum (UV detectors) **(Figure 14.35, p. 634)**
- Those that detect light in the infrared wave spectrum (IR detectors) **(Figure 14.36, p. 634)**
- Those that detect light in both UV and IR waves

An infrared detector is effective in monitoring large areas, such as an aircraft hangar or computer room. UV detectors are easily activated, they are also easily activated by such nonfire conditions as welding, sunlight, and other bright light sources. They must only be placed in areas where these triggers can be avoided or limited. They must also be positioned so that they have an unobstructed view of the protected area. If they are blocked, they cannot activate.

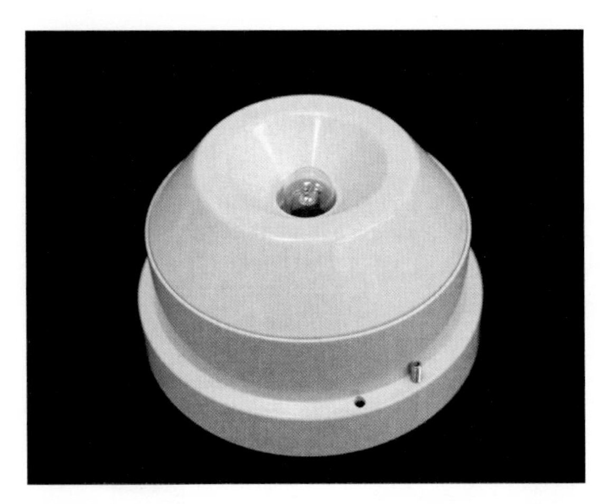

Figure 14.35 A UV flame detector detects light in the untraviolet wave spectrum.

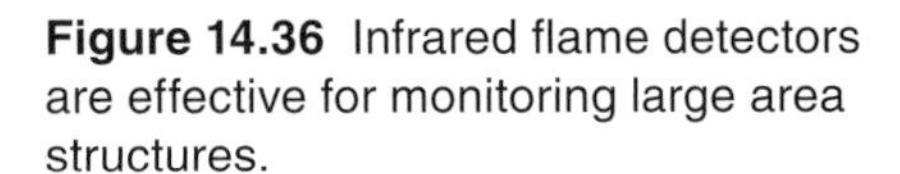

Figure 14.36 Infrared flame detectors are effective for monitoring large area structures.

To prevent accidental activation, an infrared detector requires the flickering action of a flame before it activates to send an alarm. This detector is typically designed to respond to 1 square foot (0.09 m^2) of fire from a distance of 50 feet (15 m).

There are also video-based flame detectors that work on the same principle as the video-based smoke detectors. The images from the closed-circuit televisions are sent to a computer with software designed to detect the characteristics of a flame. This type of flame detection system is used in certain chemical or petroleum facilities.

Combination Detectors

Depending on the design of the system, various combinations of the previously described detection devices may be used in a single device. These combinations include fixed-rate/rate-of-rise detectors, heat/smoke detectors, and smoke/fire-gas detectors (**Figure 14.37**). These combinations give the detector the benefit of both services and increase their responsiveness to fire conditions.

Sprinkler Waterflow Alarm-Initating Devices

An automatic initiating device is designed to activate an audible alarm (horn/strobe) when water begins to flow through the sprinkler system **(Figure 14.38)**. The notification appliance is usually located on the exterior of the building near the sprinkler riser. In addition, many alarm systems are connected to fire alarm systems that monitor fire sprinkler water flow through the use of electronic flow switches that notify the FACU when water is flowing through the system. This action may, in turn, cause the notification and signaling devices to function.

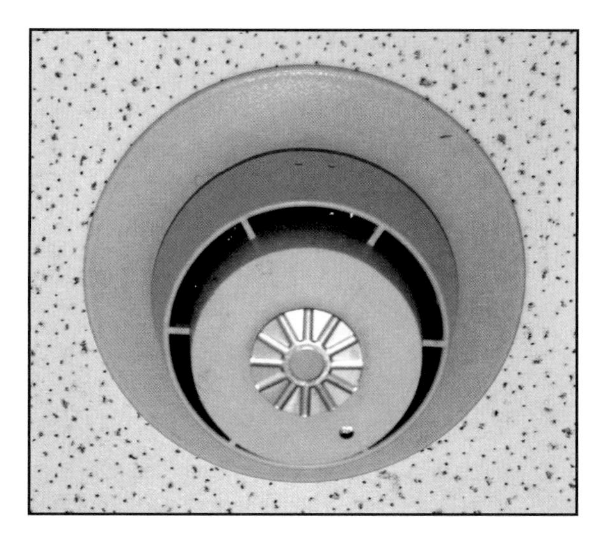

Figure 14.37 A combination heat/smoke detector.

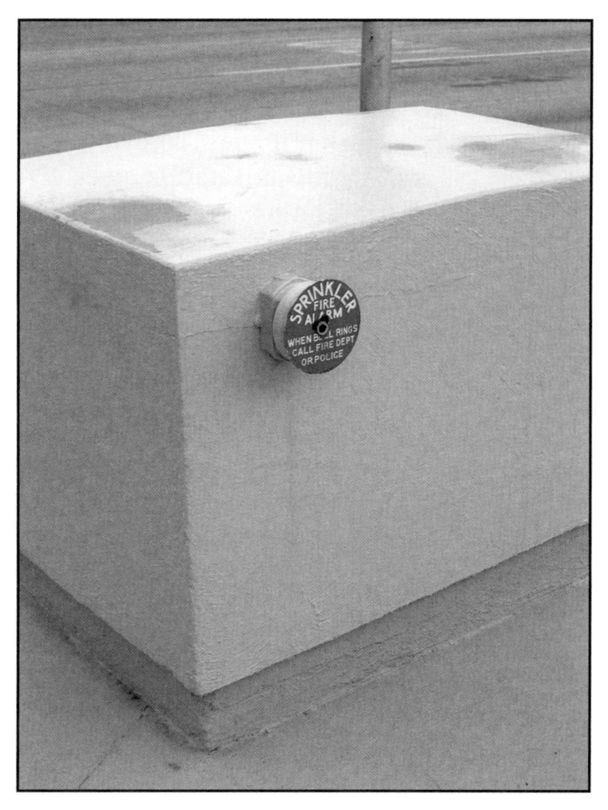

Figure 14.38 A waterflow alarm indicates the flow of water through the system, and may be accompanied by a system drain valve.

Manually Actuated Alarm-Initiating Devices

Manually actuated fire alarm boxes, commonly called manual pull stations, allow occupants to manually initiate the fire alarm signaling system. Manual pull stations may be connected to systems that sound local alarms, off-premise alarm signals, or both.

Although manual pull stations come in a variety of shapes and sizes, they are usually red in color with white lettering that specifies what they are and how they are to be used **(Figures 14.39 a-c)**. The manual pull station should only be used for fire-signaling purposes unless it is designed for other uses or to activate a fixed fire suppression system.

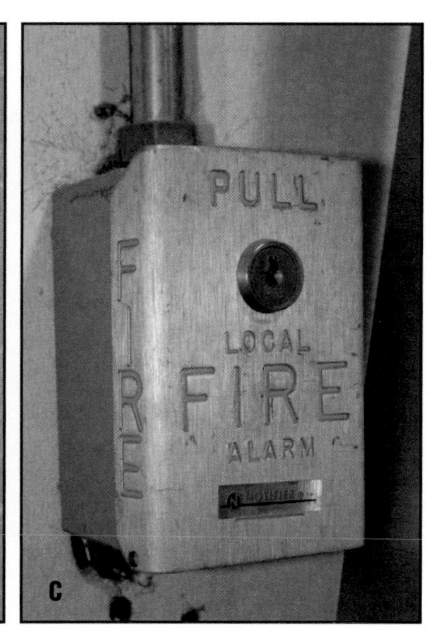

Figures 14.39 a-c Manually operated alarm-initiating devices may look slightly different, but all are important tools for signaling emergencies.

According to NFPA® 72, the pull station should be mounted on walls or columns so that the operable part is not less than 42 inches and not more than 48 inches above the floor. The manual pull station should be positioned so that it is in plain sight and unobstructed. Multistory facilities should have at least one pull station on each floor. In all cases, travel distances to the manual pull station should not exceed 200 feet (60 m).

Manual pull stations can be single-action or double-action, which are described as follows:

- **Single-action** — Operates after a single motion is made by the user. When the station lever is pulled, a lever or other movable part is moved into the alarm position and a corresponding signal is sent to the FACU (**Figure 14.40**).
- **Double-action** — Requires the operator to perform two steps to initiate an alarm. First, the operator must lift a cover or open a door to access the alarm control **(Figure 14.41)**. Then the operator must manipulate an alarm lever, switch, or button to send the signal to the fire alarm control panel. Double-action manual pull stations may be confusing to certain occupant/operators due to the need to perform two separate steps before an alarm is initiated.

A manual pull station may be protected by a listed protective cover in areas where it would be subject to damage or accidental activation (**Figure 14.42**). These protective devices are used in gymnasiums, materials handling areas, or in other locations where accidental activation is possible. Listed protective covers can only be installed over single-action manual pull stations.

Figure 14.40 Single-action pull stations require only a single motion to trigger an alarm.

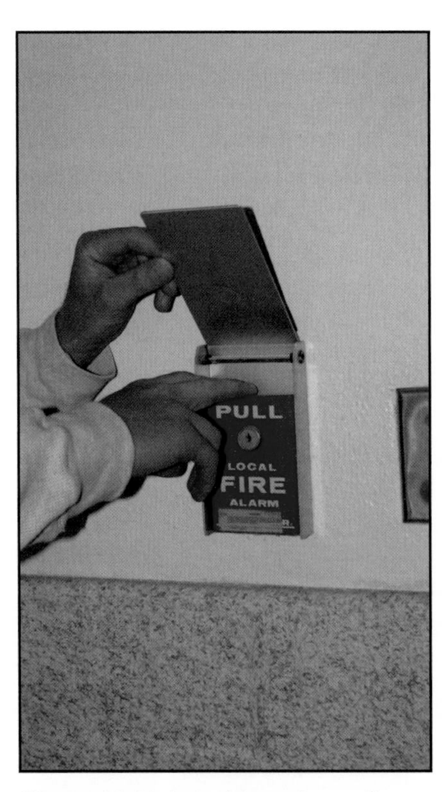

Figure 14.41 A double-action pull station has a panel that must be lifted so the operator can access and operate the pull station.

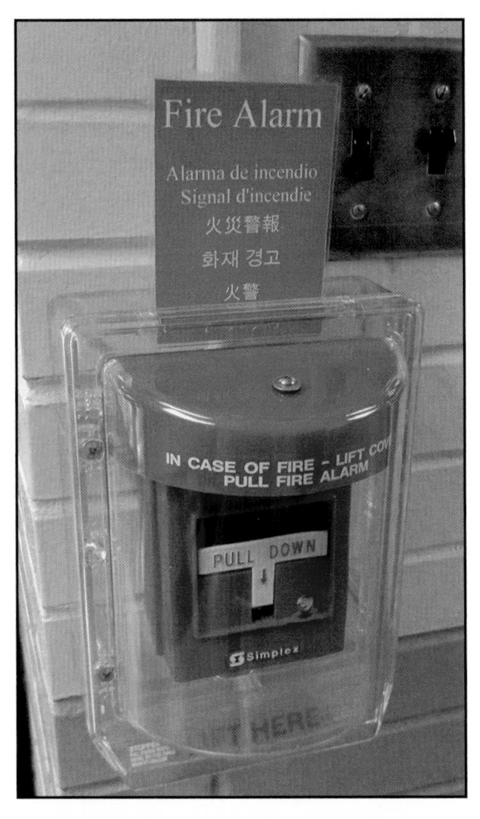

Figure 14.42 A manual alarm may require a cover to prevent damage. *Courtesy of Rich Mahaney.*

Manual pull stations that require the operator to break a small piece of glass with a mallet are no longer recommended. These devices were designed to discourage false alarms and were somewhat effective for that purpose. However, broken glass presents an injury hazard to the operator at a time when an untrained operator is least capable of clear thinking. Polycarbonate covers have taken the place of glass; however, these older types of pull stations may still be found in many old structures. Some pull stations leave a dye or ultraviolet residue on the activator that will transfer to the fingers of someone pulling the alarm to discourage malicious false alarms.

Inspection and Testing

To verify operational readiness and proper performance, fire detection and alarm signaling systems must be tested when they are installed and again on a continuing basis. Periodic testing is often referred to as a **service test**. Inspectors should be familiar with the service testing and inspection requirements for detection and alarm systems.

Service Test — Series of tests performed on fire protection, detection, and/or suppression systems in order to verify operational readiness. These tests should be performed at least yearly or whenever the system has undergone extensive repair or modification.

During inspections, an inspector should note the functional aspects of the fire detection and alarm systems. An inspector should be able to recognize physical and environmental conditions that may negatively affect system operation or even render the system inoperative. An inspector should also recognize conditions that may trigger an unwanted alarm and recommend corrective action to reduce or eliminate the number of fire department responses to possible nuisance alarms.

Inspectors who routinely conduct inspections need to have a working knowledge of these systems. Inspectors are generally limited to visual inspections and supervision of system tests **(Figure 14.43)**. Often they will not have the authority or responsibility for operating or maintaining these systems. In most cases, representatives of the owner/occupant or alarm system contractors perform system tests and maintenance. These individuals should be qualified and experienced in the various types of devices and systems. Some general inspection considerations for a fire inspector include the following:

Figure 14.43 Fire department personnel who conduct inspections must have a working knowledge of detection and alarm systems.

- Check for changes to the building or use of rooms that may result in different requirements for detection systems, audio/visual alarms, or that cause a coverage issue.
- Verify that all equipment, especially initiating and signaling devices, are free of dust, dirt, paint, and other foreign materials.
- Verify that manual pull stations, audible or visual warning devices, and any other components are not blocked or obstructed in any way.
- Verify that the monitoring system is operational, if applicable.

Fire detection and alarm systems must be tested and inspected regularly if the systems are to work reliably during an emergency. The actual performance of the tests is the responsibility of the owner/occupant or the fire alarm monitoring company. Locally adopted fire codes usually mandate that inspectors witness system tests. Periodic tests and inspections are performed on all components of the fire detection and alarm systems.

NOTE: Testing and inspection intervals may vary in accordance with local codes and ordinances.

Because it is time-consuming to test fire alarm systems in a jurisdiction, it is not always possible for fire department or inspection personnel to witness every test. Most of the time, occupants have to test the systems on their own and document the results. Inspectors should review the inspection/testing carried out by qualified third parties. The inspector should review the documentation for any deficiencies found or repairs needed, and orders sent for correction of deficiencies.

Inspectors must be familiar with their jurisdiction's inspection procedures. The following sections will provide more detail into inspecting and testing various alarm system components and types of alarm signaling systems.

Inspection Considerations for Fire Alarm Control Units

Inspectors should check the FACU to verify that all parts are operating properly. All switches should perform their intended functions and all indicators should light or sound when tested (**Figure 14.44**). When individual detectors are triggered, the FACU should indicate the proper location and warning lamps should light. Remember that the indicated location could very well be out of date due to renovations.

Verify that access to FACUs, recording instruments, and other devices are not obstructed and do not have objects stored on, in, or around them. Many FACUs have storage areas with locking doors for extra relays, light bulbs, and test equipment. Store items somewhere else if the unit does not have a storage area designed into it. Otherwise, they may foul moving parts or cause electrical shorts that can result in system failure.

Figure 14.44 Inspectors must verify that all the switches and detector lights on annunciator panels work when tested.

Auxiliary devices can also be checked at this time. The auxiliary devices include the following: local evacuation alarms and HVAC functions, such as air-handling system shutdown controls, and fire dampers. All devices must be restored to proper operation after testing.

The receiving signals should also be checked during this time. The proper signal and/or number of signals should be received and recorded. Signal impulses should be definite, clear, and evenly spaced to identify each coded signal. There should be no sticking, binding, or other irregularities.

At least one complete round of printed signals should be clearly visible and unobstructed by the receiver at the end of the test. The time stamp should clearly indicate the time of the signal and should not interfere in any way with the recording device.

Inspection Considerations for Alarm-Initiating Devices

Any fire detection and alarm-signaling system will be ineffective unless the alarm-initiating devices are in proper working order and send the appropriate signal to the system control unit. Inspectors may wish to witness the activation of selected devices to verify that the device and the system are operational.

Automatic alarm-initiating devices should be checked after installation, after a fire, and at scheduled times based on guidelines established by the AHJ or the manufacturer. Often these guidelines are found in the locally adopted fire code. All detector testing should be in accordance with local guidelines, manufacturer's specifications, and NFPA® 72.

Periodic testing procedures are included in NFPA® 72 and in the fire alarm manufacturer's literature for the system and its components. The following periodic tests are recommended test procedures for the listed device:

- **Restorable heat detection device** — Test one detector on each signal circuit semiannually. A different detector should be selected each time and so noted on the inspection report. Subsequent inspections should include a copy of the previous report so that the same detector is not tested each time.

Restorable heat detectors should be checked by following approved testing procedures described by the manufacturer . Hair dryers or electric heat guns can be used to test restorable heat detectors. Nonrestorable heat detectors must not be tested during field inspections.

Remember that some combination detectors have both restorable and nonrestorable elements. Exercise caution to avoid tripping the nonrestorable element. Nonrestorable pneumatic detectors should be tested mechanically. Those detectors equipped with replaceable fusible links should have the links removed to see whether the contacts touch and send an alarm signal. The links can then be replaced.

- **Fusible-link detector with replaceable links** — Check semiannually by removing the link and observing whether or not the contacts close. After the test, the fusible link must be reinstalled. It is recommended that the links be replaced at 5-year intervals.
- **Pneumatic detector** — Test semiannually with a heating device or a pressure pump. If a pressure pump is used, the manufacturer's instructions must be followed.

- **Smoke detector** — Test semiannually in accordance with the manufacturer's recommendations. The instruments required for performance and sensitivity testing are usually provided by the manufacturer. Sensitivity testing should be performed after the detector's first year of service and every two years after that.
- **Flame and gas detection devices** — Require testing by highly trained individuals because they are very complicated devices. Testing is typically performed by professional alarm service technicians on a contract basis.

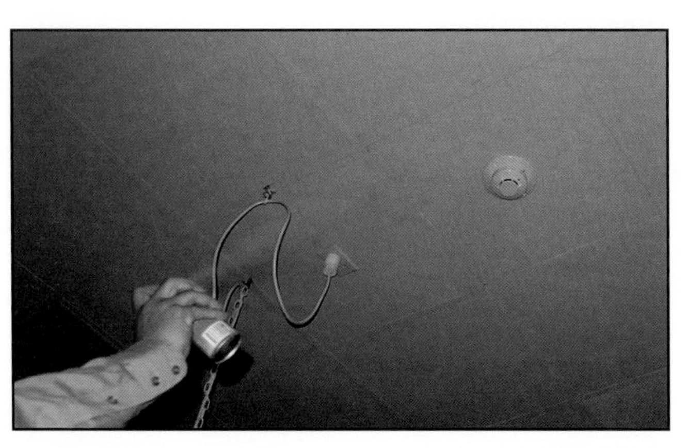

Figure 14.45 It is important to use approved testing procedures for smoke heat detectors.

NOTE: Inspectors are only responsible for witnessing the tests, not performing them.

The manufacturers of smoke and flame detectors usually have specific instructions for testing their detectors. These instructions must be followed on both the acceptance and service tests. They may include the use of smoke-generating devices, aerosol sprays, or magnets **(Figure 14.45)**. Using nonapproved testing devices may result in the manufacturer's warranty being voided.

For manual alarm-initiating devices, check the following:

- Verify that access to the device is unobstructed.
- Observe that each unit is easy to operate.

Regardless of the type of detector in use, detectors found in the following conditions should be replaced or sent to a recognized testing laboratory for testing:

- Restored to service after a period of disuse
- Obviously corroded
- Painted even if attempts were made to clean them
- Mechanically damaged or abused
- Subjected to current surges, overvoltages, or lightning strikes
- Subjected to foreign substances that might affect their operation
- Subjected to direct flame, excessive heat, or smoke damage

A permanent record of all detector tests must be maintained for at least 5 years. The following minimum information should be included in the record:

- Test date **(Figure 14.46)**
- Detector type
- Location
- Type of test
- Test results

A nonrestorable fixed-temperature detector cannot be tested periodically. Testing would destroy the detector and require the system to be rendered inoperable until a replacement detector can be located and installed. For this reason, tests are not required until 15 years after the detector has been installed.

Figure 14.46 Check that inspection records are up-to-date.

At that time, two percent of the detectors must be removed and laboratory tested. If a failure occurs in one of the detectors, additional detectors must be removed and laboratory tested. These tests are designed to determine if there is a problem with failure of the product in general or a localized failure involving just one or two detectors.

Inspection Considerations for Alarm Signaling Systems

The following list gives a brief synopsis of the inspection and testing requirements for various types of systems and timetable guidelines. If any of these systems use backup electrical generators for emergency power, those generators should be run under load monthly for at least 30 minutes.

- **Local alarm systems** — Test in accordance with guidelines established in NFPA® 72 and the manufacturer's recommendations.
- **Central station systems** — Test signaling equipment on a monthly basis. Check waterflow indicators, automatic fire detection systems, and supervisory equipment bimonthly. Check manual fire alarm devices, water tank level devices, and other automatic sprinkler system supervisory devices semiannually. In scheduling these tests, both facility/building supervisory personnel and central station personnel must be notified before the test to prevent them from evacuating occupants or dispatching fire units.
- **Auxiliary fire alarm systems** — The occupant should visually inspect and actively test these systems monthly to verify that all parts are in working order and the operation of the system results in a signal being sent to the fire department telecommunications center. Test noncoded manual fire alarm boxes semiannually.
- **Remote station and proprietary systems** — Test according to the testing requirements established by the AHJ. Test fire detection components of these systems monthly. Test water-flow indicators semiannually; however, the frequency of testing may depend upon the type of indicator.

- **Emergency voice/alarm systems** — Conduct functional test of the various components in these systems quarterly (by the owner/occupant). Include selected parts of the system that are likely to be used during an incident. Check all components at least annually.

NOTE: In all cases of testing and inspection frequencies, check with the most current edition of NFPA® 72.

Inspecting, Testing, and Evaluating Fire Detection and Alarm Systems

An Inspector II must be familiar with the service testing and inspection requirements for detection and alarm systems, initiating devices, and FACUs, along with operation of fire detection and alarm systems and the building code requirements for their installation, testing, and monitoring. During plan reviews for new buildings, an inspector may be required to evaluate and approve the design of fire detection and alarm systems.

Detection devices were covered earlier in this chapter for Fire Inspector I. This portion of the chapter will deal in greater detail with acceptance tests and occupant notification devices typically found in a modern fire detection and alarm system.

Acceptance Test — Preservice test on fire protection, detection, and/or suppression systems after installation to ensure that the system operates as intended.

Acceptance Testing

Acceptance testing is performed soon after the system has been installed and prior to occupancy to ensure it meets design criteria and functions properly. Representatives of the building owner/occupant, the fire department, and the system installer/manufacturer should witness acceptance tests. The fire department representative may be a fire inspector, a staff fire protection engineer, or in some cases the fire marshal.

Some jurisdictions require a record of completion from the system installer/manufacturer demonstrating that the system has been thoroughly tested before the fire department inspection. This record prevents the inspector from checking a system that may not have been properly installed. **See Appendix E** for a commercial fire alarm acceptance test checklist.

Activities during an acceptance test include the inspector witnessing the following:

- Inspect all wiring for proper support.
- Look for wear, damage, or any other defects that may render the insulation ineffective.
- Inspect conduit for solid connections and proper support wherever circuits are enclosed in conduit.
- Check batteries that are used as an emergency power source for clean contacts and proper charge. Immediately replace batteries that fail inspection and testing procedures. Many batteries have floating-ball indicators that show whether they are properly charged.

All of the following functions of the fire detection and alarm-signaling system should be operated during the acceptance tests:

- **Alarm, supervisory, and trouble signals** — Check actual wiring and circuitry against the system drawing to verify that all are connected properly **(Figure 14.47)**.
- **Fire alarm control unit (FACU)** — Operate all interactive controls at the FACU to verify that they control the system as designed. Inspect thoroughly to verify that the FACU is in proper working order.
- **Alarm-initiating and occupant notification devices and circuits** — Check all items for proper operation. Test pull stations, detectors, bells, and strobe lights to verify that they are operational. Test each initiating device to note that it sends an appropriate signal and causes the system to send the prescribed alarm, supervisory or trouble signal as prescribed on the approved plans.
- **Power supplies** — Operate the system on both the primary and secondary power supplies to verify that both will supply the system adequately **(Figure 14.48)**.

Check the ability of supervising station services to respond to an alarm. The alarm-receiving capability and response of those involved must be verified, and a telecommunications facility must receive the request for emergency service.

The results of all tests must be documented to the satisfaction of the AHJ. Only after all parts of the system have successfully passed the tests should a system certification be issued. Issuing the alarm system certificate is typically a preliminary step toward the issuance of a certificate of occupancy. NFPA® 72 contains complete information on system acceptance tests.

Occupant Notification Devices

The locally adopted building code, other governing code or standard, or system installation standard will determine the system requirements for occupant notification. Generally, notification is the activation of horns, bells, and strobes that are used in the occupancy to alert occupants of a fire. Notification may

Figure 14.47 Verify that the circuitry and wiring match the system drawing.

Figure 14.48 The alarm system must contain a storage battery and charger.

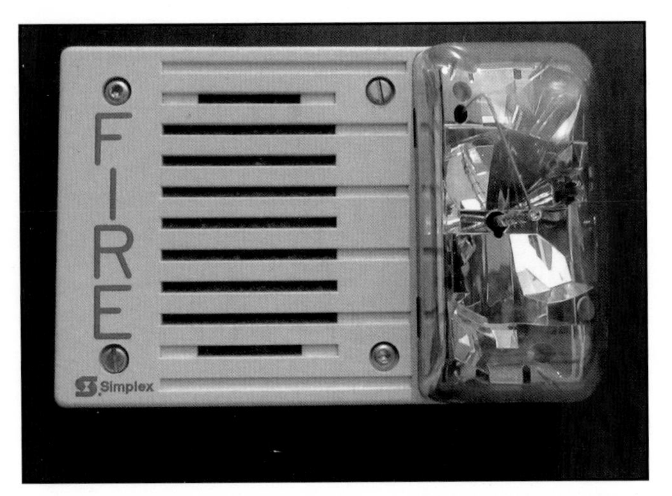

Figure 14.49 Verify that occupant notification devices are consistent with the requirements of the locally adopted code. *Courtesy of Rich Mahaney.*

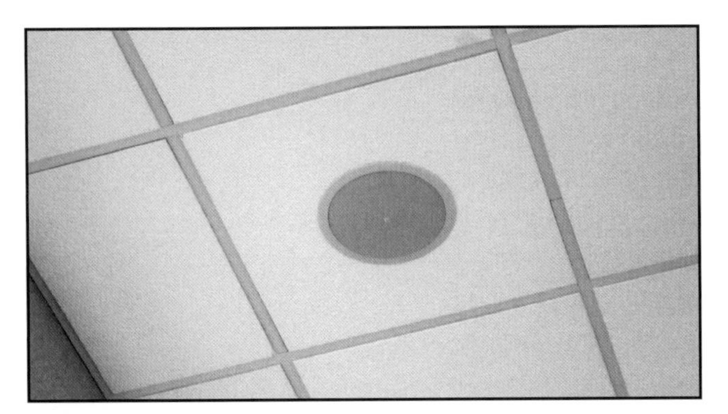
Figure 14.50 A recessed voice evacuation system. *Courtesy of Rich Mahaney.*

also include voice evacuation systems, alarm printers, annunciators, textual displays, and graphic displays that are included as a part of the installed system **(Figure 14.49)**.

The local building code usually dictates the installation of voice evacuation systems based on the size and type of occupancy. NFPA® 72 dictates where and how these devices are installed **(Figure 14.50)**. The locally adopted building and fire codes typically specify requirements for the annunciators and graphic displays. Often, the graphic displays are located near the front entrance to the building and provide a complete view of the building with alarm zones and detection points. In some installations, remote annunciators will be located near the entrance where the emergency responders will enter the building.

Almost all fire detection and alarm system installations require the use of audible signals through horns, bells, or chimes. Building code requirements for disability access will require visual occupant notification appliances. Visual alarm indicators are accomplished through the use of flashing strobe lights. In some instances, a rotating beacon may also be used for outdoor installation, or for warning on large industrial and commercial complexes.

Use alarms with integrated audible and visual signals to accommodate the hearing and visually impaired, and for areas where a person may be working alone. This includes areas such as restrooms, storage areas, offices, and similar areas.

Audible Notification

Audible notification is a method of providing occupant notification of fire. In order for audible devices to alert the building occupants, the device must be loud enough to be heard. The level of loudness is a measurement of sound pressure and that measurement is in decibels. The decibels produced by a notification device are expressed as dBA.

NFPA® 72 requires the sound level to be at least 15 dBA above the average or normal sound level or 5 dBA above the maximum sound level that lasts at least one minute in the protected occupancy. The total sound pressure produced by the audible devices must not exceed 120 dBA because permanent hearing damage may occur above this level.

If the ambient sound in a building is above 105 dBA, the building must have visible notification. NFPA® 72 allows the audible notification in noisy areas if the ambient sound level can be reduced (such as a night club).

This measurement is required to be made at 5 feet (1.5 m) above the floor. The measurement in sleeping areas is required to be measured at the pillow level. The Average Ambient Sounds levels for various types of occupancies are given in **Appendix H** and should only be considered a guide. Each installation will require specific evaluation.

Visual Notification

The requirement for specific visual notification appliances comes from the adopted building code. The strobe requirements of the ADA apply to new construction and renovations to portions of buildings open to the general public **(Figure 14.51)**. Likewise, the ADA requires visual alarm notification devices in portions of any building accessible to a hearing impaired person. In commercial facilities, visual alarm notification appliances would be located in areas accessible to the public and to occupants of the facility who may have a hearing disability. Areas such as conference rooms, restrooms, hallways, routes of tours, and the private office of someone with a hearing disability are examples. When visual alarm notification appliances are required, the installation, operation and location requirements are the same for both the ADA and NFPA® 72.

Figure 14.51 Strobe requirements apply to areas that are open to the public. *Courtesy of Rich Mahaney.*

Candela Information

The light intensity of a visual device is measured in candela (cd). The appliances listed for evacuation have specific light output requirements that must be complied with for the listing. Power is applied to these devices and the light output is measured to ensure the proper light output. The minimum light directly in front of the device is 15 cd.

Visual alarm notification appliances are installed in one of two orientations: wall mount and ceiling mount. Visual notification appliances are listed for a particular orientation and are required to be installed in that orientation.

Requirements for Wall-Mount Devices

Wall-mount devices cannot be mounted on ceilings for visual notification. Wall-mount devices are required to be mounted between 80 and 96 inches from the finished floor level. The spacing requirements for the visual devices are based on the tables in NFPA® 72. The spacing is based on the square area covered by a single appliance. The area of notification is determined when the device that entirely covers that area is used. As an example, a room is 40 feet wide by 20 feet deep. The room would be required to have a minimum of a single 60-candela strobe or two 30-candela strobes on the shorter sidewalls opposite of each other.

When visual appliances are mounted on the walls, the strobe configuration is one, two, or four devices per area. NFPA® 72 has tables that define the minimum required light output. Generally, the largest room area covered by a single wall-mounted device is 70 feet by 70 feet. The maximum room area covered by a ceiling mounted strobe is 50 feet by 50 feet. In addition, the strobe must be mounted in the center of the room to achieve the light levels as specified in the tables in NFPA® 72.

Figure 14.52 The inspector should verify correct placement for visual devices.

Where required, visual notification appliances must be installed in corridors that are less than 20 feet wide **(Figure 14.52)**. If the corridor is wider than 20 feet, the requirements for room spacing are applied. The minimum candela rating for the visual appliances mounted in corridors is 15 cd. The visual devices must be mounted within 15 feet of the end of the corridor and must not be spaced more than 100 feet apart on center. They must be mounted in accordance with NFPA® 72 for the proper height and placement. In addition, if there are any interruptions in the corridor, such as fire doors, partitions, or changes in elevation, the areas are to be viewed as separate areas.

The fire inspector must be observant with how visual appliances, and strobes in particular, are installed. When more than two strobes are in the field of vision, the strobes must be synchronized to flash at the same time. Some people are prone to photosensitive epileptic seizures when exposed to random flashing lights and synchronization of audible signals and strobe flashes prevent a potential seizure in these individuals.

The inspector should also recognize that certain visual notification appliances are field adjustable for a range of candela values. The inspector should verify that the correct candela setting is made. This information should be shown on the approved plans.

Required Detection and Alarm System Documentation

When fire alarm systems are installed, plans for the complete fire detection and alarm system should be submitted for review and approval by local authorities. The submitted plans should include building dimensions to scale with partition walls, duct work, and separation barriers. In addition, a point-by-point initiating device detail should be submitted showing:

- Detector placement
- Notification device placement
- Sequence of operation

- Voltage drop calculations
- Battery calculations

Manuals and manufacturers' cut sheets should be also be submitted for review and approval. When the system installation is completed, an "as-built" drawing should also be provided to at least the property owner and a Record of Completion should be provided to the local authorities.

On newer fire detection and alarm systems where the system installer has used computers to program the panels, additional design details should be submitted. The design configuration of the system in the building should be given to the building owner and remain on site. This is to verify that the panel can be reconfigured to the exact way it was designed and built if a catastrophic failure occurs in the panel. In addition, the system should be completely tested anytime a change in the software is made to verify that it is operating as designed and originally intended.

Chapter Summary

The inspection of fire detection and alarm-signaling systems is an almost daily activity for the inspector. This chapter described the fundamental components and operation of these systems. The procedures that must be used while conducting inspections and tests of these systems have also been provided. Many unique situations may demand a particular type of system or initiating device within the community's jurisdiction. The fire inspector must be familiar with these conditions and the systems that have been installed.

Review Questions

1. What is the basic function of the fire alarm control unit (FACU)?
2. List types of notification appliances.
3. Name types of alarm-signaling systems.
4. List types of heat detectors.
5. Names types of smoke detectors.
6. List types of flame detectors.
7. What are the placement requirements for manual pull boxes?
8. Name some of the general inspection considerations for a fire inspector.

1. What should an inspector witness during an acceptance test?
2. List inspection requirements for different types of occupant notification devices.

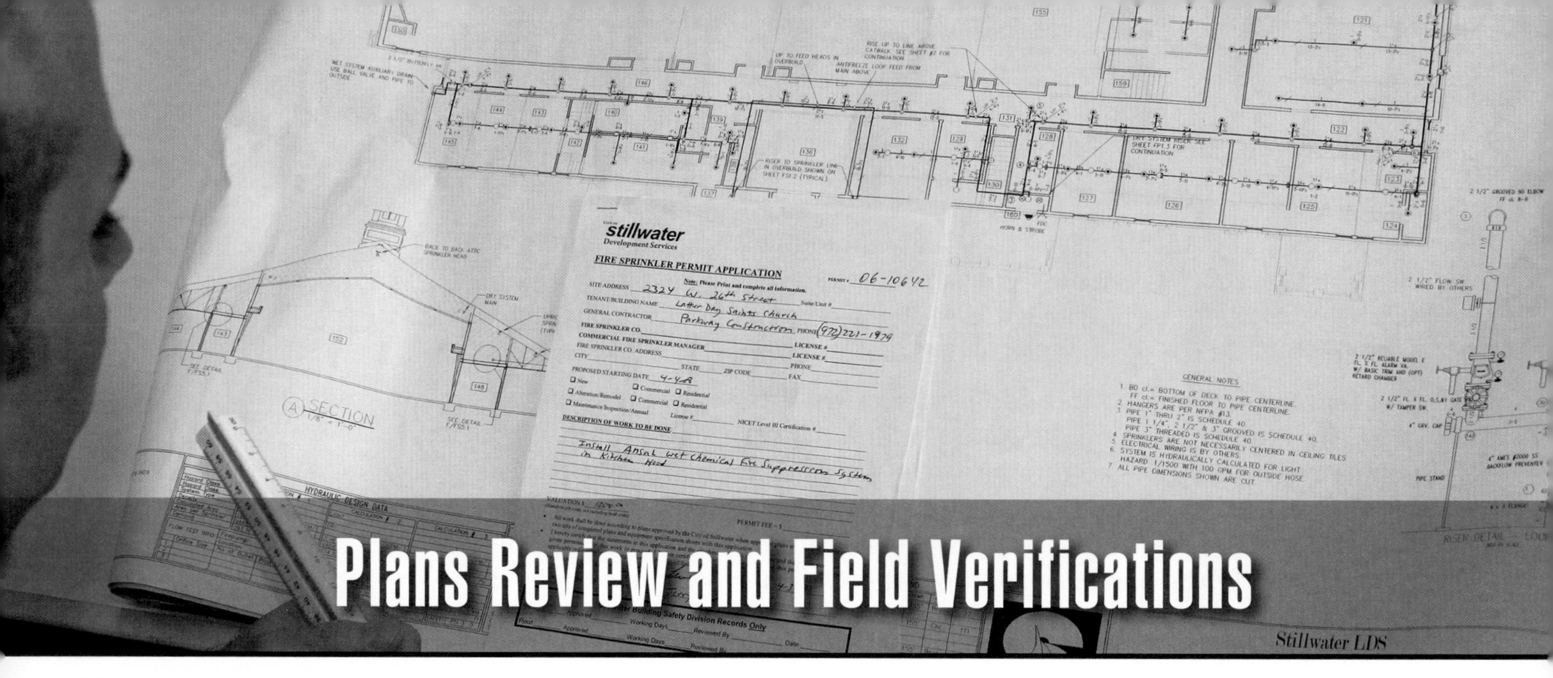

Plans Review and Field Verifications

Chapter Contents

Case History **651**

I Overview of Plans Review **651**

Benefits of Plans Review 652

Field Verification 653

Building Plans and Construction Drawings............. 654

II Plans Review Process **666**

Sequence 658

Fees.. 658

Construction and Support Documents.................... 658

Building Construction Plans **662**

Plan Views 663

Elevation View 666

Sectional View 668

Detailed View.. 669

System Plans 669

Systematic Plans Review **679**

Overall Size of Building... 680

Occupancy and Construction Classification............ 680

Occupant Load.. 681

Means of Egress 681

Exit Capacity... 682

Building Compartmentation 682

Additional Concerns 682

Chapter Summary **683**

Review Questions **685**

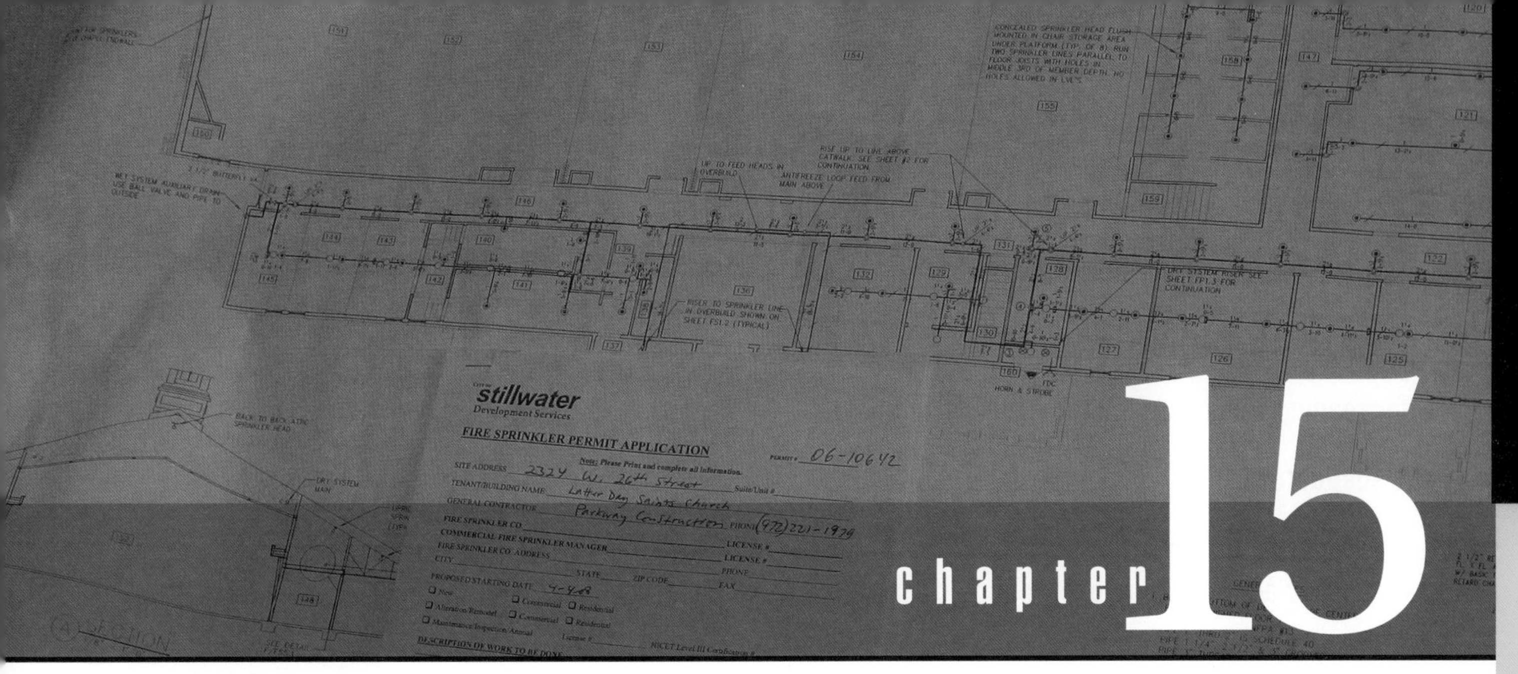

Key Terms

Benchmark 668
Change Order 661
Clear Dimensions 670
Detailed View 673
Dimensioning 669
Elevation View 670
Orientation 670
Plan View 667
Sectional View 672

NFPA® Job Performance Requirements

This chapter provides information that addresses the following job performance requirements of NFPA® 1031, *Standard for Professional Qualifications for Fire Inspector and Plan Examiner* (2014).

Fire Inspector I	Fire Inspector II
4.2.3	5.2.2
4.3.9	5.4.1
	5.4.2
	5.4.3
	5.4.4
	5.4.5
	5.4.6

Plans Review and Field Verifications

Learning Objectives

After reading this chapter, students will be able to:

Inspector I

1. Recognize the need for a plans review. (4.2.3, 4.3.9)
2. Identify actions an inspector should take during a plans review (4.3.9)

Inspector II

1. Explain the impact of local codes and ordinances on the plans review process. (5.2.2)
2. Describe the different views for building construction plans. (5.4.1, 5.4.6)
3. Describe building system plans. (5.4.2, 5.4.3, 5.4.4, 5.4.5)
4. Explain the process of systematic plans review. (5.4.4)

Chapter 15
Plans Review and Field Verifications

Case History

Largo, Florida, 2015: The city building official resigned after allegations that he knew the city's plans examiners were signing off on plans they were not qualified to review. The allegations stated that several of the plans examiners signed off on large construction plans that went beyond their licensing. Major construction problems were discovered at several large projects, necessitating removal and reinstallation so that the construction would meet code requirements.

Earlier in the year, a group of local contractors began meeting with the city to complain about long wait times for permits, lack of communication, and lack of response to inquiries. As a result of these meetings, the city is looking to make a number of improvements:

- Investigating ways to evaluate and approve smaller projects more quickly.
- Shortening the time for resubmittals.
- Hiring a senior plans examiner to relieve the workload.
- Updating the city permitting software, which would allow contractors quicker access to inspection results.

This chapter provides a general overview of the plans review and permitting process, descriptions of construction plans and supporting documents, and on-site field verification procedures used to confirm the implementation of the approved plans. A systematic approach to plans review also addresses fire and life safety concerns. The sections that follow describe the general plans review process for an Inspector I and the plans review process for an Inspector II.

NOTE: For additional information on plans examination, consult the IFSTA **Plans Examiner for Fire and Emergency Services** manual.

Overview of Plans Review

To verify that structures that are new, remodeled, or that have a change in occupancy classification meet the requirements of locally adopted building and fire codes, the AHJ establishes a plans review and permitting process **(Figure**

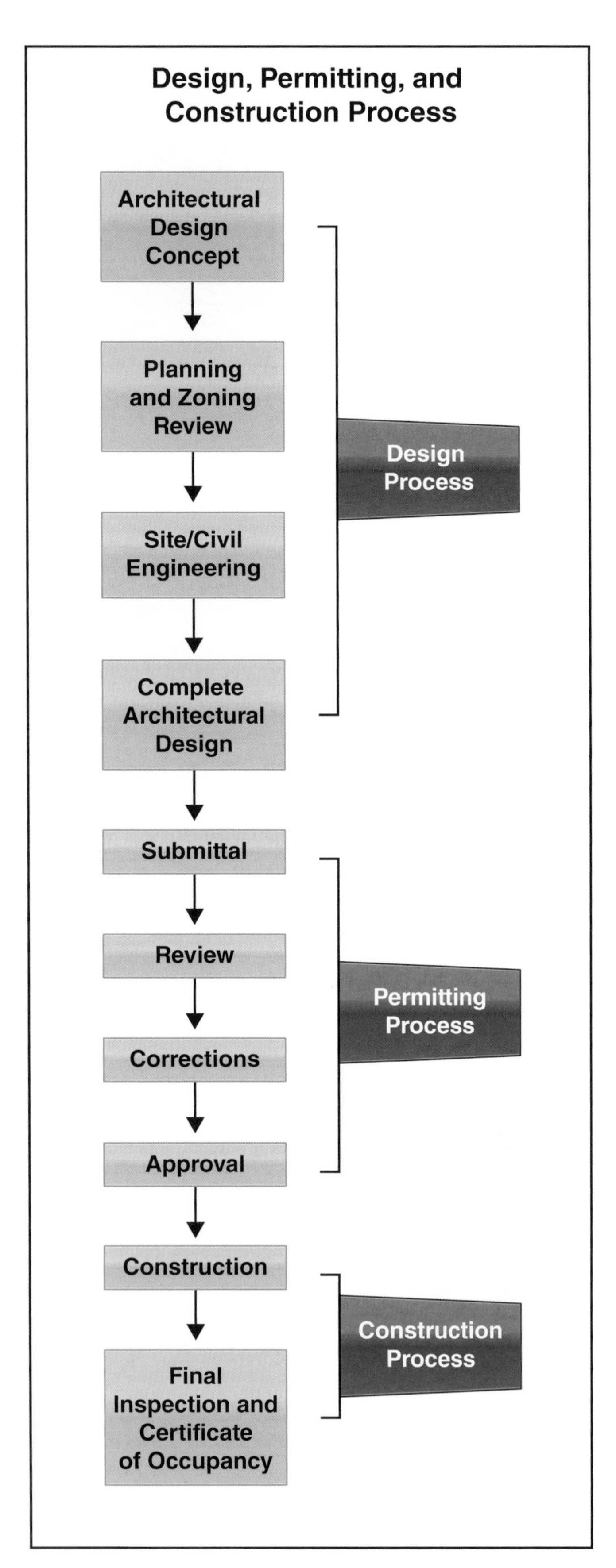

Figure 15.1 This flowchart reflects the major steps in the design, permitting, and construction processes.

15.1). This process is commonly administered by the AHJ and involves representatives of the building and fire departments and other agencies. Members of the zoning department may also review plans to verify that they meet zoning requirements.

Building department plans examiners include specialists in electrical, plumbing, structural, and mechanical (HVAC) design and installation. Fire department plans examiners are trained in fire protection systems, including sprinklers, standpipes, fire detection and alarm devices, and the fire and life safety requirements of the locally adopted codes.

Inspectors are responsible for verifying that construction activities conform to the plans that were approved by the plans examiner and issued as part of the permit documents. A Fire Inspector I is not expected to perform plans review, but must be familiar with the plans review and permitting processes. In some jurisdictions, the inspector or plans examiner who reviews the plans is also responsible for field verification. This method is most efficient because the reviewer is very familiar with the code requirements for the project. However, this can also cause problems because there is only one person looking at the documents.

The inspector should talk with the plans examiner to get an orientation on the project and review any areas that may be confusing or that require particular care during inspection. Some contractors or owners may state that permission to alter the accepted plans has already been granted. Inspectors must not accept this statement as fact unless proper verification is made with an appropriate building or fire official. A team approach also helps to prevent the owner/occupant or contractor from attempting to omit a requirement or use nonapproved alternative materials or methods.

Policies established by the AHJ will dictate the procedures for accepting field changes. Inspectors may not have the authority to approve or make any changes to the approved plans. The authority to approve changes may rest solely with the fire marshal or building official. Inspectors need to recognize when unapproved changes have been made and determine whether construction should be stopped.

Benefits of Plans Review

The main reason local governments require construction permits and plans review is community safety. Building design and construction not only affect fire and life safety but also structural integrity, electrical safety, plumbing, and ventilation. The plans review process enables reviewers to verify or improve the following areas:

- **Construction standards** — Enforcing building construction standards protects the public from shoddy construction and inferior workmanship. Adhering to these standards also ensures that construction or renovation methods conform to existing infrastructure and other off-site considerations.
- **Firefighter and emergency responder safety** — Information learned in the areas of building remodeling and renovation helps improve firefighter and emergency responder safety. If building departments did not require permits for such work, firefighters and emergency responders would have to deal with numerous unknown and potentially dangerous structural conditions **(Figure 15.2)**.

Figure 15.2 Workers (working without a permit) were attempting to enlarge a nightclub by removing part of a load-bearing wall, resulting in the collapse of the three-story building.

- **Input in design process** — Input in the design process enables fire and emergency services to verify that fire code requirements are met for the following:
  - — Hazardous materials storage
  - — Use and handling operations
  - — Installation of fire protection systems and/or assemblies
  - — Emergency vehicle access
- **Information about construction projects** — Information learned about construction projects within the jurisdiction becomes very useful to the organization for preincident planning. It also provides ongoing information to the organization on evolving construction technology.
- **Correcting errors** — Conducting plans reviews enables plans examiners to discover any errors or omissions in a building plan. Correcting fire and life safety issues before construction begins ultimately saves money and time.

Field Verification

During construction, an inspector should visit the site often for two reasons **(Figure 15.3)**:

- To verify that each code requirement is being met based on the approved plan. At the initial visit, an inspector meets the general contractor or project manager and explains the purpose of the field verification inspections. The contractor can supply a list of target dates when specific items will be completed and ready for inspection. By coordinating inspections with the completion dates, an inspector's efforts will be more efficient and effective.

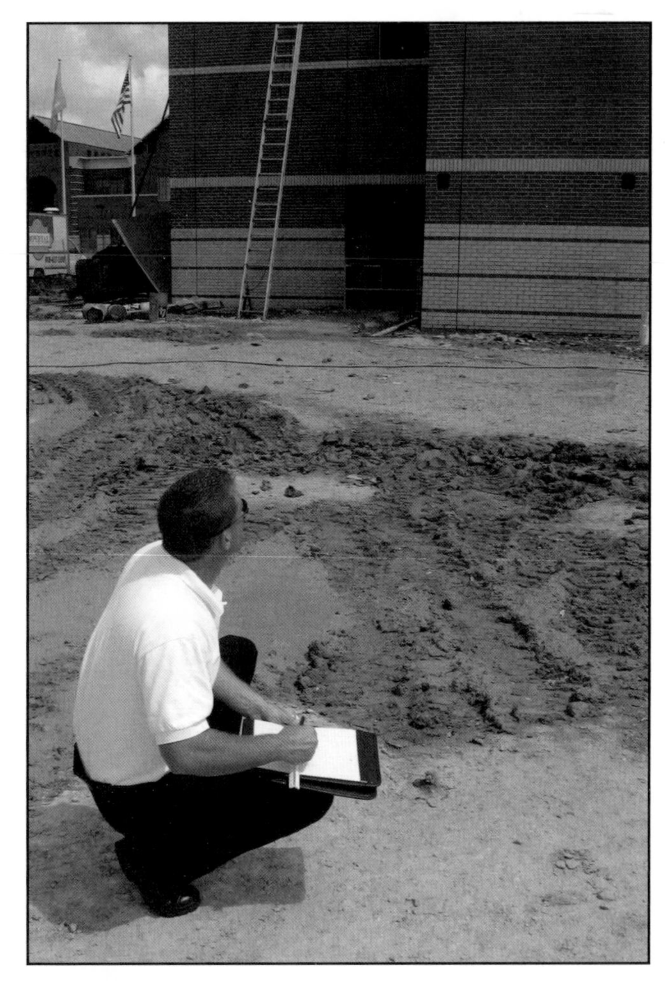

Figure 15.3 Inspectors should continuously monitor changes in new construction and remodeling to verify that work conforms to the approved plans.

- To verify that fire and life safety requirements needed during construction are met. These requirements include site access, a reliable water supply, and inspection of temporary fire protection systems.

At a construction site, the inspector must comply with Occupational Safety and Health Administration (OSHA) regulations or locally adopted requirements for the use of personal protective clothing and equipment. An inspector should perform a number of actions:

- Check in with the general contractor, project manager, or superintendent before entering the site.
- Explain the specific items or activities that must be verified.
- Compare the approved construction plans that are required to be kept on site with the ones maintained by the building or fire prevention office.
- Use a checklist to record items as they are inspected **(Figure 15.4)**.
- Notify the general contractor when violations are found and verify that corrections be made. Depending on the jurisdiction's fire code, construction may not resume until the violations are corrected.
- Develop a report noting the items inspected as part of the field verification inspection.
- Contact the general contractor or other responsible party before leaving the site.

NOTE: An inspector should follow the same general procedures outlined in Chapter 16, Inspection Procedures, during each field verification inspection.

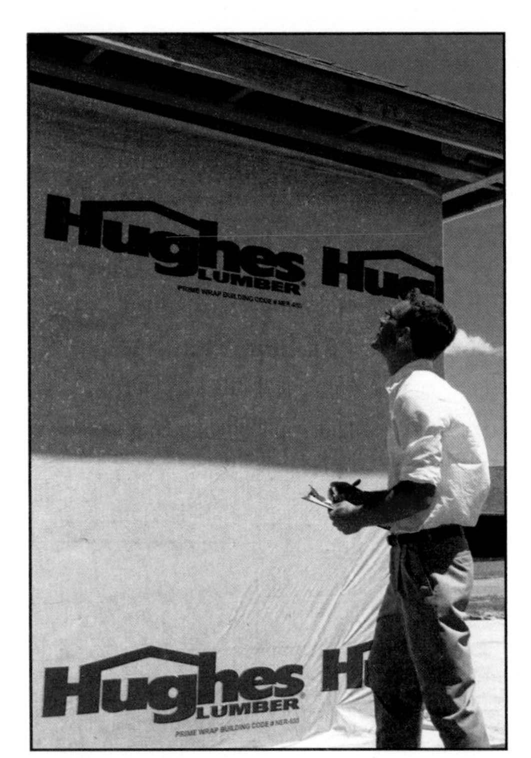

Figure 15.4 Using a checklist will help the inspector in remembering construction elements to verify.

Accurate and complete record keeping during plans review and field verification visits establishes the basis for all future inspections of the building. At the same time, records may be required for use in any legal actions should the owner or contractor fail to follow the requirements of the building or fire code.

Occupants and their belongings may remain in the building during construction. Their possessions, in addition to accumulated debris, construction materials, and equipment, can increase the fire load. Inspectors must verify that all means of egress are clear. Fire detection or alarm systems should remain intact for the duration of construction, and they may be covered to protect them if necessary. If a system will be nonfunctional for any length of time, a fire watch should be posted for the duration.

NOTE: The same precautions should be followed for renovations.

Building Plans and Construction Drawings

In order to expedite the plans review process, building plans, construction drawings, and support documents should be submitted to building officials in a professional and uniform manner. The AHJ usually establishes the exact manner in which plans must be submitted for review. Local ordinances may require that a registered design professional submit plans. Plans submitted to an AHJ are official documents. Sketches, incomplete drawings or plans, or documents with inaccurate calculations are unacceptable.

On construction drawings, abbreviations and symbols are widely used to increase the efficiency with which drawings convey information and reduce the clutter that would result from describing every detail. A large number of abbreviations can be used on construction drawings. It is not necessary to memorize all of them; however, plans reviewers should be familiar with some of the more commonly used abbreviations and symbols. **Table 15.1** shows some of the commonly used architectural and fire safety symbols.

Table 15.1
Common Architectural and Fire Safety Symbols

Symbol	Meaning	Symbol	Meaning
	Fire Hydrant		Concrete Wall
E OR	Electric Service		Concrete Block Wall
	Utility Meter or Valve		Brick Wall
G OR	Natural Gas Line		Sliding Door
FE	Fire Extinguisher		Bifold Door
	Door		Double Door
	Rebars		Pocket Door
	Building Line		Double Fixed Window
	Point of Beginning (POB)		Double Casement Window
	Fluorescent Light		Fixed Center Casement Window
	Wall	UP	Stairs

Plans Review Process

Reviewing documents and issuing permits are official government acts regardless of whether they concern a building permit or a fire code permit. These acts include reviewing documents or issuing permits for new building construction, alterations, or repairs **(Figure 15.5)**. Usually, the building department is the government agency responsible for reviewing construction documents and issuing building permits.

Many agencies use third-party agencies to assist with construction document review. If this is the case, the agency should coordinate document review with the fire department to make sure that locally adopted statutes and codes are addressed.

Many plans review agencies publish a process guide that outlines the plans review process. Individuals can review a copy and ask questions before they submit applications. While the actual sequence of events may vary between jurisdictions, the general steps of the process are:

1. **Application and Submittal** — Building owner, design professional, or contractor acquires an application form from the jurisdiction and completes it. The completed application, with complete construction documents including drawings and specifications, is submitted.
2. **Review** — Plans examiners from the building and fire departments review the documents **(Figure 15.6)**. They note corrections or alterations needed to meet code requirements.
3. **Resubmittal** — The plans review report is given to the design professional in charge. The owner/designer professional/contractor may meet with the plans examiners or building official to discuss any needed changes. The documents are resubmitted for final review after corrections are made.
4. **Accept** — Final set of plans and documents are stamped and signed to indicate official acceptance. The building department retains one copy; the fire department retains one copy (usually the fire protection system plans and documents); and one copy is returned to the owner, design professional, or contractor. This last set of plans must be kept at the construction site until the project is completed. These plans are the official construction set that will be used as the basis of inspection on the project.
6. **Authorize** — Permit to begin construction is issued for the specific site and project. The permit is usually required to be displayed at the construction site, similar to a business license.

The administrative chapter of the model building code contains specific provisions for making a building permit application. An application may require such information as the legal address of the property, its intended use and occupancy, a description of the work to be done, and the estimated value of the work. The application is normally a local document provided by the AHJ in a format suitable to the jurisdiction.

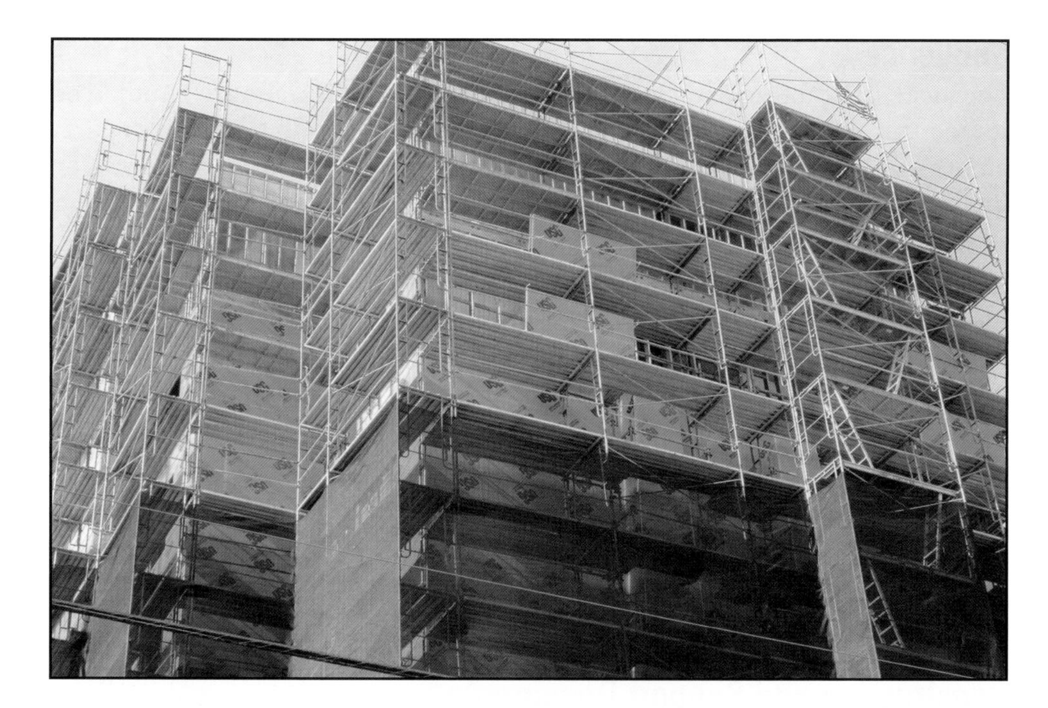

Figure 15.5 Several agencies may need to review large construction plans. *Courtesy of Rich Mahaney.*

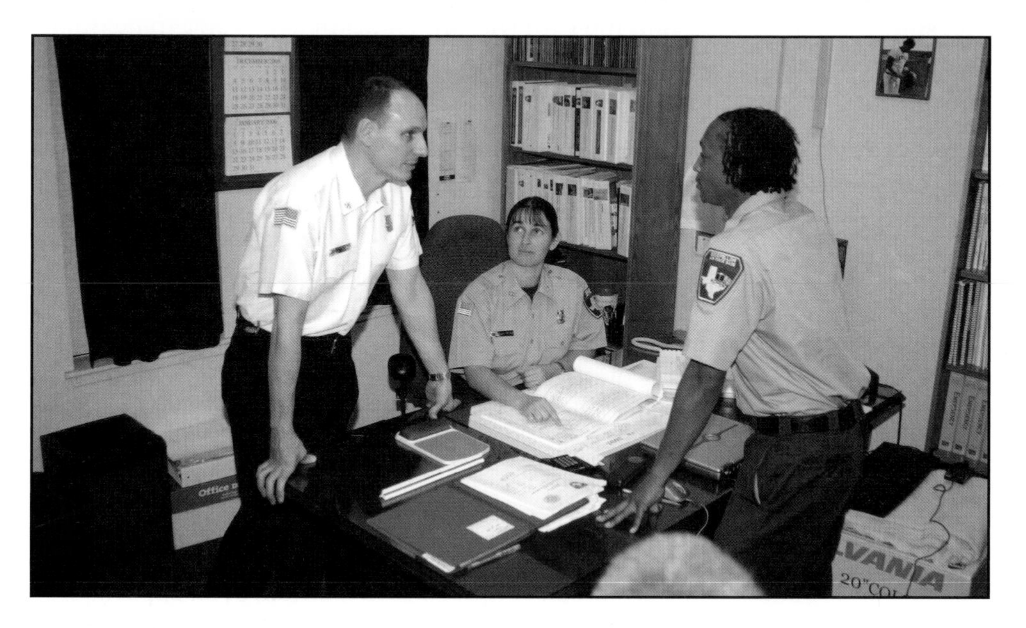

Figure 15.6 Plans examiners review the submitted documents.

The application may request other information required by the building official, such as the identifications and licenses of the various contractors (plumbers or electricians, for example). The local jurisdiction may require that contractors be bonded and licensed.

During the plans review and construction processes, the client may make minor revisions, called **change orders**, to the plans. Plans examiners must maintain records of these communications for possible reference. Inspectors should be aware of all changes to the construction plans and inspect the project to verify that the changes are made as approved. In some municipalities, all the changes made throughout the project must be collated into a single set of documents and submitted as a record set of construction documents for the project. This set of drawings is known as an "as-built" set of documents.

Change Order — Client's written order to a contractor, issued after execution of the construction contract, which authorizes a change in the construction work, project completion time, and/or cost of the project.

Sequence

The local jurisdiction determines the manner in which plans are handled or routed through the review process. In some jurisdictions, building department examiners review the plans before plumbing, electrical, mechanical, or fire protection reviewers because this order follows the general construction sequence.

Ideally, all plans for a project are submitted to the reviewing authority at the same time. There are circumstances, however, where this procedure may not be the best approach. Examples are very large structures like shopping malls that may take years to complete.

Fees

Agencies that review construction documents may charge a fee to defray the administrative costs associated with the review. These fees are established by ordinances and are similar to other municipal fees. The manner in which fees are determined varies with the jurisdiction.

Construction and Support Documents

The submitted plans need to have all the information necessary to evaluate compliance with the locally adopted code. The plans also help an inspector know what to expect during construction.

All of the information needed for the plans review must be clearly visible and not obscured because of the size or print quality of the drawing. Details such as sprinkler-riser components or other structural elements may be required to be in a larger scale than the floor plan or elevation drawings **(Figure 15.7)**.

Construction documents that are submitted for review require supporting documents in addition to drawings. For example, submitting plans for a hydraulically designed automatic sprinkler system would not be complete without hydraulic calculations. Because the calculations are usually computer generated, they cannot be conveniently included on the actual drawings and are provided as separate, supporting documents. Other supporting documents can include material specifications, safety data sheets (SDS), and manufacturers' product submittal information sheets (also called manufacturers' cut sheets, catalog sheets, or cut sheets) **(Figures 15.8 a and b, p. 664)**.

Manufacturers' product submittal information sheets are basically illustrations of individual components taken from a manufacturer's catalog. When components are unique, such as a fire alarm system annunciator unit, cut sheets can be very helpful. Cut sheets often include certification from nationally recognized laboratories stating that the proposed product has been tested and evaluated as suitable for the installation.

Performance-Based Design

Performance-based design is a formalized engineering process that involves technical analysis, materials testing, human-factor studies, and the application of the principles of fire protection engineering. Qualified individuals must develop performance-based designs. Those involved in the development of the final design must be licensed design professionals. Products or processes that are developed using performance-based designs must be accompanied by complete documentation.

Because of its complexity, performance-based design is usually applied only in large projects where standard code application is not practical. In most projects that plans examiners review, designers find it more practical to use prescriptive (required) codes. However, validated performance-based criteria may be useful in addressing elements of a project requiring alternate methods. NFPA® 1, *Fire Code,* and the *International Fire Code* have chapters that address performance-based design issues.

Dealing with supporting documents in the plans review process can become cumbersome. Project specifications, for example, are part of the contractual agreements between owners and contractors and are often given in a specifications book. In fact, the term *project manual* is sometimes used in place of specifications book. Because of the bulk of a specifications book and the fact that much of it is legal in nature, some jurisdictions do not review specifications or require their submittal. In these jurisdictions, all technical data required for approval of plans must be placed on the plans.

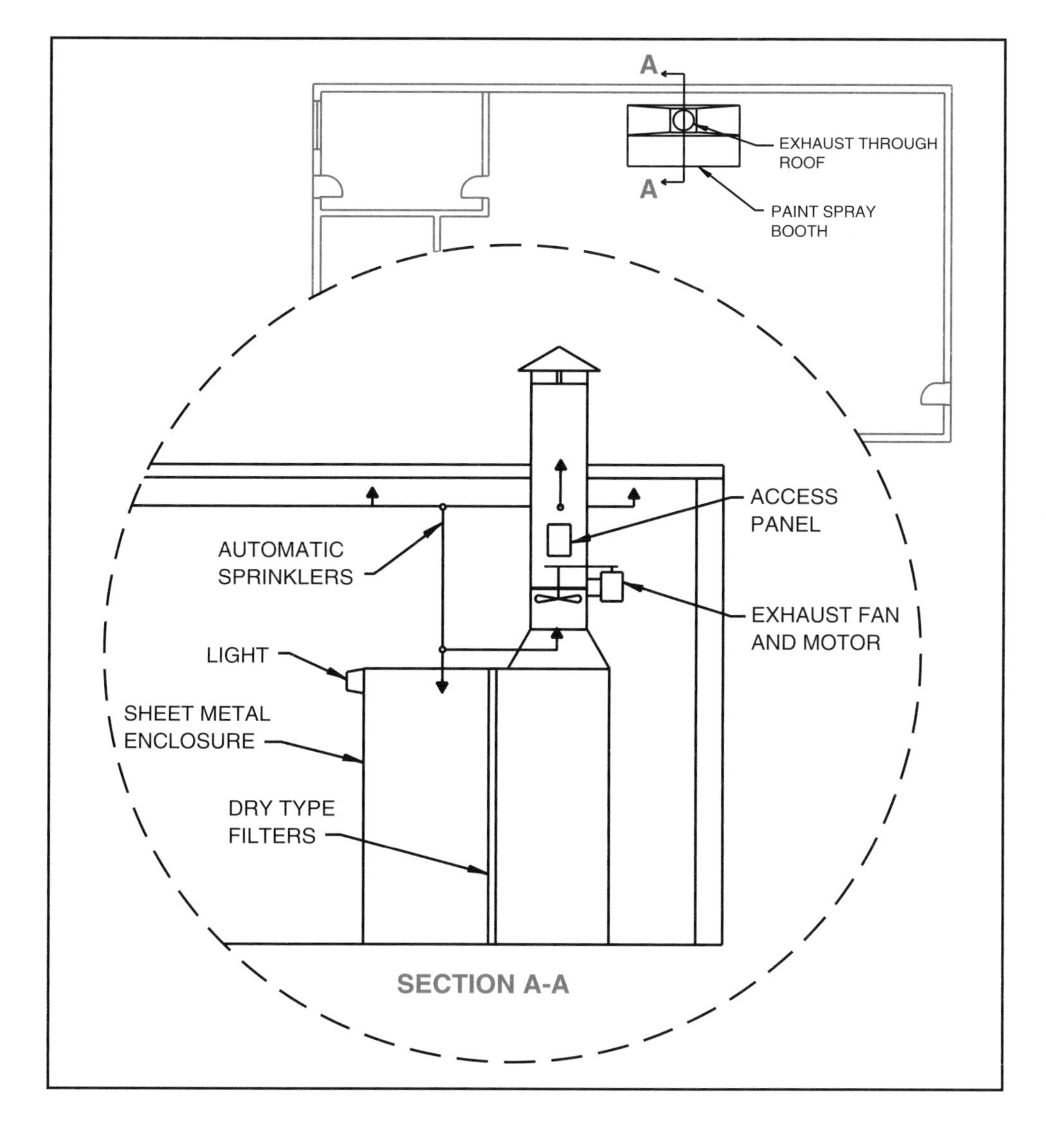

Figure 15.7 Floor plan showing details of a paint spray booth.

CENTRAL 570 & 580 4-01

Butterfly Valve

Figures 570 and 580

Tyco Fire Products --- www.tyco-grooved.com
451 North Cannon Avenue, Lansdale, Pennsylvania 19446 --- USA
Customer Service/Sales: Tel: (215) 362-0700 / Fax: (215) 362-5385
Technical Services: Tel: (800) 381-9312 / Fax: (800) 791-5500

General Description

See Fire Protection Submittal Sheet for pressure rating and Listing and Approval information

Figures 570 and 580 Butterfly Valves are specifically designed to provide for efficient control of fire protection water supplies. The Figure 570 is designed to operate in systems up to and including 175 psi. The Figure 580 is designed to meet the increasing pressure requirements of the Fire Protection Industry with a maximum operating pressure of 300 psi. Flow may be from either direction, and the valves may be positioned in any orientation. The valves are furnished with grooved ends and can be easily adapted to flanged components utilizing our Figure 71 Class 150 flange adapters. The body and disc construction provides for increased strength and durability. The Figure 570 and 580 Butterfly Valves are provided with 2 sets of SPDT Supervisory Switches for use in outdoor and indoor applications. A high strength conventional stainless steel upper stem is provided for dependability as compared to iron upper stems provided by others. The surfaces at the upper stem and lower trunnion areas incorporate a reduced dynamic torque and anti-compression set design to ensure low operating torque and increased seal longevity.

This innovative feature prevents elastomeric failure of the disc encapsulation that is commonly experienced with other manufacturers. This is accomplished by providing uniform compression throughout the opening and closing operation of the disc. Other manufacturers have non-uniform compression, which results in permanent compression set and subsequent splitting of the material and undermining of the seal.

Technical Data

Figure: 570 and 580

Sizes:
570 - 2½", 3", 4", 5", 6", 8", 10" and 12"
580 - 2½", 3", 4", 5", 6", 8", 10" and 12"

Max. Working Pressure:
Figure 570 = 175 psi *(1,207 kPa)*
Figure 580 = 300 psi *(2,068 kPa)*

Approvals:
UL, FM, and ULC (indoor and outdoor use) Note:The 8" - 12" Figure 580 are FM Approved only.

Manufacturing Source: Domestic

Materials of Construction:

Body:
Ductile iron in accordance with ASTM A-536, Grade 65-45-12

Body Coating: Orange Epoxy

Disc:
Ductile Iron in accordance with ASTM A-536, Grade 65-45-12.

Disc Seal:
Grade E EPDM encapsulated rubber in accordance with ASTM D-2000
See Gasket Grades and Recommendations Sheet for aid in selecting the proper seal.

Upper Stem: (Figures 570 & 580)
Type 440 Stainless Steel (2½" - 8")
Type 17-4 Stainless Steel (10" - 12")

Lower Plug/Stem:
Figure 570: C360 Brass (2½" - 8")
Type 17-4 Stainless Steel (10" - 12")

Lower Stem:
Figure 580:
Type 17-4 Stainless Steel (2½" - 12")

Operator: Gear operator with Iron housing - coated with Black Epoxy.

Bracket: Steel - Black Zinc Plated

Factory Hydro Test:
100% @ two times pressure rating in accordance with MSS-SP-67, UL, ULC, and FM requirements.

Installation

In piping systems, butterfly valves should be located where they will be readily accessible for operation, inspection, and maintenance.

When a valve "closes hard", it may be due to debris lodged in the sealing area. This may be corrected by backing-off the handwheel and closing it again, several times if necessary. The valve should never be forced to seat by applying a wrench to the handwheel as it may distort the valve components or score the sealing surfaces.

Before installing conduit at the actuator, remove the rubber grommet at the connection.

Important: All replacement parts must be obtained from the manufacturer to assure proper operation of the valve.

To prevent rotation of the valve, it is recommended that the Butterfly Valve be installed with rigid type couplings, such as the Figure 572 or 772 Coupling. If flexible couplings are used, additional support may be needed to prevent rotation.

Ordering Procedure

Ordering Information: When placing an order, indicate the full product name. Please specify the quantity, figure number, type of disc seal, EPDM "E", and size.

Availability and Service: Grooved Piping Products, valves, accessories and other products are available throughout the U.S., Canada, and internationally, through a network of distribution centers. You may write directly or call 215-362-0700 for the distributor nearest you.

a

No. 15-3.0

Figures 15.8 a and b Manufacturer's cut sheets providing detailed information for a butterfly valve. *Courtesy of Tyco Fire Suppression & Building Products.*

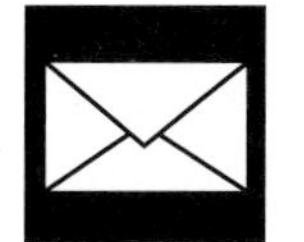

Limited Warranty

Products manufactured by Tyco Fire Products are warranted solely to the original Buyer for ten (10) years against defects in material and workmanship when paid for and properly installed and maintained under normal use and service. This warranty will expire ten (10) years from date of shipment by Tyco Fire Products. No warranty is given for products or components manufactured by companies not affiliated by ownership with Tyco Fire Products or for products and components which have been subject to misuse, improper installation, corrosion, or which have not been installed, maintained, modified or repaired in accordance with applicable Standards of the National Fire Protection Association (NFPA), and/or the standards of any other Authorities Having Jurisdiction. Materials found by Tyco Fire Products to be defective shall be either repaired or replaced, at Tyco Fire Products' sole option. Tyco Fire Products neither assumes, nor authorizes any person to assume for it, any other obligation in connection with the sale of products or parts of products. Tyco Fire Products shall not be responsible for sprinkler system design errors or inaccurate or incomplete information supplied by Buyer or Buyer's representatives.

IN NO EVENT SHALL TYCO FIRE PRODUCTS BE LIABLE, IN CONTRACT, TORT, STRICT LIABILITY OR UNDER ANY OTHER LEGAL THEORY, FOR INCIDENTAL, INDIRECT, SPECIAL OR CONSEQUENTIAL DAMAGES, INCLUDING BUT NOT LIMITED TO LABOR CHARGES, REGARDLESS OF WHETHER TYCO FIRE PRODUCTS WAS INFORMED ABOUT THE POSSIBILITY OF SUCH DAMAGES, AND IN NO EVENT SHALL TYCO FIRE PRODUCTS' LIABILITY EXCEED AN AMOUNT EQUAL TO THE SALES PRICE.

THE FOREGOING WARRANTY IS MADE IN LIEU OF ANY AND ALL OTHER WARRANTIES EXPRESS OR IMPLIED, INCLUDING WARRANTIES OF MERCHANTABILITY AND FITNESS FOR A PARTICULAR PURPOSE.

Care and Maintenance

The owner is responsible for the inspection, testing, and maintenance of their fire protection system and devices in accordance with the applicable standards of the National Fire Protection Association (e.g., NFPA 25), in addition to the standards of any authority having jurisdiction. The installing contractor or product manufacturer should be contacted relative to any questions. Any impairment must be immediately corrected.

It is recommended that automatic sprinkler systems be inspected, tested, and maintained by a qualified Inspection Service.

Figures 570 and 580

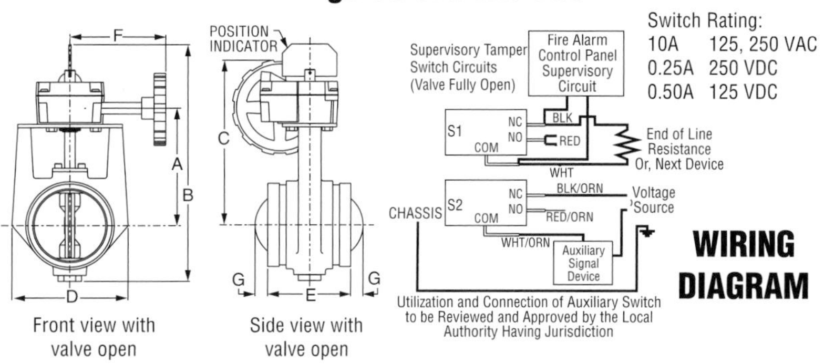

Front view with valve open

Side view with valve open

Switch Rating:
10A 125, 250 VAC
0.25A 250 VDC
0.50A 125 VDC

WIRING DIAGRAM

Friction Resistance	
Size	Equiv. Length in Feet
2½"	6'
3"	7'
4"	6'
5"	10'
6"	13'
8"	14'
10"	16'
12"	18'

Size	A Inches mm	B Inches mm	C Inches mm	D Inches mm	E Inches mm	F Inches mm	G Inches mm	Weight Lbs. Kg.
2½"	5.34 135.6	10.41 264.4	7.22 183.4	2.88 73.0	3.81 96.8	5.72 145.3	N/A*	12.0 5.4
3"	5.67 144.0	11.38 289.1	8.75 222.3	3.50 88.9	3.81 96.8	5.72 145.3	N/A*	14.0 6.4
4"	6.76 171.7	12.70 322.6	9.65 245.1	4.75 120.7	4.56 115.8	5.72 145.3	N/A*	22.0 10.0
5"	7.15 181.6	14.56 369.8	10.23 259.8	6.25 158.8	5.81 147.6	6.18 157.0	N/A*	31.0 14.1
6"	7.88 200.2	15.23 386.8	10.96 278.4	6.75 171.5	5.81 147.6	6.18 157.0	N/A*	36.0 16.3
8"	9.24 234.7	17.50 444.5	13.37 339.6	10.00 254.0	5.25 133.4	6.43 163.3	1.22 31.0	52.0 23.6
10"	11.81 299.9	21.78 553.2	16.93 430.0	12.00 304.3	6.25 158.8	7.96 202.2	1.75 44.5	75.0 34.1
12"	12.97 329.4	24.11 612.4	19.10 485.1	14.27 362.5	6.50 165.1	8.83 224.3	2.60 66.0	90.0 40.9

* END OF DISC DOES NOT EXTEND BEYOND VALVE BODY.
Note: Contact Tyco Fire Products for 10" and 12" availability.

General Notes: It is the Designer's responsibility to select products suitable for the intended service and to ensure that pressure ratings and performance data is not exceeded. Always read and understand the installation instructions (IH-1000). Never remove any piping component nor correct or modify any piping deficiencies without first depressurizing and draining the system. Material and gasket selection should be verified to be compatible for the specific application.

Certified Company

b

tyco / Flow Control / Tyco Fire Products

Printed in U.S.A. 4-01

Figure 15.8 b

Building Construction Plans

One key part of a set of construction documents are the graphical representation of the project known as working drawings. The working drawings detail the design and building's construction using four main views:

- Plan
- Elevation
- Sectional
- Detailed

Each of these drawings should have a title block that contains specific information about the drawing and the project **(Figure 15.9)**. The originators of the plan determine the format or location of the information. The title block generally contains the following information:

- Title of the drawing, such as Basement Floor Plan or Site Plan
- Description of the project
- Scale of the drawing; for example, ⅛" = 1'0" (⅛-inch drawn on the paper is equal to 1 foot of actual construction)

NOTE: Scaling the drawing provides a view in exact proportion to the actual size of the building. Sometimes the scale is placed on the drawing rather than in the title block. If a drawing is not to scale, there will be a note indicating that it is not. Copies of old drawings are often not drawn to scale.

- Date the drawing was completed, any revisions (if applicable) made to the drawing, and the date of those revisions
- Name of the firm producing the drawings
- Name (usually initials) of the person who made the drawings and the person responsible for checking the final drawings
- Sheet number in the set
- The design professional's official stamp or seal as issued by a state architectural/engineering registration board when a licensed architect or engineer supervised the design

The working drawings on a commercial construction project are typically divided into groups by individual discipline, such as architectural, structural, mechanical, and electrical. Sheets containing a specific type of information are marked accordingly. For example, architectural information would be marked A1, A2, A3; structural sheets would be marked S1, S2, S3; mechanical sheets

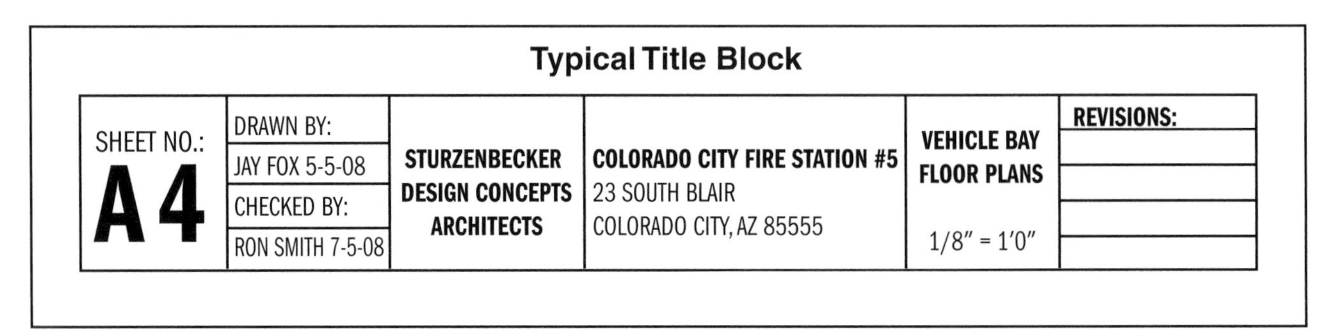

Figure 15.9 The title block identifies the type of information found in the plans.

would be marked M1, M2, M3; and electrical sheets would be marked E1, E2, E3. Although architectural, structural, mechanical, and electrical sections are typical, many larger plans include sections for plumbing, fire protection, site plans, and landscaping.

Plan Views

Plan View — Drawing containing the two-dimensional view of a building as seen from directly above the area.

A **plan view** is a two-dimensional view of the site or the building as seen from directly above the area. A plan view provides information concerning the overall layout of the site or building. An inspector or plans examiner will commonly see two basic plan views: site and floor. Occasionally, ceiling and roof plans will also be submitted.

Site Plan

The site (or plot) plan is usually one of the first sheets of a set of construction drawings. It identifies conditions currently existing on the site and relates information needed to locate the building **(Figure 15.10)**. The site plan usually includes the following pieces of information:

- **North direction symbol** — Usually points toward the top of the sheet, which conforms to universal standards used in the drawing of all maps **(Figure 15.11)**.
- **Lot dimensions** — Often shown as a broken line with the dimensions expressed in feet (meters) and decimal fractions of a foot (meter); property lines are legal instruments that define the land boundaries for a particular site. The location of the building in relation to the property line determines the fire-resistance rating of the exterior wall of the building.

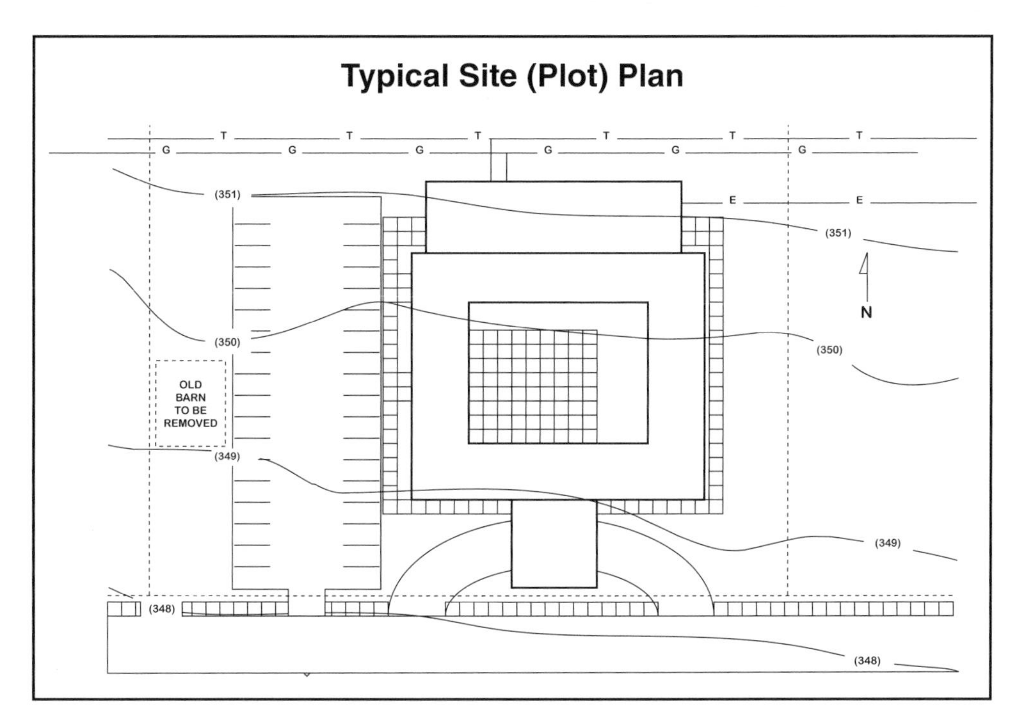

Figure 15.10 The site plan identifies conditions currently existing at the site.

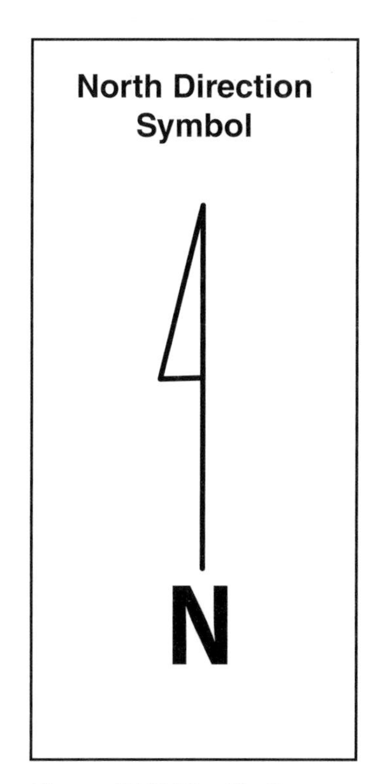

Figure 15.11 The North direction symbol points to the top of the plan.

- **Utility lines** — Shown as broken lines marked at intervals with a letter (gas, water, sewer, power, and communication); in addition to the character in the broken line, a legend on the map also identifies the type of utility service line **(Figure 15.12)**.

- **Structures to be removed** — Drawn with a light broken line **(Figure 15.13)**. A brief identification of these structures and objects is also provided on the plan.

Utility Service Line Locations

Water lines:---w---w---w---w---w---w---w---w

Sewer lines:---s---s---s---s---s---s---s---s

Power lines:---p---p---p---p---p---p---p---p

Gas lines:---g---g---g---g---g---g---g---g---g

Figure 15.12 Utility lines on site plans feature a letter corresponding to the type of utility that follows the line of the site.

Benchmark — (1) Permanently affixed mark such as a stake driven into the ground that establishes the exact elevation of a place; used by surveyors in measuring site elevations or as a starting point for surveys. (2) Standard measurement that forms the basis for comparison.

- **Contour lines** — Display the elevation of the property above sea level (ASL) or based on the elevation of the **benchmark**. Proposed contour lines show the planned elevations after grading is completed. This information is important when determining grade elevations for fire apparatus access to the site. The information is also useful in determining the ability of the fire department to place aerial apparatus and ground ladders for access to upper stories of a structure.

- **Fire apparatus access** — Illustrate the private road network serving one or more buildings on a property; it will include width, turning radius, and load-bearing capacity. Locations for fire department connections (FDCs), hydrants, or aerial apparatus placement are shown.

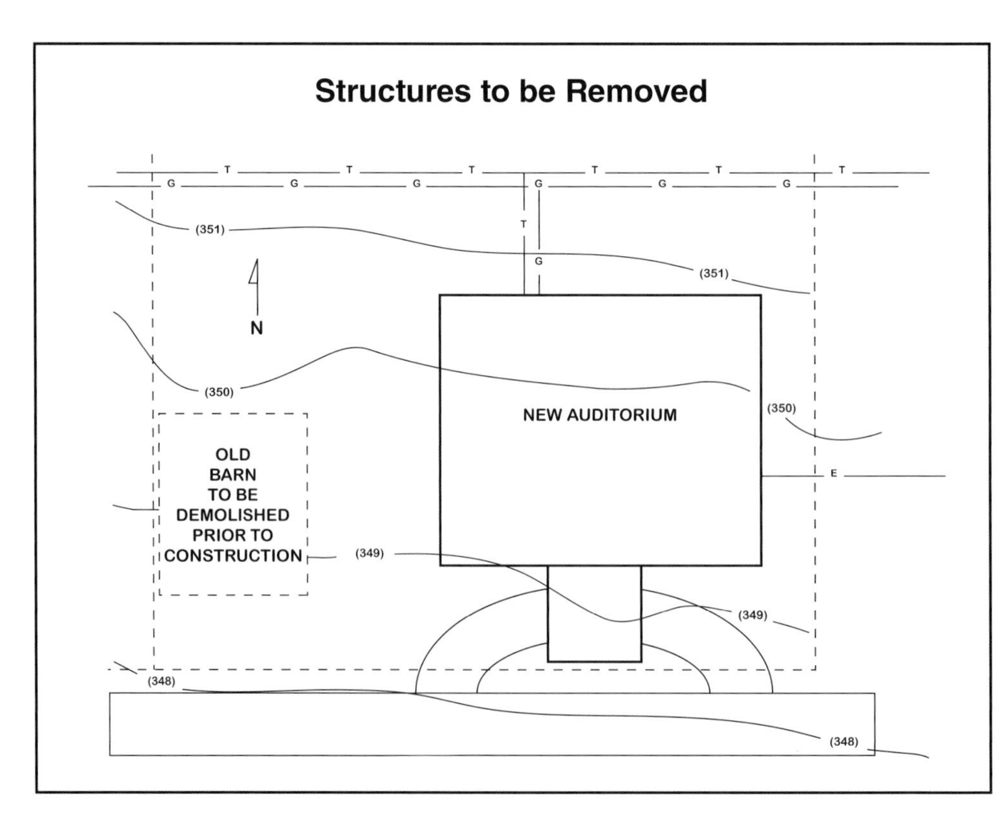

Figure 15.13 A light broken line identifies structures to be removed during construction.

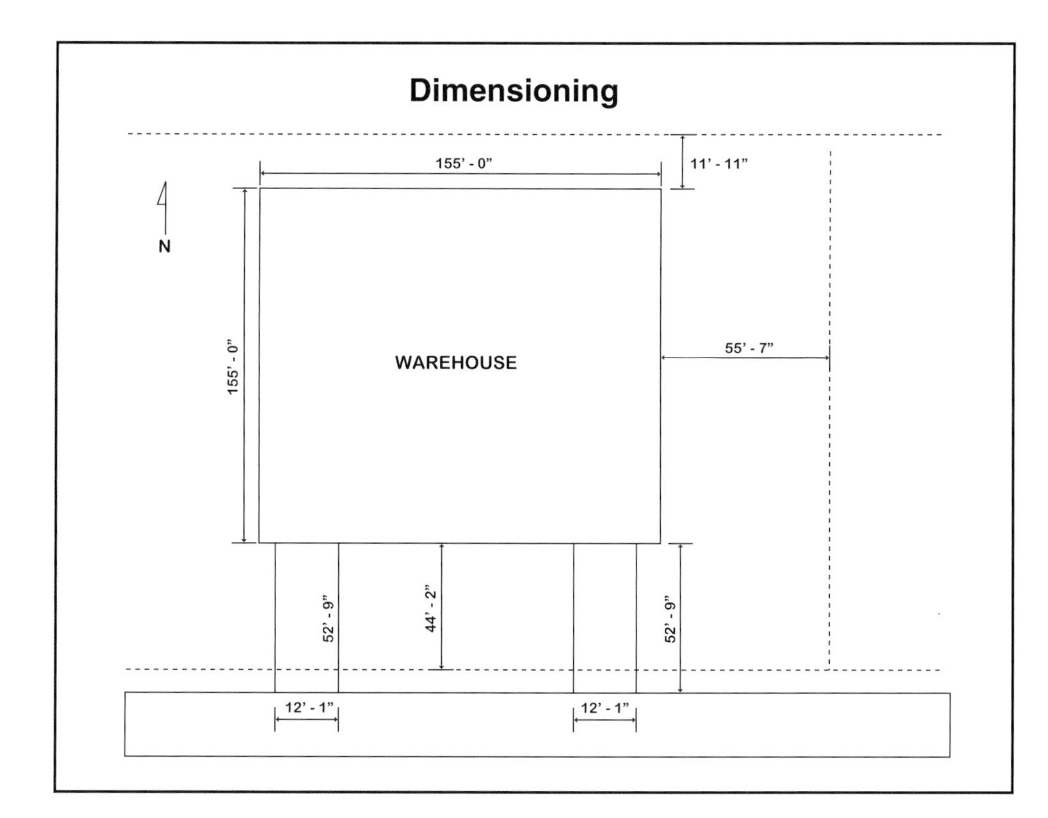

Figure 15.14 Dimensioning indicates the size and location of the building on the site.

Dimensioning — Indicating or determining size and position in space.

The placing of the building on a site plan is called **dimensioning (Figure 15.14)**. The dimensions are usually placed some distance away from the shapes on the drawing through the use of extension lines. Dimensions are expressed in feet and inches (meters and millimeters).

Floor Plan

A floor plan is a horizontal drawing of a structure at a given level, such as the ground floor, basement, or roof. The plan provides general information on the types of compartments (rooms or work spaces) and their intended uses as well as the specific details about the following building components:

- External walls
- Internal walls and partitions
- Fire separation walls
- Doors
- Windows
- Ceiling and roof joists or trusses
- Stairs
- Ramps
- Building services (such as ventilation shafts and mechanical spaces)
- Elevators or escalators

Depending on the intended use of the structure, the floor plan may include only the location of permanent walls, doors, lighting and plumbing fixtures, and electrical outlets. However, the plan may also reflect the final appearance of the structure with the location of movable walls or partitions, cabinets, shelves, and work spaces **(Figure 15.15, p. 666)**.

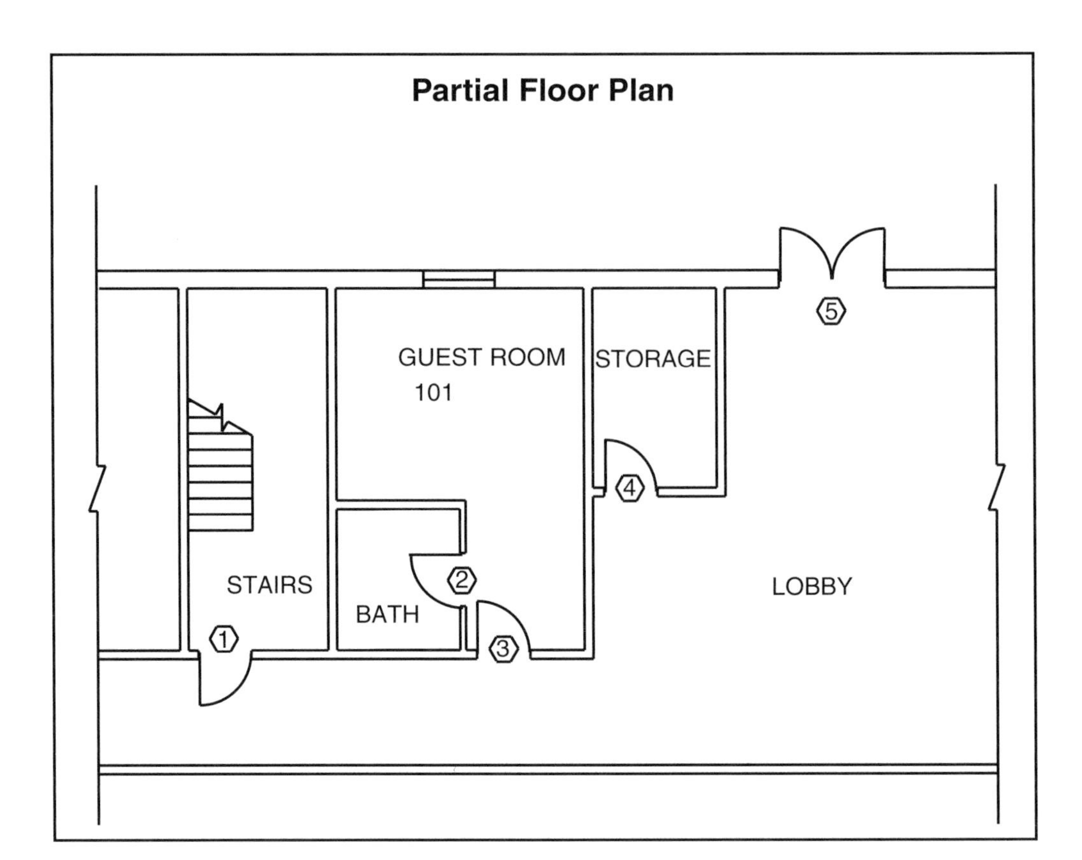

Figure 15.15 The floor plan shows the relative location of walls, doors, windows, and other features.

The usual symbol for exterior walls is a pair of parallel lines. The materials from which the wall is to be constructed may be listed on a floor plan. However, construction materials may also be listed in a sectional view of the wall. This information is also contained in the construction documents.

Clear Dimensions — Interior compartment measurements made from the inside surface of one wall to the inside surface of the opposite wall.

The interior walls or partitions are also drawn with parallel lines. The function of each room may be listed on the sheet where the room is located or the rooms may just be numbered as with proposed offices. Room dimensions should be shown as **clear dimensions** (wall surface to wall surface) or they can be shown centerline to centerline of the walls.

The windows and doors on the drawings are coded by letters or numbers enclosed in a circle or other geometric shape. Window sizes are found on some floor plans but usually are found on window schedules located in the construction documents. Door and window schedules are also used to indicate the type and installation of the fixture and its hardware.

Elevation View — Architectural drawing used to show the number of floors of a building, ceiling heights, and the grade of surrounding ground.

The symbol for a standard hinged door is a single line drawn from the hinge side of the doorway. An arc indicates the door's direction of swing. An inspector or plans examiner must be able to determine whether the location and swing of the door meet the exit requirements of the local building and fire codes **(Figure 15.16)**.

Orientation — Location or position relative to the points of the compass.

Elevation View

The **elevation view** is a two-dimensional view of a building as seen from the exterior **(Figure 15.17)**. The labeling and **orientation** of the drawing on the elevation view are usually defined by the AHJ within its plan submittal requirements. There are four exterior elevation views on any rectangular building, and they can be labeled using one of the following two methods:

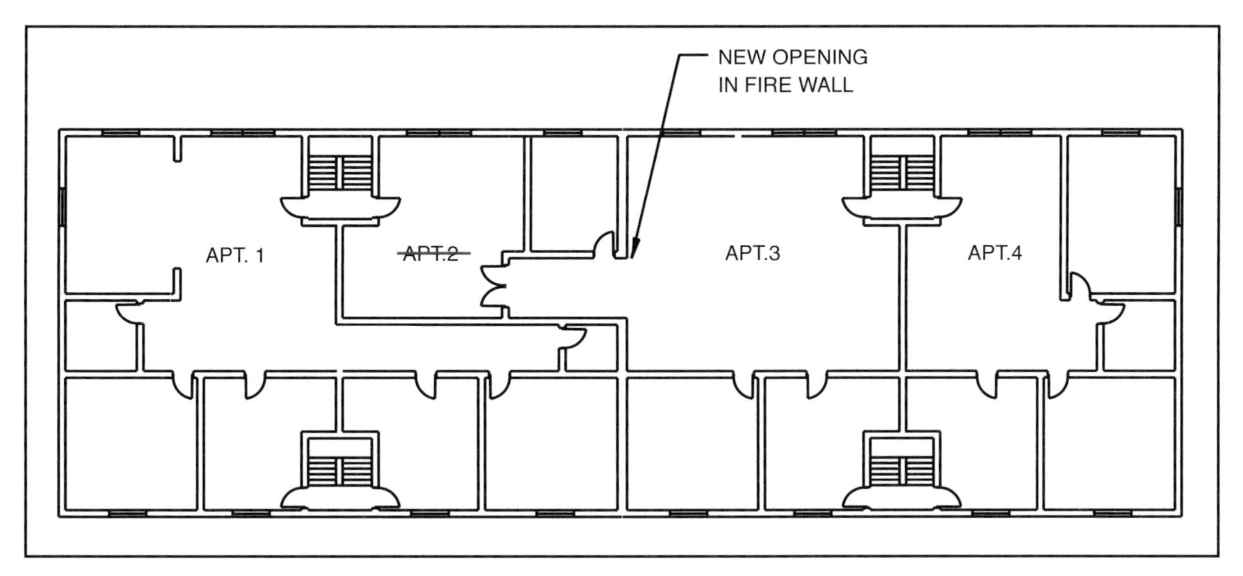

Figure 15.16 Proposed layout for three apartments.

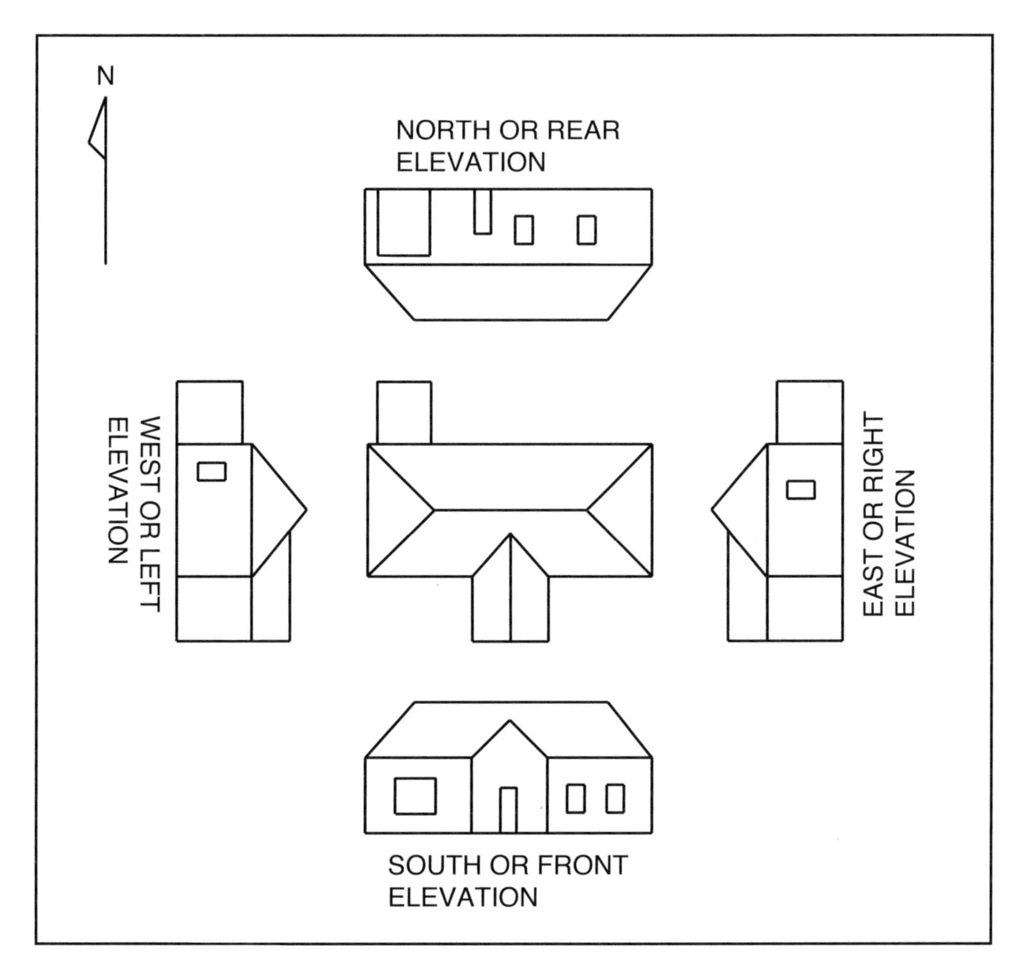

Figure 15.17 Exterior elevation views of a building.

1. If a wall is facing north, it is labeled the north elevation. The other views are then labeled according to the direction they face.
2. The front elevation faces the front of the street, and the rear elevation faces the back. When viewed from the front of the building, the left elevation is on the left side of the building and the right elevation is on the right side of the building.

Elevation views contain information about the exterior components of the building, such as type(s) of exterior finish(es) and locations, sizes, and types of doors and windows. On an elevation view as well as the site plan, the grade level is shown as a dark line. Utility service line connection points and FDCs will also appear on an elevation drawing.

Sectional View — Vertical view of a building as if it were cut into two parts; intended to show the internal construction of each assembly.

Sectional View

A **sectional view** is a vertical view of a building as if it were cut into two parts **(Figure 15.18)**. The purpose of a sectional view is to show the internal construction of each assembly. There are three types of sectional views:

- Cross-section of the complete building from exterior wall to exterior wall and from foundation to roof
- Typical cross-section of a common construction feature such as a wall assembly
- Detailed cross-section of some construction feature such as an atrium or mezzanine to show exact detail

Sectional views show a number of construction details:

- **Exterior walls** — Exterior finish, sheathing, studs, plates, insulation (thermal), vapor barrier, interior finish, and base detail
- **Interior walls** — Interior finish, insulation (sound), studs (wood or metal), sills and plates
- **Floor** — Finished floor and subfloor as well as the size and type of floor joists, bridging and location, and floor-support beams
- **Roof** — Roof pitch, roofing material, eaves protection, sheathing, eave troughs, soffit and fascia details, roof structure (truss or rafters), and insulation **(Figure 15.19)**
- **Foundation** — Wall height and thickness, width and thickness of footings (pad and strip), location and size of weeping tile, location and depth of granular fill, thickness of floor slabs, location of expansion joints, and protective coatings.

The process of listing the construction materials is referred

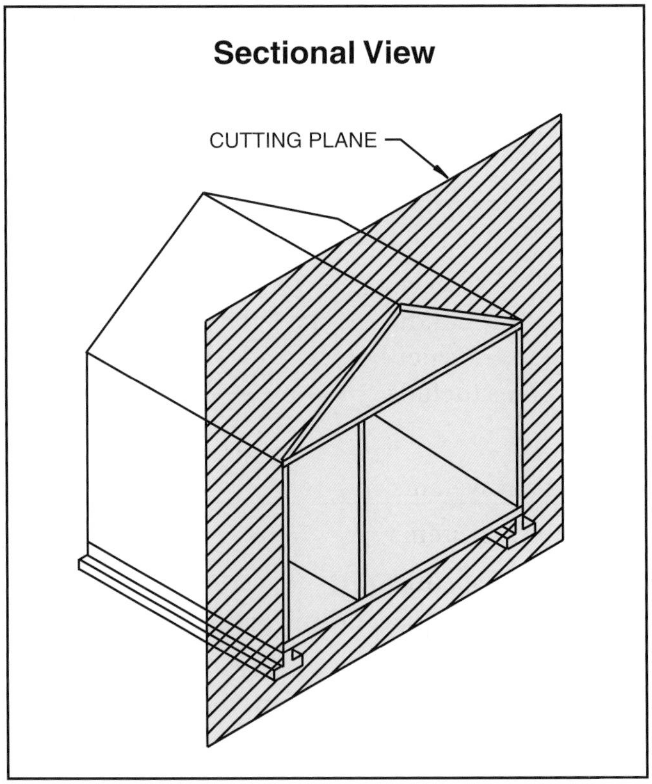

Figure 15.18 A sectional view shows the internal construction of each building assembly as if the structure was cut in two.

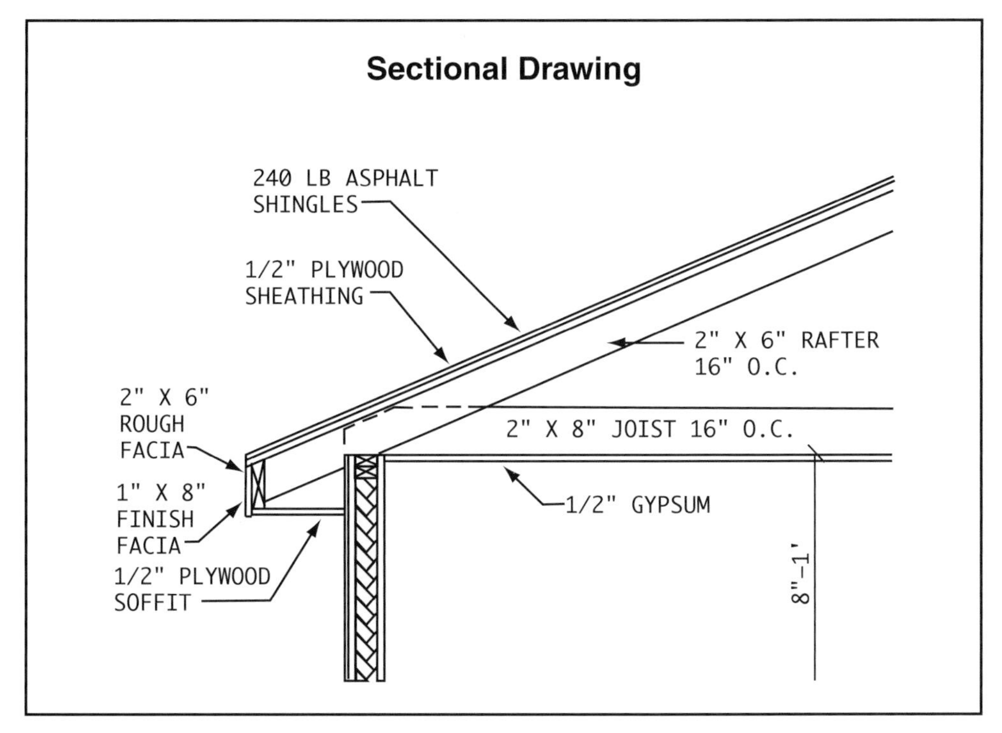

Figure 15.19 The specific design features of a roof are illustrated in a sectional drawing of that assembly.

to as calling up the materials. This procedure is used to describe wall, floor, and ceiling/roof assemblies. The materials are listed on the drawing or in the construction documents.

Detailed View

Detailed View — Additional, close-up information shown on a particular section of a larger drawing.

Detailed views show a feature in a larger size than what is used on the small-scale floor plan or elevation drawings **(Figure 15.20, p. 670)**. The additional information is provided in a larger scale on a separate sheet of the plans. For example, a floor plan drawing may show a fire door. A detailed view will show the exact construction of the fire door and its activation mechanism.

System Plans

While plan views depict the relationship between construction components and their location within the site, system plans provide details of the various systems in the building. Additional plans for building systems may be included alongside construction site, floor, elevation, sectional, and detailed plans. System plans include the following:

- Mechanical systems
- Electrical systems
- Plumbing systems
- Automatic sprinkler systems
- Standpipe and hose systems
- Special-agent fire extinguishing systems
- Fire detection and alarm systems

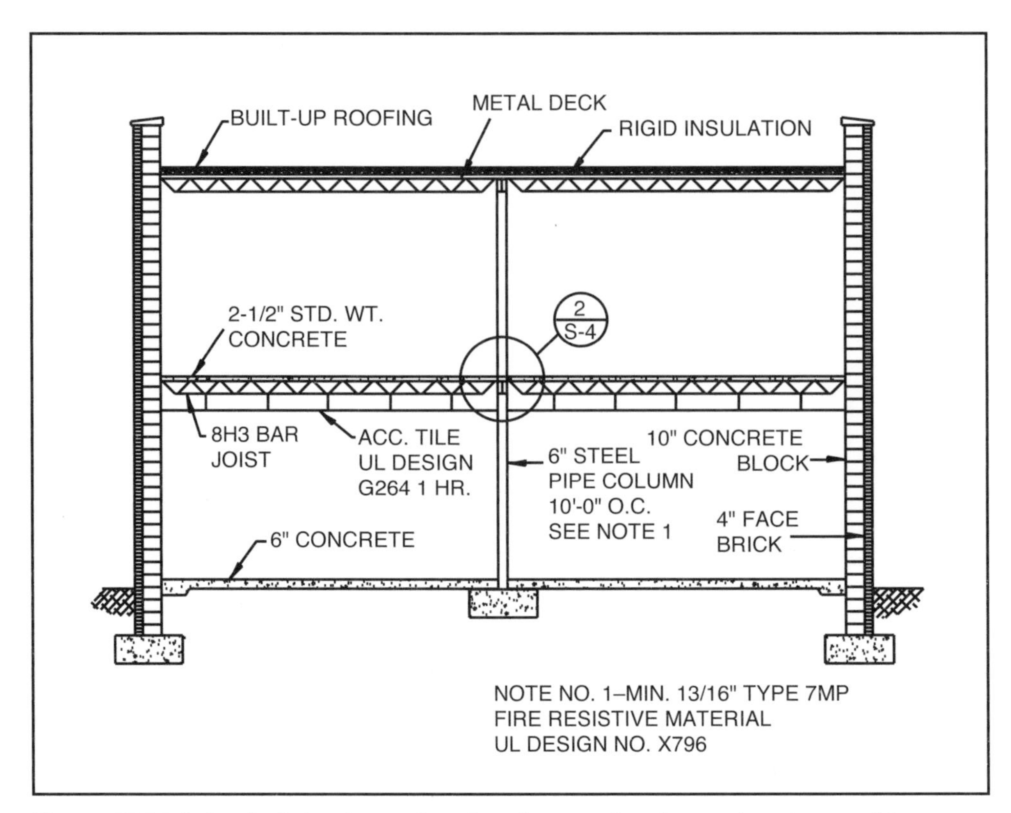

Figure 15.20 A detailed view is used to show features in a larger view that would not otherwise be visible on the small-scale floor plan.

Inspectors may only be concerned with the fire protection systems plans. However, inspectors should be aware that elements of the other plans may influence the review of the fire protection systems plans.

Mechanical Systems

Building mechanical systems provide the building's occupants with HVAC and access to upper and subgrade floors through elevators, lifts, and escalators. These systems are included on mechanical plans or drawings and should be designed by a mechanical engineer or building systems engineer. Because mechanical systems are capable of spreading fire and smoke, an inspector or plans examiner must verify that all detection and alarm systems and control dampers are included in the plans **(Figure 15.21)**.

During the review of the mechanical system plans, an inspector should verify the style and design of the HVAC system (**Figure 15.22**). The type of refrigerant used in the system and listed in the construction documents should be checked for possible health or fire hazards. Inspectors should be familiar with code requirements concerning the automatic de-energizing of heating systems to control the spread of fire and smoke. Plans must indicate the location and type of smoke detectors inside the ductwork used to de-energize air handlers.

Fire and/or smoke dampers may be required where the ductwork penetrates fire-resistive walls or floors. The system plans will indicate the location of these features. Some codes prohibit ductwork from penetrating exit areas. Exits are not permitted to be used for supply or return air. Additional references for damper placement and installation of ductwork are contained in NFPA® 90A, 90B, 92A, 92B, or the locally adopted mechanical code.

NFPA® Standards for HVAC Systems

The type of heating, ventilating, and air-conditioning (HVAC) equipment is determined by the type of fuel to be used in the system. The type of fuel also determines the type of venting needed, required clearances from combustibles, and quantities of air needed for proper combustion of the fuel. Most HVAC code requirements are based on the following standards:

- NFPA® 54, *National Fuel Gas Code,* and ANSI Z223.1, *National Fuel Gas Code*
- NFPA® 90A, *Standard for the Installation of Air-Conditioning and Ventilating Systems*
- NFPA® 90B, *Standard for the Installation of Warm Air Heating and Air-Conditioning Systems*
- NFPA® 92A, *Standard for Smoke-Control Systems Utilizing Barriers and Pressure Differences*
- NFPA® 92B, *Standard for Smoke Management Systems in Malls, Atria, and Large Spaces*
- NFPA® 96, *Standard for Ventilation Control and Fire Protection of Commercial Cooking Operations*

NOTE: The International Code Council combines these requirements into its *International Mechanical Code®*.

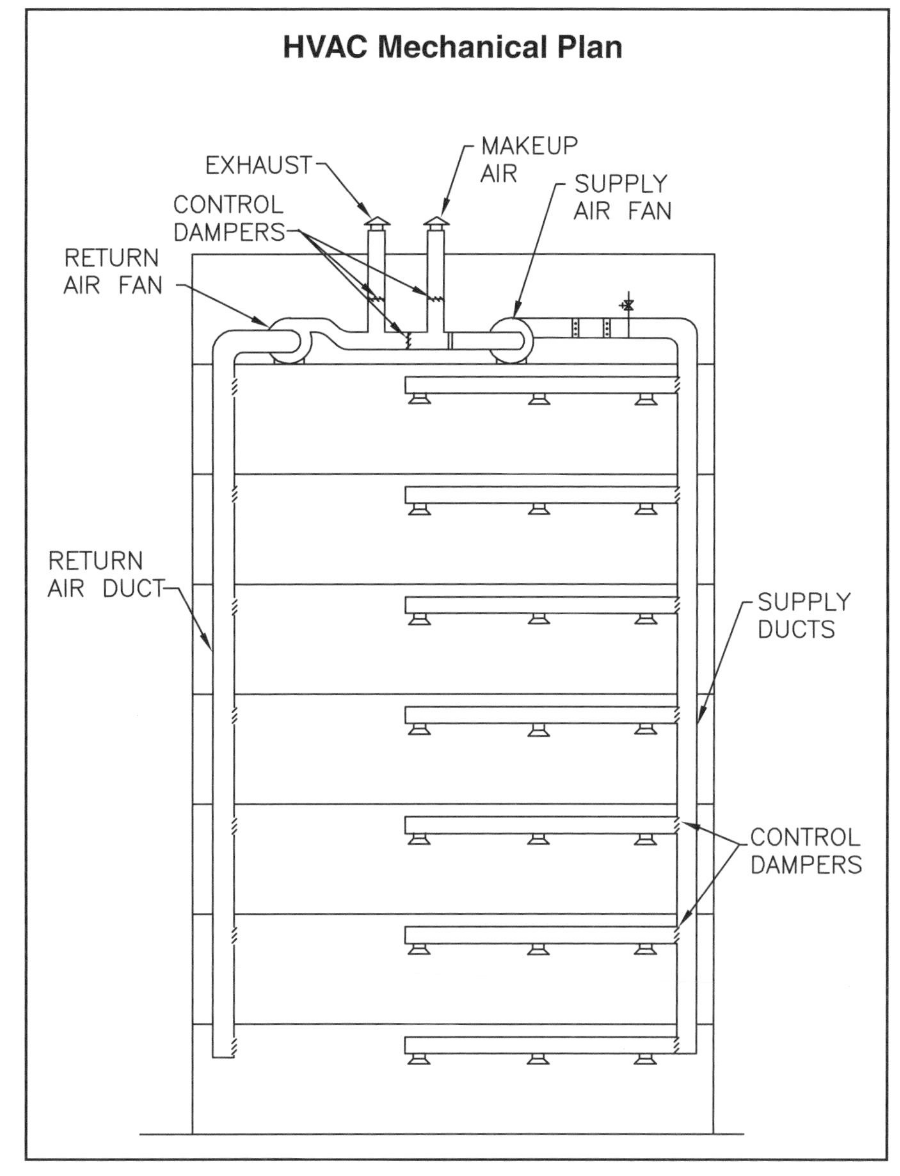

Figure 15.21 The mechanical plan shows features of the building's HVAC system.

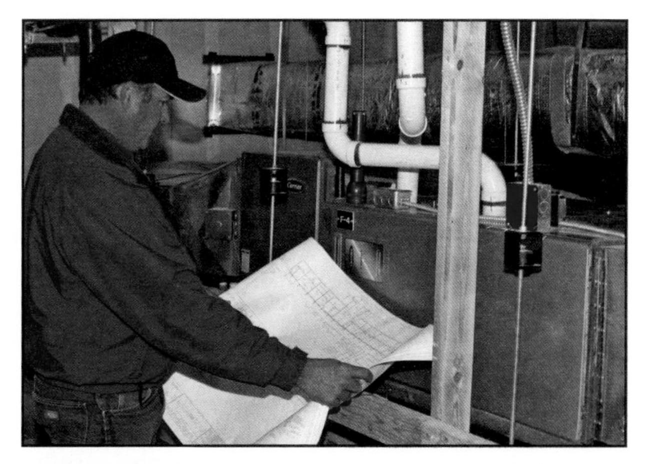

Figure 15.22 An inspector compares the details in a mechanical plan with the actual installation of a HVAC system.

In structures that contain commercial kitchens, an inspector and plans examiner must be familiar with the mechanical and fire code requirements for the following equipment and systems:

- Cooking appliances
- Range ventilation hoods
- Fuel shutoffs **(Figure 15.23, p. 672)**
- Automatic fire suppression systems
- Fire alarm systems

Figure 15.23 In facilities with hazardous fuels, fuel shutoffs are vital for employee safety.

Electrical Systems

General construction plans should include as much information about the electrical systems as possible **(Figure 15.24)**. During a review of the electrical systems, an inspector must verify the following:

- Location of exit signs
- Means of egress lighting
- Emergency and standby power
- Power supply to fire detection and alarm systems

NFPA 70®, *National Electrical Code®,* regulates the design and installation of the electrical system. There are other electrical system components that an inspector or examiner must be aware of:

- **Emergency or standby power systems** — Location of the generator or fuel cell, the source of fuel and required volume, and the location of the transfer switch and annunciator panel.
- **Hazardous location electrical equipment** — Electrical equipment designed so it does not create an ignition source around flammable liquid and gas storage, such as motor fuel dispensers.

Plumbing Systems

Plumbing system plans include the layout of the water distribution system in the structure. The drawings may also include the wastewater piping, storm drains, roof drains, and secondary containment and drainage systems for hazardous materials.

An inspector may be required to review these plans to determine whether the water supply is adequate to support the fire protection systems in the structure. Adequate water supply is important when residential automatic sprinkler systems are installed according to NFPA® 13D, *Standard for the Installation*

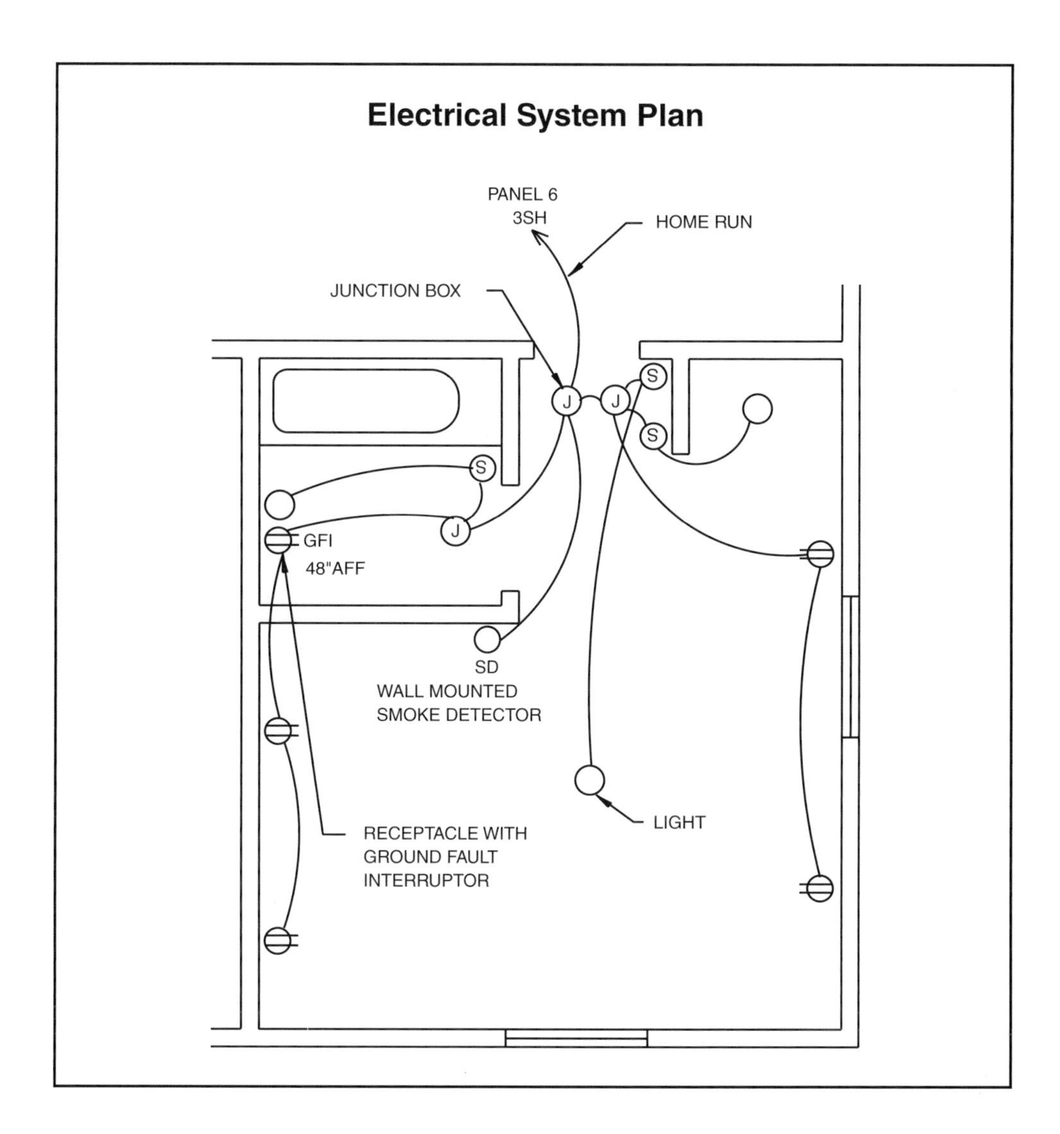

Figure 15.24 Wiring details are shown in the electrical plan.

of Sprinkler Systems in One- and Two-Family Dwellings and Manufactured Homes, and NFPA® 13R, *Standard for the Installation of Sprinkler Systems in Low-Rise Residential Occupancies*. Other fire protection system components that may appear on the plumbing plans are water supply connections to the standpipe system and fire pump.

Automatic Sprinkler Systems

A thorough, accurate review of sprinkler system plans is necessary to verify both code compliance and effectiveness of the system when it is installed **(Figure 15.25, p. 674)**. Inspectors need to be familiar with different types of automatic sprinkler systems, their operations, and specifications for each type of system. See Chapter 12, Water-Based Fire Suppression Systems, for additional information.

When required, an inspector needs to be able to develop a field inspection checklist of the specific requirements for sprinkler systems and the code sections that relate to the requirements. This checklist can be used as a guide during sprinkler plans review. See **Appendix I** for an automatic sprinkler systems acceptance test checklist.

Before the sprinkler system plans review begins, the plans reviewer verifies that the design professional has applied the proper NFPA® standard. This

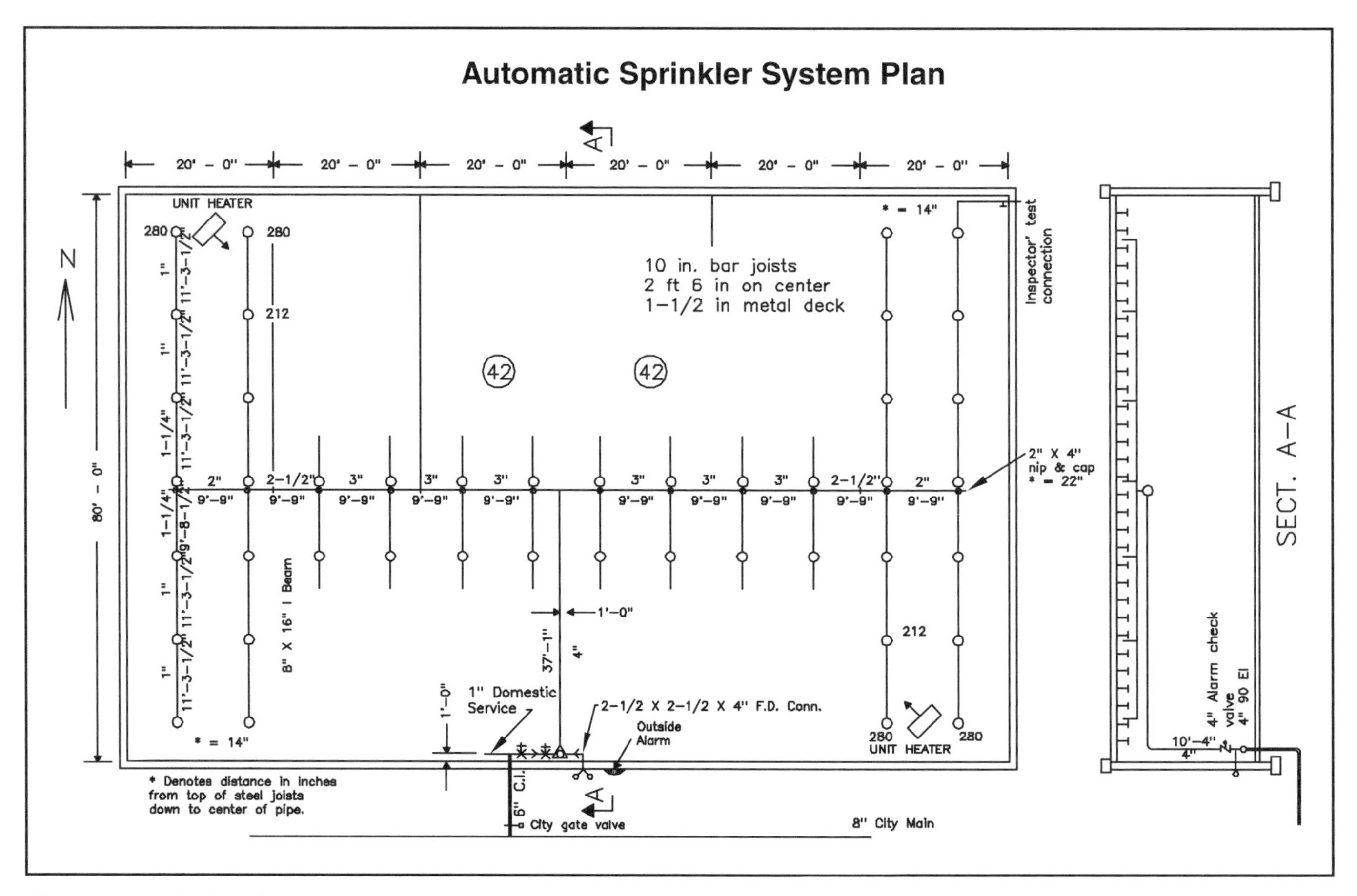

Figure 15.25 A plan of an automatic fire sprinkler system shows the proper location and size and the system components.

determination involves classifying the hazard by occupancy (light, ordinary, or extra hazard) and by storage commodity (shelf, high-piled, rack, or warehouse), or by both. The plans reviewer must evaluate mixed occupancies very carefully to verify that they adhere to the locally adopted codes or standards.

Drawings for hydraulically calculated sprinkler systems require calculations and supporting documentation proving that the design complies with the code **(Figure 15.26)**. The system designer provides all necessary information, including calculations and applicable data, on the drawings or in the specifications. All documentation must be reviewed carefully for accuracy.

Plans reviewers need to check a number of sprinkler system design factors:

- Extent of coverage
- Design area of system
- Type of system
- Size of system
- Water supply connections and valves
- Water supply capacity at site
- Fire department connection (FDC)
- Sprinkler types, temperature ratings, and locations
- Pipe sizes and lengths
- Number of pipe elbows and tees

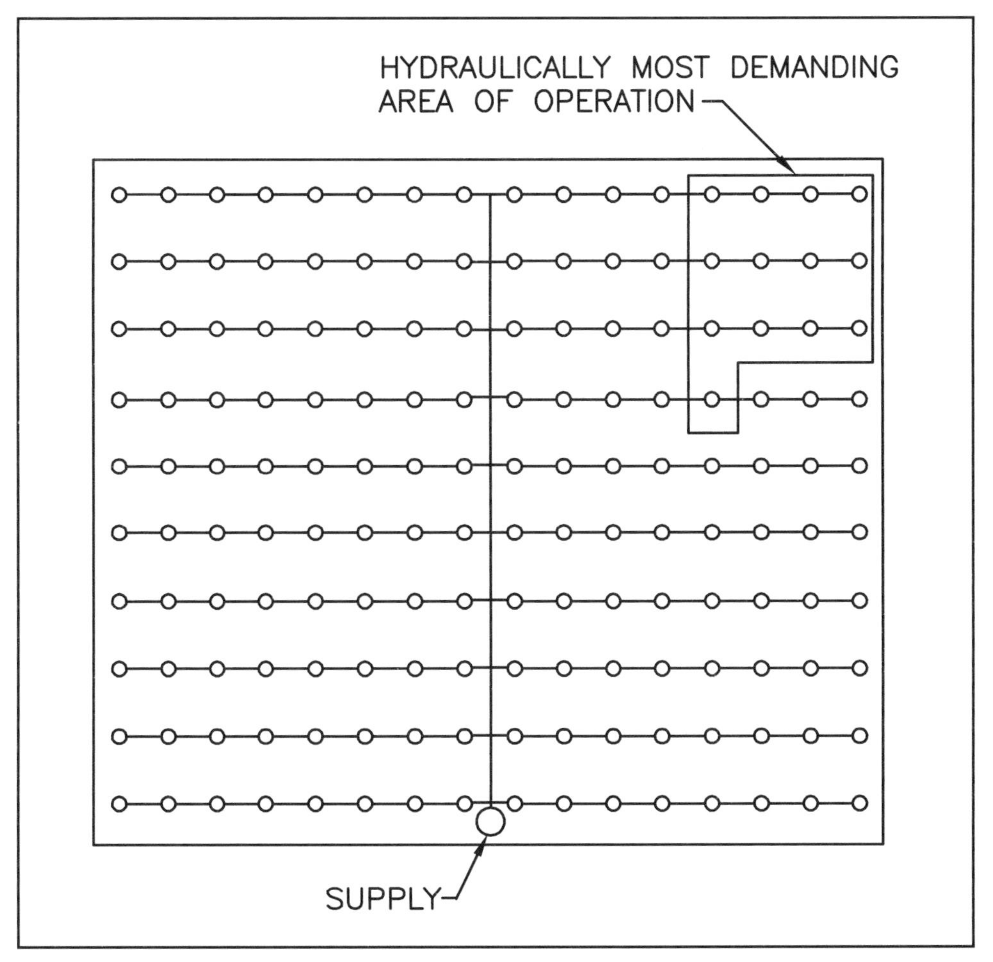

Figure 15.26 An example of the hydraulically most-demanding area for a centrally supplied sprinkler system design.

- Type and number of pipe hangers
- Requirements for seismic bracing

The inspector should review the sprinkler cut sheet to verify that the sprinkler is installed in accordance with its UL listing. Sprinklers should also be installed so that they provide complete coverage of the building. This coverage includes closets, stairwells, storage areas, walk-in freezers, and concealed spaces, such as areas above suspended ceilings with combustible construction.

The size of each sprinkler system is based on the total floor area, in square feet or square meters, that a single riser and control valve can protect. Installation standards determine the size of the systems, the number of individual systems, and the number of system control valves needed.

Plans reviewers must examine the specifications to verify the temperature rating, type, orifice size, area of coverage, use in special areas, spacing, and location of sprinklers. The drawings should clearly indicate all supports, bracing, connections, piping, valves, drains, and gauges throughout the system.

Water supplies for sprinklers must have sufficient capacity and pressure to satisfy the calculated requirements for hydraulically designed systems. The sprinkler system designer must include a graph that shows the water demand compared to the available water supply **(Figure 15.27, p. 676)**.

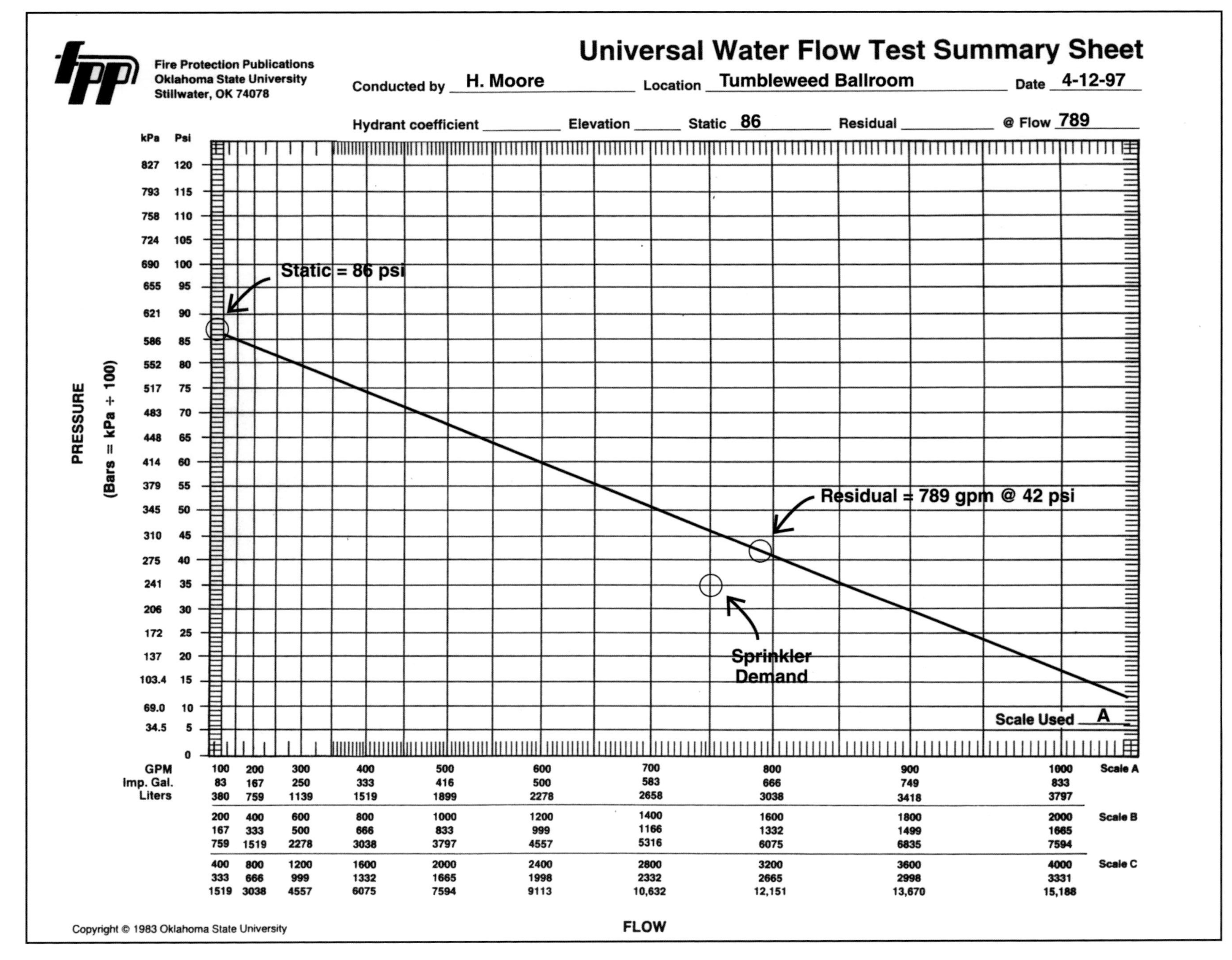

Figure 15.27 A waterflow test summary must accompany sprinkler system plans.

Standpipe and Hose Systems

An inspector or examiner must use fire protection and plumbing system plans to determine the following information about the standpipe and hose system:

- System type
- System classification
- FDC location(s)
- Water supply main location and capacity
- Riser control valve type and location
- Location of hose cabinets or discharges
- Discharge control valves locations
- Pressurized air supply type, capacity, and locations

Combined standpipe and hose systems that are part of a sprinkler system must also have waterflow calculations that indicate the effect on the operation of the sprinklers when a hose discharge is opened.

Special-Agent Fire Extinguishing Systems

Plan reviewers must be familiar with each type of special-agent fire extinguishing system and the applicable requirements of each **(Figure 15.28)**. The specifications must be very detailed and include the calculation sheets used to design the system. Additional information must be included in the documentation:

- Definition of the area or equipment to be protected
- Type of system (local application or total flooding)
- Type of extinguishing agent being used
- Amount of agent required
- Concentration of extinguishing agent to be developed
- Storage container size
- Type of expellant gas
- Rate of discharge
- Duration of flow
- Layout and type of piping included (whether engineered or pre-engineered)
- Location and type of discharge nozzles
- Method of actuation and auxiliary alarm functions such as turning off ventilation
- Type of presignaling devices used if required
- Area and volume of the protected space
- Sequence of operation

Figure 15.28 Special-agent systems are installed to meet specific fire protection situations. The inspector must be familiar with the requirements for the special-agent systems in their jurisdiction.

Fire Detection and Alarm Systems

Fire detection and alarm systems are designed to the requirements found in NFPA® 72, *National Fire Alarm*

Code®. Plans for fire detection and alarm signaling systems must contain enough information for plans reviewers to evaluate system components and functions:

Components:

- Signal initiation devices
- Signal notification appliances
- Automatic door closing devices
- Lightning protection devices
- Smoke control devices
- Damper control devices
- Fire pumps
- Power supply
- Auxiliary power supply

Functions:

- Supervision of alarm systems
- Elevator control **(Figure 15.29)**
- Stair pressurization
- Doors that unlock or close automatically when the alarm activates

NOTE: Consult NFPA® 72, Chapter 4, for information on detection and alarm system components and Chapter 10 for information about shop drawings and calculation submittals.

The following systems information may be included in various documents that the manufacturer provides or in the construction documents:

- Battery load and voltage drop calculations
- Manufacturers' cut sheets
- Point-to-point wiring diagrams

To properly review the fire detection and alarm system, plans reviewers evaluate information from the specifications, floor plans, equipment list, and symbol list. The specifications include the following information:

- Type and gauge of wire
- Protection provided for the wire
- Wiring methods
- Methods of supervision
- Sequence of operation

All fire detection and alarm systems must have electrical supervision. All fire detection and alarm equipment must be listed by a nationally recognized testing agency for their intended use. Inspectors must also be aware of specified maximum travel distances to manual pull boxes and code requirements concerning pull station placement for accessibility to persons with disabilities.

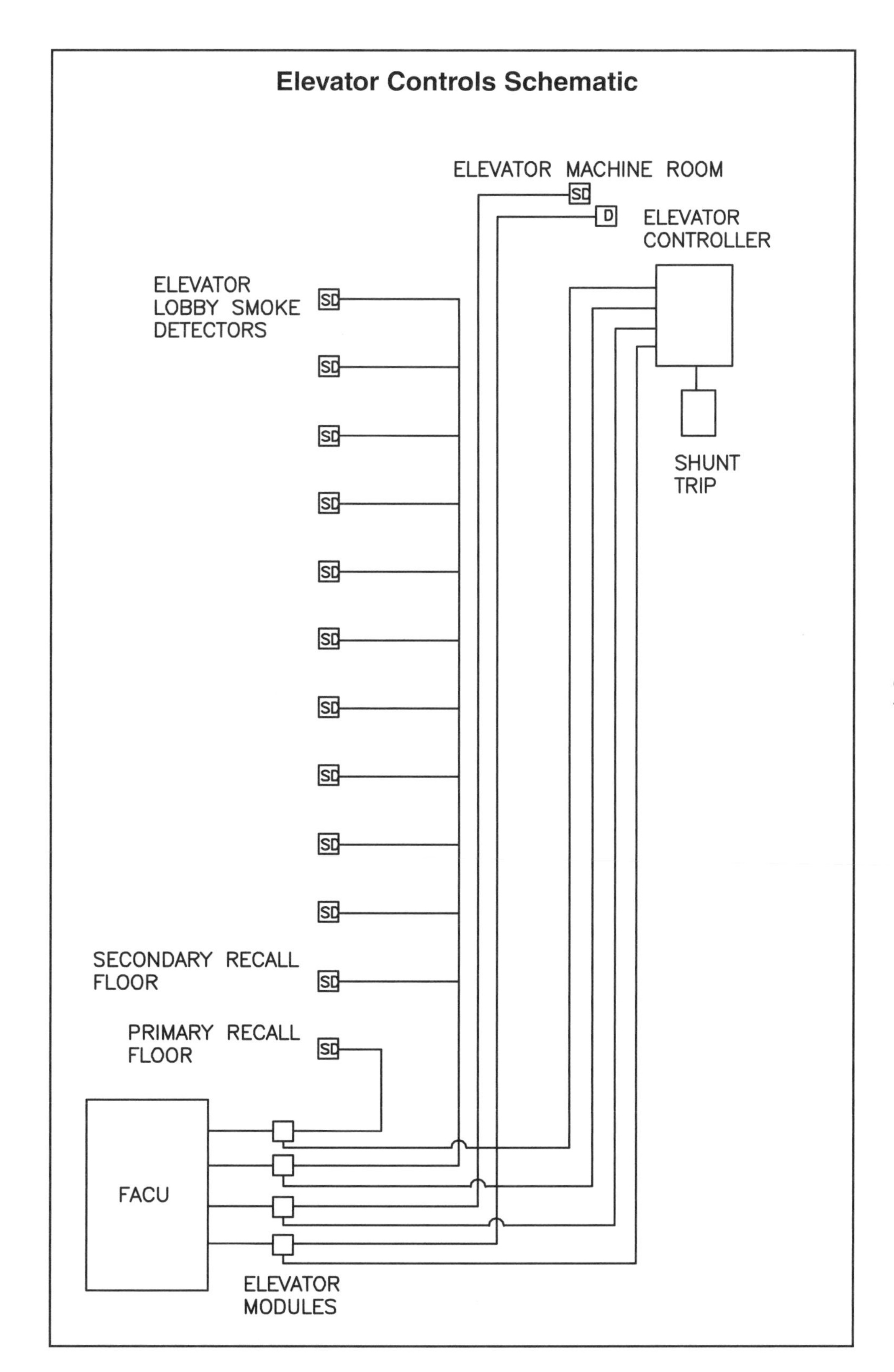

Figure 15.29 The elevator control schematic is part of the fire-protection system plan.

Systematic Plans Review

The plans review process can be long and detailed. To simplify the process, plans reviewers should take a systematic approach to reduce the chance that an important element will be overlooked. General elements that should be reviewed:

- Overall size of building (height and area)
- Occupancy classification
- Occupant load
- Means of egress
- Exit capacity
- Building compartmentation
- Additional concerns

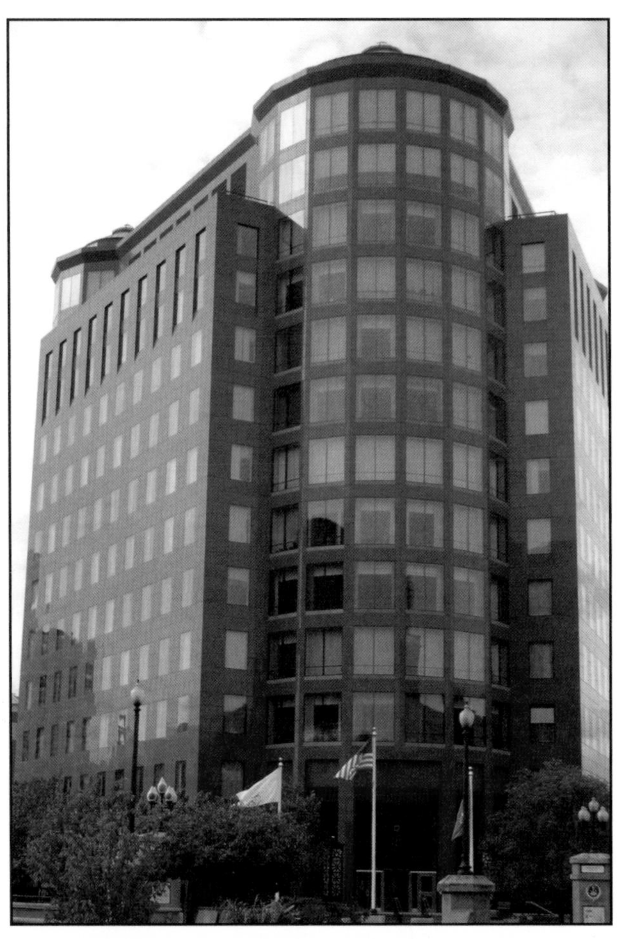

Figure 15.30 Review the overall height and area of the proposed building. *Courtesy of David DeStefano.*

Overall Size of Building

The first step in evaluating documents is to review the overall size of the building in terms of height and area. Building height and area are based on the local building code and zoning regulations, occupancy classification, and construction type of the building **(Figure 15.30)**. Building codes commonly allow an additional story if the building has complete automatic sprinkler system protection.

The model codes handle sloping grades, basements, parapet walls, and penthouses differently when determining the lowest or highest points. Inspectors should be aware of how their locally adopted codes handle these items.

Like height limitations, area limitations are intended to facilitate effective fire control and suppression operations. Such factors as the capabilities of the jurisdiction's water supply, street and road access to the site, fire risks, and climatic conditions are considered during a plans review. Each of the model building codes has different area limitations based on local requirements and resources; therefore, inspectors and plans examiners must use the definition from the locally adopted code.

Allowances for additional areas also vary among codes. Area increases or decreases are based on automatic sprinkler protection, accessible building perimeter, and the fire separation distance between buildings. These increases and decreases are usually expressed in percentages.

Occupancy and Construction Classification

It is the design professional's responsibility to determine the occupancy classification for the proposed building and identify the intended use of each room or section of the building. An inspector or plans examiner verifies that occupancy classification is consistent with the intended use of the structure. Inspectors or plans examiners evaluate the fire and life safety requirements throughout the building to verify that they comply with the required features prescribed for the occupancy. An inspector or plans examiner must be familiar with the applicable codes concerning mixed occupancies and special-use requirements.

The various classifications for building construction consider only the factors necessary to define the building types. As with occupancy classification, the intended construction is identified on the drawings. This identification is critical for required fire flow determination in some model codes. To verify the construction classification, fire inspectors refer to the appropriate fire resistance rating lists.

NOTE: Refer to Chapter 4, Construction Types and Occupancy Classifications, for more information about construction classifications.

Occupant Load

A plans examiner verifies the occupant load before evaluating the life safety and fire protection features of a building. Inspectors can use the occupant density or occupant load factors given in the codes to determine the occupant load. Design professionals are required to provide the occupant load calculations for verification.

NOTE: Chapter 7, Means of Egress, contains calculations for determining occupant load.

Means of Egress

After verifying the occupant load of the space or building, the plans examiner needs to evaluate the means of egress. Means of egress should be highlighted on the construction, mechanical, and electrical drawings so that inspectors can detect penetrations or other code violations easily.

The building plans must show details of the separation walls as well as fire door opening protectives information. The codes require that construction with a specified fire-resistance rating must separate exits from other parts of a building. Inspectors must verify the following:

- There are no unnecessary openings.
- All penetrations are sealed **(Figure 15.31)**.
- Fire-resistance ratings of the fire door and opening protectives are correct.
- Separation from the exit discharges and exit accesses is verified.

All the various components of the exit system must be in place. Each means of egress must provide a continuous path of travel to the level of exit discharge or area of refuge for mobility-impaired occupants.

Inspectors should evaluate other related code requirements applying to means of egress, such as the following:

- Interior finishes
- Headroom
- Elevation changes
- Emergency lighting
- Obstructions

NOTE: For more information about occupant load and means of egress, see Chapter 7.

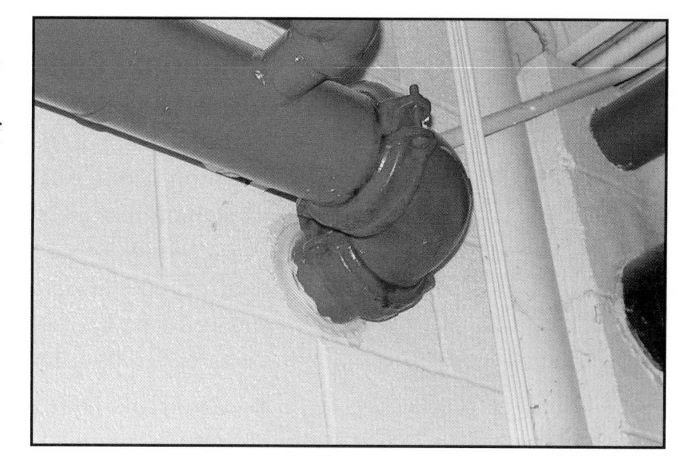

Figure 15.31 A pipe penetrating a fire wall.

Exit Capacity

After a plans examiner has identified each means of egress and verified that each component is acceptable, the capacity of the means of egress is evaluated. Plans examiners must evaluate the exit capacity according to the specifications of the appropriate codes.

Building Compartmentation

The theory behind building compartmentation is to limit any fire and resulting products of combustion to one area of a building. Building compartmentation includes the following:

- Fire walls
- Fire barriers
- Horizontal assemblies
- Shafts
- Protection of vertical openings and concealed spaces
- Opening protectives, such as fire door assemblies, fire dampers, and fire-stop assemblies

An inspector or plans examiner verifies the fire-resistance rating of the construction separating all vertical openings, such as stairways and elevator shafts, from the remaining parts of the building. Atriums and open shafts connecting two or more stories are permitted if they are provided with adequate smoke control, automatic sprinkler system protection, and fire barriers. Inspectors or plans examiner must locate and check fire and smoke barriers for continuity, fire-resistance rating, and opening protection.

During field inspections, it is often difficult to verify the existence of fire-stops in such concealed spaces and penetrations as commercial cooking equipment, HVAC systems, and the like. The design professional must provide plans that clearly show the location of required fire-stops and the building system that is being installed. When evaluating hazardous areas, inspectors or plans examiners must verify that the required separation and/or automatic sprinkler protection has been provided.

Additional Concerns

Inspectors and plans examiners may find that all the information needed to evaluate a set of plans is not contained on the plans. Such omissions may not be intentional, but they can have an effect on fire and life safety. For instance, a set of plans that is labeled *Warehouse* and shows only the exterior walls, floor, and ceiling does not provide enough information for the inspector to perform a correct plans review. In this example, contents, shelving height, and interior paths between shelves must be considered. Additional concerns that must be considered by inspectors and plans reviewers include:

- **Materials that produce smoke and toxic gases when burning** — Code requirements exclude the use of certain materials with high smoke production or smoke densities.
- **Interior finishes** — Requirements usually apply only to walls and ceilings (specific interior finishes such as cellular or foam plastic materials, incidental trim, and fire-retardant coatings) because floor coverings are tested

using a different standard. Codes often permit less stringent requirements for interior finishes when automatic sprinkler systems are installed.

- **Furnishings and decorations** — In certain occupancies, the inspector or plans examiner must be provided with flammability information for certain furnishings.
- **Portable fire extinguishers** — Inspectors must verify that extinguishers are located and installed according to the adopted code **(Figure 15.32, p. 684)**. Locations and capacities of portable fire extinguishers may be found on the floor plan. If they are missing from the plan, an inspector or plans examiner must require that the architect or designer include them based on the potential hazard, travel distances, and code requirements.
- **Insulation materials** — Both exposed (surface-mounted) and concealed (located between wall surfaces or above ceilings) insulation materials can contribute to fire spread or smoke development. Construction documents must contain the type, fire-resistance rating, and smoke-development index as determined in the building code.
- **Special hazards** — Contents as well as the processes that may occur in the completed structure must be considered during the plans review activities. The existence of special hazards such as those mentioned in Chapter 9, Fire Hazard Recognition, will mandate the types of fire and life safety protection to be installed in the structure as well as the fire flow capacity of fire hydrants in the area.

Chapter Summary

Inspectors and plans examiners from the fire department play an important role in the plans review and field verification process. While a Level II Fire Inspector may perform some plans review activities, a Level I Fire Inspector is only required to be able to read construction plans and verify that the requirements are being met during construction.

The inspector needs to be able to verify that all code requirements regarding occupant fire and life safety are met before and during the construction of the building or structure. This activity may require coordinating with inspectors from other fields, such as mechanical or electrical.

Finally, construction plans review and field verification are keys to the proactive approach to fire and life safety. Verifying that a building provides an acceptable level of fire and life safety construction and systems saves lives, reduces property loss, and reduces fire protection costs to the community.

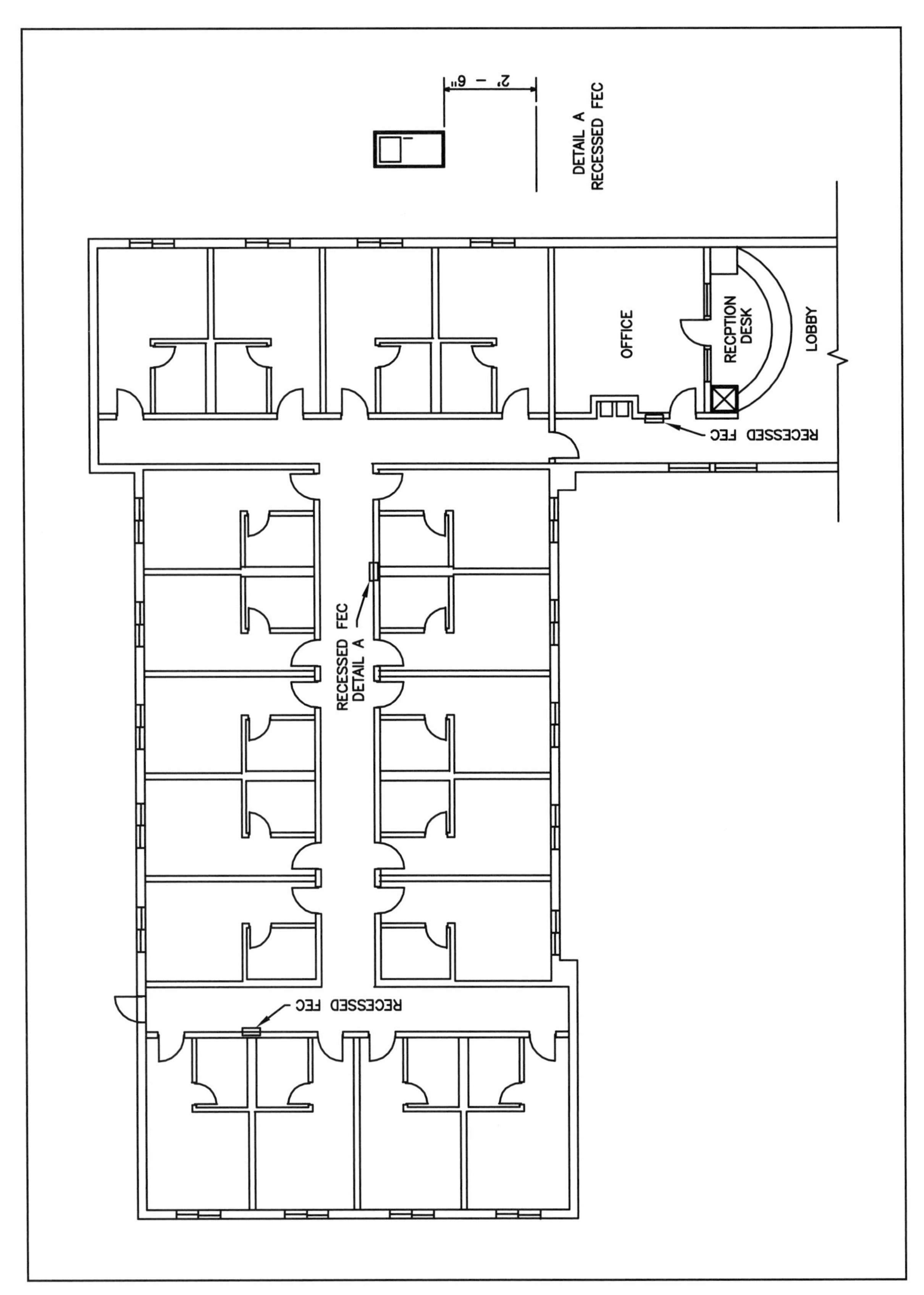

Figure 15.32 Partial floor plan drawing of a hotel showing locations of fire extinguisher cabinets (FECs).

Review Questions

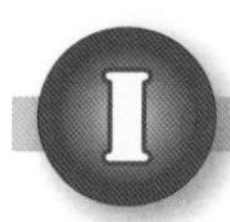

1. What is the role of the Fire Inspector I in the plans review and permitting process?
2. What is the main reason local governments require a plans review?
3. Why should an inspector visit the site often during construction?

1. What are the general steps of the plans review process?
2. What are the four main views for building construction working drawings?
3. What information is provided by a site plan?
4. What is the purpose of a sectional view?
5. List various types of system plans.
6. What are the general elements that should be reviewed in a systematic plans review?

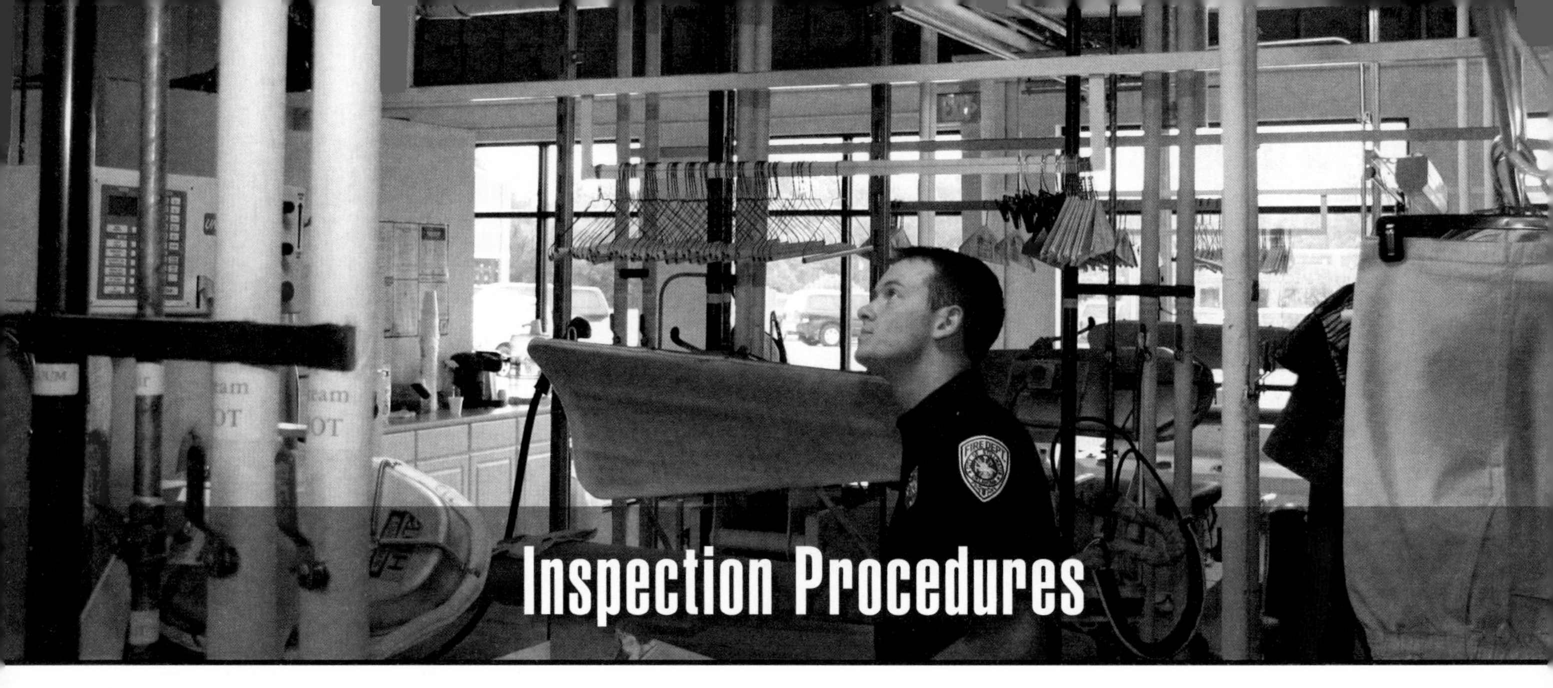

Inspection Procedures

Chapter Contents

Case History **689**

I **Interpersonal Communication** **690**

- Communication Model Elements 690
- Interpersonal Skills 692

Administrative Duties **695**

- Written Communications 695
- Files and Records 697

Inspection Preparation **698**

- Personal Appearance 700
- Equipment Lists 700
- Inspection Scheduling 701
- Inspection Records Review 702

Inspection Procedures **702**

- Guidelines for Inspections 703
- General Inspection Practices 705
- Code Requirements 707
- Photographs 708
- Inspection Checklists 708
- Building Occupancy Changes 712
- Results Interview 712
- Letters and Reports 714

Follow-Up Inspections **715**

Emergency Planning and Preparedness **716**

- Educational Facilities 718
- Health Care Facilities 719
- Correctional Facilities 721
- Hotels and Motels 721

Complaint Management **722**

II **Complex Complaint Procedures** **724**

Evaluate Emergency Planning **725**

Chapter Summary **727**

Review Questions **727**

Key Terms

Body Language..694
Cease-and Desist Order 704
Evacuation...716
Feedback...691
Monitor 726
Violation 713

NFPA® Job Performance Requirements

This chapter provides information that addresses the following job performance requirements of NFPA® 1031, *Standard for Professional Qualifications for Fire Inspector and Plan Examiner* (2014).

Fire Inspector I	Fire Inspector II
4.2.1	5.2.3
4.2.4	5.3.7
4.3.10	

Inspection Procedures

Learning Objectives

After reading this chapter, students will be able to:

Inspector I

1. Explain the duties of an Inspector I. (4.2.1)
2. Describe components of interpersonal communication. (4.2.1)
3. Describe the basic administrative duties of an Inspector I. (4.2.1)
4. Describe the preparation required before an inspection. (4.2.1)
5. Explain basic inspection procedures. (4.2.1)
6. Explain the role of an Inspector I in follow-up inspections. (4.2.1)
7. Identify ways an Inspector I will participate in emergency planning. (4.3.10)
8. Describe the complaint management process. (4.2.4)

Inspector II

1. Describe what makes a complaint complex. (5.2.3)
2. Describe how to evaluate emergency preparedness plans. (5.3.7)

Chapter 16
Inspection Procedures

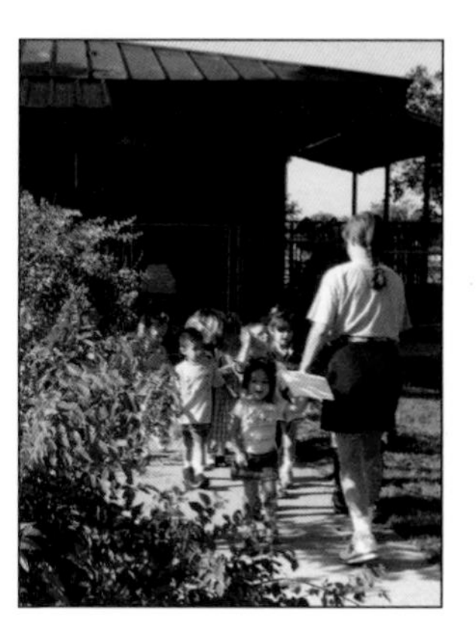

Case History

Ocean City, Maryland, 2015: The Ocean City Fire Marshal's office plans to make use of a third-party filing electronic system to enable inspectors to spend more time in the field. Currently, there are about 1,500 protection systems in the city, and the Fire Marshal's office monitors inspections done by third-party companies. Because there were no standardized formats or transmission methods for reports, the office concluded that it was spending too much time on administrative work and not enough time seeing that problems were corrected.

An additional third-party company now receives inspection reports and files them electronically, freeing the staff to concentrate more on resolving issues. Other cities nearby are using third-party organizations to issue Notices of Violations to businesses, with e-mail copies sent to the fire marshal's office. The goal in all systems is greater efficiency.

Inspectors strive to provide the safest possible construction for the citizens and emergency responders of their communities through education and the application of the locally adopted fire code. Citizens are more likely to interact with EMS responders or fire prevention personnel than they are with operations personnel. The professional attitudes displayed by inspectors go a long way in providing education and building a positive image of fire prevention activities in general.

One of the most important duties an inspector has is to communicate with building owners and occupants, design professionals, contractors, other government officials, and the general public to explain the basis and importance of the fire and life safety codes they administer. These relationships help to educate the user and establish a strong foundation for the provision of a safe built environment.

Enforcement occurs when inspectors perform the following duties:

- Inspect new and existing buildings and facilities for code compliance
- Verify occupant loads
- Investigate complaints
- Evaluate emergency preparedness plans
- Review construction and renovation plans

An inspection organization must establish systematic procedures to perform inspections, train personnel in the proper procedures, and perform administrative tasks. Preparing for an inspection, following standard inspection procedures, and using good interpersonal skills are all necessary if an inspector is to perform the job consistently and effectively.

The plans review process was described in Chapter 15, Plans Review and Field Verifications, and calculating occupant loads in Chapter 7, Means of Egress. This chapter focuses on the remainder of an inspector's duties, which include inspecting structures or facilities and investigating complaints. Inspections are opportunities for an inspector to educate and enforce the maintenance of a safe environment by applying fire and life safety code requirements, identifying unsafe behaviors, and ordering the correction of unsafe conditions.

This chapter addresses the basic principles and skills of performing an effective inspection, particularly the following:

- Interpersonal communication
- Administrative duties
- Inspection preparation
- Inspection procedures
- Follow-up inspections
- Emergency planning and preparedness
- Complaint management
- Handling complex complaint procedures
- Evaluating Emergency Planning

Interpersonal Communication

Interpersonal communication may be defined as all aspects of personal interaction and communication between individuals or members of a group. An inspector must be able to listen and communicate effectively in addition to being able to persuade citizens to correct unsafe behaviors or conditions. An inspector must be familiar with the elements of a communication model as well as having good skills in listening, conversing, and persuading.

Communication Model Elements

The basic elements of a communication model are as follows **(Figure 16.1)**:

- **Sender** — Originates a message by encoding or turning thoughts and mental images into words (often referred to as the speaker). Words are selected based on the perceived ability of the receiver (also known as the listener) to understand the message.
- **Message** — Contains a meaning, idea, or concept that a speaker is attempting to communicate; may be transmitted by sound, sight, touch, smell, taste, gesture, or any combination of these means. An effective message includes a combination of these elements.
- **Medium or channel** — Gives the path that a message takes between the sender and the receiver; usually face to face.

- **Receiver** — Accepts a message and decodes or interprets it. Education, cultural background, perception, attitude, and context all provide the receiver's frame of reference for interpretation of the message. Frequently, it is the receiver's misinterpretation, based on these frames of reference, that leads to misunderstandings and disagreements.
- **Interference** — Prevents the receiver from fully receiving a message; may be created by either internal or external sources. These include hearing impairments, excessive noise, and mixed messages that are confusing.
- **Feedback (response) to the sender** — Completes the communication process, resulting in an ongoing cycle. Can be positive, in which desired results are achieved, or negative, resulting in confrontation or misinterpretation.

Feedback — Responses that clarify and ensure that the message was received and understood.

To overcome internal interference, the receiver must focus on what the sender is saying, listen carefully, provide feedback immediately, and use nonverbal factors or clues to emphasize and acknowledge understanding of the message.

During an inspection, controlling external interference may include turning off an air-conditioner, closing a window or door, or moving the conversation to a quieter location. Conversations that involve the exchange of detailed information or complete concentration should take place in a controlled environment such as an office **(Figure 16.2)**. In areas that involve noisy operations, it may be necessary to take notes during the inspection instead of attempting to talk. These notes can be the basis for a conversation at the end of the inspection. Each inspection has to be approached as a special situation that may require special solutions.

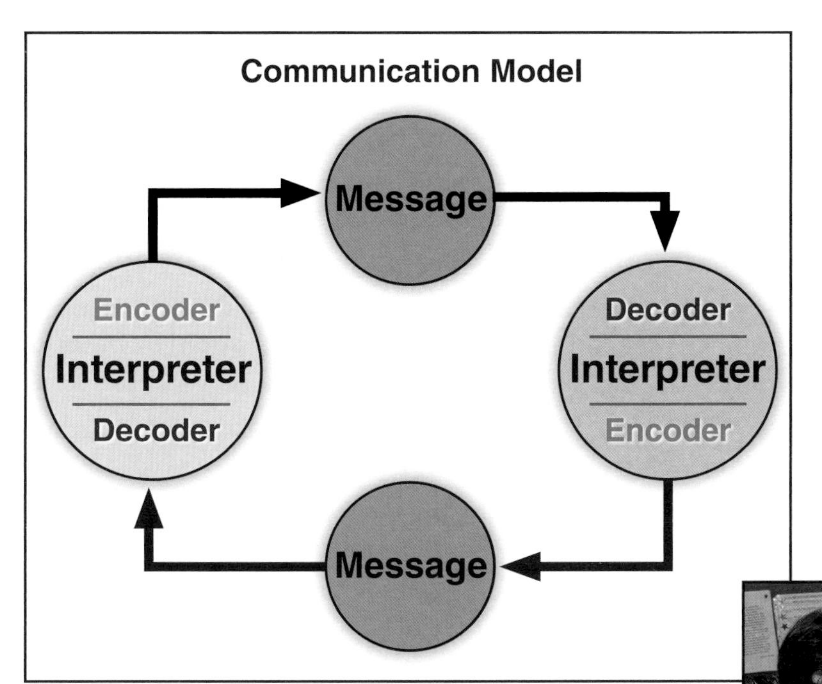

Figure 16.1 This visual representation of the communication model illustrates the circular pattern of listening and response. Understanding the communication model makes inspectors more efficient communicators.

Figure 16.2 It may be best to discuss inspection issues with citizens in a quiet location where interruptions are minimized.

Interpersonal Skills

Effective interpersonal communication skills can be broadly summarized into three areas:

1. Listening
2. Conversing
3. Persuading

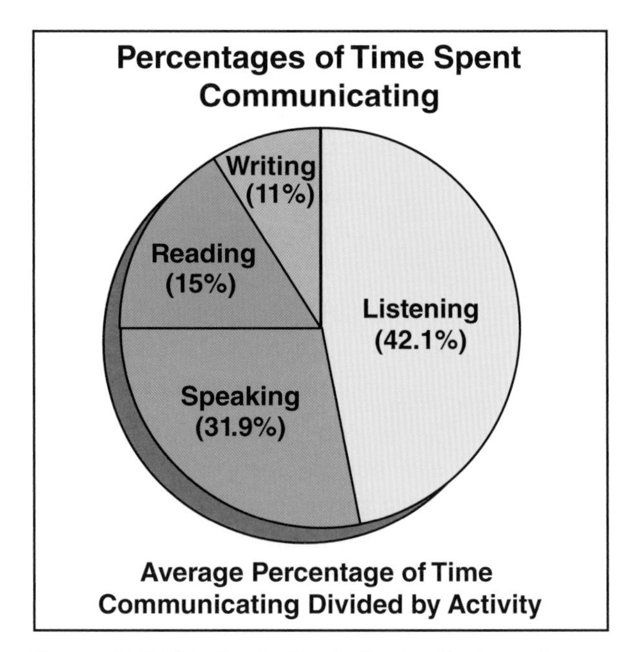

Figure 16.3 This illustration indicates the importance of listening based upon how much time per day people spend listening to others. The percentages used can vary based upon the methodology used to gather the data.

Listening Skills

Because a large portion of communication involves listening, an inspector must master good listening skills to be successful when dealing with the public **(Figure 16.3)**:

- **Attending** — Focuses on the speaker and ignores other distractions by doing the following:
  - — Look at the speaker if possible and think about what is being said.
  - — Listen to the complete message: Wait until the speaker has finished before responding.
  - — Maintain eye contact: Listen to both verbal and nonverbal messages.
  - — Remove physical barriers. For example, sit in chairs that are at right angles to one another instead of behind a desk.
  - — Listen to one speaker at a time.
- **Understanding** — Consists of decoding a message and assigning meaning to it. The receiver can repeat the message or restate it in different words to clarify the meaning.
- **Remembering** — Critical for the message to have the correct effect. To assist in remembering, repeat the information or take notes.
- **Evaluating** — Involves critically analyzing the message to determine how factual it is. The listener must be able to separate facts from opinion. Message analysis factors include:
  - — Personal experience of the listener
  - — Interpretation of the nonverbal clues from the speaker
  - — Credibility of the speaker
  - — Other available information, such as clues from the surrounding environment
- **Responding** — Involves the feedback portion of the communication process. Without any response, the speaker does not know if the message was received, understood, or will be acted upon.

Practicing good listening skills is the best way to improve them. Inspectors can listen to speeches or stories and try to repeat the key elements as a listening exercise. They can also take notes at meetings or speeches to improve note-taking skills. These actions help to overcome problems with information overload by pinpointing the essential elements of the message.

When listening to a speaker, good listeners focus on the speaker and the message. The greatest distraction is the listener's internal voice. This voice may be responding to something that was said earlier or something the listener would like to say. It could just be daydreaming. While this internal monologue is active, the words of the speaker are being ignored.

To overcome this internal barrier, a good listener should first identify it. Next, try asking questions or paraphrasing what has been said. When responding is inappropriate, such as it would be with a formal speech, good listeners take notes of the key points and ask questions later if possible.

Prejudice based on preconceived concepts of dress, voice, or attitude can be major barriers to effective listening. An inspector should work to overcome psychological barriers such as prejudice by accepting others as they are and not as people think they should be.

Other listening barriers and that may affect inspectors and ways to overcome them include:

- Information overload — Identify the essential elements of the message.
- Personal concerns — Focus on the speaker and the message rather than personal concerns or thoughts.
- Outside distractions — Take control of the environment and remove as many distractions as possible.
- Prejudice — Focus on the message and not the sender.

Conversing Skills

Learning to communicate well helps everyone develop relationships and increase organizational effectiveness and efficiency. To build strong interpersonal relationships, an inspector must develop and use the following verbal and nonverbal conversing skills:

- **Engage in dual perspective** — Be aware of the listener's frame of reference. Recognize that the listener may have a different point of view and do not try to diminish or make fun of it.
- **Take responsibility for personal feelings and thoughts** —Use language that is I-based, such as *I believe* or *I think*. Avoid phrases such as *You hurt me* or *You disappoint me* and focus instead on ownership of the feelings and the cause of those feelings.
- **Show respect for the feelings and thoughts of the other person** — Avoid trying to apply personal feelings to another person by applying good interpersonal skills.
- **Try to gain accuracy and clarity in speaking** —Be clear and accurate in all types of communication. Avoid generalizations.
- **Be aware of any special needs of the receiver** — Be patient and speak clearly while facing a person.

- **Avoid speaking or addressing a problem while angry or emotional** — Pause and place the conversation on hold until emotions are under control.

Nonverbal communications skills also require awareness and work. Inspectors should apply the following general concepts:

Body Language — Nonverbal communication including but not limited to body posture and gestures; represents a large portion of communication in human interactions.

- **Eye contact** — Learn to maintain eye contact while speaking to people. Because some cultures find direct eye contact to be disrespectful, learn to modify the use of eye contact when it is appropriate.
- **Body language** — Convey an image of self-confidence by maintaining good posture. Slouching or keeping hands in trouser pockets can be interpreted as being tired, depressed, or uninterested.
- **Facial expression** — Learn to match facial expressions to the message.
- **Gestures** — Identify and control gestures that are annoying or distracting to others.
- **Poise** — Create poise by building self-confidence and overcoming any fear associated with public speaking or dealing with strangers. Poise is gained through practice and a command of the information or topic.
- **Personal appearance** — Maintain a professional appearance by adhering to the organization's dress policies.
- **Touch** — Become conscious of the effect that touch can have on others, both positive and negative.
- **Proximity** — Be aware of the cultural differences that determine the use of space and apply it appropriately **(Figure 16.4)**

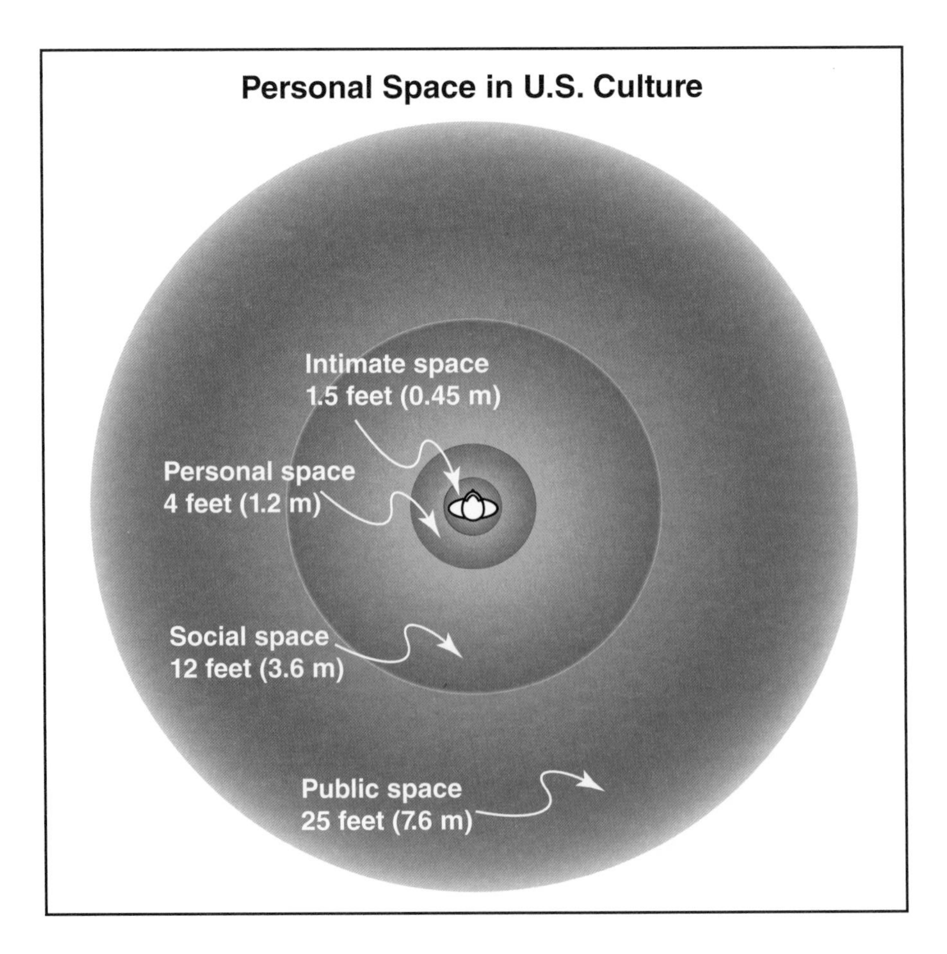

Figure 16.4 The area individuals consider their personal space varies from culture to culture. In the United States, personal space is typically considered to be 4 feet in circumference around an individual. Inspectors should respect this space in the course of their duties.

- **Use of time** — Budget your time based on the individuals or groups that are present. When it is necessary to maintain strict control over time, explain the requirements to all who are concerned.

Persuading Skills

An important part of an inspector's duty is persuading people to change unsafe behaviors or conditions. One method of persuasion is based on Monroe's Motivated Sequence Pattern developed by Alan Monroe in the 1930s. Developed properly, Monroe's sequence can be the basic outline for persuading a person to follow a desired course of action:

- **Attention** — Gain the attention of the listener.
- **Need** — Describe the problem and demonstrate a need for a change in the current situation.
- **Satisfaction** — Present the solution, providing sufficient information and evidence to allow the listener to understand how it accomplishes the goal.
- **Visualization** — Describe the solution for the problem and how it will benefit the listener.
- **Action** — State which actions that the listener will need to take to solve the problem and establish an expected date for completion.

Administrative Duties

Inspectors have additional duties and activities that fall under the classification of administrative duties. Primarily, these duties include preparing written communications, including both interoffice and external memos, e-mail messages, and letters, and maintaining files and records. The sections that follow discuss these administrative duties in greater detail.

Written Communications

An inspector needs to be proficient in the various types of written communication, such as

- E-mail messages
- Interoffice memos
- Formal business letters to owners/occupants and outside agencies
- Reports on the inspections and complaint investigations

Each of these types of written communication serves the purpose of providing rapid communication, typically on just one subject. This communication is not intended to be lengthy, but should be descriptive enough that the receiver should not need to request additional information.

Although forms of informal communications, such as e-mails and memos are designed to be brief and quickly dispensed, they must still be considered official documents. Paper memos should have copies filed, and e-mail messages should be saved in an electronic filing system. Contrary to popular belief, memos and e-mail messages are subject to the Freedom of Information Act in the U.S. and can be admitted into evidence during court proceedings.

Memos

Memos are short documents — usually only two or three paragraphs — intended to convey a single message for interoffice use. A memo is designed to serve as a permanent record of decisions, be somewhat informal, and contain one of three main communicative elements:

1. **Direction** — Assigns tasks, gives instruction, requests action.
2. **Advice** — Provides suggestions or opinions in response to a request or query.
3. **Information** — Provides details regarding a topic.

Even though a memo is brief, it is an official document. Memos should be dated, contain Heading and Subject lines, and include the inspector's signature.

E-Mail Messages

The following are considerations for using e-mails:

- Use proper formatting and grammar when sending e-mail messages. Avoid all capital or all lower case letters. Do not include catchy e-mail colloquialisms.
- Follow up e-mail messages to verify that they were received. When confirmation of an e-mail message is received, reply indicating receipt of the confirmation and save the confirmation and reply.
- Remember that people may change e-mail addresses without prior notification and leave no forwarding information. Most e-mail systems provide a means of recording other contact information such as phone numbers and addresses in a directory with the e-mail address.
- Remember that a contact person's e-mail address may be different from the contact name and e-mail address in a facility file.

Usually, all e-mail messages that are written by a public official during the performance of their duties are official documents and will be released subsequent to a request for records. Recent U.S. court rulings have declared that e-mail messages are not private or protected under a person's First Amendment rights.

Letters

Written communications reflect on an inspector's professionalism and credibility in general, along with that of the inspection organization or agency. Letters must not be misleading, threatening, rude, or poorly written, or they will give the reader a negative impression of an inspector and the organization. Letters need to be concise without sounding rude or abrupt.

All letters sent must be grammatically correct and proofread, neatly formatted, and reflect a positive professional attitude. To be consistent, all letters sent from the organization should follow the same format. Departmental policy will dictate who signs letters.

Reports

The majority of reports that inspectors write or complete consist of documenting serious or numerous fire code violations or the results of construction plans reviews. Reports are designed to provide vital and useful information to the occupant and the inspection organization or agency.

A formal report should include the following:

- Scope of the problem and the report's findings
- Introduction
- Executive summary
- Supporting documents

Each report needs to be neat in appearance and legible to the reader. However, the content of the report is even more important than its appearance. Information within the report must be delivered in a manner that is easily understood and not misleading or biased. A good, easy-to-use style manual can be consulted for correct word usage and punctuation. These procedures are also important in case the document becomes part of legal proceedings.

Files and Records

Whether an inspector is responsible for performing inspections, maintaining the organization's records management function, or simply accessing the records management system on a periodic basis, it is necessary to know what the system contains and how it is used. It is also necessary to know how long to retain files and records and the benefits of keeping them in either hard copy or electronic files.

The various files and records normally maintained by an inspection organization or agency include:

- Facility inspection files
- Inspection reports, forms, checklists, and letters
- Violation notices
- Plans review comments, approvals, and drawings
- Investigation reports
- Permits and certificates of occupancy
- Complaints

Ideally, inspection records should be maintained for all properties or facilities within a jurisdiction. For many inspection organizations, however, this may not be a realistic goal when the jurisdiction is large. Retention times for records will vary according to jurisdictional policies and procedures. At a minimum, records should be maintained on occupancies that meet certain criteria:

- Possess a permit, certificate of occupancy, or license of some type
- Contain automatic fire suppression, detection, or alarm systems
- Contain hazardous materials or perform hazardous operations on a routine basis

Each inspected property or facility should have an inspection file that contains copies of all building and inspection records for that facility. Each time an inspector has contact with the owner/occupant, records of those contacts are added to the file. It is important that this file be kept as up to date as possible **(Figure 16.5, p. 698)**.

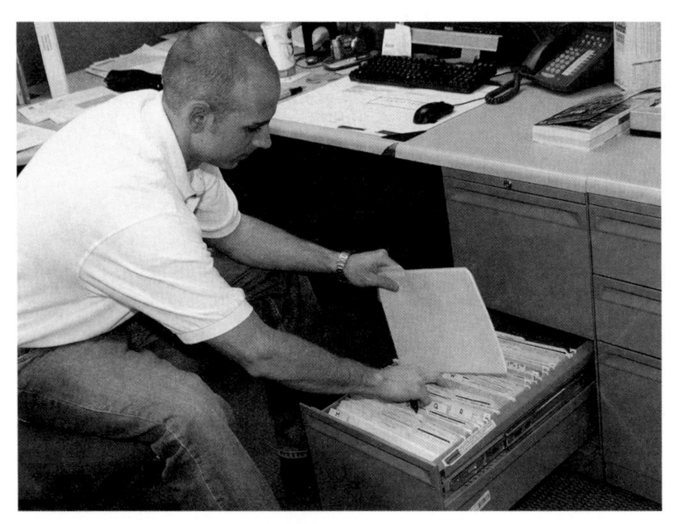

Figure 16.5 It is essential that inspectors keep files and reports organized and available as a reference on the history of the buildings they inspect.

It is recommended that records be maintained on a building or facility for its lifetime. By maintaining a record on the structure throughout its life cycle, inspectors can refer to it to learn which changes have been made to the structure and how proposed changes may be affected by the previous uses of the structure.

When a structure is demolished, the inspections organization or agency should maintain the records for a short period of time and then purge the files according to local policies and legal recommendations. All files and records maintained by the inspection organization or agency are considered public domain documents.

Computer records can reduce and/or eliminate many of the costs associated with the traditional hard-copy inspection processes. Many departments use computer or web-based systems to schedule, record, and file official inspection documents. The advances in web-based systems allow access using numerous devices such as smartphones, laptops, and tablets. These systems also enable inspectors to add supplemental details such as documents and pictures.

Having access to information and data in real and direct time greatly increases efficiency and accuracy in records management and retention. These records can also be shared within the department and provide information during an emergency incident. An electronic system can also connect with other entities such as third-party ITM companies and/or commercial property owners, or other municipal departments. Sharing information increases a community's ability for compliance.

Many aspects of computer system management must be given careful consideration when inspection records are stored:

- How will information be filed?
- How can information be retrieved?
- What portion of the information will be stored in a read-only format so that records cannot be changed without authorization?
- Which personnel will be given access to retrieve information from the system?
- Who owns the data?
- Who has access to the data?

Inspectors must receive the appropriate amount of training on their department's computer system. Not even the most well-designed computer data management system will work unless information is entered and stored in the proper manner.

Inspection Preparation

The success of an inspection visit depends upon how familiar an inspector is with the site being inspected **(Figures 16.6 a - c)**. The time spent reviewing the structure or facility inspection records file as well as the applicable fire codes and standards can determine whether the inspection is effective or not.

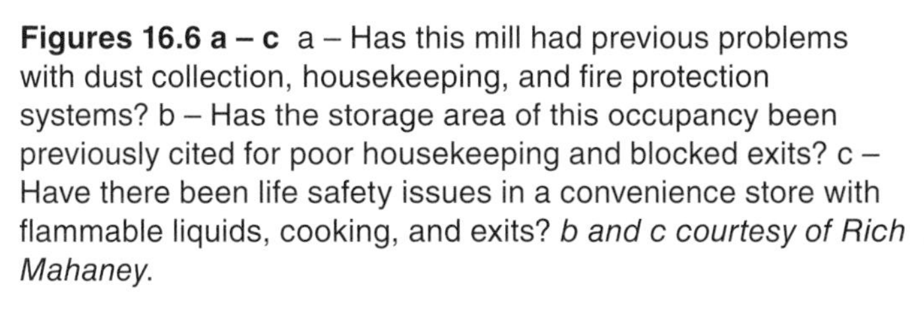

Figures 16.6 a – c a – Has this mill had previous problems with dust collection, housekeeping, and fire protection systems? b – Has the storage area of this occupancy been previously cited for poor housekeeping and blocked exits? c – Have there been life safety issues in a convenience store with flammable liquids, cooking, and exits? *b and c courtesy of Rich Mahaney.*

Inspection programs that demand proper preparation by the inspection staff are much more likely to gain community support than those that appear to be only going through the motions of an inspection. Inspectors who are unprepared often overlook fire and life safety hazards and potential code violations or cannot advise the owner/occupant with accurate information regarding the results of the inspection. Inspectors who have not reviewed inspection records will also be unaware of previous uncorrected violations or a pattern of repeated violations.

Positive Impressions

Several important considerations will help an inspector make a positive initial contact:

- Maintain a professional appearance.
- Use a friendly and professional manner with the owner/occupant.
- Display official identification.
- Create and maintain a positive atmosphere with the owner/occupant.
- Display a courteous manner.
- Avoid ethical compromises or dilemmas.
- Remain impartial.
- Maintain a positive approach to code compliance resolutions.
- Always remain calm and do not become argumentative.
- Recognize and understand potential cultural differences.

Personal Appearance

Maintaining a professional personal appearance is essential when dealing with the public. In many cases, this contact is the only one the owner/occupant has with a representative of the inspection organization or agency. When departmental policy requires inspectors to wear a uniform, they must wear the complete uniform **(Figure 16.7)**. The inspector must remember that appearance is crucial in giving a positive impression.

Figure 16.7 Wearing a uniform during inspections presents a professional image; however, a uniform is often not adequate as a sole form of identification.

Equipment Lists

An inspector assembles all of the necessary tools and equipment and personal protective equipment (PPE) that may be needed for an inspection. The actual tools and equipment will vary with each jurisdiction, the type of inspection being conducted, and an inspector's preference:

- Personal protective clothing as needed for the site:
  - Helmet
  - Hard hat
  - Gloves, safety shoes, or boots
  - Eye and hearing protection
- Flashlight
- Hand-cleaning materials to wipe off any contaminants
- Clipboard with inspection forms and map symbols
- Electronic data collection equipment when it is used by the jurisdiction
- Materials needed to make sketches and maps
- Measuring tape, 50 feet (15 m) in length
- Camera with flash equipment
- Pitot tube and gauge, hydrant wrench, and static pressure cap gauge when water supply testing will be conducted

- Reference materials, such as code books, code sections copies, building plans, and previous reports

The first consideration for an inspector is to prepare to conduct the inspection safely. A review of the inspection records file will reveal the types of hazards that may be present at the site. Reviewing this information will enable the inspector to wear the appropriate personal protective clothing and equipment. For example, coveralls with padded knees and elbows are of great benefit when inspecting areas where crawling is necessary or in confined spaces.

Figure 16.8 Inspection of hazardous locations may require the inspector to use respiratory protection.

Each inspector must be trained to recognize potential respiratory hazards as well as know proper methods for using personal respiratory protection. Inspectors who are assigned to inspect industrial facilities where respiratory hazards may exist must be able to select the appropriate respiratory equipment and operate it correctly within the organization's policies and procedures **(Figure 16.8)**.

NOTE: For more information regarding critical personal protection requirements, refer to the IFSTA **Respiratory Protection for Fire and Emergency Services** manual.

Inspection Scheduling

Inspectors must use common sense and courtesy when scheduling an inspection. The amount of time it takes to conduct an inspection varies for each facility and occupancy. A number of factors will influence the length and complexity of the inspection:

- Size of the facility or occupancy
- Complexity of operations or processes within the occupancy
- Type of inspection:
  - Regular periodic inspection
  - Follow-up inspection
- Location of the facility or occupancy
- Inspector's familiarity and experience with the occupancy

Experience and good professional judgment are essential when determining the number of inspections that can be scheduled during a normal work period. Some facilities may take an entire day to inspect, while several small occupancies can be inspected in the same amount of time. Inspection scheduling is based on department policy and the number of staff in each office.

When scheduling multiple occupancies in the same day, an inspector should consider the location of the inspection sites. Scheduling occupancies that are in the same general area and require a minimum travel time between them is desirable.

Computer-based electronic programs can be used to schedule inspections based on type of hazard, location, or cycle (annual or semiannual). Inspection schedules can be generated from the file along with reinspection dates, reminders of previous violations, permits, and complaints.

The first step in scheduling the inspection is to contact the facility owner/occupant and determine a mutually agreeable date and time for the inspection. The inspector should then record the necessary contact information on an inspection schedule form that includes phone numbers and e-mail addresses for efficient communication. Scheduling information should be available to other personnel within the inspection division in case the scheduled inspector is unavailable. In some cases, a letter can be sent to the owner/occupant confirming the date and time of the inspection. A phone call on the business day before the inspection is also a good idea **(Figure 16.9)**.

It is desirable to schedule annual inspections so both the inspector and the business owner can have enough time to perform a thorough inspection. However, the reality is that many inspections will be drop-ins or unscheduled. In some situations, inspectors will need to make unscheduled or unannounced inspections of assembly occupancies after hours and on weekends to verify compliance with the following: **(Figure 16.10)**

- Adherence to posted occupant load
- Presence of locked, blocked, or obstructed exits
- Investigation of public complaints that may be time sensitive

Inspection Records Review

The inspector should review any related files, records, codes, and documents relating to the facility before conducting an inspection. If this is the first inspection of the occupancy, the inspector should become familiar with similar types of occupancies and the fire and life safety requirements for them. Regarding the specific site, the inspection should pay special attention to the following:

- Permits that have been issued for the facility or occupancy
- Any fire and life safety complaints
- Previous inspection violations
- Specific hazards at the location that deserve special attention

This information helps identify important changes in the occupancy when the same conditions are not found during the current visit. When the occupancy contains structural elements, manufacturing processes, or operations that are not commonly encountered, an inspector should review those sections of the locally adopted code.

A review of the emergency response reports for the facility may also provide important information. These reports can indicate insight and additional information on issues that may have occurred over the course of the year.

Inspection Procedures

Effective inspections should be systematic, logically arranged, and consistent. The inspector should establish a consistent method for conducting inspections in order to better identify unsafe conditions or behaviors. Most inspection procedures include the following elements:

- General inspection practices
- Inspection concerns
- Photographs

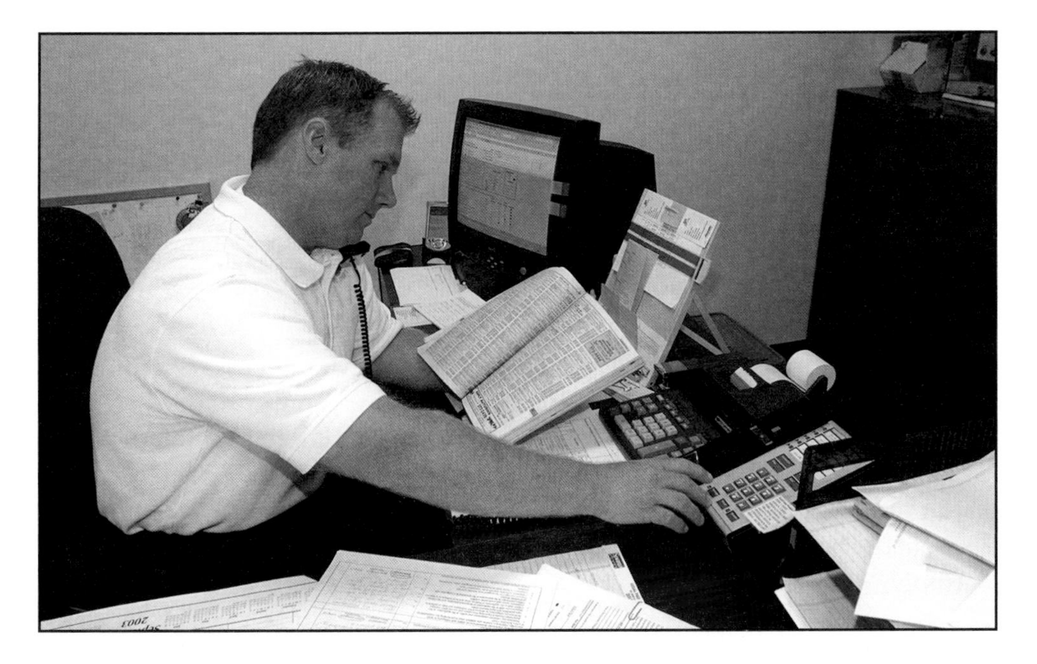

Figure 16.9 For good time management, it is a good idea to call with a reminder before a scheduled inspection.

Figure 16.10 Inspectors may need to visit some occupancies after hours if there are suspected or reported life safety issues.

- Inspection forms and checklists
- Inspection drawings
- Building occupancy changes
- Results interview
- Inspection-related letters and reports

Guidelines for Inspections

Many AHJs have guidelines and procedures for inspection staff, including but not limited to the following:

- Display identification — Inspectors should display current official credentials or have them immediately available for review. In many places, credentials include a photo.
- State the reason for the visit — The inspector must meet with the business representative and clearly state the reason for requesting entry. Often an oral request is sufficient. Occasionally, however, the owner/occupant

may ask for a written request before allowing entry. If the owner/occupant denies access, the inspector will need to document this action and follow departmental protocols for securing permission to obtain legal access.

NOTE: See Chapter One of this manual for more information about legal procedures regarding entry.

- If the business representative grants permission to inspect the property, the inspector should invite the building owner/occupant or a representative to accompany the inspector during the inspection.
- Coordinate inspections — In some cases, it may be necessary to include other agencies in an inspection. Coordination is especially useful when these inspections are part of an annual inspection, investigating complaints, new construction inspection, or when a certificate of occupancy is requested. Joint inspections reduce interruptions to the owner/occupant. They also avoid the appearance of harassment that multiple inspection requests might imply.
- Follow a written inspection procedure — Inspect within the authority of your certification. Inspectors should be aware of their authority and limitations given to them by the AHJ, their place of employment, and written policies.
- Prepare to issue a *stop work* or **cease-and-desist order** for extremely hazardous conditions or when it has been identified that work is being conducted without legally obtained permits. Always follow local protocols in regard to cease-and-desist orders **(Figure 16.11)**.
- Maintain a reliable record-keeping system of inspections.

Cease-and-Desist Order — Order prohibiting a person or business from continuing a particular course of conduct.

Owner/occupants may feel that they can improve their chances of avoiding a citation for violations by offering a drink or other friendly inducement. Because even simple courtesy gestures can be ethically misinterpreted, it is often the

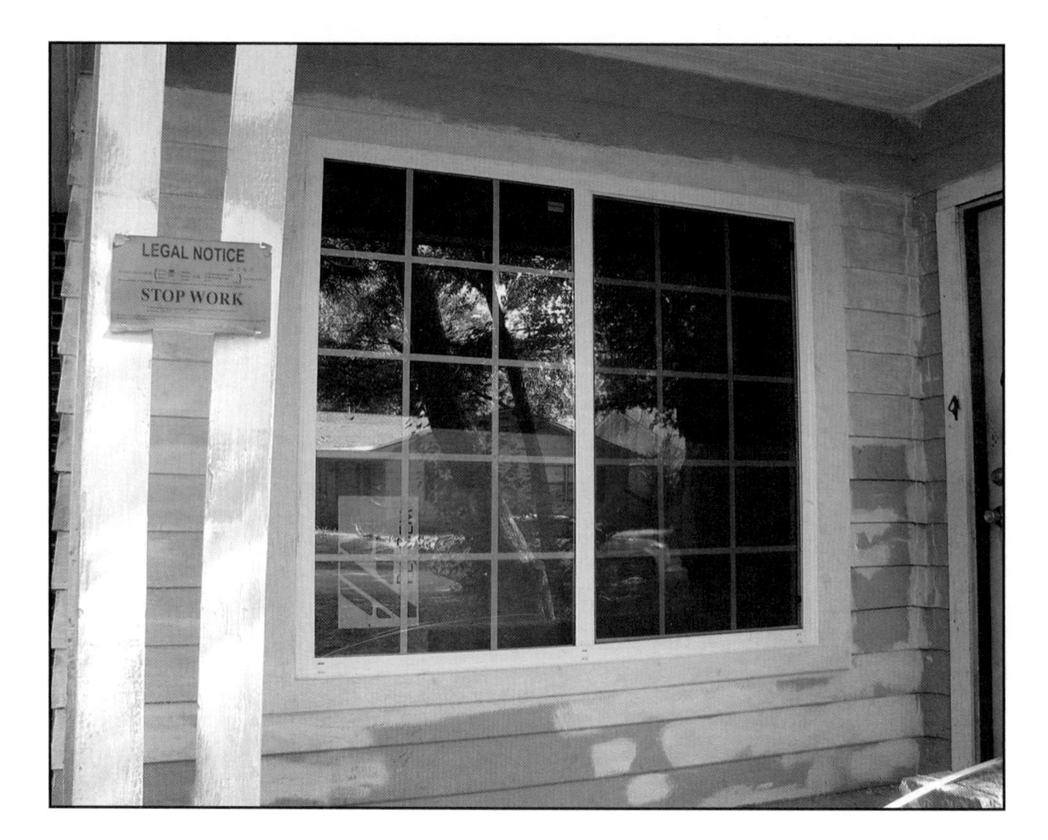

Figure 16.11 Know the procedures for issuing stop work orders.

best policy for inspectors to refuse any gratuities while in the performance of their duties. Most jurisdictions have codes of conduct/ethics that address what is appropriate or acceptable behavior and what is prohibited.

General Inspection Practices

Inspections are generally performed in a certain order to completely examine a facility or occupancy. The inspector should compare any photographs, sketches, site plans, or elevation drawings of the structure in the building inspection file to verify that they match existing conditions and to note any changes since the previous inspection. Following the same sequence for similar types of occupancies increases the consistency of the inspection process.

An inspector should make some observations regarding the general vicinity and exterior of the premises before entering a structure or facility:

- Conditions, barriers, or obstructions that may affect fire department access to the building (**Figures 16.12 a and b**)
- Locations of all hydrants, alternative water sources, and fire department connections (FDCs). These should match previous inspection notes.
- Building name and address match those listed in the building/facility inspection file
- Address numbers clearly visible on the outside of the building or facility and conform to fire/building code requirements
- Exterior maintenance or housekeeping problems, such as trash or dry vegetation, that may create a fire hazard
- Type of occupancy. Has it changed since previous inspection?
- Number of stories of the structure – should match previous inspection
- Proximity to other structures or potential exposures such as common or shared walls; distance to and type of exposures
- Construction, renovation, or demolition activities that create special or temporary fire and life safety conditions

Figures 16.12 a and b Make note of any potential or actual barriers to safe entry or access to fire protection systems. *Both photos courtesy of Rich Mahaney.*

- Exterior signage identifying any materials or dictating that certain practices are prohibited, depending on occupancy type **(Figures 16.13 a and b)**

The exterior inspection may progress clockwise or counterclockwise from the front of the structure. If the site has multiple structures, it may be necessary to complete one structure before proceeding to the next. During the exterior inspection, the inspector should draw a site plan if one does not exist in the inspection file. Site plans can be used to note the location of violations or fire protection equipment. When taking photographs of the site, be sure that the location and date of the photograph are noted on the plan.

Once the exterior examination of the structure has been completed, the inspection of the interior begins. The inspection focuses on factors that could affect the life safety of occupants or emergency responders as well as any potential hazards that are present.

The inspector should follow a systematic pattern during the inspection so that no portion of the facility is overlooked **(Figure 16.14)**. The inspector should not complicate the inspection sequence by going back and forth through an area inspecting individual items. Instead, the inspector should walk through an area noting inspection concerns on a checklist that can be referred to later when preparing the report.

There are several common methods for performing an inspection:

- Start from the roof and work downward.
- Start from the basement and work upward.
- Follow an established manufacturing process within the structure from the raw materials to the finished product and include all storage areas.

Regardless of the method used, each level or floor of the structure is inspected in a systematic manner. Use either a clockwise or counterclockwise pattern so that all areas are addressed. If a basic floor plan is available, it can

Figures 16.13 a and b Exterior signage identifying hazards must be in place and accurately describe the current conditions. *Both photos courtesy of Rich Mahaney.*

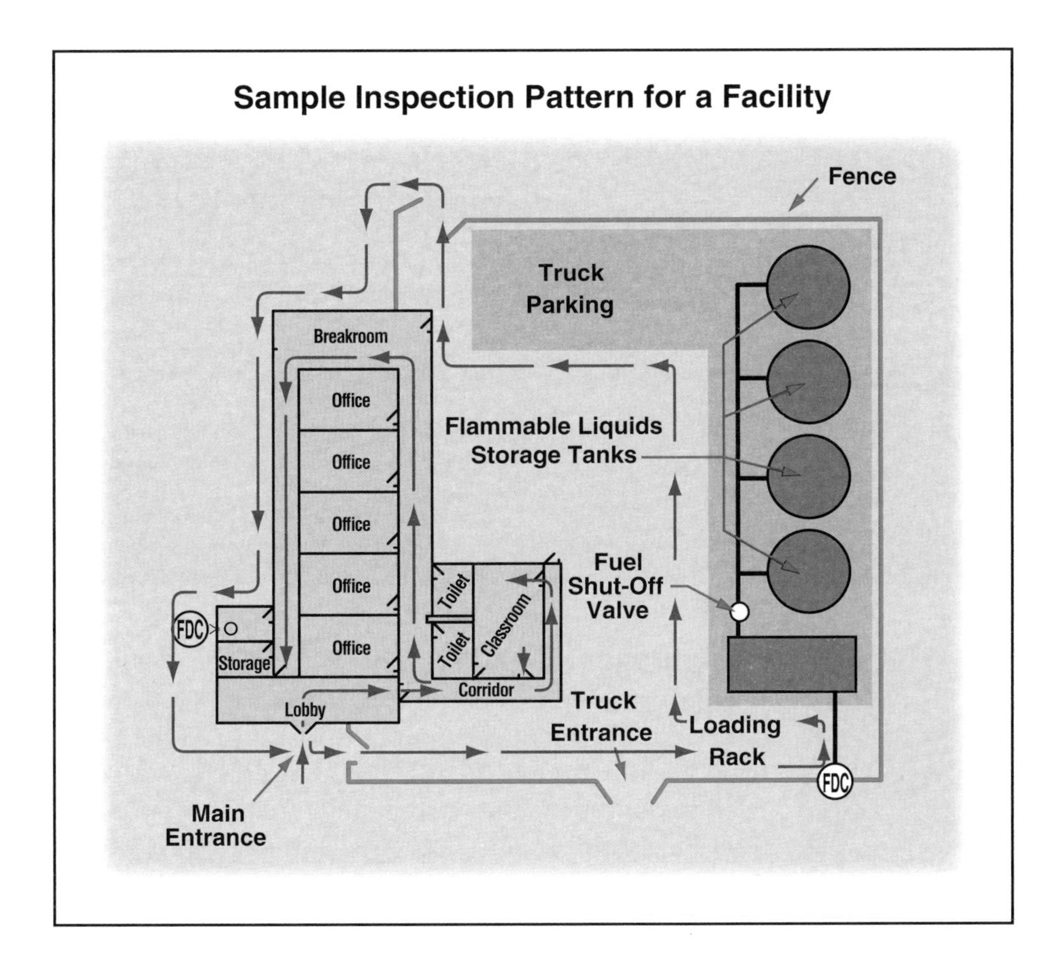

Figure 16.14 Inspectors should follow a systematic approach to make sure all areas of an occupancy are inspected.

be used to note the location of violations and can also be used to verify that all areas have been inspected.

When an area is locked, the inspector can request that the building representative unlock it. If admission is refused because of confidentiality reasons, the inspector should note the denial and continue to the next item. An inspector should not demand entry or argue with the representative. An inspector needs to respect the confidentiality of any trade secrets that exist in a facility.

The inspector should not sign any documents without consulting with a supervisor and/or legal counsel in accordance with departmental policies. After the inspection is concluded, the inspector should consult with his or her supervisor regarding any areas where entry was denied and determine if any additional steps will be taken to gain entry to these spaces.

Other facilities may have entry restrictions based on security concerns. These facilities include banks, data and computer system centers, law enforcement complexes, and detention/correction centers **(Figure 16.15, p. 708)**. In some instances, an inspector may be able to gain entry at a more appropriate time, such as during a maintenance period.

Code Requirements

Many occupancies have code requirements that are specific to that type of occupancy classification. At the same time, all occupancies share some similar fire and life safety code requirements of which an inspector must be aware. These general requirements are grouped into the similar categories for ease of application in **Table 16.1, p. 709.**

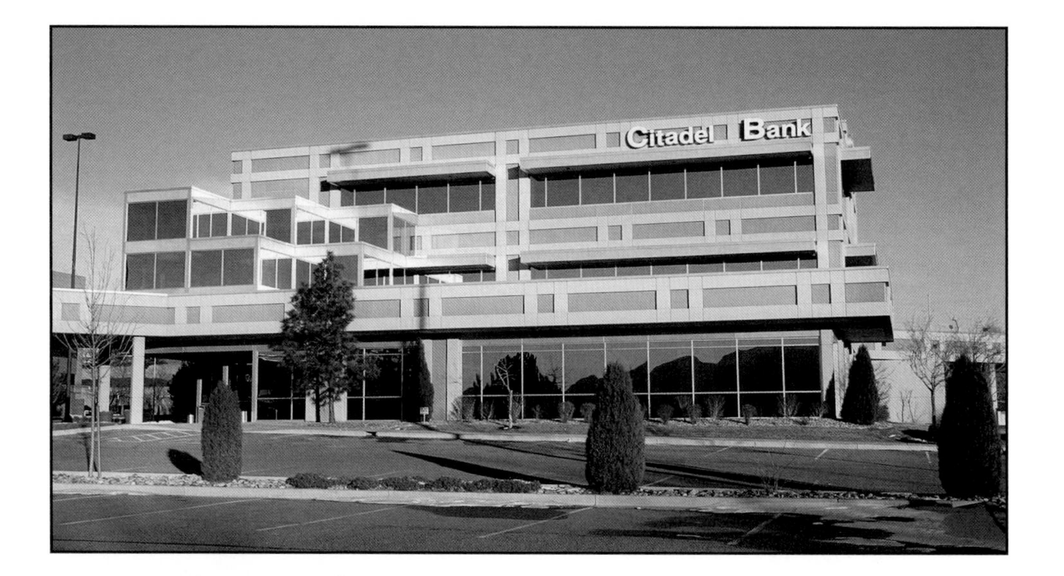

Figure 16.15 Inspectors need to work with management to facilitate inspections in occupancies that have entry restrictions.

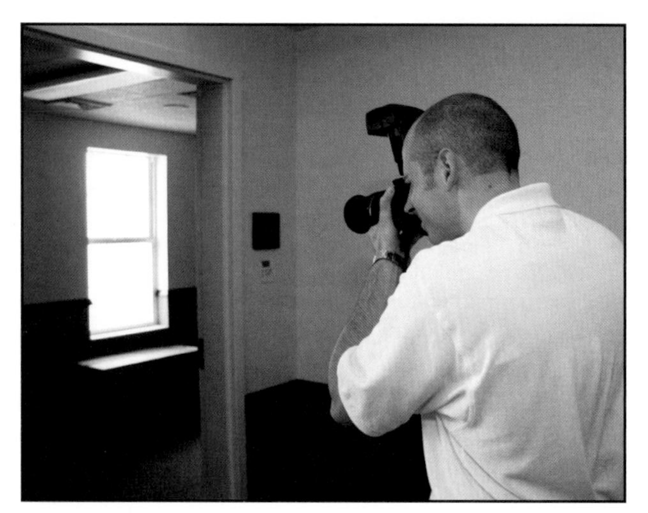

Figure 16.16 Photographing code violations creates a visual record that may be helpful in resolving the issue.

Photographs

Photographic documentation is particularly helpful when documenting violations that may result in a citation or legal action **(Figure 16.16)**. Photographs may also be used to document the current state of a feature or system. Fire suppression personnel can use these photos during preincident planning. Before and after photos will also be valuable to inspectors by providing a means of comparison when code violations are found during future inspections.

In certain facilities, the facility management and legal counsel may want to refuse the inspector taking photos or review the photos or take them for the inspector. The inspector should follow department policies for record-keeping.

Inspection Checklists

An inspection checklist not only serves as an aid during the inspection, it also serves as a record of the inspection itself. Although a checklist is a valuable tool, an inspector must be cautious not to overlook conditions or violations that are present but may not be included on the checklist.

Inspection organizations can purchase commercially developed checklists or develop their own to reflect the types of occupancies that are common in their response areas. If the organization does not use a standardized checklist, inspectors may wish to develop their own. **Appendix J** contains an inspection report checklist and a self-inspection checklist.

Handheld data collection units, such as tablets and smartphones, provide an efficient means of recording and transferring information. A number of important items should be contained on the inspection checklist:

- Inspection agency's name and address
- Inspector's name
- Date of inspection

Table 16.1
Occupancy Category Requirements

Occupancy Category	Code Requirements
General	• Posted exit plans for the facility • Posted occupant load for areas of assembly • Good housekeeping practices evident • All waste containers constructed of noncombustible materials • All trash collection areas separated by approved partitions and protected • Trash stored in trash collection areas • Trash storage room or bin within building • Chute closures fit closely and latch tightly
Egress	• Exit doors unlocked or unblocked • Exit signage indicate direction to exits • Aisles and egress paths clear of storage items • Exit doors swing in the direction of egress travel • Exit signs illuminated or readily visible • Battery-powered exit lights test properly • Exit discharges unobstructed on the exterior of the building • Handrails, stair treads, and landing areas are secured in place • Two remotely located exits provided for each employee
Auxiliary/Emergency Lighting	• Emergency lighting meets code requirements • Auxiliary lighting appropriately located for illumination of exit ways • Battery-powered emergency lighting test properly • Periodic backup power tests performed as required
Fire-Resistant Assemblies	• Openings (penetrations) protected • Pipe chases fireproofed and protected • Fire door assemblies approved and properly installed • Door closures operable and code-compliant
Vertical Openings	• Stairwells clear of storage • Wall and ceiling finishes conform to code requirements • Vertical openings enclosed and maintained with fire-rated walls and doors • Glass in exit doors comply with code (look for fire-rated glazing) • Stairwells in high-rises labeled at each floor level giving the following information: — Stairwell number — Floor number — Next available access floor — Level of exit discharge — Roof access signage • Trash and linen chute doors fire-rated and maintained in proper working condition
Interior Finishes	• Materials and finishes meet fire code requirements • Decorations and furniture meet code requirements

Continued

Table 16.1 (Continued)

Occupancy Category	Code Requirements
Fire-Protection Systems	• Fire-protection system meets building code requirements • Fire-protection system valves and operating conditions inspected and functioning properly • Fire alarm control panel (FACP) free of trouble or supervisory alarms • Records of annual, semiannual, and quarterly tests accessible and reviewed • Control valves open
Ignition Sources	• Ashtrays or noncombustible ash containers provided in authorized smoking areas • Electrical connections approved and extension cords *not* used
Portable Fire Extinguishers	• Extinguishers present, approved, and in working order • Extinguishers mounted on wall in approved manner • Extinguisher type consistent with fire hazards present • Required travel distances to extinguishers met • Extinguishers labeled correctly • Extinguishers located and accessible as required by code • Current extinguisher inspection date documented • Marking signs posted at extinguisher locations
Hazard Areas	• Hazardous materials present within the facility checked, listed, and categorized • Quantities of hazardous materials permitted within the facility at allowable levels • Hazardous processes comply with established safety and code requirements • Hazardous materials properly labeled and stored • Combustible or hazardous materials secured from all except authorized personnel
Smoking Safety Enforcement	• Prohibited in areas that contain flammable liquids, gases, or processes • *No Smoking* signs posted in nondesignated smoking areas • Noncombustible ashtrays and disposal devices available in designated areas • Metal trash containers used for disposal of ashtrays and other devices
Exterior	• Air intakes located away from combustible materials • Fire department connections (FDCs) maintained and accessible • Emergency vehicle access roads unobstructed • No parking allowed near exits or in fire lanes • Means of egress paths clear of landscaping overgrowth

Continued

Table 16.1 (Concluded)	
Occupancy Category	**Code Requirements**
Fire Alarms	• Manual alarm stations accessible • Alarm system passed periodic test • Smoke alarms present in required locations • Smoke detectors present in required locations • Fire alarm control panel (FACP) operational • Auxiliary power supply connected to the system and operational
Kitchens	• Range, filters, and ducts clean • Range hood extinguishing system inspected for operability • Proper extinguishing agent for the type of risk
Utilities	• Adequate electrical circuits for safe operation • Cover plates in place on electrical service boxes • Electrical extension cords meet code requirements • Electrical circuits and outlets *not* overloaded • Appliances in safe and operable condition • Clear area around heating appliances • All heating appliances listed by a testing laboratory
Fireplaces and Stoves	• Fireplaces and stoves meet provisions of the code • Minimum clearance of 36 inches (914.4 mm) between free-standing stoves and walls and ceilings or per manufacturer's listing • Chimney checked for serviceability • Proper fuel used in fireplace or stove
Storage	• Orderly and *not* impeding exit ways of residence • Flammable and combustible materials stored away from sources of ignition • Flammable and combustible liquids stored in approved containers and cabinets
Industrial Trucks	• Battery and recharging location according to code • Refueling location according to code • Equipped with portable fire extinguishers • Maintenance adequate and recorded

- Name and address of occupancy being inspected
- Name, phone number, and e-mail of person in charge of occupancy and/or the owner of the building if not the same person
- Name and title of building representative who accompanied inspector
- Occupancy classification
- Height of building and number of stories above and below grade
- Interior and exterior finishes

- Number, location, and size of exits
- Fire protection features
- All other areas that must be checked to verify code compliance
- Space to note items that are not contained on checklists

For most routine inspections, a comprehensive inspection checklist combined with appropriate photographs, drawings, and sketches is the only documentation needed to record the inspection. Formal inspection reports (discussed later in this chapter) are only required on major inspections or when serious code violations have been encountered.

Building Occupancy Changes

An inspector needs to be aware of the permitted occupancy classification before conducting the inspection. When it is apparent that the occupancy has changed, an inspector must then consider the following:

- What code requirements now apply to the new occupancy
- How they differ from the previous requirements
- Whether the requirements have been applied
- Whether there are any exemptions to the requirements

In particular, changes in occupancy can affect a number of fire and life safety code requirements, such as:

- Means of egress
- Fire protection equipment design or installation
- Fire detection and alarm system design or installation
- Emergency lighting
- Occupant load calculations

Even if no physical changes have occurred, the new occupancy may have increased the fuel load or amount of hazardous materials in the structure. An increase in the fuel load may require the installation of a fire protection system or changes to an existing system. Changes in the physical layout and contents of a structure can affect the location and type of sprinklers or discharge devices (**Figure 16.17**).

A change in occupancy classification typically requires a building permit and may require facility upgrades to meet current adopted codes. Being vigilant to these changes to occupancy classification is one tool that allows the AHJ to keep buildings up to date with important life safety changes.

Results Interview

When an inspection is complete, an inspector discusses the results with the owner/occupant or property manager. The purpose of this final or closing discussion is to review the entire inspection process and note areas that are in compliance as well as situations that need correction. Review possible violations and answer any questions that the owner/occupant have or note them for follow-up if research is needed to provide an answer to the owner/occupant. This is also the time to schedule a reinspection if it is needed.

Figure 16.17 Structures like strip malls have frequent changes in individual occupants and in potential fire hazards.

Violation Discussions

The most difficult aspect of dealing with people while conducting an inspection is the need to discuss unsafe conditions, behaviors, or **violations** that have been encountered. When handled in a tactful, professional manner, an inspector has an opportunity to make these discussions very positive. The first thing the inspector must do is to verify that all code-related points are properly added and are accurate. Nothing erodes an owner/occupant's confidence in an inspector more than to find a mistake. Never assume that an owner/occupant is ignorant of the codes and can be easily fooled. An inspector should always be ready to provide the backup information for a specific violation if requested.

Violation — Infringement of existing rules, codes, or laws.

To be effective, an inspector must make every effort to listen to and understand the owner/occupant's concerns. Even though the decision remains the same, being listened to helps owners/occupants feel that they have been treated in a fair manner. An inspector must fully explain the problems that have been noted and offer possible solutions for their correction if appropriate. The inspector should remember that while the owner/occupant has caused the situation, the correction will cost the owner/occupant money, down time, or both. In some instances, it may be better to refer a hostile individual to the inspector's supervisor or the head of the inspection organization. At all times, an inspector must remain polite but firm.

An inspector must not attempt to design a solution for the owner/occupant. It is better to refer the person to the appropriate qualified design professional. For example, advising about the need for sprinklers based on code requirements is acceptable. Recommending the number, location, and spacing of the sprinklers is inappropriate and can create a liability for an inspector and the organization.

When violations exist, the notice of violation is a notification and written expectation that a violation will be corrected as soon as possible. The follow-up inspection date is notification that a reinspection will be performed on or before the stated date to verify that the violation has been corrected.

If a follow-up inspection will be necessary, the time allotted between regular or follow-up inspections is usually defined by local code or policy. Additionally, in some instances, an inspector should explain the enforcement procedures when violations have been noted. The owner/occupant should also be made aware of the appeals process.

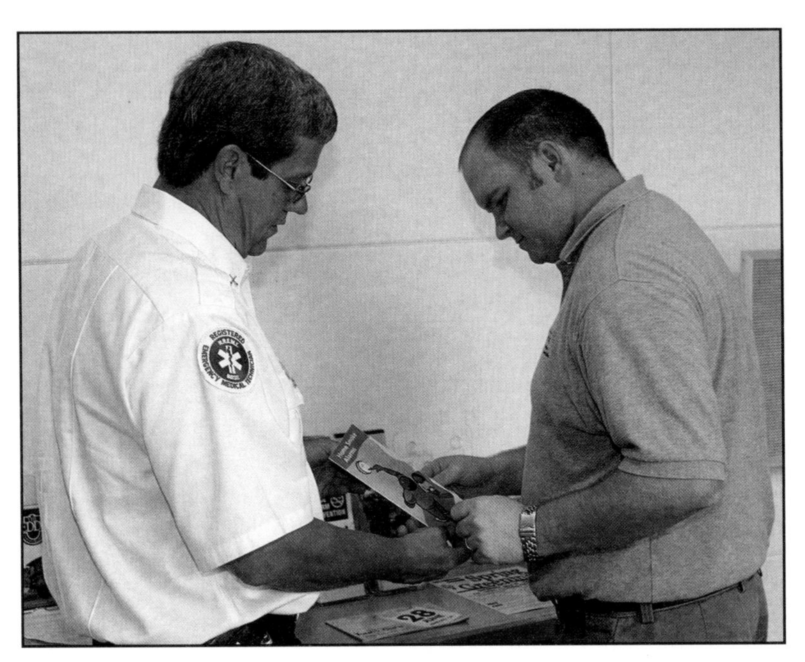

Figure 16.18 One of the inspector's most important duties is to provide education in life safety issues.

Educational Opportunities

An inspector should take advantage of the fire and life safety educational opportunities provided by an inspection to help positively influence behaviors in the future and reduce unsafe conditions **(Figure 16.18)**. A prepared educational package that contains general safety messages as well as more specific code-compliance-related messages can be very useful and may be provided to the owner/occupant during both the inspection and the results discussion.

Long-Term Relationships

One of the goals of the inspection is to build positive long-term relationships with the owner/occupant. This relationship makes it easier for an inspector to gain compliance when necessary and gives the owner/occupant the security of knowing that an inspector is a partner in providing a safe facility. The owner/occupant is more likely to comply when an inspector has educated the owner/occupant about the benefits that occur from adhering to fire and life safety code requirements.

Provide written documentation such as a follow-up letter describing the specific details of the inspection to the owner/occupant or property manager. This documentation is provided even if no violations or hazards are found.

Letters and Reports

Written letters and reports serve as official records of an inspection and form the basis for legal action when required. Without a written notice of violation that describes unsafe conditions or behaviors, citing the code section(s) violated, and giving a deadline for correction, there is no evidence that requirements were made and that the owner/occupant was notified of them.

An inspection report is made for every inspection and placed in the facility or site inspection file maintained by the inspections organization or agency. The majority of inspections are easily recorded on an inspection checklist. Some situations, however, require a formal letter:

- Code violations that create life-threatening hazards
- Code violations that require major renovations to comply

- A number of minor violations that taken together constitute a life safety issue for the occupancy

The inspection letter and the inspection report must clearly present the results of the inspection. Because these documents are used not only as official records but also to initiate corrective legal action, their formats or styles must be consistent with acceptable legal procedures. In addition to outlining inspection results and advising the owner/occupant of those results, each correspondence must include the following information:

- Name of business
- Type of occupancy
- Date of regular or periodic inspection
- Name of inspector
- Name of business owner/occupant
- Name of facility representative present during the inspection
- Name of property owner
- Edition of applicable code as a reference for future inspections
- List of violations and their locations, referencing code section numbers and titles
- Date of follow-up inspection

In the final portion of any inspection report or letter, the owner/occupant must be given an explanation of the violations noted. If the inspector uses a tablet computer and portable printer, in many instances, the formal letter and file update can be completed and signed by the occupancy owner or occupant before an inspector returns to the office.

Follow-Up Inspections

Follow-up inspections are made to verify that corrections noted in the inspection report have been completed. Inspectors should confirm the time and date of the follow-up inspection with the owner/occupant before arrival. During a follow-up inspection, the inspector focuses only on the problem areas included in the inspection report. This information should also be noted in the final inspection report.

If all code violations have been corrected, an inspector should be sure to compliment the owner/occupant for taking the appropriate actions. In order to close the file, the inspection office must send a letter to the owner/occupant stating that the violation was corrected. This letter also gives an inspector a second opportunity to thank the owner/occupant for cooperating.

If some code violations have not been corrected but the owner/occupant is making a conscientious effort to comply, the inspector should compliment the owner/occupant for the progress that has been made and schedule another follow-up inspection. The inspection files are updated and the original copy of the inspection form remains with the owner/occupant.

If the code violations have not been corrected and it is apparent that the owner/occupant has made no effort to correct them, the inspector issues a final notice with a date for another inspection. The final notice informs the owner/occupant exactly what legal action can be taken if full compliance is not attained by the date specified. At the same time, the owner/occupant's right of appeal and the appeal process is reemphasized. See Chapter 2, Standards, Codes, and Permits, for a description of the appeals process.

Emergency Planning and Preparedness

Evacuation — Controlled process of leaving or being removed from a potentially hazardous location, typically involving relocating people from an area of danger or potential risk to a safer place.

No matter what type of occupancy is involved, the key to successfully handling any emergency is the effective execution of an emergency plan. Each of the model codes establishes specific emergency planning requirements for various occupancy classifications, the training and implementation of the plan, and the frequency and participation level of emergency **evacuation** drills **(Table 16.2)**. The Inspector I is responsible for verifying that emergency plans are in place and in compliance with locally adopted codes.

In addition to verifying code compliance, the inspector also should verify that the emergency plan is updated as personnel leave or move within the organization. This is to verify that personnel contact information, as well as roles and responsibilities are clearly defined.

Emergency evacuation drills should be conducted in all types of occupancies. All occupants must be instructed on the actions to take during an emergency evacuation drill or any type of emergency. For some occupancies, such as schools, businesses, or high-rise structures, the instruction and emergency evacuation drill may be very basic and involve nothing more than an efficient evacuation of the structure.

Emergency evacuation drills for occupancies such as health care, correctional, and industrial facilities are more complex. These drills involve the movement of critically ill or bedridden patients, perimeter security of inmates, or manufacturing process shutdowns and industrial fire brigade response. All of these factors have to be evaluated before, during, and after the emergency evacuation drill. In occupancies such as hospitals, hotels, assembly occupancies, and stores, emergency evacuation drills usually involve only staff personnel to avoid alarming patients, guests, and customers.

Emergency evacuation drills need to be conducted at different times of the day so that all staff shifts have the benefit of a drill. The inspector keeps a copy of emergency evacuation drills in the facility inspection file for future reference.

The effectiveness of evacuation plans should be tested on a regular basis by conducting emergency evacuation drills. The occupancy's loss control or safety staff is responsible for planning and executing these drills. Plans for emergency evacuation drills need to be discussed with all levels of management and supervisory staff to verify that they understand and cooperate with the drill. Following an emergency evacuation drill, all members of management and the loss control staff meet to critique the effectiveness of the evacuation plan and the emergency evacuation drill.

The following sections will provide more detail on emergency evacuation drills in the different occupancies:

- Educational facilities
- Health care facilities
- Correctional facilities
- Hotels and motels

Table 16.2
Emergency Evacuation Plans and Drills per Occupancy Type

Occupancy Type	Evacuation Drill Frequency	Participation in Drill	Notes
Assembly	Quarterly	Employees	Not required for places of worship that have an occupant load of less than 2,000
Business	Annually	Employees	Applies to buildings having a total occupant load of 500 or more or more than 100 persons above or below the lowest exit discharge
Education	Monthly	All occupants	Frequency may be subject to changes based on severe climates
Hazardous	Not required	Not required	None
Institution	Quarterly on each work shift	Employees	In residential care facilities, drills shall include all occupants
Residential-1	Quarterly on each work shift	Employees	None
Residential-2	Four annually	All occupants	Applies to college and university residential quarters
Residential-4	Quarterly on each work shift	Employees	None
Mercantile	Not required	Not required	Applies to buildings having a total occupant load of 500 or more or more than 100 persons above or below the lowest exit discharge
High-rise buildings	Annually	Employees	None
Covered malls	Not required	Not required	When greater than 50,000 square feet (4 645 m^2)
Underground buildings	Not required	Not required	None
Buildings with atriums and used for Groups A, E, and M	Not required	Not required	None

Based on information provided in International Fire Code, 2006 edition.

Educational Facilities

Emergency evacuation drills in educational occupancies require that all persons in the building participate. Emphasis should be placed upon orderly evacuation under proper discipline rather than upon speed. Inspectors must verify that the frequency of emergency evacuation drills meets code requirements. In addition, inspectors should evaluate the time needed to evacuate educational occupancies and recommend alternative methods if the time is determined to be too long.

Codes that require monthly emergency evacuation drills may be modified to account for severe weather conditions. Such modifications may consist of conducting more drills during mild weather to account for ones missed due to extreme weather.

Emergency evacuation requirements apply to all types of educational occupancies, both public and private. Some portions of this latter category may be classified as places of assembly, which will require more frequent emergency evacuation drills. Emergency evacuation drills should occur at different times of the school day, whether students are in class or in other activities (**Figure 16.19**). School officials should be encouraged to develop an emergency evacuation plan that provides direction on where and how students should report if the fire alarm sounds during times other than normal class periods.

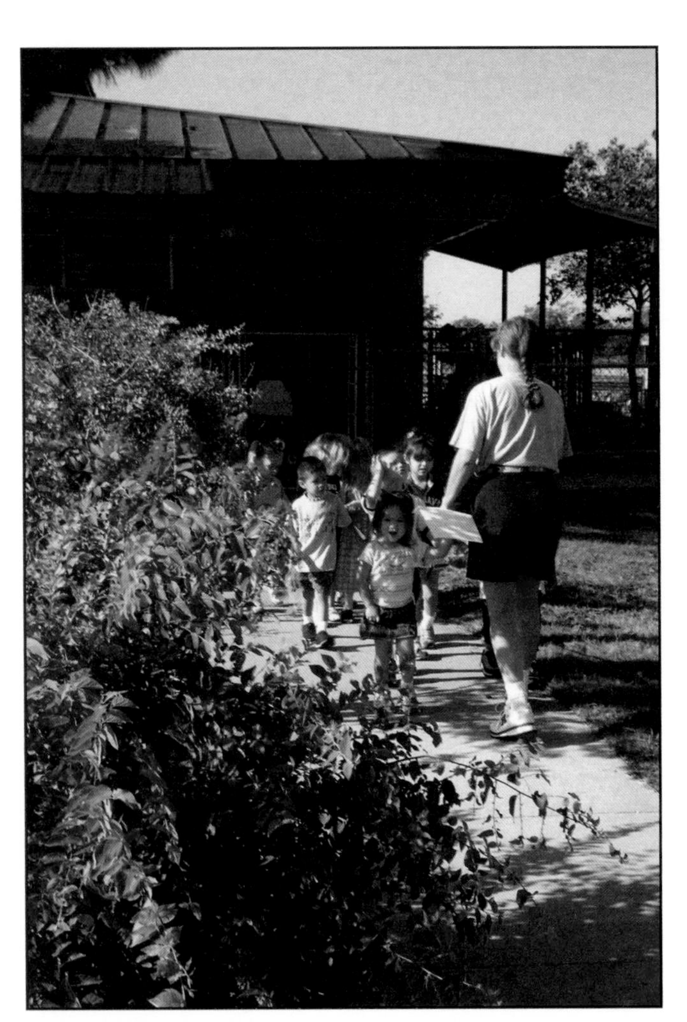

Figure 16.19 Evacuation drills must be based on the evaluation plan.

When an emergency evacuation drill occurs when classes are in session, students are instructed to form a line and immediately proceed to the nearest exit in an orderly manner. Teachers must control the class so that they can quickly and calmly contain the students, form them into lines, and direct them as necessary. Students who are incapable of holding their places in a line moving at a reasonable speed should move independently of the regular line. Students who have physical limitations may require special assistance from another student or teacher.

Teachers, teacher assistants, staff members, and older students can all function as monitors and assist in the proper execution of emergency evacuation drills. Monitors should either hold doors open in the line of procession or close doors when necessary to prevent the spread of smoke and fire. Teachers and other members of the staff have responsibilities for searching classrooms, restrooms, or other spaces.

Each class or group proceeds to a predetermined assembly point outside the building and remains there while a check is made to see that all students are accounted for. Assembly points should be far enough from the building to avoid danger from fire or smoke, interference with fire department operations, or confusion among classes and groups. No one may reenter the building until directed. Alternate exit routes must also be designated and practiced during emergency evacuation drills.

Lockdown Procedures

A heightened awareness of school security as a result of violent attacks or the need to protect students and faculty from hazardous materials accidents has led to an increased emphasis on the use of shelter-in-place or lockdown procedures. In the current security climate, many administrators and other officials are focusing on ways to secure the facility instead of emergency egress concerns. Although this concept is very important with regard to the protection of students and faculty, it is not appropriate when a fire or smoke hazard occurs in the school.

New potential threats should not erode older, proven safety practices. Shelter-in-place, commonly used during tornadoes and other severe weather, should be part of the emergency preparedness plan and practiced along with the emergency evacuation drills. The inspector must educate the administration on the importance of egress and the requirements of the code and to work with the team to come to an acceptable solution.

Health Care Facilities

In health care facilities, emergency evacuation drills are conducted quarterly on each shift, and every member of the staff must participate. Emergency evacuation drills must include all emergency notification signals that are used by the facility. Staff members must perform all assigned duties that are required when responding to a particular emergency (**Figure 16.20**).

The written health care fire safety emergency evacuation plan is followed during the drill. When the emergency evacuation drill reveals that the plan needs modification, the written plan should be updated.

Figure 16.20 Health care workers have specific duties during emergency evacuation drills.

The facility must maintain a full record of all emergency evacuation drills and names of facility personnel participating in them. An inspector receives a copy of these reports and files them in the facility inspection file as a permanent record.

Hospitals and nursing homes require special emergency evacuation procedures. Because a total evacuation of a facility is not warranted during every emergency, the extent of an evacuation is determined by the severity of the emergency. Evacuation plans proceed in progressive steps or phases.

Because emergency evacuation events can include tornadoes, hurricanes, floods, earthquakes, or terrorism as well as fire, each of the four activity phases described may differ, depending upon the nature of the emergency. However, these emergency evacuation plans must be consistent with a community's overall emergency preparedness plan. Phase activities are generally described as follows:

- **Phase 1 activities** — Involve evacuating a single room. A common procedure used in many facilities involves following a system that uses the acronym REACT:
  - **R**: Remove those in immediate danger.
  - **E**: Ensure that the room door is closed.
  - **A**: Activate the fire alarm or notify law enforcement agency, depending on the nature of the emergency if it has not already been done.
  - **C**: Call the fire department or local law enforcement agency.
  - **T**: Try to extinguish or control the fire or remove those at risk without endangering additional lives.

  Another acronym that is used is RACE:
  - **R:** Remove (patients)
  - **A:** Alert (call 9-1-1)
  - **C:** Confine (close off the area of origin)
  - **E:** Extinguish (if able to do so)
- **Phase 2 activities** — Involve evacuating an entire zone of a building. Close all room and smoke barrier doors first. Continuing actions:
  - Evacuate rooms adjacent to the fire or emergency areas first.
  - Concentrate on evacuating the remaining rooms within that zone.
  - Move occupants to predetermined safe areas on the affected floor.
  - Take medical charts/records with patients anytime they are moved from their rooms.
  - Place the facility in a lockdown situation to limit the movement of unwanted persons into patient areas when law enforcement agencies require that action.
- **Phase 3 activities** — Involve evacuating an entire floor and zones on the floor above the incident floor. Evacuate the fire floor first. Once this is done, evacuate the necessary zones on the floor above the incident. Move occupants down stairwells or elevators if they are deemed secure, safe, and operable.

- **Phase 4 activities** — Require that the entire building be evacuated. The activities work best when preincident plans are in place and practiced on a regular basis.

  Actions for fire incident:

  — Evacuate the fire floor first, followed by all floors above the incident.

  — Evacuate the floor(s) below the fire when the fire floor and floors above it are clear.

  — Transport patients to other buildings or sites.

  Actions for law enforcement incident:

  — Establish direct coordination with the local law enforcement agency.

  — Remove patients to predetermined sites that are ready to provide adequate shelter and care for special/medical needs.

Considerations for large-scale evacuations due to tornadoes, hurricanes, floods, earthquakes, or wildland/urban interface fires:

— Many close evacuation sites that would normally be used may be unsuitable because of their proximity to the incident.

— Evacuation areas may be many miles (kilometers) away during these incidents.

— It may be necessary to move patients to an intermediate location while arranging transportation to a distant evacuation site **(Figure 16.21)**.

Figure 16.21 Preplanning and coordination will be needed if large numbers of evacuees need to be transported from the emergency scene. *FEMA/Jacinta Quesada.*

Correctional Facilities

Emergency evacuation drills and procedures for correctional facilities provide a unique problem due to the physical security features of the building and the need to maintain control of inmates. The lives of inmates, visitors, and security personnel depend on the quick and effective actions of staff members during an emergency.

Correctional facilities must maintain an effective fire detection system, a key-control system, and a written emergency preparedness plan. An emergency evacuation plan must be available to inmates, and plans must be regularly reviewed and upgraded depending on changing conditions, especially inmate population.

Inmates are either protected where they are or evacuated to secure areas of refuge. Inmates should be released by a reliable means, whether it is electrical, mechanical, or using a set of master keys. There should be at least two means of access to each main cell block.

Hotels and Motels

The temporary nature of hotel and motel occupants presents a challenge with regard to evacuations. In case of an emergency, none of the occupants are familiar with the layout of the building. For this reason, local, state, or provincial fire codes require the installation of a reliable fire detection and

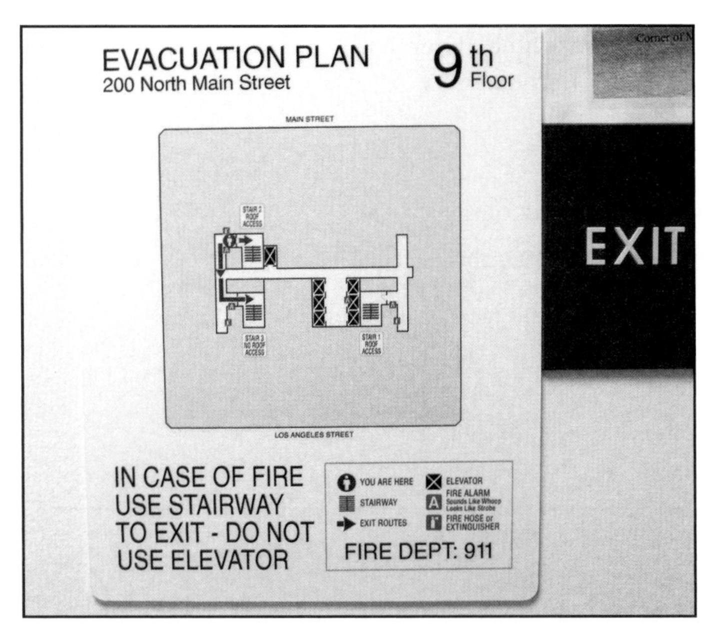

Figure 16.22 Inspectors should make sure appropriate evacuations signs are in place. *Courtesy of Tanya Hoover.*

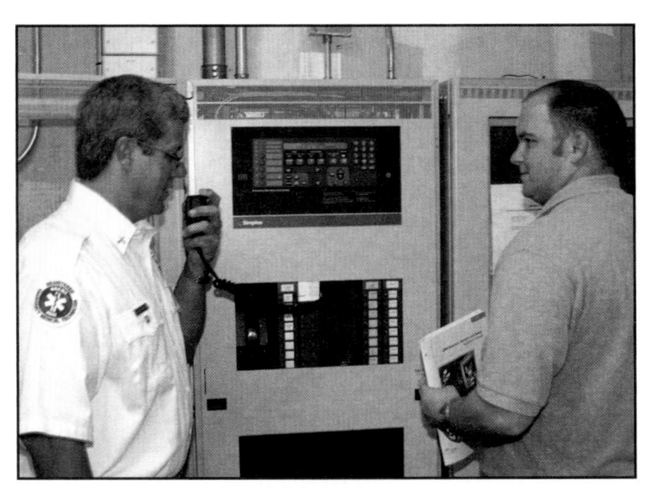

Figure 16.23 Inspectors and building owners or administrators must work together to train employees in emergency evacuation procedures.

alarm system. Many systems also include an occupant notification feature. Evacuation procedures are usually posted in each room on the back of the door, at all fire alarm manual pull stations, and next to exits **(Figure 16.22)**.

The hotel/motel owner or administrator is responsible for training employees in emergency evacuation procedures. Employees may be designated as fire wardens to direct the flow of occupants during an evacuation. Employees should attempt to account for guests and match the number evacuated against the guest register to determine if any guests are missing. It is very difficult to be 100 percent accurate in these situations. Emergency evacuation drills must be held frequently to keep all personnel familiar with the emergency preparedness plan and their assigned duties **(Figure 16.23)**.

Complaint Management

Allowing the public to identify and report serious fire and life safety hazards is an integral part of maintaining a successful fire prevention and life safety program. Complaints that citizens file should be met with courtesy and genuine interest. Prompt investigation of these complaints encourages continued support of fire prevention measures within a community.

The inspection organization should have a procedure for receiving and processing complaints in the organization's standard operating procedures (SOPs). An inspector should process and act upon each complaint in a consistent manner. An inspector taking a complaint should record all pertinent information. A standardized complaint form is useful to record all necessary information **(Figure 16.24)**.

Complaints that do not require immediate attention can be routinely assigned to inspection staff or qualified company officers for research and investigation. Complaints that involve a serious life safety threat require immediate action.

City of Ansonia
Fire Department
Office of the Fire Marshal

P.O. Box 421
Ansonia, Connecticut 06401
Voice: 203-734-3525
Fax 203:736-6537

Ralph E. Tingley
Fire Marshal

Fire Code Complaint

Complainant: ____ **Anonymous** ____ **Male** ____ **Female**

Name: ______________________________

Address: ______________________ Telephone: ______________

Involvement: ____ **Employee** ____ **Occupant** ____ **Other** ______________

How Received: ____ **In Person** ____ **Telephone** ____ **Mail** ____ **Other** ______________

Building Information

Building Name: ______________________________

Address: ______________________________

Owner: ______________________ Telephone: ______________

Owners Address: ______________________________

Use of Building: ____ **Apartment, No. of Units** ____; ____ **Mercantile;** ____ **Industrial;** ____ **Storage**

____ **Health Care;** ____ **Assembly;** ____ **Educational;** ____ **Residential Board;** ____ **1&2 Family;** ____ **Other** ____

Nature of Complaint:

____

____

____

____

____

____

____

____

Complaint received by: ______________ Date: __________ Time: __________

____ **Comments on reverse**

Figure 16.24 Typical fire code complaint form. *Courtesy of Ralph E. Tingley, Ansonia, Connecticut.*

When investigating a code complaint, inspectors may or may not need to give the owner/occupant advance notice. It may be best not to give advance notice of a complaint inspection concerning serious fire and life safety violations. The type of occupancy, location within the occupancy, and severity of the nature of the complaint often determine whether an inspector needs to obtain an administrative warrant to enter the location.

Upon finding a code violation, the inspector must initiate the process that leads to corrective action. Inspectors must be prepared to deal with negative attitudes by the business owner/occupant or representative when they act on complaints. An inspector should maintain a professional demeanor even if the other person does not.

When a complaint has been resolved, the person who filed the complaint may need to be formally notified and thanked. This acknowledgement provides confirmation that the person has contributed to the safety of the community.

All complaints are documented and maintained in the organization's record-keeping system. The complaint form may be filed and cross-referenced by location, date, or type. The following information should be kept on the complaint:

- Date/time complaint received
- Location of perceived violation
- Name of person reporting complaint
- Name of owner/occupant
- Nature of complaint or perceived violation
- Date/time of inspection
- Date/time of follow-up inspection
- Resolution of complaint

Accurate record keeping is essential. Records can confirm when a pattern of violations begins to appear. An inspector should always refer to the complaint records before making any inspection to determine whether violations have been reported about the property previously. It is also possible that the owner of several properties may have a record of code violations involving more than one address. Cross-referencing code violations, property ownership, and property addresses are important means for tracking repeat offenders.

Complex Complaint Procedures

Complex complaint procedures involve considerably more time and investigation on the part of the inspection office. Such complaints can involve the following:

- Size of facility
- How much research is involved in addressing the issue(s)
- Number of codes or standards that need to be addressed
- Whether the inspector has experience in the questions asked

Complex complaints do not involve single, straightforward issues. An example of a complex issue may be the condition of an existing apartment building where fire, building, electrical, plumbing and/or health issues are present. The inspector may require additional time to conduct research and inspections, or may require expert assistance or assistance from multiple departments or agencies.

Complex complaints may require an interpretation on a particular code and research into additional codes. The Inspector II will need to be able to determine which official has the ultimate authority over the questions. The inspector may determine that the complaint is multi-jurisdictional and therefore involves a number of agencies. At this level, complaint resolution will rely on additional technical expertise. In addition, experts may recommend several courses of action to reach a solution.

Evaluate Emergency Planning

A further responsibility of an inspector is evaluating the emergency planning and preparedness procedures in the jurisdiction. These duties may require reviewing the facility's current emergency plan and determining its compliance with the applicable codes and standards. Things that an Inspector II will look for when evaluating an emergency plan include:

- Emergency-reporting procedures for facility personnel
- Duties and responsibilities of facility personnel in the event of an emergency
- Procedures for personnel to evacuate the facility, relocate to another area, and/or shelter-in-place
- Procedures for conducting emergency fire drills
- Information about facility's fire protection systems

Inspectors should be aware of the facility's occupancy and hazards when evaluating the emergency plan because each facility has its own specific issues. One part of the emergency plan can be fairly standardized regardless of the occupancy; all emergency evacuation plans should have the following three key elements:

1. **Evacuation routes** — Primary and secondary routes of exit from a structure are given for each compartment in an occupancy. Maps showing these routes are given to each employee and posted in conspicuous locations (**Figure 16.25**). Maps are used for the following:
   — Indicate the present location of the person reading the map.
   — Indicate the routes to the exits.
   — Show the location of the meeting point for occupants once they have left the structure, which allows for accounting of evacuees and gives information on the evacuation to the Incident Commander (IC).

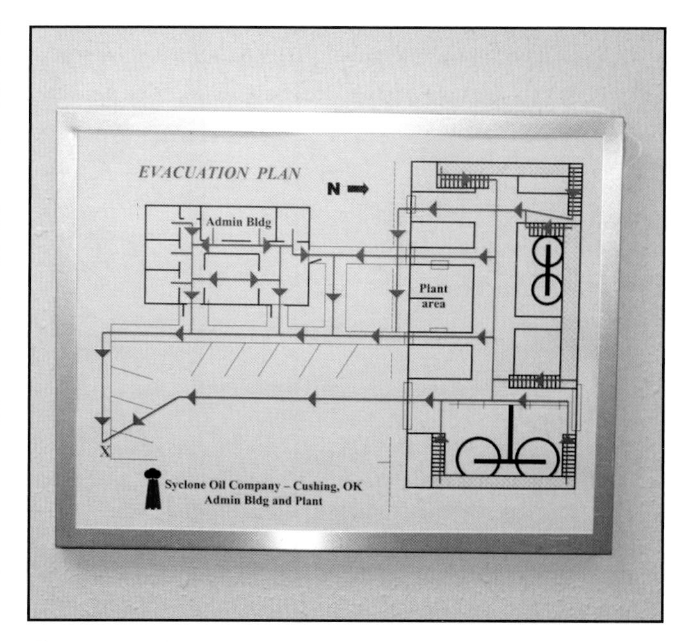

Figure 16.25 Evacuation plans like this one should be clearly posted throughout large complexes so that occupants can easily find a safe means of egress in case of an emergency.

Monitor — Individual assigned to supervise an evacuation process for a specified area within a structure, such as a ward monitor or floor monitor.

2. **Monitor duties** — Each compartment has an assigned primary **monitor** to supervise the evacuation process **(Figure 16.26)**. This monitor is responsible for seeing that all people in the particular area know the emergency evacuation plan and their individual duties and responsibilities. Alternate monitors are appointed to handle these duties and responsibilities if the primary monitor is not available. Names of both monitors may be listed on the evacuation route maps and in the emergency plan. The monitor's duties include:
   — Accounting for the people who are evacuated from the area.
   — Making a thorough search of the area if possible to verify that everyone has left before leaving the area.
3. **Employee/occupant duties** — Everyone is responsible for verifying that their area is completely evacuated during an emergency. Duties include:
   — Turning off all equipment.
   — Closing doors and windows.
   — Checking the performance of other essential safety shutdown functions if conditions and time allow.

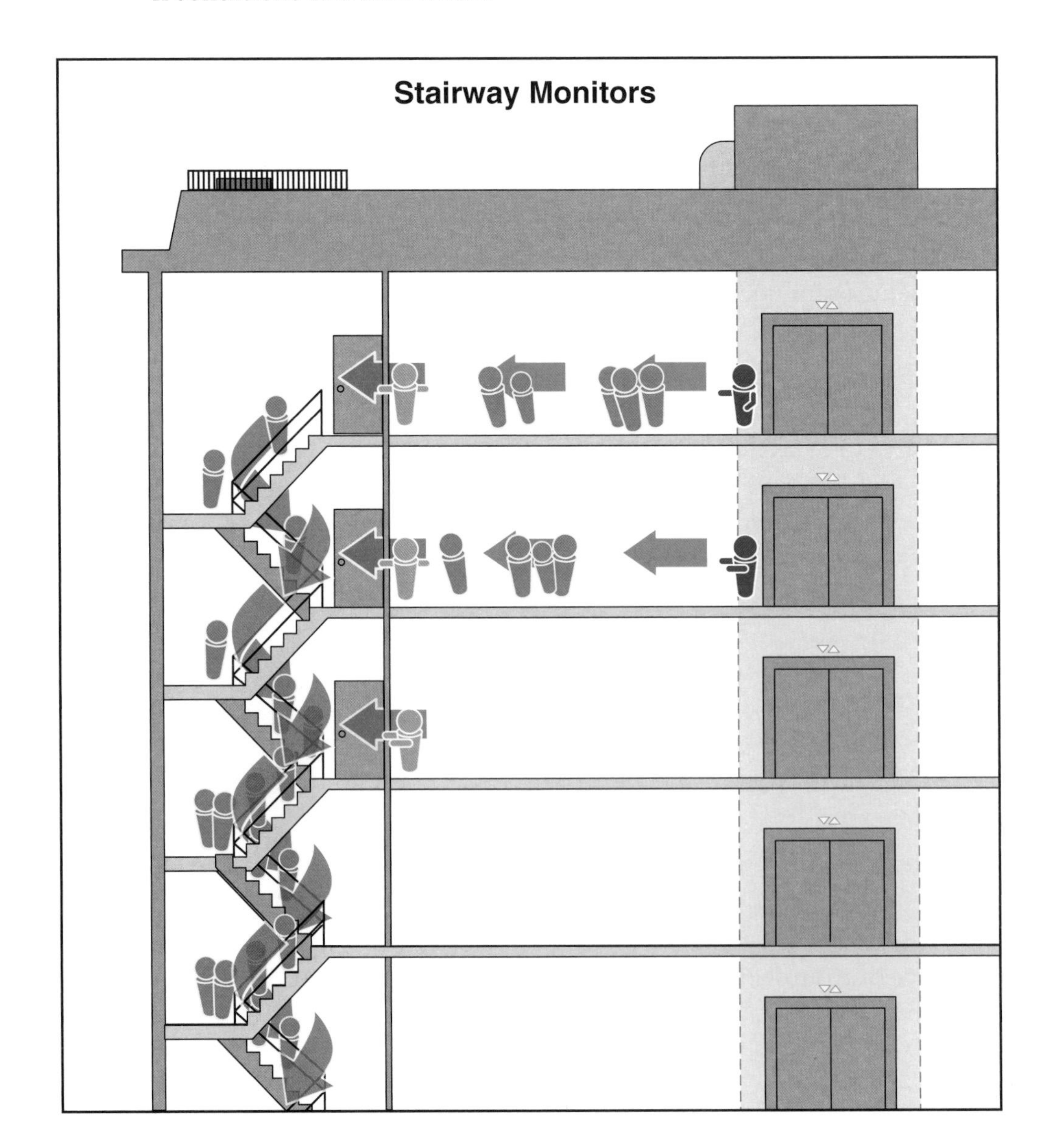

Figure 16.26 Elevator monitors can help direct occupants to the nearest stairway. Coordinating stair monitors can help ensure a smooth flow of traffic.

— Providing information to responding firefighters on the types of actions taken.

— Verifying that visitors are accounted for and evacuated from the building.

The complexity of an emergency plan depends on the type of occupancy and the level of occupant preparedness. An emergency plan for a hospital, for example, is significantly more complex than a plan for a warehouse. It is beyond the scope of this manual to address all of the information needed to develop detailed emergency plans for complex occupancies or possible risk options that might be confronted. However, an element of an emergency plan that is common to all types of occupancies is the need to evacuate the structure in a timely and efficient manner.

NOTE: Inspection personnel may be requested to assist building administrators and occupants in the development, review, and testing of emergency plans for occupancies within their jurisdictions.

Chapter Summary

The process of conducting an inspection must follow a systematic approach from preparation through the filing of the final reports. The inspector needs to prepare by researching information about the facility and applicable codes. A successful inspection involves more than research, though. Inspectors need to present a professional appearance, communicate in a respectful manner, and follow up with any issues uncovered during an inspection.

An important aspect of this chapter has been the discussion of the necessary customer-service skills that the inspector must display when working with the public. The thorough knowledge and application of these skills will most often serve the best interests of the community, the inspector, and the inspection organization or agency.

This chapter presented the methods for conducting an inspection, handling complaints, as well as fundamental communication skills that are essential to the successful performance and completion of the inspection process. Taking the time to educate citizens on safety matters will increase cooperation and help to enlist owners/occupants as allies in working toward safety.

Review Questions

1. List the enforcement duties of an inspector.
2. List the basic elements of a communication model.
3. What skills are needed to be an effective listener?
4. What skills are needed in order to converse successfully?
5. Explain Monroe's Motivated Sequence Pattern.
6. What types of written communications might an inspector use in the course of business?

6. List types of files and records normally maintained by an inspection organization or agency.
7. What does an inspector need to do to prepare for an inspection?
8. Describe common inspection procedures.
9. Why are follow-up inspections conducted?
10. Which tasks in emergency planning fall under the jurisdiction of an Inspector I?
11. What are some of the emergency evacuation requirements for educational facilities?
12. What are some of the emergency evacuation requirements for health care facilities?
13. What unique conditions impact evacuation drills for correctional facilities and hotels?
14. What complaint information should be kept by an inspector?

1. What makes a complaint complex?
2. What does an Inspector II look for when evaluating an emergency plan?

Appendices

Appendix A
NFPA® 1031 Job Performance Requirements (JPRS) with Chapter and Page References

NFPA® 1031 JPR Numbers	Chapter References	Page References
4.2.1	16	714 - 715
4.2.2	2	52 -55
4.2.3	15	651 - 653
4.2.4	2, 16	51 – 52, 722 - 724
4.2.5	2	41 - 50
4.2.6	1	19 - 24
4.3.1	4, 6	125 – 145, 223 - 235
4.3.2	7	265 – 267, 276
4.3.3	7	421 - 265
4.3.4	4, 5, 6	116 – 125, 153 – 168, 194 - 216
4.3.5	12, 13	504 – 538, 564 - 584
4.3.6	14	608 - 641
4.3.7	13	588 - 600
4.3.8	3, 9, 10	73 - 108, 308 - 334, 379 - 430, 449
4.3.9	15	655 - 659
4.3.10	16	720 - 725
4.3.11	8	285 - 299
4.3.12	9	308, 313 - 317
4.3.13	9, 10	308, 319 – 334, 379 - 438
4.3.14	3, 9	99 – 108, 319 - 334
4.3.15	9	308 - 334
4.3.16	11	466 - 492
5.2.1	2	65 - 67
5.2.2	15	660 - 665
5.2.3	16	728 - 729
5.2.4	1, 2	34 – 36, 62 - 65
5.2.5	1	34 - 36
5.3.1	7	268 – 279, 277 - 279
5.3.2	4, 7	146 – 148, 268 - 273

NFPA 1031 (cont.)

NFPA® 1031 JPR Numbers	Chapter References	Page References
5.3.3	5, 7	171 – 187, 268 - 273
5.3.4	12, 14	538 – 556, 642 - 647
5.3.5	7	268 - 275
5.3.6	3, 9, 10	48 – 51, 335 – 374, 449 – 451, 455 - 459
5.3.7	16	729 - 731
5.3.8	3, 9, 12	48 – 51, 335 – 374, 538 - 557
5.3.9	3, 9, 12	48 – 51, 352 – 373, 451 - 458
5.3.10	3, 5	48 – 51, 171 - 187
5.3.11	7, 9, 12	268 – 274, 335 – 373, 540 - 556
5.3.12	6, 9	228 – 236, 335 - 351
5.4.1	7, 15	268 – 274, 666 - 686
5.4.2	7, 15	277 – 279, 683 - 687
5.4.3	11, 12, 13, 14, 15	477 – 492, 538 – 556, 594 – 597, 642 – 646, 666 - 686
5.4.4	12, 13, 14, 15	538 – 556, 594 – 597, 642 – 646, 566 - 586
5.4.5	7, 15	268 – 274, 683 - 686
5.4.6	5, 6	171 – 187, 220 - 235

Appendix B
Example of a Citation Program

Code Citation Policy and Procedures

Section I. Purpose

1.1 To gain compliance with the *International Fire Code®*, State Administrative Code, and Title 19 when all reasonable efforts have been unsuccessful.

1.2 A course of legal action to be taken when a condition exists that causes a threat to life or property from fire or explosion.

Section II. Background

2.1 During 2007, the Fire Department wrote 2,426 violations, achieved 1,698 corrections, and had 728 outstanding violations.

2.2 Our present process of enforcement (City Attorney, District Attorney, Office Hearings, filing of complaint, etc.) does not lend itself to providing uniformity of compliance within the community. The majority of fire violations written are characteristic of the following three conditions:

2.2.1 *Transient* problems such as overcrowding of public entertainment facilities, illegal parking in fire lanes, or mischievous fire setting.

2.2.2 *Changeable* or portable situations such as illegal locking devices on public exit doors and obstructions to aisles or exitways.

2.2.3 *Maintenance* of fire extinguishing and alarm systems, portable situations such as electrical violations, or housekeeping including outdoor fire hazards.

2.3 Transient violations are specific occurrences, which should be acted upon immediately through a citation process. Changeable violations are corrected by the person responsible (in most cases) on a temporary basis but are changed back after the inspector leaves the premises. The same situations are encountered yearly and are not being permanently abated.

2.4 Citizen awareness of the Fire Department's ability to cite for violations would create an effective deterrent in maintaining corrective abatements on a more permanent basis. The citation process would be used on a discretionary basis and would be very cost-effective from the standpoint of available manpower utilization and steadily increasing workload demand.

Section III. Policy

3.1 Members of the Fire Prevention Division who are authorized by the Division Chief in charge shall have the authority to issue citations for fire and life safety violations of the *International Fire Code®* and the State Administrative Code, Title 19, Ordinance No. 1088, Resolution No. 1296.7.

3.2 It is the intent of the Fire Prevention Division to achieve compliance of the majority of code violations by traditional means of inspection, notification, the granting of reasonable time limits to comply, and reinspection.

3.3 Citations shall be issued by Fire Prevention Personnel when the following conditions exist:

3.3.1 Failure to gain reasonable compliance for the *International Fire Code®* violations.

3.3.2 Deliberate or mischievous fire setting not involving property loss.

3.3.3 Justification is evident that the violation was restored after inspection (Inspection records must verify the facts of violations).

3.34 Obstruction of fire lanes.

3.35 Upon direction of the City Attorney's office.

Section IV. Definitions

4.1 Reasonable time to comply.

4.1.1 Generally means after the third visit or second reinspection, with proper time intervals between visits for responsible party to make corrections of conditions.

4.2 Justification is evident that the violation was restored after inspection.

4.2.1 When the same violation has been written two times during any one-year period of a facility inspected two times or more annually.

4.2.2 When the same violation has been written two times during any two-year period of a facility inspected one time per year.

Section V. Procedure

5.1 Citations issued under the conditions as set forth in policy should have previous notifications history as set forth below.

5.1.1 Exiting or overcrowding of public assembly facilities requires two previous notices of violation.

5.1.2 Engine company referrals to the Prevention Division require engine company survey and reinspection, and notice of violation written by an inspector.

5.1.3 Prevention inspection (originated by Fire Prevention) requires survey report and notice of violation.

5.1.4 Repeated violation, which is corrected during inspection process and restored to violation condition afterwards, requires previous history of specific abatement process within the previous two years with the same responsible party.

5.1.5 Illegal controlled burning requires a warning notice of violation issued. Second offense by the same party requires citation to be issued.

5.1.6 Deliberate mischievous fire setting by persons of responsible age requires a citation to be issued to person or persons committing the act.

5.2 **Citing for Misdemeanor Violations.** The citation is a release stating the defendant will appear in court or post bail in lieu of physical arrest. If not sure whether you have the owner, check the liquor license or cite whoever is listed. (It may be a corporation.) If there is no liquor license, cite whoever is listed on the business license. If this is not available, check the City Business License Division for the correct owner.

5.2.1 Try to give the citation directly to the owner. You may have to return at a later time to catch him or her in the place of business. If the owner is not available, mail the citation (unsigned and certified mail) to the owner's address listed on the liquor or business license.

5.2.2 The citation must be signed by the responsible party for the premises or the person committing the act in cases of transient fire setting or illegal burning.

5.2.3 All misdemeanors by the bail schedule require a court appearance by the violator. At that time the defendant appears and either (a) pleads guilty, in which case we hear nothing more about it; (b) has the matter continued, which is not our concern; or (c) pleads not guilty, in which case the court sets a date for trial and pretrial conference and notifies the city attorney's office or the acting agency. It is the latter hearing date set by the court that requires the citing officer appearance.

5.3 **Citing for Infractions.** The procedure for writing a citation for an infraction is the same as for a misdemeanor violation.

5.3.1 All infractions can be paid at the court clerk's office or by mail (after phoning the clerk to find out what their final assessment will be). Do not indicate to the recipient of the citation what you think the fine will be.

5.4 **Failure to Appear in Court.** When a violator fails to appear in court on the appointed date, the citation is forwarded to the District Attorney's office, which will then notify the agency issuing the citation. A declaration must be filled out by the issuing officer at which time a warrant will be issued to the violator. A maximum of three weeks from the due date on the citation will be allowed for the officer to complete the paper work and the District Attorney to act.

5.5 **Failure to Abate Violation**. After the date of appearance has expired, a reinspection of the violation is warranted. Failure to abate a violation requires that a second citation be issued.

5.6 Attach copies of previous notices issued with your copy of the citation. Submit citation copy and notices to the Fire Prevention Officer, who will review them and forward to the proper authorities.

5.7 The department citation log shall be filled out and kept current. The log will be posted on the wall behind inspector's desk. The Prevention Officer and/or Division Chief shall be notified the following day of any citations issued by Fire Prevention personnel.

5.8 The Division Chief shall maintain a citation book record indicating which citation book numbers have been assigned to officers.

5.9 Attached is a schedule of recommended bails set for violations. Bail costs are required if a violator is physically arrested and may be used by a judge to determine first-offense fines. (**NOTE:** The bail schedule is lengthy and is not printed here).

Section VI. Exceptions

6.1 All exceptions or deviations shall be discussed with the Division Chief and City Attorney prior to writing of a citation.

Appendix C
Adams vs. State of Alaska

Case Summary

Thomas L. Adams et al., Appellants, v. State of Alaska et al., Appellees
No. 2326
Supreme Court of Alaska

October 1, 1976

Procedural Posture: Appellants, survivors and representatives of decedents killed in a hotel fire, sought review of the order of the trial court (Alaska), which rendered judgment on the pleadings to appellee State of Alaska in appellants' tort action alleging that the State breached its duty of care by failing to abate fire hazards it uncovered in a fire inspection of the hotel.

Overview: The State's fire inspection of the hotel uncovered several hazards, some of which were extreme life hazards. The State advised the hotel, promising to provide written details, but no further action was taken and the hazards remained unabated. Several months later, a fire in the hotel killed a number of people and injured others. The injured and decedents' representatives filed an action against the State. The State admitted as true the facts alleged, but argued successfully for judgment on the pleadings. On appeal, the court reversed and remanded, finding that the State, having undertaken to conduct the fire inspection, owed a duty to appellants to exercise reasonable care to abate the hazards it uncovered and breached that duty by its inaction. The court rejected the State's "duty to all, duty to no one" theory of immunity, holding that such theory was a form of sovereign immunity that was treated by statute and not to be amplified by court-created doctrine. The court held that appellants were in the class of persons for whose protection the fire inspection provisions were written and it was irrelevant that the State was not in control of the instrument causing the harm.

Outcome: The court reversed the judgment on the pleadings granted to the State, finding that once the State undertook to conduct a fire inspection of the hotel and uncovered several hazards, the State had a duty to exercise reasonable care to abate the hazards, and that the State breached that duty by its inaction. The court remanded for further action consistent with its opinion.

Appendix D

Sample Fire Lane Sign Requirements

APPROVED PHOENIX FIRE DEPARTMENT
FIRE LANE SIGNS
(FIGURE 5-4)

STANDARD FIRE LANE SIGN DETAIL

UNIVERSAL SYMBOL FIRE LANE SIGN DETAIL

1. THE SIGN PLATE SHALL BE A MINIMUM OF 12" X 18" WITH A THICKNESS OF .080 ALUMINUM CONSTRUCTION.

2. THE SIGN FACE SHALL HAVE A WHITE REFLECTIVE BACKGROUND WITH A RED LEGEND. USE THE STANDARD 3M SCOTCHLITE SIGN FACE NUMBER R7-32 OR EQUIVALENT, WITH RED SCREEN PRINTED LETTERING AS SHOWN ABOVE.

FILE NAME: Nopk.dwg
CALL FIRE PREVENTION-ACCESS OFFICER
FOR QUESTIONS OR ASSISTANCE
602-256-3382 OR 602-262-6771

PHOENIX FIRE DEPARTMENT

FIRE LANE SIGNS

L. HERTZ DETAIL 1.1 APRIL 2003

APPROVED PHOENIX FIRE DEPARTMENT FIRE LANE SIGNS

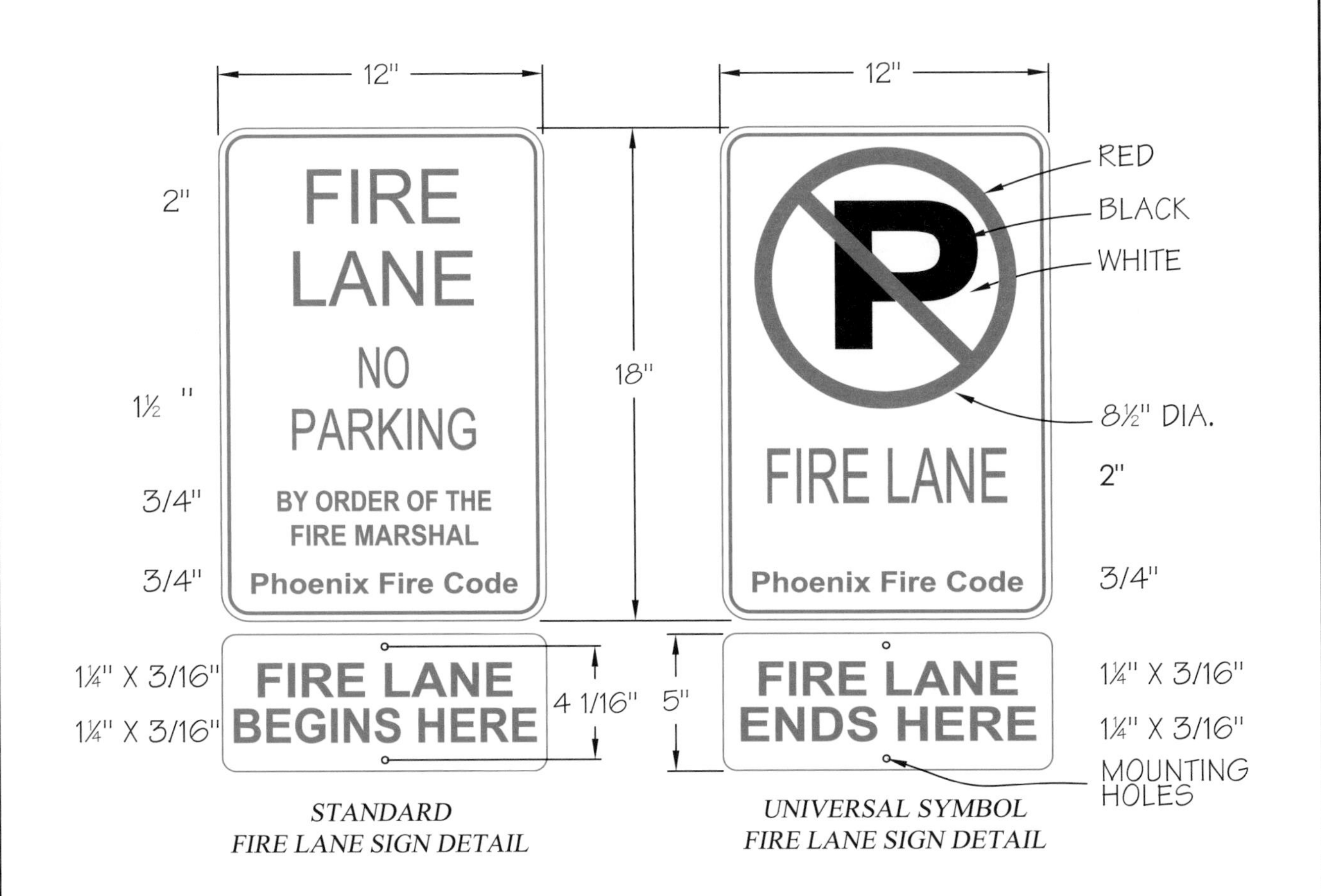

STANDARD
FIRE LANE SIGN DETAIL

UNIVERSAL SYMBOL
FIRE LANE SIGN DETAIL

1. THE SIGN PLATE SHALL BE A MINIMUM OF 12" X 18" WITH A THICKNESS OF .080 ALUMINUM CONSTRUCTION.

2. THE SIGN FACE SHALL HAVE A WHITE REFLECTIVE BACKGROUND WITH A RED LEGEND. USE THE STANDARD 3M SCOTCHLITE SIGN FACE NUMBER R7-32 OR EQUIVALENT, WITH RED SCREEN PRINTED LETTERING AS SHOWN ABOVE.

3. SPECIAL APPLICATION DETAILS MUST BE PRE-APPROVED BY PFD-EMERGENCY ACCESS OFFICER AT 602-256-3382

SPECIAL APPLICATION

FILE NAME: Nopk_begins.dwg
CALL FIRE PREVENTION-ACCESS OFFICER
FOR QUESTIONS OR ASSISTANCE
602-256-3382 OR 602-262-6771

FIRE LANE STRIPING, STENCILING AND SIGN INSTALLATION
(FIGURE 5-6)

15'
12"-18" (TYP.)
75'
75'
15'

STANDARD DETAIL 1
(WITHOUT STENCIL)

15'
STENCIL SHALL BE CENTERED BETWEEN SIGNS ON CURB FACE
FIRE LANE NO PARKING
100'
12"-18" (TYP.)
15'

DETAIL 2*
(WITH STENCIL)
* ACCESS OFFICER MUST PRE-APPROVE THIS APPLICATION

15'
SIGN SHALL BE CENTERED
65'
12"-18" (TYP.)
STENCIL SHALL START 15 FT IN
FIRE LANE NO PARKING
15'

DETAIL 3*
(65 FT OR LESS)
* MUST BE PRE-APPROVED

2 APPROVED FIRE LANE SIGNS-BACK TO BACK
35'
CARPORT PARKING

DETAIL 4
(TYPICAL BULLNOSE)

NOTES:

1. APPROVED FIRE LANE SIGNS SHALL BE INSTALLED 12" TO 18" FROM BACK OF CURB OR BACK OF SIDEWALK.

2. SIGN MUST FACE THE ONCOMING TRAFFIC.

3. STENCILS SHALL BE IN WHITE LETTERING (3" HT. ½" STROKE) ON RED PAINTED CURB.

4. STENCIL SHALL READ "FIRE LANE NO PARKING".

FILE NAME: PLANVIEW.dwg
CALL FIRE PREVENTION-ACCESS OFFICER
FOR QUESTIONS OR ASSISTANCE
602-256-3382 OR 602-262-6771

FIRE LANE STRIPING FOR HYDRANTS
(FIGURE 5-12)

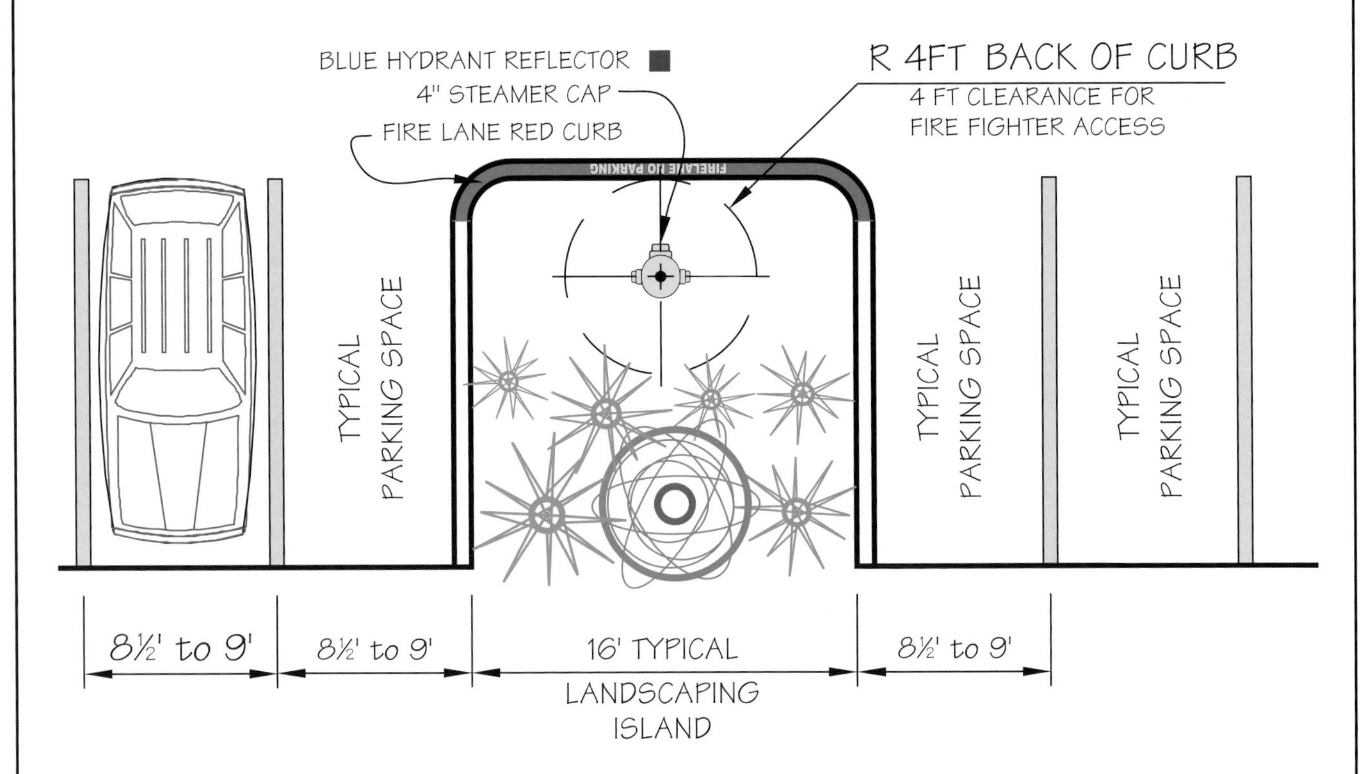

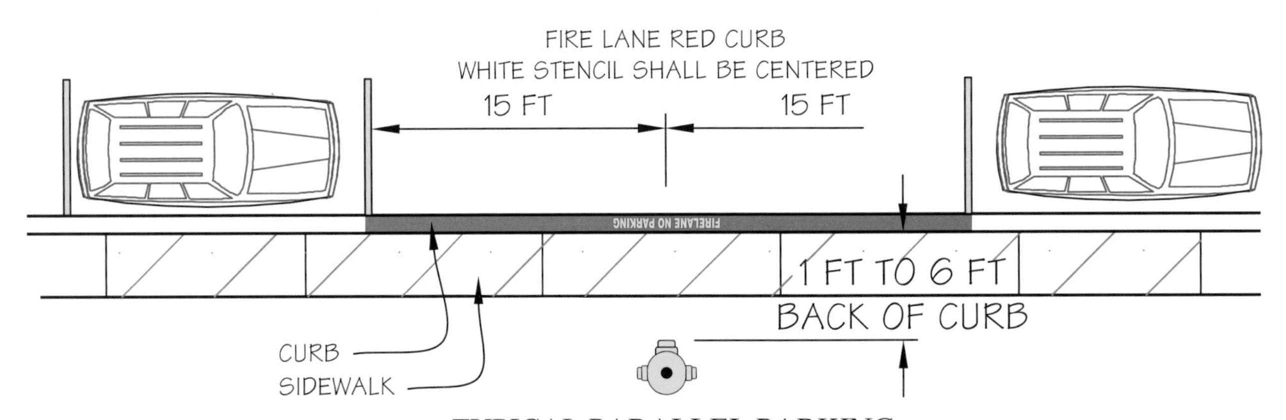

TYPICAL PARALLEL PARKING
STREET DETAIL
FOR HYDRANT INSTALLATION

NOTES:

- BLUE HYDRANT REFLECTOR SHALL BE PLACED 1 FT OFF OF CENTERLINE TOWARD HYDRANT SIDE.
- PUBLIC HYDRANT YELLOW
- PRIVATE HYDRANT YELLOW WITH WHITE REFLECTIVE BONNET. (SEE DETAIL 1.8B)
- OVER ***100 PSI*** WILL BE SIGNED HIGH PRESSURE.

(FIGURE 5-11)

FILE NAME: HYDRO_INSTALL.dwg
CALL FIRE PREVENTION-ACCESS OFFICER
FOR QUESTIONS OR ASSISTANCE
602-256-3382 OR 602-262-6771

Appendix E
Tent Guidlines From

CITY OF MADISON FIRE DEPARTMENT

Fire Prevention Division
325 W. Johnson St., Madison, WI 53703-2295
Phone: 608-266-4484 ◆ FAX: 608-267-1153

TENTS AND TEMPORARY MEMBRANE STRUCTURES

Compliance Guidelines

This list of guidelines does not include all requirements for the safe use of tents. This list of guidelines applies to all tents not just those requiring a license or notification. It is the license/permit holder's responsibility to comply with the applicable sections of the Madison General Ordinance, the International Fire Code and the Wisconsin Administrative Code.

1. **Flame-resistant treatment** required for all tents and membrane structures.
2. **Use period.** Temporary tents, air-supported, air-inflated or tensioned membrane structures shall not be erected for a period of more than 180 days within a 12-month period on a single premise.
3. **Smoking shall not be permitted** in tents or membrane structures. Approved "No Smoking" signs shall be conspicuously posted.
4. **Fireworks shall not be used** within 100 feet of tents or membrane structures.
5. **No open flame** or other devices emitting flame, fire or heat or any flammable or combustible liquids, gas, charcoal or other cooking device or any other unapproved devices shall be permitted inside or located within 20 feet of the tent or membrane structures while open to the public unless approved by the fire code official.
6. **Heating or cooking equipment**, tanks, piping, hoses, fittings, valves, tubing and other related components shall be installed as specified in the *International Mechanical Code* and the *International Fuel Gas Code,* and shall be approved by the fire code official.

   Outdoor cooking that produces sparks or grease-laden vapors shall not be performed within 20 feet from a tent.

   Operations such as warming of foods, cooking demonstrations and similar operations that use solid flammables, butane or other similar devices which do not pose an ignition hazard, shall be approved.
7. **Housekeeping.** All weeds and combustible vegetation shall be removed from the premises adjacent of or within 30 feet of any tent. Hay, straw, shavings, trash and other combustible material may not be stored less than 30 feet from any tent, except upon special permission from the fire chief.

   The grounds both inside and outside of tents shall be kept free and clear of combustible waste. The waste shall be stored in approved containers or removed from the premises.
8. **Flammable and combustible liquids shall be stored outside** in an approved manner not less than 50 feet from tents or membrane structures. The storage and dispensing of flammable or combustible liquids shall be in accordance with COMM 10.

   Flammable and combustible liquids shall not be used in tents or membrane structures.
9. **LP-gas containers shall be located outside.** Containers 500 gallons or less, shall have a minimum separation of 10 feet between container and tents or membrane structures.
10. **Gas cylinders** - Adequately secure all compressed gas cylinders.
11. **Electrical Installations.** Electrical systems in all tents that accommodate more than 50 persons shall be installed in accordance with the requirements of Wisconsin Administrative Code Chapter Comm 16 Electrical. All such systems shall be maintained and operated in a safe and workmanlike manner.

    The electrical system and equipment shall be isolated from the public by proper elevation and guarding. All electrical fuses and switches shall be installed in approved enclosures. Cables laid on the ground or in areas traversed by the public shall be placed in trenches or protected by approved covers.

12. **Exits.**

   a. **Distribution of exits.**

   - Exits shall be spaced at approximately equal intervals around the perimeter of the tent and shall be located such that all points are 100 feet (30 480 mm) or less from an exit.

   b. **Number of exits.**

   - Tents or usable portion thereof shall have at least one exit and not less than the number of exits required by Table 2403.12.2 The widths of means of egress required by Table 2403.12.2 shall be divided approximately equally among the separate means of egress. The total width of means of egress in inches (mm) shall not be less than the total occupant load served by a means of egress multiplied by 0.2 inches per person.
   - Every tent occupied by the public shall have at least 2 standard exits located at or near opposite ends of the structure.
   - Exits shall be uniformly distributed but in no case shall the line of travel to an exit be greater than 100 feet.

13. **Exit lights and signs.** All exits, aisles and passageways leading to exits in tents used as places of outdoor assembly shall be kept adequately lighted at all times when the structure is occupied by the public. Artificial illumination having an intensity of not less than 1-foot candles at the floor line shall be provided when natural light is inadequate.

    Exit lights and signs complying with the requirements of IFC 2403.12.6 shall be provided in all tents used as places of outdoor assembly where more than 50 persons can be accommodated.

14. **Portable Fire extinguishers** with a minimum rating of 2A-10B:C shall be provided. The travel distance to extinguisher(s) shall not exceed 75 feet. Extinguishers shall be readily accessible and shall not be obstructed or obscured from view.

15. **An Unobstructed Fire Break** passageway or fire road not less than 12 feet wide and free from guy ropes or other obstructions shall be maintained on all sides of all tents unless otherwise approved by the code official.

16. **Fire Access Roads** shall extend to within 150 feet of all portions of the event site and shall have an unobstructed width of not less than 20 feet and an unobstructed vertical clearance of not less than 13 feet 6 inches.

17. **Generators and other internal combustion power sources shall be separated from tents** or membrane structures by a minimum of 20 feet and shall be isolated from contact with the public by fencing, enclosure or other approved means.

18. **Tents or membrane structures shall not be located within 20 feet** of lot lines, buildings, other tents or membrane structures, parked vehicles or internal combustion engines. For the purpose of determining required distances, support ropes and guy wires shall be considered as part of the temporary membrane structure or tent.

    Exceptions:

    1. Separation distance between membrane structures and tents not used for cooking, is not required when the aggregate floor area does not exceed 15,000 sq. ft.

    2. Membrane structures and tents need not be separated from buildings when all of the following conditions are met:

       - The aggregate floor area of the membrane structure and tent shall not exceed 10,000 square feet.
       - The aggregate floor area of the building and membrane structure and tent shall not exceed the allowable floor area including increases as indicated in the *International Building Code*.
       - Required means of egress provisions are provided for both the building and the membrane structure and tent, including travel distances.
       - Fire apparatus access roads are provided in accordance with the IFC.

    Membrane structures having an area of 15,000 sq. ft. or more shall be located not less than 50 feet from any other tent or structure as measured from the sidewall of the tent or membrane structure unless joined together by a corridor.

REFERENCES: Madison General Ordinance (MGO) 29.355 - Use of Tents for Merchandising and Display; Wisconsin Administrative Code Chapter COMM 62; and 14, International Fire Code Chapter 24.

If you have any questions regarding the above requirements telephone the Fire Prevention Division at 266-4484.

Appendix F
Hot Work Program

HOT WORK PERMIT

STOP!

Avoid hot work or seek an alternative/safer method, if possible.

This Hot Work Permit is required for any temporary operation involving open flames or producing heat and/or sparks. This includes, but is not limited to: brazing, cutting, grinding, soldering, torch-applied roofing and welding.

Part 1

Instructions for Fire Safety Supervisor:

A. Specify the precautions to take.
B. Fill out and keep **Part 1** during the hot work process.
C. Issue **Part 2** to the person doing the job.
D. Keep **Part 2** on file for future reference, including signed confirmation that the one-hour fire watch and three-hour monitoring have been completed.
E. Final signoff is on **Part 2**.

HOT WORK BY
☐ Employee
☐ Contractor ______________________

DATE | JOB NUMBER

SPECIFIC LOCATION/BUILDING AND FLOOR

NATURE OF JOB

NAME (PRINT) AND SIGNATURE OF PERSON PERFORMING HOT WORK

NAME (PRINT) AND SIGNATURE OF PERSON PERFORMING FIRE WATCH

I verify the above location has been examined, the precautions checked on the Required Precautions Checklist have been taken to prevent fire, and permission is authorized for this work.

NAME (PRINT) AND SIGNATURE OF FIRE SAFETY SUPERVISOR/OPERATIONS SUPERVISOR

TIME STARTED: a.m. p.m. TIME FINISHED: a.m. p.m.

Permit Expires DATE | TIME a.m. p.m.

Note: Emergency notification on back of form.
Use as appropriate for your facility.

To order additional hot work permits or other FM Global resources, order online 24 hours a day, seven days a week, at fmglobalcatalog.com.

FM Global

Required Precautions Checklist

Y NA

☐☐ The fire pump is in operation and switched to automatic.
☐☐ Control valves to water supply for sprinkler system are open.
☐☐ Hose streams and extinguishers are in service/operable.
☐☐ Hot work equipment is in good working condition.

Requirements within 35 ft. (11 m) of hot work

☐☐ Ignitable liquid, dust, lint and oily deposits removed.
☐☐ Explosive atmosphere in area eliminated.
☐☐ Floors swept clean.
☐☐ Combustible floors wet down, covered with damp sand or fire-resistive sheets.
☐☐ Remove other combustible material where possible. Otherwise, protect with FM Approved welding pads, blankets and curtains, fire-resistive tarpaulins or metal shields.
☐☐ All wall and floor openings covered.
☐☐ FM Approved welding pads, blankets and curtains installed under and around work.
☐☐ Protect or shut down ducts and conveyors that might carry sparks to distant combustible material.

Hot work on walls, ceilings or roofs

☐☐ Construction is noncombustible and without combustible covering or insulation.
☐☐ Combustible material on other side of walls, ceilings or roofs is moved away.

Hot work on enclosed equipment

☐☐ Enclosed equipment cleaned of all combustible material.
☐☐ Containers purged of ignitable liquid/vapor.
☐☐ Pressurized vessels, piping and equipment removed from service, isolated and vented.

Fire watch/hot work area monitoring

☐☐ Fire watch will be provided during and for one (1) hour after work, including any break activity.
☐☐ Fire watch is supplied with suitable extinguishers, and where practical, a charged small hose.
☐☐ Fire watch is trained in use of equipment and in sounding alarm.
☐☐ Fire watch may be required in adjoining areas, above and below.
☐☐ Monitor hot work area for up to an additional three (3) hours after the one (1) hour fire watch.

☐☐ **Other precautions taken:** ______________________

WARNING

HOT WORK IN PROGRESS! Watch for fire!

Part 2

Instructions for Fire Safety Supervisor:

A. Specify the precautions to take.
B. Fill out and keep **Part 1** during the hot work process.
C. Issue **Part 2** to the person doing the job.
D. Keep **Part 2** on file for future reference, including signed confirmation that the one-hour fire watch and three-hour monitoring have been completed.

HOT WORK BY
☐ Employee
☐ Contractor ____________________

DATE	JOB NUMBER

SPECIFIC LOCATION/BUILDING AND FLOOR

NATURE OF JOB

NAME (PRINT) AND SIGNATURE OF PERSON PERFORMING HOT WORK

NAME (PRINT) AND SIGNATURE OF PERSON PERFORMING FIRE WATCH

I verify the above location has been examined, the precautions checked on the Required Precautions Checklist have been taken to prevent fire, and permission is authorized for this work.

NAME (PRINT) AND SIGNATURE OF FIRE SAFETY SUPERVISOR/OPERATIONS SUPERVISOR

TIME STARTED: a.m. p.m. TIME FINISHED: a.m. p.m.

Permit Expires	DATE	TIME a.m. p.m.

Fire watch signoff: Work area and all adjacent areas to which sparks and heat might have spread were inspected during the watch period and were found fire-safe.

Signed:

Final checkup: Work area was monitored for three (3) hours following completion of the one (1) hour fire watch and found fire-safe.

Signed:

Required Precautions Checklist

Y NA

☐☐ The fire pump is in operation and switched to automatic.
☐☐ Control valves to water supply for sprinkler system are open.
☐☐ Hose streams and extinguishers are in service/operable.
☐☐ Hot work equipment is in good working condition.

Requirements within 35 ft. (11 m) of hot work

☐☐ Ignitable liquid, dust, lint and oily deposits removed.
☐☐ Explosive atmosphere in area eliminated.
☐☐ Floors swept clean.
☐☐ Combustible floors wet down, covered with damp sand or fire-resistive sheets.
☐☐ Remove other combustible material where possible. Otherwise, protect with FM Approved welding pads, blankets and curtains, fire-resistive tarpaulins or metal shields.
☐☐ All wall and floor openings covered.
☐☐ FM Approved welding pads, blankets and curtains installed under and around work.
☐☐ Protect or shut down ducts and conveyors that might carry sparks to distant combustible material.

Hot work on walls, ceilings or roofs

☐☐ Construction is noncombustible and without combustible covering or insulation.
☐☐ Combustible material on other side of walls, ceilings or roofs is moved away.

Hot work on enclosed equipment

☐☐ Enclosed equipment cleaned of all combustible material.
☐☐ Containers purged of ignitable liquid/vapor.
☐☐ Pressurized vessels, piping and equipment removed from service, isolated and vented.

Fire watch/hot work area monitoring

☐☐ Fire watch will be provided during and for one (1) hour after work, including any break activity.
☐☐ Fire watch is supplied with suitable extinguishers, and where practical, a charged small hose.
☐☐ Fire watch is trained in use of equipment and in sounding alarm.
☐☐ Fire watch may be required in adjoining areas, above and below.
☐☐ Monitor hot work area for up to an additional three (3) hours after the one (1) hour fire watch.

☐☐ **Other precautions taken:** ____________________

WARNING!

HOT WORK IN PROGRESS
Watch for fire!

In case of emergency:

Call: ______________________________

At: ______________________________

____

____

WARNING!

Appendix G
Floating Roof Tanks

Figure G.1

Figure G.2

Figure G.3

Floating roof tanks are constructed with either external or internal roofs. An external floating roof tank consists of an open-topped cylindrical steel shell equipped with a roof that floats on the surface of the liquid, rising and falling with the liquid level. External floating roof tanks can be identified by stairs that lead to an open platform of a tank with no roof **(Figure G.1)**.

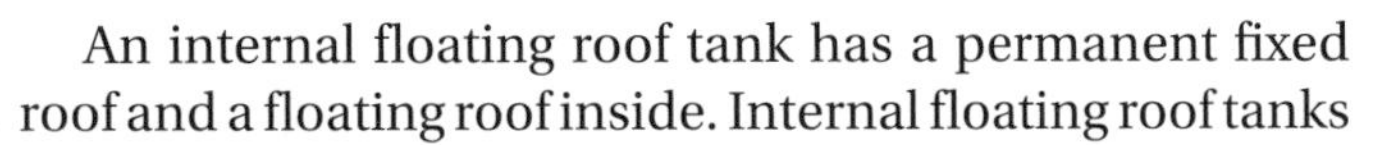

An internal floating roof tank has a permanent fixed roof and a floating roof inside. Internal floating roof tanks are identified by equally spaced openings at the top of the tank shell that inhales and exhausts air as the roof rises and falls as the liquid level changes. Floating roof tanks are constructed to API 650 **(Figure G.2)**.

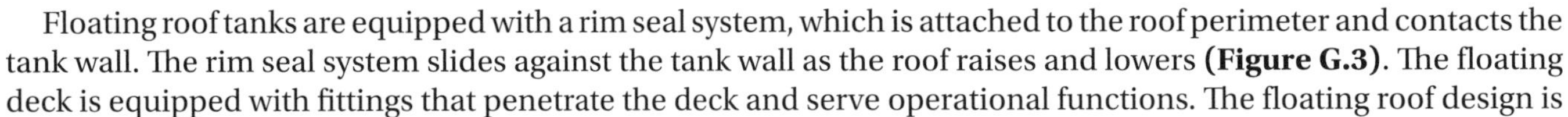

Floating roof tanks are equipped with a rim seal system, which is attached to the roof perimeter and contacts the tank wall. The rim seal system slides against the tank wall as the roof raises and lowers **(Figure G.3)**. The floating deck is equipped with fittings that penetrate the deck and serve operational functions. The floating roof design is such that evaporative losses from the stored liquid are limited to losses from the rim seal system and deck fittings and any exposed liquid on the tank walls.

Floating roof tanks offer better fire safety compared to cone roof tanks because the roof sits directly on the stored liquid in the tank. By design, the only area where vapor exists is between the tank shell and the rim seal **(Figure 6.4)**. The typical width of a rim seal is 12-24 inches. If the seal is compromised and an ignition source is introduced, only the area between the tank shell and the floating roof will burn. The remainder of the surface cannot be ignited because no air space exists.

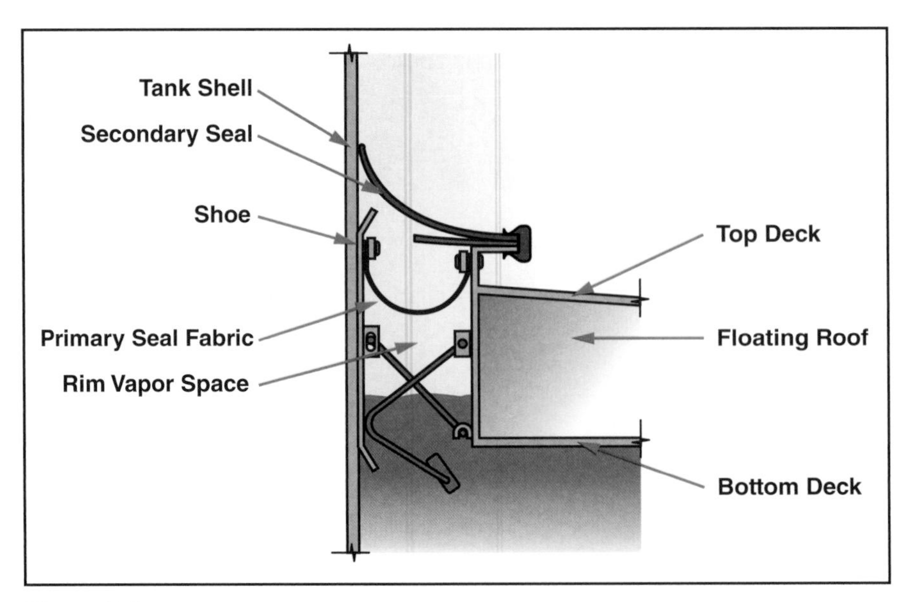

Figure G.4

Appendix H
Audible Warning Devices

Audible Alarms

Audible notification signaling appliances are the most common types of alarm-signaling systems used for signaling a fire alarm in a structure. During an inspection, the inspector must remember that there are two important factors that affect the performance of the audible signaling system: its ability to alert and its ability to convey information.

Keep in mind that NFPA® 72, *National Fire Alarm and Signaling Code*, does not require any notification appliances to be installed. Notification devices are required by the applicable building, fire, or life safety code. For example, horns are not required if a fire alarm system is installed only to monitor a sprinkler system. In that case, horns would only be required if the sprinkler sections of the applicable code required occupant notification.

Audible alarms include bells, horns, sirens, voice announcement systems, and other devices that can be distinguished above and apart from the normal sound level within the workplace. Temporal and voice signals are the most effective means of providing audible warnings. Depending on the design and programming of the system, the local alarm may sound only in the area of the activated detector, or it may sound in the entire facility.

When it becomes necessary to evacuate a building, the Standard Audible Emergency Evacuation Signal should be used. This signal consists of the following 'three-pulse' temporal pattern:

- Three successive "on" phases, lasting 0.5 second each, separated by 0.5 second of "off" time.
- At the completion of the third "on" phase there must be 1.5 seconds of "off" time before the full cycle is repeated.

Therefore, the total cycle shall last 4.0 seconds (0.5 second "on," 0.5 second "off," 0.5 second "on," 0.5 second "off," 0.5 second "on," 1.5 seconds "off").

The emergency evacuation signal should only be used to notify occupants of the need to immediately evacuate the building. Total evacuation is not always desirable or necessary during an emergency; therefore, there should be a different signal pattern assigned to signify the need for occupants to relocate to a safe area within the building or for a "shelter in place."

In some occupancies, audible devices may be ineffective or are inappropriate. Restrooms, elevators, and stairwells are examples of locations where audible devices are inappropriate. In elevators, the occupants have nowhere to exit until the car stops and the doors open. Stairwells are already in the path of egress.

Audible notification devices are listed for wall or ceiling installation. When ceiling heights allow and other installations are not permitted, wall-mounted devices are required to be mounted so that the top of the device is at least 6 inches (300 mm) below the ceiling and the top is at least 90 inches (5 400 mm) above the finished floor level.

The level of sound made by the audible device is reduced as the distance from the source is increased. As a rule of thumb, double the distance from the source and the sound pressure (dB) drops by 6 dB. Double the distance again and the sound pressure is decreased another 6 dB. When the sound must travel through walls and doors, the sound loss may be more than 6 dB.

Average Sound Levels in Occupancies

Location	Average Sound Level (dba)
Business Occupancies	55
Educational Occupancies	45
Industrial Occupancies	80
Institutional Occupancies	50
Mercantile Occupancies	40
Mechanical Rooms	85
Piers and Water Surrounded Structures	40
Places of Assembly	55
Residential Occupancies	35
Storage Occupancies	30
Thoroughfares, high-density urban	70
Thoroughfares, medium-density urban	55
Thoroughfares, rural and suburban	40
Tower Occupancies	35
Underground and Windowless Structures	40
Vehicles and Vessels	50

Automatic Sprinkler Systems Acceptance Checklist

Colorado Springs Fire Department
Automatic Sprinkler Systems Acceptance Test
NFPA 13 and UFC Section 1003

Business Address:	
Business Name:	
CSFD Project #:	Installer Name:
CSFD Permit #:	Installer License #:
Area of Protection:	
Location of Main Sprinkler Riser:	
System Pressure: PSI	
FDC Location:	
Type of System: ☐ wet ☐ dry ☐ preaction ☐ deluge ☐ water spray ☐ water mist	
Fire Pump Present: ☐ yes ☐ no	

Documentation

Yes	No	N/A	
☐	☐	☐	Installer's certificate provided to CSFD
☐	☐	☐	Certified installer has permit and plans approved by CSFD on site
☐	☐	☐	All components located same as approved drawings or "As Builts" submitted
☐	☐	☐	Copy of system instruction manuals on site

COMPONENTS

Yes	No	N/A	
☐	☐	☐	Proper Signage
☐	☐	☐	Inspector's test valve is smoothbore, corrosion resistant orifice same diameter as smallest sprinkler head
☐	☐	☐	Relief valve is present if pressure is 175 psi or greater, or 10 psi above maximum system pressure in a gridded wet pipe system
☐	☐	☐	Double check valve or back flow prevention valve installed
☐	☐	☐	Pressure gauges on both sides of alarm check valve or back flow preventer
☐	☐	☐	Valve room light and heat provided
☐	☐	☐	Extra heads and correct head wrench for each sprinkler head type in cabinet in riser room as specified in plan review
☐	☐	☐	Extra sprinkler heads are of proper type and temperature

INSTALLATION

Permanent system identification signs for each:

Yes	No	N/A	
☐	☐	☐	Control valve (if multiple valves present)
☐	☐	☐	Riser valves
☐	☐	☐	Hydraulic calculation label attached to riser
☐	☐	☐	FDC
☐	☐	☐	Water supply valves, test valves, and flow switch electrically monitored for I-1 occupancies (20 heads or greater) and all others (100 or greater)

INSTALLATION *(continued)*

Yes	No	N/A	
☐	☐	☐	Water supply valves are of indicating type and supervised electronically or locked
☐	☐	☐	Main drain to exterior with turned down elbow
☐	☐	☐	Riser supported by hanger or attachment, according to plans
☐	☐	☐	Horn and strobe water flow alarm is located above the FDC (or within 20' on same side of building)
☐	☐	☐	FDC piping sized to system according to approved plans
☐	☐	☐	FDC is installed according to plans
☐	☐	☐	FDC has national standard hose thread internal threaded swivel fittings
☐	☐	☐	FDC readily accessible and free from obstructions
☐	☐	☐	All FDC caps securely in place
☐	☐	☐	FDC clapper operates easily
☐	☐	☐	FDC has check valve and drip valve
☐	☐	☐	4' of pipe installed between FDC and check valve provided in a heated area
☐	☐	☐	Sprinkler heads are properly spaced according to approved plans
☐	☐	☐	If subject to damage, sprinkler heads have guards in place
☐	☐	☐	Sprinkler heads free from coatings not applied by manufacturer
☐	☐	☐	Sprinkler heads parallel to slope of ceiling , roof, and /or stairs
☐	☐	☐	Piping penetrations have proper clearance: 1" clearance for 1" to 3 ½ " pipe 2" clearance for 4" and larger pipe
☐	☐	☐	Escutcheon plates installed

OPERATIONAL TESTS

Pass	Fail	N/A	
☐	☐	☐	Hydrostatic testing of system @ 200 psi for 2 hours (50psi above static, if static is above 150psi) including FDC
☐	☐	☐	Main drain flow test completed per flow curve Results using attached discharge chart: _______ psi ________ gpm
☐	☐	☐	Flow test through inspector's test valve
☐	☐	☐	Activation of flow alarm for all types of systems within 90 seconds per NFPA 72

DRY & PRE-ACTION SYSTEMS ONLY

Yes	No	N/A	
☐	☐	☐	If system volume is greater than 750 gallons, then multiple dry pipe valves are required, unless system can provide water at the most remote sprinkler in less than 60 seconds
☐	☐	☐	Paddle type water flow devices has not been installed in dry, preaction or deluge systems
☐	☐	☐	Dry system compressor with minimum ½ " fill line and pressure gauges
☐	☐	☐	Air pressure maintained above the water pressure as per manufacturer's specifications
☐	☐	☐	Auxiliary drains (drum drip valve) at lowest point in the system from inspector's test valve
☐	☐	☐	Dry system relief valve functions automatically
☐	☐	☐	Low air alarm monitored
☐	☐	☐	System to refill with air from compressor within 30 minutes
☐	☐	☐	Operational test of dry pipe valve performed
☐	☐	☐	Quick opening device tested

Comments:

____

System tested by:	Print name here	Sign here

☐ **System Approved**
☐ **System Disapproved** Date: _______________ Inspector's Initials ________

Automatic Sprinkler Systems
NFPA 13 and UFC Section 1003

s:forms/acceptance tests/automatic sprinklers

3

Appendix J
Inspection Report Checklist

Page ________ of ________

COLORADO SPRINGS FIRE DEPARTMENT

INSPECTION REPORT

Permit Number ______________________ Inspection Number ______________________

Street Name | D | Number | Unit Type | Unit #

Occupant Name | Occup Class | Business Owner or Manager

Business Phone | Responsible Party After Hours Phone

INSPECTION TYPE

☐ Office of the Fire Marshal ☐ CSFD Company

☐ Regular
☐ Plan/Number ______________________
☐ Referral
☐ Haz Mat
☐ High Pile
☐ Revocable Permit
☐ State License (OFM only)
☐ Other ______________________

INSPECTION STATUS

☐ Initial Complete
☐ Initial Incomplete
☐ 1st Reinspection
☐ 1st Reinspection Complete
☐ 2nd Reinspection
☐ 2nd Reinspection Complete
☐ ____ Reinspection
☐ ____ Reinspection Complete
☐ Order Notice
☐ Order Notice/Complete
☐ Summons
☐ Summons/Complete

☐ Disapproved *☐ Approved* *☐ Close Inspection* *☐ Reinspection Fee*

FIRE CODE VIOLATIONS

Item #	Fire Code	Remarks	Approx Reinspect Date	OK

Fire department inspections are intended for your safety and that of our city's citizens and visitors. This document is an official notice of violation(s) of the Fire Code, as amended. All corrections shall be completed by the date noted above. Violation of the Fire Code is a misdemeanor punishable by a fine and/or imprisonment; therefore, failure to comply with these requirements may lead to legal action. Your cooperation is greatly appreciated. For information concerning this inspection, please call (719) 385-5978.

Copy Received by (print and sign)	Title	Date Received

Date of Inspection	# Insp	Inspected by/Title (print)	State Certification #	Span of Time

White: Occupant Yellow: Data Entry/File

(See back for Self Inspection Checklist)

SELF INSPECTION PROGRAM

The Colorado Springs Fire Department prides itself on its proactive approach to life safety. The Office of the Fire Marshal's mission is based on three principles – Education, Engineering, and Enforcement. Our goal is to provide a safe environment for our city's citizens and visitors. Unfortunately, we do not have the necessary staff to regularly inspect every commercial property. We are, therefore, asking for your help.

The following Self Inspection Checklist is an educational tool for business owners and managers to provide a fire safe business for their patrons, visitors, and staff. It is not, however, intended to take the place of an official Fire Department inspection. The items are general requirements based on common Fire Code violations typically found in most businesses.

Building Exterior

- ☐ Address numbers are posted, contrast with the background, are at least five (5) inches tall, and visible from the road.
- ☐ Fire lane signs are posted for driveways and access roads less than 34 feet wide.
- ☐ Gas meters and attached piping are protected from vehicular damage by concrete or steel posts.
- ☐ If the building is equipped with a fire sprinkler or standpipe system, the fire department connection (FDC) has a clear space of at least three (3) feet around connections and caps are in place.
- ☐ There is at least three (3) feet of clearance around all hydrants.
- ☐ Dumpster or trash containers are least five (5) feet from building openings or roof overhangs.
- ☐ Up-to-date keys are within the Knox Rapid Entry System Box (where applicable).

Building Interior

- ☐ All exit doors are free of obstructions and unlocked during business hours.
- ☐ Aisles and exit paths are at least 36 inches wide when storage and/or equipment are on one side; 44 inches wide where storage and/or equipment are on both sides.
- ☐ Aisles and exits are free of storage or obstructions.
- ☐ Exit signs and emergency lights are operational with primary and emergency power supplies.
- ☐ Each floor has one fire extinguisher for every 3,000 square feet.
- ☐ Fire extinguishers have been inspected and tagged by a licensed contractor within the last 12 months.
- ☐ Fire extinguishers have a minimum 2A:10BC rating (rating on the label).
- ☐ Combustible materials are not stored in exit paths, under stairs, under floors, above ceilings, or in mechanical rooms.
- ☐ Storage is maintained at least two (2) feet from the ceiling.
- ☐ A clear space of 30 inches is maintained in front of all electrical panels.
- ☐ Extension cords are not used in place of permanent wiring.
- ☐ There are no multi-plug adapters. Listed power taps or trip outlets with over-current protection are used.
- ☐ Signs identifying electrical rooms, mechanical rooms, and roof access are installed.
- ☐ Fire alarm control panels and fire sprinkler valve locations are identified.
- ☐ Fire-resistive construction such as drywall is free of holes and maintained in good condition.
- ☐ Fire doors are not propped or blocked open and are self closing or self latching.
- ☐ Lint traps and areas behind clothes dryers are free of lint buildup and other combustible debris.

Special Systems

- ☐ Fire sprinkler systems are inspected annually by a licensed Fire Suppression Contractor A. To verify licensure, go to www.pprbd.org or call the Pikes Peak Regional Building Department at 327-2884.
- ☐ Fire alarm systems are inspected annually by a licensed Fire Alarm A Contractor. To verify licensure, go to www.pprbd.org or call the Pikes Peak Regional Building Department at 327-2884.
- ☐ Kitchen fire suppression systems are inspected every six months by a licensed Fire Suppression Contractor B. To verify licensure, go to www.pprbd.org or call the Pikes Peak Regional Building Department at 327-2884.
- ☐ Sprinkler valves have an unobstructed clear space of at least three (3) feet.

Congratulations, you have just taken steps to maintain a fire safe business. If you have questions or concerns, please call our office at (719) 385-5978. Thank you for your interest in this extremely important program.

Glossary

EXIT
5 GALLON
GASOLINE
PEAK
HANDLE WITH CARE
TIC BOTTLES
WIN THE UNIQUE
Mobil
SAE 15W-50
POWER SERVICE PS

5 GALLON
GASOLINE
NEVER MIX
EXIT
PEAK
HANDLE WITH CARE
TIC BOTTLES
WIN THE UNIQUE
Mobil 1
Fully Synthetic Motor Oil
Extended Performance
SAE 15W-50
6/1QT BOTTLES
POWER SERVICE PS
DIESEL FUEL SUPPLEMENT
PART NO. 1025-12
12/1 32-OUNCE
CETANE BOOST

Glossary

A

Acceptance Test — Preservice test on fire protection, detection, and/or suppression systems after installation to ensure that the system operates as intended.

Administrative Warrant — Written order that authorizes an inspector to enter a property for the purpose of conducting an inspection.

Americans with Disabilities Act (ADA) of 1990 - Public Law 101-336 — Federal statute intended to remove barriers, physical and otherwise, that limit access by individuals with disabilities.

Angle of Approach/Departure — Relationship described in degrees that is created by an incline from or to a road surface.

Arc — A luminous discharge of electricity across a gap. Arcs produce very high temperature.

Area of Refuge — An area where persons unable to use stairways can remain temporarily to await instructions or assistance during emergency evacuation.

Asphyxiant — Any substance that prevents oxygen from combining in sufficient quantities with the blood or from being used by body tissues.

Assembly — Occupancy classification of buildings, structures, or compartments (rooms) that are used for the gathering of 50 or more persons.

Authority Having Jurisdiction (AHJ) — Term used in codes and standards to identify the legal entity, such as a building or fire official, that has the statutory authority to enforce a code and to approve or require equipment; may be a unit of a local, state, or federal government, depending on where the work occurs. In the insurance industry it may refer to an insurance rating bureau or an insurance company inspection department.

Autoignition — Initiation of combustion by heat but without a spark or flame. (NFPA® 921)

Autoignition Temperature — The lowest temperature at which a combustible material ignites in air without a spark or flame. (NFPA® 921)

Balloon-Frame Construction — Type of structural framing used in some wood-frame buildings; studs are continuous from the foundation to the roof, and there may be no fire stops between the studs.

B

Bearing Wall Structure — Common type of structure that uses the walls of a building to support spanning elements such as beams, trusses, and pre-cast concrete slabs.

Benchmark — (1) Permanently affixed mark such as a stake driven into the ground that establishes the exact elevation of a place; used by surveyors in measuring site elevations or as a starting point for surveys. (2) Standard measurement that forms the basis for comparison.

Body Language — Nonverbal communication including but not limited to body posture and gestures; represents a large portion of communication in human interactions.

Buoyant — The tendency or capacity to remain afloat in a liquid or rise in air or gas.

Butterfly Valve — Control valve that uses a flat circular plate in a pipe that rotates 90 degrees across the cross section of the pipe to control the flow of water.

Calcined — Process that heats a substance to a high temperature but below the melting or fusing point, causing loss of moisture, reduction or oxidation, and decomposition of carbonates and other compounds.

C

Capital — Broad top surface of a column or pilaster, designed to spread the load held by a column.

Cavitation — Condition in which vacuum pockets form due to localized regions of low pressure at the vanes in the impeller of a centrifugal pump and cause vibrations, loss of efficiency, and possible damage to the impeller.

Cease-and-Desist Order — Court order prohibiting a person or business from continuing a particular course of conduct.

Ceiling Jet — A relatively thin layer of flowing hot gases that develops under a horizontal surface (e.g., ceiling) as a result of plume impingement and the flowing gas being forced to move horizontally. (NFPA® 921)

Central Station System — Alarm system that functions through a constantly attended location (central station) operated by an alarm company. Alarm signals from the protected property are received in the central station and are then retransmitted by trained personnel to the fire department alarm communications center.

Centrifugal Pump — Pump with one or more impellers that rotate and utilize centrifugal force to move the water. Most modern fire pumps are of this type.

Certificate of Occupancy — Issued by a building official after all required electrical, gas, mechanical, plumbing, and fire protection systems have been inspected for compliance with the technical codes and other applicable laws and ordinances.

Change Order — Client's written order to a contractor, issued after execution of the construction contract, which authorizes a change in the construction work, project completion time, and/or cost of the project.

Chemical Flame Inhibition — Extinguishment of a fire by interruption of the chemical chain reaction.

Clear Dimensions — Interior compartment measurements made from the inside surface of one wall to the inside surface of the opposite wall.

Clear Width — Actual unobstructed opening size of an exit.

Code — A collection of rules and regulations that has been enacted by law in a particular jurisdiction. Codes typically address a single subject area; examples include a mechanical, electrical, building, or fire code.

Combination System — Water supply system that is a combination of both gravity and direct pumping systems. It is the most common type of municipal water supply system.

Combustion — A chemical process of oxidation that occurs at a rate fast enough to produce heat and usually light in the form of either a glow or flame. (NFPA® 921)

Commodity — Combination of material and packaging that is for sale.

Common Path of Travel — The route of travel used to determine measured egress distances in code enforcement. The common path of travel is considered to be down the center of a straight corridor and a 1-foot (0.3 m) radius around each corner. *Also called* the normal path of travel.

Compartmentation System — Series of barriers designed to keep flames, smoke, and heat from spreading from one room or floor to another; barriers may be doors, extra walls or partitions, fire-stopping materials inside walls or other concealed spaces, or floors.

Compressive Loads — Vertical and/or horizontal forces that tend to push the mass of a material together; for example, the force exerted on the top chord of a truss.

Concrete Block — Large rectangular brick used in construction; the most common type is the hollow concrete block. *Also known as* Concrete Masonry Units (CMU).

Consensus Standard — Rules, principles, or measures that are established though agreement of the members of the standards-setting organization.

Conduction — Transfer of heat through or between solids that are in direct contact.

Convection — Heat transfer by circulation within a medium such as a gas or a liquid. (NFPA® 921)

Course — Horizontal layer of individual masonry units.

Cryogen — Gas that is converted into a liquid by being cooled below -150°F (-100°C).

Curtain Wall — Nonload-bearing exterior wall attached to the outside of a building.

D

Dead Load — Weight of the structure, structural members, building components, and any other features permanently attached to the building that are constant and immobile.

Deflagration — (1) Chemical reaction producing vigorous heat and sparks or flame and moving through the material (as black or smokeless powder) at less than the speed of sound. A major difference among explosives is the speed of this reaction. (2) Intense burning, a characteristic of Class B explosives.

Detailed View — Additional, close-up information shown on a particular section of a larger drawing.

Detonation — (1) Supersonic thermal decomposition, which is accompanied by a shock wave in the decomposing material. (2) Explosion with an energy front that travels faster than the speed of sound. (3) High explosive that decomposes extremely rapidly, almost instantaneously.

Dimensional Lumber - Lumber with standard, nominal measurements for use in building construction. Dimensional lumber is also available in rough, green components with actual dimensions that match the nominal dimensions.

Dimensioning — Indicating or determining size and position in space.

Direct Pumping System — Water supply system supplied directly by a system of pumps rather than elevated storage tanks.

Distribution System — That part of an overall water supply system that receives the water from the pumping station and delivers it throughout the area to be served.

Drop Panel — Type of concrete floor construction in which the portion of the floor above each column is dropped below the bottom level of the rest of the slab, increasing the floor thickness at the column.

Due Process — Conduct of legal proceedings according to established rules and principles for the protection and enforcement of private rights, including notice and the right to a fair hearing before a tribunal with the power to decide the case; *also called* due process of law or due course of law.

Due Process Clause — Constitutional provision that prohibits the government from unfairly or arbitrarily depriving a person of life, liberty, or property.

E

Elevation View — Architectural drawing used to show the number of floors of a building, ceiling heights, and the grade of surrounding ground.

Enabling Legislation — Legislation that gives appropriate officials the authority to implement or enforce the law.

Encapsulating — Completely enclosed or surrounded as in a capsule.

Endothermic Reaction — Chemical reaction that absorbs thermal energy or heat.

Energy — Capacity to perform work; occurs when a force is applied to an object over a distance, or when a chemical, biological, or physical transformation is made in a substance.

Evacuation — Controlled process of leaving or being removed from a potentially hazardous location, typically involving relocating people from an area of danger or potential risk to a safer place.

Exit — Portion of a means of egress that is separated from all other spaces of the building structure by construction or equipment, and provides a protected way of travel to the exit discharge.

Exit Access — Portion of a means of egress that leads from any occupied portion of a building or structure to an exit.

Exit Discharge — That portion of the exit that is between the exit and a public way.

Exothermic Reaction— Chemical reaction that releases thermal energy or heat.

Expellant Gas — Inert gas that is compressed and used to force extinguishing agents from a portable fire extinguisher; nitrogen is a commonly used expellant gas.

Explosion — Physical or chemical process that results in the rapid release of high pressure gas into the environment.

Exterior Insulation and Finish Systems (EIFS) — Exterior cladding or covering systems composed of an adhesively or mechanically fastened foam insulation board, reinforcing mesh, a base coat, and an outer finish coat; *also known as* synthetic stucco.

F

Fire Alarm Control Unit (FACU) — The main fire alarm system component that monitors equipment and circuits, receives input signals from initiating devices, activates notification appliances ,and transmits signals off-site.

False Front — Additional facade on the front of a building applied after the original construction for decoration; often creating a concealed space.

Feedback — Responses that clarify and ensure that the message was received and understood.

Finger Joint — Connection between two parts made by cutting complementary mating parts, and then securing the joint with glue.

Fire — A rapid oxidation process, which is a chemical reaction resulting in the evolution of light and heat in varying intensities. (NFPA® 921)

Fire Alarm Control Unit (FACU) — The main fire alarm system component that monitors equipment and circuits, receives input signals from initiating devices, activates notification appliances ,and transmits signals off-site. Formerly called the *fire alarm control panel (FACP).*

Fire Damper — Device installed in air ducts that penetrate fire-resistant-rated vertical or horizontal assemblies; prohibits the transfer of heat or flames through the ducts at the point where the duct passes through the assembly.

Fire Department Connection (FDC) — Point at which the fire department can connect into a sprinkler or standpipe system to boost the water flow in the system.

Fire Load — The amount of fuel within a compartment expressed in pounds per square foot obtained by dividing the amount of fuel present by the floor area. Fire load is used as a measure of the potential heat release of a fire within a compartment.

Fire Point — Temperature at which a liquid fuel produces sufficient vapors to support combustion once the fuel is ignited. Fire point must exceed 5 seconds of burning duration during the test. The fire point is usually a few degrees above the flash point.

Fire Retardant — Chemical applied to a material or another substance that is designed to slow ignition or the spread of fire.

Fire Retardant — Chemical applied to material or another substance that is designed to retard ignition or the spread of fire.

Fire-Stop — Materials used to prevent or limit the spread of fire in hollow walls or floors, above false ceilings, in penetrations for plumbing or electrical installations, or in cocklofts and crawl spaces.

Fire Tetrahedron — Model of the four elements/conditions required to have a fire. The four sides of the tetrahedron represent fuel, heat, oxygen, and self-sustaining chemical chain reaction.

Fire Triangle — Model used to explain the elements/conditions necessary for combustion. The sides of the triangle represent heat, oxygen, and fuel.

Fire Wall — Wall with a specified degree of fire resistance that is designed to prevent the spread of fire within a structure or between adjacent structures.

Flame — Visible, luminous body of a burning gas emitting radiant energy including light of various colors given off by burning gases or vapors during the combustion process.

Flame Spread Rating — Numerical rating assigned to a material based on the speed and extent to which flame travels over its surface.

Flammable Liquid — Any liquid having a flash point below 100°F (37.8°C) and a vapor pressure not exceeding 40 psi absolute (276 kPa) {2.76 bar}.

Flammable Range — The range between the upper flammable limit and lower flammable limit in which a substance can be ignited.

Flash Point — Minimum temperature at which a liquid gives off enough vapors to form an ignitable mixture with air near the liquid's surface.

Flashover — A rapid transition from the growth stage to the fully developed stage.

Flow Hydrant — Fire hydrant from which water is discharged during a hydrant fire flow test.

Flow Path — Composed of at least one inlet opening, one exhaust opening, and the connecting volume between the openings. The direction of the flow is determined by difference in pressure. Heat and smoke in a high pressure area will flow towards areas of lower pressure (NIST).

Frame — Internal system of structural supports within a building.

Free Radicals — Molecular fragments that are highly reactive.

Fuel — A material that will maintain combustion under specified environmental conditions. (NFPA® 921)

Fuel Load — The total quantity of combustible contents of a building, space, or fire area, including interior finish and trim, expressed in heat units of the equivalent weight in wood.

Fuel Controlled — A fire with adequate oxygen in which the heat release rate and growth rate are determined by the characteristics of the fuel, such as quantity and geometry.

Fusible Link — (1) Connecting link device that fuses or melts when exposed to heat; used in sprinklers, fire doors, dampers, and ventilators. (2) Two-piece link held together with a metal that melts or fuses at a specific temperature.

G

Gate Valve — Control valve with a solid plate operated by a handle and screw mechanism; rotating the handle moves the plate into or out of the waterway.

Glazing — Glass or thermoplastic panel in a window that allows light to pass.

Global Positioning System (GPS) — System for determining position on the earth's surface by calculating the difference in time for the signal from a number of satellites to reach a receiver on the ground.

Glued-Laminated Beam — Term used to describe wood members produced by joining small, flat strips of wood with glue. *Also known as* Glulam.

Gravity System — Water supply system that relies entirely on the force of gravity to create pressure and cause water to flow through the system. The water supply, which is often an elevated tank, is at a higher level than the system.

Grounding — Reducing the difference in electrical potential between an object and the ground by the use of various conductors; similar to *bonding*.

H

Head Pressure — Pressure exerted by a stationary column of water, directly proportional to the height of the column.

Header Course — Course of bricks with the ends of the bricks facing outward.

Heat — A form of energy characterized by vibration of molecules and capable of initiating and supporting chemical changes and changes of state. (NFPA® 921)

Heat Flux — The measure of the rate of heat transfer to a surface, expressed in kilowatts/m^2, kilojoules/$m^2 \cdot$ sec, or Btu/$ft^2 \cdot$ sec. (NFPA® 921)

Heat of Combustion — Total amount of thermal energy (heat) that could be generated by the combustion (oxidation) reaction if a fuel were completely burned. The heat of combustion is measured in British Thermal Units (Btu) per pound or Megajoules per kilogram.

Heat Release Rate (HRR) — Total amount of heat released per unit time. The HRR is measured in kilowatts (kW) and megawatts (MW) of output.

Horizontal Exit — A path of egress travel from one building to an area in another building on approximately the same level, or a path of egress travel through or around a wall or partition to an area on approximately the same level in the same building.

Hot Work — Any operation that requires the use of tools or machines that may produce a source of ignition.

Hydrostatic Test — Test method that uses water under pressure to check the integrity of pressure containment components such as pipes or cylinders.

I

Ignition — The process of initiating self-sustained combustion. (NFPA® 921)

Ignition Temperature — Minimum temperature to which a fuel, other than a liquid, in air must be heated in order to start self-sustained combustion independent of the heating source.

Incipient Stage — First stage of the burning process in a compartment in which the substance being oxidized is producing some heat, but the heat has not spread to other substances nearby. During this phase, the oxygen content of the air has not been significantly reduced and the temperature within the compartment is not significantly higher than ambient temperature.

Indemnify — To agree that that one party will compensate another party for losses or damages that are incurred if specific actions or events occur.

Indict — To present a formal, written accusation charging a defendant with a crime.

Industry Standard — Set of published procedures and criteria that peer, professional, or accrediting organizations recognize as acceptable practice.

Initiating Device — Alarm system component that transmits a signal when a change occurs; change may be the result of an action such as the activation of a manual fire alarm box, the presence of products of combustion in the atmosphere, or the automatic activation of a supervisory switch.

Inspector — Person who is trained and certified to perform fire prevention and life safety inspections of all types of new construction and existing occupancies; *also called* Code Enforcement Officer and Fire and Life Safety Inspector.

Interior Finish — Exposed interior surfaces of buildings including, but not limited to, fixed or movable walls and partitions, columns, and ceilings. Commonly refers to finish on walls and ceilings and not floor coverings.

Intermediate Bulk Container (IBC) — Rigid (RIBC) or flexible (FIBC) portable packaging, other than a cylinder or portable tank, that is designed for mechanical handling with a maximum capacity of not more than three 3 cubic meters and a minimum capacity of not less than 0.45 cubic meters or a maximum net mass of not less than 400 kilograms.

Interstitial Space — In building construction, refers to generally inaccessible spaces between layers of building materials. May be large enough to provide a potential space for fire to spread unseen to other parts of the building.

Intrinsically Safe Equipment — Equipment designed and approved for use in flammable atmospheres that is incapable of releasing sufficient electrical energy to cause the ignition of a flammable atmospheric mixture.

Intumescent Coating — Coating or paintlike product that expands when exposed to the heat of a fire to create an insulating barrier.

Ionization Smoke Detector — Type of smoke detector that uses a small amount of radioactive material to make the air within a sensing chamber conduct electricity.

K

Kinetic Energy — The energy possessed by a body because of its motion.

L

Large Diameter Hose (LDH) — Relay-supply hose of 3½ to 6 inches (90 mm to 150 mm) in diameter; used to move large volumes of water quickly with a minimum number of pumpers and personnel.

Liability — State of being legally obliged and responsible.

Lightweight Steel Truss — Structural support made from a long steel bar that is bent at a 90-degree angle with flat or angular pieces welded to the top and bottom.

Lightweight Wood Truss — Structural supports constructed of 2 x 3-inch or 2 x 4-inch (50 mm by 75 mm or 50 mm by 100 mm) members that are connected by gusset plates.

Live Load — (1) Items within a building that are movable but not included as a permanent part of the structure; (2) force placed upon a structure by the addition of people, objects, or weather.

Load-Bearing Wall — Walls of a building that by design carry at least some part of the structural load of the building in the direction of the ground or base.

Lower Flammable Limit (LFL) — Lower limit at which a flammable gas or vapor will ignite and support combustion; below this limit the gas or vapor is too *lean* or *thin* to burn (lacks the proper quantity of fuel). *Also known as* Lower Explosive Limit (LEL).

M

Malfeasance — Commission of an unlawful act; committed by a public official.

Masonry — Bricks, blocks, stones, and unreinforced and reinforced concrete products.

Mass Notification System (MNS) — System that notifies occupants of a dangerous situation and provides information and instructions.

Matter — Anything that occupies space and has mass.

Maximum Allowable Quantity Per Control Area — Maximum amount of a hazardous material to be stored or used within a control area inside a building or an outdoor control area; maximum allowable quantity per control area is based on the material state (solid, liquid, or gas) and the material storage or use conditions (*International Fire Code®*, 2015 edition).

Means of Egress — Safe, continuous path of travel from any point in a structure to a public way; the means of egress is composed of three parts: the exit access, the exit, and the exit discharge.

Membrane Structure — Weather-resistant, flexible, or semiflexible covering consisting of layers of materials over a supporting framework.

Miscible — Materials that are capable of being mixed in all proportions.

Misfeasance — Improper performance of a legal or lawful act.

Mitigate — To make less harsh or intense; to alleviate.

Mixture — Substance containing two or more materials not chemically united.

Monitor — Individual assigned to supervise an evacuation process for a specified area within a structure, such as a ward monitor or floor monitor.

N

National Fire Incident Reporting System (NFIRS) — One of the main sources of information (data, statistics) about fires in the United States; under NFIRS, local fire departments collect fire incident data and send these to a state coordinator; the state coordinator develops statewide fire incident data and also forwards information to the USFA; begun by FEMA.

Negligence — Failure to exercise the same care that a reasonable, prudent, and careful person would under the same or similar circumstances.

Neutral Plane — The level at a compartment opening where the difference in pressure exerted by expansion and buoyancy of hot smoke flowing out of the opening and the inward pressure of cooler, ambient temperature air flowing in through the opening is equal.

Noncombustible — Incapable of supporting combustion under normal circumstances.

Nonfeasance — Failing to perform a required duty.

Nonload-Bearing Wall — Wall, usually interior, that supports only its own weight. These walls can be breached or removed without compromising the structural integrity of the building.

Normal Operating Pressure — Amount of pressure that is expected to be available from a hydrant, prior to pumping.

O

Occupant Load — Total number of people for which the means of egress of a building or portion thereof is designed.

Opening Protective — A device installed over an opening to protect it against smoke, flame, and heated gases.

Ordinance — Local or municipal law that applies to persons and things of the local jurisdiction; a local agency act that has the force of a statute; different from law that is enacted by federal or state/provincial legislatures.

Orientation — Location or position relative to the points of the compass.

Oriented Strand Board (OSB) — A wooden structural panel formed by gluing and compressing wood strands together under pressure.

OS&Y Valve — Outside stem and yoke valve; a type of control valve for a sprinkler system in which the position of the center screw indicates whether the valve is open or closed. *Also known as* Outside Screw and Yoke Valve.

Overfiring — Overheating of a solid-fuel room heater by heating the unit and its connections until they glow a dull red, thus subjecting the unit and surrounding materials to higher-than-expected heat.

Oxidation — Chemical process that occurs when a substance combines with an oxidizer such as oxygen in the air; a common example is the formation of rust on metal.

Oxidizing Agent — Substance that oxidizes another substance; can cause other materials to combust more readily or make fires burn more strongly. *Also known as* Oxidizer.

P

Panic Hardware — Hardware mounted on exit doors in public buildings that unlocks from the inside and enables doors to be opened when pressure is applied to the release mechanism.

Parapet — (1) Portion of the exterior walls of a building that extends above the roof. A low wall at the edge of a roof. (2) Any required fire walls surrounding or dividing a roof or surrounding roof openings such as light/ventilation shafts.

Passive Agents — Materials that absorb heat but do not participate actively in the combustion process.

Photoelectric Smoke Detector — Type of smoke detector that uses a small light source, either an incandescent bulb or a light-emitting diode (LED), to detect smoke by shining light through the detector's chamber: smoke particles reflect the light into a light-sensitive device called a photocell.

Piloted Ignition — Moment when a mixture of fuel and oxygen encounters an external heat (ignition) source with sufficient heat or thermal energy to start the combustion reaction.

Pitot Tube — Instrument that is inserted into a flowing fluid (such as a stream of water) to measure the velocity pressure of the stream; commonly used to measure flow. A pitot tube functions by converting the velocity energy to pressure energy that can then be measured by a pressure gauge. The gauge reads in units of pounds per square inch (psi) or kilopascals (kPa).

Plan View — Drawing containing the two-dimensional view of a building as seen from directly above the area.

Plume — The column of hot gases, flames, and smoke rising above a fire; *also called* convection column, thermal updraft, or thermal column. (NFPA® 921)

Police Power — Constitutional right of the state government to make laws and regulations to protect the safety, health, welfare, and morals of the community. Police power is used as the basis for enacting a variety of laws including land use (zoning), fire and building codes, gambling, vehicle registration and parking, nuisances, schools and sanitation.

Post and Beam Construction — Construction style using vertical elements to support horizontal elements. Historically associated with heavy wood beams and columns; modern construction may use materials including heavy timber, steel, and precast concrete.

Post Indicator Valve (PIV) — Type of valve used to control underground water mains that provides a visual means for indicating "open" or "shut" positions; found on the supply main of installed fire protection systems. The operating stem of the valve extends above ground through a "post," and a visual means is provided at the top of the post for indicating "open" or "shut."

Post-Tensioned Reinforcement — Concrete reinforcement method. Reinforcing steel strands in the concrete are tensioned after the concrete has hardened.

Potential Energy — Stored energy possessed by an object that can be released in the future to perform work once released.

Pressure Maintenance Pump — Pump used to maintain pressure on a fire protection system in order to prevent false starts at the fire pump.

Pressure-Reducing Valve — Valve installed at standpipe connection that is designed to reduce the amount of water pressure at that discharge to a specific pressure, usually 100 psi (700 kPa).

Pretensioned Reinforcement (Concrete) — Concrete reinforcement method. Steel strands are stretched, producing a tensile force in the steel. Concrete is then placed around the steel strands and allowed to harden.

Probable Cause — Sufficient information or facts to believe that it is probable (more likely than not) that a certain party is responsible for committing a felony (indictable offense).

Products of Combustion — Materials produced and released during burning.

Proportioning — Mixing of water with an appropriate amount of foam concentrate in order to form a foam solution.

Proprietary Alarm System — Fire alarm system owned and operated by the property owner.

Protected Premises System — Alarm system that alerts and notifies only occupants on the premises of the existence of a fire so that they can safely exit the building and call the fire department. If a response by a public safety agency (police or fire department) is required, an occupant hearing the alarm must notify the agency.

Protected Steel — Steel beams that are covered with either spray-on fireproofing (an insulating barrier) or fully encased in an Underwriters Laboratories Inc. (UL) designed system.

Public-Duty Doctrine — States that a government entity (such as a state or municipality) cannot be held liable for an individual plaintiff's injury resulting from a governmental officer's or employee's breach of a duty owed to the general public rather than to the individual plaintiff.

Public Emergency Alarm Reporting System — System that connects the protected property with the fire department alarm communications center by a municipal master fire alarm box or over a dedicated telephone line.

Public Way — Parcel of land such as a street or sidewalk that is essentially open to the outside and is used by the public to move from one location to another.

Pyrolysis — The chemical decomposition of a solid material by heating. Pyrolysis often precedes combustion.

Pyrophoric — Material that ignites spontaneously when exposed to air. *Also known as* Air-Reactive Material.

R

Radiation — Heat transfer by way of electromagnetic energy. (NFPA® 921)

Reactivity — Ability of two or more chemicals to react and release energy and the ease with which this reaction takes place.

Rebar — Short for reinforcing bar. These steel bars are placed in concrete forms before the cement is poured. When the concrete sets (hardens), the rebar within it adds considerable strength and reinforement.

Reducing Agent — The fuel that is being oxidized or burned during combustion.

Regulations — Rules or directives of administrative agencies that have authorization to issue them.

Remote Receiving System — System in which alarm signals from the protected premises are transmitted over a leased telephone line to a remote receiving station with a 24-hour staff; usually the municipal fire department's alarm communications center sending an alarm signal to the FACU.

Residual Pressure — Pressure remaining at a given point in a water supply system while water is flowing.

Return-Air Plenum — Unoccupied space within a building through which air flows back to the heating, ventilating, and air-conditioning (HVAC) system; normally is immediately above a ceiling and below an insulated roof or the floor above.

Rigid Frame — Load-bearing system constructed with a skeletal frame and reinforcement between a column and beam.

Riser — Vertical water pipe used to carry water for fire protection systems above ground, such as a standpipe riser or sprinkler riser.

Roof Deck — Bottom components of the roof assembly that support the roof covering; may be constructed of plywood, wood studs (2 by 4 inches or larger), lath strips, or other materials.

Right of Entry — Rights set forth by the administrative powers that allow the inspector to inspect buildings to ensure compliance with applicable codes.

S

Safety Data Sheet (SDS) — Form provided by the manufacturer and blender of chemicals that contains information about chemical composition, physical and chemical properties, health and safety hazards, emergency response procedures, and waste disposal procedures of the specified material.

Salamander — Portable heating device generally found on construction sites.

Sanction — Notice or punishment attached to a violation for the purpose of enforcing a law or regulation.

Saponification — A phenomenon that occurs when mixtures of alkaline-based chemicals and certain cooking oils come into contact resulting in the formation of a soapy film.

Scarf Joint — Connection between two parts made by the cutting of overlapping mating parts and securing them by glue or fasteners so that the joint is not enlarged and the patterns are complementary.

Search Warrant — Written order, in the name of the People, State, Province, Territory, or Commonwealth, signed by a magistrate, commanding a peace officer to search for personal property or other evidence and return it to the magistrate.

Sectional View — Vertical view of a building as if it were cut into two parts; intended to show the internal construction of each assembly.

Self-Heating — The result of exothermic reactions, occurring spontaneously in some materials under certain conditions, whereby heat is generated at a rate sufficient to raise the temperature of the material. (NFPA® 921)

Semiconductor — Material that is neither a good conductor nor a good insulator, and therefore may be used as either in some applications.

Service Test — Series of tests performed on fire protection, detection, and/or suppression systems in order to ensure operational readiness. These tests should be performed at least yearly or whenever the system has undergone extensive repair or modification.

Site Plan — Drawing that provides a view of the proposed construction in relation to existing conditions; includes survey information and information on contours and grades; generally the first sheet on a set of drawings.

Smoke Damper — Device that restricts the flow of smoke through an air-handling system; usually activated by the building's fire alarm signaling system.

Smoke-Developed Index — A measure of the concentration of smoke a material emits as it burns.

Smokeproof Enclosures — Stairways that are designed to limit the penetration of smoke, heat, and toxic gases from a fire on a floor of a building into the stairway and that serve as part of a means of egress.

Solubility — Degree to which a solid, liquid, or gas dissolves in a solvent (usually water).

Special Duty — Type of obligation that an inspector assumes by providing expert advice or assistance to a person; this obligation may make the inspector liable if it creates a situation in which a person moves from a position of safety to a position of danger by relying upon the expertise of the inspector.

Specific Gravity — Mass (weight) of a substance compared to the mass of an equal volume of water at a given temperature. A specific gravity less than 1 indicates a substance lighter than water; a specific gravity greater than 1 indicates a substance heavier than water.

Spontaneous Ignition — Initiation of combustion of a material by an internal chemical or biological reaction that has produced sufficient heat to ignite the material. (NFPA® 921)

Sprinkler — Waterflow device in a sprinkler system. The sprinkler consists of a threaded nipple that connects to the water pipe, a discharge orifice, a heat-actuated plug that drops out when a certain temperature is reached, and a deflector that creates a stream pattern that is suitable for fire control. Formerly referred to as a sprinkler head.

Standard — A set of principles, protocols, or procedures that explain how to do something or provide a set of minimum standards to be followed. Adhering to a standard is not required by law, although standards may be incorporated in codes, which are legally enforceable.

Standardization — Process of making or creating things that meet established criteria.

Static Pressure — Pressure at a given point in a water system when no water is flowing.

Statute or Statutory Law — Federal or state/provincial legislative act that becomes law; prescribes conduct, defines crimes, and promotes public good and welfare.

Steamer Connection — Large-diameter outlet, usually 4½ inches (115 mm), at a hydrant or at the base of an elevated water storage container.

Steiner Tunnel Test — Unofficial name for the test used to determine the flame spread ratings of various materials.

Stud — Vertical structural member within a wall (most often made of wood but some are made of light-gauge metal) that make up the walls and partitions in a frame building.

Sunset Provision — Clause in a law or ordinance that stipulates the periodic review of government agencies and programs in order to continue their existence.

Supervisory Signal — Signal given by a fixed fire protection system when there is a problem with the system.

T

Temperature — Measure of a material's ability to transfer heat energy to other objects; the greater the energy, the higher the temperature. Measure of the average kinetic energy of the particles in a sample of matter, expressed in terms of units or degrees designated on a standard scale.

Tensile — Force of pulling apart or stretching.

Test Hydrant — Fire hydrant used during a fire flow test to read the static and residual pressures.

Thermal Energy — The kinetic energy associated with the random motions of the molecules of a material or object; often used interchangeably with the terms heat and heat energy. Measured in joules or Btu.

Thermal Layering — Outcome of combustion in a confined space in which gases tend to form into layers, according to temperature, with the hottest gases found at the ceiling and the coolest gases at floor level.

Thermistor — Semiconductor made of substances whose resistance varies rapidly and predictably with temperature.

Thermocouple — Device for measuring temperature in which two electrical conductors of dissimilar metals such as copper and iron are joined at the point where heat is applied.

Threshold Limit Value (TLV) — Concentration of a given material that may be tolerated for an 8-hour exposure during a regular workweek without ill effects.

Thrust Plate — Steel plate located on the exterior of a masonry building to which a tension rod is anchored.

Topographical — Pertaining to configuration of the land or terrain.

Total Flooding System — Fire-suppression system designed to protect hazards within enclosed structures; foam is released into a compartment or area and fills it completely, extinguishing the fire.

Toxicity — Ability of a substance to do harm within the body.

Travel Distance — Distance from any given area in a structure to the nearest exit or to a fire extinguisher.

Trouble Signal — Signal given by a fixed fire protection alerting system when a power failure or other system malfunction occurs.

U

Upper Flammable Limit (UFL) — Upper limit at which a flammable gas or vapor will ignite; above this limit the gas or vapor is too *rich* to burn (lacks the proper quantity of oxygen). *Also known as* Upper Explosive Limit (UEL).

Upper Layer — Buoyant layer of hot gases and smoke produced by a fire in a compartment.

V

Vapor Density — Weight of a given volume of pure vapor or gas compared to the weight of an equal volume of dry air at the same temperature and pressure. A vapor density less than 1 indicates a vapor lighter than air; a vapor density greater than 1 indicates a vapor heavier than air.

Vapor Pressure — (1) Measure of the tendency of a substance to evaporate. (2) The pressure at which a vapor is in equilibrium with its liquid phase for a given temperature; liquids that have a greater tendency to evaporate have higher vapor pressures for a given temperature.

Vaporization — Physical process that changes a liquid into a gaseous state; the rate of vaporization depends on the substance involved, heat, pressure, and exposed surface area.

Vessel — Tank or container used to store a commodity that may or may not be pressurized.

Violation — Infringement of existing rules, codes, or laws.

Water Main — A principal pipe in a system of pipes for conveying water, especially one installed underground.

Watt — A unit of measure of power or rate of work equal to one joule per second (J/s).

Work Session — Formal, open meeting by a legislative body to study the merits of the proposed legislation and ask questions of the fire and life safety code official and the public regarding the provisions of the proposed code.

W

Wythe — Single vertical row of multiple rows of masonry units in a wall, usually brick.

Index

Index

A

Abatement Order, 21
Aboveground storage tanks (ASTs), 440–446
  field-erected, 442–446
  shop-fabricated, 440–442
Absorption (contact) as toxic material entry into the body, 396
Acceptance testing
  fire detection and alarm systems, 642–643
  water-based fire extinguishing systems, 540, 541–545
Access to sites. *See* Site access
Access-controlled egress, 254–255
ACEA (American Code Enforcement Association), 18
Acetylene, 388
ACGIH (American Conference of Governmental Industrial Hygienists), 397
Acronyms
  RACE, 720
  REACT, 720
ADA (Americans with Disabilities Act), 245
Adams vs. State of Alaska, 32
Addressable alarm systems, 617
Administrative duties of inspectors, 695–698
  files and records, 697–698
  written communications, 695–697
Administrative warrant, 31
Adoption by reference, 27
Aeration of foam solution, 581–582
AFFF (aqueous film forming foam), 589
Agents. *See* Extinguishing agents
AHJ. *See* Authority having jurisdiction (AHJ)
Air conditioning systems
  purpose of, 343
  refrigerants, 343–344
Air-cooled transformers, 234
Air-Reactive Material, 393
Air-sampling smoke detector, 632, 633
AIT (autoignition temperature), 80
Alarm signaling systems, 613–623
  alarm signal, defined, 613
  coded fire alarm systems, 617
  emergency communications systems, 621–623
    functions of, 621
    mass notification system, 614, 622, 623
    two-way communication systems, 622, 623
    voice notification systems, 622, 623
  factors for selection of, 614
  inspection and testing, 641–642, 643
  protected premises system, 614–617
    addressable alarm systems, 617
    conventional alarm systems, 615–616
    presignal alarms, 615
    zoned conventional alarm systems, 616–617
  public emergency alarm reporting systems, 621
  supervising station alarm systems, 617–621
    central station system, 618–619
    multiplexing systems, 620
    parallel telephone system, 614, 620
    proprietary system, 618, 619, 620
    remote, 618
    remote receiving system, 619, 621
  supervisory signal, 613
  tamper switches, 614
  trouble signal, 613–614
Alarm systems. *See* Fire detection and alarm systems
Alarm-test valve, 510
Alley dock, 288, 289
Ambulatory health care occupancies, 136–137
American Code Enforcement Association (ACEA), 18
American Conference of Governmental Industrial Hygienists (ACGIH), 397
American National Standards Institute (ANSI)
  ANSI Z223.1, *National Fuel Gas Code*, 671
  ASME/ANSI Standard A17.1 *Safety Code for Elevators and Escalators*, 47
  consensus standards, 48
  cross-reference with NFPA® and OSHA, 50
  ICC/ANSI 117.1, Standard on Accessible and Usable Buildings and Facilities, 245
  purpose of, 49–50
  refrigerants, 343–344
  Z26.1-1996, *Safety Code for Safety Glazing Materials for Glazing Motor Vehicles Operating on Land Highways*, 50
American Petroleum Institute (API)
  API 620 *Design and Construction of Large, Welded, Low-Pressure Storage Tanks*, 384, 443–444
  API 650 *Welded Steel Tanks for Oil Storage*, 384
  API 650 *Welded Tanks for Oil Storage*, 443
  API 653 *Tank Inspection, Repair, Alteration and Reconstruction*, 384, 444, 446
  field-erected aboveground storage tanks, 444
American Society for Testing and Materials (ASTM). *See also* ASTM International
  ASTM D-3175, *Standard Test Method for Volatile Matter in the Analysis Sample of Coal and Coke*, 367
  consensus standard, 48
  fire door testing, 224
  flash point tests, 384
American Society of Heating, Refrigeration and Air Conditioning Engineers (ASHRAE)
  refrigerant classifications, 231
  Standard 34, *Designation and Safety Classifications of Refrigerants*, 231–232, 343–344
American Society of Mechanical Engineers (ASME)
  ASME B31.3, 451
  ASME/ANSI Standard A17.1 *Safety Code for Elevators and Escalators*, 47
  B31.3 *Process Piping*, 384
  Boiler and Pressure Vessel Codes, 338, 384
American Society of Testing and Materials (ASTM). *See also* ASTM International
American Water Works Association, 477
Americans with Disabilities Act (ADA), 245
Angle of approach and departure, 286, 287
Anhydrous ammonia for air conditioning systems, 343
Anhydrous chlorine, 385
ANSI. *See* American National Standards Institute (ANSI)
Antifreeze sprinkler system, 507
Apartment occupancy, 131, 143–144
API. *See* American Petroleum Institute (API)
Appeals of codes, 62–65

Aqueous film forming foam (AFFF), 589
Architectural symbols, 655
Arcing, 81, 320
Area of refuge, 244-245
Around-the-pump proportioner, 583, 584
Arterial mains, 472
Asbestos, 160
ASHRAE. *See* American Society of Heating, Refrigeration and Air Conditioning Engineers (ASHRAE)
ASME. *See* American Society of Mechanical Engineers (ASME)
Asphalt kettle fire hazards, 372–373
Asphalt shingles, 199, 200
Asphyxiant, 575
Assembly occupancies, 132
ASTM. *See* American Society for Testing and Materials (ASTM)
ASTM International
  ASTM D-3175, *Standard Test Method for Volatile Matter in the Analysis Sample of Coal and Coke*, 367
  consensus standards, 48
  E84 *Standard Test Method for Surface Burning Characteristics of Building Materials*, 49, 225
  E108 *Standard Test Methods for Fire Tests of Roof Coverings*, 49
  E119 *Standard Test Methods for Fire Tests of Building Construction*, 49, 167, 222, 432
  E-152 *Standard Methods of Fire Tests of Door Assemblies*, 224
  flash point tests, 384
  purpose of, 49
ASTs (aboveground storage tanks), 440–446
Atriums, 202, 682
Audible notification, 644
Audible notification appliances, 611
Auditorium occupancies, 146
Authority
  enabling legislation, 24–25, 27
  federal laws, 25–26
  Inspector II, 24–33
  legal status of inspectors, 27–29
    private sector, 29
    public sector, 28, 29
  liability, 29–33
  local laws and ordinances, 27
  minimum/maximum laws, 26
  state/provincial laws, 26
Authority having jurisdiction (AHJ)
  access barriers, knowledge of, 294–295
  alarm signaling system testing, 641
  asbestos, handling of, 160
  automatic alarm-initiating device inspections, 639
  bridge weight capacity, knowledge of, 296
  building plans for review, 654
  certification requirements, 17
  code modification, 63
  defined, 17
  elevation view drawings, 666
  enabling legislation, 24–25
  fire alarm system design and installation, 608
  fire detection and alarm system testing, 643
  fire lane signs, 291
  hot work program approval, 354
  indemnification from liability, 32
  inspection guidelines and procedures, 703–704
  inspector professional development, 17, 18
  interpreting code provisions, 47
  model codes, 43, 46
  performance-based codes, 44
  personnel training and certification, 13
  plans review and permitting, 651–652
  preacceptance inspections of fire suppression systems, 539
  right of entry, 30
  site access responsibilities, 285
  training level for inspectors, 17
  two-way communication systems, 622
  warrants, 30
  water-based fire suppression system testing and maintenance, 538
Autoignition, 79–80
Autoignition of pyrophorics, 394
Autoignition temperature (AIT), 80
Automatic alarm-initiating devices, 623–635
  combination detectors, 634, 635
  fixed-temperature heat detectors, 624–627
  flame detectors, 633–634
  inspection and testing, 639–641
  nuisance alarms, 623
  rate-of-rise heat detectors, 627–629
  smoke detectors, 629–633
  sprinkler waterflow alarm-initiating devices, 634, 635
Automatic fire suppression systems. *See* Fire suppression systems
Automatic sprinkler systems. *See also* Sprinkler
  acceptance testing procedures, 541
  antifreeze, 507
  in big-box stores, 323
  circulating closed-loop, 507
  combined dry pipe and preaction, 507
  components, 508–518
    detection and activation devices, 516–518
    operating valves, 510
    sprinklers, 512–516
    water distribution pipes, 510–512
    water supplies, 508
    waterflow control valves, 508–509
  deluge, 506, 517
  detection and activation devices, 516–518
  dry-pipe, 505
  effectiveness of, 504
  for fire temperature reduction, 109
  high-hazard occupancies, 458–459
  hydraulically calculated, 674, 675
  for industrial furnaces and ovens, 348–349
  inspections and testing, 546–552, 553
  in material storage facilities, 323
  maximum allowable quantity per control area, 432, 433
  mechanics of, 504
  multicycle, 507
  plans review, 673–676
  preaction, 506, 517
  purpose of, 504
  in-rack sprinklers, 325, 382
  residential systems, 518–519
  in warehouses, 327–328
  waterflow alarms, 517–518, 634, 635
  wet-pipe, 505
Automatic-dry standpipe and hose system, 527, 528, 529
Automatic-wet standpipe and hose system, 527, 528, 529
Auxiliary fire alarm system inspection and testing, 641
Auxiliary power for exit signs, 264–265
Average Ambient Sounds, 644

B

B31.3 *Process Piping*, 384
B306-M, *Portable Fuel Tanks for Marine Use*, 436
B376-M, *Portable Containers for Gasoline and Other Petroleum Fuels*, 436
B620, *Highway Tanks and Portable Tanks for the Transportation of Dangerous Goods*, 436
Backflow preventers, 475, 508, 510
Balanced pressure proportioner, 583
Balloon-frame construction, 180
Barn occupancies, 145
Baseboard heaters, 341
Beams, glued-laminated (glulam), 122, 123, 155
Bearing wall structures, 172, 173
Behaviors, unsafe. *See* Unsafe behaviors
Bench trial, 22
Benchmark, 664
Big-box store storage, 322, 323
Bill of Rights, 29
Bimetallic heat detector, 625–626
Bladder tank proportioner, 555
Blasting agent hazard classifications, 395
BLEVE (Boiling Liquid Expanding Vapor Explosion), 75, 379
BNQ (Bureau de normalisation du Quebec), 50
Board of Appeals for codes, 65
Boarding house occupancies, 140–141
Body language, 694
Boiler and Pressure Vessel Code (BPVC)
- cylinders, 437
- nameplate marking, 447–448
- pressure vessel capacities, 446–447
- purpose of, 338
- storage, handling and use standards, 384

Boilers, 338–339
Boiling Liquid Expanding Vapor Explosion (BLEVE), 75, 379
Bolt locks, 254
Bonding dispensing stations, 315–317
BPVC. *See* Boiler and Pressure Vessel Code (BPVC)
Brick building materials
- floor construction materials, 200, 201
- nonreinforced bearing wall, 183
- production method, 161

Bromochlorodifluoromethane (Halon 1211), 573, 590
Bromotrifluoromethane (Halon 1301), 573, 590
Btu, 78, 87
Building codes, purpose of, 193
Building compartmentation, 682
Building components
- building services
  - electrical systems, 232–234
  - elevator hoistways and doors, 217–218, 234–235
  - evaluating, 228–236
  - HVAC systems, 229–230
  - refrigeration systems, 231–232
  - utility chases and vertical shafts, 218–219, 235–236
- ceilings, 203, 223
- doors, 204–213
  - fire doors, 207–213, 223–224
  - styles and construction materials, 206–207
  - types of, 204–206
- floors, 200–202, 222
- interior finishes, 216–217
  - evaluating, 222–223
  - fire-retardant coatings, 228
  - flame spread ratings, 225–226, 227
  - smoke-developed index, 226
  - systematic plans review, 682–683
- roofs, 199–200
- rooftop photovoltaic systems, 221
- stairs, 203–204, 223
- structural stability, evaluating, 222
- walls, 194–198
  - curtain walls, 197–198
  - enclosure and shaft walls, 196–197
  - fire partitions and fire barriers, 196
  - fire walls, 194–195, 220
  - party walls, 195
- windows, 213–216

Building construction. *See* Construction
Building Construction and Safety Code®. *See* NFPA® 5000, *Building Construction and Safety Code®*
Building department
- fire and life safety inspection programs, 15–16
- plans reviews, 15, 16

Building icing, 159
Building plans review, 654–655
Building services, 217–219
- elevator hoistways and doors, 217–218
- evaluating, 228–236
  - electrical systems, 232–234
  - elevator hoistways and doors, 234–235
  - HVAC systems, 229–231
  - refrigeration systems, 231–232
  - utility chases, 218, 235–236
  - vertical shafts, 218, 235–236
- grease ducts, 218, 219
- refuse and linen chutes, 218, 219

Building system fire hazard recognition, 335–352
- cooking equipment, 344–347
- fire suppression systems, 347–348
- furnaces and ovens, 348–349
- HVAC systems, 335–344
  - air conditioning systems, 343–344
  - boilers, 338–339
  - filtering devices, 344
  - furnaces, 336–338
  - room heaters, 340–342
  - temporary/portable heating equipment, 342–343
  - unit heaters, 340
  - ventilation systems, 344
- powered industrial trucks, 349–350
- tents, 350–352

Bulk packaging. *See also* Packaging
- aboveground storage tanks, 440–446
- atmospheric AST, 439, 441
- defined, 438
- field-erected aboveground storage tanks, 442–446
- hazardous material storage and use, 435
- intermediate bulk containers, 439
- pressure vessels, 446–448
- shop-fabricated aboveground storage tanks, 440–442
- stationary tanks, 438–440
- storage tank types by pressure rating, 441
- underground storage tanks, 446

Bullet-resisting fire doors, 213
Buoyant, 84
Bureau de normalisation du Quebec (BNQ), 50
Business occupancies, 133
Butterfly valve, 474

C

Camara vs. City and County of San Francisco, 29
Canada
  assembly occupancies, 132
  Bureau de normalisation du Quebec, 50
  combustible construction, 125
  construction types, 125
  Controlled Product Regulations, 422
  dangerous goods
    Health Canada, 422
    placards, labels, and markings, 423–427
    regulations system, 421–427
    waste material disposal, 333
    Workplace Hazardous Materials Information System, 422, 427, 428
  educational occupancies, 134
  Environment Canada, 333
  factory/industrial occupancies, 134, 135
  Hazardous Products Act, 422
  health care and ambulatory health care occupancies, 136
  hotel occupancies, 142
  Human Resources and Skills Development, 422
  lodging (boarding) or rooming house occupancies, 140
  National Building Code of Canada, 43, 125
  residential board and care occupancies, 137
  right of entry, 30
  Standards Council of Canada, 48
  standpipe and hose systems, 529
  Transport Canada
    clean-agent storage containers, 596
    corrosives, 398
    health hazard materials, 396
    nonbulk and bulk packaging, 435
    standards development, 50
    waste material disposal, 333
  Transportation of Dangerous Goods (TDG) Regulations, 333, 436
  ULC. *See* Underwriters Laboratories of Canada (ULC)
Canadian Commission on Building and Fire Codes® (CCBFC®)
  construction types, 116
  National Building Code of Canada, 43
  National Fire Code of Canada, 43
Canadian Environmental Protection Act (CEPA), 333
Canadian General Standards Board (CGSB), 50
Canadian Labour Code (CLC), 422
Canadian Standards Association (CSA)
  B306-M, *Portable Fuel Tanks for Marine Use*, 436
  B376-M, *Portable Containers for Gasoline and Other Petroleum Fuels*, 436
  B620, *Highway Tanks and Portable Tanks for the Transportation of Dangerous Goods*, 436
  standards development, 50
Candela of visual devices, 645
Canopies as access barrier, 297
Carbon dioxide extinguishing systems, 574–578
  applications for, 575
  components, 577–578
  hazardous facilities protected by, 575
  high-pressure, 577–578
  inspection and testing, 596
  low-pressure, 577–578
  means of actuation, 577
  personnel safety, 575–576
  portable fire extinguisher agent, 587–588
Carbon steel underground storage tanks, 446, 447
Cartridge-operated fire extinguisher, 590, 591
Casement window, 214, 215
Cast iron, 163
Cast-in-place concrete, 186
Catalog sheets, 658, 660–661
Cavitation, 485
CCBFC. *See* Canadian Commission on Building and Fire Codes (CCBFC)
Cease-and-desist order, 704
Ceiling jet, 104
Ceilings
  construction materials, 203
  evaluating, 223
  functions of, 203
  height as factor in fire development, 101
  interstitial space, 223
  means of egress enclosure, 257
  membrane ceilings, 165
Cellulose fiber insulation, 159
Celsius temperature scale, 78
Cement tile roofs, 199, 200
Central station system, 618–619, 641
Centrifugal pump, 532
CEPA (Canadian Environmental Protection Act), 333
Certificate of Occupancy, 385
Certification, inspector requirements, 17
CFR (Code of Federal Regulations), Title 49 (Transportation), 409
CGA. *See* Compressed Gas Association (CGA)
CGSB (Canadian General Standards Board), 50
Change order, 657
Check valve (backflow preventer), 475, 508, 510
Chemical change of matter, 74–76
Chemical energy, 80–81
Chemical flame inhibition, 96, 111
Chemical reaction, self-sustained, 95–96
Chemical Safety and Hazard Investigation Board (CSB), 362–363
Chemical Transportation Emergency Center (CHEMTREC®), 402
CHEMTREC® (Chemical Transportation Emergency Center), 402
Chimney maintenance, 341
Chlorinated fluorocarbons (Freon®) for air conditioning systems, 343
Church occupancy, 132
Chute opening fire doors, 213
Circuit breakers, 233
Circulating closed-loop sprinkler system, 507
Civil proceedings, 20
Civil rights, 31–32
Cladding, 197
Class A assembly occupancy, 132
Class A fire door, 209
Class A fires, 96, 565
Class A fuel, 99, 599
Class A interior wall finishes, 256
Class A ovens and furnaces, 348
Class A portable fire extinguishers, 585
Class B assembly occupancy, 132
Class B fire door, 209
Class B fires, 96, 565
Class B fuel, 99, 599
Class B interior wall finishes, 256
Class B ovens and furnaces, 348
Class B portable fire extinguishers, 585
Class C assembly occupancy, 132

Class C fire door, 209
Class C fires, 98, 565
Class C fuel, 99, 599
Class C ovens and furnaces, 348
Class C portable fire extinguishers, 585
Class D fire door, 209
Class D fires, 98, 565
Class D fuel, 600
Class D ovens and furnaces, 348
Class D portable fire extinguishers, 585
Class E fire door, 209
Class E fires (Europe), 565
Class I floor finishes, 222
Class I standpipe systems: Firefighters, 526, 527, 530
Class II, Division 1 combustible dust, 367
Class II, Division 2 combustible dust, 368
Class II floor finishes, 222
Class II standpipe systems: Trained Building Occupants, 526, 527
Class III standpipe systems: Combination, 526, 527, 530
Class K fires, 98, 565
Class K fuel, 600
Class K portable fire extinguishers, 585
Classification of hazardous materials, 385–401
  compressed and liquefied gases, 388
  corrosives, 398
  cryogenic fluids, 388–390
  evaluating hazards, 385
  explosives and blasting agents, 395
  flammable and combustible liquids, 386–387
  flammable solids or gases, 390–391
  health hazard materials, 396–401
  highly toxic and toxic materials, 396–397
  incompatible materials, 399–401
  mixtures, 399
  organic peroxides, 391–392
  overview, 385–386
  oxidizers and oxidizing gases, 392–393
  physical hazard materials, 386–395
  pyrophorics, 391, 393–394
  unstable (reactive) materials, 394, 395
  water-reactive materials, 394–395
Clay tile block, 162
Clay tile floors, 200, 201
Clay tile roofs, 199, 200
CLC (Canadian Labour Code), 422
Clean-agent extinguishing systems, 573–574
  agents, 573–574, 590
  applications for, 572
  components, 574
  inspection and testing, 596
Clear dimensions, 666
Clear width of exits, 244
CMSA (control mode specific application) sprinkler, 513
CMUs (concrete masonry units), 161
Coal boilers, 339
Coatings, fire-retardant, 228
Cocoanut Grove, Boston, MA, fire (1942), 262
Code enforcement department, fire and life safety inspection programs, 16
Code of Federal Regulations (CFR), Title 49 (Transportation), 409
Coded fire alarm systems, 617
Codes
  adoption of, 59, 61–62
    formal resolution preparation, 59, 61
    introducing the new code, 61–62
    sunset provisions, 61
    work session for discussion, 61
  appeals, 62–65
  application of, 44–46
  building codes, purpose of, 193
  consistency, 46–47
  defined, 42
  development of local codes, 56–62
    amendments, 56, 59, 60
    drafting the proposed code, 58–59
    formally adopting the code, 59, 61–62
    problem identification, 56–57
    stakeholder identification, 57
    submit code for legal review, 59
    task force, 57–58
  Inspector II responsibilities, 56–67
  life cycle and time line, 46
  modifications, 62–65
  occupancy classifications, 707, 709–711
  performance-based, 43, 44
  permit or license requirements, 52, 53–55
  prescriptive-based, 43
  purpose of, 41
  retention of old fire codes, 62
  standards vs., 42
Cold flow of sprinklers, 552
Combination heat-smoke detector, 634, 635
Combination of hour and letter fire door classifications, 209
Combination water supply system, 470, 472
Combined dry pipe and preaction sprinkler system, 507
Combustible construction (Canada), 125
Combustible dust, defined, 363
Combustible liquids. *See* Flammable and combustible liquids use and storage
Combustion
  building materials, 154, 155
  defined, 75
  fire tetrahedron, 76, 77
  fire triangle, 76
  flaming, 76, 77
  heat of combustion, 87
  nonflaming, 76–77
  products of combustion, 84
Commercial cooking
  equipment, 344–345
  fire extinguishing systems, 49, 347–348
  meat smokers, 346, 347
  solid-fuel cooking equipment, 346–347
Commercial inspection, 17
Commodity, 324
Common path of travel
  defined, 246, 272
  example, 273
  limits by occupancy, 274
Communications, interpersonal, 690–695
Compactors in recycling facilities, 332
Compartment
  compartmentalized floor plan, 100
  thermal properties, 102
  volume as factor in fire development, 101
Compartmentation system, 220
Complaints

complex complaints, 724–725
procedures for handling, 722–725
response to, 18
Compliance procedures, 19–20
Compressed Gas Association (CGA)
CGA C-7, *Guide to Preparation of Precautionary Labeling and Marking of Compressed Gas Containers*, 384, 430
CGA G-13, *Storage and Handling of Silane and Silane Mixtures*, 394
CGA P-18, *Standard for Bulk Inert Gas Systems*, 384
CGA P-20, *Standard for Classification of Toxic Gas Mixtures*, 384
CGA P-23, *Standard for Categorizing Gas Mixtures Containing Flammable & Nonflammable Components*, 384
CGA S-1.1, S-1.2, S-1.3, *Pressure Relief Devices - Parts 1-3*, 384
Compressed Gases and Cryogenic Fluids Code. See NFPA® 55, *Compressed Gases and Cryogenic Fluids Code*
Compressed/liquefied gases
cylinder storage and use, 317–319
dispensing of, 316
dissolved gases, 388
hazard classifications, 388
safety guidelines, 317–318
Compressive loads, 172
Concealed sprinkler, 512, 513
Concrete
concrete block, 161, 162
floor construction materials, 200, 201
grasscrete, 290, 291
structural systems, 186–188
cast-in-place, 186
ordinary reinforcing, 187
post-tensioned reinforcement, 188
precast, 186–187
pretensioned reinforcement, 187–188
Concrete masonry units (CMUs), 161
Conditions, unsafe. *See* Unsafe conditions
Conduction, 83–84
Cone roof storage tanks, 444–445
Consensus standard, 48, 49–50
Constitution. *See* U.S. Constitution
Construction, 193–237
building element fire-resistance rating requirements, 118
design, plans review of, 653
error corrections during plans, 653
exterior wall fire-resistance rating requirements, 118
materials. *See* Construction materials
new construction
building codes, 45–46
inspections, 18, 19
plans review, 662–679
preincident planning, 653
structures to be removed, 664
types of, 115–125
based on fire resistance, 115
Canadian construction, 125
factors determining, 116–117
United States construction, 117–125
basic elements, 117–118
Type I, 118–120
Type II, 120, 121
Type III, 120–122
Type IV, 122, 123
Type V, 122–125
Construction materials, 153–171
asbestos, 160
concrete, 163
fire protection rating, 154
fire-resistant materials, 154
fire-retardant coatings, 154
glazing, 166–167
gypsum board, 167–168
ignition-resistant construction, 154
lath and plaster wall construction, 168
masonry, 160–162
plastics, 168–171
common construction plastics, 170
Exterior Insulation and Finish Systems, 170–171
fire hazards, 169–170
flammability, 169
thermal barriers, 170
steel, 163–165
systematic plans review, 682
wood, 154–160
engineered wood products, 155–157
exterior wall materials, 157–160
fire-retardant treated wood, 157
solid lumber, 155
Construction permit, 52, 55
Construction sites, 291–294
Construction standards, plans review of, 653
Consumer Product Safety Commission (CPSC), 321
Containers for hazardous materials, 435–438
code requirements, 435
construction requirements, 435, 436
cylinders, 437–438. *See also* Cylinders
intermediate bulk containers, 437, 439
safety cans, 436
Containment of product, 381
Contents of buildings as factor in fire development, 100–101
Continuous-line heat detector, 626–627
Contour lines, 664
Control areas, 434
Control valves in water distribution systems, 474–475
Controlled Product Regulations, 422
Convection, 83, 84, 85
Convection column, 103
Conventional alarm systems, 615–616
Conventional sprinkler, 514
Conversing skills, 693–695
Cooking equipment
fire extinguisher standards, 573
grease traps, 345
solid-fuel cooking equipment, 346–347
uses for, 344
ventilation-hood systems, 345, 346, 347
Coreboard, 168
Correctional facility
evacuation drills, 721
occupancy, 137
Corridor door, 210
Corrosion-resistant sprinkler, 515
Corrosive
characteristics, 398
hazard classifications, 398
uses for, 398
Coupled water motor-pump proportioner, 584
Course of masonry units, 181, 182

Court proceedings, 22–23
Court rulings
  Adams vs. State of Alaska, 32
  Camara vs. City and County of San Francisco, 29
  Sea vs. City of Seattle, 29
CPSC (Consumer Product Safety Commission), 321
Criminal proceedings, 20–22
Crown corporation, 50
Cryogenic fluids
  containers for, 435
  defined, 388
  examples, 388–389
  explosion hazards, 390
  hazard classifications, 388–390
  liquid hydrogen tank, 389
  uses for, 389
CSA. *See* Canadian Standards Association (CSA)
CSB (U.S. Chemical Safety and Hazard Investigation Board), 362–363
Cul-de-sac, 288, 289
Curtain wall
  construction materials, 198
  defined, 197
  firestopping materials, 197
  installation method, 197, 198
Cut sheets, 658, 660–661
Cyclotrimethylenetriamine (RDX), 454
Cylinders
  features, 437–438
  labeling, 316
  markings, 430, 431
  valve fittings, 438, 439
  water capacity, 435

D

Dangerous goods (Canada)
  Health Canada, 422
  placards, labels, and markings, 423–427
  regulations system, 421–427
  waste material disposal, 333
  Workplace Hazardous Materials Information System, 422, 427, 428
Dangerous-when-wet materials, 391
Day care occupancies, 138
Dead load, 172–173
Deadbolt locks, 254
Dead-end access roads, 288–289
Dead-end corridor, 272–274
Decay stage of fire development, 108
Decorations, flammability of, 683
Decorative sprinkler, 516
Deflagration, 386, 453
Deflectors on sprinklers, 512
Delayed egress, 255
Deluge sprinkler system, 506, 517
Demolition sites, 291–294
Department of Defense. *See* U.S. Department of Defense (DoD)
Department of Transportation. *See* U.S. Department of Transportation (DOT)
Department store occupancy, 131
Design and Construction of Large, Welded, Low-Pressure Storage Tanks (API 620), 384, 443–444
Designation and Safety Classifications of Refrigerants (Standard 34), 231–232, 343–344
Detailed view, 669
Detectors
  automatic devices, 623. *See also* Automatic alarm-initiating devices
  flame detectors, 633–634
  smoke detectors. *See* Smoke detector
Detention facility occupancies, 137
Detonation, 386, 453
Diesel engine driver, 536, 537
Diesel motor controller, 538
Dimensional lumber, 155
Dimensioning, 665
Direct pumping system, 470, 471
Dissolved gases, 388
Distillery fire hazards, 373–374
Distribution system for water, 466–467
Distributors, 473
Documents, construction and support, 658–661. *See also* Records
DoD. *See* U.S. Department of Defense (DoD)
Doors, 204–213
  construction materials, 206–207
  corridor, 210
  elevator hoistways and doors, 234–235
  fire doors, 207–213
  flush doors, 206
  folding, 204, 205
  glass, 206–207
  hollow core, 206
  as means of egress, 251–256
    floor or landing, 252–253
    handrails, 259
    locked or blocked doors, 241, 243, 301
    locking hardware, 253–256, 257
    maximum door swing, 253
    minimum door width, 251, 252
    openings in exit stairways, 258, 259
    self-closing doors, 257
  pocket sliding, 205
  revolving, 204, 206
  sliding, 204, 205, 209
  solid core, 206
  styles, 206–207
  swinging, 204, 205
  vertical, 204, 205
Dormitory occupancies, 142
DOT. *See* U.S. Department of Transportation (DOT)
Double interlock sprinkler system, 507
Double preaction sprinkler system, 507
Double-action pull station, 636
Double-hung window, 214, 215
Drain valves, 510
Drainage systems, 456–457
Drip check or drip ball valves, 510
Drop panel, 176
Dry Chemical Fire Extinguishers (UL 299), 49
Dry cleaning operations, 361–362
Dry powder extinguishing systems, 569–571
Dry-barrel fire hydrant, 476
Dry-chemical extinguishing system, 565–569
  agents, 567–568
  application methods, 566–567
  applications for, 565–566
  components, 568–569

inspection and testing, 566, 594–595
maintenance of, 569
warning signals, 570
Dry-pipe sprinkler system, 505
Drywall. *See* Gypsum board
Duct smoke detectors, 632
Ductwork
construction of, 335
in high hazard occupancies, 458
smoke dampers, 230
Due process clauses, 19
Due process of law, 19
Dumbwaiter fire doors, 213
DuPont FE-36™, 574, 590
Dust hazards, 362–372
classifications, 367–368
combustible dust, defined, 363
control of dust, 368, 369
examples, 363, 365
explosion
conditions for occurrence of, 363
defined, 362
explosive conditions, 363, 365
primary, 363, 366
secondary, 363, 366
fire protection, 368–369
grain facilities, 369–370
inspection procedures, 371–372
manufacturing facilities, 371
process hazard analysis, 365–367
Duties, Inspector I, 14–19
Duty to inspect, 31

E

E84 *Standard Test Method for Surface Burning Characteristics of Building Materials*, 49
E108 *Standard Test Methods for Fire Tests of Roof Coverings*, 49
E119 *Standard Test Methods for Fire Tests of Building Construction*, 49, 167, 222
E-152, *Standard Methods of Fire Tests of Door Assemblies*, 224
Early suppression fast-response (ESFR) sprinkler, 514
EC (extended coverage) sprinkler, 514
ECARO-25™, 574
Educational facility evacuation drills, 718–719
Educational occupancies, 133–134
Educational opportunities from inspections, 714
Egress. *See* Means of Egress (MOE)
EIFS (Exterior Insulation and Finish Systems), 170–171
Electric motor controller, 537–538
Electric motor driver, 535–536
Electrical energy, 81–82
Electrical equipment, improper use as unsafe behavior, 311–313
Electrical hazards
arcing, 81, 320
defective or improper electrical installations, 321–322
improper use of equipment, 320–321
overloaded circuits, 320–321
worn electrical equipment, 320
Electrical systems
alterations, 312, 313
construction plans, 672, 673
electrical service panels, 233
generators, 234
purpose of, 232
switch gear, 233
transformers, 234
wiring details, 673
Electricity generated from waste incineration, 334
Electromagnetic door holders, 208–209
Electromagnetic locks, 255
Electronic air cleaners, 344, 345
Electronic heat detector for sprinkler activation, 517
Electronic spot-type heat detector, 628–629
Elevated water tanks as water supply source, 470
Elevation view, 666–668
Elevators
control schematic in system plan, 678, 679
detector activating controls, 612, 613
fire doors, 213
hoistways and doors, 234–235
as means of egress, 251
E-mail messages, 696
Emergency
defined, 29
lighting, 263
right of entry and, 29–30
switches, 232
Emergency communications systems, 621–623
Emergency planning and preparedness, 716–722
Emergency Response Guidebook (ERG)
commodity identification numbers, 404
contents, 419–420
placards and container profiles, 420
purpose of, 418
Enabling legislation
authority of inspectors, 24–25
defined, 14
local laws and ordinances, 27
Encapsulating, 323
Enclosure walls, 196–197
Encrustations in fire hydrants, 489
End suction fire pump, 534–535
Endothermic reaction, 75–76
Energy
chemical energy, 80–81
defined, 74, 78
electrical energy, 81–82
kinetic energy, 75, 78
mechanical energy, 82
potential energy, 75, 78, 79
thermal energy, 78
Engineered wood products, 155–157
Engineering controls for hazardous materials, 455–459
automatic sprinkler protection, 458–459
mechanical ventilation system, 458
spill control and secondary containment, 456–457
Environment Canada, 333
Environmental Protection Agency (EPA). *See* U.S. Environmental Protection Agency (EPA)
EPA. *See* U.S. Environmental Protection Agency (EPA)
Equipment lists for inspections, 700–701
ERG. *See Emergency Response Guidebook (ERG)*
Escalators as a means of egress, 251
ESFR (early suppression fast-response) sprinkler, 514
Ethanol in distilleries, 373
Evacuation of residential board and care occupancies, 137–138. *See also* Exits
Evacuation drills, 716–722

Evidence, inspection records, 22
Executive summary, 61
Exit access
  defined, 246
  as means of egress, 242, 243, 246
Exit capacity, systematic plans review of, 682
Exit discharge, 242, 243, 250
Exits. *See also* Evacuation
  auxiliary power, 264–265
  defined, 246
  egress capacity, 268–270
  emergency lighting, 263
  examples, 246–247
  exit discharge, 250
  exit passageway, 247–248
  horizontal exits, 248
  illumination, 262
  location of, 271
  markings, 263–264
  maximum travel distance to, 272–273, 274
  means of egress
    blocked exits, 308, 309
    capacity, 269–270
    elements, 242, 243, 246–249
  remodeling and adequate exits, 141
  required number of, 270
  smokeproof enclosures, 249
Exothermic reaction, 75
Expellant gas, 568
Expert testimony, 24
Explosion
  cryogenic fluids, 390
  defined, 362
  explosive conditions, 363, 365
  Imperial Sugar Company, Wentworth, GA (2008), 365
  Lakeland Mills, Prince George, BC (2012), 365
  pentagon, 365
  primary, 363, 366
  secondary, 363, 366
  West Pharmaceutical Services, Kinston, NC (2003), 363
Explosives
  hazard classifications, 395, 455
  hazardous/high-hazard occupancies, 453–454
Extended coverage (EC) sprinkler, 514
Extension cords, 311, 312
Exterior access barriers, 294–299
Exterior Insulation and Finish Systems (EIFS), 170–171
Extinguishers. *See* Fire extinguishers
Extinguishing agents
  clean-agent extinguishing systems, 573–574
    characteristics, 590
    ECARO-25™, 574
    FE-36™, 574, 590
    FM-200™, 573
    halon, 573
    Halotron®, 573
    Inergen®, 574
  dry-chemical extinguishing systems, 567–568
  portable fire extinguishers, 586–590
    carbon dioxide, 587–588
    dry chemical, 589
    foam, 589
    water, 586–587
    wet chemical, 590
  wet-chemical extinguishing systems, 572, 590
Extra-hazard occupancy for locating fire extinguishers, 598, 599

F

Facial expression, 694
FACP (fire alarm control panel), 608
Fact witness, 24
Factory/Industrial High Hazard Group H Occupancy classification, 148
Factory/industrial occupancies, 134, 135, 148
FACU. *See* Fire alarm control unit (FACU)
False front, 302
Fan belts in HVAC systems, 336
FDC. *See* Fire department connections (FDC)
FE-36™, 574
Federal Hazardous Substances Act (FHSA), 414
Federal Insecticide, Fungicide, and Rodenticide Act (FIFRA), 383
Federal laws for fire inspections, 25–26
Feedback, 691
Fees for plans review, 658
Felonies, 21
Felony trial, 22–23
Fence occupancy classification, 145
FFFP (film forming fluoroprotein) foam, 589
FHSA (Federal Hazardous Substances Act), 414
FIBC (flexible intermediate bulk container), 437
Fiberglass insulation, 158
Field verification, 653–654
Field-erected aboveground storage tanks, 442–446
FIFRA (Federal Insecticide, Fungicide, and Rodenticide Act), 383
Fifth Amendment, 19
Files and records of inspectors, 697–698
Film forming fluoroprotein (FFFP) foam, 589
Filtering devices on HVAC systems, 344, 345
Finger joints, 156
Finished foam, 579
Fire
  Class A fires, 96
  Class B fires, 96
  Class C fires, 98
  Class D fires, 98
  Class K fires, 98
  defined, 75
Fire alarm control panel (FACP), 608
Fire alarm control unit (FACU)
  addressable alarm systems, 617
  components, 608, 609
  conventional alarm systems, 615–616
  defined, 608
  functions of, 608
  inspection and testing, 638–639, 643
  notification appliances, 610–611
  primary power supply control, 609, 610
  smoke detector signals to, 629
  zoned conventional alarm systems, 616–617
Fire alarms. *See* Fire detection and alarm systems
Fire and life safety inspector, 14
Fire apparatus access roads. *See* Fire lanes and fire apparatus access roads
Fire barriers, 196
Fire behavior, 73–111

fire development, 98–108
decay stage, 108
factors affecting, 99–103
fully developed stage, 107–108
growth stage, 104–107
incipient stage, 103–104
interior finishes, impact on, 217
science of fire, 74–98
classification of fires, 96, 98
combustion modes, 76–77
energy, 78
forms of ignition, 79–80
fuel, 87–92, 93
heat, 78
heat transfer, 83–87
oxidizer, 92–95
passive agents, 87
physical and chemical changes, 74–76
products of combustion, 97
self-sustained chemical reaction, 95–96
temperature, 78
thermal energy, 80–82
Fire Code™. *See* NFPA® 1, *Fire Code*™
Fire control, 108–111
Fire damper, 165, 230
Fire department connections (FDC)
automatic sprinkler systems, 508
defined, 298
inspections and testing, 546, 547, 556
locking intake cap, 532
standpipe and hose systems, 524, 525, 531–532
Fire department, fire and life safety inspection programs, 15
Fire detection and alarm systems, 607–647
alarm signaling systems, 613–623
alarm-initiating devices
inspection and testing, 639–641, 643
overview, 610, 611
automatic alarm-initiating devices, 623–635
combination detectors, 634, 635
fixed-temperature heat detectors, 624–627
flame detectors, 633–634
nuisance alarms, 623
rate-of-rise heat detectors, 627–629
smoke detectors, 629–633
sprinkler waterflow alarm-initiating devices, 634, 635
components, 608–613
additional functions, 612–613
fire alarm control units, 608–609
initiating devices, 610
notification appliances, 610–612
primary power supply, 609, 610
secondary power supply, 609–610, 611
fire alarm control units, 608–609, 638–639
inspection and testing, 637–647
acceptance testing, 642–643
alarm signaling systems, 641–642, 643
alarm-initiating devices, 639–641
fire alarm control units, 638–639
occupant notification devices, 643–647
service test, 637
manually actuated alarm-initiating devices, 635–637
plans review, 677–679
Fire development, 98–108
availability and location of additional fuels, 99–101
decay stage, 108
factors affecting, 99–103
availability and location of additional fuels, 99–101
ceiling height, 101
compartment volume, 101
fuel load, 102–103
fuel type, 99
thermal properties of the compartment, 102
ventilation, 102
fully developed stage, 107–108
growth stage, 104–107
incipient stage, 103–104
Fire doors, 207–213
alarm system opening of, 612, 613
classifications, 209–210
closures, 208–209
detection devices, 207
evaluating, 223–224
frames and hardware, 210–211
glass panels and louvers, 213
horizontal sliding, 212
inspector liability for inspections, 32
open doors as factor in fire development, 102
propped open, 207, 208
purpose of, 207
rolling steel, 211–212
special types, 213
swinging, 213
testing, 224
Fire escape ladders, 223, 224
Fire escape slides, 223, 224, 261
Fire extinguishers
portable
agents, 586–590
cartridge-operated extinguisher, 590, 591
classifications, 585
effectiveness of, 584–585
inspection and maintenance, 600–602
installation and placement, 592–594
obsolete extinguishers, 592
pictorial and letter systems, 586, 587
pump-operated extinguisher, 592
ratings, 585–586
stored-pressure extinguisher, 590
systematic plans review, 683, 684
training, 600
types of, 590–592
in process industry buildings, 368
special-agent. *See* Special-agent fire extinguishing systems
Fire flow
test computations, 472–484
test results computations, 490–493
graphical analysis, 490–493
mathematical method, 493
Universal Water Flow Test Summary Sheet, 491, 492, 676
testing, 479–480
water flow vs., 477
Fire hazard recognition, 307–374
building systems, 335–352
cooking equipment, 344–347
fire suppression systems, 347–348
furnaces and ovens, 348–349
HVAC systems, 335–344
powered industrial trucks, 349–350
tents, 350–352

hazardous processes, 352–374
asphalt and tar kettles, 372–373
distilleries, 373–374
dry cleaning operations, 361–362
dust hazards, 362–372
flammable finishing operations, 356–359
quenching operations, 359–361
torch-applied roofing materials, 372
welding and thermal cutting operations, 352–356
nonresidential fires, 307
systematic plans review, 683
unsafe behaviors, 308–319
compressed/liquefied gases, 317–319
electrical equipment, 311–313
flammable and combustible liquids, 313–317
ignition sources, 308–309
inadequate housekeeping, 308, 309
open burning, 309–311
unsafe conditions, 319–334
electrical hazards, 320–322
material storage facilities, 322–331
recycling facilities, 331–334
Fire hydrants
color designations, 477, 479
discharge outlets, 477, 483
dry-barrel, 476
fire flow test computations, 482–484
fire flow test procedures, 485–490
flow hydrant, 485
inspections, 480–481
obstructions and clearance, 480, 481, 489–490
pitot tube and gauge, 480, 481–482, 483
pressure gauge, 480, 481
obstructions, 489–490
records of inspections, 477, 478
steamer connection, 477
test hydrant, 485
wet-barrel, 476
Fire lanes and fire apparatus access roads, 285–291
angle of approach and departure, 286, 287
dead-end access roads, 288–289
defined, 286
requirements, 287–288
road markings and signs, 289–291
site plans showing, 664
Fire load
building contents, 100–101
defined, 125
ventilation, 102
Fire partitions, 196
Fire point, 90, 91
Fire Prevention Officer, 30
Fire protection engineer (FPE), 33
Fire Protection Engineering for Facilities (UFC 3-600-01), 288
Fire Protection Guide to Hazardous Materials, 94–95
Fire Protection Handbook®, 49
Fire protection rating, 154
Fire protection systems
field-erected aboveground storage tanks, 445
hazardous materials, 382
labeling, 292
material storage facilities, 323
water distribution systems and, 467
Fire pumps. *See* Stationary fire pumps
Fire Safety Standard for Powered Industrial Trucks Including Type Designations, Areas of Use, Conversions, Maintenance, and Operations (NFPA® 505), 349
Fire suppression systems
for commercial cooking equipment, 347–348
dry-chemical, 348
extinguishing nozzles, 347
inspection items, 347
water-based. *See* Fire suppression systems, water-based
wet-chemical, 347
Fire suppression systems, water-based, 503–557
automatic sprinkler systems, 504–519
acceptance testing procedures, 541
detection and activation devices, 516–518
inspections and testing, 546–552
operating valves, 510
residential systems, 518–519
sprinklers, 512–516
types and design, 505–507
water distribution pipes, 510–512
water supplies, 508
waterflow control valves, 508–509
evaluate system components and equipment, 538–557
acceptance testing, 540, 541–545
inspections and testing, 540, 545–557
plans reviews, 539
foam-water systems
acceptance testing procedures, 542
overview, 523–524
proportioner, 555
private water supply system acceptance testing procedures, 545
residential systems, 518–519, 552–554
standpipe and hose systems, 524–532
acceptance testing procedures, 543
classifications, 526–527
components, 526
fire department connections, 531–532
hose valves, 530
inspections and testing, 555–557
pressure-regulating devices, 530–531
types, 527–529
water supplies and residual pressure, 529
stationary fire pumps, 532–538
acceptance testing procedures, 543–545, 557
controllers, 536–538
end suction, 534–535
horizontal split-case, 532–533
inspections and testing, 557
pressure-maintenance, 535
pump drivers, 535–536, 537
vertical inline, 534
vertical split-case, 533
vertical turbine, 534
water-mist systems
acceptance testing procedures, 542
overview, 521–523
water-spray fixed system
acceptance testing procedures, 541
inspection and testing, 554–555
purpose of, 520–521
Fire Testing of Fire Extinguishing Systems for Protection of Restaurant Cooking Areas (UL Standard 300), 347, 573
Fire tetrahedron, 76, 77

Fire triangle, 76
Fire walls
apartment buildings, 143
construction types, 194
defined, 194
freestanding, 220
opening protective, 194–195
with a parapet, 194, 195
purpose of, 220
rating, 195
Fire watch after welding or thermal cutting operations, 355–356
Fire-rated glass, 166, 167
Fire-Resistant Aboveground Storage Tanks (UL 2080), 385
Fire-resistant materials, 154
Fire-retardant coatings, 154, 228
Fire-retardant treated wood, 157
Fires
Cocoanut Grove, Boston, MA (1942), 262
factors in loss-of-life fires, 242
Hartford, Connecticut, tent fire, 350
Hotel Tap fire, 13
Imperial Sugar Company, Wentworth, GA (2008) explosion, 365
Lakeland Mills, Prince George, BC explosion (2012), 365
Madden Textile Plant, Georgia (1995), 538
Polyethylene Wax Processing Facility, 448
Station Nightclub fire, 92
West Pharmaceutical Services, Kinston, NC (2003) explosion, 363
Winecoff Hotel, Atlanta, GA (1946), 262
Fire-stop, 122, 682
Firestopping, 179
Fittings on pipes for hazardous materials, 451
Fixed foam fire extinguishing system, 578–579
Fixed system application of dry-chemical extinguishing systems, 566–567
Fixed-temperature heat detectors, 624–627
bimetallic heat detector, 625–626
color coding, 624
continuous-line heat detector, 626–627
frangible bulbs, 625
fusible links, 625
temperature classification, 624
temperature rating, 624
Flag Protection Act (1989), 310
Flame, defined, 76
Flame detectors, 633–634, 640
Flame spread ratings of interior finishes, 225–226
Flaming combustion, 76, 77
Flaming foods as open burning, 310, 311
Flammability of plastics, 169
Flammable & Combustible Liquid Code, Motor Vehicle Fuel Dispensing and Repair Garages Code (NFPA® 30 and 30A), 384
Flammable and Combustible Liquids Code. See NFPA® 30, *Flammable and Combustible Liquids Code*
Flammable and combustible liquids use and storage, 313–317
dispensing, 314–317
guidelines for safe handling, 314
hazard classifications, 386–387
Lower Flammable Limit, 387
transporting, 317
unsafe behaviors, 313
Flammable finishing operations, 356–359
Flammable liquid
defined, 90
hazard classifications, 386–387
Lower Flammable Limit, 387
Flammable range of substances, 94
Flammable solids or gases, hazard classifications, 390–391
Flash point, 90, 91
Flashover, 106–107
Flexible intermediate bulk container (FIBC), 437
Floods as access barriers, 299, 300
Floor plan
building components shown, 665
clear dimensions, 666
defined, 665
as factor in fire development, 99–100
information included, 665–666
plan review, 658, 659
symbols and codes, 666, 667
Floor proximity exit signs, 264
Floors, 200–202
construction materials, 200, 201
evaluating finishes, 222
as means of egress, 257
penetrations and openings, 202
sectional view, 668
supports, 200–201
Flow hydrant, 485
Flow path, 104
Flue maintenance, 341
Flush doors, 206
Flush sprinkler, 512, 513
FM Global
heaters listed, 340
hot work permit, 355
Loss Prevention Data Sheets, 354
safety cans, 436
FM-200™, 573
Foam
aqueous film forming foam (AFFF), 589
defined, 579, 589
effectiveness of, 589
film forming fluoroprotein (FFFP) foam, 589
Foam core building panels, 157
Foam fire extinguishing systems, 578–584
foam concentrate types, 582
foam expansion rates, 581–582
foam generation, 579–580
foam proportioning rates, 581–582
inspection and testing, 596–597
proportioners, 583–584
types of, 578–579
Foam insulation, 158
Foam-water sprinkler systems
acceptance testing procedures, 542
applications for, 523
inspection and testing, 555
proportioner, 524, 555
Foil-backed gypsum board, 168
Folding doors, 204, 205
Follow-up inspections
compliance procedures, 20
procedures, 715–716
to verify corrections, 13
Forms, inspection, 36
Foundation, sectional view of, 668

Fourteenth Amendment of the U.S. Constitution, 19
Fourth Amendment of the U.S. Constitution, 29
FPE (fire protection engineer), 33
Frame structural systems, 176–178
Frangible bulbs, on fixed-temperature heat detectors, 625
Free radicals, 95–96
Freedom of Information Act, 695
Freight elevator fire doors, 213
Freon®, 343
Fuel
  consumption of, and decay stage of fire development, 108
  defined, 74, 87
  fire control and removal of, 110
  fire development and type of, 99
  gaseous fuel, 89–90
  heat of combustion, 87
  heat release rate, 87–89
  liquid fuel, 90–91
  reducing agent, 87
  shutoffs, 671, 672
  solid fuel, 91–92, 93
Fuel load
  defined, 102
  as factor in fire development, 102–103
Fuel-controlled fires, 105
Fully developed stage of fire development, 107–108
Furnaces
  fire hazards, 337
  floor furnace, 337, 338
  forced-air, 336–337
  gravity, 336
  industrial, 348–349
  wall furnace, 338
  warm-air, 336, 337
Furnishings, flammability of, 683
Fuse boxes, 233
Fusible link
  defined, 515
  dry-chemical extinguishing systems, 568, 569
  on fixed-temperature heat detectors, 625
  inspection and testing, 639
  for sprinkler activation, 516
  upright sprinkler with, 514

G

Gas detection systems, 232, 640
Gaseous fuel, 89–90
Gasoline containers, 436
Gate valve, 474
Generators
  auxiliary power for exit signs, 264–265
  purpose of, 234
  uses for, 235
Gestures as nonverbal communications, 694
Glass block, 166
Glass bulbs for sprinkler activation, 516
Glass doors, 206–207
Glass fire door panels and louvers, 213
Glass wool insulation, 158, 159
Glass-fiber reinforced underground storage tanks, 446
Glass-Fiber-Reinforced Plastic Underground Storage Tanks for Petroleum Products, Alcohols, and Alcohol-Gasoline Mixtures (UL 1316), 384, 446
Glazing (glass)
  for enclosure wall construction, 197
  fire-rated, 166, 167
  types, 166
Glued-laminated beam, 122, 123, 155
Grain facility dust hazards, 369–370
Graphical analysis of fire flow test results, 490–493
Gravity system as means of moving water, 470, 471
Gravity tanks as water supply source, 470
Grease ducts
  evaluating, 236
  extractors, 236
  fire suppression system, 218
  inspection, 219, 236
  purpose of, 218
Grease traps, 345
Gridded sprinkler system, 507
Grounding dispensing stations, 315–317
Group B Business Occupancies, 137
Group B Care of Detention Occupancies, 136, 137
Group C Residential Occupancies, 137, 142
Group E combustible dust, 367
Group E Educational occupancies, 138, 147, 148
Group F combustible dust, 367
Group F Factory/Industrial Occupancies, 134, 135
Group G combustible dust, 367
Group H Hazardous classification, 144
Group H High Hazard occupancies, 147, 148
Group H-1/High Hazard Level 1 occupancy, 452, 454
Group H-2/High Hazard Level 2 occupancy, 452
Group H-3/High Hazard Level 3 occupancy, 452–453
Group H-4/High Hazard Level 4 occupancy, 453
Group H/High-Hazard Level occupancy, 455–459
Group I Institutional occupancy, 136
Group I Institutional Subdivision I-2 (medical) occupancy, 147
Group I Institutional Subdivision I-3 occupancy, 137
Group I Institutional Subdivision I-4 occupancy, 138
Group R Residential Occupancies, 140, 142
Group S Storage classification, 144
Growth stage of fire development, 104–107
Guide for Fire and Explosion Investigations. See NFPA® 921, *Guide for Fire and Explosion Investigations*
Guide on Alternative Approaches to Life Safety (NFPA® 101A), 138
Guide to Preparation of Precautionary Labeling and Marking of Compressed Gas Containers (CGA C-7), 384, 430
Gypsum block, 162
Gypsum board
  for ceilings, 203
  for enclosure wall construction, 196–197
  foil-backed, 168
  passive agents, 87
  regular gypsum board, 167
  Type C, 168
  Type X, 167, 168
  uses for, 167
  water-resistant, 168

H

Halocarbon agents, 573
Halon 1211 (bromochlorodifluoromethane), 573, 590
Halon 1301 (bromotrifluoromethane), 573, 590
Halotron®, 573
Hammerhead turnaround, 288, 289
Handheld hoseline application of dry-chemical extinguishing

systems, 567
Handrails on exit stairways, 258, 259
Hartford, Connecticut, tent fire, 350
Hazard communications (HAZCOM), 421
Hazard Communications Standard (HCS), 411, 412
Hazard identification signs, refrigeration mechanical room, 232
Hazardous Material Identification Guide (HMIG), 427
Hazardous materials, 379–459
classification of, 385–401
compressed and liquefied gases, 388
corrosives, 398
cryogenic fluids, 388–390
explosives and blasting agents, 395
flammable and combustible liquids, 386–387
flammable solids or gases, 390–391
health hazard materials, 396–401
highly toxic and toxic materials, 396–397
incompatible materials, 399–401
mixtures, 399
organic peroxides, 391–392
overview, 385–386
oxidizers and oxidizing gases, 392–393
physical hazard materials, 386–395
pyrophorics, 391, 393–394
unstable (reactive) materials, 394, 395
water-reactive materials, 394–395
engineering controls, 455–459
automatic sprinkler protection, 458–459
mechanical ventilation system, 458
spill control and secondary containment, 456–457
hazardous/high-hazard occupancies, 452–455
identification of, 401–430
Canadian dangerous goods system, 421–427, 428
cylinder markings, 430, 431
manufacturers' warnings, 412, 414
Mexican hazard communication system, 427, 429–430
military markings, 413, 414, 415–416
NFPA® 704 system, 413, 416–418
pipeline markings, 413, 414, 416, 417
piping identification, 430, 431
resource guidebooks, 418–421
safety data sheets, 402, 403
transportation placards, labels, and markings, 402, 404–411, 412–413
nonbulk and bulk packaging, 435–448
containers, 435–438
cylinders, 437–438
field-erected aboveground storage tanks, 442–446
intermediate bulk containers, 437
pressure vessels, 446–448
safety cans, 436
shop-fabricated aboveground storage tanks, 440–442
stationary tanks, 438–440, 441
underground storage tanks, 446
permissible amounts in buildings, 430–434
control areas, 434
maximum allowable quantity per control area, 431–433
piping, valves, and fittings, 451
process control, 449–450
regulations, application of, 379–385
exemptions, 382–385
fire protection, 382
pressure relief, 381–382
product containment, 381
testing, maintenance, and operations, 449
unauthorized discharge, 450
Hazardous Materials Code. See NFPA® 400, *Hazardous Materials Code*
Hazardous Materials Guide for First Responders, 421
Hazardous Materials Information Resource System (HMIRS), 421
Hazardous Materials Inventory Statement, 380
Hazardous Materials Management Plan, 380
Hazardous processes, 352–374
asphalt and tar kettles, 372–373
distilleries, 373–374
dry cleaning operations, 361–362
dust hazards, 362–372
classifications, 367–368
combustible dust, defined, 363
control of dust, 368, 369
explosive conditions, 363
fire protection, 368–369
grain facilities, 369–370
groups, 367
inspection procedures, 371–372
machine shops, 371
manufacturing facilities, 371
primary explosion, 363, 366
process hazard analysis, 365–367
secondary explosion, 363, 366
woodworking and processing facilities, 371
flammable finishing operations, 356–359
quenching operations, 359–361
torch-applied roofing materials, 372
welding and thermal cutting operations, 352–356
combustible material clearance, 354
fire safety issues, 353
fire watch, 355–356
hot work program, 354–355
permits, 355
rag or towel storage, 353, 354
situations not allowed, 353
Hazardous Products Act, 422
Hazardous/high-hazard occupancies, 452–455
HAZCOM (hazard communications), 421
HCFC (hydrochlorofluorocarbon), 573
HCS (Hazard Communications Standard), 411, 412
Head pressure, 533
Header course, 181, 182
Health care and ambulatory health care occupancies, 136–137
Health care facility evacuation drills, 719–721
Health hazard materials, 396–401
corrosives, 398
highly toxic and toxic materials, 396–397
incompatible materials, 399–401
mixtures, 399
Healthcare occupancy, 126
Heat detectors
fixed-temperature, 624–627
rate-of-rise, 627–629
Heat flux, 83
Heat of combustion, 87
Heat reflectivity and thermal properties of the compartment, 102
Heat release rate (HRR)
defined, 87
fire development and fuel type, 99

measure of, 88, 89
reducing agent, 87
Heat transfer, 83–87
conduction, 83–84
convection, 83, 84, 85
radiation, 83, 84–87
Heaters
baseboard heaters, 341
room heaters, 340–341
temporary/portable, 342–343
unit heaters, 340, 341
Heating, ventilating, and air conditioning (HVAC) system
components, 229, 335
duct smoke detectors, 632
ductwork, 230
evaluating, 229–231
fire hazard recognition, 335–344
air conditioning systems, 343–344
boilers, 338–339
fan belts, 336
filtering devices, 344
furnaces, 336–338
room heaters, 340–342
temporary/portable heating equipment, 342–343
unit heaters, 340
ventilation systems, 344
fire-resistant fire walls, 195
mechanical systems plans, 670–671
NFPA® standards, 671
passive smoke control, 230
purpose of, 229
return-air plenum, 223
smoke detectors, 230
smoke or fire dampers, 230
ventilation, 102
Heat-strengthened glazing, 166
Heavy-timber construction, 122, 125, 181
HFC (Hydrofluorocarbon), 573
Hidden sprinkler, 512, 513
High-expansion foam, 581, 582
Highly toxic materials hazard classifications, 396–397
High-Piled Combustible Storage Application Guide, 323
High-rise structures
active smoke control systems, 230–231
building codes during renovations, 45
defined, 532
fire pumps installed in, 532
Highway Tanks and Portable Tanks for the Transportation of Dangerous Goods (CSA B620), 436
HMIG (*Hazardous Material Identification Guide)*, 427
HMIRS (Hazardous Materials Information Resource System), 421
Horizontal exits, 248
Horizontal sliding fire doors, 212
Horizontal sliding window, 214, 215
Horizontal split-case fire pump, 532–533
Hot work. *See also* Welding and thermal cutting operations
defined, 352
hot work program, 354–355
permits, 355
Hotel occupancies, 141–142, 721–722
Hotel Tap fire, 13
Hourly fire protection rating, 209
Housekeeping unsafe behaviors, 308, 309
HRR. *See* Heat release rate (HRR)
Human Resources and Skills Development, Canada, 422
HVAC systems. *See* Heating, ventilating, and air conditioning (HVAC) system
Hydrants. *See* Fire hydrants
Hydrochlorofluorocarbon (HCFC), 573
Hydrofluorocarbon (HFC), 573
Hydrogen chloride, 169
Hydrostatic test, 595

I

IBC (Intermediate Bulk Container), 437, 439
IBC®. See International Building Code® (IBC®)
ICC. *See* International Code Council® (ICC®)
ICC Electrical Code™, 321
Identification of hazardous materials, 401–430
Canadian dangerous goods system, 421–427, 428
cylinder markings, 430, 431
manufacturers' warnings, 412, 414
Mexican hazard communication system, 427, 429–430
military markings, 413, 414, 415–416
NFPA® 704 system, 413, 416–418
pipeline markings, 413, 414, 416, 417
piping identification, 430, 431
resource guidebooks, 418–421
safety data sheets, 402, 403
transportation placards, labels, and markings, 402, 404–411, 412–413
IFC. See International Fire Code® (IFC®)
Ignition
autoignition (nonpiloted), 79–80
defined, 79
piloted, 79–80
Ignition sources
defined, 308
examples of, 308
sparks, 308, 309
static electricity, 317
unsafe behaviors, 309
Ignition temperature, 320
Ignition-resistant construction, 154
I-joists, 124–125
Illegal parking as access barrier, 296–297
IMC. *See International Mechanical Code* (IMC)
Immersion coating, 356, 358–359
Imminent hazard, 18
Imperial Sugar Company, Wentworth, GA, explosion (2008), 365
Inadequate housekeeping as unsafe behavior, 308, 309
Incidental use occupancies, 147
Incinerators, 333–334
Incipient stage of fire development, 103–104
Incompatible materials
chemical compatibility matrix, 400
defined, 399
hazard classifications, 398–401
separation of, 401
Indemnification, 32–33
Indicating control valve, 474, 475, 546, 547
Indict, defined, 21
Industrial occupancies, 134, 135, 148
Inergen®, 574
Infrared wave spectrum (IR) detectors, 633, 634
Initiating devices for fire detection systems, 610, 611
Inline balanced-pressure proportioner, 555

Insecticide regulations, 383
Inspection procedures, 689–727
administrative duties, 695–698
building occupancy changes, 712, 713
checklists, 708, 711–712
code enforcement, 689
code requirements, 707, 709–711
complaint management, 722–725
emergency planning and preparedness, 716–722
correctional facilities, 721
educational facilities, 718–719
evacuation drills, 716, 717
health care facilities, 719–721
hotels and motels, 721–722
emergency planning evaluation, 725–727
follow-up inspections, 715–716
general practices, 705–707, 708
guidelines for inspections, 703–705
interpersonal communication, 690–695
preparation for inspections, 698–702
equipment lists, 700–701
personal appearance, 700
records review, 702
scheduling, 701–702, 703
results interview, 712–714
educational opportunities, 714
long-term relationships, 714
violation discussions, 713–714
Inspections
categories, 18–19
checklist, 654
complaints, 51–52
legal guidelines, 19–24
material storage facilities, 325–327
Inspector
defined, 14
inspection categories, 18–19
legal status, 27–28
private fire safety, 16–17
professional development, 17–18
third-party, 16
training and certification, 17
Inspector I
administrative duties, 695–698
building components knowledge, 194–219
codes, 41–47
complaint management, 722–724
complaint procedures, 51–52
construction materials, 153–171
construction types, 116–125
duties, 14–19
emergency planning and preparedness, 716–722
fire detection and alarm systems, 608–642
alarm signaling systems, 613–623
automatic devices, 623–635
components, 608–613
inspection and testing, 637–642
manually actuated devices, 635–637
fire development stages, knowledge of, 98–108
fire hazard recognition, 307–334
follow-up inspections, 715–716
hazardous materials, 379–449
application of regulations, 379–385
classification of, 385–401
identification of, 401–430
nonbulk and bulk packaging, 435–448
permissible amounts in buildings, 430–434
testing, maintenance, and operations, 449
inspection preparation, 698–702
inspection procedures, 702–715
interpersonal communication, 690–695
legal guidelines for inspections, 19–24
means of egress, 241–267
occupancy classifications, 125–145
permits, 52–55
plans review, 651–655
portable fire extinguishers, 594–602
science of fire, knowledge of, 74–98
site access, 285–302
special-agent fire extinguishing systems, 564–594
standards, 41–42, 47–50
water supply systems, 465–493
water-based fire suppression systems, 503–538
Inspector II
appeals, 64–65
authority, 24–33
building components
fire doors, 223–224
fire walls, 220
knowledge required, 220–236
rooftop photovoltaic systems, 221
code development, 56–62
code modification and appeals, 62–65
code requirements, 47
complaint procedures, 724–725
emergency planning evaluation, 725–727
fire control, knowledge of, 108–111
fire detection and alarm systems, 642–647
fire extinguisher inspection and maintenance, 594–602
fire hazard recognition, 335–374
hazardous materials, 449–459
engineering controls, 455–459
hazardous/high-hazard occupancies, 452–455
piping, valves, and fittings, 451
process control, 449–450
unauthorized discharge, 450
means of egress, 268–275
multiple-use occupancies, 146–148, 149
permits, 65–67
plans review, 656–684
policy recommendations and modifications, 34–36
special-agent fire extinguishing systems, 564–587
structural systems, building, 171–188
water supply systems, 465–493
water-based fire suppression systems, 538–557
Institutional occupancies
Canadian codes, 134, 136
day care, 138
detention and correctional facilities, 137
health care and ambulatory health care, 136–137
residential board and care, 137–138
sprinklers, 515
Insulation
conduction characteristics, 84
Exterior Insulation and Finish Systems, 170–171
fiberglass, 158
foam, 158
loose-fill material, 159

noncombustible materials, 158
shell and membrane structures, 175
systematic plans review, 683
thermal barriers, 170, 175
thermal properties of the compartment, 102
Insurance company inspection department, 17
Insurance rating bureau, 17
Interference of communications, 691
Interior access barriers
interior access, 299–302
rapid entry systems, 300
Interior finishes
approved materials, 216
ceilings, 223
combustibility and fire behavior, 217
defined, 216
evaluating, 225–228
ceiling finishes, 227
fire-retardant coatings, 228
flame spread ratings, 225–226
interior wall finishes, 227
smoke-developed index, 226
toxicity of materials, 226
as factor in fire development, 101
floor finishes, 222
stairs, 223, 224
Intermediate Bulk Container (IBC), 437, 439
Intermediate level sprinkler, 515
International Building Code® (IBC®)
application of codes, 44
building element fire-resistance rating requirements, 118
CO detector requirements, 342
construction types, 117
dual hazardous occupancy classification, 453
exterior wall fire-resistance rating requirements, 119
hazardous materials storage, handling, and use, 384
hazardous/high-hazard occupancies, 452, 454
high-rise buildings, 532
live loads, 173
model codes, 43
physical hazard materials, 386
wood building materials, 154
International Code Council® (ICC®)
accessory occupancies, 148
assembly occupancies, 132
business occupancies, 133
codes and standards, 42
construction types, 116
day care occupancies, 138
detention and correctional occupancies, 137
dormitory occupancies, 142
educational occupancies, 133, 134
exit capacity, 270
exit requirements, 92
fire lanes and fire apparatus access roads, 286
health care and ambulatory health care occupancies, 136, 137
High-Piled Combustible Storage Application Guide, 323
hotel occupancies, 142
IBC. See International Building Code® (IBC®)
ICC Electrical Code™, 321
ICC/ANSI 117.1, Standard on Accessible and Usable Buildings and Facilities, 245
IFC®. See International Fire Code® (IFC®)
IMC®. *See International Mechanical Code* (IMC)
incidental use occupancies, 147, 148
industrial occupancies, 134
institutional occupancies, 134
lodging (boarding) or rooming house occupancies, 140
model codes, 43
multiple-use occupancies, 147
nonseparated occupancies, 148
occupancy category comparison, 127–130
residential board and care occupancies, 137
separated occupancies, 148, 149
storage occupancies, 144
training for inspectors, 17
utility/miscellaneous occupancies, 145
International Fire Code® (IFC®)
CO detector requirements, 342
construction permit, 52, 55
emergency communication systems, 622
exemptions, 382–385
explosives and blasting agents, 395
fire lanes and fire apparatus access roads, 286
fire protection systems, 382
fire watch, 356
flammable gases, 391
hazardous materials, 380
hazardous materials classification, 385
hazardous materials storage, handling, and use, 384
lodging (boarding) or rooming house occupancies, 140
maximum allowable quantity per control area, 431
military bases, 26
operational permit, 52–55
performance-based design, 659
piping systems, 451
prescriptive/model codes, 43
process control, 449
pyrophorics, 393, 394
Section 604.5, auxiliary power testing, 265
spill control, 456
standpipe requirements, 556
unauthorized discharge, 450
International Mechanical Code (IMC)
hazardous materials storage, handling, and use, 384
hazardous mechanical exhaust systems, 458
HVAC standards, 671
refrigerant classifications, 231
International Organization for Standardization (ISO)
hydrant inspection guidelines, 477
ISO-3864, "Safety Colors and Safety Signs," 429
safety symbols, 429
Interpersonal communication, 690–695
Interstitial space, 223
Intrinsically safe equipment, 312
Intumescent coatings, 157
Ionization smoke detector, 631–632, 633
IR (infrared wave spectrum) detectors, 633, 634
IRC P 2904 residential sprinkler system standards, 518
ISO. *See* International Organization for Standardization (ISO)
Isolated flames, 106

J
Jalousie window, 215
Job aids, 36
Joists
  I-joists, 124–125
  open web bar joist, 185
Joules, 78, 88

K
Kilojoules, 88
Kilowatts (kW), 88, 89
Kinetic energy, 75, 78
Knox rapid entry system, 300

L
Lab Safety Supply Inc., 427
Labels
  Canadian transportation, 423–427
  cylinders, 316
  fire protection systems, 292
  manufacturers' warnings, 412
  Mexican, 429
  piping, 430, 431
  on tanks, 381
  U.S. DOT, 405, 408–409, 411
Ladders, fire escape, 261
Lakeland Mills, Prince George, BC explosion (2012), 365
Laminated glazing, 166
Laminated members, 155–156
Landscaping as access barrier, 298
Large-diameter hose (LDH), 532
Lath and plaster wall construction, 168
Laws
  federal laws, 25–26
  local laws and ordinances, 27
  minimum/maximum laws, 26
  state/provincial laws, 26
Lay testimony, 24
LC_{50}, 397
LD_{50}, 397
LDH (large-diameter hose), 532
Legal guidelines for inspections, 19–24
  civil proceedings, 20
  compliance procedures, 19–20
  court proceedings, 22–23
  criminal proceedings, 20–22
  due process, 19
  testimony, 23–24
Legal review of new codes, 59
Legislation
  enabling legislation, 14, 24–25, 27
  local code adoption work session, 61
LEL (lower explosive limit), 94
Letters from inspectors, 696, 714–715
LFL (Lower Flammable Limit), 94, 387
Liability of inspectors, 29–33
  civil rights, 31–32
  defined, 29
  duty to inspect, 31
  indemnification, 32–33
  malfeasance, 33
  negligence, 33
  right of entry, 29–30
  warrants, 30–31
Life Safety Code®. *See* NFPA 101 *Life Safety Code®*
Light-hazard occupancy for locating fire extinguishers, 597, 598, 599
Lighting, emergency, 263
Line proportioner, 555
Linen chutes, 218, 219, 235
Liquefied compressed gases, 388
Liquefied gases. *See* Compressed/liquefied gases
Liquefied Petroleum Gas Code (NFPA® 58), 384
Liquid fuel, 90–91
Listening skills, 692–693
LITH-X®, 571
Live load, 172–174
Load-bearing wall, 117
Loads, 172–174
Local alarm system inspection and testing, 641
Local codes
  adoption of, 59, 61–62
    formal resolution preparation, 59, 61
    introducing the new code, 61–62
    sunset provisions, 61
    work session for discussion, 61
  development of, 56–62
    amendments, 56, 59, 60
    drafting the proposed code, 58–59
    formally adopting the code, 59, 61–62
    problem identification, 56–57
    stakeholder identification, 57
    submit code for legal review, 59
    task force, 57–58
Local laws and ordinances, 27
Lock boxes as access barrier, 300
Lockdowns at schools, 719
Lodging (boarding) house occupancies, 140–141
Looped sprinkler system, 507
Loss Prevention Data Sheets, 354
Lot dimensions, 663
Lower explosive limit (LEL), 94
Lower Flammable Limit (LFL), 94, 387
Low-expansion foam, 581–582
Lumber building materials, 154. *See also* Wood construction materials
Lumberyards
  factors in fire growth, 329
  fuel loads, 328
  open-yard storage, 329–330

M
Machine shop dust hazards, 371
Madden Textile Plant, Georgia (1995), 538
Magnesium phosphide, 391
Main drain for sprinkler systems, 548, 549
Malfeasance, 33
Mall occupancy
  fire load, 126
  occupancy changes, 713
Manual emergency switches, 232
Manual-dry standpipe and hose system, 529
Manually activated sprinkler system, 518
Manually actuated alarm-initiating devices, 635–637
Manual-wet standpipe and hose systems, 529
Manufactured wood products, 155
Manufacturers' cut sheets, 658, 660–661
Manufacturers' product submittal information sheets, 658, 660–661

Manufacturers' warnings, 412, 414
Manufacturing facility dust hazards, 371
MAQ (maximum allowable quantity) per control area, 431–434
Markings
  Canadian transportation, 423–427
  exit, 263–264
  illegal shipments, 411
  manufacturers' warnings, 412, 414
  Mexican, 429
  military services, 413, 414
  NFPA® 704 System, 413, 416–418
  pipeline companies, 413, 414, 416, 417
  pressure vessels, 447–448
  U.S. DOT, 405, 409, 411, 412–413
Masonry
  building materials, 160–162
  floor, 201
  structural systems, 181–184
Mass notification system (MNS), 614, 622, 623
Master switches, 233
Material safety data sheet (MSDS), 402. *See also* Safety data sheet (SDS)
Material storage facilities, 322–331
  big-box stores, 322, 323
  changes in inventories and location, 322
  fire protection systems, 323
  hazardous conditions, 322
  inspection guidelines, 325–327
  lumberyards, 328–330
  nonbulk storage containers, 327
  pallet storage, 323–324, 330–331
  rack storage, 324, 325
  solid piling, 324–325
  tires, 322, 330
  warehouses, 327–328
Mathematical method to determine fire flow, 493
Matter, defined, 74
Maximum allowable quantity (MAQ) per control area, 431–434
Maximum standards for inspections, 26
Means of Egress (MOE), 241–275
  areas of refuge, 245
  arrangement, 270–273
    common path of travel, 272, 273, 274
    dead-end corridor, 272, 273, 274
    exit location, 271
    maximum travel distance to an exit, 272–273, 274
  building components, 251–261
    ceilings, 257
    doors, 251–256, 257
    fire escape stairs, ladders, and slides, 260–261
    floors, 257
    ramps, 260, 261
    stairs, 258–260
    walls, 256–257
  components, 244
  defined, 242
  determinations, 268–275
    arrangement, 270–273, 274
    capacity, 268–270
    effectiveness, 273, 275
    exit capacity, 269–270
    required number of exits, 270
    width per occupant served, 269
  doors, 205–206
  elements
    exit, 242, 243, 246–249
    exit access, 242, 243, 246
    exit discharge, 242, 243, 250
  exit illumination and markings, 262–265
  factors in loss-of-life fires, 241–242
  locked exits, 241, 243
  occupant loads, 265–268
  passage areas, 244
  performance-based design during remodeling, 44
  stairs, 223
  systematic plans review, 681
  termination areas, 244
Mechanical energy, 82
Mechanical systems, 670–672
Mechanical ventilation systems for high hazard occupancies, 458
Medium of communications, 690
Medium-expansion foam, 581–582
Megawatts (MW), 87, 88, 89
Membrane ceilings, 165
Membrane structures, 174–175
Memorandum of Understanding (MOU), 19
Memos, 696
Mercantile occupancies, 138–139
Message of communications, 690
Metal doors, 207
Metal roof coverings, 199, 200
MET-L-X®, 571
Mexican hazard communication system, 427, 429–430
Mexican Regulation for the Land Transport of Hazardous Materials and Wastes, 427
Military base inspection standards, 26
Military markings, 413, 414, 415–416
Minimum standards for inspections, 26
Miscible materials, 91
Misdemeanors, 21
Misfeasance, 33
Mitigate risks, 16
Mixed occupancy, 131
Mixed-use occupancies, 147–149
Mixture of chemicals, 399
MNS (mass notification system), 614, 622, 623
Mobile foam fire extinguishing system, 579
Model code
  activities requiring permits, 53–55
  AHJ adoption of, 43
  defined, 43
  life cycle and time line, 46
  occupancy classifications, 125, 127–131
  organizations, 43
Modification of codes, 62–65
MOE. *See* Means of Egress (MOE)
Monitoring evacuations, 726
Monoammonium phosphate extinguishing agent, 568
Montreal Protocol, 573
Motel evacuation drills, 721–722
Motor Vehicle Safety Act, 50
MOU (Memorandum of Understanding), 19
Moving walkways as a means of egress, 251
MSDS (material safety data sheet), 402. *See also* Safety data sheet (SDS)
Multiagency meetings, 47
Multicycle sprinkler system, 507

Multiple-use occupancies, 146–149
Multiplexing systems, 620
Multiuse occupancy, 268

N

NAFTA (North American Free Trade Agreement), 402
Nameplates on storage tanks, 440, 442, 447–448
National Building Code of Canada (NBC)
  occupancy category comparison, 127–130
  prescriptive/model codes, 43
  smoke control in buildings, 231
  types of building construction, 125
National Electrical Code®. See NFPA® 70, *National Electrical Code®*
National Electrical Manufacturers Association (NEMA), 535
National Fire Academy (NFA), 17
National Fire Alarm and Signaling Code®. See NFPA® 72, *National Fire Alarm and Signaling Code®*
National Fire Code of Canada (NFC), 30, 43
National Fire Protection Association® (NFPA®). *See also specific NFPA*
  assembly occupancies, 132
  codes and standards, 41–42
  consensus standards, 48
  construction types, 116
  cross-reference with ANSI and OSHA, 50
  educational occupancies, 134
  exit illumination, 262
  exit requirements, 92
  fire death rates in sprinkler-equipped buildings, 504
  Fire Protection Guide to Hazardous Materials, 94–95
  handbooks, 49
  industrial occupancies, 134
  military base inspection standards, 26
  model codes, 43
  occupancy category comparison, 127–130
  purpose of, 48
  residential occupancies, 139
  storage occupancies, 144
  training for inspectors, 17
National Fire Sprinkler Association (NFSA), 504
National Fuel Gas Code (ANSI Z223.1), 671
National Fuel Gas Code (NFPA® 54), 671
National Institute for Standards and Technology (NIST), 73, 217
National Institute of Occupational Safety and Health (NIOSH)
  NIOSH Pocket Guide to Chemical Hazards, 420–421
  threshold limit values, 397
National Oceanic and Atmospheric Administration (NOAA), 399, 457
Natural gas boilers, 339
NA-X®, 570
NBC. See National Building Code of Canada (NBC)
Negligence, 33
NEMA (National Electrical Manufacturers Association), 535
Neutral plane, 105
NFA (U.S. National Fire Academy), 17
NFC (National Fire Code of Canada), 30, 43
NFPA®. *See* National Fire Protection Association® (NFPA®)
NFPA® 1, *Fire Code*™
  detention and correctional occupancies, 137
  fire lanes and fire apparatus access roads, 286
  fire watch after welding or thermal cutting operations, 356
  handbook for inspectors, 49
  hazardous materials, 380
  hazardous materials classification, 385
  hazardous materials storage, handling, and use, 384
  health care and ambulatory health care occupancies, 136
  heaters, 340
  inspector familiarity with codes, 48
  institutional occupancies, 134
  lodging (boarding) or rooming house occupancies, 140
  model codes, 43
  occupancy classifications, 131
  performance-based design, 659
  permits, 52
  pyrophorics, 394
  residential board and care occupancies, 137
NFPA® 10, *Standard for Portable Fire Extinguishers*
  depth of flammable liquids, 599
  extinguisher size requirements, 599
  extinguishers for flammable solids, 391
  fire distance requirements, 600
  inspection and maintenance, 600, 602
  installation and placement, 585
  locating or distributing fire extinguishers, 597
  pictorial and letter systems, 586
  process industry fire extinguishers, 368
  replacing obsolete extinguishers, 592
NFPA® 11, *Standard for Low-, Medium-, and High-Expansion Foam*, 578, 582
NFPA® 12, *Standard on Carbon Dioxide Extinguishing Systems*, 574
NFPA® 12A, *Standard on Halon 1301 Fire Extinguishing Systems*, 596
NFPA® 13, *Standard for the Installation of Sprinkler Systems*
  acceptance testing procedures, 541, 542
  automatic sprinkler systems, 504
  building code requirements, 44
  handbooks to interpret the standards, 49
  hazardous/high-hazard occupancies, 454, 459
  inspector familiarity with codes, 48
  maximum number of sprinklers for pipes, 511
  pallet storage, 330
  rack storage, 324
  sprinkler area requirements, 433
  sprinkler types, 513–515
  types of automatic sprinkler systems, 505–507
NFPA® 13D, *Standard for the Installation of Sprinkler Systems in One- and Two-Family Dwellings and Manufactured Homes*
  plumbing system plans, 672–673
  property conservation and life safety concerns, 518
  residential (RES) sprinklers, 515
  water supply requirements, 552–553
NFPA® 13R, *Standard for the Installation of Sprinkler Systems in Low-Rise Residential Occupancies*
  control valves in water distribution systems, 519
  inspections and testing, 553
  plumbing system plans, 673
  property conservation and life safety concerns, 518
  residential (RES) sprinklers, 515
NFPA® 14, *Standard for the Installation of Standpipe and Hose Systems*
  acceptance testing procedures, 543
  classifications of systems, 526–527
  inspector familiarity with codes, 48
  pressure-regulating devices, 530
  residual pressure, 529
  types of systems, 527–529

NFPA® 15, *Standard for Water-spray Fixed Systems for Fire Protection*, 521, 541
NFPA® 16, *Standard for the Installation of Foam-Water Sprinkler and Foam-Water-Spray Systems*, 523, 542
NFPA® 17, *Standard for Dry Chemical Extinguishing Systems*, 566, 568, 594
NFPA® 17A, *Standard for Wet Chemical Extinguishing Systems*, 347, 571, 595
NFPA® 20, *Standard for the Installation of Stationary Pumps for Fire Protection*
acceptance testing procedures, 541, 543, 544
fire pump controllers, 536–537
pressure-maintenance pumps, 535
requirements, 532
NFPA® 22, *Standard for Water Tanks for Private Fire Protection*, 545
NFPA® 24, *Standard for the Installation of Private Fire Service Mains and Their Appurtenances*, 541, 545
NFPA® 25, *Standard for the Inspection, Testing, and Maintenance of Water-Based Fire Protection Systems*
foam systems, 596
foam-water proportioner testing, 524
inspector familiarity with codes, 48
lists of requirements, 538
main drain tests, 548
private hydrants, 477
requirements, 546
standpipe and hose system pressure-regulating devices, 531
standpipe and hose systems, 525
warehouses, 328
water-mist system testing, 523
water-spray fixed system testing, 521
water-spray fixed systems, 554
NFPA® 30 and 30A, *Flammable & Combustible Liquid Code, Motor Vehicle Fuel Dispensing and Repair Garages Code*, 384
NFPA® 30, *Flammable and Combustible Liquids Code*
aboveground storage tanks, 440
fuel oil tanks, 401
hazardous materials storage and use, 435
Intermediate Bulk Containers, 437
sprinkler systems, 459
unauthorized discharge, 450
vents, 440, 442, 458, 459
NFPA® 32, *Standard for Drycleaning Plants*, 362
NFPA® 33, *Standard for Spray Application Using Flammable or Combustible Material*, 358
NFPA® 51, *Standard for the Design and Installation of Oxygen-Fuel Gas Systems for Welding, Cutting, and Allied Processes*, 354
NFPA® 51B, *Standard for Fire Prevention During Welding, Cutting, and Other Hot Work*, 354, 356
NFPA® 54, *National Fuel Gas Code*, 671
NFPA® 55, *Compressed Gases and Cryogenic Fluids Code*
flammable gases, 391
multiple hazards, 385
pyrophorics, 394
storage, handling and use standards, 384
NFPA® 58, *Liquefied Petroleum Gas Code*, 384
NFPA® 61, *Standard for the Prevention of Fires and Dust Explosions in Agricultural and Food Processing Facilities*, 369
NFPA® 68, *Standard on Explosion Protection by Deflagration Venting*, 368
NFPA® 69, *Standard on Explosion Prevention Systems*, 368
NFPA® 70, *National Electrical Code®*
combustible dusts, 367–368
distillery electrical systems, 374
dust controls, 368
electrical equipment used in explosive areas, 312, 313
electrical installation requirements, 321
electrical system construction plans, 672
equipment in flammable finishing operations, 358
fire detection and alarm system design and installation, 608
inspector familiarity with codes, 48, 49
NFPA® 72, *National Fire Alarm and Signaling Code®*
acceptance testing, 643
automatic alarm-initiating device inspections, 639
central station system, 618
heat detector color coding and marking, 624
inspector familiarity with codes, 48, 49
local alarm system testing, 641
mass notification system, 622
occupant notification devices, 644
parallel telephone system, 620
plans review, 677–678
pull stations, 636
requirements, 608, 614
secondary power supply, 609–610
shunt systems, 621
supervising station alarm systems, 618
two-way communication systems, 622
visual notification, 645, 646
NFPA® 77, *Recommended Practice on Static Electricity*, 360
NFPA® 80, *Standard for Fire Doors and Other Opening Protectives*, 147
NFPA® 82, *Standard on Incinerators and Waste and Linen Handling Systems and Equipment*, 334
NFPA® 86, *Standard for Ovens and Furnaces*, 348
NFPA® 90A, *Standard for the Installation of Air-Conditioning and Ventilating Systems*, 671
NFPA® 90B, *Standard for the Installation of Warm Air Heating and Air-Conditioning Systems*, 671
NFPA® 91, *Standard for Exhaust Systems for Air Conveying of Vapors, Gases, Mists, and Noncombustible Particulate Solids*, 384, 458
NFPA® 92A, *Standard for Smoke-Control Systems Utilizing Barriers and Pressure Differences*, 671
NFPA® 92B, *Standard for Smoke Management Systems in Malls, Atria, and Large Spaces*, 671
NFPA® 96, *Standard for Ventilation Control and Fire Protection of Commercial Cooking Operations*, 671
NFPA 101 *Life Safety Code®*
application of codes, 45
automatic sprinkler exemptions, 251
day care occupancies, 138
detention and correctional occupancies, 137
educational occupancies, 133
exit widths, 252
floor proximity exit signs, 264
handbook for inspectors, 49
hazardous/high-hazard occupancies, 452
health care and ambulatory health care occupancies, 136
inspector familiarity with codes, 48
institutional occupancies, 134
lodging (boarding) or rooming house occupancies, 140
means of egress, 244
model codes, 43
multiple-use occupancies, 146
occupancy classifications, 131
residential board and care occupancies, 137

NFPA® 101A, *Guide on Alternative Approaches to Life Safety*, 138
NFPA® 102, *Standard for Grandstands, Folding and Telescoping Seating, Tents, and Membrane Structures*, 350
NFPA® 241, *Standard for Safeguarding Construction, Alteration, and Demolition Operations*, 48, 291–292
NFPA® 252, *Standard Methods of Fire Tests of Door Assemblies*, 224
NFPA® 255, *Standard Method of Test of Surface Burning Characteristics of Building Materials*, 225
NFPA® 291, *Recommended Practice for Fire Flow Testing and Marking of Hydrants*, 477
NFPA® 318, *Standard for the Protection of Semiconductor Fabrication Facilities*, 394
NFPA® 400, *Hazardous Materials Code*
  dual hazardous occupancy classification, 453
  exemptions, 382–385
  fire protection systems, 382
  hazardous materials classification, 385
  hazardous materials storage, handling, and use, 384, 435
  hazardous/high-hazard occupancies, 452
  maximum allowable quantity per control area, 431
  permits, 380
  physical hazard materials, 386
  piping systems, 451
  process control, 449
  pyrophorics, 393
  spill control, 456
  unauthorized discharge, 450
NFPA® 495, explosives and blasting agents, 395
NFPA® 499, *Recommended Practice for the Classification of Combustible Dusts and of Hazardous (Classified) Locations for Electrical Installations in Chemical Process Areas*, 367
NFPA® 505, *Fire Safety Standard for Powered Industrial Trucks Including Type Designations, Areas of Use, Conversions, Maintenance, and Operations*, 349
NFPA® 654, *Standard for the Prevention of Fire and Dust Explosions from the Manufacturing, Processing, and Handling of Combustible Particulate Solids*, 365–366
NFPA® 701, *Standard Methods of Fire Tests for Flame Propagation of Textiles and Films*, 351
NFPA® 704, *Standard System for the Identification of the Hazards of Materials for Emergency Response*
  hazards not covered, 416–417
  inspector familiarity with codes, 48
  limitations, 418
  model codes, 232
  Official Mexican Standards adoption of, 427
  purpose of, 413, 416
  rating system, 417
NFPA® 720, *Standard for the Installation of Carbon Monoxide (CO) Detection and Warning Equipment*, 342–343
NFPA® 750, *Standard on Water-mist Fire Protection Systems*, 523, 542
NFPA® 921, *Guide for Fire and Explosion Investigations*
  autoignition (nonpiloted), 79
  autoignition temperature, 80
  ceiling jet, 104
  combustion, 75
  convection, 83
  fire, defined, 75
  fuel, 74
  heat flux, 83
  ignition, 79
  plume, 103
  radiation, 83
  self-heating, 80
  spontaneous ignition, 80
NFPA® 1031, *Standard for Professional Qualifications for Fire Inspector and Plan Examiner*
  inspector familiarity with codes, 49
  professional development, 17
  training and certification standards, 17
NFPA® 1141, *Standard for Fire Protection Infrastructure for Land Development in Wildland, Rural, and Suburban Areas*, 286
NFPA® 1851, *Standard of Selection, Care and Maintenance of Protective Ensembles, Structural Firefighting and Proximity Firefighting*, 160
NFPA® 1963, *Standard for Fire Hose Connections*, 532
NFPA® 2001, *Standard on Clean Agent Fire Extinguishing Systems*, 572, 576, 596
NFPA® 5000, *Building Construction and Safety Code®*
  educational occupancies, 133
  fire lanes and fire apparatus access roads, 286
  inspector familiarity with codes, 49
  prescriptive/model codes, 43
NFSA (National Fire Sprinkler Association), 504
NIOSH. *See* National Institute of Occupational Safety and Health (NIOSH)
NIST (National Institute for Standards and Technology), 73, 217
NOAA (National Oceanic and Atmospheric Administration), 399, 457
NOM-018-STPS-2000, 427
NOM-026-STPS-1998, "Signs and Colors for Safety and Health," 429
Nomex®, 94
Nonbulk packaging, 435. *See also* Packaging
Noncoded alarm system, 616
Noncombustible, defined, 118
Noncombustible construction (Canada), 125
Nonfeasance, 33
Nonflaming combustion, 76–77
Nonindicating control valve, 474, 475
Nonliquefied compressed gases, 388
Nonload-bearing wall, 117
Nonseparated occupancies, 148
Nonverbal communications skills, 694
Normal operating pressure, 486
Normal path of travel, 246
North American Free Trade Agreement (NAFTA), 402
Notification appliances for fire detection systems, 610–612, 643
Notification of fire code infractions, 20
Nozzle
  discharge patterns, 515
  dry-chemical extinguishing systems, 568, 570
  fire hydrant fire flow tests, 486, 487
NPG (NIOSH Pocket Guide to Chemical Hazards), 420–421
Nuisance alarms, 623

O

Occupancy
  based on use, 115
  building changes, 712, 713
  Certificate of Occupancy, 385
  changes and risks, 115, 116, 131
  classifications, 125–149
    assembly, 132
    business, 133

changes and risks, 131
comparative overview, 127–130
educational, 133–134
factory/industrial, 134, 135
institutional, 134–138
mercantile, 138–139
model codes, 125, 127–131
residential, 139–144
storage, 144–145
utility/miscellaneous, 145
code requirements, 707, 709–711
hazardous/high-hazard, 452–455
inspection of changes, 18
Inspector II responsibilities, 146–148, 149
for locating or distributing fire extinguishers, 597–598
mixed, 131
systematic plans review, 680–681
using current codes for changes in, 45
Occupant load
calculating, 265–266
changes in classification and, 265
code requirements, 265
defined, 265
maximum floor area per occupant, 266, 267
multiuse, 268
posting of maximum occupancy, 265
systematic plans review, 681
Occupant notification devices
acceptance testing, 643
audible notification, 644
candela information, 645
documentation, 646–647
inspection and testing, 643–647
visual notification, 645–646
wall-mount devices, 645
Occupational Safety and Health Administration (OSHA)
authority of inspectors, 24
cross-reference with NFPA® and OSHA, 50
field verification, 654
Hazard Communications Standard, 411, 412
obsolete fire extinguishers, 592
safety data sheets, 334, 402
Official Mexican Standards (*Normas Oficiales Mexicanas, NOMs*), 427
Oil boilers, 339
Oil-cooled transformers, 234
Oil-filled transformers, 234
One-half diagonal rule, 271
One-way voice notification system, 622, 623
Open burning, 309–311
Open floor plan, 100
Open sprinkler, 515
Opening protectives, 194–195
Openings, floor assembly, 202
Operational permit, 52–55
Ordinance, defined, 14
Ordinary-hazard occupancy for locating fire extinguishers, 598, 599
Organic peroxides, 391–392
Orientation
elevation view drawings, 666–667
of sprinklers, 512–513
Oriented strand board (OSB), 124, 156–157
Orifice plate proportioner, 555
Ornamental sprinkler, 516
OS&Y (outside stem and yoke) valve, 474, 509
OSB (oriented strand board), 124, 156–157
OSHA. *See* Occupational Safety and Health Administration (OSHA)
Outside screw and yoke valve, 509
Outside stem and yoke (OS&Y) valve, 474, 509
Outside technical assistance, 33
Ovens, industrial, 348–349
Overcurrent, 81, 82
Overhead obstructions as access barrier, 297
Overhead rolling steel fire door, 211–212
Overload, 81, 82
Owner/occupant request for inspection, 18
Oxidation, 75
Oxidizer, 92–95
common substances, 94
defined, 74
examples, 392–393
flammable range, 94, 95
hazard classifications, 392–393
lower flammable limit, 94
reactions, 392
upper flammable limit, 94
Oxidizing agent, 74
Oxidizing gases, 392–393
Oxyacetylene tanks, 353
Oxygen exclusion as fire control, 110

P

Packaging
bulk. *See* Bulk packaging
hazardous materials, 435–448
containers, 435–438
cylinders, 437–438
field-erected aboveground storage tanks, 442–446
intermediate bulk containers, 437
pressure vessels, 446–448
safety cans, 436
shop-fabricated aboveground storage tanks, 440–442
stationary tanks, 438–440, 441
underground storage tanks, 446
impact of loss of integrity, 430, 431
Paint spray booth extinguishing systems, 565, 566
Painted curbs for fire lanes, 289–290
Pallet storage, 323–324, 330–331
Panic hardware, 254
Parallel telephone system, 614, 620
Parapet, 194, 195
Party walls, 195
Passive agents, 87
PBD (performance-based design), 43, 658–659
Pellet heat sensors for sprinkler activation, 516
Pendant sprinkler, 512, 513, 514
Penetrations, floor assembly, 202
Pentaerthyritol Trinatrate (PETN), 454
Perchlorethylene, 361
Performance-based codes/standards, 43
Performance-based design (PBD), 44, 658–659
Permits, 52–55
automatic notification of changes, 46
construction, 52, 55
defined, 52
hazardous materials storage and use, 380

issuance of, 18
model code activities requiring permits, 52, 53–55
open burning, 310, 311
operational, 52–55
process, 65–67
reasons for, 52
welding and thermal cutting operations, 355
Personal appearance of inspectors, 694, 700
Personal space in U.S. culture, 694
Persuading skills, 695
PETN (Pentaerthyritol Trinatrate), 454
Petroleum storage tank fire, 380
Pharmacy occupancies, 138, 139
PHMSA (Pipeline and Hazardous Materials Safety Administration), 383
Photoelectric smoke detector, 630
Photographic documentation, 708
Photovoltaic systems, 221
Physical change of matter, 74–76
Physical hazard materials, 386–395
compressed and liquefied gases, 388
cryogenic fluids, 388–390
explosives and blasting agents, 395
flammable and combustible liquids, 386–387
flammable solids or gases, 390–391
organic peroxides, 391–392
oxidizers and oxidizing gases, 392–393
pyrophorics, 391, 393–394
unstable (reactive) materials, 394, 395
water-reactive materials, 394–395
Physical science, 74. *See also* Science of fire
Pilot line detector, 516
Piloted ignition, 79–80
Pipeline and Hazardous Materials Safety Administration (PHMSA), 383
Pipes
dry-chemical extinguishing systems, 568, 569
fire hydrant malfunctions, 489
for hazardous materials, 451
identification, 430, 431
pipeline markings, 413, 414, 416, 417
water distribution for automatic sprinkler systems, 510–512
water distribution systems, 472–474
Pitot tube and gauge, 480, 481–482, 483
PIV (post indicator valve), 474, 509, 546, 547
PIVA (post indicator valve assembly), 509
Placards
Canadian transportation, 423–427
Mexican, 429
U.S. DOT, 405, 406–408
Plan view
defined, 663
floor plan, 665–666
site plan, 663–665
Plans review, 651–684
architectural and fire safety symbols, 655
benefits of, 652–653
building construction plans, 662–679
detailed view, 669
elevation view, 666–668
floor plan, 665–666
plan views, 663–666
sectional view, 668–669
site plan, 663–665
system plans, 669–679
title block, 662
building department, 15, 16
building plans and construction drawings, 654–655
field verification, 653–654
fire suppression systems, 539
flowchart of design, permitting, and construction process, 652
overview, 651–655
performance-based design, 658–659
process of, 656–661
change orders, 657
construction and support documents, 658–661
fees, 658
sequence, 658
system plans, 669–679
automatic sprinkler systems, 673–676
electrical systems, 672, 673
fire detection and alarm systems, 677–679
mechanical systems, 670–672
plumbing systems, 672–673
special-agent fire extinguishing systems, 677
standpipe and hose systems, 677
systematic plans review, 679–684
additional concerns, 682–683, 684
building compartmentation, 682
exit capacity, 682
means of egress, 681
occupancy and construction classification, 680–681
occupant load, 681
overall building size, 680
Plastic construction materials, 168–171
common construction plastics, 170
Exterior Insulation and Finish Systems, 170–171
fire hazards, 169–170
thermal barriers, 170
Platform framing, 179, 180
Plenum, 357
Plumbing systems, 672–673
Plume, 103
Pneumatic rate-of-rise line heat detector, 627, 628, 639
Pneumatic rate-of-rise spot heat detector, 628
Pocket sliding door, 205
Police power, 28
Policies and procedures, 34–36
Polyethylene Wax Processing Facility fire, 448
Portable Containers for Gasoline and Other Petroleum Fuels (CSA B-376-M), 436
Portable fire extinguishers
agents, 586–590
classifications, 585
effectiveness of, 584–585
fixed systems vs., 585
inspection and maintenance, 594–602
installation and placement, 592–594
obsolete extinguishers, 592
pictorial and letter systems, 586, 587
ratings, 585–586
selection and location
extinguisher size and travel distance, 599–600
installation and placement, 592–594
nature of the hazard, 597–598
signage, 602, 603
systematic plans review, 683, 684

training, 600
types of, 590–592
Portable foam fire extinguishing system, 579
Portable Fuel Tanks for Marine Use (CSA B306-M), 436
Portable heating equipment, 342–343
Post and beam construction, 176, 177
Post indicator valve assembly (PIVA), 509
Post indicator valve (PIV), 474, 509, 546, 547
Posted signs for fire lanes, 289–290
Post-tensioned reinforcement, 188
Potassium bicarbonate (Purple-K) extinguishing agent, 568
Potential energy, 75, 78, 79
Powder coating, 356, 358
Power and energy, 78
Power supply inspection and testing, 643
Powered industrial trucks
fire hazard recognition, 350
as ignition source, 313
scissor lifts, 351
types of, 349
uses for, 350
Preacceptance inspections, 539–540
Preaction sprinkler system, 506, 517
Precast concrete, 186–187
Preincident planning, plans review, 653
Preliminary hearing, 21, 22
Preparation for inspections, 698–702
equipment lists, 700–701
personal appearance, 700
positive impressions, 699
records review, 702
scheduling, 701–702, 703
Prescriptive-based codes/standards, 43
Presignal alarms, 615
Pressure proportioning tank system, 584
Pressure Relief Devices - Parts 1-3 (CGA S-1.1, S-1.2, S-1.3), 384
Pressure relief of hazardous materials
cryogenic fluids, 390
hazardous materials, 381–382
Pressure relief valve, 448
Pressure tanks as water supply source, 469
Pressure vessels, 446–448
Pressure-maintenance pump, 535
Pressure-reducing valve, 531
Pressure-regulating devices on standpipe and hose systems, 530–531
Pressure/vacuum vent, 440, 443
Pretensioned reinforcement, 187–188
Primary explosives, 453–454
Primary feeders, 472
Primary power supply for fire alarm systems, 609, 610
Private fire safety inspectors, 16–17
Private sector inspector status, 29
Private water supply systems, 467, 468, 545
Probable cause hearing, 21
Procedures. *See* Policies and procedures
Process control, 449–450
Process hazard analysis, 365–367
Process Piping (B31.3), 384
Product containment, 381
Products of combustion, 84
Professional development of inspectors, 17–18
Project manual, 659
Project specifications, 659
Projected-beam photoelectric detector, 630, 631
Projecting window, 215
Proportioner
around-the-pump, 583, 584
balanced pressure, 583
bladder tank proportioner, 555
coupled water motor-pump proportioner, 584
foam-water sprinkler systems, 524, 555
line proportioner, 555
pressure proportioning tank system, 584
standard and inline balanced-pressure, 555
standard pressure proportioner, 555
Proportioning, 579
Proprietary alarm system
capabilities, 619
defined, 619
inspection and testing, 641
Prosecution for fire code violations, 20
Protected Aboveground Storage Tanks (UL 2085), 385
Protected premises system, 614–617
activation devices, 615
addressable alarm systems, 617
conventional alarm systems, 615–616
defined, 614
presignal alarms, 615
zoned conventional alarm systems, 616–617
Protected steel, 118
Provincial laws for fire inspections, 26
Public emergency alarm reporting system, 621
Public organization safety inspection programs, 15–16
Public sector inspector status, 28, 29
Public way, 244
Public-duty doctrine, 33
Pull stations, 635–637
Pump malfunctions in fire hydrants, 490
Pumping stations to facilitate movement of water, 469
Pump-operated fire extinguisher, 592
Purple-K (potassium bicarbonate) extinguishing agent, 568
Pyrolysis
conditions for occurrence of, 91
defined, 79
Pyrophorics
autoignition, 394
defined, 393
examples, 393
hazard classifications, 391, 393–394
Pyrotechnics, 92

Q

QREC (quick-response extended coverage) sprinkler, 515
QRES (quick-response early suppression) sprinkler, 515
Quenching operations, 359–361
Quick-response early suppression (QRES) sprinkler, 515
Quick-response extended coverage (QREC) sprinkler, 515

R

RACE acronym, 720
Rack storage, 324, 325
Rack storage sprinkler, 515
Radiation, 83, 84–87
Ramps
as means of egress, 260, 261
specifications, 260, 261
Rapid entry systems, 300, 301

Rate-compensation heat detector, 628
Rate-of-rise heat detectors, 627–629
RDX (Cyclotrimethylenetriamine), 454
REACT acronym, 720
Reactive (unstable) materials, 394, 395
Reactivity of chemicals, 382
Reasonable doubt standard, 23
Rebar, 163, 187
Receiver of communications, 691
Recessed sprinkler, 512
Recognition of fire hazards. *See* Fire hazard recognition
Recommended Practice for Fire Flow Testing and Marking of Hydrants (NFPA® 291), 477
Recommended Practice for the Classification of Combustible Dusts and of Hazardous (Classified) Locations for Electrical Installations in Chemical Process Areas (NFPA® 499), 367
Recommended Practice on Static Electricity (NFPA® 77), 360
Records
  complaints, 51–52
  detection and alarm system, 646–647
  detector testing, 640, 641
  extinguisher inspections, 600, 602
  inspection files and records, 22, 697–698
  inspection records review, 702
Recreational fires, 310
Recycling facilities
  compactors, 332
  equipment inspections, 332
  hazardous conditions, 331
  incinerators, 333–334
  inside operations, 332
  types of materials, 331–332
  waste-handling facilities, 332–333
Red oak standard for smoke-developed index, 226
Reducing agent, 87
Refractory photoelectric smoke detector, 630, 631
Refrigerants, 343–344
Refrigeration plant hazardous material requirements, 383
Refrigeration systems, 231–232
Refuse chutes, 218, 219, 235–236
Registry of Toxic Effects of Chemical Substances (RTECS), 397
Regulations, defined, 14
Remodeling
  occupancy classification changes, 131
  performance-based design during, 44
  using current codes during, 45
Remote receiving system, 619, 621
Remote station system inspection and testing, 641
Renovations. *See* Remodeling
Reports from inspectors, 696–697, 714–715
Reservoirs as water supply source, 469
Residential (RES) sprinkler, 515
Residential board and care occupancies, 137–138
Residential occupancies, 139–144
  apartment buildings, 143–144
  dormitories, 142
  hotels, 141–142
  lodging (boarding) or rooming houses, 139–140
  one- and two-family dwellings, 139–140
Residential sprinkler systems, 518–519
  flow rate requirements, 519
  inspections and testing, 552–554
  NFPA® standards, 518
  spacing, 519
  water supply requirements, 519
Residual pressure
  for standpipe and hose systems, 529
  in water supply system, 479
Resistance heating, 81, 82
Restaurant occupancy, 134, 135
Restaurants. *See* Commercial cooking
Restorable heat detection device testing, 639
Retention of heat in compartments, 102
Return-air plenum, 223
Revolving doors, 204, 206
RIBC (rigid intermediate bulk container), 437
Right of entry, 29–30
Rigid frame, 178
Rigid intermediate bulk container (RIBC), 437
Riser, 511
Road extensions, 286, 287
Road markings and signs, 289–291
Rodenticide regulations, 383
Rolled paper storage foam fire extinguishing systems, 578
Rolling steel fire doors, 211–212
Roof coverings, 199–200
Roof deck, 186
Roofing material fire hazards, 372
Roofs, sectional view of, 668, 669
Rooftop photovoltaic systems, 221
Room heaters, 340–341
Rooming house occupancies, 140–141
Routine inspections, 18
Row house party walls, 195
RTECS (Registry of Toxic Effects of Chemical Substances), 397
Rust as oxidation, 75

S

Safety
  carbon dioxide extinguishing systems, 575–576
  plans review and safety of responders, 653
  symbols, 655
Safety cans, 436
Safety Code for Elevators and Escalators (ASME/ANSI Standard A17.1), 47
Safety Code for Safety Glazing Materials for Glazing Motor Vehicles Operating on Land Highways (Z26.1-1996), 50
Safety Colors and Safety Signs (ISO-3864), 429
Safety Containers (ULC/ORD C30), 436
Safety data sheet (SDS)
  defined, 334, 402
  hazardous chemicals, 367
  hazardous materials classification, 385
  information disclosed on, 403
  uses for, 402
Sanction, defined, 20
Saponification, 98
Sax's Dangerous Properties of Industrial Materials, 397
Scarf joints, 156
SCBA (self-contained breathing apparatus), 82
SCC. *See* Standards Council of Canada (SCC)
Scheduling inspections, 701–702, 703
School auditorium occupancies, 146
Science of fire
  classification of fires, 96, 98
  combustion modes, 76–77
  energy, 78
  forms of ignition, 79–80
  fuel, 87–92, 93

heat, 78
heat transfer, 83–87
oxidizer, 92–95
passive agents, 87
physical and chemical changes, 74–76
products of combustion, 97
self-sustained chemical reaction, 95–96
temperature, 78
thermal energy, 80–82
SDS. *See* Safety data sheet (SDS)
Sea vs. City of Seattle, 29
Search warrant, 31
Seasonal attractions, 180
Seasonal climate conditions as access barrier, 298–299
Secondary containment for hazardous materials, 456–457
Secondary explosives, 454
Secondary feeders, 472
Secondary power supply for fire alarm systems, 609–610, 611
Sectional view, 668–669
Security
causing access problems, 327
door bars and grills, 256, 257, 302
entry restrictions, 707, 708
fire doors, 213
on windows, 215–216
Sedimentation deposits in fire hydrants, 489
Self-closing doors, 256, 257
Self-contained breathing apparatus (SCBA), 82
Self-heating, 80–81
Self-heating materials, 391
Self-sustained chemical reaction, 95–96
Semiautomatic-dry standpipe and hose system, 529
Semifixed foam fire extinguishing system, 579
Sender of communications, 690
Separated occupancies, 148, 149
Service counter opening fire doors, 213
Service test, 637
Services in buildings. *See* Building services
Shaft walls, 196–197
Shakes, 199
Sheathing, 158
Sheetrock®, 167. *See also* Gypsum board
Shell structures, 174–175
Shingles, 199–200
Shop-fabricated aboveground storage tanks, 440–442
Shopping mall occupancy, 126
Shredded wood insulation, 159
Shunt systems, 621
SI (standard international) system of units, 78, 88
Sidewall sprinkler, 512, 513, 514
Siding, 159–160
Signs
hazard identification, 706
posted signs for fire lanes, 289–290
road markings and signs, 289–291
"Signs and Colors for Safety and Health" (NOM-026-STPS-1998), 429
Single-action pull station, 636
Single-hung window, 214, 215
Single-strength annealed glazing, 166
Site access, 285–302
construction and demolition sites, 291–294
entry restrictions, 707, 708
fire lanes and fire apparatus access roads, 285–291
angle of approach and departure, 286, 287
dead-end access roads, 288–289
defined, 286
road markings and signs, 289–291
inspection observations, 705
requirements, 286–287
security and. *See* Security
structure access barriers, 294–302
exterior access, 294–299, 300
gates and security barriers, 295
illegal parking, 296–297
interior access, 299–302
landscaping, 298
rapid entry systems, 300, 301
seasonal climate conditions, 298–299
topographical conditions, 298, 299
weight requirements, 295–296
Site plan, 292, 663–665
Slab and column frames, 176, 178
Slab door, 206
Slate tile roofs, 199, 200
Sliding doors, 204, 205, 209
Sliding window, 214, 215
Smoke, ventilation from roofs with photovoltaic systems, 221
Smoke alarm vs. smoke detector, 629
Smoke dampers
defined, 612
in ductwork, 230
Smoke detector
defined, 629
duct smoke detectors, 632
for fire doors, 207–208
in HVAC systems, 230
inspection and testing, 640
ionization, 631–632, 633
performance factors, 630
photoelectric, 630–631
smoke alarm vs., 629
types, 629
video-based detectors, 633
Smoke Detectors for Fire Alarm Systems (UL 268), 49
Smoke-developed index, 225, 226
Smokeproof enclosures, 249
Smokers for cooking, 346, 347
Sodium bicarbonate extinguishing agent, 567–568
Sofa Super Store fire, 73
SOG (standard operating guidelines), 34
Solar panels, 221
Soldier course, 181, 182
Solid fuel, 91–92
Solid fuel appliance hood and ducts, 236
Solid piling, 324–325
Solid-fuel cooking equipment, 346, 347
Solubility, 90–91
SOP. *See* Standard operating procedures (SOPs)
Sparks as ignition sources
sparking, 82
unintentional ignition sources, 308, 309
welding, 328
from welding and thermal cutting operations, 353
Special duty, 33
Special-agent fire extinguishing systems, 564–594
carbon dioxide systems, 574–578
applications for, 575

components, 577–578
inspection and testing, 596
personnel safety, 575–576
clean-agent systems, 573–574
agents, 573–574
applications for, 572
inspection and testing, 596
dry powder systems, 569–571
dry-chemical systems, 565–569
agents, 567–568
application methods, 566–567
applications for, 565
components, 568–569
inspection and testing, 566, 594–595
fire hazard classification, 565
foam systems, 578–584
applications for, 578
foam concentrate types, 582
foam expansion rates, 581–582
foam generation, 579–580
foam proportioning rates, 581–582
inspection and testing, 596–597
methods of fire extinguishment, 578
proportioners, 583–584
types of, 578–579
locations, 564
plans review, 677
uses for, 564
wet-chemical systems, 571–572
Specific gravity, 90, 91
Specifications book, 659
SPFE Code Official's Guide to Performance-Based Design Review, 44
Spill control for hazardous materials, 456–457
Spontaneous ignition, 80–81
Spray finishing, 356, 357–358
Sprinkler
cold flow, 552
color coding, 517
defined, 512
deflector, 512
fusible link, 514, 515
hangers and piping, 551
NFPA® 13 sprinkler types, 513–515
orientation, 512–513
protective cage, 551
spare sprinklers, 513, 514
systems. *See* Automatic sprinkler systems
temperature classification, 517
variations of basic types, 515–516
SS (standard spray) sprinkler, 515
Stairs
components, 203–204
fire code requirements, 204
fire escape ladders, 224
as means of egress, 258–260
protected or enclosed, 223
purpose of, 203
rise and run, 203–204
stairwell egress capacity, 270
Stakeholders in code development process, 57
Standard 34, *Designation and Safety Classifications of Refrigerants*, 231–232, 343–344
Standard 300, *Fire Testing of Fire Extinguishing Systems for Protection of Restaurant Cooking Areas*, 347, 573
Standard balanced-pressure proportioner, 555
Standard for Bulk Inert Gas Systems (CGA P-18), 384
Standard for Categorizing Gas Mixtures Containing Flammable & Nonflammable Components (CGA P-23), 384
Standard for Classification of Toxic Gas Mixtures (CGA P-20), 384
Standard for Dry Chemical Extinguishing Systems (NFPA® 17), 566, 568, 594
Standard for Dry Pipe and Deluge Valves for Fire-Protection Service (UL 260), 49
Standard for Drycleaning Plants (NFPA® 32), 362
Standard for Exhaust Systems for Air Conveying of Vapors, Gases, Mists, and Noncombustible Particulate Solids (NFPA® 91), 384, 458
Standard for Fire Doors and Other Opening Protectives (NFPA® 80), 147
Standard for Fire Hose Connections (NFPA® 1963), 532
Standard for Fire Prevention During Welding, Cutting, and Other Hot Work (NFPA® 51B), 354, 356
Standard for Fire Protection Infrastructure for Land Development in Wildland, Rural, and Suburban Areas (NFPA® 1141), 286
Standard for Fire Testing of Fire Extinguishing Systems for Protection of Commercial Cooking Equipment (UL 300), 49
"Standard for Fire Tests of Joint Systems (UL Standard 2049), 222
Standard for Grandstands, Folding and Telescoping Seating, Tents, and Membrane Structures (NFPA® 102), 350
Standard for Low-, Medium-, and High-Expansion Foam (NFPA® 11), 578, 582
Standard for Ovens and Furnaces (NFPA® 86), 348
Standard for Portable Fire Extinguishers. See NFPA® 10, *Standard for Portable Fire Extinguishers*
Standard for Professional Qualifications for Fire Inspector and Plan Examiner. See NFPA® 1031, *Standard for Professional Qualifications for Fire Inspector and Plan Examiner*
Standard for Residential Sprinklers for Fire-Protection Service (UL 1626), 49
Standard for Safeguarding Construction, Alteration, and Demolition Operations (NFPA® 241), 48, 291–292
Standard for Smoke Management Systems in Malls, Atria, and Large Spaces (NFPA® 92B), 671
Standard for Smoke-Control Systems Utilizing Barriers and Pressure Differences (NFPA® 92A), 671
Standard for Spray Application Using Flammable or Combustible Material (NFPA® 33), 358
Standard for Steel Underground Tanks for Flammable and Combustible Liquids (UL 58), 446
Standard for the Design and Installation of Oxygen-Fuel Gas Systems for Welding, Cutting, and Allied Processes (NFPA® 51), 354
Standard for the Inspection, Testing, and Maintenance of Water-Based Fire Protection Systems. See NFPA® 25, *Standard for the Inspection, Testing, and Maintenance of Water-Based Fire Protection Systems*
Standard for the Installation of Air-Conditioning and Ventilating Systems (NFPA® 90A), 671
Standard for the Installation of Carbon Monoxide (CO) Detection and Warning Equipment (NFPA® 720), 342–343
Standard for the Installation of Foam-Water Sprinkler and Foam-Water-Spray Systems (NFPA® 16), 523, 542
Standard for the Installation of Private Fire Service Mains and Their Appurtenances (NFPA® 24), 541, 545
Standard for the Installation of Sprinkler Systems. See NFPA® 13, *Standard for the Installation of Sprinkler Systems*

Standard for the Installation of Sprinkler Systems in Low-Rise Residential Occupancies. See NFPA® 13R, *Standard for the Installation of Sprinkler Systems in Low-Rise Residential Occupancies*
Standard for the Installation of Sprinkler Systems in One- and Two-Family Dwellings and Manufactured Homes. See NFPA® 13D, *Standard for the Installation of Sprinkler Systems in One- and Two-Family Dwellings and Manufactured Homes*
Standard for the Installation of Sprinkler Systems (NFPA® 13), 48
Standard for the Installation of Standpipe and Hose Systems. See NFPA® 14, *Standard for the Installation of Standpipe and Hose Systems*
Standard for the Installation of Stationary Pumps for Fire Protection. See NFPA® 20, *Standard for the Installation of Stationary Pumps for Fire Protection*
Standard for the Installation of Warm Air Heating and Air-Conditioning Systems (NFPA® 90B), 671
Standard for the Prevention of Fire and Dust Explosions from the Manufacturing, Processing, and Handling of Combustible Particulate Solids (NFPA® 654), 365–366
Standard for the Prevention of Fires and Dust Explosions in Agricultural and Food Processing Facilities (NFPA® 61), 369
Standard for the Protection of Semiconductor Fabrication Facilities (NFPA® 318), 394
Standard for Ventilation Control and Fire Protection of Commercial Cooking Operations (NFPA® 96), 671
Standard for Water Tanks for Private Fire Protection (NFPA® 22), 545
Standard for Water-spray Fixed Systems for Fire Protection (NFPA® 15), 521, 541
Standard for Wet Chemical Extinguishing Systems (NFPA® 17A), 347, 573, 595
Standard international (SI) system of units, 78, 88
Standard Method of Test of Surface Burning Characteristics of Building Materials (NFPA® 255), 225
Standard Methods of Fire Tests for Flame Propagation of Textiles and Films (NFPA® 701), 351
Standard Methods of Fire Tests of Door Assemblies (E-152), 224
Standard Methods of Fire Tests of Door Assemblies (NFPA® 252), 224
Standard of Selection, Care and Maintenance of Protective Ensembles, Structural Firefighting and Proximity Firefighting (NFPA® 1851), 160
Standard on Carbon Dioxide Extinguishing Systems (NFPA® 12), 574
Standard on Clean Agent Fire Extinguishing Systems (NFPA® 2001), 572, 576, 596
Standard on Explosion Prevention Systems (NFPA® 69), 368
Standard on Explosion Protection by Deflagration Venting (NFPA® 68), 368
Standard on Halon 1301 Fire Extinguishing Systems (NFPA® 12A), 596
Standard on Incinerators and Waste and Linen Handling Systems and Equipment (NFPA® 82), 334
Standard on Water-mist Fire Protection Systems (NFPA® 750), 523, 542
Standard operating guidelines (SOGs), 34
Standard operating procedures (SOPs)
  code enforcement departments, 34
  compressed/liquefied gas, dispensing of, 316
Standard pressure proportioner, 555
Standard System for the Identification of the Hazards of Materials for Emergency Response. See NFPA® 704, *Standard System for the Identification of the Hazards of Materials for Emergency Response*
Standard Test Method for Surface Burning Characteristics of Building Materials (E84), 49
Standard Test Method for Volatile Matter in the Analysis Sample of Coal and Coke (ASTM D-3175), 367
Standard Test Methods for Fire Tests of Building Construction (E119), 49, 167, 222
Standard Test Methods for Fire Tests of Roof Coverings (E108), 49
Standardization, defined, 49
Standards
  American National Standards Institute, 49–50. *See also* American National Standards Institute (ANSI)
  ASTM International, 49. *See also* ASTM International
  codes vs., 42
  consensus standards, 48, 49–50
  consistency, 46–47
  defined, 42
  industry standards, 48
  life cycle and time line, 46
  National Fire Protection Association®, 48–49. *See also* National Fire Protection Association® (NFPA®)
  performance-based, 43
  prescriptive-based, 43
  Standards Council of Canada, 50
  Underwriters Laboratories, 49
Standards Council of Canada (SCC)
  consensus standards, 48
  purpose of, 50
Standpipe and hose systems, 524–532
  acceptance testing procedures, 543
  classifications, 526–527
  components, 526
  fire department connections, 531–532
  hose valves, 530
  inspections and testing, 555–557
  location of standpipes, 530
  plans review, 677
  pressure-regulating devices, 530–531
  types, 527–529
  water supplies and residual pressure, 529
State laws for fire inspections, 26
Static discharge, 317
Static pressure in a water supply system, 479
Station Nightclub fire, 92
Stationary fire pumps, 532–538
  acceptance testing procedures, 543–545
  centrifugal pump, 532
  controllers, 536–538
  drivers, 535–536
  end suction, 534–535
  functions of, 532
  horizontal split-case, 532–533
  inspections and testing, 557
  pressure ratings, 533
  pressure-maintenance, 535
  vertical inline, 534
  vertical split-case, 533
  vertical turbine, 534
Stationary tanks, 438–440
Statute, defined, 14
Steam turbine, 536
Steamer connection, 477
Steel, 163–165
  characteristics, 164
  fire protection, 164–165

floor joist systems, 200, 201
material content, 163
membrane ceilings, 165
protected, 118
rolling steel fire doors, 211–212
spray-on coatings, 165
structural systems, 184–185
stud wall framing, 176, 177
truss frames, 177
trusses, 185
Steel Aboveground Storage Tanks for Flammable and Combustible Liquids (UL 142), 384, 440, 458
Steel Underground Storage Tanks for Flammable and Combustible Liquids (UL 58), 384
Steiner Tunnel Test, 225–226
Stone building materials, 161–162
Stop work order, 67, 704
Storage and Handling of Silane and Silane Mixtures (CGA G-13), 394
Storage occupancies, 131, 144–145. *See also* Material storage facilities
Storage tanks
field-erected aboveground, 442–446
product reconciliation, 446, 447
shop-fabricated aboveground, 440–442
underground, 446, 447
water distribution system, 474
Stored-pressure fire extinguisher, 590
Storm data from NOAA, 457
Stretcher course, 181, 182
Strobe visual notification, 645
Structural stability evaluations, 222
Structural systems, 171–188
bearing wall structures, 172, 173
concrete, 186–188
construction type relationships with, 171
frame, 176–178
inspection of, 172
loads, 172–174
masonry, 181–184
shell and membrane, 174–175
steel, 184–185
wood, 178–181
Structure access barriers, 294–302
exterior access, 294–299, 300
illegal parking, 296–297
interior access, 299–302
landscaping, 298
rapid entry systems, 300
seasonal climate conditions, 298–299
topographical conditions, 298, 299
weight requirements, 295–296
Suction tanks, 469
Summary offenses, 21
Sunset provision, 61
Supervising station alarm systems, 617–621
central station system, 618–619
multiplexing systems, 620
parallel telephone system, 614, 620
proprietary system, 618, 619, 620
remote, 618
remote receiving system, 619, 621
Supervisory signal
defined, 613
inspection and testing, 643
Surface-burning characteristic, 225
Surface-to-mass ratio, 92, 93
Swinging doors, 204, 205
Swinging fire doors, 213
Switch gear, 233
Symbols
architectural and fire safety, 655
floor plan, 666, 667
ISO safety symbols, 429
north directional symbol, 663
Synthetic Stucco, 170
System plans, 669–679
automatic sprinkler systems, 673–676
electrical systems, 672, 673
fire detection and alarm systems, 677–679
mechanical systems, 670–672
plumbing systems, 672–673
special-agent fire extinguishing systems, 677
standpipe and hose systems, 677
Systematic plans review, 679–684
additional concerns, 682–683, 684
building compartmentation, 682
exit capacity, 682
means of egress, 681
occupancy and construction classification, 680–681
occupant load, 681
overall building size, 680

T

T turnaround, 288, 289
Tactile notification appliances, 611
Tamper seals on extinguishers, 594, 602
Tamper switches, 614
Tank Inspection, Repair, Alteration, and Reconstruction (API 653), 384, 444, 446
Tar kettle fire hazards, 372–373
Task force for code development, 57–58
TC. *See* Transport Canada (TC)
TDG (Transportation of Dangerous Goods) Regulations, 333, 436
Telecommunications facilities, 572
Temperature
autoignition temperature, 80
defined, 78
fire control and reduction of, 109–110
Tensile force, 163
Tents, 350–352
Test hydrant, 485
Testimony, 23–24
Testing of Restaurant Cooking Area Fire Extinguishing System Units (ULC/ORD-C1254.6), 573
Testing water supplies, 477, 479–493
fire flow test computations, 482–484
fire flow test procedures, 485–490
fire flow test results computations, 490–493
fire flow vs. water flow, 477
fire hydrant inspections, 480–481
pitot tube and gauge, 480, 481–482, 483
residual pressure, 479, 484–485
static pressure, 479
Textual notification appliances, 611
Thermal barrier, 170, 175
Thermal column, 103

Thermal energy
chemical energy, 80–81
defined, 78
electrical energy, 81–82
mechanical energy, 82
Thermal layering, 104–105
Thermal properties of the compartment, 102
Thermal updraft, 103
Thermistor, 628
Third-party inspectors, 16
Third-party technical assistance, 33
Threshold limit value (TLV), 397
Thrust plate, 183
Time management, 695, 702, 703
Tire storage, 322, 330
Title block, 662
TLV (threshold limit value), 397
Topographical conditions as access barrier, 298, 299
Torch-applied roofing material fire hazards, 372
Total flooding system, 573
Totes, 437
Toxic or highly toxic materials hazard classifications, 396–397
Toxicity of products of combustion, 226
Training, inspector training levels, 17
Transformers, 234, 235
Transport Canada (TC)
clean-agent storage containers, 596
corrosives, 398
health hazard materials, 396
nonbulk and bulk packaging, 435
standards development, 50
waste material disposal, 333
Transportation of Dangerous Goods (TDG) Regulations, 333, 436
Transportation placards, labels, and markings
DOT labels, 405, 408–409
DOT markings, 405, 409, 411, 412–413
DOT placards, 405, 406–408
Mexican transportation placards, labels, and markings, 429
UN Recommendations
background of, 402
commodity identification numbers, 404, 406
hazard classes, 404, 405
Trash as a health hazard, 309
Travel distance, 272, 274
Treated wood, 157
Trials, 22–23
Trim valves, 510
Trouble signal
defined, 613
inspection and testing, 643
Trusses
lightweight wooden, 124
steel, 185
truss frames, 176, 177
wood ceiling trusses, 183
Tubing-type continuous line heat detector, 626
Tunnel test, 225–226
Two-way communication system, 622, 623
Type A refrigerants, 232
Type B refrigerants, 232
Type C gypsum board, 168
Type I construction
characteristics, 118–120
structural systems, 171
Type II construction
building materials, 121
characteristics, 120
structural systems, 171
Type III construction
characteristics, 120–122
interior structural framing, 183
structural systems, 171
Type IV construction
building materials, 123
characteristics, 122
interior structural framing, 183
structural systems, 171
wood structural systems, 178, 179
Type V construction
characteristics, 122–125
structural systems, 171
wood structural systems, 178, 179
Type X gypsum board, 167, 168

U

UEL (upper explosive limit), 94
UFC. *See Unified Facilities Criteria (UFC)*
UFL (upper flammable limit), 94
UL. *See* Underwriters Laboratories Inc. (UL)
ULC. *See* Underwriters' Laboratories of Canada (ULC)
Ultraviolet wave spectrum (UV) detectors, 633, 634
UN Recommendations
background of, 402
commodity identification numbers, 404, 406
hazard classes, 404, 405
Mexican transportation placards, labels, and markings, 429
Unauthorized discharge, 450
Underground storage tanks (USTs), 446, 447
Underwriters Laboratories Inc. (UL)
consensus standards, 48
cooking equipment, 573
fire-rated door assemblies, 206
foam fire extinguishing system, 582
heaters listed, 340
interior finishes, evaluating, 225
NA-X® listing, 570
protected steel, 118
purpose of, 49
safety cans, 436
smoke detectors, 630
UL 58 *Standard for Steel Underground Tanks for Flammable and Combustible Liquids*, 446
UL 58 *Steel Underground Storage Tanks for Flammable and Combustible Liquids*, 384
UL 142 *Steel Aboveground Storage Tanks for Flammable and Combustible Liquids*, 384, 440, 458
UL 260 *Standard for Dry Pipe and Deluge Valves for Fire-Protection Service*, 49
UL 268 *Smoke Detectors for Fire Alarm Systems*, 49
UL 299 *Dry Chemical Fire Extinguishers*, 49
UL 300, *Fire Testing of Fire Extinguishing Systems for Protection of Restaurant Cooking Areas*, 347, 573
UL 300 *Standard for Fire Testing of Fire Extinguishing Systems for Protection of Commercial Cooking Equipment*, 49
UL 753, Steiner Tunnel Test, 225–226
UL 1316, *Glass-Fiber-Reinforced Plastic Underground Storage Tanks for Petroleum Products, Alcohols, and Alcohol-Gasoline Mixtures*, 384, 446

UL 1626 *Standard for Residential Sprinklers for Fire-Protection Service*, 49
UL 2080 *Fire-Resistant Aboveground Storage Tanks*, 385
UL 2085 *Protected Aboveground Storage Tanks*, 385
UL 2245 *Vaulted Storage Tank Systems*, 385
UL Standard 2049, "Standard for Fire Tests of Joint Systems," 222
Underwriters' Laboratories of Canada (ULC)
heaters listed, 340
safety cans, 436
standards development, 50
ULC/ORD C30, *Safety Containers*, 436
ULC/ORD-C1254.6, *Testing of Restaurant Cooking Area Fire Extinguishing System Units*, 573
UNDMGC
corrosives, 398
explosives and blasting agent regulations, 395
explosives hazard classification, 455
Unified Facilities Criteria (UFC)
military base fire departments, 26
model codes, 43
UFC 3-600-01, *Fire Protection Engineering for Facilities*, 288
Unit heaters, 340, 341
Universal Water Flow Test Summary Sheet, 491, 492, 676
Unsafe behaviors, 308–319
compressed/liquefied gases, 317–319
electrical equipment, 311–313
flammable and combustible liquids, 313–317
ignition sources, 308–309
inadequate housekeeping, 308, 309
open burning, 309–311
Unsafe conditions, 319–334
electrical hazards, 320–322
material storage facilities, 322–331
big box stores, 322, 323
changes in inventories and location, 322
fire protection systems, 323
inspection guidelines, 325–327
lumberyards, 328–330
pallet storage, 323–324, 330–331
rack storage, 324, 325
solid piling, 324–325
tires, 322, 330
warehouses, 327–328
recycling facilities, 331–334
Unstable (reactive) materials, 394, 395
Upper explosive limit (UEL), 94
Upper flammable limit (UFL), 94
Upper layer, 87
Upright sprinkler, 512, 513, 514
Urgent care facilities, 137
U.S. Chemical Safety and Hazard Investigation Board (CSB), 362–363
U.S. Constitution
Fifth Amendment, 19
Fourteenth Amendment, 19
Fourth Amendment, 29
U.S. Department of Defense (DoD)
emergency service access road requirements, 288
Hazardous Materials Information Resource System, 421
Unified Facilities Criteria, 26, 43, 288
U.S. Department of Transportation (DOT)
clean-agent storage container standards, 596
corrosives, 398
cylinders, 437
explosives and blasting agents, 395
explosives hazard classification, 455
health hazard materials, 396
Intermediate Bulk Containers, 437
labels, 405, 408–409
markings, 405, 409, 411, 412–413
natural gas pipeline regulations, 383
nonbulk and bulk packaging, 435
placards, 405, 406–408
U.S. Environmental Protection Agency (EPA)
clean-agent extinguishing systems, 572
Halotron®, 573
hazardous materials transportation, 333
pesticides, fungicides, and rodenticides, 383
water residual pressure requirements, 484–485
U.S. Federal Hazardous Substances Act (FHSA), 414
U.S. National Fire Academy (NFA), 17
Use-closed system, 430–431, 432
Use-open system, 430–431, 432
USFA Fire Incident Survey, 307
UST (underground storage tank), 446, 447
Utility chase, 218, 235–236
Utility lines shown in site plans, 664
Utility/miscellaneous occupancies, 145
UV (ultraviolet wave spectrum) detectors, 633, 634

V

Valves
alarm-test, 510
automatic sprinkler systems, 508–510
butterfly valve, 474
check valve (backflow preventer), 475, 508, 510
control valves in water distribution systems, 474–475
on cylinders, 438, 439
drain valve, 510
drip check or drip ball, 510
fire hydrant malfunctions, 489
gate valve, 474
for hazardous materials, 451
indicating, 474, 475, 546, 547
nonindicating, 474, 475
outside stem and yoke (OS&Y), 474, 509
post indicator valve, 474, 509, 546, 547
post indicator valve assembly, 509
pressure relief, 448
pressure-reducing, 531
residential control valves, 519
trim valve, 510
wall post indicator valve, 509
Vapor density, 89
Vapor pressure, 90
Vaporization, 79
Variances of codes, 63
Vaulted Storage Tank Systems (UL 2245), 385
Ventilation
cooking hoods, 345, 346, 347
during decay stage of fire development, 108
as factor in fire development, 102
HVAC systems, 102
mechanical systems in high hazard occupancies, 458
purpose of ventilation systems, 344, 345
smoke ventilation from roofs with photovoltaic systems, 221
vents on aboveground storage tanks, 442, 443

Vertical clearance over fire lanes or fire apparatus access roads, 286
Vertical doors, 204, 205
Vertical inline fire pump, 534
Vertical shafts, 218, 235–236
Vertical split-case fire pump, 533
Vertical turbine fire pump, 534
Vessel, defined, 338
Video-based detectors, 633
Violations found during inspections, 713–714
Visual notification appliances, 611, 645–646
Voice notification systems, 622, 623, 642

W

Wall post indicator valve (WPIC), 509
Walls
  bearing wall structures, 172, 173
  curtain walls, 185, 197–198
  enclosure and shaft walls, 196–197
  exterior materials, 157–160
  exterior wall fire-resistance rating requirements, 119
  fire partitions and fire barriers, 196
  fire walls, 143, 194–195
  fire-resistance rating, 256–257
  lath and plaster construction, 168
  load-bearing, 117
  masonry, 160, 181–182
  nonload-bearing, 117
  nonreinforced brick bearing wall, 183
  party walls, 195
  reinforced masonry, 182
  sectional view, 668
  steel stud wall framing, 176, 177
Warehouses
  exposure protection steps, 328
  fire load, 126
  inspection of fire suppression systems, 327–328
  inspections, 145
  storage occupancies, 144
  welding operations in, 328
Warrants, 30–31
Waste-handling facilities, 332–333
Water
  capacity of compressed gas cylinders, 435
  as extinguishing agent, 586–587
  heat capacity of, 110
Water flow, 477
Water flowmeter, 475
Water main, 472
Water supply, 465–493
  adequacy for fire fighting operations, 465, 467
  automatic sprinkler systems, 508
  distribution systems, 472–477, 478, 479
    backflow preventers, 475
    control valves, 474–475
    fire hydrants, 476–477, 478, 479
    piping, 472–474
    private water supply, 467, 468
    storage tanks, 474
    system design, 466–468
    water towers, 473
  fire hydrants. *See also* Fire hydrants
    classifications and markings, 477, 479
    dry-barrel, 476
    fire flow test computations, 482–484
    inspections, 480–481
    outlets, 477, 483
    records of inspections, 477, 478
    wet-barrel, 476
  means of moving water, 466, 470–472
  primary, 508
  processing or treatment facilities, 466
  residential sprinkler systems, 519
  secondary, 508
  sources, 466, 468–470
  standpipe and hose systems, 529
  system design, 466–468
  testing, 477, 479–493
    fire flow test computations, 482–484
    fire flow test procedures, 485–490
    fire flow test results computations, 490–493
    fire flow vs. water flow, 477
    fire hydrant inspections, 480–481
    pitot tube and gauge, 480, 481–482, 483
    residual pressure, 479, 484–485
    static pressure, 479
Water-based fire suppression systems. *See* Fire suppression systems, water-based
Waterflow alarm for sprinkler activation, 517–518, 634, 635
Water-mist sprinkler systems, 521–523, 542
Water-reactive materials, 394–395
Water-resistant gypsum board, 168
Water-spray fixed sprinkler system
  acceptance testing procedures, 541
  components, 520
  inspections and testing, 554–555
  mechanics of, 520
  purpose of, 520–521
Watt (W), 78, 89
Welded Tanks for Oil Storage (API 650), 384, 443
Welding and thermal cutting operations, 352–356
  combustible material clearance, 354
  fire safety issues, 353
  fire watch, 355–356
  hot work program, 354–355
  ignition sources, 353
  oxyacetylene tanks, 353
  permits, 355
  rag or towel storage, 353, 354
  situations not allowed, 353
  3D welding, 363
West Pharmaceutical Services, Kinston, NC, explosion (2003), 363
Wet-barrel fire hydrant, 476
Wet-chemical extinguishing system
  agents, 572, 590
  purpose of, 571–572
Wet-chemical fire suppression system, 347
Wet-pipe sprinkler system, 505
WHMIS (Workplace Hazardous Materials Information System), 422, 427, 428
Wind as live load, 173
Windows, 213–216
Winecoff Hotel, Atlanta, GA, fire (1946), 262
Wired glass, 166
Wireless Information System For Emergency Responders (WISER), 386
Wire-type continuous line heat detector, 627

Wiring, 311, 312
WISER (Wireless Information System For Emergency Responders), 386
Wood construction materials, 154–160
  dimensional lumber, 155
  engineered wood products, 155–157
  exterior wall materials, 157–160
  fire-retardant treated wood, 157
  for floors, 200, 201
  lumber, 154
  shingles, 199
  solid lumber, 155
  wood frame construction (Type V), 122–125
  wood panel doors, 206
  wood structural systems, 178–181
  wood-frame structures as source of fire fuel, 101
Woodworking and processing facility dust hazards, 371
Work and energy, 78
Work session, 61
Workplace Hazardous Materials Information System (WHMIS), 422, 427, 428
WPIV (wall post indicator valve), 509
Written communications from inspectors, 695–697
Wythe of masonry units, 181, 182

Z

Z26.1-1996, *Safety Code for Safety Glazing Materials for Glazing Motor Vehicles Operating on Land Highways*, 50
Zoned conventional alarm system, 616–617

NOTES

NOTES